1989

Probability and Mathematical Statistics (Continued)

PURI, VILAPLANA, and WERTZ • New Perspectives in Theoretical and Applied Statistics

RANDLES and WOLFE • Introduction to the Theory of Nonparametric Statistics

RAO • Linear Statistical Inference and Its Applications, *Second Edition*

RAO • Real and Stochastic Analysis

RAO and SEDRANSK • W.G. Cochran's Impact on Statistics

RAO • Asymptotic Theory of Statistical Inference

ROHATGI • An Introduction to Probability Theory and Mathematical Statistics

ROHATGI • Statistical Inference

ROSS • Stochastic Processes

RUBINSTEIN • Simulation and The Monte Carlo Method

SCHEFFE • The Analysis of Variance

SEBER • Linear Regression Analysis

SEBER • Multivariate Observations

SEN • Sequential Nonparametrics: Invariance Principles and Statistical Inference

SERFLING • Approximation Theorems of Mathematical Statistics

SHORACK and WELLNER • Empirical Processes with Applications to Statistics

TJUR • Probability Based on Radon Measures

Applied Probability and Statistics

ABRAHAM and LEDOLTER • Statistical Methods for Forecasting

AGRESTI • Analysis of Ordinal Categorical Data

AICKIN • Linear Statistical Analysis of Discrete Data

ANDERSON, AUQUIER, HAUCK, OAKES, VANDAELE, and WEISBERG • Statistical Methods for Comparative Studies

ARTHANARI and DODGE • Mathematical Programming in Statistics

BAILEY • The Elements of Stochastic Processes with Applications to the Natural Sciences

BAILEY • Mathematics, Statistics and Systems for Health

BARNETT • Interpreting Multivariate Data

BARNETT and LEWIS • Outliers in Statistical Data, *Second Edition*

BARTHOLOMEW • Stochastic Models for Social Processes, *Third Edition*

BARTHOLOMEW and FORBES • Statistical Techniques for Manpower Planning

BECK and ARNOLD • Parameter Estimation in Engineering and Science

BELSLEY, KUH, and WELSCH • Regression Diagnostics: Identifying Influential Data and Sources of Collinearity

BHAT • Elements of Applied Stochastic Processes, *Second Edition*

BLOOMFIELD • Fourier Analysis of Time Series: An Introduction

BOX • R. A. Fisher, The Life of a Scientist

BOX and DRAPER • Empirical Model-Building and Response Surfaces

BOX and DRAPER • Evolutionary Operation: A Statistical Method for Process Improvement

BOX, HUNTER, and HUNTER • Statistics for Experimenters: An Introduction to Design, Data Analysis, and Model Building

BROWN and HOLLANDER • Statistics: A Biomedical Introduction

BUNKE and BUNKE • Statistical Inference in Linear Models, Volume I

CHAMBERS • Computational Methods for Data Analysis

CHATTERJEE and PRICE • Regression Analysis by Example

CHOW • Econometric Analysis by Control Methods

CLARKE and DISNEY • Probability and Random Processes: A First Course with Applications, *Second Edition*

COCHRAN • Sampling Techniques, *Third Edition*

COCHRAN and COX • Experimental Designs, *Second Edition*

CONOVER • Practical Nonparametric Statistics, *Second Edition*

CONOVER and IMAN • Introduction to Modern Business Statistics

CORNELL • Experiments with Mixtures: Designs, Models and The Analysis of Mixture Data

(*continued on back*)

Biostatistics:
A Foundation for Analysis
in the Health Sciences

Fourth Edition

Wayne W. Daniel
Professor of Decision Sciences
Georgia State University

John Wiley & Sons

New York   Chichester   Brisbane   Toronto   Singapore

Library of Congress Cataloging in Publication Data:

Daniel, Wayne W., 1929–
  Biostatics, a foundation for analysis in the health
sciences.

  (Wiley series in probability and mathematical
statistics. Applied probability and statistics)
  Includes bibliographies and index.
  1. Medical statistics. 2. Biometry. I. Title.
II. Series. [DNLM: 1. Biometry. 2. Statistics.
WA 950 D184b]
RA409.D35 1987   519.5′024574   86-24578
ISBN 0-471-85264-3

Printed in the United States of America

10 9 8 7 6 5 4

To my wife, Mary, and
my children, Jean, Carolyn, and John

Preface

This fourth edition of *Biostatistics*: *A Foundation for Analysis in the Health Sciences* should appeal to the same audience for which the first three editions were written: the advanced undergraduate and beginning graduate student and the health professional in need of a reference book on statistical methodology.

The present edition improves on the previous editions in three important areas. I have (1) included several large data sets; (2) expanded, revised, and further clarified much of the discussion appearing in the first three editions; and (3) included additional student exercises.

Like the previous editions, this edition requires few mathematical prerequisites. Only reasonable proficiency in algebra is required for an understanding of the concepts and methods underlying the calculations. The emphasis continues to be on an intuitive understanding of principles rather than an understanding based on mathematical sophistication.

Since the publication of the first edition, the widespread use of microcomputers has had a tremendous impact on the teaching of statistics, as it is the rare student who does not now own or have access to a personal computer. Now, more than ever, the statistics instructor can concentrate on teaching concepts and principles and devote less class time to tracking down computational errors made by students. Relieved of the tedium and labor associated with lengthy hand calculations, today's students have reason to view their course in statistics as a much more enjoyable experience than their predecessors. Consequently this edition contains a greater emphasis on computer applications.

The instructor who makes use of the computer may be able to cover more material during a term than previously. However, the material in the present edition should be sufficient for a one- or two-term course.

The following specific improvements have been incorporated in this fourth edition:

1. Several new drawings to help clarify difficult concepts.

2. Several examples of computer printouts.

3. More than 100 new exercises—some at the ends of sections and some at the ends of chapters.

4. Twenty large data sets, each containing 1000 or more observations. Exercises require the student to select from these data sets simple random samples to be analyzed by the different techniques presented in the text—descriptive techniques, estimation, hypothesis testing, analysis of variance, regression and correlation, chi-square, and nonparametric methods.

5. The nine-step hypothesis testing procedure introduced in Chapter 6 is used with other hypothesis testing procedures such as analysis of variance, regression and correlation analysis, chi-square, and nonparametric techniques. This feature helps to reinforce the hypothesis-testing concepts and provides students with an organized approach to the solution of exercises involving hypothesis testing.

6. The chapter on multiple regression contains a discussion of the use of qualitative, or categorical, variables in regression analysis.

I would like to express my gratitude to the many readers and users of my book as a textbook for their helpful suggestions for this fourth edition. In particular I thank the following people who made detailed recommendations for this revision: Dr. Gary Brager, Towson State University; Dr. John E. Hewett, University of Missouri–Columbia; and Dr. Kirby Jackson, University of South Carolina School of Public Health.

I also thank my colleagues at Georgia State University, Professors Geoffrey Churchill and Brian Schott who wrote computer programs for generating some of the appendix tables. For any remaining deficiencies in the book, however, I must accept full responsibility.

Finally, I am grateful for the help of my wife Mary, without whose skills as a typist, proofreader, and critic this book could not have been published.

Wayne W. Daniel

Contents

CHAPTER 1

Organizing and Summarizing Data

1.1 INTRODUCTION

The objectives of this book are twofold: (1) to teach the student to organize and summarize data and (2) to teach the student how to reach decisions about a large body of data by examining only a small part of the data. The concepts and methods necessary for achieving the first objective are presented under the heading of *descriptive statistics*, and the second objective is reached through the study of what is called *inferential statistics*. This chapter discusses descriptive statistics. Chapters 2 to 5 discuss topics that form the foundation of statistical inference, and most of the remainder of the book deals with inferential statistics.

Since this volume is designed for persons preparing for or already pursuing a career in the health field, the illustrative material and exercises reflect the problems and activities that these persons are likely to encounter in the performance of their duties.

1.2 SOME BASIC CONCEPTS

Let us first define some basic terms that will be encountered here.

Statistics The meaning of *statistics* is implicit in the previous section. More concretely, however, we may say that *statistics is a field of study concerned with (1) the collection, organization, and summarization of data, and (2) the drawing of inferences about a body of data when only a part of the data are observed.*

Biostatistics The tools of statistics are employed in many fields—business, education, psychology, agriculture, and economics, to mention only a few. When the data being analyzed are derived from the biological sciences and medicine, we use the term biostatistics to distinguish this particular application of statistical tools and concepts. This area of application is the concern of this book.

1

Variable If, as we observe a characteristic, we find that it takes on different values in different persons, places, or things, we label the characteristic a *variable*. We do this for the simple reason that the characteristic is not the same when observed in different possessors of it. Some examples of variables include the heights of adult males, the weights of preschool children, and the ages of patients seen in a dental clinic.

Quantitative Variables A *quantitative variable* is one that can be measured in the usual sense. We can, for example, obtain measurements on the heights of adult males, the weights of preschool children, and the ages of patients seen in a dental clinic. These are examples of *quantitative variables*. Measurements made on quantitative variables convey the concept of amount.

Qualitative Variables Some characteristics are not capable of being measured in the sense that height, weight, and age are measured. Many characteristics can be categorized only, as for example, when an ill person is given a medical diagnosis, a person is designated as belonging to an ethnic group, or a person, place, or object is said to possess or not to possess some characteristic of interest. In such cases measuring consists of categorizing. We refer to variables of this kind as *qualitative variables*. Measurements made on qualitative variables convey the concept of attribute.

Although, in the case of qualitative variables, measurement in the usual sense of the word is not achieved, we can count the number of persons, places, or things belonging to various categories. A hospital administrator, for example, can count the number of patients admitted during a day under each of the various admitting diagnoses.

Random Variable Whenever we determine the height, weight, or age of an individual, the result is frequently referred to as a *value* of the respective variable. When the values obtained arise as a result of chance factors, the variable is called a *random variable*. Values resulting from measurement procedures are often referred to as *observations* or, simply, *measurements*.

Discrete Random Variable Variables may be characterized further as to whether they are *discrete* or *continuous*. Since mathematically rigorous definitions of discrete and continuous variables are beyond the level of this book, we offer, instead, nonrigorous definitions and give an example of each.

A discrete random variable is characterized by gaps or interruptions in the values that it can assume. These gaps or interruptions indicate the absence of values between particular values that the variable can assume. Some examples illustrate the point. The number of daily admissions to a general hospital is a discrete random variable since the number of admissions each day must be represented by a whole number, such as 0, 1, 2, 3, and so on. The number of admissions on a given day cannot be a number such as 1.5, 2.997, or 3.3333. The number of decayed, missing, or filled teeth per child in an elementary school is another example of a discrete variable.

Continuous Random Variable   *A continuous random variable does not possess the gaps or interruptions characteristic of a discrete random variable.* A continuous random variable can assume any value within a specified interval of values assumed by the variable. Examples of continuous variables include the various measurements that can be made on individuals such as height, weight, and skull circumference. No matter how close together the observed heights of two people, for example, we can, theoretically, find another person whose height falls somewhere in between.

Because of the limitations of available measuring instruments, however, observations on variables that are inherently continuous are recorded as if they were discrete. Height, for example, is usually recorded to the nearest one-quarter, one-half, or whole inch, whereas, with a perfect measuring device, such a measurement could be made as precise as desired.

Population   The average person thinks of a population as a collection of entities, usually people. A population or collection of entities may, however, consist of animals, machines, plants, or cells. For our purposes, we define a *population of entities as the largest collection of entities for which we have an interest at a particular time.* If we take a measurement of some variable on each of the entities in a population, we generate a population of values of that variable. We may, therefore, define a *population of values as the largest collection of values of a random variable for which we have an interest at a particular time.* If, for example, we are interested in the weights of all the children enrolled in a certain county elementary school system, our population consists of all these weights. If our interest lies only in the weights of first grade students in the system, we have a different population—weights of first grade students enrolled in the school system. Hence, populations are determined or defined by our sphere of interest. Populations may be *finite* or *infinite.* If a population of values consists of a fixed number of these values, the population is said to be *finite.* If, on the other hand, a population consists of an endless succession of values, the population is an *infinite* one.

Sample   A sample may be defined simply as *a part of a population.* Suppose our population consists of the weights of all the elementary school children enrolled in a certain county school system. If we collect for analysis the weights of only a fraction of these children, we have only a part of our population of weights, that is, we have a *sample.* There are many kinds of samples that can be selected from a population. They are discussed in succeeding chapters.

1.3   COMPUTERS AND BIOSTATISTICAL ANALYSIS

The relatively recent widespread use of computers has had a tremendous impact on health sciences research in general and biostatistical analysis in particular. The necessity to perform long and tedious arithmetic computations as part of the statistical analysis of data lives only in the memory of those researchers and practitioners whose careers antedate the so-called "computer revolution." Computers can perform more calculations faster and far more accurately than can human technicians. The use of computers makes it possible for investigators to devote more time to the improvement of the quality of raw data and the interpretation of the results.

Canned computer programs are available for performing most of the descriptive and inferential statistical procedures that the average investigator is likely to need. Some widely used "packages" of statistical procedures are *BMDP: Biomedical Computer Programs* (1), *SPSS Statistical Package for the Social Sciences* (2). *The IMSL Library* (3), *Minitab* (4), and *SAS* (5). Dixon and Jennrich (6), in a review article, describe 38 different packaged statistical programs. They provide information regarding the machines on which the programs may be run, core memory requirements, program languages, available documentation, and sources of additional information. The calculations for many of the exercises in this book can be performed by the programs in these and other statistical program packages.

Statistical programs differ with respect to their input requirements, their output formats, and the specific calculations they will perform. The reader who wishes to use the computer to obtain solutions to the exercises in this book should become familiar with the programs that can be used at his or her computer installation to determine, first of all, if there is an existing program that will do the required calculations. Once an appropriate program has been located, its input requirements should be studied carefully preparatory to entering the data of the exercises into the computer. Finally, the program's output format should be studied so that the proper interpretation of the results can be made. Readers who have studied a computer language may, in some instances, wish to write their own computer programs for use with the exercises.

The reader who uses a microcomputer will find an abundance of statistical software packages from which to choose. For information on those that are currently available one may consult some of the many microcomputer magazines that are on the market. These magazines not only carry advertisements for available statistical software packages for use on microcomputers, but also, from time to time, feature review articles that compare the strengths and weaknesses of the packages.

The usefulness of the computer in the health sciences is not limited to statistical analysis. The reader interested in learning more about the use of computers in biology, medicine, and the other health sciences will find the books by Krasnoff (7), Ledley (8), Lindberg (9), Sterling and Pollack (10), and Taylor (11) helpful.

Current developments in the use of computers in biology, medicine, and related fields are reported in several periodicals devoted to the subject. A few such periodicals are *Computers in Biology and Medicine*, *Computers and Biomedical Research*, *International Journal of Bio-Medical Computing*, *Computer Programs in Biomedicine*, and *Computers and Medicine*.

1.4 THE ORDERED ARRAY

When measurements of a random variable are taken on the entities of a population, the resulting values usually come to the researcher or statistician as a mass of unordered data. Unless the number of observations is extremely small it will be unlikely that these data will impart much information until they have been put in some kind of order. If the number of observations is not too large, a first step in organizing these data is the preparation of an *ordered array*. An *ordered array* is a listing of the values of a collection (either population or sample) in order of magnitude from the smallest value to the largest value.

Example 1.4.1 Table 1.4.1 consists of a list of the ages of patients admitted to a chronic disease hospital during a certain month. The table reflects the order in which the patients were admitted. As can be seen, it requires considerable

TABLE 1.4.1

Ages (in Years) of Patients Admitted to a Chronic Disease Hospital During a Certain Month

Number	Age	Number	Age	Number	Age	Number	Age
1	10	26	48	51	63	76	53
2	22	27	39	52	53	77	33
3	24	28	6	53	88	78	3
4	42	29	72	54	48	79	85
5	37	30	14	55	52	80	8
6	77	31	36	56	87	81	51
7	89	32	69	57	71	82	60
8	85	33	40	58	51	83	58
9	28	34	61	59	52	84	9
10	63	35	12	60	33	85	14
11	9	36	21	61	46	86	74
12	10	37	54	62	33	87	24
13	7	38	53	63	85	88	87
14	51	39	58	64	22	89	7
15	2	40	32	65	5	90	81
16	1	41	27	66	87	91	30
17	52	42	33	67	28	92	76
18	7	43	1	68	2	93	7
19	48	44	25	69	85	94	6
20	54	45	22	70	61	95	27
21	32	46	6	71	16	96	18
22	29	47	81	72	42	97	17
23	2	48	11	73	69	98	53
24	15	49	56	74	7	99	70
25	46	50	5	75	10	100	49

TABLE 1.4.2

An Ordered Array of the Ages of Patients Admitted to a Chronic Disease Hospital During a Certain Month

1	14	37	58
1	15	39	60
2	16	40	61
2	17	42	61
2	18	42	63
3	21	46	63
5	22	46	69
5	22	48	69
6	22	48	70
6	24	48	71
6	24	49	72
7	25	51	74
7	27	51	76
7	27	51	77
7	28	52	81
7	28	52	81
8	29	52	85
9	30	53	85
9	32	53	85
10	32	53	85
10	33	53	87
10	33	54	87
11	33	54	87
12	33	56	88
14	36	58	89

searching to ascertain such elementary information as the age of the youngest and oldest patient admitted. Table 1.4.2 presents the same data in the form of an ordered array. If additional computations and organization of the data have to be done by hand, the work may be facilitated by working from the ordered array. If the data are to be analyzed by a computer, it may be undesirable to prepare an ordered array.

1.5 GROUPED DATA—THE FREQUENCY DISTRIBUTION

Although a set of observations can be made more comprehensible and meaningful by means of an ordered array, further useful summarization may be achieved by grouping the data. To group a set of observations we select a set of contiguous, nonoverlapping intervals such that each value in the set of observations can be placed in one, and only one, of the intervals. These intervals are usually referred to as *class intervals*.

One of the first things to consider, when data are to be grouped, is how many intervals to include. Too few intervals are undesirable because of the resulting loss of information. On the other hand, if too many intervals are used, the objective of summarization will not be met. The best guide to this, as well as to other decisions to be made in grouping data, is your knowledge of the data. It may be that class intervals have been determined by precedent, as in the case of annual tabulations, when the class intervals of previous years are maintained for comparative purposes. Those who wish guidance in this matter may consult a formula given by Sturges (12). This formula gives $k = 1 + 3.322(\log_{10} n)$, where k stands for the number of class intervals and n is the number of values in the data set under consideration. The answer obtained by applying *Sturges' rule* should not be regarded as final, but should be considered as a guide only. The number of class intervals specified by the rule should be increased or decreased for convenience and clear presentation.

Suppose, for example, that we have a sample of 275 observations that we want to group. The logarithm to the base 10 of 275 is 2.4393. Applying Sturges' formula gives $k = 1 + 3.322(2.4393) \approx 9$. In practice, other considerations might cause us to use 8 or fewer or perhaps 10 or more class intervals.

Another question that must be decided regards the width of the class intervals. Although this is sometimes impossible, class intervals generally should be of the same width. This width may be determined by dividing the range by k, the number of class intervals. Symbolically, the class interval width is given by

$$w = \frac{R}{k} \tag{1.5.1}$$

where R (the range) is the difference between the smallest and the largest observation in the data set. As a rule this procedure yields a width that is inconvenient for use. Again, we may exercise our good judgment and select a width (usually close to the one given by Equation 1.5.1) that is more convenient.

Example 1.5.1   Table 1.5.1 shows the weights in ounces of malignant tumors removed from the abdomens of 57 subjects. To get an idea as to the number of class intervals to use, we can apply Sturges' rule to obtain

$$k = 1 + 3.322(\log 57)$$
$$= 1 + 3.322(1.7559)$$
$$\approx 7$$

Now let us divide the range by 7 to get some idea about the class interval width. We have

$$\frac{R}{k} = \frac{79 - 12}{7} = \frac{67}{7} = 9.6$$

TABLE 1.5.1

Weights in Ounces of Malignant Tumors
Removed from the Abdomens
of 57 Subjects

68	65	12	22
63	43	32	43
42	25	49	27
27	74	38	49
30	51	42	28
36	36	27	23
28	42	31	19
32	28	50	46
79	31	38	30
27	28	21	43
22	25	16	49
23	45	24	12
24	12	69	
25	57	47	
44	51	23	

It is apparent that a class interval width of 10 will be more convenient to use, as well as more meaningful to the reader. We may now construct our intervals. Since the smallest value in Table 1.5.1 is 12 and the largest value is 79, we may begin our intervals with 10 and end with 79. This gives the following intervals:

10–19

20–29

30–39

40–49

50–59

60–69

70–79

We see that there are seven of these intervals, the number suggested by Sturges' rule.

Determining the number of values falling into each class interval is merely a matter of looking at the values one by one and of placing a tally mark beside the appropriate interval. When we do this, we have Table 1.5.2.

A table of this kind is called a *frequency distribution*. This table shows the way in which the values of the variable are distributed among the specified class intervals. By consulting it, we can determine the frequency of occurrence of values within any one of the class intervals shown.

TABLE 1.5.2

Frequency Distribution of Weights (in Ounces) of Malignant Tumors Removed from the Abdomens of 57 Subjects

Class interval		Frequency
10–19	ⅬⱵꝷ	5
20–29	ⅬⱵꝷ ⅬⱵꝷ ⅬⱵꝷ ////	19
30–39	ⅬⱵꝷ ⅬⱵꝷ	10
40–49	ⅬⱵꝷ ⅬⱵꝷ ///	13
50–59	////	4
60–69	////	4
70–79	//	2
Total		57

It may be useful at times to know the proportion, rather than the number, of values falling within a particular class interval. We obtain this information by dividing the number of values in the particular class interval by the total number of values. If, in our example, we wish to know the proportion of values between 30 and 39, inclusive, we divide 10 by 57, obtaining .18. Thus we say that 10 out of 57, or 10/57ths, or .18, of the values are between 30 and 39. Multiplying .18 by 100 gives us the percentage of values between 30 and 39. We can say, then, that 18 percent of the 57 values are between 30 and 39. We may refer to the proportion of values falling within a class interval as the *relative frequency of occurrence* of values in that interval.

In determining the frequency of values falling within two or more class intervals, we obtain the sum of the number of values falling within the class intervals of interest. Similarly, if we want to know the relative frequency of

TABLE 1.5.3

Frequency, Cumulative Frequency, Relative Frequency, and Cumulative Relative Frequency Distributions for Example 1.5.1

Class intervals	Frequency	Cumulative frequency	Relative frequency	Cumulative relative frequency
10–19	5	5	.0877	.0877
20–29	19	24	.3333	.4210
30–39	10	34	.1754	.5964
40–49	13	47	.2281	.8245
50–59	4	51	.0702	.8947
60–69	4	55	.0702	.9649
70–79	2	57	.0351	1.0000
Total	57		1.0000	

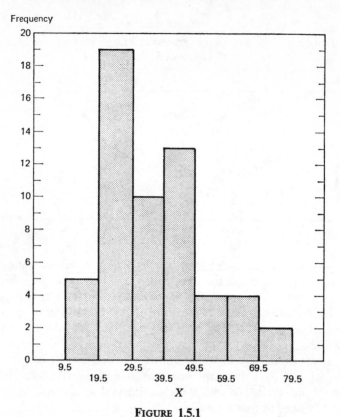

FIGURE 1.5.1

Histogram of Weights (in Ounces) of Malignant Tumors Removed from the Abdomens of 57 Subjects.

occurrence of values falling within two or more class intervals, we add the respective relative frequencies. We may *cumulate* the frequencies and relative frequencies to facilitate obtaining information regarding the frequency or relative frequency of values within two or more contiguous class intervals. Table 1.5.3 shows the data of Table 1.5.2 along with the *cumulative frequencies*, the *relative frequencies*, and *cumulative relative frequencies*.

Suppose that we are interested in the relative frequency of values between 40 and 69. We use the cumulative relative frequency column of Table 1.5.3 and subtract .5964 from .9649, obtaining .3685.

The Histogram We may display a frequency distribution (or a relative frequency distribution) graphically in the form of a *histogram* as shown in Figure 1.5.1. In constructing a histogram the values of the variable under consideration make up the horizontal axis, while the vertical axis has as its scale the frequency (or relative frequency if desired) of occurrence. Above each class interval on the horizontal axis a rectangular bar, or cell, as it is sometimes called, is erected so

TABLE 1.5.4

The Data of Table 1.5.2 Showing True Class Limits

True class limits	Frequency
9.5–19.5	5
19.5–29.5	19
29.5–39.5	10
39.5–49.5	13
49.5–59.5	4
59.5–69.5	4
69.5–79.5	2

that the height corresponds to the respective frequency. The cells of a histogram must be joined and, to accomplish this, we must take into account the true boundaries of the class intervals to prevent gaps from occurring between the cells of our graph.

The level of precision observed in reported data that are measured on a continuous scale indicates some order of rounding. The order of rounding reflects either the reporter's personal preference or the limitations of the measuring instrument employed. When a frequency distribution is constructed from the data, the class interval limits usually reflect the degree of precision of the raw data. This has been done in our illustrative example. We know, however, that some of the values falling in the second class interval, for example, when measured precisely, would probably be a little less than 20 and some would be a little greater than 29. Considering the underlying continuity of our variable, and assuming that the data were rounded to the nearest whole number, we find it convenient to think of 19.5 and 29.5 as the *true limits* of this second interval. The true limits for each of the class intervals, then, we take to be as shown in Table 1.5.4.

If we draw a graph using these class limits as the base of our rectangles, no gaps will result, and we will have the histogram shown in Figure 1.5.1.

Note that each observation is allotted one unit of the area of the histogram. Since we have 57 observations, the histogram consists of a total of 57 units. Each cell contains a certain proportion of the total area, depending on the frequency. The second cell, for example, contains nineteen fifty-sevenths of the area. This, as we have learned, is the relative frequency of occurrence of values between 19.5 and 29.5. From this we see that subareas of the histogram defined by the cells correspond to the frequencies of occurrence of values between the horizontal scale boundaries of the areas. The ratio of a particular subarea to the total area of the histogram is equal to the relative frequency of occurrence of values between the corresponding points on the horizontal axis.

Computer Analysis Many computer software packages contain programs for the construction of histograms. One such package is MINITAB. Figure 1.5.2 shows

```
MIDDLE OF      NUMBER OF
INTERVAL       OBSERVATIONS
   14.5            5        *****
   24.5           19        ******************
   34.5           10        **********
   44.5           13        *************
   54.5            4        ****
   64.5            4        ****
   74.5            2        **
```

FIGURE 1.5.2

Computer-constructed Histogram Using the Data of Example 1.5.1 and the MINITAB Software Package.

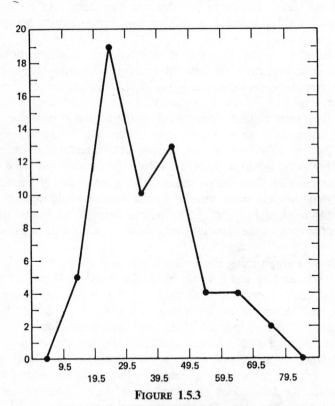

FIGURE 1.5.3

A frequency Polygon. Weights (in Ounces) of Malignant Tumors Removed from the Abdomens of 57 Subjects.

the histogram constructed from the data of Example 1.5.1 by the MINITAB program. After the data were entered into the computer, the computer was instructed to construct a histogram with a first midpoint of 14.5 and an interval width of 10.

The Frequency Polygon   A frequency distribution can be portrayed graphically in yet another way by means of a *frequency polygon*. To draw a frequency polygon we first place a dot above the midpoint of each class interval represented on the horizontal axis of a graph like the one shown in Figure 1.5.1. The height above the horizontal axis of a given dot corresponds to the frequency of the relevant class interval. Connecting the dots by straight lines produces the frequency polygon. Figure 1.5.3 is the frequency polygon for the data of Example 1.5.1.

Note that the polygon is brought down to the horizontal axis at the ends at points that would be the midpoints if there were an additional cell at each end of the corresponding histogram. This allows for the total area to be enclosed. The total area under the frequency polygon is equal to the area under the histogram. Figure 1.5.4 shows the frequency polygon of Fig. 1.5.3 superimposed on the histogram of Figure 1.5.1. This figure allows you to see, for the same set of data, the relationship between the two graphic forms.

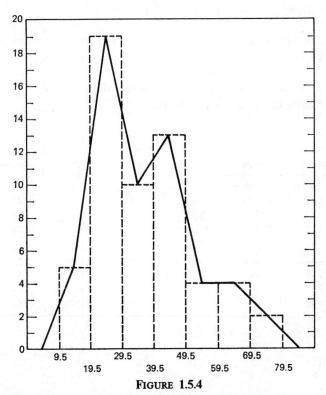

FIGURE 1.5.4

Frequency Polygon and Histogram for Example 1.5.1.

EXERCISES

1.5.1 The following are the fasting blood glucose levels of 100 children:

56	61	57	77	62	75	63	55	64	60
60	57	61	57	67	62	69	67	68	59
65	72	65	61	68	73	65	62	75	80
66	61	69	76	72	57	75	68	81	64
69	64	66	65	65	76	65	58	65	64
68	71	72	58	73	55	73	79	81	56
65	60	65	80	66	80	68	55	66	71
72	73	73	75	75	74	66	68	73	65
73	74	68	59	69	55	67	65	67	63
67	56	67	62	65	75	62	63	63	59

Prepare:

(a) A frequency distribution.   (b) A histogram.
(c) A relative frequency distribution. (d) A frequency polygon.

1.5.2 Using the data of Table 1.4.1, prepare:

(a) A frequency distribution.   (b) A relative frequency distribution.
(c) A histogram.   (d) A frequency polygon.

1.5.3 The following are scores made on an intelligence test by a group of children who participated in an experiment:

Child number	Score	Child number	Score	Child number	Score	Child number	Score
1	114	16	90	31	137	46	118
2	115	17	89	32	120	47	110
3	113	18	106	33	138	48	108
4	112	19	104	34	111	49	134
5	113	20	126	35	100	50	118
6	132	21	127	36	116	51	114
7	130	22	115	37	101	52	142
8	128	23	116	38	111	53	120
9	122	24	109	39	110	54	119
10	121	25	108	40	137	55	143
11	126	26	122	41	119	56	133
12	117	27	123	42	115	57	85
13	115	28	149	43	83	58	117
14	88	29	140	44	109	59	147
15	113	30	121	45	117	60	102

Construct:

(a) A frequency distribution.   (b) A relative frequency distribution.
(c) A histogram.   (d) A frequency polygon.

1.5.4 Seventy-five employees of a general hospital were asked to perform a certain task. The time in minutes required for each employee to complete the task was recorded. The results were as shown below.

Employee number	Time	Employee number	Time	Employee number	Time
1	1.3	26	2.2	51	3.2
2	1.5	27	2.3	52	3.0
3	1.4	28	2.6	53	3.4
4	1.5	29	2.8	54	3.4
5	1.7	30	2.1	55	3.2
6	1.0	31	2.3	56	4.5
7	1.3	32	2.4	57	4.6
8	1.7	33	2.0	58	4.9
9	1.2	34	2.8	59	4.1
10	1.8	35	2.2	60	4.6
11	1.1	36	2.5	61	4.2
12	1.0	37	2.9	62	4.0
13	1.8	38	2.0	63	4.3
14	1.6	39	2.9	64	4.8
15	2.1	40	2.5	65	4.5
16	2.1	41	3.6	66	5.1
17	2.1	42	3.1	67	5.7
18	2.1	43	3.5	68	5.1
19	2.4	44	3.7	69	5.4
20	2.9	45	3.7	70	5.7
21	2.7	46	3.4	71	6.7
22	2.3	47	3.1	72	6.8
23	2.8	48	3.5	73	6.6
24	2.0	49	3.6	74	6.0
25	2.7	50	3.5	75	6.1

From these data construct:

(a) A frequency distribution.   (b) A relative frequency distribution.
(c) A histogram.   (d) A frequency polygon.

1.5.5 The following table shows the number of hours 45 hospital patients slept following the administration of a certain anesthetic.

7	10	12	4	8	7	3	8	5
12	11	3	8	1	1	13	10	4
4	5	5	8	7	7	3	2	3
8	13	1	7	17	3	4	5	5
3	1	17	10	4	7	7	11	8

From these data construct:

(a) A frequency distribution.   (b) A relative frequency distribution.
(c) A histogram.   (d) A frequency polygon.

1.5.6 The following are the number of babies born during a year in 60 community hospitals.

30	55	27	45	56	48	45	49	32	57	47	56
37	55	52	34	54	42	32	59	35	46	24	57
32	26	40	28	53	54	29	42	42	54	53	59
39	56	59	58	49	53	30	53	21	34	28	50
52	57	43	46	54	31	22	31	24	24	57	29

From these data construct:

(a) A frequency distribution.   (b) A relative frequency distribution.
(c) A frequency polygon.

1.5.7 In a study of physical endurance levels of male college freshmen the following composite endurance scores based on several exercise routines were collected.

254	281	192	260	212	179	225	179	181	149
182	210	235	239	258	166	159	223	186	190
180	188	135	233	220	204	219	211	245	151
198	190	151	157	204	238	205	229	191	200
222	187	134	193	264	312	214	227	190	212
165	194	206	193	218	198	241	149	164	225
265	222	264	249	175	205	252	210	178	159
220	201	203	172	234	198	173	187	189	237
272	195	227	230	168	232	217	249	196	223
232	191	175	236	152	258	155	215	197	210
214	278	252	283	205	184	172	228	193	130
218	213	172	159	203	212	117	197	206	198
169	187	204	180	261	236	217	205	212	218
191	124	199	235	139	231	116	182	243	217
251	206	173	236	215	228	183	204	186	134
188	195	240	163	208					

From these data construct:

(a) A frequency distribution.   (b) A relative frequency distribution.
(c) A frequency polygon.   (d) A histogram.

1.6 MEASURES OF CENTRAL TENDENCY

Although frequency distributions serve useful purposes, there are many situations that require other types of data summarization. What we need in many instances is the ability to summarize data by means of just a few *descriptive measures*. Descriptive measures may be computed from the data of a sample or the data of a population. To distinguish between them we have the following definitions.

1. A descriptive measure computed from the data of a sample is called a *statistic*.
2. A descriptive measure computed from the data of a population is called a *parameter*.

Several types of descriptive measures can be computed from a set of data. In this chapter, however, we limit discussion to *measures of central tendency* and *measures of dispersion*. We consider measures of central tendency in this section and measures of dispersion in the following one.

In each of the measures of central tendency, of which we discuss three, we have a single value that is considered to be typical of the set of data as a whole. The three most commonly used measures of central tendency are the *mean*, the *median*, and the *mode*.

Arithmetic Mean The most familiar measure of central tendency is the arithmetic mean. It is the descriptive measure most people have in mind when they speak of the "average." The adjective, arithmetic, distinguishes this mean from other means that can be computed. Since we are not covering these other means in this book, there should be no cause for confusion if we refer to the arithmetic mean simply as the *mean*. The mean is obtained by adding all the values in a population or sample and dividing by the number of values that are added. To obtain the mean age of the population of 100 patients represented in Table 1.4.1, we proceed as follows:

$$\text{mean age} = \frac{10 + 22 + \cdots + 70 + 49}{100} = \frac{3920}{100} = 39.20$$

The three dots in the numerator represent the values not shown to save space.

It will be convenient if we can generalize the procedure for obtaining the mean and, also, represent the procedure in a more compact notational form. Let us begin by designating the random variable of interest by the capital letter X. In our present illustration we let X represent the random variable, age. Specific values of a random variable will be designated by the lowercase letter x. To distinguish one value from another we attach a subscript to the x and let the subscript refer to the first, the second, the third value, and so on. For example, from Table 1.4.1 we have

$$x_1 = 10, \qquad x_2 = 22, \ldots, \qquad \text{and} \qquad x_{100} = 49$$

In general, a typical value of a random variable will be designated by x_i and the final value, in a finite population of values, by x_N. Finally, we will use the Greek letter μ to stand for the population mean. We may now write the general formula for a finite population mean as follows:

$$\mu = \frac{\sum_{i=1}^{N} x_i}{N} \tag{1.6.1}$$

The symbol $\sum_{i=1}^{N}$ instructs us to add all values of the variable from the first to the last. This symbol, $\sum$, called the *summation sign*, will be used extensively in this book. When from the context it is obvious which values are to be added, the symbols above and below $\sum$ will be omitted.

When we compute the mean for a sample of values, the procedure just outlined is followed with some modifications in notation. We use $\bar{x}$ to designate the sample mean and n to indicate the number of values in the sample. The sample mean then is expressed as

$$\bar{x} = \frac{\sum_{i=1}^{n} x_i}{n} \tag{1.6.2}$$

Suppose we have available a sample consisting of the following five ($n = 5$) observations from Table 1.4.1. Substitution of our sample data into

POPULATION OBSERVATION	SAMPLE OBSERVATION	
Number	Number	Value
x_{12}	x_1	10
x_{20}	x_2	54
x_{36}	x_3	21
x_{62}	x_4	33
x_{98}	x_5	53

Equation 1.6.2 gives

$$\bar{x} = \frac{\sum_{i=1}^{n} x_i}{n} = \frac{10 + 54 + 21 + 33 + 53}{5} = \frac{171}{5} = 34.2$$

The arithmetic mean possesses certain properties, some desirable and some not so desirable. These properties include the following.

1. Uniqueness. For a given set of data there is one and only one arithmetic mean.

2. Simplicity. The arithmetic mean is easily understood and easy to compute.

3. Since each and every value in a set of data enters into the computation of the mean, it is affected by each value. Extreme values, therefore, have an influence on the mean and, in some cases, can so distort it that it becomes undesirable as a measure of central tendency.

As an example of how extreme values may affect the mean, consider the following situation. Suppose the five physicians who practice in an area are surveyed to determine their charges for a certain procedure. Assume that they report these charges: $15, $15, $15.50, $15.50, and $80. The mean charge for the five physicians is found to be $28.20, a value that is not very representative of the set of data as a whole. The single atypical value had the effect of inflating the mean.

Median   The median of a finite set of values is that value which divides the set into two equal parts such that the number of values equal to or greater than the median is equal to the number of values equal to or less than the median. If the number of values is odd, the median will be the middle value when all values have been arranged in order of magnitude. When the number of observations is even, there is no single middle observation, but two middle observations. In this case the median is taken to be the mean of these two middle observations, when all observations have been arranged in the order of their magnitude.

Let us illustrate this by finding the median of the data in Table 1.4.2. Here the values are already ordered so we need only to find the two middle values. These are observation numbers 50 and 51. The values are 36 and 37, and the median, therefore, is $(36 + 37)/2 = 36.5$.

Let us now obtain the median of the sample consisting of the values 10, 54, 21, 33, and 53. Arraying these in the order of their magnitude we have 10, 21, 33, 53, and 54. Since this is an odd number of values, the median is the middle value, or 33.

Properties of the median include the following:

1. Uniqueness. As is true with the mean, there is only one median for a given set of data.

2. Simplicity. The median is easy to calculate.

3. It is not as drastically affected by extreme values as is the mean.

The Mode   The mode of a set of values is that value which occurs most frequently. If all the values are different there is no mode; on the other hand, a set of values may have more than one mode. Again, looking at the data in Table 1.4.2 we find that 7, which occurs five times, is the most frequently occurring value, and, therefore, is the mode.

For an example of a set of values that has more than one mode, let us consider a laboratory with 10 employees whose ages are 20, 21, 20, 20, 34, 22, 24, 27, 27, and 27. We could say that these data have two modes, 20 and 27. The sample consisting of the values 10, 21, 33, 53, and 54 has no mode since all the values are different.

The mode may be used for describing qualitative data. For example, suppose the patients seen in a mental health clinic during a given year received one of the following diagnoses: mental retardation, organic brain syndrome, psychosis, neurosis, and personality disorder. The diagnosis occurring most frequently in the group of patients would be called the modal diagnosis.

EXERCISES

1.6.1 The following are the fasting blood glucose levels of a sample of 10 children.

Number	Value	Number	Value
1	56	6	65
2	62	7	65
3	63	8	68
4	65	9	70
5	65	10	72

Compute:

(a) The mean.   (b) The median.
(c) The mode.

1.6.2 The following are the postsurgery weights of a sample of 10 experimental animals.

Number	Weight (kg)	Number	Weight (kg)
1	13.2	6	14.4
2	15.4	7	13.6
3	13.0	8	15.0
4	16.6	9	14.6
5	16.9	10	13.1

Find:

(a) The mean.   (b) The median.

1.6.3 A sample of 15 patients making initial visits to a county health department traveled these distances.

Patient	Distance (miles)	Patient	Distance (miles)
1	5	8	6
2	9	9	13
3	11	10	7
4	3	11	3
5	12	12	15
6	13	13	12
7	12	14	15
		15	5

Find:

(a) The mean distance traveled by these patients.
(b) The median distance traveled.

1.6.4 A sample of 11 patients admitted for diagnosis and evaluation to a newly opened psychiatric ward of a general hospital experienced the following lengths of stay.

Number	Length of stay (days)	Number	Length of stay (days)
1	29	6	14
2	14	7	28
3	11	8	14
4	24	9	18
5	14	10	22
		11	14

Find:

(a) The mean length of stay for these patients.
(b) The median.
(c) The mode.

1.6.5 Twenty patients in the convalescent wing of a general hospital were offered a choice of four meats for lunch. Their choices were as follows: chicken, fish, fish, liver, chicken, chicken, meat loaf, chicken, fish, meat loaf, liver, fish, meat loaf, chicken, chicken, chicken, liver, fish, meat loaf, chicken.
What was the modal choice?

1.6.6 The following are the numbers of minutes required for 15 male college students to run 1 mile.

5.77	6.86	5.00	7.33	7.67
5.48	7.24	6.05	7.65	6.75
6.66	8.31	6.43	5.75	5.73

From these data compute:

(a) The mean.   (b) The median.

1.7 MEASURES OF DISPERSION

The *dispersion* of a set of observations refers to the variety that they exhibit. If all the values are the same, there is no dispersion; if they are not all the same, dispersion is present in the data. The amount of dispersion may be small, when the values, though different, are close together. Figure 1.7.1 shows the frequency polygons for two populations that have equal means but different amounts of variability. Population B, which is more variable than population A, is more spread out. If the values are widely scattered, the dispersion is greater. Other terms used synonymously with dispersion include *variation*, *spread*, and *scatter*.

The Range   One way to measure the variation in a set of values is to compute the *range*. The range is the difference between the smallest and largest value in a set of observations. If we denote the range by R, the largest value by x_L, and the

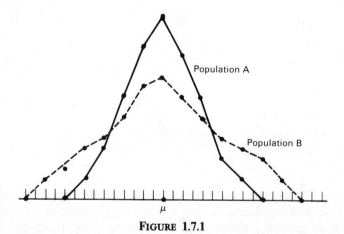

FIGURE 1.7.1

Two Frequency Distributions with Equal Means But Different Amounts of Dispersion.

smallest value by x_S, we compute the range as follows:

$$R = x_L - x_S \qquad (1.7.1)$$

Using the data in Table 1.4.2 we have

$$R = 89 - 1 = 88$$

The usefulness of the range is limited. The fact that it takes into account only two values causes it to be a poor measure of dispersion. The main advantage in using the range is the simplicity of its computation.

The Variance   When the values of a set of observations lie close to their mean, the dispersion is less than when they are scattered over a wide range. Since this is true, it would be intuitively appealing if we could measure dispersion relative to the scatter of the values about their mean. Such a measure is realized in what is known as the *variance*. In computing the variance, we subtract the mean from each of the values, square the differences, and then add them up. This sum of the squared deviations of the values from their mean is divided by the sample size, minus 1, to obtain the sample variance. Letting s^2 stand for the sample variance, the procedure may be written in notational form as follows:

$$s^2 = \frac{\sum_{i=1}^{n} (x_i - \bar{x})^2}{n - 1} \qquad (1.7.2)$$

Let us illustrate by computing the variance of our now familiar sample that we drew from Table 1.4.1.

$$s^2 = \frac{(10 - 34.2)^2 + (54 - 34.2)^2 + (21 - 34.2)^2 + (33 - 34.2)^2 + (53 - 34.2)^2}{4}$$

$$= \frac{1506.8}{4} = 376.7$$

The reason for dividing by $n - 1$ rather than n, as we might have expected, is the theoretical consideration referred to as *degrees of freedom*. In computing the variance, we say that we have $n - 1$ *degrees of freedom*. We reason as follows. The sum of the deviations of the values from their mean is equal to zero, as can be shown. If, then, we know the values of $n - 1$ of the deviations from the mean, we know the nth one, since it is automatically determined because of the necessity for all n values to add to zero. From a practical point of view, dividing the squared differences by $n - 1$ rather than n is necessary in order to use the sample variance in the inference procedures discussed later. The concept of degrees of freedom will be discussed again later. Students interested in pursuing the matter further at this time should refer to the article by Walker (13).

When the number of observations is large, the use of Equation 1.7.2 can be tedious. The following formula may prove to be less troublesome, especially when a desk calculator or hand-held calculator is used.

$$s^2 = \frac{n \sum_{i=1}^{n} x_i^2 - \left(\sum_{i=1}^{n} x_i \right)^2}{n(n-1)} \tag{1.7.3}$$

When we compute the variance from a population of values, the procedures outlined above are followed except that we divide by N rather than $N - 1$. If we let σ^2 stand for the finite population variance, the definitional and computational formulas, respectively, are as follows:

$$\sigma^2 = \frac{\sum_{i=1}^{N} (x_i - \mu)^2}{N} \tag{1.7.4}$$

$$\sigma^2 = \frac{N \sum_{i=1}^{N} x_i^2 - \left(\sum_{i=1}^{N} x_i \right)^2}{N \cdot N} \tag{1.7.5}$$

The variance represents squared units and, therefore, is not an appropriate measure of dispersion when we wish to express this concept in terms of the original units. To obtain a measure of dispersion in original units, we merely take the square root of the variance. The result is called the *standard deviation*. In general, the standard deviation of a sample is given by

$$s = \sqrt{s^2} = \sqrt{\frac{\sum_{i=1}^{n} (x_i - \bar{x})^2}{n-1}} \tag{1.7.6}$$

The standard deviation of a finite population is obtained by taking the square root of the quantity obtained by Equation 1.7.4.

The Coefficient of Variation   The standard deviation is useful as a measure of variation within a given set of data. When one desires to compare the dispersion in two sets of data, however, comparing the two standard deviations may lead to fallacious results. It may be that the two variables involved are measured in different units. For example, we may wish to know, for a certain population, whether serum cholesterol levels, measured in milligrams per 100 ml, are more variable than body weight, measured in pounds.

Furthermore, although the same unit of measurement is used, the two means may be quite different. If we compare the standard deviation of weights of first grade children with the standard deviation of weights of high school freshmen, we may find that the latter standard deviation is numerically larger than the former, because the weights themselves are larger, not because the dispersion is greater.

What is needed in situations like these is a measure of relative variation rather than absolute variation. Such a measure is found in the *coefficient of variation*, which expresses the standard deviation as a percentage of the mean. The formula is given by

$$\text{C.V.} = \frac{s}{\bar{x}}(100) \qquad\qquad (1.7.7)$$

We see that, since the mean and standard deviations are expressed in the same unit of measurement, the unit of measurement cancels out in computing the coefficient of variation. What we have, then, is a measure that is independent of the unit of measurement.

Suppose two samples of human males yield the following results.

	Sample 1	Sample 2
Age	25 years	11 years
Mean weight	145 pounds	80 pounds
Standard deviation	10 pounds	10 pounds

A comparison of the standard deviations might lead one to conclude that the two samples possess equal variability. If we compute the coefficients of variation, however, we have for the 25-year-olds

$$\text{C.V.} = \tfrac{10}{145}(100) = 6.9$$

and for the 11-year-olds

$$\text{C.V.} = \tfrac{10}{80}(100) = 12.5$$

If we compare these results we get quite a different impression.

The coefficient of variation is also useful in comparing the results obtained by different persons who are conducting investigations involving the same variable.

EXERCISES

1.7.1   Refer to Exercise 1.6.1 and compute:

(a)   The range.   (b)   The variance, s^2.
(c)   The standard deviation, s.

133,280

1.7.2   Refer to Exercise 1.6.2. Compute:

(a)   The range.   (b)   The variance, s^2.
(c)   The standard deviation, s.   (d)   The coefficient of variation.

1.7.3   Refer to Exercise 1.6.3. Compute:

(a)   The range.   (b)   The variance, s^2.
(c)   The standard deviation, s.   (d)   The coefficient of variation.

1.7.4   Refer to Exercise 1.6.4. Compute:

(a)   The range.   (b)   The variance, s^2.
(c)   The standard deviation, s.   (d)   The coefficient of variation.

1.7.5   The following are the strength of right grip scores (kilograms) of 12 postpubescent adolescent males: 30, 21, 62, 18, 53, 29, 65, 19, 62, 56, 63, 64. Compute:

(a)   The mean.   (b)   The median.
(c)   The range.   (d)   The variance.
(e)   The standard deviation.

1.8   MEASURES OF CENTRAL TENDENCY COMPUTED FROM GROUPED DATA

After data have been grouped, it may be desirable to compute some of the descriptive measures, such as the mean and variance. Frequently an investigator does not have access to the raw data in which he or she is interested, but does have a frequency distribution. Data frequently are published in the form of a frequency distribution without an accompanying list of individual values or descriptive measures. Readers interested in a measure of central tendency or a measure of dispersion for these data must compute their own.

When data are grouped the individual observations lose their identity. By looking at a frequency distribution we are able to determine the number of observations falling into the various class intervals, but the actual values cannot be determined. Because of this we must make certain assumptions about the values when we compute a descriptive measure from grouped data. As a consequence of making these assumptions our results are only approximations to the true values.

The Mean Computed from Grouped Data   In calculating the mean from grouped data, we assume that all values falling into a particular class interval are located

TABLE 1.8.1

**Work Table for Computing the Mean from
the Grouped Data of Table 1.5.3.**

Class interval	Class midpoint m_i	Class frequency f_i	$m_i f_i$
10–19	14.5	5	72.5
20–29	24.5	19	465.5
30–39	34.5	10	345.0
40–49	44.5	13	578.5
50–59	54.5	4	218.0
60–69	64.5	4	258.0
70–79	74.5	2	149.0
Total		57	2086.5

at the *midpoint* of the interval. The midpoint of a class interval is obtained by computing the mean of the upper and lower limits of the interval. The midpoint of the first class interval of the distribution shown in Table 1.5.3, is equal to $(10 + 19)/2 = 29/2 = 14.5$. The midpoints of successive class intervals may be found by adding the class interval width to the previous midpoint. The midpoint of the second class interval in Table 1.5.3, for example, is equal to $14.5 + 10 = 24.5$.

To find the mean we multiply each midpoint by the corresponding frequency, sum these products, and divide by the sum of the frequencies. If the data represent a sample of observations, the computation of the mean may be shown symbolically as

$$\bar{x} = \frac{\sum_{i=1}^{k} m_i f_i}{\sum_{i=1}^{k} f_i} \qquad (1.8.1)$$

where k = the number of class intervals, m_i = the midpoint of the ith class interval, and f_i = the frequency of the ith class interval.

When we compute the mean from grouped data, it is convenient to prepare a work table such as Table 1.8.1, which has been prepared for the data of Table 1.5.3.

We may now compute the mean.

$$\bar{x} = \frac{\sum_{i=1}^{k} m_i f_i}{\sum_{i=1}^{k} f_i} = \frac{2086.5}{57} = 36.6$$

Computing a mean from a population of values grouped into a finite number of classes is carried out in exactly the same manner.

The Median–Grouped Data When computing the mean from grouped data, we assume that the values within a class interval are located at the midpoint; however, in computing the median, we assume that they are evenly distributed through the interval.

The first step in computing the median from grouped data is to locate the class interval in which it is located. We do this by finding the interval containing the $n/2$ value. Referring to the data of Example 1.5.1 again for illustrative purposes, we recall that there are 57 observations. The $n/2$ value is 28.5. Looking at Table 1.5.3 we see that the first two class intervals account for 24 of the observations and that 34 observations are accounted for by the first three class intervals. The median value, therefore, is in the third class interval. It is somewhere between 29.5 and 39.5 if we consider the true class limits. The question now is: How far must we proceed into this interval before reaching the median? Under the assumption that the values are evenly distributed through the interval, it seems reasonable that we should move a distance equal to $(28.5 - 24)/10$ of the total distance of the class interval, since after reaching the lower limit of the class interval containing the median we need $4\frac{1}{2}$ more observations, and there are a total of 10 observations in the interval. The value of the median then is equal to the value of the lower limit of the interval containing the median plus 4.5/10 of the interval width. For the data of Table 1.5.3 we compute the median to be $29.5 + (4.5/10)(10) = 34$.

In general, the median may be computed from grouped data by the following formula:

$$\text{median} = L_i + \frac{j}{f_i}(U_i - L_i) \tag{1.8.2}$$

where L_i = the true lower limit of the interval containing the median, U_i = the true upper limit of the interval containing the median, j = the number of observations still lacking to reach the median, after the lower limit of the interval containing the median has been reached, and f_i = the frequency of the interval containing the median.

The Mode—Grouped Data We have defined the mode of a set of values as the value that occurs most frequently. When designating the mode of grouped data, we usually refer to the *modal class*, where the modal class is the class interval with the highest frequency. In Example 1.5.1 the modal class would be the second class, $20 - 29$, or $19.5 - 29.5$, using the true class limits. If a single value for the mode of grouped data must be specified, it is taken as the midpoint of the modal class. In the present example this is 24.5. The assumption is made that all values in the interval fall at the midpoint.

The reader may wonder why we use $(n + 1)/2$ rather than $n/2$ to locate the median when we compute this measure from grouped data. In our example it

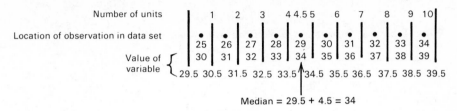

$$\text{Median} = 29.5 + 4.5 = 34$$

FIGURE 1.8.1

Diagram Showing Calculation of Median from Grouped Data in Table 1.5.3.

may appear that $57/2 = 28.5$ does not locate an observation that is halfway between the 28th and the 30th. If we were to use $(n + 1)/2$ with our example, we would compute $(57 + 1)/2 = 29$, which is halfway between the 28th and the 30th. In other words, we would say that the median observation is the 29th observation after the data have been ordered. This finding agrees with our rule for determining the median observation when we have an odd number of measurements.

To show that $57/2 = 28.5$ does locate an observation halfway between the 28th and 30th, we recall that when calculating the median from grouped data, we assume that the observations in any interval are evenly distributed over the interval. In our example, the 10 observations in the third class interval (see Table 1.5.3) are assumed to be evenly distributed over that internal. That is, we think of the interval as being divided into 10 equal units, each with a width of 1 ounce. One observation is thought of as being located at the midpoint of each of these units. This concept is illustrated in Figure 1.8.1.

When we move into the third interval a distance of $(28.5 - 24) = 4.5$ units we are at the 29th observation, or halfway between the 28th and 30th observations. We see in Figure 1.8.1 that the value of the variable at the point is 34, and we take this to be the median of the 57 observations.

1.9 THE VARIANCE AND STANDARD DEVIATION—GROUPED DATA

In calculating the variance and standard deviation from grouped data we assume that all values falling into a particular class interval are located at the midpoint of the interval. This, it will be recalled, is the assumption made in computing the mean and the mode. The variance of a sample, then, is given by

$$s^2 = \frac{\sum\limits_{i=1}^{k} (m_i - \bar{x})^2 f_i}{\sum\limits_{i=1}^{k} f_i - 1} \tag{1.9.1}$$

where the symbols have the definitions given in Equation 1.8.1.

TABLE 1.9.1

Work Table for Computing the Variance and Standard Deviation from the Grouped Data of Table 1.5.3

(1)	(2)	(3)	(4)	(5)	(6)	(7)	(8)
	Class	Class					
Class	midpoint	frequency					
interval	m_i	f_i	$(m_i - \bar{x})$	$(m_i - \bar{x})^2$	$(m_i - \bar{x})^2 f_i$	m_i^2	$m_i^2 f_i$
10–19	14.5	5	−22.1	488.41	2442.05	210.25	1051.25
20–29	24.5	19	−12.1	146.41	2781.79	600.25	11404.75
30–39	34.5	10	−2.1	4.41	44.10	1190.25	11902.50
40–49	44.5	13	7.9	62.41	811.33	1980.25	25743.25
50–59	54.5	4	17.9	320.41	1281.64	2970.25	11881.00
60–69	64.5	4	27.9	778.41	3113.64	4160.25	16641.00
70–79	74.5	2	37.9	1436.41	2872.82	5550.25	11100.50
Total		57		3236.87	13347.37		89724.25
$\bar{x} = 36.6$							

The following computing formula for the sample variance on occasion may be preferred:

$$s^2 = \frac{n \sum_{i=1}^{k} m_i^2 f_i - \left(\sum_{i=1}^{k} m_i f_i \right)^2}{n(n-1)} \qquad (1.9.2)$$

where

$$n = \sum_{i=1}^{k} f_i$$

The definitional formula for σ^2 is the same as for s^2 except that μ replaces $\bar{x}$ and the denominator is $\sum_{i=1}^{k} f_i$. The computational formula for σ^2 has $N \cdot N$ in the denominator rather than $n(n-1)$.

Let us now illustrate the computation of the variance and standard deviation, by both the definitional and the computational formula, using the data of Table 1.5.3. To do this, another work table such as Table 1.9.1 will be useful.

Dividing the total of column 6 by the total of column 3, less 1, we have

$$s^2 = \frac{\sum_{i=1}^{k} (m_i - \bar{x})^2 f_i}{\sum_{i=1}^{k} f_i - 1} = \frac{13347.37}{56} = 238.3459$$

The standard deviation is

$$s = \sqrt{238.3459} = 15.44$$

If we use the computing formula of Equation 1.9.2, we have

$$s^2 = \frac{57(89724.25) - (2086.5)^2}{57(56)} = 238.3459$$

EXERCISES

In the following exercises, treat the data sets as samples.

1.9.1 See Exercise 1.5.1. Find:

(a) The mean.

(b) The median.

(c) The modal class.

(d) The variance.

(e) The standard deviation.

1.9.2 See Exercise 1.5.2. Find:

(a) The mean.

(b) The median.

(c) The modal class.

(d) The variance.

(e) The standard deviation.

1.9.3 See Exercise 1.5.3. Find:

(a) The mean.

(b) The median.

(c) The modal class.

(d) The variance.

(e) The standard deviation.

1.9.4 See Exercise 1.5.4. Find:

(a) The mean.

(b) The median.

(c) The modal class.

(d) The variance.

(e) The standard deviation.

1.9.5 See Exercises 1.5.5. Find:

(a) The mean.

(b) The median.

(c) The variance.

(d) The standard deviation.

1.9.6   See Exercise 1.5.6. Find:

(a)   The mean.   (b)   The median.
(c)   The variance.   (d)   The standard deviation.

1.9.7   Refer to Exercise 1.5.7.
Find:

(a)   The mean.   (b)   The median.
(c)   The variance.   (d)   The standard deviation.

1.10   SUMMARY

In this chapter we define statistics as an area of study that is concerned with collecting and describing data and making inferences. Early in the chapter a basic statistics vocabulary is provided. Various descriptive statistical procedures are explained. These include the organization of data by means of the ordered array, the frequency distribution, the relative frequency distribution, the histogram, and the frequency polygon. The concepts of central tendency and variation are described, along with methods for computing their more common measures: the mean, median, mode, range, variance, and standard deviation. The concepts and methods are presented in a way that makes possible the handling of both grouped and ungrouped data.

In particular, the computer is a useful tool for calculating descriptive measures and constructing various distributions from large sets of data. The availability of a computer obviates the need for the grouped-data formulas of Sections 1.8 and 1.9 when the raw data are also available.

REVIEW QUESTIONS AND EXERCISES

1.   Explain what is meant by descriptive statistics.
2.   Explain what is meant by inferential statistics.
3.   Define:
   (a)   Statistics   (b)   Biostatistics
   (c)   Variable   (d)   Quantitative variable
   (e)   Qualitative variable   (f)   Random variable
   (g)   Population   (h)   Finite population
   (i)   Infinite population   (j)   Sample
   (k)   Discrete variable   (l)   Continuous variable
4.   What is an ordered array?

5. Define and compare the characteristics of the mean, the median, and the mode.

6. What are the advantages and limitations of the range as a measure of dispersion?

7. What is a frequency distribution?

8. What is a relative frequency distribution?

9. Explain the difference between a statistic and a parameter.

10. Explain the rationale for using $n - 1$ to compute the sample variance.

11. What is the purpose of the coefficient of variation?

12. What is the purpose of Sturges' rule?

13. What is a histogram?

14. What is a frequency polygon?

15. What assumptions does one make when computing the mean from grouped data? The median? The variance?

16. What is meant by the term *true class limits*?

17. Describe from your field of study a population of data where knowledge of the central tendency and dispersion would be useful. Obtain real or realistic synthetic values from this population and compute the mean, median, mode, variance, and standard deviation, using the techniques for ungrouped data.

18. Collect a set of real, or realistic, data from your field of study and construct a frequency distribution, a relative frequency distribution, a histogram and a frequency polygon.

19. Compute the mean, median, modal class, variance, and standard deviation for the data in Exercise 18, using the techniques for grouped data.

20. Find an article in a journal from your field of study in which some measure of central tendency and dispersion have been computed.

21. In a study designed to investigate the effectiveness of a potential local anesthetic, various doses were administered to 15 laboratory animals. A record was made of the duration (in minutes) of response. The results were as follows.

Animal number	Duration of response	Animal number	Duration of response
1	31	9	22
2	14	10	20
3	19	11	32
4	17	12	19
5	34	13	27
6	25	14	11
7	17	15	23
8	35		

Compute the mean, median, variance, and standard deviation for these sample data.

22. The following table shows the daily fat intake (grams) of a sample of 150 adult males in a developing country. Prepare a frequency distribution and histogram for these data. Compute the mean, median, variance, and standard deviation.

22	62	77	84	91	102	117	129	137	141
42	56	78	73	96	105	117	125	135	143
37	69	82	93	93	100	114	124	135	142
30	77	81	94	97	102	119	125	138	142
46	89	88	99	95	100	116	121	131	152
63	85	81	94	93	106	114	127	133	155
51	80	88	98	97	106	119	122	134	151
52	70	76	95	107	105	117	128	144	150
68	79	82	96	109	108	117	120	147	153
67	75	76	92	105	104	117	129	148	164
62	85	77	96	103	105	116	132	146	168
53	72	72	91	102	101	128	136	143	164
65	73	83	92	103	118	127	132	140	167
68	75	89	95	107	111	128	139	148	168
68	79	82	96	109	108	117	130	147	153

23. The following are the hemoglobin values (g/100 ml) of 10 children receiving treatment for hemolytic anemia:

9.1	8.3
10.0	9.9
11.4	9.1
12.4	7.5
9.8	6.7

Compute the sample mean, median, variance, and standard deviation.

24. Twenty postmenopausal women who had undergone a hysterectomy during their premenopausal years received synthetic estrogen therapy daily for four months. After treatment the following estrogen values were recorded:

61	58	54	54
81	56	81	75
61	80	92	59
63	83	71	58
82	92	69	94

Compute the sample mean, median, variance, and standard deviation.

25. The following table shows the age distribution of cases of a certain disease reported during a year in a particular state.

Age	Number of cases
5–14	5
15–24	10
25–34	20
35–44	22
45–54	13
55–64	5
Total	75

Compute the sample mean, median, variance, and standard deviation.

26. Give three synonyms for variation (variability).

27. As part of a research project, investigators obtained the following data on serum lipid peroxide (SLP) levels from laboratory reports of a sample of 10 adult subjects undergoing treatment for diabetes mellitus: 5.85, 6.17, 6.09, 7.70, 3.17, 3.83, 5.17, 4.31, 3.09, 5.24. Compute the mean, median, variance, and standard deviation.

28. The following are the SLP values obtained from a sample of 10 apparently healthy adults: 4.07, 2.71, 3.64, 3.37, 3.84, 3.83, 3.82, 4.21, 4.04, 4.50. For these data compute the mean, the variance, and the standard deviation.

REFERENCES

References cited
1. W. J. Dixon and M. B. Brown, Editors, *BMDP: Biomedical Computer Programs P-Series*, University of California Press, Berkeley, 1979.
2. Norman H. Nie, C. Hadlai Hull, Jean G. Jenkins, Karin Steinbrenner, and Dale H. Bent, *SPSS Statistical Package for the Social Sciences*, Second Edition, McGraw-Hill, New York, 1975.
3. *The IMSL Library*, *Vols*. 1–3, International Mathematical and Statistical Libraries, Inc., Dallas, Texas, 1979.
4. Thomas A. Ryan Jr., Brian L. Joiner, and Barbara F. Ryan, *Minitab Student Handbook*, Duxbury Press, North Scituate, Mass., 1976.
5. Anthony J. Barr, James H. Goodnight, John P. Sall, and June T. Helwig, *A User's Guide to SAS 79*, SAS Institute, Inc., Raleigh, N. C., 1979.
6. W. J. Dixon, and R. L. Jennrich, "Scope, Impact, and Status of Packaged Statistical Programs," *Annual Review of Biophysics and Bioengineering*, *1* (1972), 505–528.
7. Sidney O. Krasnoff, *Computers in Medicine*, Charles C. Thomas, Springfield, Ill., 1967.

8. Robert Steven Ledley, *Use of Computers in Biology and Medicine*, McGraw-Hill, New York, 1965.

9. Donald A. B. Lindberg, *The Computer and Medical Care*, Charles C. Thomas, Springfield, Ill., 1968.

10. Theodor D. Sterling and Seymour V. Pollack, *Computers and the Life Sciences*, Columbia University Press, New York, 1965.

11. Thomas R. Taylor, *The Principles of Medical Computing*, Blackwell Scientific Publications, Oxford, 1967.

12. H. A. Sturges, "The Choice of a Class Interval," *Journal of the American Statistical Association, 21* (1926), 65–66.

13. Helen M. Walker, "Degrees of Freedom," *The Journal of Educational Psychology, 31* (1040), 253–269.

Other References, Books
1. Wilfred J. Dixon and Frank J. Massey, Jr., *Introduction to Statistical Analysis*, Third Edition, McGraw-Hill, New York, 1969.

2. A. Bradford Hill, *Principles of Medical Statistics*, Eighth Edition, Oxford University Press, New York, 1967.

3. George W. Snedecor and William G. Cochran, *Statistical Methods*, Seventh Edition, The Iowa State University Press, Ames, 1967.

4. George H. Weinberg and John A. Schumaker, *Statistics: An Intuitive Approach*, Second Edition, Wadsworth, Belmont, Cal., 1980.

5. Bernard G. Greenberg, "Biostatistics," in Hugh Rodman Leavell, and E. Gurney Clark, *Preventive Medicine*, McGraw-Hill, New York, 1965.

Other References, Journal Articles
1. I. Altman and A. Ciocco, "Introduction to Occupational Health Statistics I," *Journal of Occupational Medicine, 6* (1964), 297–301.

2. A. R. Feinstein, "Clinical Biostatistics I, A New Name and Some Other Changes of the Guard," *Clinical Pharmacology and Therapeutics*, (1970), 135–138.

3. Alva R. Feinstein, "Clinical Biostatistics VI, Statistical Malpractice—and the Responsibility of a Consultant," *Clinical Pharmacology and Therapeutics, 11* (1970), 898–914.

4. Lyon Hyams, "The Practical Psychology of Biostatistical Consultation," *Biometrics, 27* (1971), 201–211.

5. Johannes Ipsen, "Statistical Hurdles in the Medical Career," *American Statistician, 19* (June 1965), 22–24.

6. Richard K. Means, "Interpreting Statistics: An Art," *Nursing Outlook 13* (May 1965), 34–37.

7. E. S. Pearson, "Studies in the History of Probability and Statistics. XIV Some Incidents in the Early History of Biometry and Statistics, 1890–94," *Biometrika*, *52* (1965), 3–18.

8. Harold M. Schoolman, "Statistics in Medical Research," *The New England Journal of Medicine*, *280* (1969), 218–19.

9. Stanley Schor and Irving Karten, "Statistical Evaluation of Medical Journal Manuscripts," *Journal of the American Medical Association*, *195* (1966), 1123–1128.

10. H. C. Selvin and A. Stuart, "Data—Dredging Procedures in Surgery Analysis," *American Statistician*, *20* (June 1966), 20–22.

11. Robert L. Stearman, "Statistical Concepts in Microbiology," *Bacteriological Reviews*, *19* (1955), 160–215.

12. Harry E. Ungerleider and Courtland C. Smith, "Use and Abuse of Statistics," *Geriatrics*, *22* (Feb. 1967), 112–120.

13. James P. Zimmerman, "Statistical Data and Their Use," *Physical Therapy*, *49* (1969), 301–302.

14. Robert I. Rollwagen, "Statistical Methodology in Medicine," *Canadian Medical Association Journal*, *112* (1975), 677.

15. Editorial, "Limitations of Computers in Medicine," *Canadian Medical Association Journal*, *104* (1971), 234–235.

16. Carol M. Newton, "Biostatistical Computing," *Federation Proceeding, Federation of American Societies for Experimental Biology*, *33* (1974), 2317–2319.

CHAPTER 2

Some Basic Probability Concepts

2.1 INTRODUCTION

The theory of probability provides the foundation for statistical inference. However, this theory, which is a branch of mathematics, is not the main concern of this book, and, consequently, only its fundamental concepts are discussed here. Students who desire to pursue this subject should refer to the books on probability by Bates (1), Dixon (2), Mosteller et al. (3), Earl et al. (4), Berman (5), Hausner (6), and Mullins and Rosen (7). They will also find helpful the books on mathematical statistics by Freund (8), Hogg and Craig (9), and Mood, Graybill and Boes (10). For those interested in the history of probability, the books by Todhunter (11) and David (12) are recommended. From the latter, for example, we learn that the first mathematician to calculate a theoretical probability correctly was Girolamo Cardano, an Italian who lived from 1501 to 1576. The objectives of this chapter are to help students gain some mathematical ability in the area of probability and to assist them in developing an understanding of the more important concepts. Progress along these lines will contribute immensely to their success in understanding the statistical inference procedures presented later in this book.

The concept of probability is not foreign to health workers and is frequently encountered in everyday communication. For example, we may hear a physician say that a patient has a 50–50 chance of surviving a certain operation. Another physician may say that she is 95 percent certain that a patient has a particular disease. A public health nurse may say that nine times out of ten a certain client will break an appointment. Thus we are accustomed to measuring the probability of the occurrence of some event by a number between zero and one. The more likely the event, the closer the number is to one; and the more unlikely the event, the closer the number is to zero. An event that cannot occur has a probability of zero, and an event that is certain to occur has a probability of one.

2.2   TWO VIEWS OF PROBABILITY—
OBJECTIVE AND SUBJECTIVE

Until fairly recently, probability was thought of by statisticians and mathematicians only as an *objective* phenomenon derived from objective processes.

The concept of *objective probability* may be categorized further under the headings of (1) *classical*, or *a priori*, *probability* and (2) the *relative frequency*, or *a posteriori*, concept of probability.

The classical treatment of probability dates back to the seventeenth century and the work of two mathematicians, Pascal and Fermat (11, 12). Much of this theory developed out of attempts to solve problems related to games of chance, such as the rolling of dice. Examples from games of chance illustrate very well the principles involved in classical probability. For example, if a fair six-sided die is rolled, the probability that a 1 will be observed is equal to 1/6 and is the same for the other five faces. If a card is picked at random from a well-shuffled deck of ordinary playing cards, the probability of picking a heart is 13/52. Probabilities such as these are calculated by the processes of abstract reasoning. It is not necessary to roll a die or draw a card to compute the above probabilities. In the rolling of the die we say that each of the six sides is *equally likely* to be observed if there is no reason to favor any one of the six sides. Similarly, if there is no reason to favor the drawing of a particular card from a deck of cards we say that each of the 52 cards is equally likely to be drawn. We may define probability in the classical sense as follows.

Definition   If an event can occur in N mutually exclusive and equally likely ways, and if m of these possess a characteristic, E, the probability of the occurrence of E is equal to m/N.

If we read $P(E)$ as "the probability of E," we may express the above definition as

$$P(E) = \frac{m}{N} \qquad (2.2.1)$$

The relative frequency approach to probability depends on the repeatability of some process and the ability to count the number of repetitions, as well as the number of times that some event of interest occurs. In this context we may define the probability of observing some characteristic, E, of an event as follows.

Definition   If some process is repeated a large number of times, n, and if some resulting event with the characteristic, E, occurs m times, the relative frequency of occurrence of E, m/n, will be approximately equal to the probability of E.

To express this definition in compact form we write

$$P(E) = \frac{m}{n} \tag{2.2.2}$$

We must keep in mind, however, that strictly speaking, m/n is only an estimate of $P(E)$.

In the early 1950s L. J. Savage (13) gave considerable impetus to what is called the "personalistic" or subjective concept of probability. This view holds that probability measures the confidence that a particular individual has in the truth of a particular proposition. This concept does not rely on the repeatability of any process. In fact, by applying this concept of probability, one may evaluate the probability of an event that can only happen once, for example, the probability that a cure for cancer will be discovered within the next 10 years.

Although the subjective view of probability has enjoyed increased attention over the years, it has not been fully accepted by statisticians who have traditional orientations.

2.3 ELEMENTARY PROPERTIES OF PROBABILITY

In 1933 the axiomatic approach to probability was formalized by the Russian mathematician A. N. Kolmogorov (14). The basis of this approach is embodied in three properties from which a whole system of probability theory is constructed through the use of mathematical logic. The three properties are as follows.

1. Given some process (or experiment) with n mutually exclusive outcomes, (called events), $E_1, E_2, \ldots, E_n$, the probability of any event, E_i, is assigned a nonnegative number. That is,

$$P(E_i) \geqq 0 \tag{2.3.1}$$

In other words, all events must have a probability greater than or equal to zero, a reasonable requirement in view of the difficulty of conceiving of negative probability. A key concept in the statement of this property is the concept of *mutually exclusive* outcomes. Two events are said to be mutually exclusive if they cannot occur simultaneously.

2. The sum of the probabilities of all mutually exclusive outcome is equal to 1.

$$P(E_1) + P(E_2) + \cdots + P(E_n) = 1 \tag{2.3.2}$$

This is the property of *exhaustiveness* and refers to the fact that the observer of a probabilistic process must allow for all possible events, and when all are taken together, their total probability is 1. The requirement that the events be mutually exclusive is specifying that the events $E_1, E_2, \ldots, E_n$ do not overlap.

3. Consider any two mutually exclusive events, E_i and E_j. The probability of the occurrence of either E_i or E_j is equal to the sum of their individual probabilities.

$$P(E_i \text{ or } E_j) = P(E_i) + P(E_j) \qquad (2.3.3)$$

Suppose the two events were not mutually exclusive; that is, suppose they could occur at the same time. In attempting to compute the probability of the occurrence of either E_i or E_j the problem of overlapping would be discovered, and the procedure could become quite complicated.

Before we use these properties to calculate the probability of an event, it is useful to review some basic ideas of set theory and counting techniques. These topics are covered in the next two sections.

2.4 SET THEORY AND SET NOTATION (BASIC NOTIONS)

George Cantor (1845–1918) introduced set theory in the latter part of the nineteenth century. It is an important mathematical tool that is useful in many branches of mathematics, including probability. For this reason, set theory is included in the present chapter. However, only a minimum of its basic concepts will be covered. The books by Breuer (15), Stoll (16), and Maher (17), among others, may be consulted for a more complete treatment.

A *set* is a collection of definite, distinct *objects*. The objects that constitute a set are called the *elements* or *members* of the set. We will use capital letters to designate a set.

A set may be described in either one of the following ways.

1. By listing all the elements in the set.
    Examples:

Set	All Elements in the Set
A = {patient number 1, patient number 2, patient number 3}	
B = {drug A, drug B, drug C, drug D}	
C = {animal 1, animal 2, ..., animal n}	
D = {Mrs. Smith, Ms. Jones, Mr. Johnson, Ms. King}	

2. By describing the type of element of which the set is composed.
    Examples:

Set	Type of Element
A = {all patients in critical condition on fourth floor}	
B = {all drugs used in a certain experiment}	
C = {all animals used in a certain experiment}	
D = {all public health nurses employed in a certain clinic}	

Some additional concepts related to sets are given below.

1. A *unit set* is a set composed of only one element.

2. A set that contains no elements is called the *empty set*, or *null set*, and is designated by the symbol $\emptyset$.

3. The set of all elements for which there is interest in a given discussion is called the *universal set*. We will designate the universal set by the capital letter U.

4. If the set A contains one or more elements from the set B, and if every element of A is an element of B, then A is said to be a *subset* of B.

5. The null set is a subset of every other set, by definition.

6. Two sets are equal if, and only if, they contain the same elements.

The following are some useful set operations. Where appropriate the various relationships among the sets will be portrayed visually by a device called a *Venn diagram*, which represents a set as a portion of a plane.

1. The *union* of two sets, A and B, is another set and consists of all elements belonging to either A or B or both A and B. The symbol $\cup$ will be used to designate the union of two sets.

Example: Suppose that in a mental health clinic a social worker's case load consists of the set of patients A, where

$$A = \{\text{patients } 1, 2, 3, 4, 5, 6\} = \text{all assigned patients}$$
who are receiving drug therapy

and the set of patients B, where

$$B = \{\text{patients } 2, 4, 7, 8, 9, 10, 11\} = \text{all assigned patients}$$
who are receiving group psychotherapy

The union of these two sets may be written

$$A \cup B = \{\text{patients } 1, 2, 3, 4, 5, 6, 7, 8, 9, 10, 11\} = \text{all assigned}$$
patients who are receiving drug therapy, group psychotherapy, or both

The two sets, A and B, are said to be *conjoint* because they have at least one element in common. In this case, the common elements are patients 2 and 4.

The three sets are shown with the use of Venn diagrams in Fig. 2.4.1. Note that $A \cup B$ is the total shaded area in the rectangle on the right

If two sets have no elements in common they are said to be *disjoint*. For example, suppose 15 patients, of whom 10 were over 30, and 5 were 30 years of

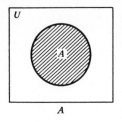 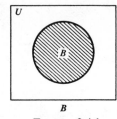 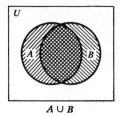

A   B   A ∪ B

FIGURE 2.4.1

The Union of Two Conjoint Sets

age or younger, were admitted to a general hospital on March 10, 1983. Let us define the following sets:

$$A = \{\text{all patients over 30 admitted on March 10, 1983}\}$$

$$B = \{\text{all patients 30 or younger admitted on March 10, 1983}\}$$

$$A \cup B = \{\text{all patients admitted on March 10, 1983}\}$$

The sets A and B here are disjoint, since a patient cannot be both 30 years old or younger and over 30 years of age at the same time. The union of two disjoint sets is shown in Fig. 2.4.2.

2. The *intersection* of two sets, A and B, is another set, and consists of all elements that are in both A and B. The symbol ∩ will be used to designate the intersection of two sets.

In the previous example, which referred to patients in a mental health clinic, the intersection of sets A and B would consist of all patients receiving both drug and group psychotherapy, that is, patients 2 and 4. In set notation we write this as $A \cap B = \{\text{all patients receiving both drug and group psychotherapy}\}$. In Figure 2.4.1, $A \cap B$ is shown as the double shaded area, representing the overlapping of sets A and B.

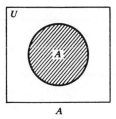 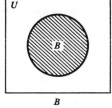 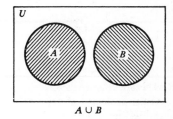

A   B   A ∪ B

FIGURE 2.4.2

The Union of Two Disjoint Sets

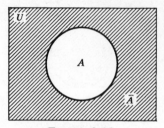

FIGURE 2.4.3
Venn Diagram Showing Sets A and $\overline{A}$

The example of the patients admitted to a general hospital on March 10, 1983, illustrates the fact that the intersection of two disjoint sets is the null set.

3. If the set A is a subset of the universal set, U, the *complement* of A is another subset of U and consists of the elements in U that are not in A. Let us designate the complement of A as $\overline{A}$.

Suppose, of 50 women receiving private prenatal care, 4 have type AB blood. If we designate the 50 women as the universal set, U, and the subset of 4 with type AB blood as the set A, then the complement of A is the set $\overline{A}$ consisting of the 46 women who have some other blood type. The complement of a set is illustrated in Figure 2.4.3.

Example 2.4.1 It is frequently useful to be able to identify sets and subsets represented by cross-tabulated data, as in Table 2.4.1 which shows the profes-

TABLE 2.4.1
Professional and Technical Employees of a Group of Hospitals
Classified by Age and Job Category

Job category	A_1 ≤ 25	A_2 $26-30$	A_3 $31-35$	A_4 > 35	Total
B_1 physicians	0	5	25	75	105
B_2 Clinical laboratory services	20	30	35	35	120
B_3 Dietary services	3	6	6	10	25
B_4 Medical record services	7	15	8	12	42
B_5 Nursing services	200	375	442	203	1220
B_6 Pharmacy	1	12	8	3	24
B_7 Radiologic technology	4	10	19	12	45
B_8 Therapeutic services	5	25	15	10	55
B_9 Other professional and technical services	20	35	50	25	130
Total	260	513	608	385	1766

sional and technical personnel of a group of hospitals tabulated by age and job category. Let us denote the number of elements in a set, say set A, as $n(A)$ and use set notation to identify some of the subsets defined in the table.

In Table 2.4.1, sets A_1 to A_4 consist of personnel falling into the specified age groups and sets B_1 to B_9 consist of personnel falling into the specified job categories. We may specify other sets by employing the concepts of intersection, union, and complement. For example, the set $B_1 \cap A_4$ consists of physicians who are over 35 years of age, and $n(B_1 \cap A_4) = 75$. The set $B_2 \cup A_2$ consists of clinical laboratory personnel, or personnel between the ages of 26 and 30, or both, and $n(B_2 \cup A_2) = 120 + 513 - 30 = 603$. In calculating $n(B_2 \cup A_2)$, the number (30) who are both clinical laboratory personnel and between the ages of 26 and 30 has to be subtracted because it has been counted twice, that is, it is included in both the figure 120 and the figure 513. The complement of A_4, $\overline{A}_4$, consists of all personnel 35 years old or younger, and $n(\overline{A}_4) = 1766 - 385 = 1381$.

EXERCISES

2.4.1 The following table shows the patients admitted to a psychiatric hospital during a year. The data are cross-tabulated by diagnosis and age.

	AGE (YEARS)							
Diagnosis	A_1 < 15	A_2 15 to 24	A_3 25 to 34	A_4 35 to 44	A_5 45 to 54	A_6 55 to 64	A_7 65 and older	Total
B_1 Involutional psychotic reaction	0	0	0	7	27	20	4	58
B_2 Manic depressive reaction	0	1	1	4	9	5	4	24
B_3 Schizophrenia	5	90	140	160	103	44	7	549
B_4 Psychoneurotic reactions	0	26	44	47	29	13	3	162
B_5 Alcohol addiction	0	7	41	77	68	26	5	224
B_6 Drug addiction	0	2	2	4	2	2	1	13
Total	5	126	228	299	238	110	24	1030

Based on the above table, explain in words the following sets and give the number of patients in each:

(a) $A_4 \cap B_3$

(b) $B_5 \cap A_6$

(c) $B_3 \cup A_4$

(d) $A_6 \cup B_5$

(e) A_1

(f) $(A_4 \cup A_5) \cap B_3$

2.4.2 Of 50 patients on the third floor of a hospital, 35 are female and 12 are over 70 years of age. Among those over 70 years of age, eight are female. How many of the 50 patients are female *or* not over 70 years of age? (Use a Venn diagram to help you answer the question.)

2.5 COUNTING TECHNIQUES— PERMUTATIONS AND COMBINATIONS

In this section we present some useful techniques for counting the number of events satisfying some set of conditions. These techniques are helpful in computing the probability of an event when the total number of possible events is large.

Factorials   Given the positive integer n, the product of all the whole numbers from n down through 1 is called n *factorial* and is written $n!$

The following are some examples of factorials:

$$10! = 10 \cdot 9 \cdot 8 \cdot 7 \cdot 6 \cdot 5 \cdot 4 \cdot 3 \cdot 2 \cdot 1$$
$$5! = 5 \cdot 4 \cdot 3 \cdot 2 \cdot 1$$
$$2! = 2 \cdot 1$$

and, in general,

$$n! = n(n-1)(n-2)(n-3) \cdot \ldots \cdot 1$$

By definition, $0! = 1$. It should also be noted that

$$10! = 10 \cdot 9!$$
$$5! = 5 \cdot 4!$$
$$n! = n(n-1)!$$

Example 2.5.1   By means of factorials we are able to answer questions regarding the number of ways objects can be arranged in a line. For example, suppose we have four containers of culture media, each of which is inoculated with a different organism. In how many ways can they be placed in a line on a shelf? The answer is $4! = 4 \cdot 3 \cdot 2 \cdot 1 = 24$ ways. A graphic aid, called a tree diagram, is helpful in visualizing the possibilities. Let us designate the positions as the first, second, third, and fourth positions and the four media as A, B, C, and D. The tree diagram in Figure 2.5.1 represents the possible arrangements.

Permutations   A *permutation* is an ordered arrangement of objects.

The 24 arrangements of culture media shown in Fig. 2.5.1 are the possible permutations of four objects taken four at a time. At times we may have more objects than we have positions to fill. Suppose, in the preceding example, we have only two positions available on the shelf. In how many ways can we fill these two

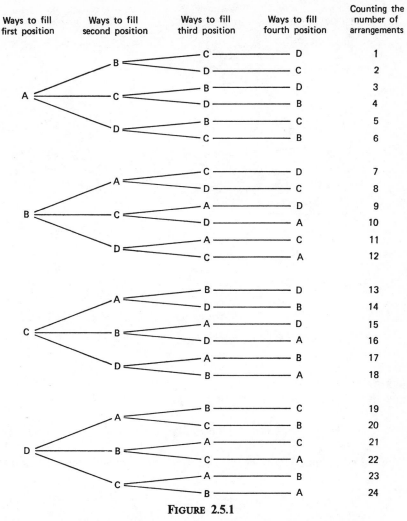

Ways to fill first position	Ways to fill second position	Ways to fill third position	Ways to fill fourth position	Counting the number of arrangements

FIGURE 2.5.1

Tree Diagram Showing Possible Arrangements of Four Objects Placed in a Line

positions using the four culture media? To answer this question we must determine the number of possible permutations of four things taken two at a time. We have four objects, A, B, C, or D, with which to fill the first position. Once that position has been filled we have three objects, only, with which to fill the second position. Again, let us use a tree diagram (Figure 2.5.2) to illustrate.

We see from Figure 2.5.2 that there are $4 \cdot 3 = 12$ possible permutations of four things taken two at a time. Let us designate by n the number of distinct objects from which an ordered arrangement is to be derived, and by r the number of objects in the arrangement. The number of possible such ordered

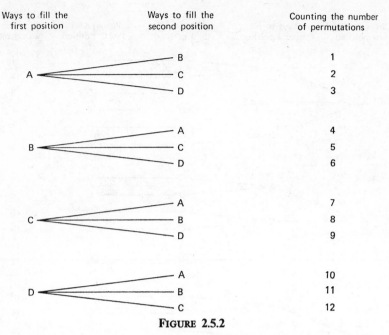

FIGURE 2.5.2

Tree Diagram Showing the Permutations of Four Things Taken Two at a Time

arrangements is referred to as the number of permutations of n things taken r at a time and may be written as $_nP_r$. In general,

$$_nP_r = n(n - 1)(n - 2) \cdots (n - r + 1) \qquad (2.5.1)$$

We may also evaluate $_nP_r$ by means of a fraction involving factorials as follows:

$$_nP_r = \frac{n!}{(n - r)!} \qquad (2.5.2)$$

Expanding Equation 2.5.2 gives

$$_nP_r = \frac{n(n - 1)(n - 2) \cdots (n - r + 1)(n - r)(n - r - 1) \cdots 1}{(n - r)(n - r - 1) \cdots 1}$$

The denominator and terms in the numerator beyond $(n - r + 1)$ cancel, leaving the right side of Equation 2.5.1.

Let us illustrate the evaluation of permutations by means of another example.

Example 2.5.2 In a county health department there are five adjacent offices to be occupied by five nurses, A, B, C, D, and E. In how many ways can the five nurses be assigned to the offices? The answer is obtained by evaluating $_5P_5$, which is

$$_5P_5 = \frac{5!}{(5-5)!} = 5 \cdot 4 \cdot 3 \cdot 2 \cdot 1 = 120$$

Suppose there are six nurses of whom four are to be assigned to four adjacent offices. Here we need to determine the number of permutations of six things taken four at a time, which is

$$_6P_4 = \frac{6!}{(6-4)!} = \frac{6 \cdot 5 \cdot 4 \cdot 3 \cdot 2!}{2!} = 360$$

Combinations   A *combination* is an arrangement of objects without regard to order. We write the number of combinations of n things taken r at a time as $\binom{n}{r}$.

In Figure 2.5.2 the permutations of four things taken two at a time consisted of the following 12 arrangements:

AB

AC

AD

BA

BC

BD

CA

CB

CD

DA

DB

DC

We note that in the list above exactly half of the pairs of arrangements are alike except for the order in which the letters occur. We may match these pairs to obtain

AB	AC	AD	BC	BD	CD
BA	CA	DA	CB	DB	DC

In certain cases we may not want to make a distinction between arrangement AB and arrangement BA, for example. We may want to consider them as the same subset, in which case we say that *order does not count* and refer to the arrangements as combinations. Whereas in the culture media example we have 12 permutations, we have only six combinations. In other words, we have two permutations of each combination. In general, we will have $r!$ permutations for each combination of n things taken r at a time, or putting it another way, we will, in general, have $r!$ times as many permutations as combinations. That is,

$$_nP_r = r!\binom{n}{r} \tag{2.5.3}$$

If we solve Equation 2.5.3 for $\binom{n}{r}$, we have

$$\binom{n}{r} = \frac{_nP_r}{r!}$$

By rewriting the numerator in accordance with Equation 2.5.2, we have the formula for the number of combinations of n things taken r at a time:

$$\binom{n}{r} = \frac{n!}{r!(n-r)!} \tag{2.5.4}$$

Now let us show that when we use Equation 2.5.4 to obtain the number of combinations of four things taken two at a time we get six:

$$\binom{4}{2} = \frac{4!}{2!2!} = \frac{4 \cdot 3 \cdot 2!}{2 \cdot 1 \cdot 2!} = 6$$

Example 2.5.3 As a further example, suppose that a group therapy leader in a mental health clinic has 10 patients from whom to form a group of 6. How many combinations of patients are possible? We find the answer to be

$$\binom{10}{6} = \frac{10!}{6!4!} = \frac{10 \cdot 9 \cdot 8 \cdot 7 \cdot 6!}{6!4 \cdot 3 \cdot 2 \cdot 1} = 210$$

Permutations of Objects That Are Not All Different In discussing permutations, we have considered the case where all objects being permuted are different. In some situations we may have, in a set of objects to be permuted, some one or more subsets of items that are indistinguishable. The question of how many permutations are possible under these circumstances now arises. It seems logical that the number would be smaller than when all objects are different.

Example 2.5.4 Let us illustrate by referring to the case of five nurses, A, B, C, D, and E, who are to be assigned adjacent offices. Suppose two of the nurses want their offices painted white, two yellow, and the remaining nurse wants the office painted green. How many sequences of colors are possible? We assume that with respect to color the two white offices are indistinguishable and the two yellow offices are indistinguishable. The possible color sequences for the five adjacent offices are as follows:

WWYYG	YYWWG	YWWYG
WWYGY	YYWGW	YWWGY
WWGYY	YYGWW	YWGWY
WGWYY	YGYWW	YGWWY
GWWYY	GYYWW	GYWWY
WYWYG	YWYWG	WYYWG
WYWGY	YWYGW	WYYGW
WYGWY	YWGYW	WYGYW
WGYWY	YGWYW	WGYYW
GWYWY	GYWYW	GWYYW

As we can see, there are 30 possible sequences. If each nurse had wanted a different color, so that with respect to color all offices were different, there would be $_5P_5 = 5! = 120$ possible color sequences.

Suppose the two whites are distinguishable, say one is an off-white; and the yellows are distinguishable, say dark and light. Let us indicate the differences by subscripts as follows: W_1, W_2, Y_1, Y_2. We can take any one of the previously given sequences and obtain three additional sequences by permuting the subscripts. We do this by permuting the two subscripts for white, while we leave the yellows unchanged. There are 2! such permutations. We can obtain 2! additional sequences by permuting the subscripts of the yellows, leaving the whites unchanged. Since there is only one green, we do not have to be concerned with its effect on the number of permutations, simply noting that there are 1! permutations of the single green. If, however, there were two distinguishable greens we would have to take into account the resulting 2! possible permutations. Let us take the first sequence and illustrate. The four possible sequences, when we distinguish between the whites and yellows, are

$$W_1W_2Y_1Y_2G$$
$$W_2W_1Y_1Y_2G$$
$$W_1W_2Y_2Y_1G$$
$$W_2W_1Y_2Y_1G$$

Since we can do this for any of the 30 previous sequences, we see that there are $30 \cdot 2! \cdot 2! \cdot 1! = 120$ sequences when all objects are different. This, of course, is equal to $_5P_5 = 120$ which was found previously.

If we let $_nP_{n_1, n_2, \ldots, n_k}$ equal the number of distinguishable sequences that can be formed from n objects taken n at a time, when n_1 are of one type, n_2 are of a second type, $\ldots$, and n_k are of a kth type, and $n = n_1 + n_2 + \cdots + n_k$, we can generalize the above results by writing

$$n! = \left(_nP_{n_1, n_2, \ldots, n_k}\right)n_1!n_2! \cdots n_k!$$

If we solve for $_nP_{n_1, n_2, \ldots, n_k}$ we obtain

$$_nP_{n_1, n_2, \ldots, n_k} = \frac{n!}{n_1! n_2! \cdots n_k!} \tag{2.5.5}$$

Using the current example to illustrate, we have

$$_5P_{2,2,1} = \frac{5!}{2! 2! 1!} = \frac{120}{4} = 30$$

which is the number of sequences listed previously.

Example 2.5.5   Let us illustrate with another example. Suppose the food service department of a hospital on a certain day is serving two white, two green, and two yellow vegetables. How many distinguishable arrangements of these vegetables can be made on the serving line if we are interested only in distinguishing between vegetables on the basis of color? By Equation 2.5.5 we find the answer to be

$$_6P_{2,2,2} = \frac{6!}{2! 2! 2!} = \frac{720}{8} = 90$$

An important special case of Equation 2.5.5 occurs when only two types of objects are present, that is, when r are of one type and $n - r$ are of another type. In this case we have

$$_nP_{r, n-r} = \frac{n!}{r!(n - r)!} = \binom{n}{r} \tag{2.5.6}$$

From this we see that the number of distinct permutations of n things of which r are of one type and $n - r$ are of another type is equal to the number of combinations of n different things, taken r at a time.

EXERCISES

2.5.1   Evaluate the following:

(a) $_6P_2$

(b) $_7P_3$

(c) $_{10}P_5$

(d) $\binom{6}{2}$

(e) $\binom{7}{3}$

(f) $\binom{10}{5}$

(g) $\binom{8}{5}$

(h) $\binom{9}{5}$

(i) $\binom{5}{2}$

2.5.2 A public health nurse is preparing a program for a meeting with expectant mothers. She has four topics to cover, and they may be covered in any order.

(a) How many different programs is it possible for her to prepare?
(b) Suppose at the last minute she finds that she has time for only three topics. How many different programs is it possible for her to present if she considers all four of equal importance?

2.5.3 For lunch a patient in a hospital may choose one of four meats, two of five vegetables, and one of three desserts. How many different meals does the patient have to choose from if he selects the specified number from each group?

2.5.4 A physical therapist planning his day's schedule, finds that he has seven activities to perform that day.

(a) If he may perform these activities in any order he wishes, how many different schedules can he prepare?
(b) If he decides to take the afternoon off so that he has time for only three of his activities, how many schedules can he prepare?

2.5.5 A health educator has three posters to display in a row on the lobby wall of a health center. How many different displays can he prepare?

2.5.6 Suppose in a certain laboratory there are four jobs to be done on a particular afternoon and five people to do the jobs. In how many ways can four of the five people be assigned to the four jobs?

2.5.7 A researcher has four drugs she wishes to test, but she has only enough experimental animals to test three of the drugs. How many combinations of drugs can she test?

2.5.8 A health educator has engaged four community leaders, two men and two women, to appear on a panel discussion program. In how many ways can she arrange the panel in a row facing the audience if she wishes to distinguish among the panelists only on the basis of sex?

2.5.9 Eight experimental animals have been inoculated with a certain drug; three with type A, three with type B, and two with type C. Each animal is to be placed in one of eight adjacent cages for observation. If the animals are distinguishable only on the basis of the type of drug received, how many different arrangements are possible?

2.5.10 Ten people are candidates to fill three vacancies on the board of directors of a hospital. In how many ways can the three positions be filled?

2.5.11 In how many ways can seven people enter a cafeteria line, single file?

2.6 CALCULATING THE PROBABILITY OF AN EVENT

We now make use of the concepts and techniques of the previous sections in calculating the probabilities of specific events. Additional ideas will be introduced as needed.

Example 2.6.1   For illustrative purposes let us refer to Example 2.4.1 and the data given in Table 2.4.1. Suppose an employee is picked at random from all employees represented; what is the probability that this person will be 25 years of age or younger? We consider the likelihood of selecting any one employee to be equal to the likelihood of selecting any other employee and define the probability of interest as the number of favorable outcomes divided by the total number of possible outcomes. The answer to our question then is

$$P(A_1) = \frac{n(A_1)}{n(U)} = \frac{260}{1766} = .15$$

Conditional Probability   On occasion, the set of "all possible outcomes" may constitute a subset of the universal set. In other words, the population of interest may be reduced by some set of conditions not applicable to the total population. When probabilities are calculated with a subset of the universal set as the denominator, the result is a *conditional probability*.

We may illustrate the concept of conditional probability by referring again to Table 2.4.1. Suppose we are interested in calculating the probability that an employee selected at random from the 1766 employees is a physician. This is an unconditional probability since no conditions have been placed on the set of all possible outcomes, and we compute it as follows.

$$P(B_1) = \frac{n(B_1)}{n(U)} = \frac{105}{1766} = .06$$

Suppose, however, that we reduce the set of all possible outcomes to those employees who are over 35 years of age (set A_4). Now, what is the probability that an employee is a physician, given that he or she is picked at random from the set of employees over 35 years old? We see in Table 2.4.1 that there are 385 members of set A_4, employees over 35 years of age. We also see that, of these, 75 are physicians and that these 75 are the number of members of the set $B_1 \cap A_4$. Thus we find the answer to our question to be

$$P(B_1|A_4) = \frac{n(B_1 \cap A_4)}{n(A_4)}$$

$$= \tfrac{75}{385} = .19$$

The vertical line in $P(B_1|A_4)$ is read "given."

We may obtain the conditional probability, $P(B_1|A_4)$ in another way. Suppose that we divide the numerator and denominator by $n(U)$, the number in the universal set, the total number of employees. This gives

$$P(B_1|A_4) = \frac{\dfrac{n(B_1 \cap A_4)}{n(U)}}{\dfrac{n(A_4)}{n(U)}}$$

The numerator of this last expression is the probability that an employee picked at random from all employees will be both a physician and over 35 years of age, and may be written $P(B_1 \cap A_4)$. The denominator is the probability that an employee picked at random will be over 35 years of age and may be written as $P(A_4)$. The entire expression may be written as

$$P(B_1|A_4) = \frac{P(B_1 \cap A_4)}{P(A_4)} = \frac{\frac{75}{1766}}{\frac{385}{1766}} = \frac{.0425}{.2190} = .19$$

This result suggests the following general definition of a conditional probability.

Definition   The conditional probability of A given B is equal to the probability of $A \cap B$ divided by the probability of B, provided the probability of B is not zero.

That is,

$$P(A|B) = \frac{P(A \cap B)}{P(B)}, \qquad P(B) \neq 0 \qquad \textbf{(2.6.1)}$$

When we ask the question, what is the probability that an employee picked at random from all employees (Table 2.4.1) is over 35 years old, we are asking for a *marginal probability*. This marginal probability provides the denominator for our conditional probability, $P(B_1|A_4)$. Our interest centers on a probability associated with a marginal total, and we disregard any other criterion of classification. In computing

$$P(A_4) = \tfrac{385}{1766} = .2180$$

a lack of interest in the B classification is implied. Similarly, if we are interested in the probability that an employee picked at random is a physician, for example, we compute this probability using the marginal total, 105, and ignore the age classifications.

In general, then, we may say that when one or more criteria of classification are ignored in computing a probability, the resulting probability is called a *marginal probability*.

Now let us look at the numerator of $P(B_1|A_4)$. This numerator is the probability of the *joint* occurrence of B_1 and A_4, that is, it is the condition of being simultaneously a physician and over 35 years of age. It has already been pointed out that if two events, A and B, cannot occur simultaneously, that is, if $A \cap B = \varnothing$, then A and B are mutually exclusive. When we refer to Table 2.4.1 we see that for this particular set of employees the events, being a physician and being 25 years of age or under are mutually exclusive, since $B_1 \cap A_1 = \varnothing$.

The third property of probability given previously states that the probability of the occurrence of either one or the other of two mutually exclusive events is equal to the sum of their individual probabilities. In our example, then, the probability of being either a physician or 25 years of age or under is found to be

$$P(B_1) + P(A_1) = \tfrac{105}{1766} + \tfrac{260}{1766} = .06 + .15 = .21$$

What if two events are not mutually exclusive? This case is covered by what is known as the *addition rule*, which may be stated as follows.

Definition Given two events A and B, the probability that event A, or event B, or both occur is equal to the probability that event A occurs, plus the probability that event B occurs, minus the probability that both events occur simultaneously.

The addition rule may be written

$$P(A \cup B) = P(A) + P(B) - P(A \cap B) \qquad (2.6.2)$$

Let us look once again at Table 2.4.1 and compute the probability that an employee picked at random will be either a physician, over 35 years of age, or both.

$$
\begin{aligned}
P(B_1 \cup A_4) &= P(B_1) + P(A_4) - P(B_1 \cap A_4) \\
&= \tfrac{105}{1766} + \tfrac{385}{1766} - \tfrac{75}{1766} \\
&= .23
\end{aligned}
$$

Note that the 75 employees who are *both* physicians *and* over 35 years of age are included in the 105 who are physicians as well as in the 385 who are over 35 years of age. Since, in computing the probability, these 75 have been added into the numerator twice, they have to be subtracted out once to overcome the effect of duplication, or overlapping.

Another useful rule for computing the probability of an event is the *multiplication rule*. This rule is suggested by our definition of conditional probability and Equation 2.6.1:

$$P(A|B) = \frac{P(A \cap B)}{P(B)}, \qquad P(B) \neq 0$$

We can rewrite Equation 2.6.1 to obtain

$$P(A \cap B) = P(B)P(A|B) \tag{2.6.3}$$

Equation 2.6.3 tells us that the probability of the joint occurrence of events A and B is equal to the conditional probability of A given B times the marginal probability of B.

To illustrate, let us use Equation 2.6.3 to find the probability that an employee picked at random from all employees is both a physician and over 35 years of age.

$$P(B_1 \cap A_4) = P(A_4)P(B_1|A_4)$$
$$= \frac{385}{1766} \cdot \frac{75}{385}$$
$$= \frac{75}{1766}$$

This, of course, agrees with what we may compute directly from Table 2.4.1 by forming the ratio of the number of favorable outcomes, 75, to the total number of possible outcomes, 1766.

Let us look again at Equation 2.6.3.

$$P(A \cap B) = P(B)P(A|B)$$

Suppose the fact that we are given event B has no effect on the probability of A. In other words, suppose the probability of event A is the same whether event B occurs or does not occur. In such a case, $P(A|B) = P(A)$, and we say that A and B are *independent events*.

Example 2.6.2 Let us illustrate the concept of independence by means of the following example. In a certain high school class, consisting of 60 girls and 40 boys, it is observed that 24 girls and 16 boys wear eyeglasses. If a student is picked at random from this class, the probability that the student wears eyeglasses, $P(E)$, is 40/100, or .4. What is the probability that a student picked at random wears eyeglasses, given that the student is a boy. By using the formula for computing a conditional probability we find this to be

$$P(E|B) = \frac{P(E \cap B)}{P(B)} = \frac{16/100}{40/100} = .4$$

Thus the additional information that a student is a boy does not alter the probability that the student wears eyeglasses, and $P(E) = P(E|B)$. We say that the events being a boy, and wearing eyeglasses, for this group are independent. We may also show that the event of wearing eyeglasses, E, and *not* being a boy, $\bar{B}$, are also independent as follows:

$$P(E|\bar{B}) = \frac{P(E \cap \bar{B})}{P(\bar{B})} = \frac{24/100}{60/100} = \frac{24}{60} = .4$$

What is the probability of the joint occurrence of the events wearing eyeglasses and being a boy? Using the rule given in Equation 2.6.3, we have

$$P(E \cap B) = P(B)P(E|B)$$

but, since we have shown that events E and B are independent we may replace $P(E|B)$ by $P(E)$ to obtain

$$P(E \cap B) = P(B)P(E)$$
$$= \left(\tfrac{40}{100}\right)\left(\tfrac{40}{100}\right)$$
$$= .16$$

In general, then, when two events, A and B, are independent, and neither $P(A)$ nor $P(B)$ is equal to zero, the probability of their joint occurrence is equal to the product of their individual probabilities. That is

$$P(A \cap B) = P(A)P(B) \tag{2.6.4}$$

if

$$P(A) \neq 0, \qquad P(B) \neq 0$$

Before we conclude this section, a reminder is appropriate. The probability of an event A is equal to 1 minus the probability of its complement, $\overline{A}$, and

$$P(\overline{A}) = 1 - P(A) \tag{2.6.5}$$

This follows from the third property of probability since the event, A, and its complement, $\overline{A}$, are mutually exclusive.

Example 2.6.3 Suppose that of 1200 admissions to a general hospital during a certain period of time, 750 are private admissions. If we designate these as set A, then $\overline{A}$ is equal to 1200 minus 750, or 450. We may compute

$$P(A) = 750/1200 = .625$$

and

$$P(\overline{A}) = 450/1200 = .375$$

and see that

$$P(\overline{A}) = 1 - P(A)$$
$$.375 = 1 - .625$$
$$.375 = .375$$

EXERCISES

2.6.1 Refer to Exercise 2.4.1 and compute:

(a) $P(A_4 \cap B_3)$ (b) $P(B_5 \cup A_6)$ (c) $P(B_3 \cup A_4)$
(d) $P(A_6 \cup B_5)$ (e) $P(\bar{A_1})$ (f) $P[(A_4 \cup A_5) \cap B_3]$

2.6.2 The following table shows the first 1000 patients admitted to a clinic for retarded children by diagnostic classification and level of intelligence. For this group find:

(a) $P(A_3 \cap B_4)$.
(b) The probability that a patient picked at random is severely retarded.
(c) The probability that a patient picked at random either is not retarded or is borderline.
(d) The probability that a patient picked at random is profoundly retarded and has Down's syndrome.
(e) The probability that a patient is profoundly retarded, given that he or she has Down's syndrome.

	LEVEL OF RETARDATION						
Diagnostic classification	A_1 *Not retarded*	A_2 *Pro- found*	A_3 *Severe*	A_4 *Moderate*	A_5 *Mild*	A_6 *Border- line*	*Total*
B_1 Encephalopathies	33	38	57	114	103	55	400
B_2 Down's syndrome	2	4	34	88	27	5	160
B_3 Congenital cerebral defect	10	2	6	6	6	0	30
B_4 Mental retardation of unknown cause	0	0	9	36	62	35	142
B_5 Other	161	0	8	16	8	75	268
Total	206	44	114	260	206	170	1000

2.6.3 In a group of 502 persons it was determined that the distribution of blood groups was as follows:

Blood group	*Number*
O	226
A	206
B	50
AB	20
Total	502

If a person is picked at random from this group, what is the probability that he or she will have blood group:

(a) O?   (b) A?   (c) B?   (d) AB?

2.6.4 Suppose that the data of Exercise 2.6.3 when classified by sex are as follows:

Blood group	SEX Male	SEX Female	Total
O	113	113	226
A	103	103	206
B	25	25	50
AB	10	10	20
Total	251	251	502

For this group of persons would you say that sex and blood group are independent? Demonstrate by computing appropriate probabilities.

2.6.5 If the probability of left-handedness in a certain group of people is .05, what is the probability of right-handedness (assuming no ambidexterity)?

2.6.6 The probability is .6 that a patient selected at random from the current residents of a certain hospital will be a male. The probability that the patient will be a male who is in for surgery is .2. A patient randomly selected from current residents is found to be a male; what is the probability that the patient is in the hospital for surgery?

2.6.7 In a certain population of hospital patients the probability is .35 that a randomly selected patient will have heart disease. The probability is .86 that a patient with heart disease is a smoker. What is the probability that a patient randomly selected from the population will be a smoker *and* have heart disease?

2.7 SUMMARY

In this chapter some of the basic ideas and concepts of probability are presented. The objective has been to provide enough of a "feel" for the subject so that the probabilistic aspects of statistical inference can be more readily understood and appreciated.

REVIEW QUESTIONS AND EXERCISES

1. Define the following:
    (a) Probability
    (b) Objective probability
    (c) Subjective probability
    (d) Classical probability
    (e) The relative frequency concept of probability
    (f) Mutually exclusive events
    (g) Independence
    (h) Marginal probability
    (i) Joint probability
    (j) Conditional probability

2. Name and explain the three properties of probability.

3. What is a set? Give three examples of sets.

4. Define the following:
    (a) Unit set
    (b) Empty set
    (c) Universal set
    (d) Subset
    (e) Disjoint set
    (f) Complement

5. Under what condition are two sets considered equal?

6. Define and give an example of the union of two sets.

7. Define and give an example of the intersection of two sets.

8. What is a Venn diagram?

9. Define and illustrate the following:
    (a) Factorial
    (b) Permutation
    (c) Combination
    (d) The addition rule
    (e) The multiplication rule

10. Set C consists of the citizens of a certain town who voted "yes" for water fluoridation. Set D consists of the citizens of the same town who have preschool children. Define:
    (a) $C \cup D$
    (b) $A \cap D$
    (c) $\overline{C}$
    (d) $\overline{D}$

11. One hundred married women were asked to specify which type of birth control method they preferred. The following table shows the 100 responses cross-classified by educational level of the respondent.

Birth control method	EDUCATIONAL LEVEL			
	High school (A)	College (B)	Graduate school (C)	Total
S	15	8	7	30
T	3	7	20	30
V	5	5	15	25
W	10	3	2	15
Total	33	23	44	100

Specify the number of members of each of the following sets:

(a) S (b) $V \cup C$ (c) A (d) $\overline{W}$

(e) U (f) $\overline{B}$ (g) $T \cap B$ (h) $\overline{(T \cap C)}$

12. A psychiatric social worker has ten clients to be visited in a week. The clients consist of three alcoholics, four drug addicts, two schizophrenic patients, and one manic-depressive. How many distinguishable visiting arrangements can the social worker prepare if she wishes to distinguish among the clients only on the basis of diagnosis?

13. A certain county health department has received 25 applications for an opening that exists for a public health nurse. Of these applicants ten are over 30 and fifteen are under 30. Seventeen hold bachelor's degrees only, and eight have master's degrees. Of those under 30, six have master's degrees. If a selection from among these 25 applicants is made at random, what is the probability that a person over 30 *or* a person with a master's degree will be selected?

14. The following table shows 1000 nursing school applicants classified according to scores made on a college entrance examination and the quality of the high school from which they graduated, as rated by a group of educators.

	QUALITY OF HIGH SCHOOL			
	Poor	Average	Superior	
Score	(P)	(A)	(S)	Total
Low (L)	105	60	55	220
Medium (M)	70	175	145	390
High (H)	25	65	300	390
Total	200	300	500	1000

(a) Calculate the probability that an applicant picked at random from this group:
   1. Made a low score on the examination.
   2. Graduated from a superior high school.
   3. Made a low score on the examination and graduated from a superior high school.
   4. Made a low score on the examination given that he or she graduated from a superior high school.
   5. Made a high score or graduated from a superior high school.
(b) Calculate the following probabilities:

   1. $P(A)$ 3. $P(M)$ 5. $P(M \cap P)$
   2. $P(H)$ 4. $P(A|H)$ 6. $P(H|S)$

15. If the probability that a public health nurse will find a client at home is .7, what is the probability (assuming independence) that on two home visits made in a day both clients will be home?

16. The following table shows the outcome of 500 interviews completed during a survey to study the opinions of residents of a certain city about legalized abortion. The data are also classified by the area of the city in which the questionnaire was attempted.

	OUTCOME			
Area of city	For (F)	Against (Q)	Undecided (R)	Total
A	100	20	5	125
B	115	5	5	125
D	50	60	15	125
E	35	50	40	125
Total	300	135	65	500

(a) If a questionnaire is selected at random from the 500, what is the probability that:
  1. The respondent was for legalized abortion?
  2. The respondent was against legalized abortion?
  3. The respondent was undecided?
  4. The respondent lived in area A? B? D? E?
  5. The respondent was for legalized abortion, given that he/she resided in area B?
  6. The respondent was undecided or resided in area D?

(b) Calculate the following probabilities:

  1. $P(A \cap R)$ 3. $P(\overline{D})$ 5. $P(B|R)$
  2. $P(Q \cup D)$ 4. $P(Q|D)$ 6. $P(F)$

17. In a certain population the probability that a randomly selected subject will have been exposed to a certain allergen and experience a reaction to the allergen is .60. The probability is .8 that a subject exposed to the allergen will experience an allergic reaction. If a subject is selected at random from this population, what is the probability that he or she will have been exposed to the allergen?

18. Suppose that 3 percent of the people in a population of adults have attempted suicide. It is also known that 20 percent of the population are living below the poverty level. If these two events are independent, what is the probability that a person selected at random from the population will have attempted suicide *and* be living below the poverty level?

19. In a certain population of women 4 percent have had breast cancer, 20 percent are smokers, and 3 percent are smokers and have had breast cancer. A woman is selected at random from the population. What is the probability that she has had breast cancer or smokes or both?

20. The probability that a person selected at random from a population will exhibit the classic symptom of a certain disease is .2, and the probability that a person selected at random has the disease is .23. The probability that a person who has the symptom also has the disease is .18. A person selected at random from the population does not have the symptom; what is the probability that the person has the disease?

REFERENCES

References cited

1. Grace E. Bates, *Probability*, Addison-Wesley, Reading, Mass., 1965.

2. John R. Dixon, *A Programmed Introduction to Probability*, Wiley, New York, 1964.

3. Frederick Mosteller, Robert E. K. Rourke, and George B. Thomas, Jr., *Probability With Statistical Applications*, Second Edition, Addison-Wesley, Reading, Mass., 1970.

4. Boyd Earl, William Moore, and Wendell I. Smith, *Introduction to Probability*, McGraw-Hill, New York, 1963.

5. Simeon M. Berman, *The Elements of Probability*, Addison-Wesley, Reading, Mass., 1969.

6. Melvin Hausner, *Elementary Probability Theory*, Harper and Row, New York, 1971.

7. E. R. Mullins, Jr., and David Rosen, *Concepts of Probability*, Bogden and Quigley, New York, 1972.

8. John E. Freund, *Mathematical Statistics*, Prentice-Hall, Englewood Cliffs, N.J., 1962.

9. Robert V. Hogg and Allen T. Craig, *Introduction to Mathematical Statistics*, Macmillan, New York, 1965.

10. Alexander M. Mood, Franklin A. Graybill, and Duane C. Boes, *Introduction to the Theory of Statistics*, Third Edition, McGraw-Hill, New York, 1974.

11. I. Todhunter, *A History of the Mathematical Theory of Probability*, G. E. Stechert, New York, 1931.

12. F. N. David, *Games, Gods and Gambling*, Hafner, New York, 1962.

13. L. J. Savage, *Foundations of Statistics*, Wiley, New York, 1954.

14. A. N. Kolmogorov, *Foundations of the Theory of Probability*, Chelsea, New York, 1964. (Original German edition published in 1933.)

15. Joseph Breuer, *An Introduction to the Theory of Sets*, Prentice-Hall, Englewood Cliffs, N.J., 1958 (Translated by Howard F. Fehr.)

16. Robert R. Stoll, *Set Theory and Logic*, W. H. Freeman, San Francisco, 1963.

17. Lawrence P. Maher, Jr., *Finite Sets, Theory, Counting, and Applications*, Charles E. Merrill, Columbus, Ohio, 1968.

CHAPTER 3

Probability Distributions

3.1 INTRODUCTION

In the preceding chapter we introduced the basic concepts of probability as well as methods for calculating the probability of an event. We build on these concepts in the present chapter and explore ways of calculating the probability of an event under somewhat more complex conditions.

3.2 PROBABILITY DISTRIBUTIONS OF DISCRETE VARIABLES

Let us begin our discussion of probability distributions by considering the probability distribution of a discrete variable, which we shall define as follows:

Definition The probability distribution of a discrete random variable is a table, graph, formula, or other device used to specify all possible values of a discrete random variable along with their respective probabilities.

Example 3.2.1 A public health nurse has a case load of 50 families. Let us construct the probability distribution of X, the number of children per family for this population. We can do this with a table in which we list in one column, x, the possible values that X assumes, and in another column, $P(X = x)$, the probability with which X assumes a particular value, x. This has been done in Table 3.2.1.

Alternatively, we can present this probability distribution in the form of a graph as in Figure 3.2.1.

In Figure 3.2.1 the length of each vertical bar indicates the probability for the corresponding value of x.

It will be observed in the present example that the values of $P(X = x)$ are all positive, they are all less than 1, and their sum is equal to 1. These are not phenomena peculiar to this particular example, but are characteristics of all

TABLE 3.2.1

Probability Distribution of Number of Children per Family in a Population of 50 Families

x	Frequency of occurrence of x	$P(X = x)$
0	1	1/50
1	4	4/50
2	6	6/50
3	4	4/50
4	9	9/50
5	10	10/50
6	7	7/50
7	4	4/50
8	2	2/50
9	2	2/50
10	1	1/50
	50	50/50

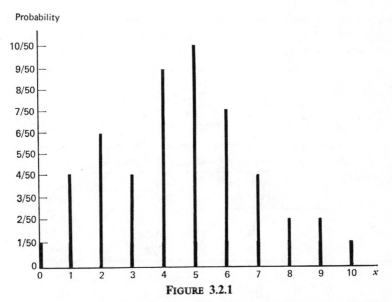

FIGURE 3.2.1

Graphical Representation of the Probability Distribution of Number of Children per Family for Population of 50 Families

probability distributions of discrete variables. We may then give the following two essential properties of a probability distribution of a discrete variable:

$$(1) \quad 0 \le P(X = x) \le 1$$
$$(2) \quad \sum P(X = x) = 1$$

The reader will also note that each of the probabilities in Table 3.2.1 is the *relative frequency of occurrence* of the corresponding value of X.

With its probability distribution available to us, we can make probability statements regarding the random variable X. Suppose the nurse, with the case load of 50 families, randomly picks a family to visit. What is the probability that the family will have three children? From Table 3.2.1 we see that the answer is $4/50 = .08$, that is, $P(X = 3) = .08$. What is the probability that a family chosen at random will have either three or four children? To answer this question we make use of the addition rule. Since the probability of three children is .08, and the probability of four children is .18, the probability of choosing a family having three or four children is $.08 + .18 = .26$. We may express this more compactly as

$$P(X = 3 \text{ or } X = 4) = P(X = 3) + P(X = 4) = .26$$

Cumulative Distributions   Sometimes it will be more convenient to work with the *cumulative probability distribution* of a random variable. The cumulative probability distribution for the discrete variable whose probability distribution is given in Table 3.2.1 may be obtained by successively adding the probabilities, $P(X = x)$,

TABLE 3.2.2

Cumulative Probability Distribution of Number of Children per Family in a Population of 50 Families

x	Frequency of occurrence of x	$P(X = x)$	$P(X \le x)$
0	1	1/50	1/50
1	4	4/50	5/50
2	6	6/50	11/50
3	4	4/50	15/50
4	9	9/50	24/50
5	10	10/50	34/50
6	7	7/50	41/50
7	4	4/50	45/50
8	2	2/50	47/50
9	2	2/50	49/50
10	1	1/50	50/50
	50	50/50	

given in the last column. The resulting cumulative probability distribution is shown in Table 3.2.2.

By consulting the cumulative probability distribution we may answer questions like the following.

1. What is the probability that a family picked at random from the 50 will have fewer than five children? What is needed to answer the question is $P(X < 5)$, and this is obtained by determining the value of the cumulative probability for values of $X = 0$ through $X = 4$. From the table we see that this is $24/50 = .48$.

2. What is the probability that a randomly picked family will have five or more children? We may answer this question by means of the concept of the complement. The set of families with five or more children is the complement set of the set of families with fewer than five children. Their sum is equal to the universal set of 50 families. Since the total probability is 1, and we have found that $P(X < 5) = .48$, the desired probability, $P(X \geq 5)$, is equal to $1 - P(X < 5) = 1 - .48 = .52$.

3. What is the probability that a randomly selected family will have between three and six children, inclusive? What is needed is $P(3 \leq X \leq 6)$, which is equal to $P(X \leq 6) - P(X < 3)$. The probability that X is less than or equal to 6 is the cumulative probability through $X = 6$, or $41/50 = .82$; and the probability that X is less than 3 is the cumulative probability through $X = 2$, or $11/50 = .22$. Thus, we see that $P(3 \leq X \leq 6) = .82 - .22 = .60$.

The probability distribution given in Table 3.2.1 was developed out of actual experience, and to find another variable following this distribution would be coincidental. The probability distributions of many variables of interest, however, can be determined or assumed on the basis of theoretical considerations. In the following sections, we study in detail three of these theoretical probability distributions, the *binomial*, the *Poisson*, and the *normal*.

3.3 THE BINOMIAL DISTRIBUTION

The *binomial distribution* is one of the most widely encountered probability distributions in applied statistics. The distribution is derived from a process known as a *Bernoulli trial*, named in honor of the Swiss mathematician James Bernoulli (1654–1705), who made significant contributions in the field of probability, including, in particular, the binomial distribution. When a single trial of some process or experiment can result in only one of two mutually exclusive outcomes, such as dead or alive, sick or well, male or female, the trial is called a Bernoulli trial.

A sequence of Bernoulli trials forms a *Bernoulli process* under the following conditions.

1. Each trial results in one of two possible, mutually exclusive, outcomes. One of the possible outcomes is denoted (arbitrarily) as a success, and the other is denoted a failure.
2. The probability of a success, denoted by p, remains constant from trial to trial. The probability of a failure, $1 - p$, is denoted by q.
3. The trials are independent; that is, the outcome of any particular trial is not affected by the outcome of any other trial.

Example 3.3.1 We are interested in being able to compute the probability of x successes in n Bernoulli trials. For example, suppose that in a certain population 52 percent of all recorded births are males. We interpret this to mean that the probability of a recorded male birth is .52. If we randomly select five birth records from this population, what is the probability that exactly three of the records will be for male births? Let us designate the occurrence of a record for a male birth as a "success," and hasten to add that this is an arbitrary designation for purposes of clarity and convenience and does not reflect an opinion regarding the relative merits of male versus female births. The occurrence of a birth record for a male will be designated a success, since we are looking for birth records of males. If we are looking for birth records of females, these would be designated successes, and birth records of males would be designated failures.

It will also be convenient to assign a value of 1 to a success (record for a male birth) and a value of 0 to a failure (record of a female birth).

The process that eventually results in a birth record we consider to be a Bernoulli process.

Suppose the five birth records selected resulted in this sequence of sexes:

MFMMF

In coded form we would write this as

10110

Since the probability of a success is denoted by p and the probability of a failure is denoted by q, the probability of the above sequence of outcomes is found by means of the multiplication rule to be

$$P(1,0,1,1,0) = pqppq = q^2p^3$$

The multiplication rule is appropriate for computing this probability since we are seeking the probability of a male, and a female, and a male, and a male, and a female, in that order or, in other words, the joint probability of the five events. For simplicity, commas, rather than intersection notation, have been used to separate the outcomes of the events in the probability statement above.

The resulting probability is that of obtaining the specific sequence of outcomes in the order shown. We are not, however, interested in the order of occurrence of records for male and female births but, instead, as has been stated already, the probability of the occurrence of exactly three records of male births out of five randomly selected records. Instead of occurring in the sequence shown above (call it sequence number 1), three successes and two failures could occur in any one of the following additional sequences as well:

Number	Sequence
2	11100
3	10011
4	11010
5	11001
6	10101
7	01110
8	00111
9	01011
10	01101

Each of these sequences has the same probability of occurring, and this probability is equal to $q^2 p^3$, the probability computed for the first sequence mentioned.

When we draw a single sample of size five from the population specified, we obtain only one sequence of successes and failures. The question now becomes, what is the probability of getting sequence number 1 or sequence number 2 $\cdots$ or sequence number 10? From the addition rule we know that this probability is equal to the sum of the individual probabilities. In the present example we need to sum the ten $q^2 p^3$'s or, equivalently, multiply $q^2 p^3$ by ten. We may now answer our original question: What is the probability, in a random sample of size 5, drawn from the specified population, of observing three successes (record of a male birth) and two failures (record of a female birth)? Since in the population, $p = .52$, $q = (1 - p) = (1 - .52) = .48$, the answer to the question is

$$10(.48)^2(.52)^3 = 10(.2304)(.140608) = .32$$

We can easily anticipate that, as the size of the sample increases, listing the number of sequences becomes more and more difficult and tedious. What is needed is an easy method of counting the number of sequences. Such a method is provided by Equation 2.5.6, since we have n things, some of which are of one type and the remainder are of another type. The number of sequences in our present example is found by Equation 2.5.6 to be $\binom{5}{2} = 5!/2!3! = 120/12 = 10$. In general, if we let n equal the total number of objects, x the number of objects of one type, and $n - x$ the number of objects of the other type, the number of

TABLE 3.3.1

The Binomial Distribution

Number of successes, x	Probability, $f(x)$
0	$\binom{n}{0} q^{n-0} p^0$
1	$\binom{n}{1} q^{n-1} p^1$
2	$\binom{n}{2} q^{n-2} p^2$
$\vdots$	$\vdots$
x	$\binom{n}{x} q^{n-x} p^x$
$\vdots$	$\vdots$
n	$\binom{n}{n} q^{n-n} p^n$
Total	1

sequences is equal to $\binom{n}{x} = n!/x!(n-x)!$ which is equal to the number of combinations of n things taken x at a time.

In our example we may let $x = 3$, the number of successes, so that $n - x = 2$, the number of failures. We then may write the probability of obtaining exactly x successes in n trials as

$$f(x) = \binom{n}{x} p^x q^{n-x} \quad \text{for } x = 0, 1, 2, \ldots, n \tag{3.3.1}$$

$$= 0, \quad \text{elsewhere}$$

This expression is called the binomial distribution. In Equation 3.3.1 $f(x) = P(X = x)$ where X is the random variable, number of successes in n trials. We use $f(x)$ rather than $P(X = x)$ because of its compactness and because of its almost universal use.

We may present the binomial distribution in tabular form as in Table 3.3.1.

We establish the fact that Equation 3.3.1 is a probability distribution by showing that

1. $f(x) \geq 0$ for all real values of x. This follows from the fact that n and p are both nonnegative and, hence, $\binom{n}{x}$, p^x, and $(1-p)^{n-x}$ are all nonnegative and, therefore, their product is greater than or equal to zero.

2. $\Sigma f(x) = 1$. This is seen to be true if we recognize that $\Sigma \binom{n}{x} q^{n-x} p^x$ is equal to $[(1-p)+p]^n = 1^n = 1$, the familiar binomial expansion. If the binomial

$(q + p)^n$ is expanded we have

$$(q + p)^n = q^n + nq^{n-1}p^1 + \frac{n(n-1)}{2}q^{n-2}p^2 + \cdots + nq^1p^{n-1} + p^n$$

If we compare the terms in the above expansion, term for term, with the $f(x)$ in Table 3.3.1 we see that they are, term for term, equivalent, since

$$f(0) = \binom{n}{0}q^{n-0}p^0 = q^n$$

$$f(1) = \binom{n}{1}q^{n-1}p^1 = nq^{n-1}p^1$$

$$f(2) = \binom{n}{2}q^{n-2}p^2 = \frac{n(n-1)}{2}q^{n-2}p^2$$

$$\vdots \qquad \vdots \qquad \vdots$$

$$f(n) = \binom{n}{n}q^{n-n}p^n = p^n$$

Example 3.3.2 As another example of the use of the binomial distribution, suppose that it is known that 30 percent of a certain population are immune to some disease. If a random sample of size 10 is selected from this population, what is the probability that it will contain exactly four immune persons? We take the probability of an immune person to be .3. Using Equation 3.3.1 we find

$$f(4) = \binom{10}{4}(.7)^6(.3)^4$$

$$= \frac{10!}{4!6!}(.117649)(.0081)$$

$$= .2001$$

The calculation of a probability using Equation 3.3.1 can be a rather tedious undertaking if the sample size is large. Fortunately, probabilities for different values of n, p, and x have been tabulated, so that we need only to consult an appropriate table to obtain the desired probability. Table A of the Appendix is one of many such tables available. It gives the probability that x is less than or equal to some specified value. That is, the table gives the cumulative probabilities from $x = 0$ up through some specified value.

Let us illustrate the use of the table by using the example above where it was desired to find the probability that $x = 4$ when $n = 10$ and $p = .3$. Drawing on our knowledge of cumulative probability distributions from the previous section, we know that $P(x = 4)$ may be found by subtracting $P(X \leq 3)$ from $P(X \leq 4)$. If in the table we locate $p = .3$ for $n = 10$ we find that $P(X \leq 4) = .8497$ and $P(X \leq 3) = .6496$. Subtracting the latter from the former gives $.8497 - .6496 = .2001$, which agrees with our hand calculation.

Let us look again at the example about the birth records. The objective was to determine the probability of observing exactly three records of male births when $n = 5$ and $p = .52$. To use the table we first locate $n = 5$. We discover, however, that the table does not contain an entry for $p = .52$. A little reflection will convince us that we can obtain the desired results if we restate the question as follows: What is the probability of observing exactly two records of female births if the probability of a record of a female birth is .48 and $n = 5$?

We can find $p = .48$ in the table for $n = 5$, and we see that $P(X \leq 2) = .5375$ and $P(X \leq 1) = .2135$. Subtraction gives $.5375 - .2135 = .324$, which agrees with the figure obtained previously.

Frequently we are interested in determining probabilities, not for specific values of X, but for intervals such as the probability that X is between, say, 5 and 10. Let us illustrate with an example.

Example 3.3.3 Suppose it is known that in a certain population 10 percent of the population is colorblind. If a random sample of 25 people is drawn from this population, use Table A in the Appendix to find the probability that:

1. Five or fewer will be colorblind.

Solution This probability is an entry in the table. No addition or subtraction is necessary. $P(X \leq 5) = .9666$.

2. Six or more will be colorblind.

Solution This set is the complement of the set specified in 1; therefore,

$$P(X \geq 6) = 1 - P(X \leq 5) = 1 - .9666 = .0334$$

3. Between six and nine inclusive will be colorblind.

Solution We find this by subtracting the probability that X is less than or equal to 5 from the probability that X is less than or equal to 9. That is,

$$P(6 \leq X \leq 9) = P(X \leq 9) - P(X \leq 5) = .9999 - .9666 = .0333$$

4. Two, three, or four will be colorblind.

Solution This is the probability that X is between 2 and 4 inclusive.

$$P(2 \leq X \leq 4) = P(X \leq 4) - P(X \leq 1) = .9020 - .2712 = .6308$$

The binomial distribution has two parameters, n and p. They are parameters in the sense that they are sufficient to specify a binomial distribution. The binomial distribution is really a family of distributions with each possible value of n and p designating a different member of the family. The mean and variance of the binomial distribution are $\mu = np$ and $\sigma^2 = np(1 - p)$, respectively.

Strictly speaking, the binomial distribution is applicable in situations where sampling is from an infinite population or from a finite population with replacement. Since in actual practice samples are usually drawn without replacement from finite populations, the question naturally arises as to the appropriateness of the binomial distribution under these circumstances. Whether or not the binomial is appropriate depends on how drastic is the effect of these conditions on the constancy of p from trial to trial. It is generally agreed that when n is small relative to N, the binomial model is appropriate. Some writers say that n is small relative to N if N is at least 10 times as large as n.

EXERCISES

In each of the following exercises, assume that N is sufficiently large relative to n that the binomial distribution may be used to find the desired probabilities.

3.3.1 Suppose that 24 percent of a certain population have blood group B. For a sample of size 20 drawn from this population find the probability that:

(a) Exactly three persons with blood group B will be found.
(b) Three or more persons with the characteristic of interest will be found.
(c) Fewer than three will be found.
(d) Exactly five will be found.

3.3.2 In a large population, 16 percent of the members are left-handed. In a random sample of size 10 find:

(a) The probability that exactly two will be left-handed.
(b) $P(X \geq 2)$.
(c) $P(X < 2)$.
(d) $P(1 \leq X \leq 4)$.

3.3.3 Suppose it is known that the probability of recovery for a certain disease is .4. If 15 people are stricken with the disease (assume this to be a random sample), what is the probability that:

(a) Three or more will recover?   (b) Four or more?
(c) Five or more?   (d) Fewer than three?

3.3.4 Suppose the mortality rate for a certain disease is .10, and suppose 10 people in a community contract the disease. What is the probability that:

(a) None will survive?   (b) Fifty percent will die?
(c) At least three will die?   (d) Exactly three will die?

3.3.5 The probability that a person suffering from migraine headache will obtain relief with a particular drug is .9. Three randomly selected sufferers from migraine headache are given the drug. Find the probability that the number obtaining relief will be:

(a) Exactly zero.   (b) Exactly one.
(c) More than one.   (d) Two or fewer.
(e) Two or three.   (f) Exactly three.

3.4 THE POISSON DISTRIBUTION

The next discrete distribution that we consider is the *Poisson distribution*, named for the French mathematician Simeon Denis Poisson (1781–1840), who is generally credited for publishing its derivation in 1837 (1, 2). This distribution has been used extensively in biology and medicine. Haight (2) has prepared a fairly extensive catalog of such applications in Chapter 7 of his book.

If x is the number of occurrences of some random event in an interval of time or space (or some volume of matter), the probability that x will occur is given by

$$f(x) = \frac{e^{-\lambda}\lambda^x}{x!}, \qquad x = 0, 1, 2, \ldots \tag{3.3.1}$$

The Greek letter λ (lambda) is called the parameter of the distribution and is the average number of occurrences of the random event in the interval (or volume). The symbol e is the constant (to four decimals) 2.7183.

It can be shown that $f(x) \geq 0$ for every x and that $\sum_x f(x) = 1$, so that the distribution satisfies the requirements for a probability distribution.

The Poisson Process We have seen that the binomial distribution results from a set of assumptions about an underlying process yielding a set of numerical observations. Such, also, is the case with the Poisson distribution. The following statements describe what is known as the *Poisson process*.

1. The occurrences of the events are independent. The occurrence of an event in an interval[1] of space or time has no effect on the probability of a second occurrence of the event in the same, or any other, interval.

2. Theoretically, an infinite number of occurrences of the event must be possible in the interval.

3. The probability of the single occurrence of the event in a given interval is proportional to the length of the interval.

---

[1]For simplicity the Poisson is discussed in terms of intervals, but other units such as a volume of matter are implied.

4. In any infinitesimally small portion of the interval the probability of more than one occurrence of the event is negligible.

An interesting feature of the Poisson distribution is the fact that the mean and variance are equal.

The Poisson distribution is employed when counts are made of events or entities that are distributed at random in space or time. One may suspect that a certain process obeys the Poisson law, and under this assumption probabilities of the occurrence of events or entities within some unit of space or time may be calculated. For example, under the assumption that the distribution of some parasite among individual host members follows the Poisson law, one may, with knowledge of the parameter λ, calculate the probability that a randomly selected individual host will yield x number of parasites. In a later chapter we will learn how to decide whether the assumption that a specified process obeys the Poisson law is plausible.

To illustrate the use of the Poisson distribution for computing probabilities, let us consider the following example.

Example 3.4.1 A hospital administrator, who has been studying daily emergency admissions over a period of several years, has concluded that they are distributed according to the Poisson law. Hospital records reveal that emergency admissions have averaged three per day during this period. If the administrator is correct in assuming a Poisson distribution find the probability that

1. Exactly two emergency admissions will occur on a given day.

Solution We let λ be 3 and X be a random variable denoting the number of daily emergency admissions. Then, if X follows the Poisson distribution

$$P(X = 2) = f(2) = \frac{e^{-3}3^2}{2!}$$

$$= \frac{.050(9)}{2 \cdot 1}$$

$$= .225$$

Values of e^x are available from most hand-held calculators.

2. No emergency admissions will occur on a particular day.

Solution

$$f(0) = \frac{e^{-3}3^0}{0!}$$

$$= \frac{.050(1)}{1}$$

$$= .05$$

3. Either three or four emergency cases will be admitted on a particular day.

Solution Since the two events are mutually exclusive, we use the addition rule to obtain

$$
\begin{aligned}
f(3) + f(4) &= \frac{e^{-3}3^3}{3!} + \frac{e^{-3}3^4}{4!} \\
&= \frac{.05(27)}{3 \cdot 2 \cdot 1} + \frac{.05(81)}{4 \cdot 3 \cdot 2 \cdot 1} \\
&= .225 + .16875 \\
&= .39
\end{aligned}
$$

In the foregoing example the probabilities were evaluated directly from the equation. We may, however, use Appendix Table B, which gives cumulative probabilities for various values of λ and X.

Example 3.4.2 In the study of a certain aquatic organism, a large number of samples were taken from a pond, and the number of organisms in each sample was counted. The average number of organisms per sample was found to be two. Assuming that the number of organisms follows a Poisson distribution, find the probability that:

1. The next sample taken will contain one or fewer organisms.

Solution In Table B we see that when $\lambda = 2$, the probability that $X \leq 1$ is .406. That is, $P(X \leq 1|2) = .406$.

2. The next sample taken will contain exactly three organisms.

Solution

$$
P(X = 3|2) = P(X \leq 3) - P(X \leq 2) = .857 - .677 = .180
$$

3. The next sample taken will contain more than five organisms.

Solution Since the set of more than five organisms does not include five we are asking for the probability that six or more organisms will be observed. This is obtained by subtracting the probability of observing five or fewer from 1. That is,

$$
P(X > 5|2) = 1 - P(X \leq 5) = 1 - .983 = .017
$$

EXERCISES

3.4.1 Suppose it is known that in a certain area of a large city the average number of rats per quarter block is five. Assuming that the number of rats follows a Poisson distribution, find the probability that in a randomly selected quarter block:

(a)   There are exactly five rats.
(b)   There are more than five rats.
(c)   There are fewer than five rats.
(d)   There are between five and seven rats, inclusive.

3.4.2   Suppose that over a period of several years the average number of deaths from a certain noncontagious disease has been 10. If the number of deaths from this disease follows the Poisson distribution, what is the probability that during the current year:

(a)   Exactly seven people will die from the disease?
(b)   Ten or more people will die from the disease?
(c)   There will be no deaths from the disease?

3.4.3   If the mean number of serious accidents per year in a large factory (where the number of employees remains constant) is five, find the probability that in the current year there will be:

(a)   Exactly seven accidents.
(b)   Ten or more accidents.
(c)   No accidents.
(d)   Fewer than five accidents.

3.4.4   In a study of the effectiveness of an insecticide against a certain insect a large area of land was sprayed. Later the area was examined for live insects by randomly selecting squares and counting the number of live insects per square. Past experience has shown the average number of live insects per square after spraying to be .5. If the number of live insects per square follows a Poisson distribution, what is the probability that a selected square will contain:

(a)   Exactly one live insect?   (b)   No live insects?
(c)   Exactly four live insects?   (d)   One or more live insects?

3.4.5   In a certain population an average of 13 new cases of esophageal cancer are diagnosed each year. If the annual incidence of esophageal cancer follows a Poisson distribution, find the probability that in a given year the number of newly diagnosed cases of esophageal cancer will be:

(a)   Exactly 10.   (b)   At least 8.
(c)   No more than 12.   (d)   Between 9 and 15, inclusive.
(e)   Fewer than 7.

3.5 CONTINUOUS PROBABILITY DISTRIBUTIONS

The probability distributions considered thus far, the binomial and the Poisson, are distributions of discrete variables. Let us now consider distributions of continuous random variables. In Chapter 1 we stated that a continuous variable is one that can assume any value within a specified interval of values assumed by the variable. Consequently, between any two values assumed by a continuous variable, there exist an infinite number of values.

To help us understand the nature of the distribution of a continuous random variable, let us consider the data presented in Table 1.5.1 and Figure 1.5.1. In the table we have 57 values of the random variable, weights in ounces of malignant tumors. The histogram of Figure 1.5.1 was constructed by locating specified points on a line representing the measurement of interest and erecting a series of rectangles, whose widths were the distances between two specified points on the line, and whose heights represented the number of values of the variable falling between the two specified points. The intervals defined by any two consecutive specified points we called class intervals. As was noted in Chapter 1, subareas of the histogram correspond to the frequencies of occurrence of values of the variable between the horizontal scale boundaries of these subareas. This provides a way whereby the relative frequency of occurrence of values between any two specified points can be calculated: merely determine the proportion of the histogram's total area falling between the specified points. This can be done more conveniently by consulting the relative frequency or cumulative relative frequency columns of Table 1.5.3.

Imagine now the situation where the number of values of our random variable is very large and the width of our class intervals is made very small. The resulting histogram could look like that shown in Figure 3.5.1.

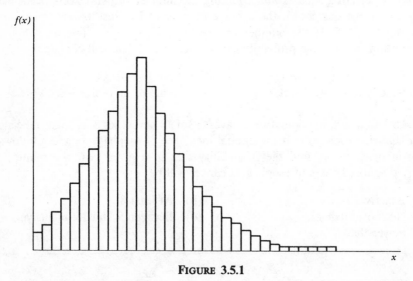

FIGURE 3.5.1

A Histogram Resulting from a Large Number of Values and Small Class Intervals

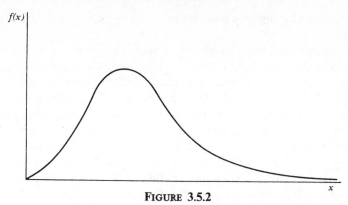

FIGURE 3.5.2

Graphical Representation of a Continuous Distribution

If we were to connect the midpoints of the cells of the histogram in Figure 3.5.1 to form a frequency polygon, clearly we would have a much smoother figure than the frequency polygon of Figure 1.5.3. It should not be difficult to believe that, in general, as the number of observations, n, approaches infinity, and the width of the class intervals approaches zero, the frequency polygon approaches a smooth curve such as is shown in Figure 3.5.2. Such smooth curves are used to represent graphically the distributions of continuous random variables. This has some important consequences when we deal with probability distributions. First, the total area under the curve is equal to one as was true with the histogram, and the relative frequency of occurrence of values between any two points on the x-axis is equal to the total area bounded by the curve, the x-axis, and perpendicular lines erected at the two points on the x-axis. See Figure 3.5.3. The probability of *any specific value* of the random variable is *zero*. This seems logical, since a

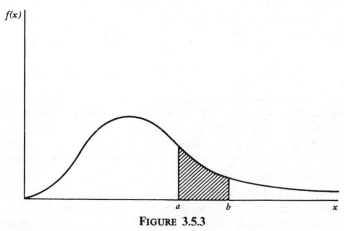

FIGURE 3.5.3

Graph of a Continuous Distribution Showing Area Between a and b

specific value is represented by a point on the x-axis and the area above a point is zero.

With a histogram, as we have seen, subareas of interest can be found by adding areas represented by the cells. We have no cells in the case of a smooth curve, so we must seek an alternative method of finding subareas. Such a method is provided by the integral calculus. To find the area under a smooth curve between any two points, a and b, the *density function* is integrated from a to b. A *density function* is a formula used to represent the distribution of a continuous random variable. Integration is the limiting case of summation, but we shall not perform any integrations, since the mathematics involved are beyond the scope of this book. As we shall see later, for all the continuous distributions we shall consider, there will be an easier way to find areas under their curves.

Although the definition of a probability distribution for a continuous random variable has been implied in the above discussion, by way of summary, we present it in a more compact form as follows.

Definition A nonnegative function $f(x)$ is called a probability distribution (sometimes called probability density function) of the continuous random variable X if the total area bounded by its curve and the x-axis is equal to 1 and if the subarea under the curve bounded by the curve, the x-axis, and perpendiculars erected at any two points a and b gives the probability that X is between the points a and b.

3.6 THE NORMAL DISTRIBUTION

We come now to the most important distribution in all of statistics—the *normal distribution*. The formula for this distribution was first published by Abraham De Moivre (1667–1754) on November 12, 1733 (3). Many other mathematicians figure prominently in the history of the normal distribution, including Carl Friedrich Gauss (1777–1855). The distribution is frequently called the *Gaussian distribution* in recognition of his contributions.

The normal density is given by

$$f(x) = \frac{1}{\sqrt{2\pi}\,\sigma} e^{-(x-\mu)^2/2\sigma^2}, \qquad -\infty < x < \infty \qquad (3.6.1)$$

In Equation 3.6.1, π and e are the familiar constants, 3.14159 and 2.71828, respectively, which are frequently encountered in mathematics. The two parameters of the distribution are μ, the mean, and σ, the standard deviation. For our purposes we may think of μ and σ of a normal distribution, respectively, as measures of central tendency and dispersion as discussed in Chapter 1. Since, however, a normally distributed random variable is continuous and takes on values between $-\infty$ and $+\infty$, its mean and standard deviation may be more rigorously defined; but such definitions cannot be given without using calculus.

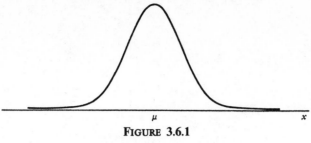

FIGURE 3.6.1

Graph of a Normal Distribution

The graph of the normal distribution produces the familiar bell-shaped curve shown in Figure 3.6.1.

The following are some important characteristics of the normal distribution.

1. It is symmetrical about its mean, μ. As is shown in Figure 3.6.1, the curve on either side of μ is a mirror image of the other side.

2. The mean, the median, and the mode are all equal.

3. The total area under the curve above the x-axis is one square unit. This characteristic follows from the fact that the normal distribution is a probability distribution. Because of the symmetry already mentioned, 50 percent of the area is to the right of a perpendicular erected at the mean, and 50 percent is to the left.

4. If we erect perpendiculars a distance of 1 standard deviation from the mean in both directions, the area enclosed by these perpendiculars, the x-axis, and the curve will be approximately 68 percent of the total area. If we extend these lateral boundaries a distance of 2 standard deviations on either side of the mean, approximately 95 percent of the area will be enclosed, and extending them a distance of 3 standard deviations will cause approximately 99.7 percent of the total area to be enclosed. These approximate areas are illustrated in Figure 3.6.2.

5. The normal distribution is completely determined by the parameters μ and σ. In other words, a different normal distribution is specified for each different value of μ and σ. Different values of μ shift the graph of the distribution along the x-axis as is shown in Figure 3.6.3. Different values of σ determine the degree of flatness or peakedness of the graph of the distribution as is shown in Figure 3.6.4.

The Unit Normal or Standard Normal Distribution The last-mentioned characteristic of the normal distribution implies that the normal distribution is really a family of distributions in which one member is distinguished from another on the basis of the values of μ and σ. The most important member of this family is the *unit normal or standard normal distribution*, so called because it has a mean of

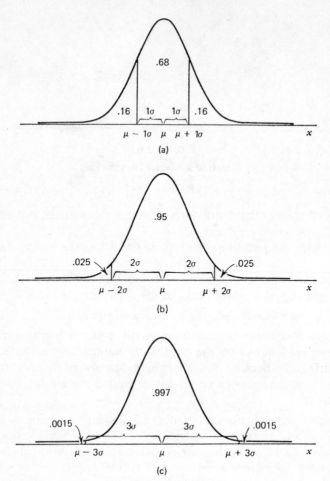

FIGURE 3.6.2

Subdivision of the Area Under the Normal Curve (Areas Are Approximate)

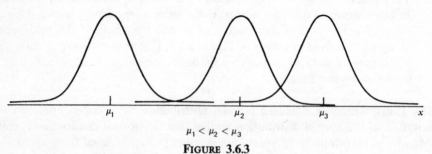

$\mu_1 < \mu_2 < \mu_3$

FIGURE 3.6.3

Three Normal Distributions with Different Means

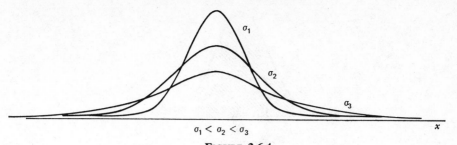

FIGURE 3.6.4

Three Normal Distributions with Different Standard Deviations

zero and a standard deviation of one. It may be obtained from Equation 3.6.1 by letting $\mu = 0$ and $\sigma = 1$. The random variable that results, $(x - \mu)/\sigma$, is usually designated by the letter z, so that the equation for the unit normal distribution is written

$$f(z) = \frac{1}{\sqrt{2\pi}} e^{-z^2/2}, \qquad -\infty < z < \infty \qquad (3.6.2)$$

The graph of the unit normal distribution is shown in Figure 3.6.5.

To find the probability that z takes on a value between any two points on the z-axis, say z_0 and z_1, we must find the area bounded by perpendiculars erected at these points, the curve, and the horizontal axis. As we mentioned previously, areas under the curve of a continuous distribution are found by integrating the function between two values of the variable. In the case of the unit normal, then, to find the area between z_0 and z_1, we need to evaluate the following integral:

$$\int_{z_0}^{z_1} \frac{1}{\sqrt{2\pi}} e^{-z^2/2} \, dz$$

Fortunately, we do not have to concern ourselves with the mathematics, since

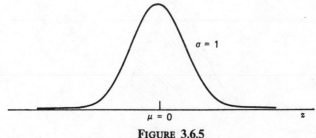

FIGURE 3.6.5

The Unit Normal Distribution

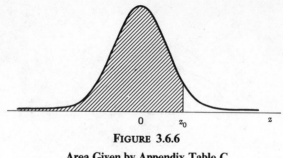

FIGURE 3.6.6

Area Given by Appendix Table C

there are tables available that provide the results of all such integrations in which we might be interested. Table C of the Appendix is an example of these tables. In the body of Table C are found the areas under the curve between $-\infty$ and the values of z shown in the leftmost column of the table. The shaded area in Figure 3.6.6 represents the area listed in the table as being between $-\infty$ and $z = z_0$.

We now illustrate the use of Table C by several examples.

Example 3.6.1   Given the unit normal distribution, find the area under the curve, above the z-axis between $-\infty$ and $z = 2$.

Solution   It will be helpful to draw a picture of the unit normal distribution and shade the desired area, as in Figure 3.6.7. If we locate $z = 2$ in Table C and read the corresponding entry in the body of the table, we find the desired area to be .9772. We may interpret this area in several ways. We may interpret it as the probability that a z picked at random from the population of z's will have a value between $-\infty$ and 2. We may also interpret it as the relative frequency of occurrence (or proportion) of values of z between $-\infty$ and 2, or we may say that 97.72 percent of the z's have a value between $-\infty$ and 2.

Example 3.6.2   What is the probability that a z picked at random from the population of z's will have a value between -2.55 and $+2.55$?

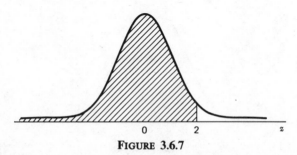

FIGURE 3.6.7

The Unit Normal Distribution Showing Area Between $-\infty$ and $z = 2$

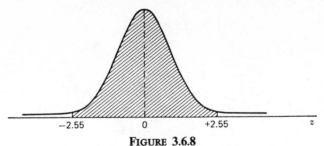

FIGURE 3.6.8

Unit Normal Curve Showing $P(-2.55 < z < 2.55)$

Solution Figure 3.6.8 shows the area desired. Table C gives us the area between $-\infty$ and 2.55, which is found by locating 2.5 in the left-most column of the table and then moving across until we come to the entry in the column headed by .05. We find this area to be .9946. If we look at the picture we draw, we see that this is more area than is desired. We need to subtract from .9946 the area to the left of -2.55. Reference to Table C shows that the area to the left of -2.55 is .0054. Thus the desired probability is

$$P(-2.55 < z < 2.55) = .9946 - .0054 = .9892.$$

Suppose we had been asked to find the probability that z is between -2.55 and 2.55 inclusive. The desired probability is expressed as $P(-2.55 \le z \le 2.55)$. Since, as we noted in Section 3.5, $P(z = z_0) = 0$, $P(-2.55 \le z \le 2.55) = P(-2.55 < z < 2.55) = .9892$.

Example 3.6.3 What proportion of z values are between -2.74 and 1.53?

Solution Figure 3.6.9 shows the area desired. We find in Table C that the area between $-\infty$ and 1.53 is .9370, and the area between $-\infty$ and -2.74 is

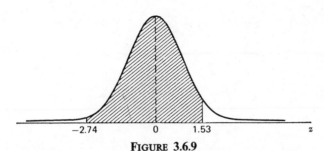

FIGURE 3.6.9

Unit Normal Curve Showing Proportion of z Values Between $z = -2.74$ and $z = 1.53$

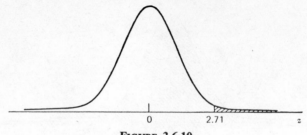

FIGURE 3.6.10
Unit Normal Distribution Showing $P(z \geq 2.71)$

.0031. To obtain the desired probability we subtract .0031 from .9370. That is,

$$P(-2.74 \leq z \leq 1.53) = .9370 - .0031 = .9339$$

Example 3.6.4   Given the unit normal distribution, find $P(z \geq 2.71)$.

Solution   The area desired is shown in Figure 3.6.10. We obtain the area to the right of $z = 2.71$ by subtracting the area between $-\infty$ and 2.71 from 1. Thus,

$$P(z \geq 2.71) = 1 - P(z \leq 2.71)$$
$$= 1 - .9966$$
$$= .0034$$

Example 3.6.5   Given the unit normal distribution, find $P(.84 \leq z \leq 2.45)$.

Solution   The area we are looking for is shown in Figure 3.6.11. We first obtain the area between $-\infty$ and 2.45 and from that subtract the area between $-\infty$ and .84. In other words,

$$P(.84 \leq z \leq 2.45) = P(z \leq 2.45) - P(z \leq .84)$$
$$= .9929 - .7995$$
$$= .1934$$

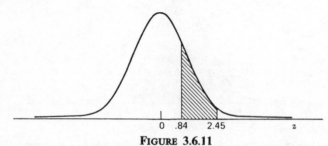

FIGURE 3.6.11
Unit Normal Curve Showing $P(.84 \leq z \leq 2.45)$

EXERCISES

Given the unit normal distribution find:
3.6.1 The area under the curve between $z = 0$ and $z = 1.43$.
3.6.2 The probability that a z picked at random will have a value between $z = -2.87$ and $z = 2.64$.
3.6.3 $P(z \geq .55)$.
3.6.4 $P(z \geq -.55)$.
3.6.5 $P(z < -2.33)$.
3.6.6 $P(z < 2.33)$.
3.6.7 $P(-1.96 \leq z \leq 1.96)$.
3.6.8 $P(-2.58 \leq z \leq 2.58)$.
3.6.9 $P(-1.65 \leq z \leq 1.65)$.
3.6.10 $P(z = .74)$.
Given the following probabilities, find z_1:
3.6.11 $P(z \leq z_1) = .0055$
3.6.12 $P(-2.67 \leq z \leq z_1) = .9718$
3.6.13 $P(z > z_1) = .0384$
3.6.14 $P(z_1 \leq z \leq 2.98) = .1117$
3.6.15 $P(-z_1 \leq z \leq z_1) = .8132$

Applications Although its importance in the field of statistics is indisputable, one should realize that the normal distribution is not a law that is adhered to by all measurable characteristics occurring in nature. It is true, however, that many of these characteristics are approximately normally distributed. Human stature and human intelligence are frequently cited as examples. On the other hand, Elveback et al. (4) and Nelson et al. (5) have pointed out that many distributions relevant to the health field cannot be described adequately by a normal distribution. Whenever it is known that a random variable is approximately normally distributed or when, in the absence of complete knowledge, it is considered reasonable to make this assumption, the statistician is aided tremendously in his or her efforts to solve practical problems relative to this variable.

There are several other reasons why the normal distribution is so important in statistics, and these will be considered in due time. For now, let us see how we may answer simple probability questions about random variables when we know, or are willing to assume, that they are, at least, approximately normally distributed.

Example 3.6.6 A physical therapist feels that scores on a certain manual dexterity test are approximately normally distributed with a mean of 10 and a standard deviation of 2.5. If a randomly selected individual takes the test, what is the probability that he or she will make a score of 15 or better?

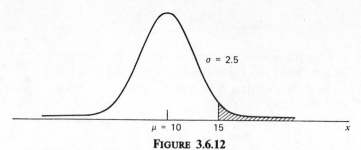

FIGURE 3.6.12

Normal Distribution to Approximate Distribution of Scores on a Manual Dexterity Test

Solution First let us draw a picture of the distribution and shade the area corresponding to the probability of interest. This has been done in Figure 3.6.12.

If our distribution were the unit normal distribution with a mean of 0 and a standard deviation of 1, we could make use of Table C and find the probability with little effort. Fortunately, it is possible for any normal distribution to be transformed easily to the unit normal. What we do is transform all values of X to corresponding values of z. This means that the mean of X must become 0, the mean of z. In Figure 3.6.13 both distributions are shown. We must determine what value of z, say z_0, corresponds to an x of 15. This is done by the following formula:

$$z = \frac{x - \mu}{\sigma} \qquad (3.6.3)$$

which transforms any value of x in any normal distribution to the corresponding

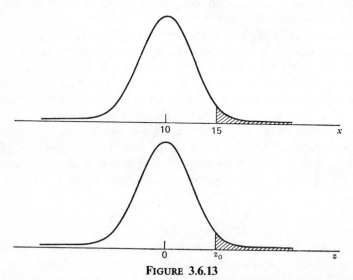

FIGURE 3.6.13

Normal Distribution of Manual Dexterity Scores (x) and the Unit Normal Distribution (z)

value of z in the unit normal distribution. For the present example we have

$$z = \frac{15 - 10}{2.5} = \frac{5}{2.5} = 2$$

The value of z_0 we seek, then, is 2.

Let us examine these relationships more closely. It is seen that the distance from the mean, 10, to the x-value of interest, 15, is $15 - 10 = 5$, which is a distance of 2 standard deviations. When we transform x values to z values, the distance of the z value of interest from its mean, 0, is equal to the distance of the corresponding x value from its mean, 10, in standard deviation units. We have seen that this latter distance is 2 standard deviations. In the z distribution a standard deviation is equal to one, and consequently the point on the z scale located a distance of 2 standard deviations from 0 is $z = 2$, the result obtained by employing the formula. By consulting Table C, we find that the area to the right of $z = 2$ is .0228. We may summarize the above discussion as follows:

$$P(x \geq 15) = P\left(z \geq \frac{15 - 10}{2.5}\right) = P(z \geq 2) = .0228$$

To answer our original question, we say that the probability that a randomly chosen individual taking the test will make a score of 15 or better is .0228.

Example 3.6.7 Suppose it is known that the weights of a certain group of individuals are approximately normally distributed with a mean of 140 pounds and a standard deviation of 25 pounds. What is the probability that a person picked at random from this group will weigh between 100 and 170 pounds?

Solution In Figure 3.6.14 are shown the distribution of weights and the z distribution to which we transform the original values to determine the desired

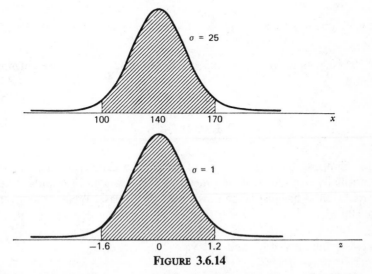

FIGURE 3.6.14

Distribution of Weights (x) and the Corresponding Unit Normal Distribution (z)

probabilities. We find the z value corresponding to an x of 100 by

$$z = \frac{100 - 140}{25} = -1.6$$

Similarly, for $x = 170$ we have

$$z = \frac{170 - 140}{25} = 1.2$$

From Table C we find the area between $-\infty$ and -1.6 to be .0548 and the area between $-\infty$ and 1.2 to be .8849. The area desired is the difference between these, $.8849 - .0548 = .8301$. To summarize

$$
\begin{aligned}
P(100 \le x \le 170) &= P\left(\frac{100 - 140}{25} \le z \le \frac{170 - 140}{25} \right) \\
&= P(-1.6 \le z \le 1.2) \\
&= P(-\infty \le z \le 1.2) - P(-\infty \le z \le -1.6) \\
&= .8849 - .0548 \\
&= .8301
\end{aligned}
$$

The probability asked for in our original question, then, is .8301.

Example 3.6.8   In a population of 10,000 of the people described in Example 3.6.7, how many would you expect to weigh more than 200 pounds?

Solution   We first find the probability that one person selected at random from the population would weigh more than 200 pounds. That is,

$$P(x > 200) = P\left(z > \frac{200 - 140}{25} \right) = P(z > 2.4) = 1 - .9918 = .0082$$

Out of 10,000 people we would expect $10,000(.0082) = 82$ to weigh more than 200 pounds.

EXERCISES

3.6.16   Suppose the ages at time of onset of a certain disease are approximately normally distributed with a mean of 11.5 years and a standard deviation of 3 years. A child has just come down with the disease. What is the probability that the child is:

(a)   Between the ages of $8\frac{1}{2}$ and $14\frac{1}{2}$ years?
(b)   Over 10 years of age?
(c)   Under 12?

3.6.17 In the study of fingerprints an important quantitative characteristic is the total ridge count for the 10 fingers of an individual. Suppose that the total ridge counts of individuals in a certain population are approximately normally distributed with a mean of 140 and a standard deviation of 50. Find the probability that an individual picked at random from this population will have a ridge count:

(a) Of 200 or more.
(b) Less than 100.
(c) Between 100 and 200.
(d) Between 200 and 250.
(e) In a population of 10,000 people how many would you expect to have a ridge count of 200 or more?

3.6.18 If the capacities of the cranial cavities of a certain population are approximately normally distributed with a mean of 1400 cc and a standard deviation of 125, find the probability that a person randomly picked from this population will have a cranial cavity capacity:

(a) Greater than 1450 cc.
(b) Less than 1350 cc.
(c) Between 1300 cc and 1500 cc.

3.6.19 Suppose the average length of stay in a chronic disease hospital of a certain type of patient is 60 days with a standard deviation of 15. If it is reasonable to assume an approximately normal distribution of lengths of stay, find the probability that a randomly selected patient from this group will have a length of stay:

(a) Greater than 50 days.
(b) Less than 30 days.
(c) Between 30 and 60 days.
(d) Greater than 90 days.

3.6.20 If the total cholesterol values for a certain population are approximately normally distributed with a mean of 200 mg/100 ml and a standard deviation of 20 mg/100 ml, find the probability that an individual picked at random from this population will have a cholesterol value:

(a) Between 180 and 200 mg/100 ml.
(b) Greater than 225 mg/100 ml.
(c) Less than 150 mg/100 ml.
(d) Between 190 and 210 mg/100 ml.

3.6.21 Given a normally distributed population with a mean of 75 and a variance of 625, find:

(a) $P(50 \leq x \leq 100)$.   (b) $P(x > 90)$.
(c) $P(x < 60)$.   (d) $P(x \geq 85)$.
(e) $P(30 \leq x \leq 110)$.

3.6.22 The weights of a certain population of young adult females are approximately normally distributed with a mean of 132 pounds and a standard deviation of 15. Find the probability that a subject selected at random from this population will weigh:

(a) More than 155 pounds.   (b) 100 pounds or less.
(c) Between 105 and 145 pounds.

3.7 SUMMARY

In the present chapter the concepts of probability described in the preceding chapter are further developed. The concepts of discrete and continuous random variables and their probability distributions are discussed. In particular, two discrete probability distributions, the binomial and the Poisson, and one continuous probability distribution, the normal, are examined in considerable detail. We have seen how these theoretical distributions allow us to make probability statements about certain random variables that are of interest to the health professional.

REVIEW QUESTIONS AND EXERCISES

1. What is a discrete random variable? Give three examples that are of interest to the health professional.
2. What is a continuous random variable? Give three examples of interest to the health professional.
3. Define the probability distribution of a discrete random variable.
4. Define the probability distribution of a continuous random variable.
5. What is a cumulative probability distribution?
6. What is a Bernoulli trial?

7. Describe the binomial distribution.

8. Give an example of a random variable which you think follows a binomial distribution.

9. Describe the Poisson distribution.

10. Give an example of a random variable that you think is distributed according to the Poisson law.

11. Describe the normal distribution.

12. Describe the unit normal distribution and tell how it is used in statistics.

13. Give an example of a random variable that you think is, at least approximately, normally distributed.

14. Using the data of your answer to question 13 above, demonstrate the use of the unit normal distribution in answering probability questions related to the variable selected.

15. The usual method for teaching a particular self-care skill to retarded persons is effective in 50 percent of the cases. A new method is tried with 10 persons. If the new method is no better than the standard, what is the probability that seven or more will learn the skill?

16. Personnel records of a large hospital show that 10 percent of housekeeping and maintenance employees quit within one year after being hired. If 10 new employees have just been hired:
    (a) What is the probability that exactly half of them will still be working after one year?
    (b) What is the probability that all will be working after one year?
    (c) What is the probability that 3 of the 10 will quit before the year is up?

17. In a certain developing country, 30 percent of the children are undernourished. In a random sample of 25 children from this area, what is the probability that the number of undernourished will be:
    (a) Exactly 10?
    (b) Less than five?
    (c) Five or more?
    (d) Between three and five inclusive?
    (e) Less than seven, but more than four?

18. On the average, two students per hour report for treatment to the first-aid room of a large elementary school.
    (a) What is the probability that during a given hour three students come to the first-aid room for treatment?
    (b) What is the probability that during a given hour two or fewer students will report to the first-aid room?
    (c) What is the probability that between three and five students, inclusive, will report to the first-aid room during a given hour?

19. On the average, five smokers pass a certain street corner every 10 minutes. What is the probability that during a given ten-minute period the number of

smokers passing will be:
(a)   Six or fewer?
(b)   Seven or more?
(c)   Exactly eight?

20.   In a certain metropolitan area there is an average of one suicide per month. What is the probability that during a given month the number of suicides will be:
(a)   Greater than one?
(b)   Less than one?
(c)   Greater than three?

21.   The IQs of individuals admitted to a state school for the mentally retarded are approximately normally distributed with a mean of 60 and a standard deviation of 10.
(a)   Find the proportion of individuals with IQs greater than 75.
(b)   What is the probability that an individual picked at random will have an IQ between 55 and 75?
(c)   Find $P(50 \leq X \leq 70)$.

22.   A nurse supervisor has found that staff nurses, on the average, complete a certain task in 10 minutes. If the times required to complete the task are approximately normally distributed with a standard deviation of 3 minutes, find:
(a)   The proportion of nurses completing the task in less than 4 minutes.
(b)   The proportion of nurses requiring more than 5 minutes to complete the task.
(c)   The probability that a nurse who has just been assigned the task will complete it within 3 minutes.

23.   Scores made on a certain aptitude test by nursing students are approximately normally distributed with a mean of 500 and a variance of 10,000.
(a)   What proportion of those taking the test score below 200?
(b)   A person is about to take the test; what is the probability that he or she will make a score of 650 or more?
(c)   What proportion of scores fall between 350 and 675?

24.   Given a binomial variable with a mean of 20 and a variance of 16, find n and p.

25.   Suppose a variable X is normally distributed with a standard deviation of 10. Given that .0985 of the values of X are greater than 70, what is the mean value of X?

26.   Given the normally distributed random variable X, find the numerical value of k such that $P(\mu - k\sigma \leq X \leq \mu + k\sigma) = .754$.

27.   Given the normally distributed random variable X with mean 100 and standard deviation 15, find the numerical value of k such that:
(a) $P(X \leq k) = .0094$
(b) $P(X \geq k) = .1093$

(c) $P(100 \le X \le k) = .4778$

(d) $P(k' \le X \le k) = .9660$, where k' and k are equidistant from μ.

28. Given the normally distributed random variable X with $\sigma = 10$ and $P(X \le 40) = .0080$, find μ.

29. Given the normally distributed random variable X with $\sigma = 15$ and $P(X \le 50) = .9904$, find μ.

30. Given the normally distributed random variable X with $\sigma = 5$ and $P(X \ge 25) = .0526$, find μ.

31. Given the normally distributed random variable X with $\mu = 25$ and $P(X \le 10) = .0778$, find σ.

32. Given the normally distributed random variable X with $\mu = 30$ and $P(X \le 50) = .9772$, find σ.

REFERENCES

References Cited
1. Normal L. Johnson and Samuel Kotz, *Discrete Distributions*, Houghton-Mifflin, Boston, 1969.

2. Frank A. Haight, *Handbook of the Poisson Distribution*, Wiley, New York, 1967.

3. Helen M. Walker, *Studies in the History of Statistical Method*, The Williams and Wilkins Company, Baltimore, 1931.

4. Lila R. Elveback, Claude L. Gulliver, and F. Raymond Deating, Jr., "Health, Normality and the Ghost of Gauss," *The Journal of the American Medical Association*, 211 (1970), 69–75.

5. Jerald C. Nelson, Elizabeth Haynes, Rodney Willard, and Jan Kuzma, "The Distribution of Euthyroid Serum Protein-Bound Iodine Levels," *The Journal of the American Medical Association*, 216 (1971), 1639–1641.

Other References
1. Sam Duker, "The Poisson Distribution in Educational Research," *Journal of Experimental Education*, 23 (1955), 265–269.

2. M. S. Lafleur, P. R. Hinrichsen, P. C. Landry, and R. B. Moore, "The Poisson Distribution: An Experimental Approach to Teaching Statistics," *Physics Teacher*, 10 (1972), 314–321.

CHAPTER 4

Some Important Sampling Distributions

4.1 INTRODUCTION

Before we examine the subject matter of the present chapter, let us review the high points of what we have covered thus far. In Chapter 1 the organization and summarization of data are emphasized. It is here that we encounter the concepts of central tendency and dispersion and learn how to compute their descriptive measures. In Chapter 2 we are introduced to the fundamental ideas of probability, and in Chapter 3 we consider the concept of a probability distribution. These concepts are fundamental to an understanding of statistical inference, the topic that comprises the major portion of this book.

The present chapter serves as a bridge between the preceding material, which is essentially descriptive in nature, and most of the remaining topics, which have been selected from the area of statistical inference.

4.2 SIMPLE RANDOM SAMPLING

Broadly speaking, there are two types of sampling, *probability sampling* and *nonprobability sampling*, but we consider only the former in this book. The reason is that only for probability sampling are there statistically sound procedures that allow us to infer, from the sample drawn, to the population of interest.

Definition   A probability sample is a sample drawn from a population in such a way that every member of the population has a known probability of being included in the sample.

From any finite population of size N, one may draw a number of different samples of size n. (This statement is made on the assumption that N is large enough to warrant sampling. As a rule, small populations, for obvious reasons, are not sampled. Instead, the entire population is examined.)

Definition If a sample of size n is drawn from a population of size N in such a way that every possible sample of size n has the same probability of being selected, the sample is called a *simple random sample*.

The mechanics of drawing a sample to satisfy the definition of a simple random sample is called *simple random sampling*.

We shall demonstrate the procedure of simple random sampling shortly, but first let us consider the problem of whether to sample *with replacement* or *without replacement*. When sampling with replacement is employed, every member of the population is available at each draw. For example, suppose that we are drawing a sample from a population of former hospital patients as part of a study of length of stay. Let us assume that the sampling involves selecting from the shelves in the medical record department a sample of charts of discharged patients. In sampling with replacement we would proceed as follows: select a chart to be in the sample, record the length of stay, and return the chart to the shelf. The chart is back in the "population" and may be drawn again on some subsequent draw, in which case the length of stay will again be recorded. In sampling without replacement, we would not return a drawn chart to the shelf after recording the length of stay, but would lay it aside until the entire sample is drawn. Following this procedure, a given chart could appear in the sample only once. As a rule, in practice, sampling is always done without replacement. The significance and consequences of this will be explained later, but first let us see how one goes about selecting a simple random sample. To ensure true randomness of selection we will need to follow some objective procedure. We certainly will want to avoid using our own judgment to decide which members of the population constitute a random sample.

Example 4.2.1 One way of selecting a simple random sample is to use a table of random numbers like that shown in Appendix Table D. Suppose our population of interest consists of the 150 fasting blood sugar values shown in Table 4.2.1.

From this population let us draw a simple random sample of size 10 using the random numbers in Table D. As the first step we locate a random starting point in the table. This can be done in a number of ways, one of which is to look away from the page while touching it with the point of a pencil. The random starting point is the digit closest to where the pencil touched the page. Let us assume that following this procedure led to a random starting point in Table D at the intersection of row 21 and column 28. The digit at this point is 5. Since we have 150 values to choose from, we can use only the random numbers 1 through 150. It will be convenient to pick three-digit numbers so that the the numbers 001 through 150 will be the only eligible numbers. The first three-digit number, beginning at our random starting point, is 532, a number we cannot use. Let us

TABLE 4.2.1

Fasting Blood Sugar Values of 150 Apparently Normal Individuals

Subject number	Value	Subject number	Value	Subject number	Value	Subject number	Value	Subject number	Value
1	91	31	107	61	87	91	91	121	90
2	94	32	94	62	104	92	104	122	105
3	115	33	101	63	109	93	109	123	100
4	85	34	95	64	93	94	92	124	89
5	89	35	80	65	95	95	85	125	90
6	107	36	104	66	107	96	108	126	106
7	94	37	94	67	88	97	99	127	94
8	105	38	102	68	107	98	103	128	100
9	94	39	89	69	113	99	81	129	92
10	103	40	98	70	95	100	96	130	91
11	104	41	106	71	102	101	105	131	87
12	105	42	85	72	94	102	91	132	105
13	88	43	93	73	99	103	115	133	102
14	107	44	103	74	87	104	108	134	101
15	90	45	119	75	102	105	102	135	111
16	95	46	90	76	105	106	101	136	91
17	104	47	82	77	80	107	94	137	92
18	93	48	90	78	90	108	93	138	98
19	109	49	113	79	108	109	102	139	81
20	87	50	104	80	105	110	119	140	117
21	92	51	97	81	90	111	96	141	103
22	117	52	101	82	115	112	104	142	96
23	98	53	90	83	82	113	85	143	101
24	89	54	88	84	90	114	108	144	88
25	105	55	108	85	102	115	103	145	100
26	101	56	95	86	91	116	90	146	100
27	81	57	100	87	103	117	105	147	95
28	108	58	103	88	107	118	99	148	103
29	94	59	108	89	107	119	88	149	101
30	104	60	85	90	97	120	103	150	90

move down past 196, 372, 654, and 928 until we come to 137, a number we can use. The 137th value from Table 4.2.1 is 92, the first value in our sample. We record the random number and the corresponding fasting blood sugar value in Table 4.2.2. We record the random number to keep track of the random numbers selected. Since we want to sample without replacement, we do not want to include the same individual's value twice. Proceeding in the manner just described leads us to the remaining nine random numbers and their corresponding blood sugar values shown in Table 4.2.2. Notice that when we get to the end of the column we simply move over three digits to 028 and proceed up the column. We could have started at the top with the number 369.

TABLE 4.2.2

Sample of 10 Fasting Blood Sugar Values Drawn from Data in Table 4.2.1

Random number	Sample subject number	Value
137	1	92
114	2	108
028	3	108
085	4	102
018	5	93
042	6	85
053	7	90
108	8	93
144	9	88
126	10	106

Thus we have drawn a simple random sample of size 10 from a population of size 150. In future discussion, whenever the term simple random sample is used, it will be understood that the sample has been drawn in this or an equivalent manner.

Many of the computers currently on the market are equipped with random number generating capabilities. As an alternative to using printed tables of random numbers, investigators may use computers to generate the random numbers they need. Actually, the "random" numbers generated by most computers are in reality *pseudorandom numbers* because they are the result of a deterministic formula. However, as Fishman (1) points out, the numbers appear to serve satisfactorily for many practical purposes.

EXERCISES

4.2.1 Using the table of random numbers, select a new random starting point, and draw another simple random sample of size 10 from the data in Table 4.2.1.

4.2.2 Draw a simple random sample of size 5 from the data in Table 1.5.1. Compute the mean, variance, and standard deviation for your sample.

4.3 SAMPLING DISTRIBUTIONS

We come now to the main topic of this chapter, *sampling distributions*. The importance of a clear understanding of sampling distributions cannot be overemphasized, as this concept is the very key to the understanding of statistical inference. Let us begin with the following definition.

Definition The distribution of all possible values which can be assumed by some statistic, computed from samples of the same size randomly drawn from the same population, is called the *sampling distribution* of that statistic.

Sampling distributions may be constructed empirically when sampling from a discrete, finite population. To construct a sampling distribution we proceed as follows.

1. From a finite population of size N, randomly draw all possible samples of size n.
2. Compute the statistic of interest for each sample.
3. List in one column the different distinct observed values of the statistic, and in another column list the corresponding frequency of occurrence of each distinct observed value of the statistic.

The actual construction of a sampling distribution is a formidable undertaking if the population is of any appreciable size and is an impossible task if the population is infinite. In such cases, sampling distributions may be approximated by taking a large number of samples of a given size.

We usually are interested in knowing three things about a given sampling distribution: its mean, its variance, and its functional form (how it looks when graphed).

We can recognize the difficulty of constructing a sampling distribution according to the steps given above when the population is of any appreciable size. We also run into a problem when considering the construction of a sampling distribution when the population is infinite. The best we can do experimentally in this case is to approximate the sampling distribution of a statistic.

Both of these problems may be obviated by means of mathematics. Although the procedures involved are not compatible with the mathematical level of this text, sampling distributions can be derived mathematically. The interested reader can consult one of many mathematical statistics textbooks, for example, Hoel (2) and Anderson and Bancroft (3).

In the sections that follow some of the more frequently encountered sampling distributions are discussed.

4.4 DISTRIBUTION OF THE SAMPLE MEAN

An important sampling distribution is the distribution of the sample mean. Let us see how we might construct this sampling distribution by following the steps outlined in the previous section.

TABLE 4.4.1

All Possible Samples of Size $n = 2$ from a Population of Size $N = 5$. Samples Above or Below the Principal Diagonal Result when Sampling Is without Replacement. Sample Means Are in Parentheses

		SECOND DRAW				
		6	8	10	12	14
	6	6, 6 (6)	6, 8 (7)	6, 10 (8)	6, 12 (9)	6, 14 (10)
	8	8, 6 (7)	8, 8 (8)	8, 10 (9)	8, 12 (10)	8, 14 (11)
First draw	10	10, 6 (8)	10, 8 (9)	10, 10 (10)	10, 12 (11)	10, 14 (12)
	12	12, 6 (9)	12, 8 (10)	12, 10 (11)	12, 12 (12)	12, 14 (13)
	14	14, 6 (10)	14, 8 (11)	14, 10 (12)	14, 12 (13)	14, 14 (14)

Example 4.4.1 Suppose we have a population of size $N = 5$, consisting of the ages of five children who are outpatients in a community mental health center. The ages are as follows: $x_1 = 6$, $x_2 = 8$, $x_3 = 10$, $x_4 = 12$, and $x_5 = 14$. The mean, μ, of this population is equal to $\sum x_i / N = 10$ and the variance

$$\sigma^2 = \frac{\sum (x_i - \mu)^2}{N} = \frac{40}{5} = 8$$

Let us compute another measure of dispersion as follows:

$$S^2 = \frac{\sum (x_i - \mu)^2}{N - 1} = \frac{40}{4} = 10$$

We will refer to this quantity again in the next chapter. Let us draw all possible samples of size $n = 2$ from this population. These samples, along with their means, are shown in Table 4.4.1.

We see in this example that when sampling is with replacement, there are 25 possible samples. In general, when sampling is with replacement, the number of possible samples is equal to N^n.

We may construct the sampling distribution of $\bar{x}$ by listing the different values of $\bar{x}$ in one column and their frequency of occurrence in another, as in Table 4.4.2.

We see that the data of Table 4.4.2 satisfy the requirements for a probability distribution. The individual probabilities are all greater than 0, and their sum is equal to 1.

TABLE 4.4.2

Sampling Distribution of $\bar{x}$ Computed from Samples in Table 4.4.1

$\bar{x}$	Frequency	Relative frequency
6	1	1/25
7	2	2/25
8	3	3/25
9	4	4/25
10	5	5/25
11	4	4/25
12	3	3/25
13	2	2/25
14	1	1/25
Total	25	25/25

It was stated earlier that we are usually interested in the functional form of a sampling distribution, its mean, and its variance.

Let us look at the distribution of $\bar{x}$ plotted as a histogram, along with the distribution of the population, both of which are shown in Fig. 4.4.1. The reader is no doubt impressed with the radical difference in appearance between the histogram of the population and the histogram of the sampling distribution of $\bar{x}$. Whereas the former is uniformly distributed, the latter gradually rises to a peak and then drops off with perfect symmetry. Now let us compute the mean, which we will call $\mu_{\bar{x}}$, of our sampling distribution. To do this we add the 25 sample means and divide by 25. Thus

$$\mu_{\bar{x}} = \frac{\sum \bar{x}_i}{N^n} = \frac{6 + 7 + 7 + 8 + \cdots + 14}{25} = \frac{250}{25} = 10$$

We note with interest that the mean of the sampling distribution of $\bar{x}$ has the same value as the mean of the original population.

Finally, we may compute the variance of $\bar{x}$, which we call $\sigma_{\bar{x}}^2$, as follows:

$$\sigma_{\bar{x}}^2 = \frac{\sum (\bar{x}_i - \mu_{\bar{x}})^2}{N^n}$$

$$= \frac{(6 - 10)^2 + (7 - 10)^2 + (7 - 10)^2 + \cdots + (14 - 10)^2}{25}$$

$$= \frac{100}{25} = 4$$

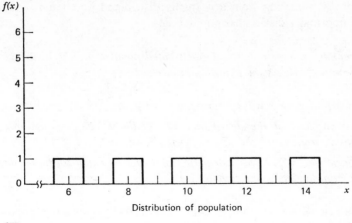

Distribution of population

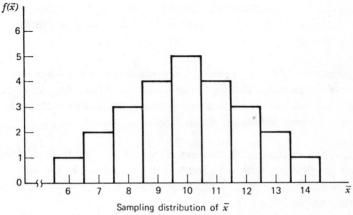

Sampling distribution of $\bar{x}$

FIGURE 4.4.1

Distribution of Population and Sampling Distribution of $\bar{x}$

We note that the variance of the sampling distribution is not equal to the population variance. It is of interest to observe, however, that the variance of the sampling distribution is equal to the population variance divided by the size of the sample used to obtain the sampling distribution. That is,

$$\sigma_{\bar{x}}^2 = \frac{\sigma^2}{n} = \frac{8}{2} = 4$$

The square root of the variance of the sampling distribution, $\sqrt{\sigma_{\bar{x}}^2} = \sigma/\sqrt{n}$, is called the *standard error of the mean* or, simply, the *standard error*.

These results are not coincidences but are examples of the characteristics of sampling distributions in general, when sampling is with replacement or when sampling is from an infinite population. To generalize, we distinguish between two situations: sampling from a normally distributed population and sampling from a nonnormally distributed population.

When sampling is from a normally distributed population, the distribution of the sample mean will possess the following properties:

1. *The distribution of $\bar{x}$ will be normal.*
2. *The mean, $\mu_{\bar{x}}$, of the distribution of $\bar{x}$ will be equal to the mean of the population from which the samples were drawn.*
3. *The variance, $\sigma_{\bar{x}}^2$, of the distribution of $\bar{x}$ will be equal to the variance of the population divided by the sample size.*

For the case where sampling is from a nonnormally distributed population, we refer to an important mathematical theorem known as the *central limit theorem*. The importance of this theorem in statistical inference may be summarized in the following statement.

Given a population of any nonnormal functional form with a mean, μ, and finite variance, σ^2, the sampling distribution of $\bar{x}$, computed from samples of size n from this population, will be approximately normally distributed with mean, μ, and variance, σ^2/n, when the sample size is large.

Note that the central limit theorem allows us to sample from nonnormally distributed populations with a guarantee of approximately the same results as would be obtained if the populations were normally distributed provided that we take a large sample.

The importance of this will become evident later when we learn that a normally distributed sampling distribution is a powerful tool in statistical inference. In the case of the sample mean, we are assured of at least an approximately normally distributed sampling distribution under three conditions: (1) when sampling from a normally distributed population; (2) when sampling from a nonnormally distributed population and our sample is large; and (3) when sampling from a population whose functional form is unknown to us so long as our sample size is large.

The logical question that arises at this point is: How large does the sample have to be in order for the central limit theorem to apply? There is no one answer, since the size of the sample needed depends on the extent of nonnormality present in the population. One rule of thumb states that, in most practical situations, a sample of size 30 is satisfactory. Let us add that, in general, the approximation to normality of the sampling distribution of $\bar{x}$ becomes better and better as the sample size increases.

The above results have been given on the assumption that sampling is either with replacement or that the samples are drawn from infinite populations. In general, we do not sample with replacement, and in most practical situations it is necessary to sample from a finite population; hence, we need to become familiar with the behavior of the sampling distribution of the sample mean under these conditions. Before making any general statements, let us again look at the data in Table 4.4.1. The sample means that result when sampling is without replacement are those above the principal diagonal, which are the same as those below the principal diagonal, if we ignore the order in which the observations were drawn. We see that there are 10 possible samples. In general, when drawing samples of size n from a finite population of size N without replacement, and ignoring the order in which the sample values are drawn, the number of possible samples is given by the combination of N things·taken n at a time. In our present example we have

$$\binom{N}{n} = \frac{N!}{n!(N-n)!} = \frac{5!}{2!3!} = \frac{5 \cdot 4 \cdot 3!}{2!3!} = 10 \text{ possible samples}$$

The mean of the 10 sample means is

$$\mu_{\bar{x}} = \frac{\sum x_i}{\binom{N}{n}} = \frac{7 + 8 + 9 + \cdots + 13}{10} = \frac{100}{10} = 10$$

We see that once again the mean of the sampling distribution is equal to the population mean.

The variance of this sampling distribution is found to be

$$\sigma_{\bar{x}}^2 = \frac{\sum(\bar{x}_i - \mu_{\bar{x}})^2}{\binom{N}{n}} = \frac{30}{10} = 3$$

and we note that this time the variance of the sampling distribution is not equal to the population variance divided by the sample size, since $\sigma_{\bar{x}}^2 = 3 \neq 8/2 = 4$. There is, however, an interesting relationship which we discover by multiplying σ^2/n by $(N-n)/(N-1)$. That is,

$$\frac{\sigma^2}{n} \cdot \frac{N-n}{N-1} = \frac{8}{2} \cdot \frac{5-2}{4} = 3$$

This result tells us that if we multiply the variance of the sampling distribution that would be obtained if sampling were with replacement, by the factor $(N-n)/(N-1)$, we obtain the value of the variance of the sampling distribution that results when sampling is without replacement. We may generalize these results with the following statement.

When sampling is without replacement from a finite population the sampling *distribution of $\bar{x}$ will have mean μ and variance*

$$\frac{\sigma^2}{n} \cdot \frac{N-n}{N-1}$$

If the sample size is large, the central limit theorem applies and the sampling distribution of $\bar{x}$ will be approximately normally distributed.

The factor $(N-n)/(N-1)$ is called the finite population correction and can be ignored when the sample size is small in comparison with the population size. When the population is much larger than the sample, the difference between σ^2/n and $(\sigma^2/n)[(N-n)/(N-1)]$ will be negligible. Suppose a population contains 10,000 observations and a sample from this population consists of 25 observations; the finite population correction would be equal to $(10,000-25)/(9999) = .9976$. To multiply .9976 times σ^2/n is almost equivalent to multiplying by 1. Most practicing statisticians do not use the finite population correction unless the sample contains more than 5 percent of the observations in the population. That is, the finite population correction is usually ignored when $n/N \leq .05$.

Applications As we shall see in succeeding chapters, a knowledge and understanding of sampling distributions will be necessary for understanding the concepts of statistical inference. The simplest application of our knowledge of the sampling distribution of the sample mean is in computing the probability of obtaining a sample with a mean of some specified magnitude. Let us illustrate with some examples.

Example 4.4.2 Suppose it is known that in a certain large human population, cranial length is approximately normally distributed with a mean of 185.6 mm and a standard deviation of 12.7 mm. What is the probability that a random sample of size 10 from this population will have a mean greater than 190?

Solution We know that the single sample under consideration is one of the all possible samples of size 10 which can be drawn from the population, so that the mean which it yields is one of the $\bar{x}$'s constituting the sampling distribution of $\bar{x}$ which, theoretically, could be derived from this population.

When we say that the population is approximately normally distributed, we assume that the sampling distribution of $\bar{x}$ will be, for all practical purposes, normally distributed. We also know that the mean and standard deviation of the sampling distribution are equal to 185.6 and $\sqrt{(12.7)^2/10} = 12.7/\sqrt{10} = 4.02$, respectively. We assume that the population is large relative to the sample so that the finite population correction can be ignored.

We learned in the preceding chapter that whenever we have a random variable that is normally distributed, we may very easily transform it to the unit normal distribution. Our random variable now is $\bar{x}$, the mean of its distribution is $\mu_{\bar{x}}$, and its standard deviation is $\sigma_{\bar{x}} = \sigma/\sqrt{n}$. By appropriately modifying the formula given previously, we arrive at the following formula for transforming the normal distribution of $\bar{x}$ to the unit normal distribution.

$$ z = \frac{\bar{x} - \mu_{\bar{x}}}{\sigma/\sqrt{n}} \tag{4.4.1} $$

The probability that answers our question is represented by the area to the right

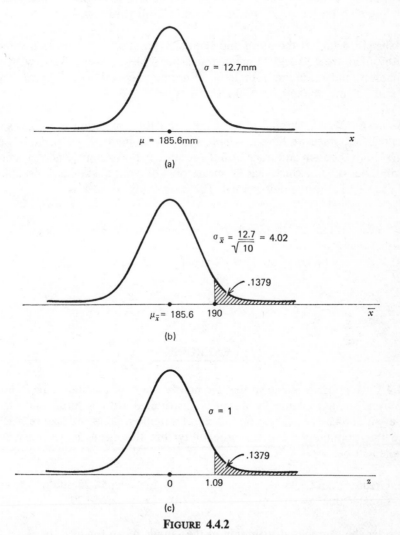

FIGURE 4.4.2

Population Distribution, Sampling Distribution, and Unit Normal Distribution, Example 4.4.2. (a) Population Distribution. (b) Sampling Distribution of $\bar{x}$ for Samples of Size 10. (c) Unit Normal Distribution

of $\bar{x} = 190$ under the curve of the sampling distribution. This area is equal to the area to the right of

$$z = \frac{190 - 185.6}{4.02} = \frac{4.4}{4.02} = 1.09$$

By consulting the unit normal table we find that the area to the right of 1.09 is .1379; hence, we say that the probability is .1379 that a sample of size 10 will have a mean greater than 190.

Figure 4.4.2 shows the relationship between the original population, the sampling distribution of $\bar{x}$, and the unit normal distribution.

Example 4.4.3 If the mean and standard deviation of serum iron values for healthy men are 120 and 15 micrograms per 100 ml, respectively, what is the probability that a random sample of 50 normal men will yield a mean between 115 and 125 micrograms per 100 ml?

Solution The functional form of the population of serum iron values is not specified, but since we have a sample size greater than 30, we make use of the central limit theorem and transform the resulting approximately normal sampling distribution of $\bar{x}$ (which has a mean of 120 and a standard deviation of $15/\sqrt{50} = 2.12$) to the unit normal. The probability we seek is

$$P(115 \le \bar{x} \le 125) = P\left(\frac{115 - 120}{2.12} \le z \le \frac{125 - 120}{2.12} \right)$$

$$= P(-2.36 \le z \le 2.36)$$

$$= .9909 - .0091$$

$$= .9818$$

EXERCISES

4.4.1 Suppose it is known that the hourly wages of a certain type of hospital employee are approximately normally distributed with a mean and standard deviation of $4.50 and $.50, respectively. If a random sample of size 16 is selected from this population, find the probability that the mean hourly wage for the sample will be:

(a) Greater than $4.25.   (b) Between $4.25 and $4.75.
(c) Greater than $4.80.   (d) Less than $4.20.

Consider the sampling distribution of the sample mean for $n = 16$.

(e) What percent of the sample means are greater than 4.65?

(f)   Suppose for this problem $\binom{N}{n} = 100,000$. How many of the sample means are greater than 4.30?

4.4.2   It has been found that, following a period of training, the mean time required for certain handicapped persons to perform a particular task is 25 seconds with a standard deviation of 5 seconds. Assuming a normal distribution of times, find the probability that a sample of 25 individuals will yield a mean:

(a)   Of 26 seconds or more.   (b)   Between 24 and 27 seconds.
(c)   26 seconds or less.   (d)   Greater than 22 seconds.

4.4.3   If the uric acid values in normal adult males are approximately normally distributed with a mean and standard deviation of 5.7 and 1 mg percent respectively, find the probability that a sample of size 9 will yield a mean:

(a)   Greater than 6.   (b)   Between 5 and 6.
(c)   Less than 5.2.

4.4.4   For a certain large segment of the population, for a particular year, suppose the mean number of days of disability is 5.4 with a standard deviation of 2.8 days. Find the probability that a random sample of size 49 from this population will have a mean:

(a)   Greater than 6 days.   (b)   Between 4 and 6 days.
(c)   Between $4\frac{1}{2}$ and $5\frac{1}{2}$ days.

4.4.5   Given a normally distributed population with a mean of 100 and a standard deviation of 20, find the following probabilities based on a sample of size 16:

(a) $P(\bar{x} \geq 100)$.   (b) $P(96 \leq \bar{x} \leq 108)$.
(c) $P(\bar{x} \leq 110)$.

4.4.6   Given: $\mu = 50$, $\sigma = 16$, $n = 64$, find:

(a) $P(45 \leq \bar{x} \leq 55)$.   (b) $P(\bar{x} > 53)$.
(c) $P(\bar{x} < 47)$.   (d) $P(49 \leq \bar{x} \leq 56)$.

4.4.7   Suppose a population consists of the following values: 1, 3, 5, 7, 9. Construct the sampling distribution of $\bar{x}$ based on samples of size two selected without replacement. Find the mean and variance of the sampling distribution.

4.4.8   Use the data of Example 4.4.1 to construct the sampling distribution of $\bar{x}$ based on samples of size three selected without replacement. Find the mean and variance of the sampling distribution.

4.4.9 For a population of 17-year-old boys, the mean subscapular skinfold thickness (in millimeters) is 9.7 and the standard deviation is 6.0. For a simple random sample of size 40 drawn from this population find the probability that the sample mean will be:

(a) Greater than 11.   (b) Less than or equal to 7.5.

(c) Between 7 and 10.5.

4.5 DISTRIBUTION OF THE DIFFERENCE BETWEEN TWO SAMPLE MEANS

Frequently the interest in an investigation is focused on two populations. Specifically, an investigator may wish to know something about the difference between two population means. In one investigation, for example, a researcher may wish to know if it is reasonable to conclude that two population means are different. In another situation, the researcher may desire knowledge about the magnitude of the difference between two population means. A medical research team, for example, may want to know whether or not the mean serum cholesterol level is higher in a population of sedentary office workers than in a population of laborers. If the researchers are able to conclude that the population means are different, they may wish to know by how much they differ. A knowledge of the sampling distribution of the difference between two means is useful in investigations of this type.

Example 4.5.1 Suppose we have two groups of individuals—one group (group 1) has experienced some condition felt to be associated with mental retardation, and the other group (group 2) has not experienced the condition. The distribution of intelligence scores in each of the two groups is believed to be approximately normally distributed with a standard deviation of 20.

Suppose, further, that we take a sample of 15 individuals from each group and compute for each sample the mean intelligence score with the following results: $\bar{x}_1 = 92$ and $\bar{x}_2 = 105$. If there is no difference between the two groups, with respect to their true mean intelligence scores, what is the probability of observing this large a difference between sample means?

To answer this question we need to know the nature of the sampling distribution of the relevant statistic, the *difference between two sample means*, $\bar{x}_1 - \bar{x}_2$. Notice that we seek a probability associated with the difference between two sample means rather than a single mean.

Although, in practice, we would not attempt to construct the desired sampling distribution, we can conceptualize the manner in which it could be done when sampling is from finite populations. We would begin by selecting from group 1 all possible samples of size 15 and computing the mean for each sample. We know

TABLE 4.5.1

Working Table for Constructing the Distribution of the Difference Between Two Sample Means

Samples from population 1	Samples from population 2	Sample means population 1	Sample means population 2	All possible differences between means
n_{11}	n_{12}	$\bar{x}_{11}$	$\bar{x}_{12}$	$\bar{x}_{11} - \bar{x}_{12}$
n_{21}	n_{22}	$\bar{x}_{21}$	$\bar{x}_{22}$	$\bar{x}_{11} - \bar{x}_{22}$
n_{31}	n_{32}	$\bar{x}_{31}$	$\bar{x}_{32}$	$\bar{x}_{11} - \bar{x}_{22}$
$\vdots$	$\vdots$	$\vdots$	$\vdots$	$\vdots$
$n_{\binom{N_1}{n_1}}^{1}$	$n_{\binom{N_2}{n_2}}^{2}$	$\bar{x}_{\binom{N_1}{n_1}}^{1}$	$\bar{x}_{\binom{N_2}{n_2}}^{2}$	$\bar{x}_{\binom{N_1}{n_1}}^{1} - \bar{x}_{\binom{N_2}{n_2}}^{2}$

that there would be $\binom{N_1}{n_1}$ such samples where N_1 is the group size and $n_1 = 15$. Similarly, we would select all possible samples of size 15 from group 2 and compute the mean for each of these samples. We would then take all possible pairs of sample means, one from group 1 and one from group 2, and take the difference. Table 4.5.1 shows the results of following this procedure. Note that the 1's and 2's in the last line of this table are not exponents, but indicators of group 1 and 2, respectively.

It is the distribution of the differences between sample means that we seek. If we plotted the sample differences against their frequency of occurrence, we would obtain a normal distribution with a mean equal to $\mu_1 - \mu_2$, the difference between the true group, or population, means, and a variance equal to $(\sigma_1^2/n_1) + (\sigma_2^2/n_2)$. That is, the standard error of the difference between sample means would be equal to $\sqrt{(\sigma_1^2/n_1) + (\sigma_2^2/n_2)}$.

For our present example we would have a normal distribution with a mean of 0 (if there is no difference between the true population means) and a variance of $[(20)^2/15] + [(20)^2/15] = 53.33$. The graph of the sampling distribution is shown in Figure 4.5.1.

We know that this normal distribution can be transformed to the unit normal distribution by means of a modification of a previously learned formula. The new formula is as follows:

$$z = \frac{(\bar{x}_1 - \bar{x}_2) - (\mu_1 - \mu_2)}{\sqrt{\dfrac{\sigma_1^2}{n_1} + \dfrac{\sigma_2^2}{n_2}}} \tag{4.5.1}$$

The area under the curve of $\bar{x}_1 - \bar{x}_2$ corresponding to the probability we seek is the area to the left of $\bar{x}_1 - \bar{x}_2 = 92 - 105 = -13$. The z value corresponding

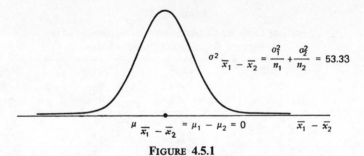

$$\sigma^2 _{\bar{x}_1 - \bar{x}_2} = \frac{\sigma_1^2}{n_1} + \frac{\sigma_2^2}{n_2} = 53.33$$

$$\mu_{\bar{x}_1 - \bar{x}_2} = \mu_1 - \mu_2 = 0 \qquad \bar{x}_1 - \bar{x}_2$$

FIGURE 4.5.1

Graph of the Sampling Distribution of $\bar{x}_1 - \bar{x}_2$ When There Is No Difference Between Population Means, Example 4.5.1

to -13, assuming there is no difference between population means, is

$$z = \frac{-13 - 0}{\sqrt{\dfrac{(20)^2}{15} + \dfrac{(20)^2}{15}}} = \frac{-13}{\sqrt{53.3}} = \frac{-13}{7.3} = -1.78$$

By consulting Table C, we find that the area under the unit normal curve to the left of -1.78 is equal to .0375. In answer to our original question, we say that if there is no difference between population means, the probability of obtaining a difference between sample means as large as or larger than 13 is .0375.

The procedure we have just followed is valid even when the sample sizes, n_1 and n_2, are different and when the population variances, σ_1^2 and σ_2^2 have different values. The theoretical results on which this procedure is based may be summarized as follows.

Given two normally distributed populations with means, μ_1 and μ_2, and variances, σ_1^2 and σ_2^2, respectively, the sampling distribution of the difference, $\bar{x}_1 - \bar{x}_2$, between the means of independent samples of size n_1 and n_2 drawn from these populations is normally distributed with mean, $\mu_1 - \mu_2$, and variance, $(\sigma_1^2/n_1) + (\sigma_2^2/n_2)$.

Sampling from Nonnormal Populations Many times a researcher is faced with one or the other of the following problems: the necessity of (1) sampling from nonnormally distributed populations, or (2) sampling from populations whose functional forms are not known. A solution to these problems is to take large samples, since when the sample sizes are large the central limit theorem applies and the distribution of the difference between two sample means is at least approximately normally distributed with a mean equal to $\mu_1 - \mu_2$ and a variance of $(\sigma_1^2/n_1) + (\sigma_2^2/n_2)$. To find probabilities associated with specific values of the statistic, then, our procedure would be the same as that given above when sampling is from normally distributed populations.

Example 4.5.2 Suppose it has been established that for a certain type of client the average length of a home visit by a public health nurse is 45 minutes with a standard deviation of 15 minutes, and that for a second type of client the average home visit is 30 minutes long with a standard deviation of 20 minutes. If a nurse randomly visits 35 clients from the first and 40 from the second group, what is the probability that the average length of home visit will differ between the two groups by 20 or more minutes?

Solution   No mention is made of the functional form of the two populations, so let us assume that this characteristic is unknown, or that the populations are not normally distributed. Since the sample sizes are large (greater than 30) in both cases, we draw on the results of the central limit theorem to answer the question posed. We know that the difference between sample means is approximately normally distributed with the following mean and variance:

$$\mu_{\bar{x}_1 - \bar{x}_2} = \mu_1 - \mu_2 = 45 - 30 = 15$$

$$\sigma_{\bar{x}_1 - \bar{x}_2}^2 = \frac{\sigma_1^2}{n_1} + \frac{\sigma_2^2}{n_2} = \frac{(15)^2}{35} + \frac{(20)^2}{40} = 16.4286$$

The area under the curve of $\bar{x}_1 - \bar{x}_2$ which we seek is that area to the right of 20. The corresponding value of z in the unit normal is

$$z = \frac{(\bar{x}_1 - \bar{x}_2) - (\mu_1 - \mu_2)}{\sqrt{\frac{\sigma_1^2}{n_1} + \frac{\sigma_2^2}{n_2}}} = \frac{20 - 15}{\sqrt{16.4286}} = \frac{5}{4.05} = 1.23$$

In Table C we find the area to the right of $z = 1.23$ is $1 - .8907 = .1093$. We say, then, that the probability of the nurse's random visits resulting in a difference between the two means as great as or greater than 20 minutes is .1093. The curve of $\bar{x}_1 - \bar{x}_2$ and the corresponding unit normal curve are shown in Figure 4.5.2.

EXERCISES

4.5.1   A researcher is willing to assume that levels of vitamin A in the liver in two human populations are each normally distributed. The variances for the two

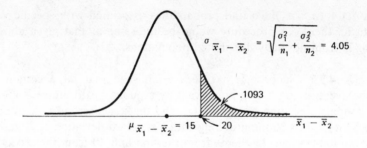

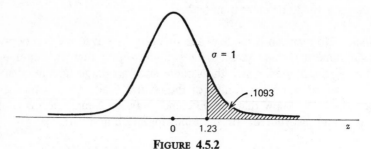

FIGURE 4.5.2

**Sampling Distribution of $\bar{x}_1 - \bar{x}_2$ and the Corresponding Unit Normal
Distribution, Home Visit Example**

populations are assumed to be as follows:

$$\text{Population 1: } \sigma_1{}^2 = 19{,}600$$

$$\text{Population 2: } \sigma_2{}^2 = 8100$$

What is the probability that a random sample of size 15 from the first population
and 10 from the second population will yield a value of $\bar{x}_1 - \bar{x}_2$ greater than or
equal to 50 if there is no difference in the population means?

4.5.2 In a study of annual family expenditures for general health care, two
populations were surveyed with the following results:

$$\text{Population 1: } n_1 = 40, \ \bar{x}_1 = \$346$$

$$\text{Population 2: } n_2 = 35, \ \bar{x}_2 = 300$$

If it is known that the population variances are $\sigma_1{}^2 = 2800$ and $\sigma_2{}^2 = 3250$, what
is the probability of obtaining sample results as extreme as those shown above if
there is no difference in the means of the two populations?

4.5.3 Given two normally distributed populations with equal means and
variances of $\sigma_1{}^2 = 100$ and $\sigma_2{}^2 = 80$, what is the probability that samples of size
$n_1 = 25$ and $n_2 = 16$ will yield a value of $\bar{x}_1 - \bar{x}_2$ greater than or equal to 8?

4.5.4 Given two nonnormally distributed populations with equal means and variances of $\sigma_1{}^2 = 240$ and $\sigma_2{}^2 = 350$, what is the probability that samples of size $n_1 = 40$ and $n_2 = 35$ will yield a value of $\bar{x}_1 - \bar{x}_2$ as large as 12?

4.5.5 For a population of 17-year-old boys and 17-year-old girls the means and standard deviations, respectively, of their subscapular skinfold thickness values are as follows: boys, 9.7 and 6.0; girls, 15.6 and 9.5. Simple random samples of 40 boys and 35 girls are selected from the populations. What is the probability that the difference between sample means $(\bar{x}_{\text{girls}} - \bar{x}_{\text{boys}})$ will be greater than 10?

4.6 DISTRIBUTION OF THE SAMPLE PROPORTION

In the previous sections we have dealt with the sampling distributions of statistics computed from measured variables. We are frequently interested, however, in the sampling distribution of statistics, such as a sample proportion, that result from counts or frequency data.

Example 4.6.1 Suppose we know that in a certain human population .08 are colorblind. If we designate a population proportion by p, we can say that in this example $p = .08$. If we randomly select 150 individuals from this population, what is the probability that the proportion in the sample who are colorblind will be as great as .15?

To answer this question we need to know the properties of the sampling distribution of the sample proportion. We will designate the sample proportion by the symbol $\hat{p}$.

The reader will recognize the similarity between the present example and those presented in Section 3.3, which dealt with the binomial distribution. Indeed, the variable colorblindness is a dichotomous variable, since an individual can be classified into one or the other of two mutually exclusive categories, colorblind or not colorblind. In Section 3.3, we were given the same information and were asked to find the number with the characteristic of interest, whereas here we are seeking the proportion in the sample possessing the characteristic of interest. We could with a sufficiently large table of binomial probabilities, such as Table A, determine the probability associated with the number corresponding to the proportion of interest. As we shall see, this will not be necessary, since there is available an alternative procedure, when sample sizes are large, that is generally more convenient.

The sampling distribution of a sample proportion would be constructed experimentally in exactly the same manner as was suggested in the case of the arithmetic mean and the difference between two means. From the population, which we assume to be finite, we would take all possible samples of a given size and for each sample compute the sample proportion, $\hat{p}$. We would then prepare a frequency distribution of $\hat{p}$ by listing the different distinct values of $\hat{p}$ along with their frequencies of occurrence. This frequency distribution (as well as the corresponding relative frequency distribution) would constitute the sampling distribution of $\hat{p}$.

When the sample size is large, the distribution of sample proportions is approximately normally distributed by virtue of the central limit theorem. The mean of the distribution, $\mu_{\hat{p}}$, that is, the average of all the possible sample proportions, will be equal to the true population proportion, p, and the variance of the distribution, $\sigma_{\hat{p}}^2$, will be equal to $p(1 - p)/n$. To answer probability questions about p, then, we use the following formula:

$$z = \frac{\hat{p} - p}{\sqrt{\dfrac{p(1 - p)}{n}}} \tag{4.6.1}$$

The question that now arises is: How large does the sample size have to be for the use of the normal approximation to be valid? A widely used criterion is that both np and $n(1 - p)$ must be greater than 5, and we will abide by that rule in this text.

We are now in a position to answer the question regarding colorblindness in the sample of 150 individuals from a population in which .08 are colorblind. Since both np and $n(1 - p)$ are greater than 5 ($150 \times .08 = 12$ and $150 \times .92 = 138$), we can say that, in this case, $\hat{p}$ is approximately normally distributed with a mean, $\mu_{\hat{p}} = p = .08$ and $\sigma_{\hat{p}}^2 = p(1 - p)/n = (.08)(.92)/150 = .00049$. The probability we seek is the area under the curve of $\hat{p}$ that is to the right of .15. This area is equal to the area under the unit normal curve to the right of

$$z = \frac{\hat{p} - p}{\sqrt{\dfrac{p(1 - p)}{n}}} = \frac{.15 - .08}{\sqrt{.00049}} = \frac{.07}{.0222} = 3.15$$

The transformation to the unit normal distribution has been accomplished in the usual manner: z is found by dividing the standard error into the difference between a value of the statistic and its mean. From Table C we find that the area to the right of $z = 3.15$ is $1 - .9992 = .0008$. We may say, then, that the probability of observing $\hat{p} \geq .15$ in a random sample of size $n = 150$ from a population in which $p = .08$ is .0008. If we should, in fact, draw such a sample most people would consider it a rather rare event.

The normal approximation may be improved by the *correction for continuity*, a device that makes an adjustment for the fact that a discrete distribution is being approximated by a continuous distribution. Suppose we let $x = n\hat{p}$, the number in the sample with the characteristic of interest when the proportion is $\hat{p}$. To apply the correction for continuity we compute

$$z_c = \frac{\dfrac{x + .5}{n} - p}{\sqrt{pq/n}}, \qquad \text{for } x > np \qquad (4.6.2)$$

or

$$z_c = \frac{\dfrac{x - .5}{n} - p}{\sqrt{pq/n}}, \qquad \text{for } x < np \qquad (4.6.3)$$

where $q = 1 - p$. The correction for continuity will not make a great deal of difference when n is large. In the above example $n\hat{p} = 150(.15) = 22.5$, and

$$z_c = \frac{\dfrac{22.5 - .5}{150} - .08}{\sqrt{.00049}} = 3.01$$

and $P(\hat{p} \geq .15) = 1 - .9987 = .0013$, a result not greatly different from that obtained without the correction for continuity.

Example 4.6.2 Suppose it is known that in a certain population of women, 90 percent entering their third trimester of pregnancy have had some prenatal care. If a random sample of size 200 is drawn from this population, what is the probability that the sample proportion who have had some prenatal care will be less than .85?

Solution We can assume that the sampling distribution of p is approximately normally distributed with $\mu_{\hat{p}} = .90$ and $\sigma_{\hat{p}}^2 = (.1)(.9)/200 = .00045$. We compute

$$z = \frac{.85 - .90}{\sqrt{.00045}} = \frac{-.05}{.0212} = -2.36$$

The area to the left of -2.36 under the unit normal curve is .0091. Therefore, $P(\hat{p} \leq .85) = P(z \leq -2.36) = .0091$.

EXERCISES

4.6.1 If, in a population of adults, .15 are on some sort of special diet, what is the probability that a random sample of size 100 will yield a proportion who are on a diet:

(a) Greater than or equal to .20?   (b) Between .10 and .20?
(c) No greater than .12?

4.6.2 In a certain city it is felt that 20 percent of the households have at least
one member who is suffering from some ill effect due to air pollution. A random
sample of 150 households yielded $\hat{p} = .27$. If the 20 percent figure is correct,
what is the probability of obtaining a sample proportion this large or larger?

4.6.3 In a random sample of 75 adults, 35 said they felt that cancer of the
breast is curable. If, in the population from which the sample was drawn, the true
proportion who feel cancer of the breast can be cured is .55, what is the
probability of obtaining a sample proportion as small as or smaller than that
obtained in this sample?

4.6.4 The standard drug used to treat a certain disease is known to prove
effective within three days in 75 percent of the cases in which it is used. In
evaluating the effectiveness of a new drug in treating the same disease, it was
given to 150 persons suffering from the disease. At the end of three days 97
persons had recovered. If the new drug is equally as effective as the standard,
what is the probability of observing this small a proportion recovering?

4.6.5 Given a population in which $p = .6$ and a random sample from this
population of size 100, find:

(a) $P(\hat{p} \ge .65)$.   (b) $P(\hat{p} \le .58)$.
(c) $P(.56 \le \hat{p} \le .63)$.

4.6.6 It is known that 35 percent of the members of a certain population
suffer from one or more chronic diseases. What is the probability that in a sample
of 200 subjects drawn at random from this population 80 or more will have at
least one chronic disease?

4.7 DISTRIBUTION OF THE DIFFERENCE
BETWEEN TWO SAMPLE PROPORTIONS

Often there are two population proportions in which we are interested and we
desire to assess the probability associated with a difference in proportions
computed from samples drawn from each of these populations. The relevant
sampling distribution is the distribution of the difference between two sample
proportions. The characteristics of this sampling distribution may be summarized
as follows:

*If independent random samples of size n_1 and n_2 are drawn from two populations of
dichotomous variables where the proportion of observations with the characteristic of*

interest in the two populations are p_1 and p_2, respectively, the distribution of the difference between sample proportions, $\hat{p}_1 - \hat{p}_2$, is approximately normal with mean

$$\mu_{\hat{p}_1 - \hat{p}_2} = p_1 - p_2$$

and variance

$$\sigma^2_{\hat{p}_1 - \hat{p}_2} = \frac{p_1(1 - p_1)}{n_1} + \frac{p_2(1 - p_2)}{n_2}$$

when n_1 and n_2 are large.

To physically construct the sampling distribution of the difference between two sample proportions, we would proceed in the manner described in Section 4.5 for constructing the sampling distribution of the difference between two means.

Given two sufficiently small populations, one would draw, from population 1, all possible simple random samples of size n_1 and compute, from each set of sample data, the sample proportion $\hat{p}_1$. From population 2, one would draw independently all possible simple random samples of size n_2 and compute, for each set of sample data, the sample proportion $\hat{p}_2$. One would compute the differences between all possible pairs of sample proportions, where one member of each pair was a value of $\hat{p}_1$ and the other a value of $\hat{p}_2$. The sampling distribution of the difference between sample proportions, then, would consist of all such distinct differences, accompanied by their frequencies (or relative frequencies) of occurrence. For large finite or infinite populations one could approximate the sampling distribution of the difference between sample proportions by drawing a large number of independent simple random samples and proceeding in the manner just described.

To answer probability questions about the difference between two sample proportions, then, we use the following formula:

$$z = \frac{(\hat{p}_1 - \hat{p}_2) - (p_1 - p_2)}{\sqrt{\dfrac{p_1(1 - p_1)}{n_1} + \dfrac{p_2(1 - p_2)}{n_2}}} \qquad (4.7.1)$$

Example 4.7.1 Suppose that the proportion of moderate to heavy users of illegal drugs in population 1 is .50 while in population 2 the proportion is .33. What is the probability that samples of size 100 drawn from each of the populations will yield a value of $\hat{p}_1 - \hat{p}_2$ as large as .30?

We assume that the sampling distribution of $\hat{p}_1 - \hat{p}_2$ is approximately normal with mean

$$\mu_{\hat{p}_1 - \hat{p}_2} = .50 - .33 = .17$$

and variance

$$\sigma^2_{\hat{p}_1 - \hat{p}_2} = \frac{(.33)(.67)}{100} + \frac{(.5)(.5)}{100}$$

$$= .004711$$

The area corresponding to the probability we seek is the area under the curve of $\hat{p}_1 - \hat{p}_2$ to the right of .30. Transforming to the unit normal distribution gives

$$z = \frac{(\hat{p}_1 - \hat{p}_2) - (p_1 - p_2)}{\sqrt{\dfrac{p_1(1 - p_1)}{n_1} + \dfrac{p_2(1 - p_2)}{n_2}}} = \frac{.30 - .17}{\sqrt{.004711}} = 1.89$$

Consulting Table C, we find that the area under the unit normal curve that lies to the right of $z = 1.89$ is $1 - .9706 = .0294$. The probability of observing a difference as large as .30 is, then, .0294.

Example 4.7.2 In a certain population of teenagers it is known that 10 percent of the boy are obese. If the same proportion of girls in the population are obese, what is the probability that a random sample of 250 boys and 200 girls will yield a value of $\hat{p}_1 - \hat{p}_2 \geq .06$?

We assume that the sampling distribution of $\hat{p}_1 - \hat{p}_2$ is approximately normal. If the proportion of obese individuals is the same in the two groups, the mean of the distribution will be 0 and the variance will be

$$\sigma^2_{\hat{p}_1 - \hat{p}_2} = \frac{p_1(1 - p_1)}{n_1} + \frac{p_2(1 - p_2)}{n_2} = \frac{(.1)(.9)}{250} + \frac{(.1)(.9)}{200}$$

$$= .00081$$

The area of interest under the curve of $\hat{p}_1 - \hat{p}_2$ is that to the right of .06. The corresponding z value is

$$z = \frac{.06 - 0}{\sqrt{.00081}} = 2.11$$

Consulting Table C, we find that the area to the right of $z = 2.11$ is $1 - .9826 = .0174$.

EXERCISES

4.7.1 In a certain population of retarded children it is known that the proportion who are hyperactive is .40. A random sample of size 120 was drawn from this population, and a random sample of size 100 was drawn from another population of retarded children. If the proportion of hyperactive children is the same in both populations, what is the probability that the sample would yield a difference, $\hat{p}_1 - \hat{p}_2$, of .16 or more?

4.7.2 In a certain area of a large city it is hypothesized that 40 percent of the houses are in a dilapidated condition. A random sample of 75 houses from this section and 90 houses from another section yielded a difference, $\hat{p}_1 - \hat{p}_2$, of .09. If there is no difference between the two areas in the proportion of dilapidated houses, what is the probability of observing a difference this large or larger?

4.7.3 The results of a health survey reveals that in population A 55 percent of the subjects and in population B 35 percent of the subjects are free of cardiovascular disease. Suppose we select a simple random sample of size 120 from population A and an independent simple random sample of size 130 from population B. What is the probability that the difference between sample proportions, $\hat{p}_A - \hat{p}_B$, will be between .30 and .40?

4.8 SUMMARY

This chapter is concerned with sampling and sampling distributions. Simple random sampling, the type of sampling that is basic to statistical inference, is defined, and a procedure for drawing this type of sample is explained. The concept of a sampling distribution is introduced and the following important sampling distributions are covered:

1. The distribution of a single sample mean.
2. The distribution of the difference between two sample means.
3. The distribution of a sample proportion.
4. The distribution of the difference between two sample proportions.

We emphasize the importance of this material and urge readers to make sure that they understand it before proceeding to the next chapter.

REVIEW QUESTIONS AND EXERCISES

1. What are the two types of sampling?
2. Why is nonprobability sampling not covered in this text?

3. Define or explain the following terms:
   (a) Probability sample.
   (b) Simple random sample.
   (c) Sampling with replacement.
   (d) Sampling without replacement.
   (e) Sampling distribution.

4. Explain how a sampling distribution may be constructed from a finite population.

5. Describe the sampling distribution of the sample mean when sampling is with replacement from a normally distributed population.

6. Explain the central limit theorem.

7. How does the sampling distribution of the sample mean, when sampling is without replacement, differ from the sampling distribution obtained when sampling is with replacement?

8. Describe the sampling distribution of the difference between two sample means.

9. Describe the sampling distribution of the sample proportion when large samples are drawn.

10. Describe the sampling distribution of the difference between two sample means when large samples are drawn.

11. Explain the procedure you would follow in constructing the sampling distribution of the difference between sample proportions based on large samples from finite populations.

12. Suppose it is known that the response time of healthy subjects to a particular stimulus is a normally distributed random variable with a mean of 15 seconds and a variance of 16. What is the probability that a random sample of 16 subjects will have a mean response time of 12 seconds or more?

13. A certain firm has 2000 employees. During a recent year, the mean amount per employee spent on personal medical expenses was $31.50, and the standard deviation was $6.00. What is the probability that a simple random sample of 36 employees will yield a mean between $30 and $33?

14. Suppose it is known that in a certain population of drug addicts the mean duration of abuse is 5 years and the standard deviation is 3 years. What is the probability that a random sample of 36 subjects from this population will yield a mean duration of abuse between 4 and 6 years?

15. Suppose the mean daily protein intake for a certain population is 125 grams, while for another population the mean is 100 grams. If daily protein intake values in the two populations are normally distributed with a standard deviation of 15 grams, what is the probability that random and independent samples of size 25 from each population will yield a difference between sample means of 12 or less?

16. Suppose that two drugs, purported to reduce the response time to a certain stimulus, are under study by a drug manufacturer. The researcher is willing to assume that response times, following administration of the two drugs, are normally distributed with equal variances of 60. As part of the evaluation of the two drugs, drug A is to be administered to 15 subjects and drug B is to be administered to 12 subjects. The researcher would like to know between what two values the central 95 percent of all differences between sample means would lie if the drugs were equally effective and the experiment were repeated a large number of times using these sample sizes.

17. Suppose it is known that the serum albumin concentration in a certain population of individuals is normally distributed with a mean of 4.2 g/100 ml and a standard deviation of .5. A random sample of nine of these individuals placed on a daily dosage of a certain oral steroid yielded a mean serum albumin concentration value of 3.8 g/100 ml. Does it appear likely from these results that the oral steroid reduces the level of serum albumin?

18. A survey conducted in a large metropolitan area revealed that among high school students, 35 percent have, at one time or another, smoked marijuana. If, in a random sample of 150 of these students, only 40 admit to having ever smoked marijuana, what would you conclude?

19. Sixty percent of the employees of a large firm were absent due to sickness three or more days last year. If a simple random sample of 150 of these employees is selected, what is the probability that the proportion in the sample who were absent three or more days due to sickness will be between .50 and .65?

20. A psychiatric social worker believes that in both community A and community B, the proportion of adolescents suffering from some emotional or mental problem is .20. In a sample of 150 adolescents from community A, 15 had an emotional or mental problem. In a sample of 100 from community B, the number was 16. If the social worker's belief is correct, what is the probability of observing a difference as great as was observed between these two samples?

21. Two drugs, A and B, are purported to be equally effective in reducing the level of anxiety in a certain type of emotionally disturbed person. The proportion of persons with whom the drugs are effective is believed to be .80. A random sample of 100 emotionally disturbed persons were given drug A and 85 experienced a reduction in anxiety level. Drug B was effective with 105 of an independent random sample of 150 emotionally disturbed subjects. If the two drugs are, in fact, equally effective as believed, what is the probability of observing a difference in sample proportions as large as or larger than that observed?

22. How many simple random samples (without replacement) of size 5 can be selected from a population of size 10?

23. It is known that 27% of the members of a certain adult population have never smoked. Consider the sampling distribution of the sample proportion based on simple random samples of size 110 drawn from this population. What is the functional form of the sampling distribution?

24. Refer to Exercise 23. Compute the mean and variance of the sampling distribution.

25. Refer to Exercise 24. What is the probability that a single simple random sample of size 110 drawn from this population will yield a sample proportion smaller than .18?

26. In a population of subjects who died from lung cancer following exposure to asbestos it was found that the mean number of years elapsing between exposure and death was 25. The standard deviation was 7 years. Consider the sampling distribution of sample means based on samples of size 35 drawn from this population. What will be the shape of the sampling distribution?

27. Refer to Exercise 26. What will be the mean and variance of the sampling distribution?

28. Refer to Exercise 26. What is the probability that a single simple random sample of size 35 drawn from this population will yield a mean between 22 and 29?

REFERENCES

References Cited
1. George S. Fishman, *Concepts and Methods in Discrete Event Digital Simulation*, John Wiley & Sons, New York, 1973.

2. Paul G. Hoel, *Introduction to Mathematical Statistics*, Third Edition, Wiley, New York, 1962.

3. R. L. Anderson and T. A. Bancroft, *Statistical Theory in Research*, McGraw-Hill, New York, 1952.

Other References, Books
1. John E. Freund and Ronald E. Walpole, *Mathematical Statistics*, Third Edition, Prentice-Hall, Englewood Cliffs, N.J., 1980.

2. Richard J. Larsen and Morris L. Marx, *An Introduction to Mathematical Statistics and Its Applications*, Prentice-Hall, Englewood Cliffs, N.J., 1981.

CHAPTER 5

Estimation

5.1 INTRODUCTION

We come now to a consideration of *estimation*, the first of the two general areas of statistical inference. The second general area, *hypothesis testing*, is examined in the next chapter.

In Chapter 1 inferential statistics is defined as the procedure whereby one reaches decisions about a large body of data by examining only a small portion of that data. More specifically, the large body of data and the small portion thereof, which are referred to in that definition, are, respectively, a population and a sample drawn from that population. This leads to the following restatement of the definition of statistical inference.

Definition Statistical inference is the procedure whereby inferences about a population are made on the basis of the results obtained from a sample drawn from that population.

The process of estimation entails calculating, from the data of a sample, some statistic that is offered as an approximation of the corresponding parameter of the population from which the sample was drawn.

The rationale behind estimation in the health sciences field rests on the assumption that workers in this field have an interest in the parameters of various populations. If this is the case, there are two good reasons why one must rely on estimating procedures to obtain information regarding these parameters. First, many populations of interest, although finite, are so large that a 100 percent examination would be prohibitive from the standpoint of cost. Second, populations that are infinite are incapable of complete examination.

Suppose the administrator of a large hospital is interested in the average age of patients admitted to his hospital during a given year. He may consider it too expensive to go through the records of all patients admitted during that particu-

127

lar year and, consequently, elects to examine a sample of the records from which he can compute an estimate of the average age of patients admitted that year.

A physician in general practice may be interested in knowing what proportion of a certain type of individual, treated with a particular drug, suffers undesirable side effects. No doubt, her concept of the population consists of all those persons who ever have been or ever will be treated with this drug. Deferring a conclusion until the entire population has been observed could have an adverse effect on her practice.

These two examples have implied an interest in estimating, respectively, a population mean and a population proportion. Other parameters, the estimation of which we shall cover in this chapter, are the difference between two means, the difference between two proportions, the population variance, and the ratio of two variances.

We shall find that for each of these parameters we can compute two types of estimate: a point estimate and an interval estimate. A *point estimate* is a single numerical value used to estimate the corresponding population parameter. An *interval estimate* consists of two numerical values defining an interval which, with a specified degree of confidence, we feel includes the parameter being estimated. These concepts will be elaborated on in the succeeding sections.

Note that a single computed value has been referred to as an *estimate*. The rule that tells us how to compute this value, or estimate, is referred to as an *estimator*. Estimators are usually presented as formulas. For example,

$$\bar{x} = \frac{\sum x_i}{n}$$

is an estimator of the population mean, μ. The single numerical value that results from evaluating this formula is called an estimate of the parameter, μ.

In many cases, a parameter may be estimated by more than one estimator. For example, we could use the sample median to estimate the population mean. How then do we decide which estimator to use for estimating a given parameter? The decision is based on criteria that reflect the "goodness" of particular estimators. When measured against these criteria, some estimators are better than others. One of these criteria is the property of *unbiasedness*.

An estimator, say T, of the parameter θ is said to be an unbiased estimator of θ if $E(T) = \theta$. $E(T)$ is read, "the expected value of T." For a finite population, $E(T)$ is obtained by taking the average value of T computed from all possible samples of a given size that may be drawn from the population. That is, $E(T) = \mu_T$. For an infinite population, $E(T)$ is defined in terms of calculus.

In the previous chapter we have seen that the sample mean, the sample proportion, the difference between two sample means, and the difference between two sample proportions are each unbiased estimates of their corresponding parameters. This property was implied when the parameters were said to be the means of the respective sampling distributions. For example, since the mean of the sampling distribution of $\bar{x}$ is equal to μ, we know that $\bar{x}$ is an unbiased estimator of μ. The other criteria of good estimators will not be discussed in this book. The interested reader will find them covered in detail by Freund and Walpole (1) and Mood, Graybill, and Boes (2), among others. A much less mathematically rigorous treatment may be found in Yamane (3).

Sampled Populations and Target Populations   The health researcher who uses statistical inference procedures must be aware of the difference between two kinds of population—the *sampled population* and the *target population*. The sampled population is the population from which one actually draws a sample. The target population is the population about which one wishes to make an inference. These two populations may or may not be the same. Statistical inference procedures allow one to make inferences about sampled populations (provided proper sampling methods have been employed). Only when the target population and the sampled population are the same is it possible for one to use statistical inference procedures to reach conclusions about the target population. If the sampled population and the target population are different, the researcher can reach conclusions about the target population only on the basis of nonstatistical considerations.

Suppose, for example, that a researcher wishes to assess the effectiveness of some method for treating rheumatoid arthritis. The target population consists of all patients suffering from the disease. It is not practical to draw a sample from this population. The researcher may, however, select a sample from all rheumatoid arthritis patients seen in some specific clinic. These patients constitute the sampled population, and, if proper sampling methods are used, inferences about this sampled population may be drawn on the basis of the information in the sample. If the researcher wishes to make inferences about all rheumatoid arthritis suffers, he or she must rely on nonstatistical means to do so. Perhaps the researcher knows that the sampled population is similar, with respect to all important characteristics, to the target population. That is, the researcher may know that the age, sex, severity of illness, duration of illness compositions, and so on are similar. And on the strength of this knowledge, the researcher may be willing to extrapolate his or her findings to the target population.

In many situations the sampled population and the target population are identical, and, when this is the case, inferences about the target population are straightforward. The researcher should, however, be aware that this is not always the case and not fall into the trap of drawing unwarranted inferences about a population that is different from the one that is sampled.

Random and Nonrandom Samples   In the examples and exercises of this book, we assume that the data available for analysis have come from random samples. The strict validity of the statistical procedures discussed depends on this assumption. In many instances in real-world applications it is impossible or impractical to use truly random samples. In animal experiments, for example, researchers usually use whatever animals are available from suppliers or their own breeding stock. If the researchers had to depend on randomly selected material, very little research of this type would be conducted. Again, nonstatistical considerations must play a part in the generalization process. Researchers may contend that the samples actually used are equivalent to simple random samples, since there is no reason to believe that the material actually used is not representative of the population about which inferences are desired.

In many health research projects, samples of convenience, rather than random samples, are employed. Researchers may have to rely on volunteer subjects or on readily available subjects such as students in their classes. Again, generalizations must be made on the basis of nonstatistical considerations. The consequences of such generalizations, however, may be useful or they may range from misleading to disastrous.

In some situations it is possible to introduce randomization into an experiment even though available subjects are not randomly selected from some well-defined population. In comparing two treatments, for example, each subject may be randomly assigned to one or the other of the treatments. Inferences in such cases apply to the treatments and not the subjects, and hence the inferences are valid.

5.2 CONFIDENCE INTERVAL FOR A POPULATION MEAN

Suppose researchers wish to estimate the mean of some normally distributed population. They draw a random sample of size n from the population, and compute $\bar{x}$, which they use as a point estimate of μ. Although this estimate of μ possesses all the qualities of a good estimator, we know that because of the vagaries of sampling, $\bar{x}$ cannot be expected to be equal to μ.

It would be much more meaningful, therefore, to estimate μ by an interval that somehow communicates the probable magnitude of μ.

To obtain such an interval estimate, we must draw on our knowledge of sampling distributions. In the present case, since we are concerned with the sample mean as an estimator of a population mean, we must recall what we know about the sampling distribution of the sample mean.

In the previous chapter we learned that if sampling is from a normally distributed population, the sampling distribution of the sample mean will be normally distributed with a mean, $\mu_{\bar{x}}$, equal to the population mean, μ, and a variance, $\sigma_{\bar{x}}^2$, equal to σ^2/n. We could plot the sampling distribution if we only knew where to locate it on the $\bar{x}$-axis From our knowledge of normal distributions, in general, we know even more about the distribution of $\bar{x}$ in this case. We know, for example, that regardless of where it is located, approximately 95 percent of the possible values of $\bar{x}$ constituting the distribution are within 2 standard deviations of the mean. The two points that are 2 standard deviations from the mean are $\mu - 2\sigma_{\bar{x}}$ and $\mu + 2\sigma_{\bar{x}}$, so that the interval, $\mu \pm 2\sigma_{\bar{x}}$, will contain approximately 95 percent of the possible values of $\bar{x}$. We know that μ and, hence, $\mu_{\bar{x}}$ are unknown, but we may arbitrarily place the sampling distribution of $\bar{x}$ on the $\bar{x}$-axis.

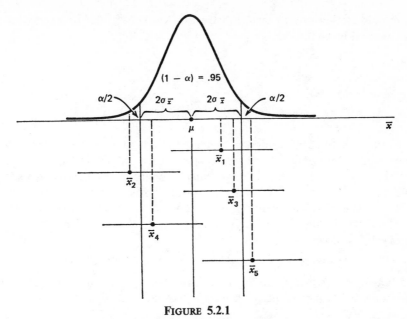

FIGURE 5.2.1

The 95 Percent Confidence Intervals for μ

Since we do not know the value of μ, not a great deal is accomplished by the expression $\mu \pm 2\sigma_{\bar{x}}$. We do, however, have a point estimate of μ, which is $\bar{x}$. Would it be useful to construct an interval about this point estimate of μ? The answer is, yes. Suppose we constructed intervals about every possible value of $\bar{x}$ computed from all possible samples of size n from the population of interest. We would have a large number of intervals of the form $\bar{x} \pm 2\sigma_{\bar{x}}$ with widths all equal to the width of the interval about the unknown μ. Approximately 95 percent of these intervals would have centers falling within the $\pm 2\sigma_{\bar{x}}$ interval about μ. Each of the intervals whose centers fall within $2\sigma_{\bar{x}}$ of μ would contain μ. These concepts are illustrated in Figure 5.2.1. In Figure 5.2.1 we see that $\bar{x}_1$, $\bar{x}_3$, and $\bar{x}_4$ all fall within the $2\sigma_{\bar{x}}$ interval about μ, and, consequently, the $2\sigma_{\bar{x}}$ intervals about these sample means include the value of μ. The sample means $\bar{x}_2$ and $\bar{x}_5$ do not fall within the $2\sigma_{\bar{x}}$ interval about μ, and the $2\sigma_{\bar{x}}$ intervals about them do not include μ.

Example 5.2.1 Suppose a researcher, interested in obtaining an estimate of the average level of some enzyme in a certain human population, takes a sample of 10 individuals, determines the level of the enzyme in each, and computes a sample mean of $\bar{x} = 22$. Suppose further it is known that the variable of interest is approximately normally distributed with a variance of 45. An approximate 95 percent confidence interval for μ is given by

$$\bar{x} \pm 2\sigma_{\bar{x}}$$

$$22 \pm 2\sqrt{\tfrac{45}{10}}$$

$$22 \pm 2(2.12)$$

$$17.76, 26.24$$

Let us examine the composition of this interval estimate. It contains in its center the point estimate of μ. The 2 we recognize as a value from the unit normal distribution that tells us within how many standard errors lie approximately 95 percent of the possible values of $\bar{x}$. This value of z is referred to as the *reliability coefficient*. The last component, $\sigma_{\bar{x}}$, is the standard error, or standard deviation of the sampling distribution of $\bar{x}$. In general, then, an interval estimate may be expressed as follows:

$$\text{estimator} \pm (\text{reliability coefficient}) \times (\text{standard error}) \qquad \textbf{(5.2.1)}$$

In particular, when sampling is from a normal distribution with known variance an interval estimate for μ may be expressed as

$$\bar{x} \pm z_{(1-\alpha/2)}\sigma_{\bar{x}} \qquad \textbf{(5.2.2)}$$

How do we interpret the interval given by Expression 5.2.2? In the present example, where the reliability coefficient is equal to 2, we say that in repeated sampling approximately 95 percent of the intervals constructed by Expression 5.2.2 will include the population mean. This interpretation is based on the probability of occurrence of different values of $\bar{x}$. We may generalize this interpretation if we designate the total area, under the curve of $\bar{x}$, which is outside the interval $\mu \pm 2\sigma_{\bar{x}}$ as α and the area within the interval as $1 - \alpha$ and give the following *probabilistic interpretation* Expression of 5.2.2.

In repeated sampling, from a normally distributed population, $100(1 - \alpha)$ *percent of all intervals of the form* $\bar{x} \pm z_{(1-\alpha/2)}\sigma_{\bar{x}}$ *will in the long run include the population mean,* μ.

The quantity $1 - \alpha$, in this case .95, is called the *confidence coefficient*, and the interval $\bar{x} \pm z_{(1-\alpha/2)}\sigma_{\bar{x}}$ is called a *confidence interval* for μ. When $(1 - \alpha) = .95$, the interval is called the 95 percent confidence interval for μ. In the present example we say that we are 95 percent confident that the population mean is between 17.76 and 26.24. This is called the *practical interpretation* of 5.2.2. In general, it may be expressed as follows.

We are $100(1 - \alpha)$ *percent confident that the single computed interval,* $\bar{x} \pm z_{(1-\alpha/2)}\sigma_{\bar{x}}$, *contains the population mean,* μ.

In the example given here we might prefer, rather than 2, the more exact value of z, 1.96, corresponding to a confidence coefficient of .95. The researcher may use any confidence coefficient he wishes; the most frequently used values are .90, .95, and .99, which have associated reliability factors, respectively, of 1.645, 1.96, and 2.58.

Example 5.2.2 A physical therapist wished to estimate, with 99 percent confidence, the mean maximal strength of a particular muscle in a certain group of individuals. He is willing to assume that strength scores are approximately normally distributed with a variance of 144. A sample of 15 subjects who participated in the experiment yielded a mean of 84.3. The z value corresponding to a confidence coefficient of .99 is found in Table C to be 2.58. This is our reliability coefficient. The standard error is $\sigma_{\bar{x}} = 12/\sqrt{15} = 3.10$. Our 99 percent confidence interval for μ, then, is

$$84.3 \pm 2.58(3.10)$$
$$84.3 \pm 8.0$$
$$76.3, 92.3$$

We say we are 99 percent confident that the population mean is between 76.3 and 92.3 since, in repeated sampling, 99 percent of all intervals that could be constructed in the manner just described would include the population mean.

Sampling from Nonnormal Populations   It will not always be possible or prudent to assume that the population of interest is normally distributed. Thanks to the central limit theorem, this will not deter us if we are able to select a large enough sample. We have learned that for large samples, the sampling distribution of $\bar{x}$ is approximately normally distributed regardless of how the parent population is distributed.

Example 5.2.3   Punctuality of patients in keeping appointments is of interest to a research team. In a study of patient flow through the offices of general practitioners, it was found that a sample of 35 patients were 17.2 minutes late for appointments, on the average. Previous research had shown the standard deviation to be about 8 minutes. The population distribution was felt to be nonnormal. What is the 90 percent confidence interval for μ, the true average amount of time late for appointments?

Since the sample size is fairly large (greater than 30), and since the population standard deviation is known, we draw on the central limit theorem and assume the sampling distribution of $\bar{x}$ to be approximately normally distributed. From Table C we find the reliability coefficient, corresponding to a confidence coefficient of .90, to be about 1.645, if we interpolate. The standard error is $\sigma_{\bar{x}} = 8/\sqrt{35} = 1.35$, so that our 90 percent confidence interval for μ is

$$17.2 \pm 1.645(1.35)$$
$$17.2 \pm 2.2$$
$$15.0, 19.4$$

Computer Analysis When confidence intervals are desired, a great deal of time can be saved if one uses a computer, which can be programmed to construct intervals from raw data.

Example 5.2.4 The following are the activity values (micromoles per minute per gram of tissue) of a certain enzyme measured in normal gastric tissue of 35 patients with gastric carcinoma.

.360	1.189	.614	.788	.273	2.464	.571
1.827	.537	.374	.449	.262	.448	.971
.372	.898	.411	.348	1.925	.550	.622
.610	.319	.406	.413	.767	.385	.674
.521	.603	.533	.662	1.177	.307	1.499

We wish to use the MINITAB computer software package to construct a 95 percent confidence interval for the population mean. Suppose we know that the population variance is .36. It is not necessary to assume that the sampled population of values is normally distributed since the sample size is sufficiently large for application of the central limit theorem. When we enter the data into the computer and issue appropriate MINITAB commands, we obtain the following printout.

```
        N =   35    MEAN =      .71643     ST.DEV.=        .511
THE ASSUMED SIGMA =       .6000
A 95.00   PERCENT C.I. FOR MU IS (       .5174,        .9155)
```

We are 95 percent confident that the population mean is somewhere between .5174 and .9155.

<hr>

EXERCISES

5.2.1 We wish to estimate the average number of heartbeats per minute for a certain population. The average number of heartbeats per minute for a sample of 49 subjects was found to be 90. If it is reasonable to assume that these 49 patients constitute a random sample, and that the population is normally distributed with a standard deviation of 10, find:

(a) The 90 percent confidence interval for μ.
(b) The 95 percent confidence interval for μ.
(c) The 99 percent confidence interval for μ.

5.2.2 We wish to estimate the mean serum indirect bilirubin level of 4-day-old infants. The mean for a sample of 16 infants was found to be 5.98 mg/100 cc. Assuming bilirubin levels in 4-day-old infants are approximately normally distributed with a standard deviation of 3.5 mg/100 cc find:

(a)   The 90 percent confidence interval for μ.
(b)   The 95 percent confidence interval for μ.
(c)   The 99 percent confidence interval for μ.

5.2.3   In a length of hospitalization study conducted by several cooperating hospitals, a random sample of 64 peptic ulcer patients was drawn from a list of all peptic ulcer patients ever admitted to the participating hospitals and the length of hospitalization per admission was determined for each. The mean length of hospitalization was found to be 8.25 days. If the population standard deviation is known to be 3 days find:

(a)   The 90 percent confidence interval for μ.
(b)   The 95 percent confidence interval for μ.
(c)   The 99 percent confidence interval for μ.

5.2.4   A sample of 100 apparently normal adult males, 25 years old, had a mean systolic blood pressure of 125. If it is felt that the population standard deviation is 15, find:

(a)   The 90 percent confidence interval for μ.
(b)   The 95 percent confidence interval for μ.

5.2.5   Some studies of Alzheimer's disease (AD) have shown an increase in $^{14}CO_2$ production in patients with the disease. In one such study (simulated) the following $^{14}CO_2$ values were obtained from 16 neocortical biopsy samples from AD patients.

$$1009, 1280, 1180, 1255, 1547, 2352, 1956, 1080,$$
$$1776, 1767, 1680, 2050, 1452, 2857, 3100, 1621$$

Assume that the population of such values is normally distributed with a standard deviation of 350 and construct a 95% confidence interval for the population mean.

5.3   THE t DISTRIBUTION

In Section 5.2 a procedure was outlined for constructing a confidence interval for a population mean. The procedure requires a knowledge of the variance of the population from which the sample is drawn. It may seem somewhat strange that one can have knowledge of the population variance and not know the value of the population mean. Indeed, it is the usual case, in situations such as have been presented, that the population variance, as well as the population mean, is unknown. This condition presents a problem with respect to constructing confidence intervals. Although, for example, the statistic

$$z = \frac{\bar{x} - \mu}{\sigma/\sqrt{n}}$$

is normally distributed when the population is normally distributed, and is at least approximately normally distributed when n is large, regardless of the functional form of the population, we cannot make use of this fact because σ is unknown. However, all is not lost, and the most logical solution to the problem is the one followed. We use the sample standard deviation, $s = \sqrt{\sum(x_i - \bar{x})^2/(n - 1)}$, to replace σ. When the sample size is large, say greater than 30, our faith in s as an approximation of σ is usually substantial, and we may feel justified in using normal distribution theory to construct a confidence interval for the population mean. In that event, we proceed as instructed in Section 5.2.

It is when we have small samples that it becomes mandatory for us to find an alternative procedure for constructing confidence intervals.

As a result of the work of W. S. Gosset (4), writing under the pseudonym of "Student," an alternative, known as *Student's t distribution*, usually shortened to *t distribution*, is available to us.

The quantity

$$t = \frac{\bar{x} - \mu}{s/\sqrt{n}}$$

follows this distribution. The t distribution has the following properties.

1. It has a mean of 0.
2. It is symmetrical about the mean.
3. In general, it has a variance greater than 1, but the variance approaches 1 as the sample size becomes large.
4. The variable t ranges from $-\infty$ to $+\infty$.
5. The t distribution is really a family of distributions, since there is a different distribution for each sample value of $n - 1$, the divisor used in computing s^2. We recall that $n - 1$ is referred to as degrees of freedom. Figure 5.3.1 shows t distributions corresponding to several degrees-of-freedom values.
6. Compared to the normal distribution the t distribution is less peaked in the center and has higher tails. Figure 5.3.2 compares the t distribution with the normal.
7. The t distribution approaches the normal distribution as $n - 1$ approaches infinity.

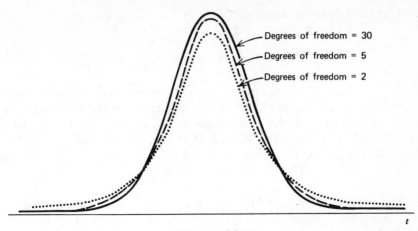

FIGURE 5.3.1

The *t*-Distribution for Different Degrees-of-Freedom Values

The *t* distribution, like the unit normal, has been extensively tabulated. One such table is given as Table E in the Appendix. As we shall see, we must take both the confidence coefficient and degrees of freedom into account when using the table of the *t* distribution.

The general procedure for constructing confidence intervals is not affected by our having to use the *t* distribution rather than the unit normal distribution. We still make use of the relationship expressed by

$$\text{estimator} \pm (\text{reliability coefficient}) \times (\text{standard error})$$

What is different is the source of the reliability coefficient. It is now obtained from the table of the *t* distribution rather than from the table of the unit normal distribution. To be more specific, *when sampling is from a normal distribution whose standard deviation, σ, is unknown, the* $100(1 - \alpha)$ *percent confidence*

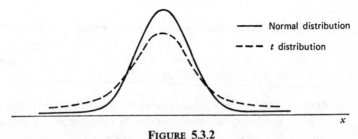

FIGURE 5.3.2

Comparison of Normal Distribution and *t*-Distribution

interval for the population mean, μ, is given by

$$\bar{x} \pm t_{(1-\alpha/2)}\frac{s}{\sqrt{n}} \qquad (5.3.1)$$

Notice that a requirement for valid use of the t distribution is that the sample must be drawn from a normal distribution. Experience has shown, however, that moderate departures from this requirement can be tolerated. As a consequence, the t distribution is used even when it is known that the parent population deviates from normality. Most researchers require that an assumption of, at least, a mound-shaped population distribution be tenable.

Example 5.3.1 We wish to estimate the mean serum amylase value in a healthy population. Determinations were made on a sample of 15 apparently healthy subjects. The sample yielded a mean of 96 units/100 ml and a standard deviation of 35 units/100 ml. The population variance was unknown. We may use the sample mean, 96, as a point estimate of the population mean but, since the population standard deviation is unknown, we must assume the population of values to be at least approximately normally distributed before constructing a confidence interval for μ. Let us assume that such an assumption is reasonable and that a 95 percent confidence interval is desired. We have our estimator, $\bar{x}$, and our standard error is $s/\sqrt{n} = 35/\sqrt{15} = 9.04$. We need now to find the reliability coefficient, the value of t associated with a confidence coefficient of .95 and $n - 1 = 14$ degrees of freedom. Since a 95 percent confidence interval leaves .05 of the area under the curve of t to be equally divided between the two tails, we need the value of t to the right of which lies .025 of the area. We locate in Table E the column headed $t_{.975}$. This is the value of t to the left of which lies .975 of the area under the curve. The area to the right of this value is equal to the desired .025. We now locate the number 14 in the degrees of freedom column. The value at the intersection of the row labeled 14 and the column labeled $t_{.975}$ is the t we seek. This value of t, which is our reliability coefficient, is found to be 2.1448. We now construct our 95 percent confidence interval as follows:

$$96 \pm 2.1448(9.04)$$
$$96 \pm 19$$
$$77, 115$$

This interval may be interpreted from both the probabilistic and practical points of view. We say we are 95 percent confident that the true population mean, μ, is between 77 and 115 because, in repeated sampling, 95 percent of intervals constructed in like manner will include μ.

Deciding Between z and t When we construct a confidence interval for a population mean, we must decide whether to use a value of z or a value of t as the reliability factor. To make an appropriate choice we must consider sample size, whether the sampled population is normally distributed, and whether the population variance is known. Figure 5.3.3 provides a flowchart that one can use to decide quickly whether the reliability factor should be z or t.

EXERCISES

5.3.1 Nine patients suffering from the same physical handicap, but otherwise comparable, were asked to perform a certain task as part of an experiment. The average time required to perform the task was 7 minutes with a standard deviation of 2 minutes. Assuming normality, construct the 90, 95, and 99 percent confidence intervals for the true mean time required to perform the task by this type patient.

5.3.2 A hospital administrator took a sample of 25 overdue accounts for the purpose of estimating the mean amount due. From the sample she computed a mean of $250 and a standard deviation of $75. Assuming that the amounts of all overdue accounts are normally distributed, find the 90, 95, and 99 percent confidence intervals for μ.

5.3.3 A sample of 25 ten-year-old boys yielded a mean weight and standard deviation of 73 and 10 pounds, respectively. Assuming a normally distributed population find the 90, 95, and 99 percent confidence intervals for the mean of the population from which the sample came.

5.3.4 A sample of 16 ten-year-old girls gave a mean weight of 71.5 and a standard deviation of 12 pounds, respectively. Assuming normality, find the 90, 95, and 99 percent confidence intervals for μ.

5.3.5 A simple random sample of 16 apparently normal subjects yielded the following values of urine excreted arsenic (milligrams per day).

Subject	Value	Subject	Value
1	.007	9	.012
2	.030	10	.006
3	.025	11	.010
4	.008	12	.032
5	.030	13	.006
6	.038	14	.009
7	.007	15	.014
8	.005	16	.011

Construct a 95% percent confidence interval for the population mean.

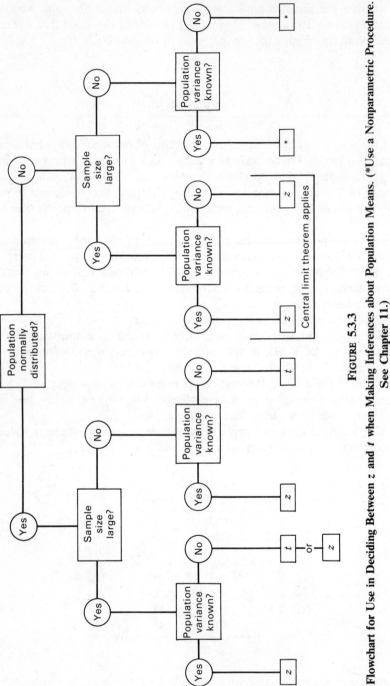

FIGURE 5.3.3

Flowchart for Use in Deciding Between z and t when Making Inferences about Population Means. (*Use a Nonparametric Procedure. See Chapter 11.)

5.4 CONFIDENCE INTERVAL FOR THE DIFFERENCE BETWEEN TWO POPULATION MEANS

Sometimes there arise cases in which we are interested in estimating the difference between two populations means. From each of the populations an independent random sample is drawn and, from the data of each, the sample means $\bar{x}_1$ and $\bar{x}_2$, respectively, are computed. We learned in the previous chapter that the estimator $\bar{x}_1 - \bar{x}_2$ yields an unbiased estimate of $\mu_1 - \mu_2$, the difference between the population means. The variance of the estimator is $(\sigma_1^2/n_1) + (\sigma_2^2/n_2)$. We also know from Chapter 4 that, depending on the conditions, the sampling distribution of $\bar{x}_1 - \bar{x}_2$ may be, at least, approximately normally distributed, so that in many cases we make use of the theory relevant to normal distributions to compute a confidence interval for $\mu_1 - \mu_2$. When the population variances are known the $100(1 - \alpha)$ percent confidence interval for $\mu_1 - \mu_2$ is given by

$$(\bar{x}_1 - \bar{x}_2) \pm z_{1-\alpha/2} \sqrt{\frac{\sigma_1^2}{n_1} + \frac{\sigma_2^2}{n_2}} \qquad (5.4.1)$$

Let us illustrate, first for the case where sampling is from a normal distribution and then for the case where the assumption of normally distributed populations cannot be made.

Example 5.4.1 A research team is interested in the difference between serum uric acid levels in patients with and without mongolism. In a large hospital for the treatment of the mentally retarded, a sample of 12 individuals with mongolism yielded a mean of $\bar{x}_1 = 4.5$ mg/100 ml. In a general hospital a sample of 15 normal individuals of the same age and sex were found to have a mean value of $\bar{x}_2 = 3.4$. If it is reasonable to assume that the two populations of values are normally distributed with variances equal to 1, find the 95 percent confidence interval for $\mu_1 - \mu_2$.

For a point estimate of $\mu_1 - \mu_2$, we use $\bar{x}_1 - \bar{x}_2 = 4.5 - 3.4 = 1.1$. The reliability coefficient corresponding to .95 is found in Table C to be 1.96. The standard error is

$$\sigma_{\bar{x}_1 - \bar{x}_2} = \sqrt{\frac{\sigma_1^2}{n_1} + \frac{\sigma_2^2}{n_2}} = \sqrt{\frac{1}{12} + \frac{1}{15}} = .39$$

The 95 percent confidence interval, then, is

$$1.1 \pm 1.96(.39)$$
$$1.1 \pm .8$$
$$.3, 1.9$$

We say that we are 95 percent confident that the true difference, $\mu_1 - \mu_2$, is between .3 and 1.9, because, in repeated sampling, 95 percent of the intervals constructed in this same manner would include the difference between the true means.

Sampling from Nonnormal Populations   The construction of a confidence interval for the difference between two population means when sampling is from nonnormal populations proceeds in the same manner as above if the sample sizes, n_1 and n_2, are large. Again, this is a result of the central limit theorem.

Example 5.4.2   Researchers wish to compare the economic status of patients treated in two hospitals. The average family income of a sample of 75 patients admitted to hospital A was $\bar{x}_1 = \$6800$, while the average based on a sample of 80 patients from hospital B was found to be $\bar{x}_2 = \$4450$. If the population standard deviations are $\sigma_1 = \$600$ and $\sigma_2 = \$500$, find the 99 percent confidence interval for $\mu_1 - \mu_2$, the difference between population means.

The point estimate of $\mu_1 - \mu_2$ is $\bar{x}_1 - \bar{x}_2 = \$6800 - \$4450 = \$2350$. From Table C we find the reliability coefficient to be 2.58. The standard error is

$$\sigma_{\bar{x}_1 - \bar{x}_2} = \sqrt{\frac{(600)^2}{75} + \frac{(500)^2}{80}} = 89$$

Our 99 percent confidence interval is

$$\$2350 \pm 2.58(89)$$
$$\$2350 \pm 230$$
$$\$2120, \$2580$$

We interpret this interval in the usual ways.

The t Distribution and the Difference Between Means   When population variances are unknown, and it is desired to estimate the difference between two population means with a confidence interval, we can use the t distribution as a source of the reliability factor if certain assumptions are met. We must know, or be willing to assume, that the two sampled populations are normally distributed. With regard to the population variances, we distinguish between two situations: (1) the situation in which the population variances are equal and (2) the situation in which they are not equal. Let us consider each situation separately.

Population Variances Equal   If the assumption of equal population variances is justified, the two sample variances which we compute from our two independent samples may be construed as estimates of the same thing, the common variance. It seems logical then that we should somehow capitalize on this in our analysis. We do just that and obtain a *pooled estimate* of the common variance. This pooled estimate is obtained by computing the weighted average of the two sample variances. Each sample variance is weighted by its degrees of freedom. If the sample sizes are equal, this weighted average is the arithmetic mean of the two sample variances. If the two sample sizes are unequal, the weighted average takes advantage of the additional information provided by the larger sample. The pooled estimate is given by the formula:

$$s_p^2 = \frac{(n_1 - 1)s_1^2 + (n_2 - 1)s_2^2}{n_1 + n_2 - 2} \tag{5.4.2}$$

The standard error of the estimate, then, is given by

$$\sqrt{\frac{s_p^2}{n_1} + \frac{s_p^2}{n_2}}$$

and the $100(1 - \alpha)$ percent confidence interval for $\mu_1 - \mu_2$ is given by

$$(\bar{x}_1 - \bar{x}_2) \pm t_{(1-\alpha/2)}\sqrt{\frac{s_p^2}{n_1} + \frac{s_p^2}{n_2}} \tag{5.4.3}$$

The number of degrees of freedom used in determining the value of t to use in constructing the interval is $n_1 + n_2 - 2$, the denominator of Equation 5.4.2. We interpret this interval in the usual manner.

Example 5.4.3   To illustrate, let us refer to Example 5.4.1 and suppose that, in addition to the apparently healthy normal subjects, serum amylase determinations were also made on an independent sample of 22 hospitalized subjects. Suppose the mean and standard deviation from this group are 120 and 40 units/ml, respectively. Let us designate the 15 normal subjects as group 2. Our point estimate of $\mu_1 - \mu_2$ is $120 - 96 = 24$.

To find a confidence interval, let us proceed under the assumption that the two populations under study are normally distributed and that the population variances are equal. Our first step is to obtain, by Equation 5.4.2, a pooled estimate of the common variance as follows:

$$s_p^2 = \frac{14(35)^2 + 21(40)^2}{15 + 22 - 2} = 1450$$

The 95 percent confidence interval for $\mu_1 - \mu_2$, by Equation 5.4.3, is as follows:

$$(120 - 96) \pm 2.0301\sqrt{\frac{1450}{15} + \frac{1450}{22}}$$

$$24 \pm (2.0301)(12.75)$$

$$24 \pm 26$$

$$-2, 50$$

Again we say we are 95 percent confident that the true difference, $\mu_1 - \mu_2$, is between -2 and 50 because, in repeated sampling, 95 percent of intervals so constructed would include $\mu_1 - \mu_2$.

Population Variances Not Equal When one is reluctant to assume that the variances of two populations of interest are equal, even though the two populations may be assumed to be normal, it is not proper to use the t distribution as just outlined in constructing confidence intervals.

A solution to the problem of unequal variances was proposed by Behrens (5) and later was verified and generalized by Fisher (6, 7). Solutions have also been proposed by Neyman (8), Scheffé (9, 10), and Welch (11, 12). The problem is discussed in detail by Aspin (13), Trickett et al. (14), and Cochran (15). Cochran's approach is also found in Snedecor and Cochran (16).

The problem revolves around the fact that the quantity

$$\frac{(\bar{x}_1 - \bar{x}_2) - (\mu_1 - \mu_2)}{\sqrt{\dfrac{s_1^2}{n_1} + \dfrac{s_2^2}{n_2}}}$$

does not follow a t distribution with $n_1 + n_2 - 2$ degrees of freedom when the population variances are not equal. Consequently, the t distribution cannot be used in the usual way to obtain the reliability factor for the confidence interval for the difference between the means of two populations that have unequal variances. The solution proposed by Cochran consists of computing the reliability factor, $t'_{1-\alpha/2}$, by the following formula:

$$t'_{1-\alpha/2} = \frac{w_1 t_1 + w_2 t_2}{w_1 + w_2} \qquad (5.4.4)$$

where $w_1 = s_1^2/n_1$, $w_2 = s_2^2/n_2$, $t_1 = t_{1-\alpha/2}$ for $n_1 - 1$ degrees of freedom, and $t_2 = t_{1-\alpha/2}$ for $n_2 - 1$ degrees of freedom. An approximate $100(1 - \alpha)$ percent confidence interval for $\mu_1 - \mu_2$ is given by

$$(\bar{x}_1 - \bar{x}_2) \pm t'_{(1-\alpha/2)} \sqrt{\frac{s_2^2}{n_1} + \frac{s_2^2}{n_2}} \qquad (5.4.5)$$

Example 5.4.3 Total serum complement activity (C_{H50}) was assayed in 20 apparently healthy subjects and 10 independent subjects with disease. The following results were obtained:

Subjects	n	$\bar{x}$	s
With disease	10	62.6	33.8
Normal	20	47.2	10.1

The investigators had reason to believe that the sampled populations are approximately normally distributed, but they were unwilling to assume that the two unknown population variances are equal. Find the 95 percent confidence interval for $\mu_1 - \mu_2$.

The point estimate for $\mu_1 - \mu_2$ is $62.6 - 47.2 = 15.4$. For the confidence interval we need to use Equation 5.4.4 to calculate the value of the reliability factor. In Table E we find that $t_1 = 2.2622$ and $t_2 = 2.0930$. We compute $w_1 = (33.8)^2/10 = 114.244$ and $w_2 = (10.1)^2/20 = 5.1005$. By Equation 5.4.4 we compute

$$t' = \frac{114.244(2.2622) + 5.1005(2.0930)}{114.244 + 5.1005} = 2.255$$

By Equation 5.4.5 the 95 percent confidence interval for $\mu_1 - \mu_2$ is

$$15.4 \pm 2.255\sqrt{\frac{33.8^2}{10} + \frac{10.1^2}{20}}$$

$$15.4 \pm 2.255(10.9245)$$

$$15.4 \pm 24.6$$

$$-9.2, 40.0$$

When constructing a confidence interval for the difference between two population means one may use Figure 5.4.1 to decide quickly whether the reliability factor should be z, t, or t'.

EXERCISES

5.4.1 In a study of educable mentally retarded children, 11 boys and 10 girls, after a year of academic schooling combined with therapy, were given achievement tests. We are interested in the difference between boys and girls with respect to their mean scores. The mean score for boys was $\bar{x}_1 = 67.0$ and for girls $\bar{x}_2 = 61.5$. If it is reasonable to assume that scores for similar children under similar circumstances are normally distributed with standard deviations, $\sigma_1 = 11$ and $\sigma_2 = 10$, find:

(a) The 90 percent confidence interval for $\mu_1 - \mu_2$.
(b) The 95 percent confidence interval for $\mu_1 - \mu_2$.
(c) The 99 percent confidence interval for $\mu_1 - \mu_2$.

5.4.2 A sample of 10 twelve-year-old girls and a sample of 10 twelve-year-old boys yielded mean heights, respectively, of $\bar{x}_1 = 59.8$ inches and $\bar{x}_2 = 58.5$ inches. Assuming normal distributions of heights with $\sigma_1 = 2$ inches and $\sigma_2 = 3$

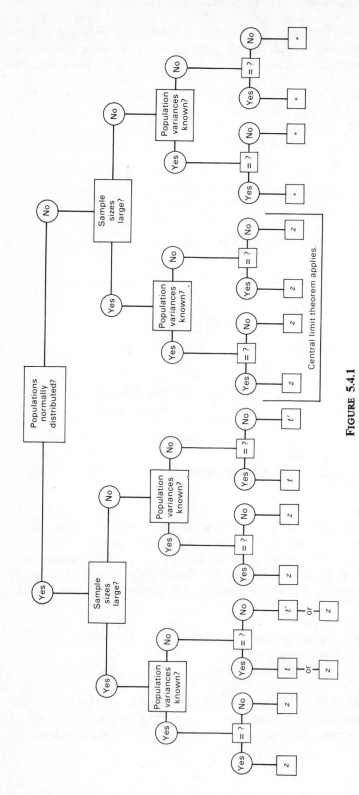

FIGURE 5.4.1

Flowchart for Use in Deciding whether the Reliability Factor Should Be z, t, or t' when Making Inferences about the Difference Between Two Population Means. (*Use a Nonparametric Procedure. See Chapter 11.)

inches, find:

(a) The 90 percent confidence interval for $\mu_1 - \mu_2$.
(b) The 95 percent confidence interval for $\mu_1 - \mu_2$.
(c) The 99 percent confidence interval for $\mu_1 - \mu_2$.

5.4.3 The difference between mean lengths of stay among patients with different diagnoses is of interest to health planners. A sample of 100 patients with disease A, admitted to a chronic disease hospital, remained in the hospital, on the average, 35 days. Another sample of 100 patients with disease B stayed, on the average, 28 days. If the population variances are 100 and 255, respectively, find:

(a) The 90 percent confidence interval for $\mu_A - \mu_B$.
(b) The 95 percent confidence interval for $\mu_A - \mu_B$.
(c) The 99 percent confidence interval for $\mu_A - \mu_B$.

5.4.4 A study was conducted to compare the high-density lipoprotein levels in adult males who were sedentary workers and those who were manual laborers. Sample data yielded the following results:

$$\text{Sedentary workers: } \bar{x} = 56.5, \ s = 14.1, \ n = 55$$
$$\text{Manual laborers: } \bar{x} = 51.3, \ s = 13.5, \ n = 50$$

Construct a 95 percent confidence interval for the difference between population means.

5.4.5 Refer to Exercises 5.3.3 and 5.3.4. Assume the population variances are equal. Construct the 90, 95, and 99 percent confidence intervals for the difference in the two population means.

5.4.6 Transverse diameter measurements on the hearts of adult males and females gave the following results:

Group	Sample size	$\bar{x}$ (cm)	s (cm)
Males	12	13.21	1.05
Females	9	11.00	1.01

Assuming normally distributed populations with equal variances, construct the 90, 95, and 99 percent confidence intervals for $\mu_1 - \mu_2$.

5.4.7 Twenty-four experimental animals with vitamin D deficiency were divided equally into two groups. Group 1 received treatment consisting of a diet that provided vitamin D. The second group was not treated. At the end of the experimental period, serum calcium determinations were made with the following results:

$$\text{Treated group: } \bar{x} = 11.1 \text{ mg/100 ml}, \ s = 1.5$$
$$\text{Untreated group: } \bar{x} = 7.8 \text{ mg/100 ml}, \ s = 2.0$$

Assuming normally distributed populations with equal variances construct the 90 95, and 99 percent confidence intervals for the difference between population means.

5.4.8 Two groups of children were given visual acuity tests. Group 1 was composed of 11 children who receive their health care from private physicians. The mean score for this group was 26 with a standard deviation of 5. The second group, consisting of 14 children who receive their health care from the health department, had an average score of 21 with a standard deviation of 6. Assuming normally distributed populations with equal variances find the .90, .95, and .99 confidence intervals for $\mu_1 - \mu_2$.

5.4.9 The average length of stay of a sample of 20 patients discharged from a general hospital was 7 days with a standard deviation of 2 days. A sample of 24 patients discharged from a chronic disease hospital has an average length of stay of 36 days with a standard deviation of 10 days. Assuming normally distributed populations with unequal variances, find the 95 percent confidence interval for the difference between population means.

5.4.10 In a study of factors thought to be responsible for the adverse effects of smoking on human reproduction, cadmium level determinations (nanograms per gram) were made on placenta tissue of a sample of 14 mothers who were smokers and an independent random sample of 18 nonsmoking mothers. The results were as follows:

Nonsmokers: 10.0, 8.4, 12.8, 25.0, 11.8, 9.8, 12.5, 15.4, 23.5,
9.4, 25.1, 19.5, 25.5, 9.8, 7.5, 11.8, 12.2, 15.0
Smokers: 30.0, 30.1, 15.0, 24.1, 30.5, 17.8, 16.8, 14.8,
13.4, 28.5, 17.5, 14.4, 12.5, 20.4

Construct a 95% confidence interval for the difference between population means. Does it appear likely that the mean cadmium level is higher among smokers than nonsmokers? Why do you reach this conclusion?

5.5 CONFIDENCE INTERVAL
FOR A POPULATION PROPORTION

Many questions of interest to the health worker relate to population proportions. What proportion of patients who receive a particular type of treatment recover? What proportion of some population has a certain disease? What proportion of a population are immune to a certain disease?

To estimate a population proportion we proceed in the same manner as when estimating a population mean. A sample is drawn from the population of interest, and the sample proportion, $\hat{p}$, is computed. This sample proportion is used as the point estimator of the population proportion. A confidence interval is obtained by the general formula:

$$\text{estimator} \pm (\text{reliability coefficient}) \times (\text{standard error})$$

In the previous chapter we saw that when both np and $n(1-p)$ are greater than 5, we may consider the sampling distribution of $\hat{p}$ to be quite close to the normal distribution. When this condition is met, our reliability coefficient is some value of z from the unit normal distribution. The standard error, we have seen, is equal to $\sigma_{\hat{p}} = \sqrt{p(1-p)/n}$. Since p, the parameter we are trying to estimate, is unknown, we must use $\hat{p}$ as an estimate. Thus we estimate $\sigma_{\hat{p}}$ by $\sqrt{\hat{p}(1-\hat{p})/n}$, and our $100(1-\alpha)$ percent confidence interval for p is given by

$$\hat{p} \pm z_{(1-\alpha/2)}\sqrt{\hat{p}(1-\hat{p})/n} \qquad (5.5.1)$$

We give this interval both the probabilistic and practical interpretations.

Example 5.5.1 A survey was conducted to study the dental health practices, and attitudes of a certain urban adult population. Of 300 adults interviewed, 123 said that they regularly had a dental check up twice a year.

The best point estimate of the population proportion is $\hat{p} = 123/300 = .41$. The size of the sample and our estimate of p are of sufficient magnitude to justify use of the unit normal distribution in constructing a confidence interval. The reliability coefficient corresponding to a confidence level of .95, for example, is 1.96 and our estimate of the standard error, $\sigma_{\hat{p}}$, is $\sqrt{\hat{p}(1-\tilde{p})/n} = \sqrt{.41(.59)/300} = .028$. The 95 percent confidence interval for p, based on these data, is

$$.41 \pm 1.96(.028)$$
$$.41 \pm .05$$
$$.36, .46$$

We say we are 95 percent confident that the true proportion, p, is between .36 and .46 since, in repeated sampling, about 95 percent of the intervals constructed in the manner of the present single interval would include the true p.

EXERCISES

5.5.1 A medical record librarian drew a random sample of 100 patients' charts and found that in 8 percent of them the face sheet had at least one item of information contradictory to other information in the record. Construct the 90, 95, and 99 percent confidence intervals for the true proportion of charts containing such discrepancies.

5.5.2 A survey, resulting in a random sample of 150 households in a certain urban community, revealed that in 87 percent of the cases at least one member of the household had some form of health insurance. Construct the 90, 95, and 99 percent confidence intervals for p, the true proportion of households in the community with the characteristic of interest.

5.5.3 In a study designed to assess the relationship between a certain drug and a certain anomaly in chick embryos, 50 fertilized eggs were injected with the drug on the fourth day of incubation. On the twentieth day of incubation the embryos were examined and in 12 the presence of the abnormality was observed. Find the 90, 95, and 99 percent confidence intervals for p.

5.5.4 In a simple random sample of 125 unemployed male high school dropouts between the ages of 16 and 21, inclusive, 88 stated that they were regular consumers of alcoholic beverages. Construct a 95% confidence interval for the population proportion.

5.6 CONFIDENCE INTERVAL FOR THE DIFFERENCE BETWEEN TWO POPULATION PROPORTIONS

The magnitude of the difference between two population proportions is often of interest. We may want to compare, for example, men and women, two age groups, two socioeconomic groups, or two diagnostic groups with respect to the proportion possessing some characteristic of interest. An unbiased point estimator of the difference in two population proportions is provided by the difference in sample proportions, $\hat{p}_1 - \hat{p}_2$. Since, as we have seen, when n_1 and n_2 are large and the population proportions are not too close to 0 or 1, the central limit theorem applies and normal distribution theory may be employed to obtain confidence intervals. The standard error of the estimate must be estimated by

$$\hat{\sigma}_{\hat{p}_1 - \hat{p}_2} = \sqrt{\frac{\hat{p}_1(1 - \hat{p}_1)}{n_1} + \frac{\hat{p}_2(1 - \hat{p}_2)}{n_2}}$$

since the population proportions are unknown. A $100(1 - \alpha)$ percent confidence interval for $p_1 - p_2$ is given by

$$(\hat{p}_1 - \hat{p}_2) \pm z_{(1 - \alpha/2)}\sqrt{\frac{\hat{p}_1(1 - \hat{p}_1)}{n_1} + \frac{\hat{p}_2(1 - \hat{p}_2)}{n_2}} \qquad (5.6.1)$$

We may interpret this interval from both the probabilistic and practical points of view.

Example 5.6.1 Researchers wish to compare the effects of two treatments on mean recovery time of patients with a certain disease. Two hundred patients were randomly divided into two equal groups. Of the first group, who received the standard treatment, 78 recovered within three days. Out of the other 100, who were treated by a new method, 90 recovered within three days. The physician wished to estimate the true difference in the proportions who would recover within three days.

Our best point estimate of the difference in the population proportions is $\hat{p}_1 - \hat{p}_2 = .78 - .90 = -.12$. The 95 percent confidence interval for $p_1 - p_2$ is

$$(.78 - .90) \pm 1.96\sqrt{\frac{(.78)(.22)}{100} + \frac{(.90)(.10)}{100}}$$

$$-.12 \pm 1.96(.05)$$
$$-.12 \pm .10$$
$$-.22, -.02$$

We say we are 95 percent confident that the true difference is between .02 and .22, because we know that, in repeated sampling, about 95 percent of the intervals that may be constructed in the manner just described will include the true difference.

Note that the negative signs merely reflect the fact that better results were obtained by using the new treatment. The interval could just as well have been constructed around $\hat{p}_2 - \hat{p}_1$, in which case the end points of the interval would have been positive.

EXERCISES

5.6.1 Out of a sample of 150, selected from patients admitted over a two-year period to a large hospital, 129 had some type of hospitalization insurance. In a sample of 160 similarly selected patients from a second hospital, 144 had some type of hospitalization insurance. Find the 90, 95, and 99 percent confidence intervals for the true difference in population proportions.

5.6.2 In a survey conducted in two sections of a large metropolitan area, the following results were obtained with respect to abnormal blood pressure.

Area	Number of persons screened	Number abnormal on screening
1	200	20
2	250	38

Construct the 90, 95, and 99 percent confidence intervals for the difference between the two population proportions.

5.6.3 In a study designed to assess the side effects of two drugs, 50 animals were given drug A and 50 animals were given drug B. Of the 50 receiving drug A, 11 showed undesirable side effects, while 8 of those receiving drug B reacted similarly. Find the 90, 95, and 99 percent confidence intervals for $p_A - p_B$.

5.6.4 In a certain community neurological examinations of 110 employees of a herbicide factory found 44 with neurological abnormalities. Among 150 residents who were not employed by the herbicide factory, 16 showed neurological abnormalities. Construct a 95 percent confidence interval for the difference between population proportions.

5.7 DETERMINATION OF
SAMPLE SIZE FOR ESTIMATING MEANS

The question of how large a sample to take arises early in the planning of any survey or experiment. This is an important question and should not be treated lightly. To take a larger sample than is needed to achieve the desired results is wasteful of resources, while very small samples often lead to results that are of no practical use. Let us consider, then, how one may go about determining the size sample that is needed in a given situation. In this section, we present a method for determining the sample size required for estimating a population mean, and in the next section we apply this method to the case of sample size determination when the parameter to be estimated is a population proportion. By straightforward extensions of these methods, sample sizes required for more complicated situations can be determined.

The objectives in interval estimation are to obtain narrow intervals with high reliability. If we look at the components of a confidence interval, we see that the width of the interval is determined by the magnitude of the quantity

$$(\text{reliability coefficient}) \times (\text{standard error})$$

since the total width of the interval is twice this amount. For a given standard error, increasing reliability means a larger reliability coefficient. But a larger reliability coefficient for a fixed standard error makes for a wider interval.

On the other hand, if we fix the reliability coefficient, the only way to reduce the width of the interval is to reduce the standard error. Since the standard error is equal to $\sigma/\sqrt{n}$, and since σ is a constant, the only way to obtain a small standard error is to take a large sample. How large a sample? That depends on the size of σ, the population standard deviation, and the desired degree of reliability and desired interval width.

Let us suppose we want an interval that extends d units on either side of the estimator. We can write

$$d = (\text{reliability coefficient}) \times (\text{standard error}) \qquad (5.7.1)$$

If sampling is to be with replacement, from an infinite population, or from a population that is sufficiently large to warrant our ignoring the finite population correction, Equation 5.7.1 becomes

$$d = z\frac{\sigma}{\sqrt{n}} \qquad (5.7.2)$$

which, when solved for n, gives

$$n = \frac{z^2\sigma^2}{d^2}$$ (5.7.3)

When sampling is without replacement from a small finite population, the finite population correction is required and Equation 5.7.1 becomes

$$d = z\frac{\sigma}{\sqrt{n}}\sqrt{\frac{N-n}{N-1}}$$ (5.7.4)

which, when solved for n, gives

$$n = \frac{Nz^2\sigma^2}{d^2(N-1) + z^2\sigma^2}$$ (5.7.5)

If the finite population correction can be ignored, Equation 5.7.5 reduces to Equation 5.7.3.

The formulas for sample size require a knowledge of σ^2 but, as has been pointed out, the population variance is, as a rule, unknown. As a result, σ^2 has to be estimated. The most frequently used sources of estimates for σ^2 are the following.

1. A *pilot* or preliminary sample may be drawn from the population and the variance computed from this sample may be used as an estimate of σ^2. Observations used in the pilot sample may be counted as part of the final sample, so that n (the computed sample size) $- n_1$ (the pilot sample size) $= n_2$ (the number of observations needed to satisfy the total sample size requirement).

2. Estimates of σ^2 may be available from previous or similar studies.

3. If it is felt that the population from which the sample is to be drawn is approximately normally distributed, one may use the fact that the range is approximately equal to 6 standard deviations and compute $\sigma \approx R/6$. This method requires some knowledge of the smallest and largest value of the variable in the population.

Example 5.7.1   A health department nutritionist, wishing to conduct a survey among a population of teenage girls to determine their average daily protein intake, is seeking the advice of a biostatistician relative to the size sample that should be taken.

What procedure does the biostatistician follow in providing assistance to the nutritionist? Before the statistician can be of help to the nutritionist, the latter must provide three items of information: the desired width of the confidence interval, the level of confidence desired, and the magnitude of the population variance.

Let us assume that the nutritionist would like an interval about 10 units wide; that is, the estimate should be within about 5 units of the true value in either direction. Let us also assume that a confidence coefficient of .95 is decided on and that, from past experience, the nutritionist feels that the population standard deviation is probably about 20 grams. The statistician now has the necessary information to compute the sample size: $z = 1.96$, $\sigma = 20$, and $d = 5$. Let us assume that the population of interest is large so that the statistician may ignore the finite population correction and use Equation 5.7.3. On making proper substitutions, the value of n is found to be

$$n = \frac{(1.96)^2(20)^2}{(5)^2}$$

$$= 61.47$$

The nutritionist is advised to take a sample of size 62. When calculating a sample size by Equation 5.7.3 or Equation 5.7.5, we round up to the next largest whole number if the calculations yield a number that is not itself an integer.

EXERCISES

5.7.1 A hospital administrator wishes to estimate the mean weight of babies born in her hospital. How large a sample of birth records should be taken if she wants a 99 percent confidence interval that is 1 pound wide. Assume that a reasonable estimate of σ is 1 pound. What size sample is required if the confidence coefficient is lowered to .95?

5.7.2 The director of the rabies control section in a city health department wishes to draw a sample from the department's records of dog bites reported during the past year in order to estimate the mean age of persons bitten. He wants a 95 percent confidence interval, he will be satisfied to let $d = 2.5$, and from previous studies he estimates the population standard deviation to be about 15 years. How large a sample should be drawn?

5.7.3 A physician would like to know the mean fasting blood glucose value (milligrams per 100 ml) of patients seen in a diabetes clinic over the past 10 years. Determine the number of records the physician should examine in order to obtain a 90 percent confidence interval for μ if the desired width of the interval is 6 units and a pilot sample yields a variance of 60.

5.7.4 For multiple sclerosis patients we wish to estimate the mean age at which the disease was first diagnosed. We want a 95 percent confidence interval that is 10 years wide. If the population variance is 90, how large should our sample be?

5.8   DETERMINATION OF SAMPLE SIZE FOR ESTIMATING PROPORTIONS

The method of sample size determination when a population proportion is to be estimated is essentially the same as that described above for estimating a population mean. We make use of the fact that one-half the desired interval, d, may be set equal to the product of the reliability coefficient and the standard error.

Assuming random sampling and conditions warranting approximate normality of the distribution of $\hat{p}$ leads to the following formula for n when sampling is with replacement, when sampling is from an infinite population, or when the sampled population is large enough to make the use of the finite population correction unnecessary:

$$n = \frac{z^2 pq}{d^2} \qquad (5.8.1)$$

where $q = 1 - p$.

If the finite population correction cannot be disregarded the proper formula for n is

$$n = \frac{Nz^2 pq}{d^2(N - 1) + z^2 pq} \qquad (5.8.2)$$

When N is large in comparison to n (that is, $n/N \leq .05$) the finite population correction may be ignored, and Equation 5.8.2 reduces to Equation 5.8.1.

As we see, both formulas require a knowledge of p, the proportion in the population possessing the characteristic of interest. Since this is the parameter we are trying to estimate, it, obviously, will be unknown. One solution to this problem is to take a pilot sample and compute an estimate to be used in place of p in the formula for n. Sometimes an investigator will have some notion of an upper bound for p which can be used in the formula. For example, if it is desired to estimate the proportion of some population who have a certain condition, we may feel that the true proportion cannot be greater than, say, .30. We then substitute .30 for p in the formula for n. If it is impossible to come up with a better estimate, one may set p equal to .5 and solve for n. Since $p = .5$ in the formula yields the maximum value of n, this procedure will give a large enough sample for the desired reliability and interval width. It may, however, be larger than needed and result in a more expensive sample than if a better estimate of p had been available. This procedure should be used only if one is unable to come up with a better estimate of p.

Example 5.8.1   A survey is being planned to determine what proportion of families in a certain area are medically indigent. It is felt that the proportion cannot be greater than .35. A 95 percent confidence interval is desired with $d = .05$. What size sample of families should be selected?

If the finite population correction can be ignored, we have

$$n = \frac{(1.96)^2(.35)(.65)}{(.05)^2} = 349.6$$

The necessary sample size, then, is 350.

EXERCISES

5.8.1 An epidemiologist wishes to know what proportion of adults living in a large metropolitan area have subtype ay hepatitis B virus. Determine the size sample that would be required to estimate the true proportion to within .03 with 95 percent confidence. In a similar metropolitan area the proportion of adults, with the characteristic is reported to be .20. If data from another metropolitan area were not available and a pilot sample could not be drawn, what size sample would be required?

5.8.2 A survey is planned to determine what proportion of the high school students in a metropolitan school system have regularly smoked marijuana. If no estimate of p is available from previous studies, a pilot sample cannot be drawn, a confidence coefficient of .95 is desired, and $d = .04$ is to be used, determine the appropriate sample size. What size sample would be required if 99 percent confidence were desired?

5.8.3 A hospital administrator wishes to know what proportion of discharged patients are unhappy with the care received during hospitalization. How large a sample should be drawn if we let $d = .05$, the confidence coefficient is .95, and no other information is available? How large should the sample be if p is approximated by .25?

5.8.4 A health planning agency wishes to know, for a certain geographic region, what proportion of patients admitted to hospitals for the treatment of trauma are discharged dead. A 95 percent confidence interval is desired, the width of the interval must be .06, and the population proportion, from other evidence, is estimated to be .20. How large a sample is needed?

5.9 CONFIDENCE INTERVAL FOR THE VARIANCE OF A NORMALLY DISTRIBUTED POPULATION

Point Estimation of the Population Variance In previous sections it has been suggested that when a population variance is unknown, the sample variance may be used as an estimator. The reader may have wondered about the quality of this estimator. We have discussed only one criterion of goodness—unbiasedness—so let us see if the sample variance is an unbiased estimator of the population variance. To be unbiased, the average value of the sample variance over all possible samples must be equal to the population variance. That is, the expression $E(s^2) = \sigma^2$ must hold. To see if this condition holds for a particular situation, let us refer to the example of constructing a sampling distribution given in Section 4.4. In Table 4.4.1 we have all possible samples of size 2 from the population consisting of the values 6, 8, 10, 12, and 14. It will be recalled that two measures of dispersion for this population were computed as follows:

$$\sigma^2 = \frac{\sum (x_i - \mu)^2}{N} = 8 \quad \text{and} \quad S^2 = \frac{\sum (x_i - \mu)^2}{N - 1} = 10$$

If we compute the sample variance $s^2 = \sum (x_i - \bar{x})^2/(n - 1)$ for each of the possible samples shown in Table 4.4.1, we obtain the sample variances shown in Table 5.9.1.

If sampling is with replacement, the expected value of s^2 is obtained by taking the mean of all sample variances in Table 5.9.1. When we do this, we have

$$E(s^2) = \frac{\sum s_i^2}{N^n} = \frac{0 + 2 + \cdots + 2 + 0}{25} = \frac{200}{25} = 8$$

and we see, for example, that when sampling is with replacement $E(s^2) = \sigma^2$, where $s^2 = \sum (x_i - \bar{x})^2/(n - 1)$ and $\sigma^2 = \sum (x_i - \mu)^2/N$.

If we consider the case where sampling is without replacement, the expected value of s^2 is obtained by taking the mean of all variances above (or below) the principal diagonal. That is,

$$E(s^2) = \frac{\sum s_i^2}{\binom{N}{n}} = \frac{2 + 8 + \cdots + 2}{10} = \frac{100}{10} = 10$$

which, we see, is not equal to σ^2, but is equal to $S^2 = \sum (x_i - \mu)^2/(N - 1)$.

TABLE 5.9.1

Variances Computed from Samples Shown in Table 4.4.1

		SECOND DRAW				
		6	8	10	12	14
	6	0	2	8	18	32
	8	2	0	2	8	18
First draw	10	8	2	0	2	8
	12	18	8	2	0	2
	14	32	18	8	2	0

These results are examples of general principles, as it can be shown that, in general,

$$E(s^2) = \sigma^2 \text{ when sampling is with replacement}$$

$$E(s^2) = S^2 \text{ when sampling is without replacement}$$

When N is large, $N - 1$ and N will be approximately equal and, consequently, σ^2 and S^2 will be approximately equal.

These results justify our use of $s^2 = \sum(x_i - \bar{x})^2/(n - 1)$ when computing the sample variance. In passing, let us note that although s^2 is an unbiased estimator of σ^2, s is not an unbiased estimator of σ. The bias, however, diminishes rapidly as n increases. Those interested in pursuing this point further are referred to the articles by Cureton (18), and Gurland and Tripathi (19).

Interval Estimation of a Population Variance   With a point estimate available to us, it is logical to inquire about the construction of a confidence interval for a population variance. Whether we are successful in constructing a confidence interval for σ^2 will depend on our ability to find an appropriate sampling distribution.

Confidence intervals for σ^2 are usually based on the sampling distribution of $(n - 1)s^2/\sigma^2$. If samples of size n are drawn from a normally distributed population, this quantity has a distribution known as the *chi-square distribution* with $n - 1$ degrees of freedom. As we shall say more about this distribution in a later chapter, we only say here that it is the distribution that the quantity $(n - 1)s^2/\sigma^2$ follows and that it is useful in finding confidence intervals for σ^2 when the assumption that the population is normally distributed holds true.

In Figure 5.9.1 are shown some chi-square distributions for several values of degrees of freedom. Percentiles of the chi-square distribution, designated by the Greek letter χ^2, are given in Table F. The column headings give the values of χ^2 to the left of which lies a proportion of the total area under the curve equal to the subscript of χ^2. The row labels are the degrees of freedom.

To obtain a $100(1 - \alpha)$ percent confidence interval for σ^2, we first obtain the $100(1 - \alpha)$ percent confidence interval for $(n - 1)s^2/\sigma^2$. To do this, we select the values of χ^2 from Table F in such a way that $\alpha/2$ is to the left of the smaller value and $\alpha/2$ is to the right of the larger value. In other words, the two values of χ^2 are selected in such a way that α is divided equally between the two tails of the distribution. We may designate these two values of χ^2 as $\chi^2_{\alpha/2}$ and $\chi^2_{(1-\alpha/2)}$, respectively. The $100(1 - \alpha)$ percent confidence interval for $(n - 1)s^2/\sigma^2$, then, is given by

$$\chi^2_{\alpha/2} < \frac{(n - 1)s^2}{\sigma^2} < \chi^2_{(1-\alpha/2)}$$

We now manipulate this expression in such a way that we obtain an expression with σ^2 alone as the middle term. First, let us divide each term by $(n - 1)s^2$

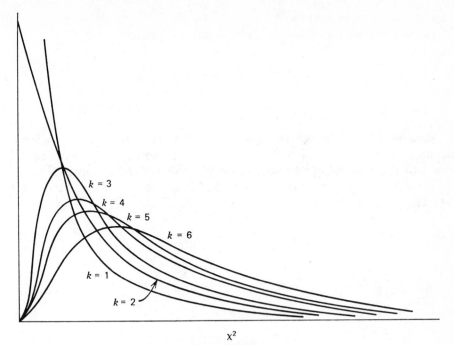

x^2

FIGURE 5.9.1

Chi-Square Distributions for Several Values of Degrees of Freedom k. (Source. Paul G. Hoel and Raymond J. Jessen. *Basic Statistics for Business and Economics*, Wiley, 1971. Used by Permission)

to get

$$\frac{\chi^2_{\alpha/2}}{(n-1)s^2} < \frac{1}{\sigma^2} < \frac{\chi^2_{(1-\alpha/2)}}{(n-1)s^2}$$

If we take the reciprocal of this expression we have

$$\frac{(n-1)s^2}{\chi^2_{\alpha/2}} > \sigma^2 > \frac{(n-1)s^2}{\chi^2_{(1-\alpha/2)}}$$

Note that the direction of the inequalities changed when we took the reciprocals. If we reverse the order of the terms we have

$$\frac{(n-1)s^2}{\chi^2_{(1-\alpha/2)}} < \sigma^2 < \frac{(n-1)s^2}{\chi^2_{\alpha/2}} \tag{5.9.1}$$

which is the $100(1-\alpha)$ percent confidence interval for σ^2. If we take the square root of each term in Expression 5.9.1, we have the following $100(1 - \alpha)$ percent confidence interval for σ, the population standard deviation:

$$\sqrt{\frac{(n-1)s^2}{\chi^2_{(1-\alpha/2)}}} < \sigma < \sqrt{\frac{(n-1)s^2}{\chi^2_{\alpha/2}}} \qquad (5.9.2)$$

Example 5.9.1 Refer to Example 5.3.1 and assume that the population of serum amylase determinations from which the sample of size 15 was drawn is normally distributed. Let us construct the 95 percent confidence interval for σ^2.

The sample yielded a value of $s^2 = 1225$. The degrees of freedom are $n - 1 = 14$. The appropriate values of χ^2 from Table F are $\chi^2_{(1-\alpha/2)} = 26.119$ and $\chi^2_{\alpha/2} = 5.629$. Our 95 percent confidence interval for σ^2 is

$$\frac{(14)(1225)}{26.119} < \sigma^2 < \frac{14(1225)}{5.629}$$
$$656.6101 < \sigma^2 < 3046.7223$$

The 95 percent confidence interval for σ is

$$25.62 < \sigma < 55.20$$

We say we are 95 percent confident that the parameters being estimated are within the specified limits, because we know that in the long run, in repeated sampling, 95 percent of intervals constructed as illustrated would include the respective parameters.

Although this method of constructing confidence intervals for σ^2 is widely used, it is not without its drawbacks. First, the assumption of the normality of the population from which the sample is drawn is crucial, and results may be misleading if the assumption is ignored.

Another difficulty with these intervals results from the fact that the estimator is not in the center of the confidence interval, as is the case with the confidence interval for μ. This is because the chi-square distribution, unlike the normal, is not symmetric. The practical implication of this is that the method for the construction of confidence intervals for σ^2, which has just been described, does not yield the shortest possible confidence intervals. Tate and Klett (20) give tables that may be used to overcome this difficulty.

EXERCISES

5.9.1 Each of a sample of 51 nursing students was given a standardized test to measure his or her level of responsibility. A value of $s^2 = 12$ was computed. Construct 95 percent confidence intervals for σ^2 and σ.

5.9.2 White blood cell counts taken on a sample of 10 adult males with some type of leukemia yielded a variance of 25,000,000. Construct 95 percent confidence intervals for σ^2 and σ.

5.9.3 Forced vital capacity determinations were made on 20 healthy adult males. The sample variance was 1,000,000. Construct 90 percent confidence intervals for σ^2 and σ.

5.9.4 In a study of myocardial transit times, appearance transit times were obtained on a sample of 30 patients with coronary artery disease. The sample variance was found to be 1.03. Construct 99 percent confidence intervals for σ^2 and σ.

5.9.5 A sample of 25 physically and mentally healthy males participated in a sleep experiment in which the percentage of each participant's total sleeping time spent in a certain stage of sleep was recorded. The variance computed from the sample data was 2.25. Construct 95 percent confidence intervals for σ^2 and σ.

5.9.6 Hemoglobin determinations were made on 16 animals exposed to a harmful chemical. The following observations were recorded: 15.6, 14.8, 14.4, 16.6, 13.8, 14.0, 17.3, 17.4, 18.6, 16.2, 14.7, 15.7, 16.4, 13.9, 14.8, 17.5. Construct 95 percent confidence intervals for σ^2 and σ.

5.9.7 Twenty air samples taken at the same site over a period of 6 months showed the following amounts of suspended particulate matter (micrograms per cubic meter of air).

68	22	36	32
42	24	28	38
30	44	28	27
28	43	45	50
79	74	57	21

Consider these measurements to be a random sample from a population of normally distributed measurements, and construct a 95 percent confidence interval for the population variance.

5.10 CONFIDENCE INTERVAL FOR THE RATIO OF THE VARIANCES OF TWO NORMALLY DISTRIBUTED POPULATIONS

It is frequently of interest to compare two variances, and one way to do this is to form their ratio, σ_1^2/σ_2^2. If two variances are equal, their ratio will be equal to 1. We usually will not know the variances of populations of interest, and, consequently, any comparisons we make will have to be based on sample variances. To be specific, we may wish to estimate the ratio of two population variances. Again, since this is a form of inference, we must rely on some sampling distribution, and this time the distribution of $(s_1^2/\sigma_1^2)/(s_2^2/\sigma_2^2)$ is utilized provided certain assumptions are met. The assumptions are that s_1^2 and s_2^2 are computed from

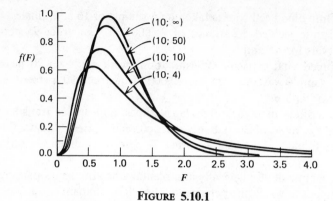

FIGURE 5.10.1

The *F*-Distribution for Various Degrees of Freedom. (From
Documenta Geigy, Scientific Tables, 7th edition, 1970.
Courtesy of Ciba-Geigy Limited, Basle, Switzerland)

independent samples of size n_1 and n_2, respectively, drawn from two normally
distributed populations.

If the assumptions are met $(s_1^2/\sigma_1^2)/(s_2^2/\sigma_2^2)$ follows a distribution known
as the *F distribution*. We defer a more complete discussion of this distribution
until a later chapter, but note that this distribution depends on two degrees of
freedom values, one corresponding to the value of $n_1 - 1$ used in computing s_1^2
and the other corresponding to the value of $n_2 - 1$ used in computing s_2^2. These
are usually referred to as the *numerator degrees of freedom* and the *denominator
degrees of freedom*. Figure 5.10.1 shows some F distributions for several numerator and denominator degrees-of-freedom combinations. Table G contains, for
specified combinations of degrees of freedom and values of α, F values to the
right of which lies $\alpha/2$ of the area under the curve of F.

To find the $100(1 - \alpha)$ percent confidence interval for σ_1^2/σ_2^2, we begin with
the expression

$$F_{\alpha/2} < \frac{s_1^2/\sigma_1^2}{s_2^2/\sigma_2^2} < F_{(1-\alpha/2)}$$

where $F_{\alpha/2}$ and $F_{(1-\alpha/2)}$ are the values from the F table to the left and right of
which, respectively, lies $\alpha/2$ of the area under the curve. The middle term of this
expression may be rewritten so that the entire expression is

$$F_{\alpha/2} < \frac{s_1^2}{s_2^2} \cdot \frac{\sigma_2^2}{\sigma_1^2} < F_{(1-\alpha/2)}$$

If we divide through by s_1^2/s_2^2, we have

$$\frac{F_{\alpha/2}}{s_1^2/s_2^2} < \frac{\sigma_2^2}{\sigma_1^2} < \frac{F_{(1-\alpha/2)}}{s_1^2/s_2^2}$$

Taking the reciprocals of the three terms gives

$$\frac{s_1^2/s_2^2}{F_{\alpha/2}} > \frac{\sigma_1^2}{\sigma_2^2} > \frac{s_1^2/s_2^2}{F_{(1-\alpha/2)}}$$

and if we reverse the order we have the following $100(1 - \alpha)$ percent confidence interval for σ_1^2/σ_2^2:

$$\frac{s_1^2/s_2^2}{F_{(1-\alpha/2)}} < \frac{\sigma_1^2}{\sigma_2^2} < \frac{s_1^2/s_2^2}{F_{\alpha/2}} \qquad \textbf{(5.10.1)}$$

Example 5.10.1 Researchers selected a simple random sample of size 21 from a population of apparently healthy adult subjects (sample 1). They selected an independent simple random sample of size 16 from a population of patients with Parkinson's disease (sample 2). The variable of interest was reaction time to a particular stimulus. The sample variances were 1600 for sample 1 and 1225 for sample 2. Let us compute the 95 percent confidence interval for σ_1^2/σ_2^2. We have the following information:

$$n_1 = 21, \qquad n_2 = 16$$
$$s_1^2 = 1600, \qquad s_2^2 = 1225$$

numerator degrees of freedom = 20
denominator degrees of freedom = 15
$$\alpha = .05$$
$$F_{.025} = .389 \qquad F_{.975} = 2.76$$

At this point we must make a cumbersome, but unavoidable, digression and explain how the values $F_{.975} = 2.76$ and $F_{.025} = .389$ were obtained. The value of $F_{.975}$ at the intersection of the column headed 20 and the row labeled 15 is 2.76. If we had a more extensive table of the F distribution, finding $F_{.025}$ would be no trouble; we would simply find $F_{.025}$ as we found $F_{.975}$. We would take the value at the intersection of the column headed 20 and the row headed 15. To include every possible percentile of F would make for a very lengthy table. Fortunately, however, there exists a relationship that enables us to compute the lower percentile values from our limited table. We proceed as follows.

Interchange the numerator and denominator degrees of freedom and locate the appropriate value of F. For the problem at hand we locate 2.57, which is at the intersection of the column headed 15 and the row labeled 20. We now take the reciprocal of this value, $1/2.57 = .389$.

We are now ready to obtain our 95 percent confidence interval for σ_1^2/σ_2^2 by substituting appropriate values into Expression 5.10.1:

$$\frac{1600/1225}{2.76} < \frac{\sigma_1^2}{\sigma_2^2} < \frac{1600/1225}{.389}$$

$$.473 < \frac{\sigma_1^2}{\sigma_2^2} < 3.36$$

We give this interval the appropriate probabilistic and practical interpretations.

Alternative procedures for making inferences about the equality of two variances when the sampled populations are not normally distributed may be found in the book by Daniel (21).

EXERCISES

5.10.1 Oxytocin concentrations (picograms per milliliter) in the amniotic fluid of two samples of women were recorded. The first sample consisted of 25 parturient women with labor contractions. The second sample consisted of 16 women with no labor contractions. The sample variances were $s_1^2 = 40,000$ and $s_2^2 = 10,000$. Construct the 95 percent confidence interval for the ratio of the two population variances.

5.10.2 A measure of the nicotine content was obtained for urine specimens collected from two groups of adult males. Group 1 consisted of 25 apparently healthy cigarette smokers, and group 2 was composed of 31 cigarette smokers with cancer of the urinary bladder. The sample variances were $s_1^2 = 1.00$ and $s_2^2 = 3.5$. Construct the 90 percent confidence interval for σ_2^2/σ_1^2.

5.10.3 Stroke index values were statistically analyzed for two samples of patients suffering from myocardial infarction. The sample variances were 12 and 10. There were 21 patients in each sample. Construct the 95 percent confidence interval for the ratio of the two population variances.

5.10.4 Thirty-two adult aphasics seeking speech therapy were divided equally into two groups. Group 1 received treatment 1, and group 2 received treatment 2. Statistical analysis of the treatment effectiveness scores yielded the following variances: $s_1^2 = 8$, $s_2^2 = 15$. Construct the 90 percent confidence interval for σ_2^2/σ_1^2.

5.10.5 Sample variances were computed for the tidal volumes (milliliters) of two groups of patients suffering from atrial septal defect. The results and sample sizes were as follows:

$$n_1 = 31, \ s_1^2 = 35,000$$

$$n_2 = 41, \ s_2^2 = 20,000$$

Construct the 95 percent confidence interval for the ratio of the two population variances.

5.10.6 Glucose responses to oral glucose were recorded for 11 patients with Huntington's disease (group 1) and 13 control subjects (group 2). Statistical analysis of the results yielded the following sample variances: $s_1^2 = 105$, $s_2^2 = 148$. Construct the 95 percent confidence interval for the ratio of the two population variances.

5.10.7 Measurements of gastric secretion of hydrochloric acid (milliequivalents per hour) in 16 normal subjects and 10 subjects with duodenal ulcer yielded the following results:

Normal subjects: 6.3, 2.0, 2.3, 0.5, 1.9, 3.2, 4.1, 4.0, 6.2, 6.1,

3.5, 1.3, 1.7, 4.5, 6.3, 6.2

Ulcer subjects: 13.7, 20.6, 15.9, 28.4, 29.4, 18.4, 21.1, 3.0,

26.2, 13.0

Construct a 95 percent confidence interval for the ratio of the two population variances. What assumptions must be met for this procedure to be valid?

5.11 SUMMARY

This chapter is concerned with one of the major areas of statistical inference—estimation. Both point and interval estimation are covered. The concepts and methods involved in the construction of confidence intervals are illustrated for the following parameters: means, the difference between two means, proportions, the difference between two proportions, variances, and the ratio of two variances.

REVIEW QUESTIONS AND EXERCISES

1. What is statistical inference?
2. Why is estimation an important type of inference?
3. What is a point estimate?
4. Explain the meaning of unbiasedness.
5. Define the following:
   (a) Reliability coefficient
   (b) Confidence coefficient
   (c) Standard error
   (d) Estimator
6. Give the general formula for a confidence interval.
7. State the probabilistic and practical interpretations of a confidence interval.
8. Of what use is the central limit theorem in estimation?
9. Describe the t distribution.

10. What are the assumptions underlying the use of the t distribution in estimating a single population mean?

11. What is the finite population correction? When can it be ignored?

12. What are the assumptions underlying the use of the t distribution in estimating the difference between two population means?

13. Arterial blood gas analyses performed on a sample of 15 physically active adult males yielded the following resting PaO_2 values:

$$75, 80, 80, 74, 84, 78, 89, 72, 83, 76, 75, 87, 78, 79, 88$$

Compute the 95 percent confidence interval for the mean of the population.

14. What proportion of asthma patients are allergic to house dust? In a sample of 140, 35 percent had positive skin reactions. Construct the 95 percent confidence interval for the population proportion.

15. An industrial hygiene survey was conducted in a large metropolitan area. Of 70 manufacturing plants of a certain type visited, 21 received a "poor" rating with respect to absence of safety hazards. Construct a 95 percent confidence interval for the population proportion deserving a "poor" rating.

16. Refer to the previous problem. How large a sample would be required to estimate the population proportion to within .05 with 95 percent confidence (.30 is the best available estimate of p):
    (a) If the finite population correction can be ignored?
    (b) If the finite population correction is not ignored and $N = 1500$?

17. In a dental survey conducted by a county dental health team, 500 adults were asked to give the reason for their last visit to a dentist. Of the 220 who had less than a high school education, 44 said they went for preventive reasons. Of the remaining 280, who had a high school education or better, 150 stated that they went for preventive reasons. Construct a 95 percent confidence interval for the difference between the two population proportions.

18. A breast cancer research team collected the following data on tumor size:

Type of tumor	n	$\bar{x}$	s
A	21	3.85 cm	1.95 cm
B	16	2.80 cm	1.70 cm

Construct a 95 percent confidence interval for the difference between population means.

19. A certain drug was found to be effective in the treatment of pulmonary disease in 180 of 200 cases treated. Construct the 90 percent confidence interval for the population proportion.

20. Seventy patients with stasis ulcers of the leg were randomly divided into two equal groups. Each group received a different treatment for edema. At the end of the experiment, treatment effectiveness was measured in terms of reduction in leg volume as determined by water displacement. The means and standard deviations for the two groups were as follows:

Group (treatment)	$\bar{x}$	s
A	95 cc	25
B	125 cc	30

Construct a 95 percent confidence interval for the difference in population means.

21. What is the average serum bilirubin level of patients admitted to a hospital for treatment of hepatitis? A sample of 10 patients yielded the following results:

$$20.5, 14.8, 21.3, 12.7, 15.2, 26.6, 23.4, 22.9, 15.7, 19.2$$

Construct a 95 percent confidence interval for the population mean.

22. Determinations of saliva pH levels were made in two independent random samples of seventh grade schoolchildren. Sample A children were caries-free while sample B children had a high incidence of carries. The results were as follows:

A: 7.14, 7.11, 7.61, 7.98, 7.21, 7.16, 7.89,
    7.24, 7.86, 7.47, 7.82, 7.37, 7.66, 7.62, 7.65

B: 7.36, 7.04, 7.19, 7.41, 7.10, 7.15, 7.36,
    7.57, 7.64, 7.00, 7.25, 7.19

Construct a 90 percent confidence interval for the difference between the population means. Assume that the population variances are equal.

23. Drug A was prescribed for a random sample of 12 patients complaining of insomnia. An independent random sample of 16 patients with the same complaint received drug B. The numbers of hours of sleep experienced during the second night after treatment began were as follows.

A: 3.5, 5.7, 3.4, 6.9, 17.8, 3.8, 3.0, 6.4, 6.8, 3.6, 6.9, 5.7

B: 4.5, 11.7, 10.8, 4.5, 6.3, 3.8, 6.2, 6.6, 7.1, 6.4, 4.5,
    5.1, 3.2, 4.7, 4.5, 3.0

Construct a 95 percent confidence interval for the difference between the population means. Assume that the population variances are equal.

24. Elevated serum cholesterol levels have long been implicated in increased risk of cardiovascular disease. Numerous studies have been conducted in an effort to gain more insight into the nature of the association. The following serum cholesterol (chol.) determinations were made on a population of apparently healthy subjects.

Subject	Chol.	Subject	Chol.	Subject	Chol.	Subject	Chol.
1.	231	2.	219	3.	216	4.	199
5.	159	6.	185	7.	292	8.	227
9.	203	10.	239	11.	196	12.	180
13.	304	14.	239	15.	186	16.	197
17.	248	18.	152	19.	226	20.	228
21.	238	22.	209	23.	221	24.	189
25.	209	26.	169	27.	216	28.	233
29.	193	30.	159	31.	181	32.	195
33.	225	34.	225	35.	227	36.	252
37.	244	38.	217	39.	220	40.	177
41.	190	42.	220	43.	218	44.	166
45.	192	46.	219	47.	180	48.	249
49.	209	50.	217	51.	229	52.	250
53.	161	54.	226	55.	206	56.	258
57.	206	58.	228	59.	282	60.	218
61.	224	62.	206	63.	244	64.	249
65.	276	66.	243	67.	212	68.	251
69.	196	70.	207	71.	177	72.	200
73.	189	74.	218	75.	257	76.	144
77.	199	78.	242	79.	221	80.	271
81.	203	82.	216	83.	236	84.	226
85.	191	86.	216	87.	214	88.	156
89.	180	90.	180	91.	198	92.	237
93.	215	94.	252	95.	214	96.	200
97.	172	98.	234	99.	218	100.	231
101.	193	102.	252	103.	200	104.	231
105.	208	106.	189	107.	223	108.	202
109.	209	110.	221	111.	186	112.	241
113.	244	114.	226	115.	265	116.	150
117.	187	118.	188	119.	180	120.	177
121.	262	122.	212	123.	197	124.	259
125.	177	126.	214	127.	250	128.	275
129.	240	130.	193	131.	228	132.	201
133.	237	134.	222	135.	169	136.	189
137.	220	138.	187	139.	248	140.	243
141.	284	142.	260	143.	220	144.	179
145.	171	146.	217	147.	286	148.	267
149.	281	150.	201	151.	217	152.	229
153.	259	154.	223	155.	292	156.	218
157.	257	158.	249	159.	263	160.	241
161.	193	162.	208	163.	295	164.	250
165.	262	166.	268	167.	215	168.	248
169.	179	170.	265	171.	244	172.	196
173.	207	174.	220	175.	196	176.	249
177.	237	178.	250	179.	199	180.	222
181.	174	182.	240	183.	218	184.	199
185.	207	186.	248	187.	180	188.	258

Subject	Chol.	Subject	Chol.	Subject	Chol.	Subject	Chol.
189.	248	190.	224	191.	221	192.	270
193.	221	194.	201	195.	231	196.	192
197.	193	198.	219	199.	238	200.	208
201.	213	202.	150	203.	248	204.	202
205.	199	206.	195	207.	244	208.	227
209.	227	210.	154	211.	176	212.	184
213.	204	214.	178	215.	200	216.	228
217.	209	218.	166	219.	211	220.	245
221.	245	222.	224	223.	242	224.	227
225.	192	226.	206	227.	234	228.	221
229.	216	230.	242	231.	201	232.	224
233.	154	234.	208	235.	172	236.	227
237.	212	238.	244	239.	219	240.	249
241.	222	242.	229	243.	247	244.	236
245.	259	246.	224	247.	186	248.	281
249.	239	250.	221	251.	257	252.	236
253.	201	254.	183	255.	222	256.	269
257.	204	258.	195	259.	208	260.	262
261.	208	262.	183	263.	170	264.	186
265.	197	266.	221	267.	202	268.	197
269.	205	270.	249	271.	221	272.	243
273.	196	274.	241	275.	236	276.	223
277.	212	278.	197	279.	248	280.	159
281.	218	282.	216	283.	186	284.	254
285.	196	286.	253	287.	259	288.	173
289.	169	290.	200	291.	218	292.	222
293.	179	294.	206	295.	208	296.	220
297.	210	298.	207	299.	226	300.	195
301.	204	302.	237	303.	160	304.	265
305.	212	306.	220	307.	171	308.	233
309.	191	310.	256	311.	238	312.	154
313.	239	314.	243	315.	175	316.	221
317.	251	318.	280	319.	208	320.	167
321.	160	322.	203	323.	239	324.	254
325.	211	326.	205	327.	255	328.	264
329.	188	330.	216	331.	221	332.	272
333.	236	334.	206	335.	160	336.	218
337.	248	338.	268	339.	224	340.	224
341.	189	342.	234	343.	156	344.	266
345.	174	346.	202	347.	230	348.	261
349.	138	350.	201	351.	262	352.	246
353.	192	354.	225	355.	162	356.	241
357.	249	358.	259	359.	207	360.	215
361.	212	362.	199	363.	268	364.	206
365.	184	366.	268	367.	213	368.	168
369.	155	370.	153	371.	220	372.	203
373.	261	374.	253	375.	283	376.	199

Subject	Chol.	Subject	Chol.	Subject	Chol.	Subject	Chol.
377.	224	378.	227	379.	229	380.	212
381.	187	382.	183	383.	230	384.	106
385.	254	386.	234	387.	193	388.	248
389.	231	390.	223	391.	220	392.	265
393.	175	394.	269	395.	211	396.	174
397.	239	398.	191	399.	219	400.	198
401.	227	402.	225	403.	204	404.	251
405.	261	406.	263	407.	195	408.	234
409.	172	410.	215	411.	257	412.	255
413.	199	414.	222	415.	224	416.	235
417.	195	418.	221	419.	289	420.	214
421.	229	422.	240	423.	239	424.	229
425.	241	426.	191	427.	181	428.	245
429.	181	430.	185	431.	220	432.	188
433.	161	434.	247	435.	232	436.	203
437.	232	438.	188	439.	222	440.	191
441.	208	442.	211	443.	177	444.	171
445.	248	446.	206	447.	230	448.	218
449.	192	450.	205	451.	183	452.	243
453.	206	454.	222	455.	191	456.	219
457.	202	458.	171	459.	194	460.	237
461.	240	462.	178	463.	242	464.	201
465.	231	466.	182	467.	246	468.	272
469.	212	470.	209	471.	235	472.	187
473.	230	474.	157	475.	221	476.	266
477.	200	478.	199	479.	240	480.	224
481.	222	482.	200	483.	187	484.	278
485.	230	486.	165	487.	243	488.	212
489.	289	490.	195	491.	241	492.	196
493.	243	494.	217	495.	256	496.	223
497.	233	498.	142	499.	245	500.	229
501.	228	502.	209	503.	172	504.	243
505.	189	506.	239	507.	187	508.	170
509.	255	510.	183	511.	260	512.	240
513.	247	514.	194	515.	175	516.	192
517.	234	518.	212	519.	246	520.	185
521.	161	522.	192	523.	236	524.	232
525.	249	526.	202	527.	213	528.	239
529.	221	530.	248	531.	281	532.	228
533.	184	534.	280	535.	264	536.	263
537.	232	538.	230	539.	205	540.	259
541.	211	542.	229	543.	277	544.	188
545.	202	546.	214	547.	238	548.	225
549.	253	550.	204	551.	185	552.	210
553.	208	554.	241	555.	214	556.	259
557.	242	558.	161	559.	191	560.	214

Subject	Chol.	Subject	Chol.	Subject	Chol.	Subject	Chol.
561.	177	562.	219	563.	163	564.	200
565.	187	566.	189	567.	236	568.	162
569.	221	570.	189	571.	213	572.	242
573.	235	574.	232	575.	231	576.	194
577.	216	578.	245	579.	235	580.	223
581.	203	582.	231	583.	253	584.	215
585.	278	586.	240	587.	263	588.	244
589.	237	590.	214	591.	266	592.	240
593.	201	594.	194	595.	200	596.	223
597.	219	598.	215	599.	200	600.	208
601.	260	602.	259	603.	223	604.	262
605.	220	606.	229	607.	155	608.	198
609.	186	610.	199	611.	201	612.	238
613.	163	614.	210	615.	234	616.	174
617.	262	618.	226	619.	263	620.	202
621.	220	622.	165	623.	233	624.	222
625.	182	626.	201	627.	223	628.	202
629.	263	630.	210	631.	198	632.	180
633.	173	634.	190	635.	177	636.	240
637.	212	638.	259	639.	197	640.	191
641.	212	642.	206	643.	221	644.	254
645.	229	646.	197	647.	220	648.	223
649.	230	650.	202	651.	231	652.	253
653.	209	654.	188	655.	222	656.	246
657.	263	658.	241	659.	200	660.	257
661.	239	662.	177	663.	225	664.	281
665.	218	666.	194	667.	279	668.	270
669.	225	670.	200	671.	183	672.	228
673.	156	674.	212	675.	258	676.	272
677.	231	678.	203	679.	253	680.	209
681.	216	682.	192	683.	234	684.	230
685.	179	686.	225	687.	276	688.	153
689.	269	690.	194	691.	265	692.	235
693.	203	694.	232	695.	221	696.	168
697.	212	698.	229	699.	234	700.	182
701.	188	702.	238	703.	218	704.	194
705.	194	706.	205	707.	207	708.	202
709.	252	710.	173	711.	205	712.	235
713.	317	714.	265	715.	232	716.	255
717.	213	718.	228	719.	271	720.	250
721.	247	722.	210	723.	262	724.	238
725.	212	726.	221	727.	257	728.	197
729.	269	730.	234	731.	201	732.	250
733.	197	734.	223	735.	174	736.	261
737.	240	738.	265	739.	212	740.	223
741.	178	742.	250	743.	246	744.	230

Subject	Chol.	Subject	Chol.	Subject	Chol.	Subject	Chol.
745.	286	746.	299	747.	177	748.	242
749.	263	750.	191	751.	196	752.	276
753.	169	754.	186	755.	220	756.	206
757.	254	758.	207	759.	222	760.	222
761.	246	762.	225	763.	241	764.	216
765.	213	766.	183	767.	241	768.	162
769.	248	770.	203	771.	200	772.	281
773.	209	774.	203	775.	193	776.	244
777.	184	778.	186	779.	226	780.	202
781.	236	782.	174	783.	243	784.	208
785.	274	786.	224	787.	215	788.	213
789.	261	790.	201	791.	198	792.	187
793.	211	794.	251	795.	211	796.	237
797.	231	798.	217	799.	238	800.	220
801.	226	802.	205	803.	187	804.	231
805.	171	806.	203	807.	204	808.	246
809.	210	810.	260	811.	240	812.	187
813.	225	814.	198	815.	229	816.	199
817.	216	818.	201	819.	253	820.	242
821.	204	822.	212	823.	231	824.	287
825.	247	826.	237	827.	161	828.	182
829.	220	830.	249	831.	171	832.	176
833.	259	834.	142	835.	222	836.	275
837.	182	838.	256	839.	246	840.	204
841.	287	842.	258	843.	226	844.	183
845.	195	846.	188	847.	174	848.	201
849.	282	850.	216	851.	215	852.	224
853.	186	854.	169	855.	195	856.	238
857.	227	858.	283	859.	209	860.	243
861.	196	862.	185	863.	258	864.	170
865.	259	866.	222	867.	211	868.	197
869.	195	870.	217	871.	252	872.	230
873.	288	874.	273	875.	244	876.	194
877.	171	878.	232	879.	182	880.	227
881.	203	882.	179	883.	146	884.	230
885.	259	886.	240	887.	251	888.	249
889.	269	890.	222	891.	277	892.	243
893.	204	894.	136	895.	184	896.	155
897.	255	898.	254	899.	262	900.	215
901.	283	902.	267	903.	231	904.	191
905.	252	906.	207	907.	164	908.	215
909.	252	910.	203	911.	286	912.	265
913.	216	914.	271	915.	201	916.	221
917.	218	918.	281	919.	220	920.	196
921.	244	922.	226	923.	230	924.	208
925.	255	926.	227	927.	216	928.	223
929.	230	930.	236	931.	211	932.	209

Subject	Chol.	Subject	Chol.	Subject	Chol.	Subject	Chol.
933.	229	934.	223	935.	205	936.	175
937.	221	938.	190	939.	172	940.	211
941.	179	942.	172	943.	267	944.	283
945.	225	946.	247	947.	294	948.	245
949.	188	950.	162	951.	195	952.	233
953.	234	954.	230	955.	227	956.	228
957.	239	958.	202	959.	208	960.	179
961.	227	962.	217	963.	228	964.	231
965.	225	966.	177	967.	206	968.	227
969.	229	970.	214	971.	172	972.	217
973.	218	974.	224	975.	243	976.	219
977.	259	978.	213	979.	208	980.	257
981.	228	982.	256	983.	216	984.	174
985.	228	986.	202	987.	191	988.	254
989.	243	990.	179	991.	284	992.	207
993.	176	994.	220	995.	219	996.	191
997.	210	998.	218	999.	220	1000.	195

Select a simple random sample of size 15 from this population and construct a 95 percent confidence interval for the population mean. Compare your results with those of your classmates. What assumptions are necessary for your estimation procedure to be valid?

25. Refer to Exercise 24. Select a simple random sample of size 50 from the population and construct a 95 percent confidence interval for the proportion of subjects in the population who have readings greater than 225. Compare your results with those of your classmates.

26. The following are the weights in grams of the babies born in a community hospital during a 3-year period.

Baby	Weight	Baby	Weight	Baby	Weight	Baby	Weight
1.	2866	2.	2661	3.	3573	4.	3639
5.	2783	6.	3596	7.	3741	8.	3555
9.	3621	10.	2917	11.	3660	12.	3627
13.	3242	14.	2864	15.	3190	16.	2496
17.	2622	18.	3827	19.	3017	20.	4157
21.	2807	22.	3526	23.	3311	24.	3086
25.	3031	26.	4162	27.	3233	28.	2386
29.	4334	30.	3240	31.	4085	32.	4468
33.	4618	34.	3110	35.	3473	36.	4796
37.	3360	38.	2546	39.	3105	40.	2654
41.	3061	42.	3763	43.	4064	44.	2829
45.	3050	46.	3452	47.	3651	48.	2436
49.	2867	50.	3024	51.	2793	52.	2441
53.	3159	54.	3976	55.	2738	56.	3534

Baby	Weight	Baby	Weight	Baby	Weight	Baby	Weight
57.	2609	58.	4792	59.	2374	60.	4137
61.	3739	62.	4042	63.	3572	64.	4589
65.	3689	66.	3463	67.	2858	68.	3287
69.	4002	70.	3604	71.	4276	72.	3186
73.	3104	74.	3178	75.	2513	76.	4380
77.	2552	78.	3544	79.	3384	80.	4661
81.	3195	82.	3446	83.	3174	84.	2597
85.	3567	86.	3297	87.	3364	88.	2961
89.	2673	90.	2286	91.	3357	92.	2947
93.	4166	94.	2949	95.	2452	96.	3587
97.	3231	98.	2052	99.	2937	100.	3197
101.	3570	102.	4180	103.	3046	104.	2752
105.	4668	106.	3490	107.	3418	108.	2958
109.	3819	110.	3937	111.	3514	112.	4268
113.	2767	114.	3276	115.	4091	116.	2717
117.	3233	118.	3477	119.	3325	120.	2170
121.	3488	122.	3851	123.	3707	124.	3038
125.	3748	126.	2779	127.	3437	128.	2729
129.	3569	130.	2938	131.	4015	132.	2810
133.	4461	134.	2409	135.	3385	136.	2750
137.	3670	138.	2451	139.	2429	140.	3506
141.	2890	142.	3024	143.	3093	144.	4888
145.	3539	146.	2803	147.	3132	148.	3338
149.	3813	150.	3898	151.	2864	152.	3769
153.	3099	154.	4188	155.	3975	156.	2676
157.	4267	158.	3332	159.	3581	160.	3512
161.	3321	162.	3347	163.	4258	164.	3866
165.	2527	166.	4092	167.	4294	168.	2925
169.	3606	170.	3113	171.	3569	172.	3083
173.	2286	174.	2564	175.	2891	176.	3773
177.	3916	178.	3510	179.	4064	180.	3462
181.	3729	182.	3903	183.	2706	184.	3631
185.	4048	186.	2364	187.	3348	188.	3107
189.	4309	190.	2600	191.	3526	192.	5108
193.	3129	194.	3814	195.	3902	196.	2649
197.	2751	198.	3031	199.	3586	200.	3650
201.	2817	202.	3694	203.	2886	204.	2638
205.	4276	206.	3374	207.	3739	208.	2493
209.	3104	210.	3033	211.	2179	212.	2839
213.	2998	214.	3967	215.	3691	216.	1810
217.	3194	218.	3407	219.	3931	220.	3336
221.	4031	222.	2751	223.	3101	224.	2194
225.	2254	226.	2553	227.	3712	228.	3615
229.	3490	230.	3403	231.	3747	232.	5522
233.	4658	234.	2858	235.	3348	236.	2196
237.	2513	238.	2649	239.	4033	240.	2542

Baby	Weight	Baby	Weight	Baby	Weight	Baby	Weight
241.	2939	242.	2524	243.	3456	244.	2885
245.	3233	246.	4195	247.	2967	248.	4278
249.	3089	250.	3458	251.	3725	252.	3007
253.	2768	254.	3103	255.	3873	256.	2348
257.	3029	258.	3161	259.	1986	260.	3037
261.	3961	262.	2871	263.	2810	264.	3756
265.	3116	266.	3399	267.	2920	268.	4314
269.	2504	270.	2613	271.	3140	272.	2529
273.	2879	274.	3306	275.	3537	276.	3324
277.	3411	278.	2652	279.	2554	280.	3249
281.	3473	282.	3354	283.	3110	284.	3373
285.	3169	286.	2981	287.	2219	288.	4262
289.	4902	290.	2983	291.	2321	292.	3169
293.	4332	294.	3390	295.	2622	296.	2833
297.	5024	298.	3550	299.	3229	300.	3558
301.	3334	302.	3741	303.	3303	304.	3845
305.	3240	306.	2281	307.	3709	308.	2903
309.	2902	310.	3830	311.	2999	312.	3981
313.	3486	314.	3990	315.	3997	316.	3307
317.	2934	318.	2649	319.	3622	320.	3011
321.	2833	322.	3589	323.	2328	324.	3945
325.	3148	326.	3791	327.	2746	328.	2620
329.	2931	330.	2959	331.	4330	332.	3254
333.	3164	334.	3774	335.	3581	336.	2818
337.	3695	338.	3027	339.	2355	340.	1689
341.	2983	342.	2832	343.	2912	344.	3014
345.	3405	346.	3791	347.	2183	348.	2604
349.	3121	350.	3269	351.	3421	352.	4629
353.	3083	354.	3696	355.	2865	356.	2578
357.	3978	358.	3143	359.	2783	360.	4049
361.	2505	362.	2379	363.	2559	364.	2318
365.	3927	366.	2478	367.	3224	368.	3575
369.	3976	370.	3483	371.	3662	372.	3399
373.	3701	374.	3160	375.	4617	376.	4031
377.	3920	378.	2710	379.	4010	380.	3709
381.	3406	382.	4016	383.	3455	384.	3920
385.	3156	386.	2309	387.	2807	388.	2287
389.	3724	390.	2894	391.	2705	392.	3438
393.	3941	394.	2348	395.	3791	396.	3951
397.	2546	398.	3319	399.	2175	400.	3395
401.	3867	402.	2817	403.	3020	404.	3212
405.	2820	406.	2557	407.	3126	408.	2483
409.	4027	410.	3091	411.	3870	412.	3568
413.	4284	414.	3736	415.	2705	416.	4078
417.	3187	418.	3049	419.	2635	420.	3040
421.	4509	422.	3968	423.	3635	424.	3691

Baby	Weight	Baby	Weight	Baby	Weight	Baby	Weight
425.	3453	426.	3116	427.	2933	428.	2143
429.	3511	430.	2301	431.	3579	432.	2952
433.	4318	434.	3118	435.	3235	436.	3607
437.	3357	438.	3354	439.	3081	440.	2414
441.	3549	442.	3008	443.	3144	444.	3716
445.	3561	446.	3576	447.	3373	448.	3750
449.	3891	450.	3749	451.	3587	452.	4128
453.	4062	454.	3395	455.	4161	456.	2640
457.	3269	458.	3393	459.	3056	460.	3139
461.	3067	462.	3248	463.	3462	464.	3775
465.	3784	466.	3603	467.	3271	468.	3309
469.	2986	470.	3008	471.	3794	472.	3741
473.	4206	474.	3530	475.	4165	476.	3035
477.	3456	478.	3133	479.	4222	480.	3133
481.	3348	482.	3022	483.	4499	484.	3836
485.	3272	486.	2559	487.	4022	488.	2552
489.	4138	490.	3535	491.	3552	492.	3506
493.	3215	494.	3818	495.	3472	496.	3190
497.	3220	498.	3286	499.	3628	500.	2967
501.	3138	502.	2899	503.	2963	504.	3453
505.	3771	506.	2041	507.	3588	508.	2655
509.	3047	510.	2625	511.	3377	512.	4111
513.	3719	514.	2585	515.	3618	516.	4202
517.	3280	518.	3787	519.	3567	520.	2910
521.	3418	522.	2826	523.	2677	524.	3673
525.	2959	526.	3160	527.	3256	528.	3697
529.	3497	530.	3557	531.	3794	532.	2408
533.	3913	534.	3340	535.	2395	536.	3126
537.	3356	538.	3547	539.	2471	540.	3179
541.	2957	542.	3903	543.	3140	544.	3082
545.	3767	546.	2913	547.	2621	548.	4972
549.	3319	550.	3524	551.	3060	552.	3117
553.	2866	554.	3131	555.	3667	556.	2133
557.	3130	558.	2369	559.	2426	560.	3204
561.	3620	562.	3774	563.	3501	564.	3412
565.	3016	566.	3497	567.	4233	568.	2266
569.	4483	570.	2552	571.	3076	572.	2919
573.	3958	574.	2671	575.	2324	576.	2571
577.	3239	578.	2189	579.	4531	580.	3526
581.	3143	582.	3332	583.	2652	584.	2425
585.	3641	586.	3051	587.	3754	588.	2134
589.	3183	590.	3585	591.	3065	592.	3233
593.	3702	594.	3351	595.	3128	596.	2745
597.	3051	598.	3606	599.	3407	600.	2909
601.	3244	602.	2891	603.	4499	604.	3384
605.	3608	606.	3264	607.	2215	608.	4229

Baby	Weight	Baby	Weight	Baby	Weight	Baby	Weight
609.	3417	610.	3894	611.	3553	612.	3236
613.	3249	614.	3279	615.	3188	616.	3084
617.	4135	618.	3702	619.	3758	620.	4325
621.	3116	622.	2526	623.	3321	624.	3136
625.	2876	626.	4727	627.	4424	628.	2425
629.	3849	630.	3681	631.	4248	632.	2630
633.	3780	634.	3454	635.	4076	636.	3465
637.	3460	638.	3387	639.	3506	640.	4441
641.	3144	642.	3019	643.	3471	644.	4134
645.	4240	646.	3520	647.	3174	648.	3310
649.	3050	650.	3356	651.	2533	652.	4115
653.	3349	654.	2983	655.	3615	656.	3260
657.	3602	658.	3592	659.	3219	660.	1746
661.	3682	662.	2817	663.	4173	664.	3054
665.	3118	666.	3285	667.	2768	668.	3597
669.	2468	670.	3149	671.	3868	672.	4161
673.	3621	674.	3744	675.	2451	676.	3234
677.	3029	678.	3859	679.	3081	680.	3220
681.	3210	682.	3782	683.	3487	684.	3809
685.	3713	686.	3177	687.	3067	688.	3719
689.	2701	690.	2443	691.	3498	692.	2696
693.	3575	694.	3608	695.	3783	696.	3656
697.	3030	698.	3363	699.	3460	700.	4175
701.	2997	702.	4347	703.	3899	704.	4371
705.	3233	706.	3177	707.	3285	708.	3126
709.	2803	710.	3434	711.	3093	712.	3718
713.	2574	714.	3122	715.	4207	716.	3634
717.	2735	718.	2075	719.	3478	720.	4118
721.	3283	722.	4082	723.	3854	724.	3303
725.	3347	726.	3343	727.	3683	728.	2915
729.	3628	730.	3345	731.	2278	732.	3345
733.	3885	734.	3134	735.	1752	736.	3180
737.	2984	738.	3604	739.	1878	740.	3752
741.	2512	742.	2979	743.	3491	744.	3487
745.	4364	746.	3656	747.	2941	748.	3466
749.	2703	750.	2361	751.	2652	752.	4512
753.	3557	754.	4114	755.	2465	756.	4072
757.	3593	758.	3053	759.	2862	760.	3359
761.	2968	762.	2227	763.	2574	764.	2949
765.	3212	766.	2382	767.	2918	768.	3508
769.	2390	770.	3508	771.	4157	772.	2607
773.	4138	774.	4338	775.	2814	776.	2810
777.	2897	778.	3146	779.	3707	780.	3046
781.	4003	782.	3173	783.	3214	784.	2418
785.	2944	786.	2619	787.	3074	788.	2591
789.	3805	790.	3005	791.	3189	792.	2542

Baby	Weight	Baby	Weight	Baby	Weight	Baby	Weight
793.	3216	794.	2914	795.	3456	796.	3579
797.	2136	798.	3973	799.	3406	800.	2577
801.	3438	802.	2581	803.	3139	804.	3276
805.	3251	806.	3468	807.	3176	808.	2899
809.	3986	810.	3579	811.	2626	812.	3976
813.	3856	814.	3273	815.	2061	816.	3965
817.	2989	818.	3505	819.	2996	820.	3721
821.	3215	822.	4115	823.	2691	824.	4429
825.	4142	826.	2743	827.	2401	828.	2050
829.	3267	830.	3703	831.	3482	832.	4369
833.	3785	834.	5281	835.	3604	836.	3219
837.	3745	838.	3640	839.	3734	840.	4106
841.	3450	842.	3765	843.	2514	844.	3344
845.	3414	846.	3470	847.	4048	848.	4325
849.	2309	850.	3651	851.	3204	852.	2477
853.	2618	854.	2491	855.	2370	856.	2511
857.	3024	858.	3744	859.	3123	860.	3560
861.	3123	862.	2522	863.	2977	864.	2722
865.	2632	866.	3496	867.	3709	868.	2901
869.	2853	870.	3315	871.	2953	872.	2152
873.	2936	874.	3526	875.	4918	876.	3210
877.	3577	878.	2910	879.	3586	880.	3968
881.	3600	882.	3657	883.	3189	884.	2518
885.	4036	886.	3312	887.	3746	888.	3842
889.	4112	890.	3425	891.	3076	892.	3470
893.	2597	894.	2951	895.	2907	896.	3767
897.	3448	898.	3119	899.	2613	900.	2976
901.	3938	902.	3672	903.	3677	904.	3935
905.	3572	906.	2997	907.	2700	908.	3382
909.	2914	910.	3668	911.	3169	912.	3782
913.	3770	914.	2770	915.	3067	916.	3226
917.	2651	918.	3194	919.	2641	920.	2808
921.	3766	922.	3353	923.	3049	924.	4055
925.	3334	926.	3815	927.	3801	928.	3461
929.	4029	930.	3165	931.	2719	932.	2721
933.	3739	934.	2303	935.	3500	936.	2725
937.	3833	938.	3466	939.	3642	940.	2549
941.	3066	942.	3341	943.	3916	944.	4064
945.	4211	946.	3082	947.	3559	948.	3962
949.	2841	950.	3403	951.	4211	952.	3347
953.	2744	954.	4130	955.	2934	956.	3572
957.	3122	958.	3933	959.	2710	960.	2206
961.	2319	962.	2066	963.	3394	964.	3475
965.	2376	966.	3146	967.	2931	968.	3845
969.	2739	970.	3141	971.	3508	972.	3667
973.	3506	974.	2845	975.	3381	976.	3717
977.	2695	978.	3188	979.	3390	980.	2330

Baby	Weight	Baby	Weight	Baby	Weight	Baby	Weight
981.	4122	982.	3732	983.	3599	984.	3266
985.	3749	986.	3370	987.	3520	988.	2371
989.	4517	990.	3052	991.	3623	992.	4857
993.	2860	994.	3927	995.	3747	996.	3590
997.	3375	998.	2366	999.	4540	1000.	2328
1001.	2816	1002.	2562	1003.	3481	1004.	3665
1005.	3957	1006.	4449	1007.	2344	1008.	4944
1009.	3346	1010.	2607	1011.	2336	1012.	3412
1013.	4142	1014.	2839	1015.	3764	1016.	3876
1017.	4045	1018.	4462	1019.	3309	1020.	2910
1021.	3717	1022.	2215	1023.	3014	1024.	3305
1025.	2985	1026.	4225	1027.	2680	1028.	2810
1029.	4254	1030.	2786	1031.	4557	1032.	3706
1033.	4104	1034.	3720	1035.	3599	1036.	3169
1037.	3091	1038.	2590	1039.	4174	1040.	3618
1041.	2761	1042.	3088	1043.	2776	1044.	3228
1045.	3254	1046.	2283	1047.	3879	1048.	3141
1049.	2516	1050.	3383	1051.	3470	1052.	3084
1053.	3387	1054.	3468	1055.	3709	1056.	2259
1057.	2456	1058.	3224	1059.	3918	1060.	2682
1061.	3498	1062.	2560	1063.	4123	1064.	3703
1065.	2861	1066.	2661	1067.	3570	1068.	4280
1069.	2189	1070.	3576	1071.	3466	1072.	3324
1073.	2562	1074.	3668	1075.	2659	1076.	3412
1077.	3199	1078.	2642	1079.	3367	1080.	3271
1081.	3312	1082.	4107	1083.	2788	1084.	3861
1085.	3215	1086.	3562	1087.	3400	1088.	2395
1089.	3372	1090.	3464	1091.	3780	1092.	2612
1093.	3954	1094.	2850	1095.	4055	1096.	4282
1097.	2766	1098.	3075	1099.	3261	1100.	2758
1101.	3457	1102.	4365	1103.	3534	1104.	3902
1105.	3049	1106.	4129	1107.	2271	1108.	4364
1109.	3556	1110.	3010	1111.	2384	1112.	3682
1113.	3115	1114.	3410	1115.	4023	1116.	2892
1117.	3507	1118.	2879	1119.	3599	1120.	2626
1121.	2932	1122.	2444	1123.	2463	1124.	3126
1125.	2421	1126.	3765	1127.	3610	1128.	3106
1129.	2937	1130.	3729	1131.	3177	1132.	4169
1133.	2291	1134.	2235	1135.	3862	1136.	3271
1137.	3888	1138.	2787	1139.	3970	1140.	2972
1141.	3374	1142.	2512	1143.	3388	1144.	4084
1145.	3356	1146.	3434	1147.	4709	1148.	4642
1149.	2761	1150.	3118	1151.	3815	1152.	3521
1153.	3289	1154.	2964	1155.	2849	1156.	3720
1157.	3344	1158.	3428	1159.	2517	1160.	2873
1161.	2996	1162.	2940	1163.	3078	1164.	2776
1165.	3049	1166.	2500	1167.	3710	1168.	3578

Baby	Weight	Baby	Weight	Baby	Weight	Baby	Weight
1169.	4439	1170.	2869	1171.	3137	1172.	2514
1173.	3165	1174.	3120	1175.	2910	1176.	3187
1177.	2381	1178.	3535	1179.	4255	1180.	3351
1181.	3155	1182.	4051	1183.	3960	1184.	3768
1185.	2605	1186.	3443	1187.	3293	1188.	3644
1189.	3596	1190.	3296	1191.	4101	1192.	3282
1193.	4076	1194.	2690	1195.	2814	1196.	3725
1197.	2507	1198.	3162	1199.	3918	1200.	2742

Draw a simple random sample of size 20 from this population and construct a 95 percent confidence interval for the population mean. Compare your results with those of your classmates. What assumptions are necessary for your estimation procedure to be valid?

27. Refer to Exercise 26. Draw a simple random sample of size 35 from the population and construct a 95 percent confidence interval for the population mean. Compare this interval with the one constructed in Exercise 26.

28. In a study of the role of dietary fats in the etiology of ischemic heart disease the subjects were 60 males between 40 and 60 years of age who had recently had a myocardial infarction and 50 apparently healthy males from the same age group and social class. One variable of interest in the study was the proportion of linoleic acid (L.A.) in the subjects' plasma triglyceride fatty acids. The data on this variable were as follows.

SUBJECTS WITH MYOCARDIAL INFARCTION

Subject	L.A.	Subject	L.A.	Subject	L.A.	Subject	L.A.
1.	18.0	2.	17.6	3.	9.6	4.	5.5
5.	16.8	6.	12.9	7.	14.0	8.	8.0
9.	8.9	10.	15.0	11.	9.3	12.	5.8
13.	8.3	14.	4.8	15.	6.9	16.	18.3
17.	24.0	18.	16.8	19.	12.1	20.	12.9
21.	16.9	22.	15.1	23.	6.1	24.	16.6
25.	8.7	26.	15.6	27.	12.3	28.	14.9
29.	16.9	30.	5.7	31.	14.3	32.	14.1
33.	14.1	34.	15.1	35.	10.6	36.	13.6
37.	16.4	38.	10.7	39.	18.1	40.	14.3
41.	6.9	42.	6.5	43.	17.7	44.	13.4
45.	15.6	46.	10.9	47.	13.0	48.	10.6
49.	7.9	50.	2.8	51.	15.2	52.	22.3
53.	9.7	54.	15.2	55.	10.1	56.	11.5
57.	15.4	58.	17.8	59.	12.6	60.	7.2

HEALTHY SUBJECTS

Subject	L.A.	Subject	L.A.	Subject	L.A.	Subject	L.A.
1.	17.1	2.	22.9	3.	10.4	4.	30.9
5.	32.7	6.	9.1	7.	20.1	8.	19.2
9.	18.9	10.	20.3	11.	35.6	12.	17.2
13.	5.8	14.	15.2	15.	22.2	16.	21.2
17.	19.3	18.	25.6	19.	42.4	20.	5.9
21.	29.6	22.	18.2	23.	21.7	24.	29.7
25.	12.4	26.	15.4	27.	21.7	28.	19.3
29.	16.4	30.	23.1	31.	19.0	32.	12.9
33.	18.5	34.	27.6	35.	25.0	36.	20.0
37.	51.7	38.	20.5	39.	25.9	40.	24.6
41.	22.4	42.	27.1	43.	11.1	44.	32.7
45.	13.2	46.	22.1	47.	13.5	48.	5.3
49.	29.0	50.	20.2				

Construct the 95 percent confidence interval for the difference between population means. What do these data suggest about the levels of linoleic acid in the two sampled populations?

29. The following are the heights (in inches) of 1000 twelve-year-old boys.

Boy	Height	Boy	Height	Boy	Height	Boy	Height
1.	60.7	2.	62.4	3.	59.7	4.	56.2
5.	55.2	6.	58.6	7.	59.1	8.	59.5
9.	56.5	10.	61.1	11.	55.8	12.	55.9
13.	61.0	14.	62.9	15.	57.2	16.	56.7
17.	61.7	18.	56.3	19.	55.8	20.	62.1
21.	60.6	22.	59.0	23.	62.3	24.	59.2
25.	57.8	26.	64.1	27.	54.3	28.	58.0
29.	62.0	30.	59.3	31.	62.8	32.	64.4
33.	57.9	34.	60.6	35.	56.5	36.	66.2
37.	60.3	38.	58.6	39.	56.8	40.	61.0
41.	59.4	42.	62.9	43.	60.6	44.	58.5
45.	61.7	46.	57.8	47.	62.2	48.	55.6
49.	62.8	50.	59.3	51.	55.7	52.	55.2
53.	59.4	54.	56.7	55.	54.6	56.	55.8
57.	59.4	58.	60.3	59.	61.5	60.	62.0
61.	57.5	62.	57.4	63.	60.6	64.	59.5
65.	57.6	66.	57.0	67.	61.2	68.	57.0
69.	60.9	70.	59.8	71.	56.3	72.	57.8
73.	58.7	74.	60.2	75.	54.8	76.	58.4
77.	56.2	78.	58.5	79.	57.8	80.	59.5
81.	53.7	82.	57.2	83.	58.3	84.	55.5
85.	63.2	86.	57.3	87.	54.3	88.	63.4
89.	58.6	90.	57.7	91.	59.2	92.	58.5
93.	59.4	94.	59.4	95.	57.7	96.	57.9
97.	60.0	98.	59.4	99.	62.8	100.	57.4

Boy	Height	Boy	Height	Boy	Height	Boy	Height
101.	59.2	102.	56.6	103.	59.7	104.	54.9
105.	62.4	106.	62.0	107.	59.6	108.	62.3
109.	60.8	110.	56.6	111.	56.5	112.	59.2
113.	56.1	114.	59.1	115.	56.8	116.	61.0
117.	58.7	118.	61.3	119.	54.2	120.	57.3
121.	62.7	122.	56.9	123.	62.7	124.	53.3
125.	62.7	126.	60.7	127.	60.8	128.	59.5
129.	55.0	130.	57.9	131.	57.0	132.	61.7
133.	54.9	134.	55.4	135.	63.1	136.	58.8
137.	62.3	138.	57.6	139.	62.4	140.	64.2
141.	60.5	142.	60.8	143.	62.2	144.	63.0
145.	57.2	146.	57.5	147.	60.1	148.	61.8
149.	59.1	150.	58.8	151.	62.8	152.	54.9
153.	61.5	154.	57.0	155.	63.6	156.	57.5
157.	57.8	158.	59.7	159.	55.5	160.	62.7
161.	60.9	162.	59.1	163.	57.6	164.	61.1
165.	58.2	166.	58.3	167.	60.8	168.	60.6
169.	58.1	170.	62.4	171.	59.6	172.	64.8
173.	64.4	174.	57.6	175.	61.7	176.	62.1
177.	57.4	178.	61.9	179.	63.3	180.	58.7
181.	56.9	182.	61.0	183.	62.0	184.	58.2
185.	61.2	186.	62.4	187.	58.7	188.	56.8
189.	55.5	190.	56.2	191.	55.2	192.	55.5
193.	63.9	194.	60.8	195.	54.8	196.	63.5
197.	62.4	198.	62.4	199.	57.4	200.	56.7
201.	58.7	202.	63.5	203.	63.2	204.	61.4
205.	58.1	206.	62.4	207.	59.4	208.	56.0
209.	55.8	210.	62.5	211.	59.8	212.	58.1
213.	55.7	214.	58.1	215.	55.6	216.	63.2
217.	55.5	218.	55.0	219.	61.4	220.	58.5
221.	56.7	222.	58.7	223.	62.8	224.	63.0
225.	59.2	226.	62.7	227.	56.6	228.	54.4
229.	55.2	230.	64.9	231.	60.1	232.	62.6
233.	60.6	234.	64.5	235.	61.7	236.	56.4
237.	59.3	238.	60.8	239.	59.3	240.	61.7
241.	58.9	242.	57.9	243.	59.3	244.	54.1
245.	63.2	246.	58.1	247.	58.0	248.	63.3
249.	59.1	250.	53.9	251.	62.6	252.	63.7
253.	60.2	254.	62.8	255.	55.8	256.	58.4
257.	59.4	258.	59.3	259.	57.8	260.	58.2
261.	63.7	262.	60.6	263.	55.7	264.	56.8
265.	55.6	266.	56.5	267.	58.0	268.	56.3
269.	58.3	270.	56.2	271.	61.4	272.	59.5
273.	54.0	274.	61.5	275.	53.8	276.	59.7
277.	58.8	278.	55.1	279.	63.2	280.	60.3
281.	59.4	282.	61.5	283.	58.6	284.	60.0
285.	58.6	286.	59.3	287.	58.2	288.	55.5
289.	61.8	290.	61.1	291.	57.2	292.	59.2
293.	56.2	294.	61.7	295.	63.7	296.	59.9

Boy	Height	Boy	Height	Boy	Height	Boy	Height
297.	59.5	298.	60.1	299.	63.5	300.	64.6
301.	56.2	302.	58.2	303.	61.4	304.	58.8
305.	63.2	306.	60.3	307.	56.9	308.	59.1
309.	56.9	310.	55.1	311.	58.5	312.	60.2
313.	59.4	314.	61.2	315.	64.7	316.	57.1
317.	59.2	318.	60.0	319.	57.8	320.	60.5
321.	59.3	322.	58.9	323.	59.2	324.	57.8
325.	55.8	326.	60.7	327.	62.9	328.	59.9
329.	57.7	330.	57.4	331.	56.9	332.	60.1
333.	58.9	334.	55.3	335.	54.4	336.	63.1
337.	58.9	338.	58.6	339.	60.6	340.	55.6
341.	56.6	342.	61.5	343.	58.3	344.	59.6
345.	56.0	346.	58.7	347.	62.2	348.	56.3
349.	54.4	350.	60.1	351.	57.1	352.	61.0
353.	56.1	354.	60.2	355.	63.6	356.	59.3
357.	60.5	358.	62.9	359.	60.7	360.	57.2
361.	59.6	362.	62.0	363.	59.3	364.	61.1
365.	58.7	366.	61.7	367.	61.1	368.	57.7
369.	54.8	370.	61.6	371.	55.8	372.	54.1
373.	58.7	374.	60.2	375.	61.4	376.	60.8
377.	61.2	378.	58.9	379.	60.7	380.	60.2
381.	54.6	382.	55.5	383.	55.1	384.	61.8
385.	58.8	386.	61.2	387.	61.4	388.	61.7
389.	60.9	390.	62.3	391.	59.3	392.	55.9
393.	58.4	394.	54.7	395.	59.6	396.	56.9
397.	58.9	398.	58.9	399.	56.6	400.	56.8
401.	60.6	402.	57.0	403.	62.5	404.	59.0
405.	54.2	406.	60.4	407.	63.8	408.	58.5
409.	58.5	410.	61.5	411.	59.5	412.	59.7
413.	57.5	414.	59.9	415.	56.4	416.	60.2
417.	62.8	418.	57.0	419.	62.1	420.	62.1
421.	55.0	422.	56.2	423.	58.3	424.	59.0
425.	52.9	426.	55.5	427.	52.8	428.	59.5
429.	62.8	430.	59.9	431.	59.3	432.	60.0
433.	56.6	434.	57.4	435.	59.6	436.	63.2
437.	58.4	438.	57.8	439.	58.1	440.	60.2
441.	64.5	442.	57.0	443.	66.1	444.	59.6
445.	63.3	446.	60.2	447.	61.9	448.	57.5
449.	57.3	450.	62.9	451.	59.5	452.	61.3
453.	57.5	454.	60.5	455.	60.9	456.	57.5
457.	59.9	458.	59.6	459.	58.8	460.	61.8
461.	57.3	462.	58.9	463.	58.8	464.	63.2
465.	59.2	466.	59.2	467.	59.9	468.	59.4
469.	61.2	470.	63.0	471.	57.2	472.	57.9
473.	59.7	474.	63.6	475.	60.3	476.	61.9
477.	63.6	478.	59.4	479.	57.2	480.	59.8
481.	59.8	482.	54.9	483.	58.4	484.	61.6
485.	58.0	486.	58.5	487.	63.8	488.	62.3
489.	58.8	490.	59.2	491.	58.7	492.	64.6

Boy	Height	Boy	Height	Boy	Height	Boy	Height
493.	64.3	494.	57.7	495.	58.0	496.	52.9
497.	54.4	498.	56.2	499.	64.1	500.	60.4
501.	60.2	502.	55.5	503.	57.4	504.	61.0
505.	56.6	506.	60.6	507.	60.3	508.	56.5
509.	57.9	510.	61.2	511.	61.6	512.	63.3
513.	62.6	514.	62.3	515.	59.5	516.	55.0
517.	58.2	518.	56.7	519.	60.0	520.	56.5
521.	58.1	522.	57.3	523.	63.2	524.	62.1
525.	58.3	526.	59.7	527.	59.6	528.	60.1
529.	58.6	530.	58.1	531.	56.0	532.	52.8
533.	58.8	534.	57.6	535.	56.8	536.	55.5
537.	60.6	538.	56.0	539.	59.5	540.	60.0
541.	61.9	542.	61.2	543.	60.9	544.	58.2
545.	58.5	546.	58.3	547.	59.9	548.	57.0
549.	55.9	550.	61.4	551.	58.9	552.	63.3
553.	56.6	554.	61.6	555.	59.0	556.	59.6
557.	63.0	558.	58.5	559.	57.0	560.	59.2
561.	58.4	562.	58.4	563.	61.0	564.	57.3
565.	60.8	566.	62.0	567.	59.5	568.	64.1
569.	57.6	570.	55.0	571.	58.4	572.	57.9
573.	62.0	574.	55.6	575.	62.3	576.	60.2
577.	55.9	578.	58.7	579.	58.5	580.	60.0
581.	58.7	582.	56.9	583.	58.3	584.	57.6
585.	58.4	586.	62.1	587.	58.7	588.	59.5
589.	59.9	590.	60.2	591.	56.2	592.	60.8
593.	55.2	594.	55.5	595.	63.2	596.	57.3
597.	61.2	598.	57.2	599.	57.9	600.	61.7
601.	55.8	602.	60.5	603.	64.6	604.	58.6
605.	57.9	606.	53.3	607.	58.1	608.	59.4
609.	58.3	610.	55.5	611.	58.0	612.	58.6
613.	54.8	614.	55.9	615.	59.8	616.	57.3
617.	55.1	618.	59.5	619.	60.9	620.	64.1
621.	59.8	622.	61.9	623.	56.6	624.	62.8
625.	58.9	626.	58.4	627.	59.6	628.	54.1
629.	57.4	630.	59.2	631.	58.2	632.	63.2
633.	57.4	634.	57.8	635.	56.4	636.	58.4
637.	60.7	638.	56.6	639.	60.3	640.	60.5
641.	58.1	642.	60.7	643.	60.5	644.	60.7
645.	60.6	646.	58.3	647.	56.5	648.	60.0
649.	59.6	650.	53.9	651.	57.4	652.	59.7
653.	60.9	654.	67.7	655.	58.6	656.	57.6
657.	60.6	658.	62.1	659.	54.9	660.	61.3
661.	57.5	662.	60.3	663.	59.0	664.	57.3
665.	61.0	666.	59.0	667.	55.7	668.	56.8
669.	61.9	670.	64.0	671.	60.9	672.	56.9
673.	61.2	674.	62.4	675.	59.4	676.	59.0
677.	56.5	678.	57.9	679.	62.7	680.	58.8
681.	53.5	682.	60.9	683.	57.4	684.	56.6
685.	58.8	686.	58.0	687.	58.5	688.	58.8

Boy	Height	Boy	Height	Boy	Height	Boy	Height
689.	57.0	690.	55.8	691.	60.7	692.	60.9
693.	57.4	694.	57.7	695.	56.9	696.	58.0
697.	58.4	698.	56.1	699.	55.3	700.	59.3
701.	59.2	702.	56.0	703.	54.2	704.	52.7
705.	53.2	706.	59.6	707.	59.3	708.	59.7
709.	58.4	710.	56.1	711.	59.1	712.	61.8
713.	62.0	714.	58.0	715.	58.3	716.	60.6
717.	58.0	718.	58.9	719.	60.2	720.	61.3
721.	56.2	722.	58.5	723.	60.3	724.	60.0
725.	63.5	726.	57.4	727.	61.0	728.	57.1
729.	54.5	730.	56.3	731.	61.2	732.	56.2
733.	59.7	734.	57.9	735.	59.3	736.	53.8
737.	57.4	738.	55.5	739.	61.2	740.	57.9
741.	52.5	742.	61.8	743.	61.7	744.	57.4
745.	63.0	746.	56.9	747.	59.8	748.	60.1
749.	58.1	750.	63.7	751.	56.0	752.	58.1
753.	61.2	754.	64.3	755.	58.7	756.	58.1
757.	56.9	758.	58.1	759.	59.0	760.	63.2
761.	60.2	762.	60.4	763.	62.8	764.	61.8
765.	61.9	766.	60.3	767.	56.4	768.	54.7
769.	61.3	770.	60.8	771.	56.3	772.	64.9
773.	59.2	774.	61.2	775.	60.1	776.	61.8
777.	62.9	778.	61.0	779.	58.2	780.	56.1
781.	61.6	782.	56.5	783.	61.1	784.	58.2
785.	64.1	786.	56.9	787.	60.6	788.	59.2
789.	54.3	790.	59.4	791.	56.9	792.	57.6
793.	58.8	794.	57.3	795.	60.1	796.	58.4
797.	60.6	798.	59.3	799.	56.6	800.	60.2
801.	62.4	802.	55.5	803.	56.8	804.	61.0
805.	56.4	806.	59.9	807.	58.7	808.	59.7
809.	57.9	810.	56.7	811.	61.1	812.	57.6
813.	60.6	814.	56.5	815.	57.3	816.	63.8
817.	61.3	818.	61.5	819.	59.4	820.	59.5
821.	63.8	822.	64.3	823.	62.0	824.	59.3
825.	61.0	826.	60.1	827.	57.0	828.	63.3
829.	59.9	830.	59.7	831.	63.9	832.	60.2
833.	52.8	834.	59.2	835.	59.6	836.	57.6
837.	55.8	838.	59.1	839.	60.1	840.	57.3
841.	58.9	842.	61.2	843.	61.3	844.	59.4
845.	53.6	846.	62.3	847.	56.7	848.	56.5
849.	58.8	850.	59.6	851.	57.7	852.	60.7
853.	57.7	854.	59.3	855.	60.5	856.	59.7
857.	59.5	858.	61.2	859.	58.0	860.	60.3
861.	61.9	862.	57.5	863.	57.6	864.	57.7
865.	64.0	866.	59.3	867.	54.9	868.	61.7
869.	58.4	870.	59.1	871.	53.3	872.	55.8
873.	55.6	874.	55.6	875.	54.9	876.	59.9
877.	62.5	878.	57.2	879.	62.3	880.	64.6
881.	59.1	882.	61.4	883.	60.1	884.	62.1

Boy	Height	Boy	Height	Boy	Height	Boy	Height
885.	59.1	886.	53.2	887.	58.1	888.	61.8
889.	57.2	890.	61.2	891.	56.2	892.	58.0
893.	58.5	894.	57.9	895.	61.4	896.	61.3
897.	56.5	898.	58.7	899.	60.4	900.	63.3
901.	57.6	902.	59.2	903.	61.9	904.	58.4
905.	62.5	906.	55.2	907.	59.7	908.	62.1
909.	59.7	910.	56.6	911.	52.8	912.	62.8
913.	60.5	914.	58.7	915.	58.8	916.	57.4
917.	54.9	918.	62.0	919.	58.3	920.	59.3
921.	62.4	922.	60.5	923.	63.3	924.	58.7
925.	62.9	926.	60.9	927.	62.5	928.	59.3
929.	59.0	930.	58.3	931.	62.5	932.	56.0
933.	56.2	934.	53.8	935.	53.9	936.	59.5
937.	59.6	938.	66.6	939.	55.1	940.	60.4
941.	58.5	942.	60.1	943.	60.1	944.	60.2
945.	59.7	946.	61.1	947.	56.9	948.	58.7
949.	58.9	950.	57.6	951.	59.0	952.	59.8
953.	56.4	954.	60.6	955.	57.6	956.	59.3
957.	58.2	958.	62.7	959.	57.1	960.	56.8
961.	59.4	962.	58.4	963.	54.5	964.	57.0
965.	61.1	966.	60.2	967.	61.0	968.	58.1
969.	58.8	970.	63.8	971.	56.3	972.	62.6
973.	56.7	974.	59.9	975.	59.2	976.	58.7
977.	62.4	978.	57.5	979.	60.1	980.	58.9
981.	55.7	982.	57.6	983.	58.8	984.	58.2
985.	63.3	986.	59.8	987.	60.4	988.	58.1
989.	57.6	990.	57.8	991.	56.8	992.	59.8
993.	57.8	994.	58.4	995.	58.6	996.	54.5
997.	59.1	998.	59.0	999.	61.5	1000.	59.2

Select a simple random sample of size 15 from this population and construct a 99 percent confidence interval for the population mean. What assumptions are necessary for this procedure to be valid?

30. Refer to Exercise 29. Select a simple random sample of size 35 from the population and construct a 99 percent confidence interval for the population mean. Compare this interval with the one constructed in Exercise 29.

REFERENCES

References Cited
1. John E. Freund and Ronald E. Walpole, *Mathematical Statistics*, Third Edition, Prentice-Hall, Englewood Cliffs, N.J., 1980.

2. Alexander M. Mood, Franklin A. Graybill and Duane C. Boes, *Introduction to the Theory of Statistics*, Third Edition, McGraw-Hill, New York, 1974.

3. Taro Yamane, *Statistics: An Introductory Analysis*, Second Edition, Harper & Row, New York, 1967.

4. W. S. Gosset, ("Student"), "The Probable Error of a Mean," *Biometrika, 6* (1908), 1–25.

5. W. V. Behrens, "Ein Beitrag zu Fehlerberechnung bei wenige Beobachtungen," *Landwirtsschaftliche Jahrbücher*, *68* (1929), 807–837.

6. R. A. Fisher, "The Comparison of Samples with Possibly Unequal Variances," *Annals of Eugenics*, *9* (1939), 174–180.

7. R. A. Fisher, "The Asymptotic Approach to Behrens' Integral with Further Tables for the *d* Test of Significance," *Annals of Eugenics*, *11* (1941), 141–172.

8. J. Neyman, "Fiducial Argument and the Theory of Confidence Intervals," *Biometrika*, *32* (1941), 128–150.

9. H. Scheffé, "On Solutions of the Behrens-Fisher Problem Based on the *t*-Distribution," *The Annals of Mathematical Statistics*, *14* (1943), 35–44.

10. H. Scheffé, "A Note on the Behrens-Fisher Problem," *The Annals of mathematical Statistics*, *15* (1944), 430–432.

11. B. L. Welch, "The Significance of the Difference Between Two Means When the Population Variances Are Unequal," *Biometrika*, *29* (1937), 350–361.

12. B. L. Welch, "The Generalization of 'Student's' Problem When Several Different Population Variances Are Involved," *Biometrika*, *34* (1947), 28–35.

13. Alice A. Aspin, "Tables for Use in Comparisons Whose Accuracy Involves Two Variances *s*, Separately Estimated," *Biometrika*, *36* (1949), 290–296.

14. W. H. Trickett, B. L. Welch, and G. S. James, "Further Critical Values for the Two-Means Problem," *Biometrika*, *43* (1956), 203–205.

15. William G. Cochran, "Approximate Significance Levels of the Behrens–Fisher Test," *Biometrics*, *20* (1964), 191–195.

16. George W. Snedecor and William G. Cochran, *Statistical Methods*, Sixth Edition, The Iowa State University Press, Ames, 1967.

17. Wilfred J. Dixon and Frank J. Massey, Jr., *Introduction to Statistical Analysis*, Third Edition, McGraw-Hill, New York, 1969.

18. Edward E. Cureton, "Unbiased Estimation of the Standard Deviation," *The American Statistician*, *22* (February 1968) 22.

19. John Gurland and Ran C. Tripathi, "A Simple Approximation for Unbiased Estimation of the Standard Deviation," *The American Statistician*, *25* (October 1971), 30–32.

20. R. F. Tate and G. W. Klett, "Optimal Confidence Intervals for the Variance of a Normal Distribution," *Journal of the American Statistical Association*, *54* (1959), 674–682.

21. Wayne W. Daniel, *Applied Nonparametric Statistics*, Houghton-Mifflin, Boston, 1978.

Other References

1. H. A. Al-Bayyati, "A Rule of Thumb for Determining a Sample Size in Comparing Two Proportions," *Technometrics*, *13* (1971), 675–677.

2. F. F. Ractliffe, "The Effect on the *t* Distribution of Nonnormality in the Sampled Population," *Applied Statistics*, *17* (1968), 42–48.

CHAPTER 6

Hypothesis Testing

6.1 INTRODUCTION

One type of statistical inference, estimation, is discussed in the preceding chapter. The other type, hypothesis testing, is the subject of the present chapter. As is true with estimation, the *purpose of hypothesis testing is to aid the clinician, researcher, or administrator in reaching a decision concerning a population by examining a sample from that population.* Estimation and hypothesis testing are not as different as they are made to appear by the fact that most textbooks devote a separate chapter to each. As we will explain later, one may use confidence intervals to arrive at the same conclusions that are reached by using the hypothesis testing procedures discussed in this chapter.

In this section some of the basic concepts essential to an understanding of hypothesis testing are presented. The specific details of particular tests will be given in succeeding sections.

A *hypothesis* may be defined simply as *a statement about one or more populations.* The hypothesis is usually concerned with the parameters of the populations about which the statement is made. A hospital administrator may hypothesize that the average length of stay of patients admitted to the hospital is five days; a public health nurse may hypothesize that a particular educational program will result in improved communication between nurse and patient; a physician may hypothesize that a certain drug will be effective in 90 percent of the cases with which it is used. By means of hypothesis testing one determines whether or not such statements are compatible with available data.

Researchers are concerned with two types of hypotheses—*research hypotheses* and *statistical hypotheses.* The research hypothesis is the conjecture or supposition that motivates the research. It may be the result of years of observation on the part of the researcher. A public health nurse, for example, may have noted that certain clients responded more readily to a particular type of health education program. A physician may recall numerous instances in which certain combinations of therapeutic measures were more effective than any one of them alone. Research projects often result from the desire of such health practitioners to determine whether or not their theories or suspicions can be supported when subjected to the rigors of scientific investigation.

Research hypotheses lead directly to statistical hypotheses. Statistical hypotheses are stated in such a way that they may be evaluated by appropriate statistical techniques. In this book the hypotheses that we shall focus on are statistical hypotheses. We shall assume that the research hypotheses for the examples and exercises have already been considered.

For convenience, hypothesis testing will be presented as a nine-step procedure. There is nothing magical or sacred about this particular format, although it does break the process down into a logical sequence of actions and decisions.

1. *Data* The nature of the data that form the basis of the testing procedures must be understood, since this determines the particular test to be employed. Whether the data consist of counts or measurements, for example, must be determined.

2. *Assumptions* As we learned in the chapter on estimation, different assumptions led to modifications of confidence intervals. The same is true in hypothesis testing: a general procedure is modified depending on the assumptions. In fact, the same assumptions that are of importance in estimation are also important in hypothesis testing. We have seen that these include, among others, assumptions about the normality of the population distribution, equality of variances, and independence of samples.

3. *Hypotheses* There are two statistical hypotheses involved in hypothesis testing and these should be explicitly stated. The first is the *hypothesis to be tested*, usually referred to as the *null hypothesis* and designated by the symbol H_0. The null hypothesis is sometimes referred to as a *hypothesis of no difference*, since it is a statement of agreement with (or no difference from) conditions presumed to be true in the population of interest. In general, the null hypothesis is set up for the express purpose of being discredited. Consequently, the complement of the conclusion that the researcher is seeking to reach becomes the statement of the null hypothesis. In the testing process the null hypothesis either is rejected or is not rejected. If the null hypothesis is not rejected, we will say that the data on which the test is based do not provide sufficient evidence to cause rejection. If the testing procedure leads to rejection, we will say that the data at hand are not compatible with the null hypothesis, but are supportive of some other hypothesis. This other hypothesis is known as the *alternative hypothesis* and may be designated by the symbol H_A.

When hypotheses are of the type considered in this chapter an indication of equality (either $=$, $\leq$, or $\geq$) must appear in the null hypothesis. Suppose, for example, that we want to answer the question: Can we conclude that a certain population mean is not 50? The null hypothesis is

$$H_0: \mu = 50$$

and the alternative is

$$H_A: \mu \neq 50$$

Suppose we want to know if we can conclude that the population mean is greater than 50. Our hypotheses are

$$H_0: \mu \leq 50 \qquad H_A: \mu > 50$$

If we want to know if we can conclude that the population mean is less than 50, the hypotheses are

$$H_0: \mu \geq 50 \qquad H_A: \mu < 50$$

It should be pointed out that neither hypothesis testing nor statistical inference, in general, leads to the proof of a hypothesis; it merely indicates whether the hypothesis is supported or is not supported by the available data. When we fail to reject a null hypothesis, therefore, we do not say that it is true, but that it may be true. When we speak of accepting a null hypothesis, we have this limitation in mind and do not wish to convey the idea that accepting implies proof.

4. *Test Statistic*   The test statistic is some statistic that may be computed from the data of the sample. As a rule, there are many possible values that the test statistic may assume, the particular value observed depending on the particular sample drawn. As we shall see, the test statistic serves as a decision maker, since the decision to reject or not to reject the null hypothesis depends on the magnitude of the test statistic. An example of a test statistic is the quantity

$$z = \frac{\bar{x} - \mu_0}{\sigma/\sqrt{n}}$$

where μ_0 is a hypothesized value of a population mean. This test statistic is related to the statistic

$$z = \frac{\bar{x} - \mu}{\sigma/\sqrt{n}}$$

with which we are already familiar.

The following is a general formula for a test statistic that will be applicable in many of the hypothesis tests discussed in this book:

$$\text{test statistic} = \frac{\text{relevant statistic} - \text{hypothesized parameter}}{\text{standard error of the relevant statistic}}$$

In the present example, $\bar{x}$ is the relevant statistic, μ_0 is the hypothesized parameter, and $\sigma/\sqrt{n}$ is the standard error of $\bar{x}$, the relevant statistic.

5. *Distribution of the Test Statistic* It has been pointed out that the key to statistical inference is the sampling distribution. We are reminded of this again when it becomes necessary to specify the probability distribution of the test statistic. The distribution of the test statistic,

$$z = \frac{\bar{x} - \mu_0}{\sigma/\sqrt{n}}$$

for example, follows the unit normal distribution if the null hypothesis is true and the assumptions are met.

6. *Decision Rule* All possible values that the test statistic can assume are points on the horizontal axis of the graph of the distribution of the test statistic and are divided into two groups; one group constitutes what is known as the *rejection region* and the other group makes up the *acceptance region*. The values of the test statistic comprising the rejection region are those values that are less likely to occur if the null hypothesis is true, while the values making up the acceptance region are more likely to occur if the null hypothesis is true. *The decision rule tells us to reject the null hypothesis if the value of the test statistic that we compute from our sample is one of the values in the rejection region and to not reject (or "accept") the null hypothesis if the computed value of the test statistic is one of the values in the acceptance region.*

The decision as to which values go into the rejection region and which ones go into the acceptance region is made on the basis of the desired *level of significance*, designated by α. The term, level of significance, reflects the fact that hypothesis tests are sometimes called significance tests, and a computed value of the test statistic that falls in the rejection region is said to be *significant*. The level of significance, α, specifies the area under the curve of the distribution of the test statistic that is above the values on the horizontal axis constituting the rejection region. It is seen, then, that α is a probability and, in fact, is the *probability of rejecting a true null hypothesis*. Since to reject a true null hypothesis would constitute an error, it seems only reasonable that we should make the probability of rejecting a true null hypothesis small and, in fact, that is what is done. We select a small value of α in order to make the probability of rejecting a true null hypothesis small. The more frequently encountered values of α are .01, .05, and .10.

Possible condition of null hypothesis

		True	False
Possible action	Fail to reject H_0	Correct action	Type II error
	Reject H_0	Type I error	Correct action

FIGURE 6.1.1 Conditions under which Type I and Type II Errors May Be Committed

The error committed when a true null hypothesis is rejected is called the *type I error*. The *type II error* is the error committed when a false null hypothesis is accepted. The probability of committing a type II error is designated by β.

Whenever we reject a null hypothesis there is always the concomitant risk of committing a type I error, rejecting a true null hypothesis. Whenever we "accept" a null hypothesis the risk of accepting a false null hypothesis is always present. We make α small, but we generally exercise no control over β, although we know that in most practical situations it is larger than α.

We never know whether we have committed one of these errors when we reject or fail to reject a null hypothesis, since the true state of affairs is unknown. If the testing procedure leads to rejection of the null hypothesis, we can take comfort from the fact that we made α small and, therefore, the probability of committing a type I error was small. If we "accept" the null hypothesis, we do not know the concurrent risk of committing a type II error, since β is usually unknown but, as has been pointed out, we do know that, in general, it is larger than α.

Figure 6.1.1 shows the possible conditions of a null hypothesis, the possible actions that an investigator may take, and the conditions under which each of the two types of error will be made.

7. *Calculation of the Test Statistic* From the data contained in the sample we compute a value of the test statistic and compare it with the acceptance and rejection regions that have already been specified.

8. *Statistical Decision* The statistical decision consists of rejecting or of not rejecting the null hypothesis. It is rejected if the computed value of the test statistic falls in the rejection region, and it is not rejected if the computed value of the test statistic falls in the acceptance region.

9. *Conclusion* If H_0 is rejected, we conclude that H_A is true. If H_0 is not rejected, we conclude that H_0 *may be true*.

Figure 6.1.2 is a flowchart of the steps that we follow when we perform a hypothesis test.

One of the purposes of hypothesis testing is to assist administrators and clinicians in making decisions. The administrative or clinical decision usually depends on the statistical decision. If the null hypothesis is rejected, the administrative or clinical decision usually reflects this, in that the decision is compatible

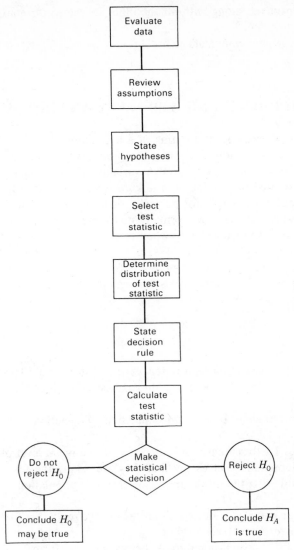

FIGURE 6.1.2

Steps in the Hypothesis Testing Procedure

with the alternative hypothesis. The reverse is usually true if the null hypothesis is not rejected. The administrative or clinical decision, however, may take other forms, such as a decision to gather more data.

We must emphasize at this point, however, that the outcome of the statistical test is only one piece of evidence that influences the administrative or clinical decision. The statistical decision should not be interpreted as definitive, but should be considered along with all the other relevant information available to the experimenter.

With these general comments as background, we now discuss specific hypothesis tests.

6.2 HYPOTHESIS TESTING: A SINGLE POPULATION MEAN

In this section we consider the testing of a hypothesis about a population mean under three different conditions: (1) when sampling is from a normally distributed population of values with known variance, (2) when sampling is from a normally distributed population with unknown variance, and (3) when sampling is from a population that is not normally distributed. Although the theory for conditions 1 and 2 depend on normally distributed populations, it is common practice to make use of the theory when relevant populations are only approximately normally distributed. This is satisfactory so long as the departure from normality is not drastic. When sampling is from a normally distributed population and the population variance is known, the test statistic for testing H_0: $\mu = \mu_0$ is

$$z = \frac{\bar{x} - \mu_0}{\sigma/\sqrt{n}} \qquad (6.2.1)$$

which, when H_0 is true, is distributed as the unit normal. The following two examples illustrate hypothesis testing under these conditions.

Sampling from Normally Distributed Populations: Population Variances Known

Example 6.2.1 Researchers are interested in the mean level of some enzyme in a certain population. Let us say that they are asking the following question: Can we conclude that the mean enzyme level in this population is different from 25?

Based on our knowledge of hypothesis testing, we reply that they can conclude that the mean enzyme level is different from 25 if they can reject the null hypothesis that the mean is equal to 25. Let us use the information from Example 5.2.1 and the nine-step hypothesis testing procedure given in the previous section to help the researchers reach a decision.

1. *Data* The data available to the researchers are the enzyme determinations made on a sample of 10 individuals from the population of interest. From this sample a mean of $\bar{x} = 22$ has been computed.

2. *Assumptions* It is assumed that the sample comes from a population of normally distributed enzyme values with a known variance, $\sigma^2 = 45$.

3. *Hypotheses* The hypothesis to be tested, or null hypothesis, is that the population mean enzyme level is equal to 25. The alternative hypothesis is

that the population mean enzyme level is not equal to 25. Notice that we are identifying with the alternative hypothesis the conclusion they wish to reach, so that if the data permit rejection of the null hypothesis, the researchers' conclusion will carry more weight, since the accompanying probability of rejecting a true null hypothesis will be small. We will make sure of this by assigning a small value to α, the probability of committing a type I error. We may present the relevant hypotheses in compact form as follows:

$$H_0: \mu = 25$$
$$H_A: \mu \neq 25$$

4. *Test Statistic*   Since we are testing a hypothesis about a population mean, since we assume that the population is normally distributed, and since the population variance is known, our test statistic is given by Equation 6.2.1.

5. *Distribution of Test Statistic*   Based on our knowledge of sampling distributions and the normal distribution, we know that the test statistic is normally distributed with a mean of 0 and a variance of 1, if H_0 is true. There are many possible values of the test statistic that the present situation can generate; one for every possible sample of size 10 that can be drawn from the population. Since we draw only one sample, we have only one of these possible values on which to base a decision.

6. *Decision Rule*   The decision rule tells us to reject H_0 if the computed value of the test statistic falls in the rejection region and to accept H_0 if it falls in the acceptance region. We must now specify the rejection and acceptance regions. We can begin by asking ourselves what magnitude of values of the test statistic will cause rejection of H_0. If the null hypothesis is false, it may be so either because the true mean is less than 25 or because the true mean is greater than 25. Therefore, either sufficiently small values or sufficiently large values of the test statistic will cause rejection of the null hypothesis. These extreme values we want to constitute the rejection region. How extreme must a possible value of the test statistic be to qualify for the rejection region? The answer depends on the significance level we choose, that is, the size of the probability of committing a type I error. Let us say that we want the probability of rejecting a true null hypothesis to be $\alpha = .05$. Since our rejection region is to consist of two parts, sufficiently small values and sufficiently large values of the test statistic, part of α will have to be associated with the large values and part with the small values. It seems reasonable that we should divide α equally and let $\alpha/2 = .025$ be associated with small values and $\alpha/2 = .025$ be associated with large values.

What value of the test statistic is so large that, when the null hypothesis is true, the probability of obtaining a value this large or larger is .025? In other words, what is the value of z to the right of which lies .025 of the area under the unit normal distribution? The value of z to the right of which lies .025 of the area

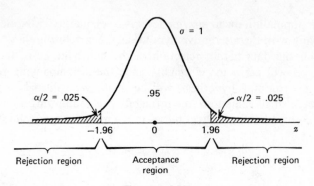

FIGURE 6.2.1

Acceptance and Rejection Regions for Example 6.2.1

is the same value between $-\infty$ and which lies .975 of the area. We look in the body of Table C until we find .975 or its closest value and read the corresponding marginal entries to obtain our z value. In the present example the value of z is 1.96. Similar reasoning will lead us to find -1.96 as the value of the test statistic so small that when the null hypothesis is true, the probability of obtaining a value this small or smaller is .025. Our rejection region, then, consists of all values of the test statistic equal to or greater than 1.96 or less than or equal to -1.96. The acceptance region consists of all values in between. We may state the decision rule for this test as follows: *Reject H_0 if the computed value of the test statistic is either ≥ 1.96 or ≤ -1.96*. Otherwise, do not reject H_0. The acceptance and rejection regions are shown in Figure 6.2.1. The values of the test statistic that separate the rejection and acceptance regions are called *critical values* of the test statistic, and the rejection region is sometimes referred to as the *critical region*.

The decision rule tells us to compute a value of the test statistic from the data of our sample and to reject H_0 if we get a value equal to or greater than 1.96 or equal to or less than -1.96 and to accept H_0 if we get any other value. The value of α and, hence, the decision rule should be decided on prior to gathering the data. This prevents our being accused of allowing the sample results to influence our decision. This condition of objectivity is highly desirable and should be preserved in all tests.

7. *Calculation of Test Statistic* From our sample we compute

$$z = \frac{22 - 25}{\sqrt{45/10}} = \frac{-3}{2.1213} = -1.41$$

8. *Statistical Decision* Abiding by the decision rule, we are unable to reject the null hypothesis since -1.41 is not in the rejection region. We can say that the computed value of the test statistic is not significant at the .05 level.

9. *Conclusion* We conclude that μ *may be* equal to 25 and let our administrative or clinical actions be in accordance with this conclusion.

p **Values** Instead of saying that an observed value of the test statistic is significant or is not significant, most writers in the research literature prefer to report the exact probability of getting a value as extreme or more extreme than that observed if the null hypothesis is true. In the present instance these writers would give the computed value of the test statistic along with the statement $p = .1586$. The statement $p = .1586$ means that the probability of getting a value as extreme as 1.41 in either direction, when the null hypothesis is true, is .1586. The value .1586 is obtained from Table C and is the probability of observing a $z \geq 1.41$ or a $z \leq -1.41$ when the null hypothesis is true. That is, when H_0 is true, the probability of obtaining a value of z as large as or larger than 1.41 is .0793, and the probability of observing a value of z as small as or smaller than -1.41 is .0793. The probability of one or the other of these events occurring, when H_0 is true, is equal to the sum of the two individual probabilities, and hence, in the present example, we say that $p = .0793 + .0793 = .1586$. The quantity p is referred to as the *p value* for the test.

Definition The p value for a hypothesis test is the probability of obtaining, when H_0 is true, a value of the test statistic as extreme as or more extreme (in the appropriate direction) than the one actually computed.

The p value for a test may be defined also as the smallest value of α for which the null hypothesis can be rejected.

The reporting of p values as part of the results of an investigation is more informative to the reader than such statements as "the null hypothesis is rejected at the .05 level of significance" or "the results were not significant at the .05 level." Reporting the p value associated with a test lets the reader know just how common or how rare is the computed value of the test statistic given that H_0 is true. Gibbons and Pratt (1), Daniel (2), and Bahn (3) may be consulted for a more extensive treatment of the subject of p-values.

Testing H_0 by Means of a Confidence Interval Earlier we stated that one can use confidence intervals to test hypotheses. In Example 6.2.1 we used a hypothesis testing procedure to test H_0: $\mu = 25$ against the alternative, H_A: $\mu \neq 25$. We were unable to reject H_0 because the computed value of the test statistic fell in the acceptance region.

Let us see how we might have arrived at this same conclusion by using a $100(1 - \alpha)$ percent confidence interval. The 95 percent confidence interval for μ is

$$22 \pm 1.96\sqrt{45/10}$$
$$22 \pm 1.96(2.1213)$$
$$22 \pm 4.16$$
$$17.84, 26.16$$

Since this interval includes 25, we say that 25 is a candidate for the mean we are estimating and, therefore, μ may be equal to 25 and H_0 is not rejected. This is the same conclusion reached by means of the hypothesis testing procedure.

If the hypothesized parameter, 25, had not been included in the 95 percent confidence interval; that is, if the lower limit of the computed interval had been larger than 25, or if the upper limit had been smaller than 25, we would have said that H_0 is rejected at the .05 level of significance. In general, *when testing a null hypothesis by means of a two-sided confidence interval, we reject H_0 at the α level of significance if the hypothesized parameter is not contained within the $100(1 - \alpha)$ percent confidence interval. If the hypothesized parameter is contained within the interval, H_0 cannot be rejected at the α level of significance.*

One-Sided Hypothesis Tests   The hypothesis test illustrated by Example 6.2.1 is an example of a *two-sided test*, so called because the rejection region is split between the two sides or tails of the distribution of the test statistic. A hypothesis test may be *one-sided*, in which case all of the rejection region is in one or the other tail of the distribution. Whether a one-sided or a two-sided test is used depends on the nature of the question being asked by the researcher.

If both large and small values will cause rejection of the null hypothesis, a two-sided test is indicated. When either sufficiently "small" values only or sufficiently "large" values only will cause rejection of the null hypothesis, a one-sided test is indicated.

Example 6.2.2   Refer to Exercise 6.2.1. Suppose, instead of asking if they could conclude that $\mu \neq 25$, the researchers had asked: Can we conclude that $\mu < 25$? To this question we would reply that they can so conclude if they can reject the null hypothesis that $\mu \geq 25$. Let us go through the nine-step procedure to reach a decision based on a one-sided test.

1.   *Data*   See the previous example.

2.   *Assumptions*   See the previous example.

3.   *Hypotheses*

$$H_0: \mu \geq 25$$

$$H_A: \mu < 25$$

The inequality in the null hypothesis implies that the null hypothesis consists of an infinite number of hypotheses. The test will be made only at the point of equality, since it can be shown that if H_0 is rejected when the test is made at the point of equality it would be rejected if the test were done for any other value of μ indicated in the null hypothesis.

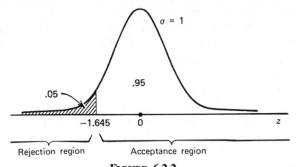

FIGURE 6.2.2

Acceptance and Rejection Regions for Example 6.2.2

4. *Test Statistic*

$$z = \frac{\bar{x} - \mu_0}{\sigma/\sqrt{n}}$$

5. *Distribution of Test Statistic* See the previous example.

6. *Decision Rule* Let us again let $\alpha = .05$. To determine where to place the rejection region, let us ask ourselves what magnitude of values would cause rejection of the null hypothesis. If we look at the hypotheses, we see that sufficiently small values would cause rejection and that large values would tend to reinforce the null hypothesis. We will want our rejection region to be where the small values are—at the lower tail of the distribution. This time, since we have a one-sided test, all of α will go in the one tail of the distribution. By consulting Table C, we find that the value of z to the left of which lies .05 of the area under the unit normal curve is -1.645 after interpolating. Our rejection and acceptance regions are now specified and are shown in Figure 6.2.2.

Our decision rule tells us to reject H_0 if the computed value of the test statistic is less than or equal to -1.645.

7. *Calculation of Test Statistic* From our data we compute

$$z = \frac{22 - 25}{\sqrt{45/10}} = -1.41$$

8. *Statistical Decision* We are unable to reject the null hypothesis since $-1.41 > -1.645$.

9. *Conclusion* We conclude that the population mean may be greater than or equal to 25 and act accordingly.

The p-value for this test is .0793, since $P(z \le -1.41)$, when H_0 is true, is .0793 as given by Table C when we determine the magnitude of the area to the left of -1.41 under the unit normal curve. One can test a one-sided null hypothesis by means of a one-sided confidence interval. However, we shall not cover the construction and interpretation of this type of confidence interval in this book. The interested reader is referred to a discussion of the topic in the book by Daniel (4).

If the researchers' question had been, "Can we conclude that the mean is greater than 25?", following the above nine-step procedure would have led to a one-sided test with all of the rejection region at the upper tail of the distribution of the test statistic and a critical value of $+1.645$.

Sampling from a Normally Distributed Population: Population Variance Unknown
As we have already noted, the population variance is usually unknown in actual situations involving statistical inference about a population mean. When sampling is from a normally distributed population with an unknown variance, the test statistic for testing H_0: $\mu = \mu_0$ is

$$t = \frac{\bar{x} - \mu_0}{s/\sqrt{n}} \tag{6.2.2}$$

which, when H_0 is true, is distributed as Student's t with $n - 1$ degrees of freedom. The following example illustrates the hypothesis testing procedure when the population is normally distributed and its variance is unknown.

Example 6.2.3 Researchers collected serum amylase values from a random sample of 15 apparently healthy subjects. They want to know whether they can conclude that the mean of the population from which the sample of serum amylase determinations came is different from 120. They can so conclude if they reject the null hypothesis that the true mean is equal to 120. This suggests a two-sided hypothesis test that we now make by using the familiar nine steps.

1. *Data* The data consist of serum amylase determinations made on 15 apparently healthy subjects. The mean and standard deviation computed from the sample are 96 and 35 units/100 ml, respectively.

2. *Assumptions* The 15 determinations constitute a random sample from a population of determinations that is normally distributed. The population variance is unknown.

3. *Hypotheses*

$$H_0: \mu = 120$$
$$H_A: \mu \neq 120$$

4. *Test Statistic* Since the population variance is unknown, our test statistic is given by Equation 6.2.2.

5. *Distribution of Test Statistic* Our test statistic is distributed as Student's t with $n - 1$ degrees of freedom if H_0 is true.

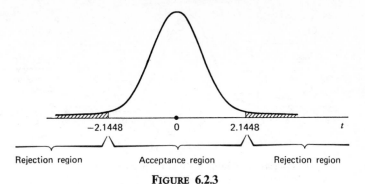

FIGURE 6.2.3

Acceptance and Rejection Regions for Example 6.2.3

6. *Decision Rule* Let $\alpha = .05$. Since we have a two-sided test, we put $\alpha/2 = .025$ in each tail of the distribution of our test statistic. The t values to the right and left of which .025 of the area lies are 2.1448 and -2.1448. These values are obtained from Table E. The acceptance and rejection regions are shown in Figure 6.2.3.

The decision rule tells us to compute a value of the test statistic and reject H_0 if the computed t is either greater than or equal to 2.1448 or less than or equal to -2.1448.

7. *Calculation of Test Statistic*

$$t = \frac{96 - 120}{35/\sqrt{15}} = \frac{-24}{9.04} = -2.65$$

8. *Statistical Decision* Reject H_0, since -2.65 falls in the rejection region.

9. *Conclusion* Our conclusion, based on these data is that the mean of the population from which the sample came is not 120.

The exact p value for this test cannot be obtained from Table E since it gives t values only for selected percentiles. The p value can be stated as an interval, however. In the present example -2.65 is less than -2.624, the value of t to the left of which lies .01 of the area under the t with 14 degrees of freedom, but greater than -2.9768, to the left of which lies .005 of the area. Consequently, when H_0 is true, the probability of obtaining a value of t as small as or smaller than -2.65 is less than .01 but greater than .005. That is, $.005 < P(t \leq -2.65) < .01$. Since the test was two-sided, we must allow for the possibility of a computed value of the test statistic as large in the opposite direction as that observed. Table E reveals that $.005 < P(t \geq 2.65) < .01$. The p value, then, is $.01 < p < .02$. Figure 6.2.4 shows the p value for this example.

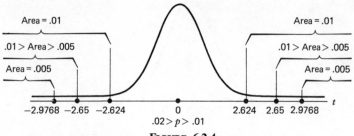

FIGURE 6.2.4

Determination of p-value for Example 6.2.3

If in the previous example the hypotheses had been

$$H_0: \mu \geq 120$$

$$H_A: \mu < 120$$

the testing procedure would have led to a one-sided test with all of the rejection region at the lower tail of the distribution, and if the hypotheses had been

$$H_0: \mu \leq 120$$

$$H_A: \mu > 120$$

we would have had a one-sided test with all of the rejection region at the upper tail of the distribution.

Sampling from a Population That Is Not Normally Distributed   If the sample on which we base our hypothesis test about a population mean comes from a population that is not normally distributed, we may, if our sample is large, take advantage of the central limit theorem and use $z = (\bar{x} - \mu_0)/(\sigma/\sqrt{n})$ as the test statistic. If the population standard deviation is not known, the usual practice is to use the sample standard deviation as an estimate. The test statistic for testing $H_0: \mu = \mu_0$, then, is

$$z = \frac{\bar{x} - \mu_0}{s/\sqrt{n}} \tag{6.2.3}$$

which, when H_0 is true, is distributed approximately as the unit normal distribution if n is large. The rationale for using s to replace σ is that the large sample, necessary for the central limit theorem to apply, will yield a sample standard deviation that closely approximates σ.

Example 6.2.4   In a health survey of a certain community 150 persons were interviewed. One of the items of information obtained was the number of prescriptions each person had had filled during the past year. The average number for the 150 people was 5.8 with a standard deviation of 3.1. The investigator wishes to know if these data provide sufficient evidence to indicate that the population mean is greater than 5. We say that the data do provide that evidence if we can reject the null hypothesis that the mean is less than or equal to 5. The following test may be carried out:

1. *Data*   The data consist of the number of prescriptions filled for 150 persons with $\bar{x} = 5.8$ and $s = 3.1$.

2. *Assumptions*   The data constitute a random sample from a population that is not normally distributed.

3. *Hypotheses*

$$H_0: \mu \le 5$$
$$H_A: \mu > 5$$

4. *Test Statistic*   The test statistic is given by Equation 6.2.3, since σ is unknown.

5. *Distribution of Test Statistic*   Because of the central limit theorem, the test statistic is approximately normally distributed with $\mu = 0$ if H_0 is true.

6. *Decision Rule*   Let $\alpha = .05$. The critical value of the test statistic is 1.645. The acceptance and rejection regions are shown in Figure 6.2.5. Reject H_0 if computed $z \ge 1.645$.

7. *Calculation of Test Statistic*

$$z = \frac{5.8 - 5.0}{3.1/\sqrt{150}} = \frac{.8}{.25} = 3.2$$

8. *Statistical Decision*   Reject H_0 since $3.2 > 1.645$.

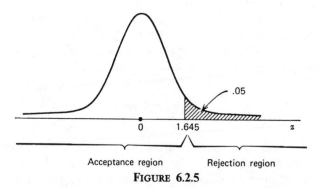

FIGURE 6.2.5
Acceptance and Rejection Regions for Example 6.2.4

9. *Conclusion* Conclude that the mean number of prescriptions filled per person per year for this population is greater than 5.

If the population variance had been known, the procedure would have been identical to the above except that the known value of σ, instead of the sample value, would have been used in the denominator of the computed test statistic. The p value for this test is .0007.

Depending on what the investigators wished to conclude, either a two-sided test or a one-sided test, with the rejection region at the lower tail of the distribution, could have been made using the above data.

When testing a hypothesis about a single population mean, we may use Figure 5.3.1 to decide quickly whether the test statistic is z or t.

Computer Analysis To illustrate the use of computers in testing hypotheses we consider the following Example.

Example 6.2.5 The following are the head circumferences (centimeters) at birth of 15 infants.

33.38	32.15	33.99	34.10	33.97
34.34	33.95	33.85	34.23	32.73
33.46	34.13	34.45	34.19	34.05

We wish to test H_0: $\mu = 34.5$ against H_A: $\mu \neq 34.5$. We assume that the assumptions for use of the t statistic are met.

After entering the data into the computer and issuing appropriate MINITAB commands, we obtain the following printout.

```
        N =  15     MEAN =      33.798     ST.DEV.=      .630
   TEST OF MU =  34.5000 VS. MU N.E.    34.5000
   T = -4.314
   THE TEST IS SIGNIFICANT AT   .0007
```

We learn from the printout that the computed value of the test statistic is -4.314 and the p value for the test is .0007.

When both the z statistic and the t statistic are inappropriate test statistics for use with the available data, one may wish to use a nonparametric technique to test a hypothesis about a single population measure of central tendency. One such procedure, the sign test, is discussed in Chapter 11.

EXERCISES

For each of the following exercises carry out the nine-step hypothesis testing procedure for the given significance level. Compute the p value for each test.

6.2.1 Administrators of a health care center want to know if the mean time spent by patients in the waiting room is greater than 20 minutes. A sample of 100 patients spent, on the average, 23 minutes in the waiting room between register-ing and being seen by a member of the health team. The sample standard deviation was 10. Let $\alpha = .05$.

6.2.2   An experiment was conducted to study the effect on rats of a certain type of brain surgery. Following surgery the rats were trained to perform a series of tasks. After the training period the rats were stimulated to perform the task and each was scored on the basis of the quality of its performance. The mean score for the 16 rats used in the experiment was 90 with a standard deviation of 10. Do these data provide sufficient evidence to indicate that the population mean is less than 95? Let $\alpha = .10$.

6.2.3   A survey of 64 medical laboratories revealed that the mean price charged for a certain test was $12 with a standard deviation of $6. Do these data provide sufficient evidence to indicate that the population mean is greater than $10? Let $\alpha = .05$.

6.2.4   A study was made of a sample of 25 records of patients seen at a chronic disease hospital on an outpatient basis. The mean number of outpatient visits per patient was 4.8, and the sample standard deviation was 2. Can it be concluded from these data that the population mean is greater than four visits per patient? Let the probability of committing a type I error be .05.

6.2.5   In a sample of 49 adolescents who served as the subjects in an immunologic study, one variable of interest was the diameter of skin test reaction to an antigen. The sample mean and standard deviation were 21 and 11 mm erythema, respectively. Can it be concluded from these data that the population mean is less than 30? Let $\alpha = .05$.

6.2.6   Nine laboratory animals were infected with a certain bacterium and then immunosuppressed. The mean number of organisms later recovered from tissue specimens was 6.5 (coded data) with a standard deviation of .6. Can one conclude from these data that the population mean is greater than 6? Let $\alpha = .05$.

6.2.7   A sample of 25 freshman nursing students made a mean score of 77 on a test designed to measure attitude toward the dying patient. The sample standard deviation was 10. Do these data provide sufficient evidence to indicate, at the .05 level of significance, that the population mean is less than 80?

6.2.8   We wish to know if we can conclude that the mean daily caloric intake in the adult rural population of a developing country is less than 2000. A sample of 500 had a mean of 1985 and a standard deviation of 210. Let $\alpha = .05$.

6.2.9   A survey of 100 similar sized hospitals revealed a mean average daily census in the pediatrics service of 27 with a standard deviation of 6.5. Do these data provide sufficient evidence to indicate that the population mean is greater than 25? Let $\alpha = .05$.

6.2.10   Following a week-long hospital supervisory training program, 16 assistant hospital administrators made a mean score of 74 on a test administered as part of the evaluation of the training program. The sample standard deviation was 12. Can it be concluded from these data that the population mean is greater than 70? Let $\alpha = .05$.

6.2.11 A random sample of 16 emergency reports was selected from the files of an ambulance service. The mean time (computed from the sample data) required for ambulances to reach their destinations was 13 minutes. Assume that the population of times is normally distributed with a variance of 9. Can we conclude at the .05 level of significance that the population mean is greater than 10 minutes?

6.2.12 The following data are the oxygen uptakes (milliliters) during incubation of a random sample of 15 cell suspensions.

$$14.0, 14.1, 14.5, 13.2, 11.2, 14.0, 14.1, 12.2$$
$$11.1, 13.7, 13.2, 16.0, 12.8, 14.4, 12.9$$

Do these data provide sufficient evidence at the .05 level of significance that the population mean is not 12 ml?

6.2.13 Can we conclude that the mean maximum voluntary ventilation value for apparently healthy college seniors is not 110 liters per minute? A sample of 20 yielded the following values.

$$132, 33, 91, 108, 67, 169, 54, 203, 190, 133$$
$$96, 30, 187, 21, 63, 166, 84, 110, 157, 138$$

Let $\alpha = .01$.

6.2.14 The following are the systolic blood pressures (mm Hg) of 12 patients undergoing drug therapy for hypertension.

$$183, 152, 178, 157, 194, 163, 144, 114, 178, 152, 118, 158$$

Can we conclude on the basis of these data that the population mean is less than 165? Let $\alpha = .05$.

6.2.15 Can we conclude that the mean age at death of patients with homozygous sickle-cell disease is less than 30 years? A sample of 50 patients yielded the following ages in years.

15.5,	2.0,	45.1,	1.7,	.8,	1.1,	18.2,	9.7,	28.1,	18.2
27.6,	45.0,	1.0,	66.4,	2.0,	67.4,	2.5,	61.7,	16.2,	31.7
6.9,	13.5,	1.9,	31.2,	9.0,	2.6,	29.7,	13.5,	2.6,	14.4,
20.7,	30.9,	36.6,	1.1,	23.6,	.9,	7.6,	23.5,	6.3,	40.2
23.7,	4.8,	33.2,	27.1,	36.7,	3.2,	38.0,	3.5,	21.8,	2.4

Let $\alpha = .05$. What assumptions are necessary?

6.2.16 The following are intraocular pressure (mm Hg) values recorded for a sample of 21 elderly subjects.

14.5,	12.9,	14.0,	16.1,	12.0,	17.5,	14.1,	12.9,	17.9,	12.0
16.4,	24.2,	12.2,	14.4,	17.0,	10.0,	18.5,	20.8,	16.2,	14.9
19.6									

Can we conclude from these data that the mean of the population from which the sample was drawn is greater than 14? Let $\alpha = .05$, What assumptions are necessary?

6.3   HYPOTHESIS TESTING: THE DIFFERENCE BETWEEN TWO POPULATION MEANS

Hypothesis testing involving the difference between two population means is most frequently employed to determine whether or not it is reasonable to conclude that the two are unequal. In such cases, one or the other of the following hypotheses may be formulated:

(1) $H_0: \mu_1 - \mu_2 = 0,$ $H_A: \mu_1 - \mu_2 \neq 0$
(2) $H_0: \mu_1 - \mu_2 \geq 0,$ $H_A: \mu_1 - \mu_2 < 0$
(3) $H_0: \mu_1 - \mu_2 \leq 0,$ $H_A: \mu_1 - \mu_2 > 0$

It is possible, however, to test the hypothesis that the difference is equal to, greater than or equal to, or less than or equal to some value other than zero.

As was done in the previous section, hypothesis testing involving the difference between two population means will be discussed in three different contexts: (1) when sampling is from normally distributed populations with known population variances, (2) when sampling is from normally distributed populations with unknown population variances, and (3) when sampling is from populations that are not normally distributed.

Sampling from Normally Distributed Populations: Population Variances Known
When each of two independent simple random samples has been drawn from a normally distributed population with a known variance, the test statistic for testing the null hypothesis of equal population means is

$$z = \frac{(\bar{x}_1 - \bar{x}_2) - (\mu_1 - \mu_2)}{\sqrt{\dfrac{\sigma_1^2}{n_1} + \dfrac{\sigma_2^2}{n_2}}}$$

(6.3.1)

The hypothesized parameter in this case would be more correctly written as $(\mu_1 - \mu_2)_0$ where the subscript 0 indicates that the difference is a hypothesized parameter. To avoid cumbersome notation, however, we are deleting the subscript in the formula for the test statistic. When H_0 is true the test statistic of Equation 6.3.1 is distributed as the unit normal.

Example 6.3.1   Researchers wish to know if the data they have collected provide sufficient evidence to indicate a difference in mean serum uric acid levels between normal individuals and individuals with mongolism. We will say that the data do provide such evidence if we can reject the null hypothesis that the means are equal. Let us reach a decision by means of the nine-step hypothesis testing procedure.

1. *Data*   The data consist of serum uric acid readings on 12 mongoloid individuals and 15 normal individuals. The means are $\bar{x}_1 = 4.5$ mg/100 ml and $\bar{x}_2 = 3.4$ mg/100 ml.

2. *Assumptions*   The data constitute two independent simple random samples each drawn from a normally distributed population with a variance equal to 1.

3. *Hypotheses*

$$H_0: \mu_1 - \mu_2 = 0$$

$$H_A: \mu_1 - \mu_2 \neq 0$$

An alternative way of stating the hypotheses is as follows:

$$H_0: \mu_1 = \mu_2$$

$$H_A: \mu_1 \neq \mu_2$$

4. *Test Statistic*   The test statistic is given by Equation 6.3.1.

5. *Distribution of Test Statistic*   When the null hypothesis is true, the test statistic follows the unit normal distribution.

6. *Decision Rule*   Let $\alpha = .05$. The critical values of z are ± 1.96. Reject H_0 unless $-1.96 < z_{\text{computed}} < 1.96$. The acceptance and rejection regions are shown in Figure 6.3.1.

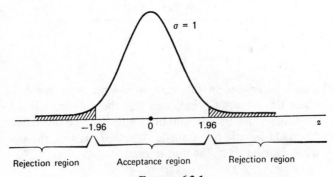

FIGURE 6.3.1

Acceptance and Rejection Regions for Example 6.3.1

7. *Calculation of Test Statistic*

$$z = \frac{(4.5 - 3.4) - 0}{\sqrt{\frac{1}{12} + \frac{1}{15}}} = \frac{1.1}{.39} = 2.82$$

8. *Statistical Decision*   Reject H_0, since $2.82 > 1.96$.

9. *Conclusion*   Conclude that, on the basis of these data, there is an indication that the two population means are not equal. For this test, $p = .0048$.

In the previous chapter the 95 percent confidence interval for $\mu_1 - \mu_2$, computed from the same data, was found to be .3 to 1.9. Since this interval does not include 0, we say that 0 is not a candidate for the true difference, and we conclude that the difference is not zero. Thus we arrive at the same conclusion by means of a confidence interval.

Although the population variances were equal in this example, this is not a requirement when population variances are known, and if they had been unequal, the same hypothesis testing procedure would have been followed.

Sampling from Normally Distributed Populations: Population Variances Unknown
As we have learned, when the population variances are unknown, two possibilities exist. The two population variances may be equal or they may be unequal. We consider first the case where it is known, or it is reasonable to assume, that they are equal.

Population Variances Equal   When the population variances are unknown, but assumed to be equal, we recall from Chapter 5 that it is appropriate to pool the sample variances by means of the following formula.

$$s_p^2 = \frac{(n_1 - 1)s_1^2 + (n_2 - 1)s_2^2}{n_1 + n_2 - 2}$$

When each of two independent simple random samples has been drawn from a normally distributed population and the two populations have equal but unknown variances, the test statistic for testing H_0: $\mu_1 = \mu_2$ is given by

$$t = \frac{(\bar{x}_1 - \bar{x}_2) - (\mu_1 - \mu_2)}{\sqrt{\dfrac{s_p^2}{n_1} + \dfrac{s_p^2}{n_2}}} \tag{6.3.2}$$

which, when H_0 is true, is distributed as Student's t with $n_1 + n_2 - 2$ degrees of freedom.

Example 6.3.2 A research team collected serum amylase data from a sample of healthy subjects and from a sample of hospitalized subjects. They wish to know if they would be justified in concluding that the population means are different.

1. *Data*   The data consist of serum amylase determinations on $n_2 = 15$ healthy subjects and $n_1 = 22$ hospitalized subjects. The sample means and standard deviations are as follows:

$$\bar{x}_1 = 120 \text{ units/ml}, \qquad s_1 = 40 \text{ units/ml}$$

$$\bar{x}_2 = 96 \text{ units/ml}, \qquad s_2 = 35 \text{ units/ml}$$

2. *Assumptions*   The data constitute two independent random samples, each drawn from a normally distributed population. The population variances are unknown but are assumed to be equal.

3. *Hypotheses* $H_0: \mu_1 - \mu_2 = 0$, $H_A: \mu_1 - \mu_2 \neq 0$.

4. *Test Statistic*   The test statistic is given by Equation 6.3.2.

5. *Distribution of Test Statistic*   When the null hypothesis is true, the test statistic follows Student's t distribution with $n_1 + n_2 - 2$ degrees of freedom.

6. *Decision Rule*   Let $\alpha = .05$. The critical values of t are ± 2.0301. Reject H_0 unless $-2.0301 < t_{computed} < 2.0301$.

7. *Calculation of Test Statistic*

$$t = \frac{(120 - 96) - 0}{\sqrt{\frac{1450}{15} + \frac{1450}{22}}} = \frac{24}{12.75} = 1.88$$

8. *Statistical Decision*   We are unable to reject H_0, since $-2.0301 < 1.88 < 2.0301$; that is, 1.88 falls in the acceptance region.

9. *Conclusion*   On the basis of these data we cannot conclude that the two population means are different. For this test $.10 > p > .05$, since $1.6896 < 1.88 < 2.0301$.

Population Variances Unequal   When two independent simple random samples have been drawn from normally distributed populations with unknown and unequal variances the test statistic for testing $H_0: \mu_1 = \mu_2$ is

$$t' = \frac{(\bar{x}_1 - \bar{x}_2) - (\mu_1 - \mu_2)}{\sqrt{\frac{s_1^2}{n_1} + \frac{s_2^2}{n_2}}} \tag{6.3.3}$$

The critical value of t' for an α level of significance and a two-sided test is approximately

$$t'_{1-\alpha/2} = \frac{w_1 t_1 + w_2 t_2}{w_1 + w_2}$$

(6.3.4)

where $w_1 = s_1^2/n_1$, $w_2 = s_2^2/n_2$, $t_1 = t_{1-\alpha/2}$ for $n_1 - 1$ degrees of freedom, and $t_2 = t_{1-\alpha/2}$ for $n_2 - 1$ degrees of freedom. The critical value of t' for a one-sided test is found by computing $t'_{1-\alpha}$ by Equation 6.3.4, using $t_1 = t_{1-\alpha}$ for $n_1 - 1$ degrees of freedom and $t_2 = t_{1-\alpha}$ for n_2 degrees of freedom.

For a two-sided test reject H_0 if the computed value of t' is either greater than or equal to the critical value given by Equation 6.3.4 or less than or equal to the negative of that value.

For a one-sided test with the rejection region in the right tail of the sampling distribution reject H_0 if the computed t' is equal to or greater than the critical t'. For a one-sided test with a left-tail rejection region reject H_0 if the computed value of t' is equal to or smaller than the negative of the critical t' computed by the indicated adaptation of Equation 6.3.4.

Example 6.3.3 Researchers wish to know if two populations differ with respect to the mean value of total serum complement activity (C_{H50}).

1. *Data* The data consist of total serum complement activity (C_{H50}) determinations on $n_2 = 20$ apparently normal subjects and $n_1 = 10$ subjects with disease. The sample means and standard deviations are

$$\bar{x}_1 = 62.6, \qquad s_1 = 33.8$$
$$\bar{x}_2 = 47.2, \qquad s_2 = 10.1$$

2. *Assumptions* The data constitute two independent random samples, each drawn from a normally distributed population. The population variances are unknown and unequal.

3. *Hypotheses*

$$H_0: \mu_1 - \mu_2 = 0$$
$$H_A: \mu_1 - \mu_2 \neq 0$$

4. *Test Statistic* The test statistic is given by Equation 6.3.3.

5. *Distribution of the Test Statistic* The statistic given by Equation 6.3.3 does not follow Student's t distribution. We, therefore, obtain its critical values by Equation 6.3.4.

6. *Decision Rule* Let $\alpha = .05$. Before computing t' we calculate $w_1 = (33.8)^2/10 = 114.244$ and $w_2 = (10.1)^2/20 = 5.1005$. In Table E we find

that $t_1 = 2.2622$ and $t_2 = 2.0930$. By Equation 6.3.4 we compute

$$t' = \frac{114.244(2.2622) + 5.1005(2.0930)}{114.244 + 5.1005} = 2.255$$

Our decision rule, then, is reject H_0 if the computed t is either ≥ 2.255 or ≤ -2.255.

7. *Calculation of the Test Statistic*   By Equation 6.3.3 we compute

$$t' = \frac{(62.6 - 47.2) - 0}{\sqrt{\dfrac{(33.8)^2}{10} + \dfrac{(10.1)^2}{20}}} = \frac{15.4}{10.92} = 1.41$$

8. *Statistical Decision*   Since $-2.255 < 1.41 < 2.255$, we cannot reject H_0.

9. *Conclusion*   On the basis of these results we cannot conclude that the two population means are different.

Sampling from Populations That Are Not Normally Distributed   When sampling is from populations that are not normally distributed, the results of the central limit theorem may be employed if sample sizes are large (say ≥ 30). This will allow the use of normal theory. When each of two large independent simple random samples has been drawn from a population that is not normally distributed, the test statistic for testing H_0: $\mu_1 = \mu_2$ is

$$z = \frac{(\bar{x}_1 - \bar{x}_2) - (\mu_1 - \mu_2)}{\sqrt{\dfrac{\sigma_1^2}{n_1} + \dfrac{\sigma_2^2}{n_2}}} \tag{6.3.5}$$

which, when H_0 is true, follows the unit normal distribution. If the population variances are known, they are used; but if they are unknown, the sample variances, which are necessarily based on large samples, are used as estimates. Sample variances are not pooled, since equality of population variances is not a necessary assumption when the z statistic is used.

Example 6.3.4   A hospital administrator wished to know if the population which patronizes hospital A has a larger mean family income than does the population which patronizes hospital B. The answer is, yes, if he can reject the null hypothesis that μ_1 is less than or equal to μ_2, where μ_1 and μ_2 are the parameters for the hospital A and hospital B populations, respectively.

1. *Data*   The data consist of the family incomes of 75 patients admitted to hospital A and of 80 patients admitted to hospital B. The sample means are $\bar{x}_1 = \$6800$ and $\bar{x}_2 = \$5450$.

2. *Assumptions*   The data constitute two independent random samples, each drawn from a nonnormally distributed population with standard deviations $\sigma_1 = \$600$ and $\sigma_2 = \$500$.

3. *Hypotheses*

$$H_0: \mu_1 - \mu_2 \leq 0$$
$$H_A: \mu_1 - \mu_2 > 0$$

or, alternatively,

$$H_0: \mu_1 \leq \mu_2$$
$$H_A: \mu_1 > \mu_2$$

4. *Test Statistic*   Since we have large samples, the central limit theorem allows us to use Equation 6.3.5 as the test statistic.

5. *Distribution of Test Statistic*   When the null hypothesis is true, the test statistic is distributed approximately as the unit normal.

6. *Decision Rule*   Let $\alpha = .01$. This is a one-sided test with a critical value of z equal to 2.33. Reject H_0 if $z_{computed} \geq 2.33$.

7. *Calculation of Test Statistic*

$$z = \frac{(6800 - 5450) - 0}{\sqrt{\dfrac{(600)^2}{75} + \dfrac{(500)^2}{80}}} = \frac{1350}{89} = 15.17$$

8. *Statistical Decision*   Reject H_0, since $z = 15.17$ is in the rejection region.

9. *Conclusion*   These data indicate that the population patronizing hospital A has a larger mean family income than does the population patronizing hospital B. For this test $p < .0001$, since $15.17 > 3.89$. When testing a hypothesis about the difference between two population means, we may use Figure 5.4.1 to decide quickly whether the test statistic should be z or t.

Sometimes neither the z statistic nor the t statistic is an appropriate test statistic for use with the available data. When such is the case, one may wish to use a nonparametric technique for testing a hypothesis about the difference between two population measures of central tendency. The Mann-Whitney test statistic and the median test, discussed in Chapter 11, are frequently used alternatives to the z and t statistics.

EXERCISES

In each of the following exercises complete the nine-step hypothesis testing procedure and compute the p value for each test.

6.3.1   Seventy patients suffering from epilepsia were divided at random into two equal groups. Group A was placed on a treatment regimen that included daily doses of vitamin D. Group B received the same treatment with the exception that for this group a placebo was given in place of vitamin D. The means of the number of seizures experienced during the period of treatment by the two groups were $\bar{x}_A = 15$ and $\bar{x}_B = 24$. The sample variances were $s_A^2 = 8$ and $s_B^2 = 12$. Do these data provide sufficient evidence to indicate that vitamin D is effective in reducing the number of seizures? Let $\alpha = .05$.

6.3.2 An epidemiologist wished to compare two rabies vaccines to see if he could conclude that they differ in effectiveness. Subjects who had previously received rabies vaccine were divided into two groups. Group 1 received a booster dose of type 1 vaccine, and group 2 received a booster dose of type 2 vaccine. Antibody responses were recorded two weeks later. The means, standard deviations, and sample sizes for the two groups were as follows:

Group	Sample size	$\bar{x}$	s
1	10	4.5	2.5
2	9	2.5	2.0

Let $\alpha = .05$.

6.3.3 Median nerve motor conduction velocity values were recorded for 10 subjects admitted to the poison control center of a metropolitan hospital with a diagnosis of methylmercury poisoning. Similar determinations also were made on 15 apparently healthy subjects. The means and standard deviations were as follows:

	$\bar{x}$	s
Poisoned group:	55	6
Normal subjects:	63	5

Do these data provide sufficient evidence to indicate that the means of the populations represented by the samples differ? Let $\alpha = .05$.

6.3.4 Can we conclude that chronically ill children tend, on the average, to be less self-confident than healthy children? A test designed to measure self-confidence was administered to 16 chronically ill and 21 healthy children. The mean scores and standard deviations were as follows:

	$\bar{x}$	s
Ill group	22.5	4.1
Well group	26.9	3.2

Let $\alpha = .05$.

6.3.5 A nurse researcher wished to know if graduates of baccalaureate nursing programs and graduates of associate degree nursing programs differ with respect to mean scores on a personality inventory. A sample of 50 associate degree graduates (sample A) and a sample of 60 baccalaureate graduates (sample B) yielded the following means and standard deviations.

Sample	$\bar{x}$	s
A	52.5	10.5
B	49.6	11.2

On the basis of these data, what should the researcher conclude? Let $\alpha = .05$.

6.3.6 A test designed to measure mothers' attitudes toward their labor and delivery experiences was given to two groups of new mothers. Sample 1 (attenders) had attended prenatal classes held at the local health department. Sample 2 (nonattenders) did not attend the classes. The sample sizes and means and standard deviations of the test scores were as follows:

Sample	n	$\bar{x}$	s
1	15	4.75	1.0
2	22	3.00	1.5

Do these data provide sufficient evidence to indicate that attenders, on the average, score higher than nonattenders? Let $\alpha = .05$.

6.3.7 Cortisol level determinations were made on two samples of women at childbirth. Group 1 subjects underwent emergency cesarean section following induced labor. Group 2 subjects delivered by either cesarean section or the vaginal route following spontaneous labor. The sample sizes, mean cortisol levels, and standard deviations were as follows:

Sample	n	$\bar{x}$	s
1	10	435	65
2	12	645	80

Do these data provide sufficient evidence to indicate a difference in the mean cortisol levels in the populations represented? Let $\alpha = .05$.

6.3.8 Protoporphyrin levels were measured in two samples of subjects. Sample 1 consisted of 50 adult male alcoholics with ring sideroblasts in the bone marrow. Sample 2 consisted of 40 apparently normal adult nonalcoholic males. The mean protoporphyrin levels and standard deviations for the two samples were as follows:

Sample	$\bar{x}$	s
1	340	250
2	45	25

Can one conclude on the basis of these data that protoporphyrin levels are higher in the represented alcoholic population than in the nonalcoholic population? Let $\alpha = .01$.

6.3.9 A researcher was interested in knowing if preterm infants with late metabolic acidosis and preterm infants without the condition differ with respect to urine levels of a certain chemical. The mean levels, standard deviations, and sample sizes for the two samples studied were as follows:

Sample	n	$\bar{x}$	s
With condition	35	8.5	5.5
Without condition	40	4.8	3.6

What should the researcher conclude on the basis of these results? Let $\alpha = .05$.

6.3.10 Researchers wished to know if they could conclude that two populations of infants differ with respect to mean age at which they walked alone. The following data (ages in months) were collected:
Sample from population A:

9.5, 10.5, 9.0, 9.75, 10.0, 13.0, 10.0, 13.5, 10.0, 9.5, 10.0, 9.75

Sample from population B:

12.5, 9.5, 13.5, 13.75, 12.0, 13.75, 12.5, 9.5, 12.0, 13.5, 12.0, 12.0

What should the researchers conclude? Let $\alpha = .05$.

6.3.11 Does sensory deprivation have an effect on a person's alpha-wave frequency? Twenty volunteer subjects were randomly divided into two groups. Subjects in group A were subjected to a 10-day period of sensory deprivation, while subjects in group B served as controls. At the end of the experimental period the alpha-wave frequency component of subjects' electroencephalograms were measured. The results were as follows:

Group A: 10.2, 9.5, 10.1, 10.0, 9.8, 10.9, 11.4, 10.8, 9.7, 10.4

Group B: 11.0, 11.2, 10.1, 11.4, 11.7, 11.2, 10.8, 11.6, 10.9, 10.9

Let $\alpha = .05$.

6.3.12 Can we conclude that, on the average, lymphocytes and tumor cells differ in size? The following are the cell diameters (μm) of 40 lymphocytes and 50 tumor cells obtained from biopsies of tissue from patients with melanoma.

Lymphocytes

9.0	9.4	4.7	4.8	8.9	4.9	8.4	5.9
6.3	5.7	5.0	3.5	7.8	10.4	8.0	8.0
8.6	7.0	6.8	7.1	5.7	7.6	6.2	7.1
7.4	8.7	4.9	7.4	6.4	7.1	6.3	8.8
8.8	5.2	7.1	5.3	4.7	8.4	6.4	8.3

Tumor cells

12.6	14.6	16.2	23.9	23.3	17.1	20.0	21.0	19.1	19.4
16.7	15.9	15.8	16.0	17.9	13.4	19.1	16.6	18.9	18.7
20.0	17.8	13.9	22.1	13.9	18.3	22.8	13.0	17.9	15.2
17.7	15.1	16.9	16.4	22.8	19.4	19.6	18.4	18.2	20.7
16.3	17.7	18.1	24.3	11.2	19.5	18.6	16.4	16.1	21.5

Let $\alpha = .05$.

6.4   PAIRED COMPARISONS

In our previous discussion involving the difference between two population means, it was assumed that the samples were independent. A method frequently employed for assessing the effectiveness of a treatment or experimental procedure is one that makes use of related observations resulting from nonindependent samples. A hypothesis test based on this type of data is known as a *paired comparisons* test.

It frequently happens that true differences do not exist between two populations with respect to the variable of interest, but the presence of extraneous sources of variation may cause rejection of the null hypothesis of no difference. On the other hand, true differences also may be masked by the presence of extraneous factors.

The objective in paired comparisons tests is to eliminate a maximum number of sources of extraneous variation by making the pairs similar with respect to as many variables as possible.

Related or paired observations may be obtained in a number of ways. The same subjects may be measured before and after receiving some treatment. Litter mates of the same sex may be randomly assigned to receive either a treatment or a placebo. Pairs of twins or siblings may be randomly assigned to two treatments in such a way that members of a single pair receive different treatments. In comparing two methods of analysis, the material to be analyzed may be equally divided so that one half is analyzed by one method and one half is analyzed by the other. Or pairs may be formed by matching individuals on some characteristic, for example, digital dexterity, which is closely related to the measurement of interest, say, posttreatment scores on some test requiring digital manipulation.

Instead of performing the analysis with individual observations, we use the difference between individual pairs of observations as the variable of interest.

When the sample differences constitute a simple random sample from a normally distributed population of differences, the test statistic for testing hypotheses about the population mean difference μ_d is

$$t = \frac{\bar{d} - \mu_d}{s_{\bar{d}}} \tag{6.4.1}$$

where $\bar{d}$ is the sample mean difference, $s_{\bar{d}} = s_d / \sqrt{n}$, and s_d is the standard deviation of the sample differences. When H_0 is true the test statistic is distributed as Student's t with $n - 1$ degrees of freedom.

Again, the hypothesized parameter in Equation 6.4.1 would be more correctly specified by a symbol such as μ_{d0}. But, again, to avoid cumbersome notation we are eliminating the zero.

Although, to begin with we have two samples—before levels and after levels—we do not have to worry about equality of variances, as with independent samples, since our variable is the difference between readings in the same individual and, hence, only one variance is involved.

Table 6.4.1

Table 6.4.1

Serum Cholesterol Levels for 12 Subjects Before and After Diet–Exercise Program

Subject	SERUM CHOLESTEROL Before (X_1)	After (X_2)	Difference (after − before) d_i
1	201	200	−1
2	231	236	+5
3	221	216	−5
4	260	233	−27
5	228	224	−4
6	237	216	−21
7	326	296	−30
8	235	195	−40
9	240	207	−33
10	267	247	−20
11	284	210	−74
12	201	209	+8

Example 6.4.1 Twelve subjects participated in an experiment to study the effectiveness of a certain diet, combined with a program of exercise, in reducing serum cholesterol levels. Table 6.4.1 shows the serum cholesterol levels for the 12 subjects at the beginning of the program (Before) and at the end of the program (After).

Concentrating on the differences, d_i, the following descriptive measures may be computed:

$$\bar{d} = \frac{\sum d_i}{n} = \frac{(-1) + (5) + (-5) + \cdots + (8)}{12} = \frac{-242}{12} = -20.17$$

$$s_d^2 = \frac{\sum (d_i - \bar{d})^2}{n-1} = \frac{n \sum d_i^2 - \left(\sum d_i \right)^2}{n(n-1)} = \frac{12(10766) - (-242)^2}{12(11)} = 535.06$$

Notice that $\bar{d} = \bar{x}_2 - \bar{x}_1$.

The question to be answered is: Do the data provide sufficient evidence for us to conclude that the diet–exercise program is effective in reducing serum cholesterol levels? We will say that such evidence is provided if we can reject the null hypothesis that the population mean change μ_d is zero or positive. We may reach a decision by means of the nine-step hypothesis testing procedure.

1. *Data* The data consist of serum cholesterol levels in 12 individuals, before and after an experimental diet–exercise program.

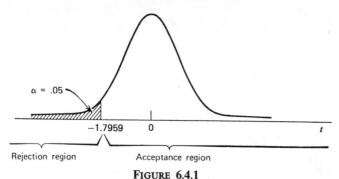

−1.7959   0 t

Rejection region   Acceptance region

FIGURE 6.4.1

Acceptance and Rejection Regions for Example 6.4.1

2. *Assumptions*   The observed differences constitute a simple random sample from a normally distributed population of differences that could be generated under the same circumstances.

3. *Hypotheses*   The null and alternate hypotheses are as follows:

$$H_0: \mu_d \geq 0$$
$$H_A: \mu_d < 0$$

4. *Test Statistic*   The appropriate test statistic is given by Equation 6.4.1.

5. *Distribution of Test Statistic*   If the null hypothesis is true, the test statistic is distributed as Student's t with $n - 1$ degrees of freedom.

6. *Decision Rule*   Let $\alpha = .05$. The critical value of t is -1.7959. Reject H_0 if computed t is less than the critical value. The acceptance and rejection regions are shown in Figure 6.4.1.

7. *Calculation of Test Statistic*

$$t = \frac{-20.17 - 0}{\sqrt{535.06/12}} = \frac{-20.17}{6.68} = -3.02$$

8. *Statistical Decision*   Reject H_0, since -3.02 is in the rejection region.

9. *Conclusion*   We may conclude that the diet–exercise program is effective. For this test, $.01 > p > .005$, since $-2.718 > -3.02 > -3.1058$.

A 95 percent confidence interval for μ_d may be obtained as follows:

$$\bar{d} \pm t_{(1-\alpha/2)} s_{\bar{d}}$$
$$-20.17 \pm 2.2010(6.68)$$
$$-20.17 \pm 14.70$$
$$-34.87, -5.47$$

If, in the analysis of paired data, the population variance of the differences is known, the appropriate test statistic is

$$z = \frac{\bar{d} - \mu_d}{\sigma_d/\sqrt{n}} \qquad (6.4.2)$$

It is unlikely that σ_d will be known in practice.

If the assumption of normally distributed d_i's cannot be made, the central limit theorem may be employed if n is large. In such cases the test statistic is Equation 6.4.2, with s_d used to estimate σ_d when, as is generally the case, the latter is unknown.

The use of the paired comparisons test is not without its problems. If different subjects are used and randomly assigned to two treatments, considerable time and expense may be involved in our trying to match individuals on one or more relevant variables. A further price we pay for using paired comparisons is a loss of degrees of freedom. If we do not use paired observations we have $2n - 2$ degrees of freedom available as compared to $n - 1$ when we use the paired comparisons procedure.

In general, in deciding whether or not to use the paired comparisons procedure, one should be guided by the economics involved as well as by a consideration of the gains to be realized in terms of controlling extraneous variation.

If neither z nor t is an appropriate test statistic for use with available data, one may wish to consider using some nonparametric technique to test a hypothesis about a median difference. The sign test, discussed in Chapter 11, is a candidate for use in such cases.

EXERCISES

In the following exercises carry out the nine-step hypothesis testing procedure at the specified significance level. Determine the p value for each test.

6.4.1 Ten experimental animals were subjected to conditions simulating disease. The number of heartbeats per minute, before and after the experiment, were recorded as follows.

Animal	HEARTBEATS PER MINUTE		Animal	HEARTBEATS PER MINUTE	
	Before	After		Before	After
1	70	115	6	100	178
2	84	148	7	110	179
3	88	176	8	67	140
4	110	191	9	79	161
5	105	158	10	86	157

Do these data provide sufficient evidence to indicate that the experimental condition increases the number of heartbeats per minute? Let $\alpha = .05$.

6.4.2 A study was conducted to see if a new therapeutic procedure is more effective than the standard treatment in improving the digital dexterity of certain handicapped persons. Twenty-four pairs of individuals were used in the study, and each pair was matched on degree of impairment, intelligence, and age. One member of each pair was randomly assigned to receive the new treatment, while the other member of the pair received the standard therapy. At the end of the experimental period each individual was given a digital dexterity test with scores as follows. Let $\alpha = .05$.

Pair	New	Standard	Pair	New	Standard
1	49	54	13	52	41
2	56	42	14	73	67
3	70	63	15	52	57
4	83	77	16	73	70
5	83	83	17	78	72
6	68	51	18	64	62
7	84	82	19	71	64
8	63	54	20	42	44
9	67	62	21	51	44
10	79	71	22	56	42
11	88	82	23	40	35
12	48	50	24	81	73

6.4.3 A group of 15 twelve-year-old boys were measured by two different nurses. The results were as follows:

Subject	Nurse 1	Nurse 2	Subject	Nurse 1	Nurse 2
1	142.9	143.0	9	142.1	142.5
2	150.9	151.5	10	159.9	160.0
3	151.9	152.1	11	141.9	142.0
4	158.1	158.0	12	140.8	141.0
5	151.2	151.5	13	147.1	148.0
6	160.2	160.5	14	143.6	144.0
7	157.8	158.0	15	139.9	141.0
8	150.1	150.0			

Do these data justify the conclusion that there is a difference in the accuracy of the two nurses? Let $\alpha = .05$. Construct a 95 percent confidence interval for μ_d.

6.4.4 Ten 16-year-old males were measured at time of arising in the morning and at time of retiring in the evening with the following results.

	HEIGHT (cm)			HEIGHT (cm)	
Subject	Morning	Evening	Subject	Morning	Evening
1	169.7	168.2	6	168.8	166.5
2	168.5	165.5	7	169.2	167.4
3	165.9	164.4	8	167.9	166.3
4	177.7	175.7	9	181.8	179.7
5	179.6	176.6	10	163.3	161.5

Do these data provide sufficient evidence to indicate that 16-year-old males are shorter in the evening than in the morning? Construct the 95 percent confidence interval for μ_d.

6.4.5 Researchers wish to know if they can conclude ($\alpha = .05$) that cranial blood flow (CBF) in healthy newborn babies differs with sleepstate. Data were collected on 20 subjects during active sleep and during quiet sleep. The results were as follows:

	CBF DURING			CBF DURING	
Subject	Active sleep	Quiet sleep	Subject	Active sleep	Quiet sleep
1	38.8	26.8	11	55.3	44.1
2	51.3	34.8	12	47.4	46.1
3	43.8	31.8	13	32.5	26.5
4	64.9	56.6	14	60.6	53.2
5	29.8	29.0	15	32.0	30.6
6	43.4	37.2	16	60.6	53.2
7	44.8	36.3	17	45.7	32.1
8	33.9	25.2	18	63.0	49.2
9	62.7	42.2	19	69.9	51.9
10	40.1	29.3	20	33.6	28.7

6.5 HYPOTHESIS TESTING:
A SINGLE POPULATION PROPORTION

Testing hypotheses about population proportions is carried out in much the same way as for means when the conditions necessary for using the normal curve are met. One-sided or two-sided tests may be made, depending on the question being asked. When a sample sufficiently large for application of the central limit theorem is available for analysis, the test statistic is

$$z = \frac{\hat{p} - p_0}{\sqrt{\dfrac{p_0 q_0}{n}}} \qquad (6.5.1)$$

which, when H_0 is true, is distributed approximately as the unit normal.

Example 6.5.1 Suppose we are interested in knowing what proportion of automobile drivers regularly wear seat belts. In a survey of 300 adult drivers, 123 said they regularly wear seat belts. Can we conclude from these data that in the sampled population the proportion who regularly wear seat belts is not .50?

1. *Data*   The data are obtained from the responses of 300 individuals of which 123 possessed the characteristic of interest, that is, $\hat{p} = .41$.

2. *Assumptions*   The sampling distribution of $\hat{p}$ is approximately normally distributed in accordance with the central limit theorem. If H_0 is true, $p = .5$ and the standard error, $\sigma_{\hat{p}} = \sqrt{(.5)(.5)/300}$. Note that we use the hypothesized value of p in computing $\sigma_{\hat{p}}$. We do this because the entire test is based on the assumption that the null hypothesis is true. To use the sample proportion, $\hat{p}$, in computing $\sigma_{\hat{p}}$ would not be consistent with this concept.

3. *Hypotheses*

$$H_0: p = .50$$
$$H_A: p \neq .50$$

4. *Test Statistic*   The test statistic is given by Equation 6.5.1.

5. *Distribution of Test Statistic*   If the null hypothesis is true the test statistic is approximately normally distributed with a mean of zero.

6. *Decision Rule*   Let $\alpha = .05$. Critical values of z are ± 1.96. Reject H_0 unless $-1.96 < z_{\text{computed}} < 1.96$.

7. *Calculation of Test Statistic*

$$z = \frac{.41 - .50}{\sqrt{\dfrac{(.5)(.5)}{300}}} = \frac{-.09}{.0289} = -3.11$$

8. *Statistical Decision*   Reject H_0 since $-3.11 < -1.96$.

9. *Conclusion*   We conclude that in the population the proportion who regularly wear seat belts is not .50. For this test, $p < .002$, since $-3.11 < -3.09$.

EXERCISES

For each of the following exercises, carry out the nine-step hypothesis testing procedure at the designated level of significance. Compute the p value for each test.

6.5.1 In a sample of 1500 residents of an inner city neighborhood who participated in a health screening program, 125 tests yielded positive results for sickle-cell anemia. Do these data provide sufficient evidence to indicate that the proportion of individuals with sickle-cell anemia in the sampled population is greater than .06? Let $\alpha = .05$.

6.5.2 A sample of 100 hospital employees who have frequent contact with blood or blood products was screened for serologic evidence of hepatitis B infection. Twenty-three were found to be positive. Can it be concluded from these data that the proportion of positives in the sampled population is greater than .15? Let $\alpha = .05$.

6.5.3 Prior to initiation of a rubella immunization program in a metropolitan county a survey revealed that 150 out of a sample of 500 elementary school children in the county had been immunized against the disease. Are these data compatible with the contention that 50 percent of the elementary school children in the county had been immunized against rubella? Let $\alpha = .05$.

6.5.4 The following questionnaire was completed by a simple random sample of 250 gynecologists. The number checking each response category is shown in the appropriate box.

(1) When you have a choice, which procedure do you prefer for obtaining samples of endometrium?

A. Dilation and curettage | 175 |

B. Vobra aspiration | 75 |

(2) Have you seen one or more pregnant women during the past year whom you knew to have elevated blood lead levels?

A. Yes | 25 |

B. No | 225 |

(3) Do you routinely acquaint your pregnant patients who smoke with the suspected hazards of smoking to the fetus?

A. Yes | 238 |

B. No | 12 |

Can we conclude from these data that in the sampled population more than 60 percent prefer dilation and curettage for obtaining samples of endometrium? Let $\alpha = .01$.

6.5.5 Refer to Exercise 6.5.4. Can we conclude from these data that in the sampled population fewer than 15 percent have seen (during the past year) one or more pregnant women with elevated blood lead levels? Let $\alpha = .05$.

6.5.6 Refer to Exercise 6.5.4. Can we conclude from these data that more than 90 percent acquaint their pregnant patients who smoke with the suspected hazards of smoking to the fetus? Let $\alpha = .05$.

6.6 HYPOTHESIS TESTING: THE DIFFERENCE BETWEEN TWO POPULATION PROPORTIONS

The most frequent test employed relative to the difference between two population proportions is that their difference is zero. It is possible, however, to test that the difference is equal to some other value. Both one-sided and two-sided tests may be made.

When the null hypothesis to be tested is $p_1 - p_2 = 0$, we are hypothesizing that the two population proportions are equal. We use this as justification for combining the results of the two samples to come up with a pooled estimate of the hypothesized common proportion. If this procedure is adopted, one computes

$$\bar{p} = \frac{x_1 + x_2}{n_1 + n_2}$$

where x_1 and x_2 are the numbers in the first and second samples, respectively, possessing the characteristic of interest. This pooled estimate of $p = p_1 = p_2$ is used in computing $\hat{\sigma}_{\hat{p}_1 - \hat{p}_2}$, the estimated standard error of the estimator, as follows

$$\hat{\sigma}_{\hat{p}_1 - \hat{p}_2} = \sqrt{\frac{\bar{p}(1 - \bar{p})}{n_1} + \frac{\bar{p}(1 - \bar{p})}{n_2}} \qquad (6.6.1)$$

The test statistic becomes

$$z = \frac{(\hat{p}_1 - \hat{p}_2) - (p_1 - p_2)}{\hat{\sigma}_{\hat{p}_1 - \hat{p}_2}} \qquad (6.6.2)$$

which is approximately distributed as the unit normal if the null hypothesis is true.

Example 6.6.1 In a study designed to compare a new treatment for migraine headache with the standard treatment, 78 of 100 subjects who received the standard treatment responded favorably. Of the 100 subjects who received the new treatment 90 responded favorably. Do these data provide sufficient evidence to indicate that the new treatment is more effective than the standard? The answer is yes if we can reject the null hypothesis that the new treatment is no more effective than the standard.

The hypothesis testing procedure for this example consists of the following steps.

1. *Data* The data consist of the responses of 100 individuals to the standard treatment and the responses of 100 individuals to the new treatment. The numbers of favorable responses were 78 and 90, respectively. Let us compute

$\hat{p}_1 = 78/100 = .78$, $\hat{p}_2 = 90/100 = .90$, and

$$\bar{p} = \frac{90 + 78}{100 + 100} = .84$$

2. *Assumptions*   It is assumed that the sampling distribution of $\hat{p}_2 - \hat{p}_1$ is approximately normally distributed with mean $p_2 - p_1 \leq 0$ and estimated standard error given by Equation 6.6.1 when the null hypothesis is true and the sample estimates are pooled.

3. *Hypotheses*

$$H_0: p_2 - p_1 \leq 0$$
$$H_A: p_2 - p_1 > 0$$

4. *Test Statistic*   The test statistic is given by Equation 6.6.2.

5. *Distribution of the Test Statistic*   If the null hypothesis is true the test statistic is distributed approximately as the unit normal.

6. *Decision Rule*   Let $\alpha = .05$. The critical value of z is 1.645. Reject H_0 if computed z is greater than 1.645.

7. *Calculation of Test Statistic*

$$z = \frac{(.90 - .78)}{\sqrt{\dfrac{(.84)(.16)}{100} + \dfrac{(.84)(.16)}{100}}} = \frac{.12}{.0518} = 2.32$$

8. *Statistical Decision*   Reject H_0 since $2.32 > 1.645$.

9. *Conclusion*   These data suggest that the new treatment is more effective than the standard ($p = .0102$).

EXERCISES

In each of the exercises below use the nine-step hypothesis testing procedure. Determine the p value for each test.

   6.6.1   A sample of patients discharged from a state mental hospital within six months after admission and a sample of those who were discharged after six months but within a year of admission were compared with respect to distance of residence from the hospital. The results were as follows:

Length of stay	n	Number who live more than 20 mi from hospital
$\leq$ 6 mo.	100	75
Longer than 6 mo.	150	90

Do these data provide sufficient evidence to indicate that the proportion who live more than 20 miles from the hospital is different in the two represented populations? Let $\alpha = .05$.

6.6.2 An epidemiologist in a developing country compared a sample of 90 adult subjects suffering from a certain neurologic disease with a sample of 100 comparable control subjects who were free of the disease. It was found that 69 of the subjects with the disease and 67 of the controls were employed in subsistence occupations. Can the epidemiologist conclude from these data that the two populations represented by the samples differ with respect to the proportion employed in subsistence occupations? Let $\alpha = .05$.

6.6.3 We want to know if children in two ethnic groups differ with respect to the proportion who are anemic. A sample of one-year-old children seen in a certain group of county health departments during a year was selected from each of the two predominant ethnic groups composing the departments' clientele. The following information regarding anemia was revealed.

Ethnic group	Number in sample	Number anemic
1	450	105
2	375	120

Do these data provide sufficient evidence to indicate a difference in the two populations with respect to the proportion who are anemic? Let $\alpha = .05$.

6.6.4 In a study of obesity the following results were obtained from samples of males and females between the ages of 20 and 75:

	n	Number overweight
Males	150	21
Females	200	48

Can we conclude from these data that in the sampled populations there is a difference in the proportions who are overweight? Let $\alpha = .05$.

6.7 HYPOTHESIS TESTING: A SINGLE POPULATION VARIANCE

In Section 5.9 we examine how it is possible to construct a confidence interval for the variance of a normally distributed population. The general principles presented in that section may be employed to test a hypothesis about a population variance. When the data available for analysis consist of a simple random sample drawn from a normally distributed population, the test statistic for testing hypotheses about a population variance is

$$\chi^2 = (n - 1)s^2/\sigma^2 \tag{6.7.1}$$

which, when H_0 is true, is distributed as χ^2 with $n - 1$ degrees of freedom.

Example 6.7.1 A simple random sample of 15 nursing students who participated in an experiment took a test to measure manual dexterity. The variance of the sample observations was 1225. We want to know if we can conclude from these data that the population variance is different from 2500.

1. *Data* The data consist of manual dexterity test scores of 15 student nurses. The sample variance was computed as $s^2 = 1225$.
2. *Assumptions* The data constitute a random sample of size 15 from a normally distributed population.
3. *Hypotheses*

$$H_0: \sigma^2 = 2500$$

$$H_A: \sigma^2 \neq 2500$$

4. *Test Statistic* The test statistic is given by Equation 6.7.1.
5. *Distribution of Test Statistic* When the null hypothesis is true the test statistic is distributed as χ^2 with $n - 1$ degrees of freedom.
6. *Decision Rule* Let $\alpha = .05$. Critical values of χ^2 are 5.629 and 26.119. Reject H_0 unless the computed value of the test statistic is between 5.629 and 26.119. The acceptance and rejection regions are shown in Figure 6.7.1.
7. *Calculation of Test Statistic*

$$\chi^2 = \frac{(14)(1225)}{2500} = 6.86$$

8. *Statistical Decision* Do not reject H_0 since $5.629 < 6.86 < 26.119$.

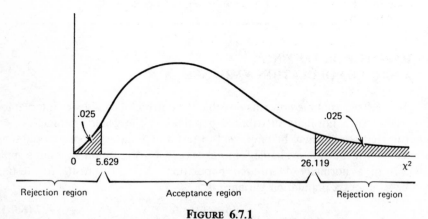

FIGURE 6.7.1

Acceptance and Rejection Regions for Example 6.7.1

9. *Conclusion*   Based on these data we are unable to conclude that the population variance is not 2500.

Although this was an example of a two-sided test, one-sided tests may also be made by logical modification of the procedure given here.

The determination of the p value for this test is complicated by the fact that we have a two-sided test and an asymmetric sampling distribution. When we have a two-sided test and a symmetric sampling distribution such as the unit normal or t, we may, as we have seen, double the one-sided p value. Problems arise when we attempt to do this with an asymmetric sampling distribution such as the chi-square distribution. Gibbons and Pratt (1) suggest that in this situation the one-sided p-value be reported along with the direction of the observed departure from the null hypothesis. In fact, this procedure may be followed in the case of symmetric sampling distributions. Precedent, however, seems to favor doubling the one-sided p value when the test is two-sided and involves a symmetric sampling distribution.

For the present example, then, we may report the p value as follows:

$p > .05$ (two-sided test). A population variance less than 2500 is suggested by the sample data, but this hypothesis is not strongly supported by the test.

If the problem is stated in terms of the population standard deviation, one may square the sample standard deviation and perform the test as indicated above.

EXERCISES

In each of the following exercises carry out the nine-step testing procedure. Determine the p value for each test.

6.7.1   As part of a research project, a sample of 25 infants was selected from those born in hospitals in a metropolitan area. The standard deviation of the weights of the infants in the sample was 150 grams. Do these data provide sufficient evidence to indicate that the population variance is greater than 10,000? Let $\alpha = .05$.

6.7.2   Each of a random sample of 30 student nurses who participated in a research project was given a test designed to measure degree of creative thinking. The standard deviation of the scores was 11. Can one conclude from these data that the population variance is less than 400? Let $\alpha = .05$.

6.7.3   Vital capacity values were recorded for a sample of 10 patients with severe chronic airway obstruction. The variance of the 10 observations was .75. Test the null hypothesis that the population variance is 1.00. Let $\alpha = .05$.

6.7.4   Hemoglobin (gm %) values were recorded for a sample of 20 children who were part of a study of acute leukemia. The variance of the observations was

5. Do these data provide sufficient evidence to indicate that the population variance is greater than 4? $\alpha = .05$.

6.7.5 A sample of 25 administrators of large hospitals participated in a study to investigate the nature and extent of frustration and emotional tension associated with the job. Each participant was given a test designed to measure the extent of emotional tension he or she experienced as a result of the duties and responsibilities associated with the job. The variance of the scores was 30. Can it be concluded from these data that the population variance is greater than 25? Let $\alpha = .05$.

6.7.6 In a study in which the subjects were 15 patients suffering from pulmonary sarcoid disease, blood gas determinations were made. The variance of the PaO_2 (mm Hg) values was 450. Test the null hypothesis that the population variance is greater than 250. Let $\alpha = .05$.

6.7.7 Analysis of the amniotic fluid from a simple random sample of 15 pregnant women yielded the following measurements on total protein (grams per 100 ml) present:

$$.69, 1.04, .39, .37, .64, .73, .69, 1.04$$

$$.83, 1.00, .19, .61, .42, .20, .79$$

Do these data provide sufficient evidence to indicate that the population variance is greater than .05? Let $\alpha = .05$. What assumptions are necessary?

6.8 HYPOTHESIS TESTING: THE RATIO OF TWO POPULATION VARIANCES

As we have seen, the use of the t distribution in constructing confidence intervals and in testing hypotheses for the difference between two population means assumes that the population variances are equal. As a rule, the only hints available about the magnitudes of the respective variances are the variances computed from samples taken from the populations. We would like to know if the difference that, undoubtedly, will exist between the sample variances is indicative of a real difference in population variances, or if the difference is of such magnitude that it could have come about as a result of chance alone when the population variances are equal.

Two methods of chemical analysis may give the same results on the average. It may be, however, that the results produced by one method are more variable than the results of the other. We would like some method of determining whether this is likely to be true.

These are two examples of the many instances in which interest focuses on the question of whether or not two population variances are equal.

Decisions regarding the comparability of two population variances are usually based on the *variance ratio test*, which is a test of the null hypothesis that two population variances are equal. When we test the hypothesis that two population variances are equal, we are, in effect, testing the hypothesis that their ratio is equal to 1.

We learned in the preceding chapter that, when certain assumptions are met, the quantity $(s_1^2/\sigma_1^2)/(s_2^2/\sigma_2^2)$ is distributed as F with $n_1 - 1$ numerator degrees of freedom and $n_2 - 1$ denominator degrees of freedom. If we are hypothesizing that $\sigma_1^2 = \sigma_2^2$, we assume that the hypothesis is true, and they cancel out in the above expression leaving s_1^2/s_2^2, which follows the same F distribution. The ratio s_1^2/s_2^2 will be designated V.R. for variance ratio.

For a two-sided test, we follow the convention of placing the larger sample variance in the numerator and obtaining the critical value of F for $\alpha/2$ and the appropriate degrees of freedom. However, for a one-sided test, which of the two sample variances is to be placed in the numerator is predetermined by the statement of the null hypothesis. For example, for the null hypothesis that $\sigma_1^2 \leq \sigma_2^2$, the appropriate test statistic is V.R. $= s_1^2/s_2^2$. The critical value of F is obtained for α (not $\alpha/2$) and the appropriate degrees of freedom. In like manner, if the null hypothesis is that $\sigma_1^2 \geq \sigma_2^2$, the appropriate test statistic is V.R. $= s_2^2/s_1^2$. In all cases, the decision rule is to reject the null hypothesis if the computed V.R. is equal to or greater than the critical value of F.

Example 6.8.1   A random sample of 22 physical therapy students took the same manual dexterity test as the nurses mentioned in Example 6.7.1. The sample variance for physical therapy students was 1600. We wish to know if these data provide sufficient evidence for us to conclude that the variance of manual dexterity test scores for the population represented by the physical therapy students is greater than the variance for the population represented by the sample of nursing students.

1. *Data*   The sample variances were 1600 and 1225, respectively.

2. *Assumptions*   The data constitute independent random samples each drawn from a normally distributed population. These are general assumptions that must be met for the following test to be valid.

3. *Hypotheses*

$$H_0: \sigma_1^2 \leq \sigma_2^2$$

$$H_A: \sigma_1^2 > \sigma_2^2$$

4. *Test Statistic*

$$V.\,R. = s_1^2/s_2^2$$

5. *Distribution of Test Statistic*   When the null hypothesis is true, the test statistic is distributed as F with $n_1 - 1$ and $n_2 - 1$ degrees of freedom.

6. *Decision Rule*   Let $\alpha = .05$. The critical value of F, from Table G, is 2.39. Note that Table G does not contain an entry for 21 numerator degrees of freedom and, therefore, 2.39 is obtained by using 20, the closest value to 21

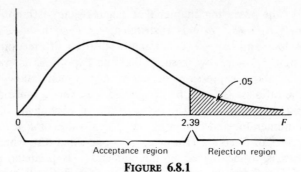

FIGURE 6.8.1

Acceptance and Rejection Regions, Example 6.8.1

in the table. Reject H_0 if V.R. ≥ 2.39. The acceptance and rejection regions are shown in Figure 6.8.1.

7. *Calculation of Test Statistic*

$$\text{V. R.} = \frac{1600}{1225} = 1.31$$

8. *Statistical Decision* We cannot reject H_0, since $1.31 < 2.39$; that is, the computed ratio falls in the acceptance region.

9. *Conclusion* We are unable to conclude that the two variances are not equal. Since the computed V.R. of 1.31 is less than 1.96, the p value for this test is greater than .10.

EXERCISES

In the following exercises perform the nine-step test. Determine the p value for each test.

6.8.1 Cardiac index values (liters/minute/M^2) were analyzed in two groups of patients following prosthetic valve replacement. The sample sizes and variances were as follows: $n_1 = 16$, $s_1^2 = 3.75$, $n_2 = 10$, $s_2^2 = 1.8$. Do these data provide sufficient evidence to indicate a difference in population variances? Let $\alpha = .05$.

6.8.2 Viable *Toxoplasma gondii* organisms were injected into 21 normal guinea pigs and 21 guinea pigs chronically infected with toxoplasma. Twenty-four hours later erythema diameter measurements were taken at the reaction site. Sample variances computed from these observations were 10 mm^2 for the infected group and 3 mm^2 for the normal group. Can it be concluded from these data that the population variance for the infected group is greater than that for the normal group? Let $\alpha = .05$.

6.8.3 A test designed to measure level of anxiety was administered to a sample of male and a sample of female patients just prior to undergoing the same surgical procedure. The sample sizes and the variances computed from the scores were as follows:

$$\text{Males:} \quad n = 16, \quad s^2 = 150$$

$$\text{Females:} \quad n = 21, \quad s^2 = 275$$

Do these data provide sufficient evidence to indicate that in the represented populations, the scores made by females are more variable than those made by males? Let $\alpha = .05$.

6.8.4 In an experiment to assess the effects on rats of exposure to cigarette smoke, 11 animals were exposed and 11 control animals were not exposed to smoke from unfiltered cigarettes. At the end of the experiment measurements were made of the frequency of the ciliary beat (beats/min at 20°C) in each animal. The variance for the exposed group was 3400 and 1200 for the unexposed group. Do these data indicate that in the populations represented the variances are different? Let $\alpha = .05$.

6.8.5 Two pain-relieving drugs were compared for effectiveness on the basis of length of time elapsing between administration of the drug and cessation of pain. Thirteen patients received drug 1 and 13 received drug 2. The sample variances were $s_1{}^2 = 64$ and $s_2{}^2 = 16$. Test the null hypothesis that the two population variances are equal. Let $\alpha = .05$.

6.8.6 Packed cell volume determinations were made on two groups of children with cyanotic congenital heart disease. The sample sizes and variances were as follows:

Group	n	s^2
1	10	40
2	16	85

Do these data provide sufficient evidence to indicate that the variance of population 2 is larger than the variance of population 1? Let $\alpha = .05$.

6.8.7 Independent simple random samples from two strains of mice used in an experiment yielded the following measurements on plasma glucose levels following a traumatic experience:

Strain A: 54, 99, 105, 46, 70, 87, 55, 58, 139, 91

Strain B: 93, 91, 93, 150, 80, 104, 128, 83, 88, 95, 94, 97

Do these data provide sufficient evidence to indicate that the variance is larger in the population of strain A mice than in the population of strain B mice? Let $\alpha = .05$. What assumptions are necessary?

6.9 SUMMARY

In this chapter the general concepts of hypothesis testing are discussed. A general procedure for carrying out a hypothesis test consisting of the following nine steps is suggested.

1. Description of data.
2. Statement of necessary assumptions.
3. Statement of null and alternate hypotheses.
4. Specification of the test statistic.
5. Specification of the distribution of the test statistic.
6. Statement of the decision rule.
7. Calculation of test statistic from sample data.
8. The statistical decision based on sample results.
9. Conclusion.

A number of specific hypothesis tests are described in detail and are illustrated with appropriate examples. These include tests concerning population means, the difference between two population means, paired comparisons, population proportions, the difference between two population proportions, a population variance, and the ratio of two population variances.

REVIEW QUESTIONS AND EXERCISES

1. What is the purpose of hypothesis testing?
2. What is a hypothesis?
3. List and explain each step in the nine-step hypothesis testing procedure.
4. What is a type I error?
5. What is a type II error?

6. Explain how one decides what statement goes into the null hypothesis and what statement goes into the alternative hypothesis.

7. What are the assumptions underlying the use of the t statistic in testing hypotheses about a single mean? The difference between two means?

8. When may the z statistic be used in testing hypotheses about
   (a) A single population mean?
   (b) The difference between two population means?
   (c) A single population proportion?
   (d) The difference between two population proportions?

9. In testing a hypothesis about the difference between two population means, what is the rationale behind pooling the sample variances?

10. Explain the rationale behind the use of the paired comparisons test.

11. Give an example from your field of interest where a paired comparisons test would be appropriate. Use real or realistic data and perform an appropriate hypothesis test.

12. Give an example from your field of interest where it would be appropriate to test a hypothesis about the difference between two population means. Use real or realistic data and carry out the nine-step hypothesis testing procedure.

13. Do Exercise 12 for a single population mean.

14. Do Exercise 12 for a single population proportion.

15. Do Exercise 12 for the difference between two population proportions.

16. Do Exercise 12 for a population variance.

17. Do Exercise 12 for the ratio of two population variances.

18. Researchers wish to know if urban and rural adult residents of a developing country differ with respect to the prevalence of blindness. A survey revealed the following information:

Group	Number in sample	Number blind
Rural	300	24
Urban	500	15

Do these data provide sufficient evidence to indicate a difference in the prevalence of blindness in the two populations? Let $\alpha = .05$. Determine the p value.

19. During an experiment using laboratory animals the following data on renal cortical blood flow during control conditions and during the administration of a certain anesthetic were recorded.

RENAL CORTICAL BLOOD FLOW
(ml / g / min)

Animal number	Control	During administration of anesthetic
1	2.35	2.00
2	2.55	1.71
3	1.95	2.22
4	2.79	2.71
5	3.21	1.83
6	2.97	2.14
7	3.44	3.72
8	2.58	2.10
9	2.66	2.58
10	2.31	1.32
11	3.43	3.70
12	2.37	1.59
13	1.82	2.07
14	2.98	2.15
15	2.53	2.05

Can one conclude on the basis of these data that the anesthetic retards renal cortical blood flow? Let $\alpha = .05$. Determine the p value.

20. An allergy research team conducted a study in which two groups of subjects were used. As part of the research, blood eosinophil determinations were made on each subject with the following results:

Group	n	Mean eosinophil value (no./ mm^3)	s
A	14	584	225
B	16	695	185

Do these data provide sufficient evidence to indicate that the population means are different? Let $\alpha = .05$. Determine the p value.

21. A survey of 90 recently delivered women on the rolls of a county welfare department revealed that 27 had a history of intrapartum or postpartum infection. Test the null hypothesis that the population proportion with a history of intrapartum or postpartum infection is less than or equal to .25. Let $\alpha = .05$. Determine the p value.

22. In a sample of 150 hospital emergency admissions with a certain diagnosis, 128 listed vomiting as a presenting symptom. Do these data provide sufficient to indicate, at the .01 level of significance, that the population proportion is less than .92? Determine the p value.

23. A research team measured tidal volume in 15 experimental animals. The mean and standard deviation were 45 and 5 cc, respectively. Do these data provide sufficient evidence to indicate that the population mean is greater than 40 cc? Let $\alpha = .05$.

24. A sample of eight patients admitted to a hospital with a diagnosis of biliary cirrhosis had a mean IgM level of 160.55 units per milliliter. The sample standard deviation was 50. Do these data provide sufficient evidence to indicate that the population mean is greater than 150? Let $\alpha = .05$. Determine the p value.

25. Some researchers have observed a greater airway resistance in smokers than in nonsmokers. Suppose a study, conducted to compare the percent tracheobronchial retention of particles in smoking-discordant monozygotic twins, yielded the following results:

PERCENT RETENTION

Smoking twin	Nonsmoking twin
60.6	47.5
12.0	13.3
56.0	33.0
75.2	55.2
12.5	21.9
29.7	27.9
57.2	54.3
62.7	13.9
28.7	8.9
66.0	46.1
25.2	29.8
40.1	36.2

Do these data support the hypothesis that tracheobronchial clearance is slower in smokers? Let $\alpha = .05$. Determine the p value for this test.

26. Researchers would like to know if psychologically stressful situations cause an increase in serum creatine phosphokinase (CPK) levels among apparently healthy individuals. The following PCK data (IU/liter) are available for a population of subjects. Determinations were made while subjects were experiencing psychological stress (A) and at a later date when subjects were enjoying a period of apparent psychological calm (B). To help the researchers reach a decision select a simple random sample from this population, perform an appropriate analysis of the sample data, and give a narrative report of your findings and conclusions. Compare your results with those of your classmates.

Subject	CONDITION A	CONDITION B	Subject	CONDITION A	CONDITION B	Subject	CONDITION A	CONDITION B
1.	193	250	2.	90	173	3.	120	135
4.	154	49	5.	149	83	6.	146	123
7.	180	126	8.	128	177	9.	180	164
10.	66	121	11.	129	200	12.	170	75
13.	121	182	14.	135	169	15.	133	234
16.	165	67	17.	125	53	18.	129	119
19.	118	241	20.	109	138	21.	89	113
22.	115	112	23.	119	104	24.	200	115
25.	158	224	26.	130	138	27.	197	223
28.	165	172	29.	80	68	30.	161	164
31.	189	232	32.	175	106	33.	120	266
34.	117	190	35.	44	153	36.	102	59
37.	174	119	38.	99	93	39.	126	162
40.	47	130	41.	132	89	42.	177	197
43.	144	189	44.	177	207	45.	161	109
46.	84	147	47.	180	193	48.	198	110
49.	158	87	50.	116	135	51.	176	115
52.	111	159	53.	109	115	54.	181	114
55.	108	202	56.	57	134	57.	155	64
58.	239	79	59.	103	96	60.	158	181
61.	240	91	62.	136	167	63.	182	200
64.	176	236	65.	113	126	66.	158	185
67.	118	141	68.	184	180	69.	116	60
70.	124	137	71.	173	195	72.	241	172
73.	141	116	74.	150	229	75.	172	142
76.	179	131	77.	146	155	78.	89	152
79.	149	246	80.	140	169	81.	100	99
82.	124	163	83.	90	124	84.	99	189
85.	208	79	86.	225	76	87.	203	154
88.	130	182	89.	99	100	90.	209	140
91.	185	115	92.	190	190	93.	148	193
94.	88	146	95.	141	143	96.	62	163
97.	256	157	98.	198	161	99.	150	192
100.	145	105	101.	201	162	102.	164	201
103.	207	203	104.	178	179	105.	145	132
106.	139	158	107.	60	147	108.	125	147
109.	140	161	110.	148	58	111.	59	91
112.	128	151	113.	135	183	114.	173	153
115.	82	130	116.	175	126	117.	140	176
118.	140	56	119.	187	160	120.	165	183
121.	126	157	122.	142	132	123.	167	183
124.	120	201	125.	154	122	126.	125	159
127.	131	115	128.	119	150	129.	129	93
130.	111	104	131.	150	228	132.	153	184
133.	55	108	134.	118	159	135.	131	134
136.	171	134	137.	137	86	138.	175	102

Subject	CONDITION		Subject	CONDITION		Subject	CONDITION	
	A	B		A	B		A	B
139.	60	164	140.	202	133	141.	175	164
142.	140	175	143.	133	93	144.	104	137
145.	129	130	146.	118	158	147.	265	160
148.	171	211	149.	110	71	150.	114	215
151.	153	116	152.	110	119	153.	190	154
154.	187	139	155.	131	169	156.	115	14
157.	130	74	158.	182	130	159.	152	105
160.	192	145	161.	64	179	162.	148	143
163.	182	220	164.	135	213	165.	202	124
166.	84	167	167.	121	121	168.	90	162
169.	187	133	170.	137	72	171.	138	88
172.	187	193	173.	137	96	174.	204	157
175.	148	102	176.	185	157	177.	120	145
178.	125	113	179.	196	185	180.	207	79
181.	145	154	182.	83	184	183.	174	90
184.	118	146	185.	136	175	186.	123	176
187.	86	172	188.	102	109	189.	102	136
190.	150	102	191.	222	148	192.	174	96
193.	179	210	194.	188	21	195.	75	116
196.	96	152	197.	152	187	198.	150	149
199.	130	168	200.	161	134	201.	77	213
202.	103	55	203.	127	152	204.	146	74
205.	130	24	206.	200	106	207.	90	192
208.	83	111	209.	122	112	210.	122	178
211.	233	176	212.	154	150	213.	125	194
214.	164	193	215.	187	143	216.	192	157
217.	99	154	218.	97	69	219.	140	69
220.	137	181	221.	178	223	222.	207	121
223.	123	187	224.	151	155	225.	206	159
226.	102	150	227.	166	55	228.	171	147
229.	154	148	230.	197	112	231.	111	103
232.	148	110	233.	78	140	234.	192	124
235.	239	101	236.	131	120	237.	67	183
238.	42	178	239.	121	80	240.	180	126
241.	70	164	242.	81	193	243.	167	182
244.	119	146	245.	109	119	246.	175	184
247.	152	162	248.	94	120	249.	64	186
250.	127	193	251.	77	210	252.	190	184
253.	169	160	254.	158	53	255.	111	122
256.	123	185	257.	39	147	258.	152	177
259.	211	159	260.	149	70	261.	137	89
262.	79	309	263.	127	132	264.	116	166
265.	150	190	266.	130	169	267.	59	122
268.	205	136	269.	149	78	270.	149	243
271.	177	162	272.	147	174	273.	153	179
274.	185	119	275.	140	171	276.	119	90

Subject	Condition A	B	Subject	Condition A	B	Subject	Condition A	B
277.	88	78	278.	137	131	279.	173	175
280.	59	113	281.	199	122	282.	160	129
283.	80	182	284.	122	107	285.	131	147
286.	73	115	287.	229	136	288.	242	124
289.	97	138	290.	127	85	291.	132	70
292.	84	151	293.	120	155	294.	161	203
295.	137	91	296.	194	143	297.	178	161
298.	205	178	299.	114	77	300.	120	195
301.	93	181	302.	125	91	303.	114	99
304.	190	128	305.	136	129	306.	114	138
307.	109	190	308.	164	178	309.	70	173
310.	56	67	311.	156	116	312.	72	159
313.	74	140	314.	74	162	315.	126	90
316.	53	154	317.	158	178	318.	182	157
319.	101	192	320.	127	153	321.	114	171
322.	201	120	323.	137	178	324.	79	240
325.	181	122	326.	86	52	327.	109	126
328.	120	187	329.	101	112	330.	184	192
331.	169	144	332.	125	82	333.	169	91
334.	123	113	335.	170	151	336.	83	155
337.	121	150	338.	121	216	339.	199	83
340.	94	138	341.	195	199	342.	92	126
343.	212	71	344.	164	80	345.	216	152
346.	178	181	347.	219	115	348.	142	179
349.	143	175	350.	104	134	351.	124	139
352.	122	165	353.	158	13	354.	129	183
355.	83	137	356.	144	101	357.	183	202
358.	132	139	359.	134	163	360.	97	114
361.	179	106	362.	141	150	363.	98	188
364.	191	62	365.	181	130	366.	165	275
367.	119	184	368.	161	138	369.	103	154
370.	159	37	371.	130	19	372.	115	216
373.	120	128	374.	189	17	375.	125	224
376.	125	169	377.	190	128	378.	127	140
379.	105	157	380.	98	118	381.	154	152
382.	113	140	383.	68	146	384.	191	138
385.	190	99	386.	191	115	387.	153	157
388.	141	178	389.	143	142	390.	71	218
391.	135	117	392.	153	212	393.	145	193
394.	117	101	395.	163	100	396.	185	163
397.	213	221	398.	96	100	399.	39	95
400.	226	132	401.	88	149	402.	151	167
403.	140	101	404.	124	116	405.	165	124
406.	50	149	407.	159	222	408.	131	101
409.	154	218	410.	140	132	411.	126	202
412.	79	76	413.	120	151	414.	163	56

Subject	**Condition** A	B	Subject	**Condition** A	B	Subject	**Condition** A	B
415.	110	70	416.	147	257	417.	153	172
418.	191	121	419.	217	155	420.	102	154
421.	242	117	422.	179	219	423.	115	66
424.	76	125	425.	156	230	426.	205	224
427.	114	151	428.	96	108	429.	99	155
430.	170	156	431.	141	196	432.	162	131
433.	179	90	434.	205	140	435.	114	149
436.	165	44	437.	121	193	438.	130	139
439.	250	188	440.	90	190	441.	95	211
442.	163	171	443.	157	164	444.	92	135
445.	173	123	446.	98	174	447.	155	123
448.	192	92	449.	152	142	450.	158	259
451.	189	142	452.	120	147	453.	131	157
454.	159	87	455.	98	164	456.	124	217
457.	83	156	458.	148	201	459.	131	140
460.	186	134	461.	151	146	462.	88	168
463.	149	130	464.	89	189	465.	104	168
466.	148	143	467.	103	83	468.	106	139
469.	69	86	470.	49	121	471.	173	160
472.	163	114	473.	207	202	474.	134	156
475.	125	204	476.	73	170	477.	160	160
478.	205	183	479.	213	57	480.	254	142
481.	67	133	482.	172	237	483.	138	82
484.	158	170	485.	221	149	486.	76	102
487.	207	164	488.	110	165	489.	196	157
490.	102	193	491.	202	141	492.	139	134
493.	87	176	494.	237	113	495.	169	140
496.	222	45	497.	181	157	498.	65	102
499.	207	68	500.	56	108	501.	221	40
502.	173	126	503.	183	129	504.	224	155
505.	97	229	506.	206	198	507.	81	178
508.	124	142	509.	136	201	510.	143	135
511.	136	99	512.	170	43	513.	118	143
514.	72	121	515.	146	110	516.	130	197
517.	178	200	518.	91	87	519.	133	171
520.	165	161	521.	176	140	522.	110	42
523.	235	192	524.	54	109	525.	160	97
526.	161	119	527.	173	75	528.	107	164
529.	132	90	530.	147	196	531.	124	193
532.	129	125	533.	273	110	534.	90	178
535.	209	146	536.	229	217	537.	105	141
538.	173	181	539.	154	225	540.	149	65
541.	92	144	542.	129	114	543.	141	191
544.	198	192	545.	114	150	546.	124	159
547.	222	139	548.	107	169	549.	147	156
550.	220	21	551.	82	156	552.	178	91

| Subject | CONDITION | | Subject | CONDITION | | Subject | CONDITION | |
	A	B		A	B		A	B
553.	196	121	554.	248	122	555.	186	111
556.	101	124	557.	116	146	558.	157	109
559.	167	173	560.	113	95	561.	135	206
562.	162	158	563.	140	134	564.	111	152
565.	240	167	566.	61	141	567.	56	158
568.	165	192	569.	160	131	570.	191	124
571.	146	192	572.	64	146	573.	148	110
574.	125	170	575.	193	201	576.	251	91
577.	154	131	578.	68	214	579.	217	153
580.	264	186	581.	115	148	582.	153	125
583.	152	155	584.	177	132	585.	117	171
586.	87	69	587.	199	93	588.	143	152
589.	219	205	590.	136	128	591.	98	165
592.	161	210	593.	201	155	594.	165	71
595.	120	177	596.	23	118	597.	180	182
598.	123	157	599.	220	196	600.	148	119
601.	113	156	602.	162	209	603.	115	154
604.	73	128	605.	116	92	606.	173	127
607.	48	93	608.	161	191	609.	183	212
610.	128	97	611.	95	141	612.	212	191
613.	119	128	614.	140	75	615.	99	166
616.	194	132	617.	178	68	618.	142	139
619.	123	73	620.	131	73	621.	205	149
622.	136	136	623.	173	52	624.	188	134
625.	172	72	626.	130	67	627.	108	96
628.	140	100	629.	193	184	630.	81	224
631.	212	58	632.	169	193	633.	175	150
634.	195	135	635.	130	151	636.	140	138
637.	130	207	638.	63	158	639.	67	223
640.	202	124	641.	127	129	642.	200	173
643.	91	179	644.	129	142	645.	116	93
646.	163	101	647.	127	166	648.	148	177
649.	123	88	650.	232	153	651.	172	74
652.	155	111	653.	50	176	654.	134	125
655.	102	122	656.	183	170	657.	94	96
658.	103	200	659.	117	170	660.	122	204
661.	48	146	662.	145	136	663.	130	123
664.	114	101	665.	159	106	666.	127	129
667.	168	169	668.	178	109	669.	135	87
670.	151	70	671.	181	160	672.	109	136
673.	184	125	674.	90	247	675.	107	131
676.	186	82	677.	207	104	678.	142	167
679.	62	99	680.	154	176	681.	196	114
682.	141	201	683.	170	148	684.	207	154
685.	109	155	686.	201	128	687.	170	192
688.	200	199	689.	76	99	690.	148	159

Subject	CONDITION A	CONDITION B	Subject	CONDITION A	CONDITION B	Subject	CONDITION A	CONDITION B
691.	111	91	692.	78	98	693.	32	158
694.	217	106	695.	166	219	696.	79	177
697.	184	143	698.	199	192	699.	178	109
700.	202	61	701.	141	229	702.	121	158
703.	169	134	704.	158	130	705.	211	106
706.	241	117	707.	120	210	708.	243	252
709.	99	119	710.	192	122	711.	115	45
712.	83	130	713.	163	99	714.	197	187
715.	134	84	716.	260	160	717.	186	103
718.	139	53	719.	64	118	720.	257	164
721.	99	128	722.	155	141	723.	150	192
724.	179	183	725.	153	222	726.	181	129
727.	142	178	728.	115	114	729.	166	157
730.	91	145	731.	216	148	732.	121	126
733.	204	194	734.	193	176	735.	113	174
736.	91	198	737.	175	121	738.	198	109
739.	185	145	740.	166	184	741.	93	201
742.	152	122	743.	170	190	744.	124	118
745.	112	124	746.	153	161	747.	179	151
748.	94	150	749.	217	96	750.	139	136
751.	176	173	752.	74	229	753.	146	192
754.	114	235	755.	213	164	756.	143	173
757.	171	145	758.	124	132	759.	83	178
760.	91	107	761.	107	136	762.	98	177
763.	130	74	764.	150	113	765.	189	72
766.	178	96	767.	177	151	768.	164	151
769.	157	117	770.	116	112	771.	165	89
772.	167	177	773.	126	109	774.	191	103
775.	253	188	776.	167	90	777.	181	162
778.	128	161	779.	192	120	780.	164	160
781.	195	160	782.	200	128	783.	132	140
784.	135	141	785.	160	89	786.	160	191
787.	50	144	788.	153	54	789.	85	147
790.	133	218	791.	172	172	792.	139	132
793.	103	125	794.	145	125	795.	143	144
796.	170	145	797.	208	79	798.	119	140
799.	100	180	800.	138	203	801.	159	148
802.	168	164	803.	250	127	804.	209	156
805.	114	116	806.	221	170	807.	217	145
808.	145	228	809.	240	137	810.	123	136
811.	144	162	812.	142	132	813.	172	119
814.	103	97	815.	143	168	816.	186	72
817.	79	116	818.	135	151	819.	124	200
820.	88	155	821.	215	139	822.	76	151
823.	166	94	824.	221	202	825.	92	108
826.	137	134	827.	201	134	828.	169	171

Subject	CONDITION		Subject	CONDITION		Subject	CONDITION	
	A	B		A	B		A	B
829.	126	91	830.	218	102	831.	144	83
832.	269	154	833.	126	221	834.	96	188
835.	119	30	836.	121	102	837.	188	207
838.	185	112	839.	165	216	840.	180	98
841.	65	166	842.	66	180	843.	109	122
844.	131	186	845.	142	110	846.	138	190
847.	178	191	848.	165	234	849.	121	184
850.	93	223	851.	134	158	852.	139	162
853.	138	121	854.	74	143	855.	34	150
856.	77	134	857.	96	48	858.	200	173
859.	139	124	860.	175	133	861.	134	103
862.	164	137	863.	106	85	864.	153	179
865.	203	176	866.	160	100	867.	68	204
868.	207	173	869.	139	141	870.	122	123
871.	128	200	872.	244	111	873.	70	143
874.	82	117	875.	163	70	876.	265	169
877.	203	206	878.	213	159	879.	75	177
880.	152	135	881.	110	103	882.	155	105
883.	181	115	884.	142	52	885.	65	161
886.	150	111	887.	236	156	888.	196	56
889.	146	105	890.	160	90	891.	148	137
892.	205	164	893.	162	156	894.	204	154
895.	128	195	896.	201	181	897.	76	163
898.	206	181	899.	157	101	900.	123	164
901.	116	129	902.	113	191	903.	79	169
904.	156	169	905.	113	196	906.	132	140
907.	141	90	908.	159	179	909.	56	171
910.	141	92	911.	260	162	912.	113	134
913.	110	117	914.	120	130	915.	143	145
916.	248	96	917.	162	115	918.	159	128
919.	155	208	920.	205	96	921.	143	173
922.	216	99	923.	164	120	924.	168	183
925.	78	193	926.	188	150	927.	193	174
928.	88	119	929.	169	154	930.	148	137
931.	111	112	932.	164	113	933.	119	64
934.	146	130	935.	112	188	936.	141	199
937.	70	186	938.	93	120	939.	137	112
940.	105	112	941.	129	134	942.	223	54
943.	135	173	944.	172	199	945.	35	129
946.	194	181	947.	185	91	948.	148	147
949.	107	131	950.	161	131	951.	127	168
952.	190	135	953.	133	132	954.	149	114
955.	118	212	956.	203	119	957.	166	109
958.	161	137	959.	193	69	960.	108	173
961.	215	116	962.	72	141	963.	168	130

Subject	CONDITION A	CONDITION B	Subject	CONDITION A	CONDITION B	Subject	CONDITION A	CONDITION B
964.	130	173	965.	99	183	966.	116	177
967.	196	161	968.	61	164	969.	59	95
970.	103	115	971.	146	189	972.	118	220
973.	128	145	974.	87	158	975.	85	177
976.	180	200	977.	120	203	978.	149	108
979.	233	101	980.	188	111	981.	135	185
982.	98	199	983.	160	154	984.	151	106
985.	121	125	986.	149	137	987.	173	150
988.	100	107	989.	99	122	990.	156	146
991.	77	197	992.	210	172	993.	160	179
994.	135	105	995.	254	73	996.	186	156
997.	100	129	998.	191	110	999.	76	165
1000.	127	208	1001.	95	150	1002.	131	92
1003.	135	135	1004.	184	126	1005.	157	155

27. A study of the blood coagulation system in newborn infants yielded the following data on the prothrombin time in seconds of blood collected from 500 full-term and 500 premature babies all of whom were free of disease. Select a simple random sample of size 16 from each of these populations and conduct an appropriate hypothesis test to determine whether one should conclude that the two populations differ with respect to mean prothrombin time. Let $\alpha = .05$. Compare your results with those of your classmates. What assumptions are necessary for the validity of the test?

FULL - TERM INFANTS

Subject	Time	Subject	Time	Subject	Time	Subject	Time	Subject	Time
1.	14	2.	14	3.	12	4.	13	5.	11
6.	11	7.	13	8.	12	9.	14	10.	12
11.	12	12.	11	13.	12	14.	12	15.	12
16.	13	17.	12	18.	12	19.	10	20.	12
21.	11	22.	10	23.	13	24.	12	25.	12
26.	13	27.	13	28.	12	29.	11	30.	12
31.	12	32.	13	33.	12	34.	12	35.	13
36.	13	37.	12	38.	12	39.	11	40.	12
41.	14	42.	11	43.	10	44.	12	45.	13
46.	11	47.	12	48.	12	49.	13	50.	12
51.	14	52.	11	53.	10	54.	11	55.	12
56.	12	57.	12	58.	14	59.	11	60.	12
61.	12	62.	12	63.	13	64.	13	65.	11

FULL - TERM INFANTS

Subject	Time	Subject	Time	Subject	Time	Subject	Time	Subject	Time
66.	13	67.	12	68.	12	69.	14	70.	13
71.	13	72.	12	73.	11	74.	13	75.	11
76.	12	77.	12	78.	12	79.	12	80.	13
81.	13	82.	12	83.	12	84.	11	85.	9
86.	13	87.	13	88.	13	89.	12	90.	11
91.	13	92.	12	93.	12	94.	14	95.	12
96.	12	97.	12	98.	11	99.	10	100.	13
101.	11	102.	11	103.	14	104.	12	105.	11
106.	11	107.	11	108.	11	109.	10	110.	14
111.	11	112.	12	113.	13	114.	12	115.	12
116.	13	117.	13	118.	12	119.	11	120.	11
121.	13	122.	11	123.	13	124.	12	125.	11
126.	12	127.	11	128.	11	129.	11	130.	14
131.	13	132.	12	133.	13	134.	11	135.	13
136.	13	137.	10	138.	12	139.	12	140.	13
141.	12	142.	11	143.	12	144.	10	145.	12
146.	13	147.	12	148.	13	149.	11	150.	12
151.	13	152.	13	153.	11	154.	11	155.	11
156.	11	157.	12	158.	11	159.	11	160.	13
161.	9	162.	13	163.	13	164.	12	165.	10
166.	11	167.	13	168.	11	169.	12	170.	13
171.	12	172.	13	173.	11	174.	13	175.	12
176.	13	177.	12	178.	12	179.	12	180.	11
181.	11	182.	13	183.	12	184.	12	185.	12
186.	11	187.	12	188.	12	189.	12	190.	11
191.	11	192.	13	193.	12	194.	12	195.	12
196.	11	197.	12	198.	13	199.	13	200.	11
201.	13	202.	13	203.	12	204.	12	205.	12
206.	11	207.	11	208.	13	209.	11	210.	11
211.	11	212.	11	213.	14	214.	13	215.	13
216.	12	217.	12	218.	11	219.	12	220.	13
221.	12	222.	11	223.	11	224.	10	225.	12
226.	13	227.	14	228.	12	229.	12	230.	12
231.	14	232.	12	233.	12	234.	11	235.	14
236.	14	237.	11	238.	13	239.	12	240.	12
241.	12	242.	12	243.	13	244.	11	245.	14
246.	10	247.	12	248.	11	249.	13	250.	11
251.	12	252.	13	253.	12	254.	11	255.	12
256.	13	257.	13	258.	12	259.	13	260.	12
261.	13	262.	12	263.	13	264.	13	265.	12
266.	13	267.	11	268.	12	269.	11	270.	11
271.	11	272.	13	273.	13	274.	13	275.	12
276.	12	277.	11	278.	12	279.	12	280.	11
281.	12	282.	13	283.	12	284.	10	285.	12

FULL - TERM INFANTS

Subject	Time	Subject	Time	Subject	Time	Subject	Time	Subject	Time
286.	12	287.	12	288.	12	289.	11	290.	12
291.	12	292.	11	293.	11	294.	13	295.	13
296.	11	297.	12	298.	12	299.	12	300.	12
301.	12	302.	12	303.	11	304.	11	305.	12
306.	12	307.	11	308.	11	309.	11	310.	12
311.	11	312.	12	313.	11	314.	13	315.	11
316.	11	317.	12	318.	10	319.	13	320.	13
321.	13	322.	12	323.	13	324.	13	325.	13
326.	12	327.	13	328.	10	329.	15	330.	10
331.	10	332.	13	333.	11	334.	12	335.	12
336.	12	337.	11	338.	12	339.	12	340.	12
341.	13	342.	13	343.	12	344.	12	345.	10
346.	11	347.	12	348.	11	349.	13	350.	10
351.	14	352.	13	353.	13	354.	10	355.	12
356.	13	357.	14	358.	13	359.	13	360.	11
361.	13	362.	10	363.	12	364.	13	365.	12
366.	10	367.	12	368.	11	369.	11	370.	14
371.	13	372.	11	373.	13	374.	12	375.	13
376.	14	377.	13	378.	11	379.	14	380.	11
381.	11	382.	12	383.	12	384.	11	385.	11
386.	13	387.	11	388.	11	389.	12	390.	13
391.	11	392.	11	393.	11	394.	13	395.	12
396.	10	397.	12	398.	12	399.	12	400.	12
401.	12	402.	11	403.	11	404.	12	405.	11
406.	12	407.	12	408.	12	409.	12	410.	12
411.	10	412.	12	413.	11	414.	13	415.	13
416.	12	417.	13	418.	12	419.	12	420.	14
421.	12	422.	11	423.	13	424.	14	425.	10
426.	12	427.	13	428.	12	429.	13	430.	12
431.	11	432.	14	433.	13	434.	12	435.	11
436.	14	437.	11	438.	11	439.	13	440.	11
441.	13	442.	11	443.	12	444.	12	445.	12
446.	11	447.	10	448.	12	449.	14	450.	11
451.	13	452.	13	453.	12	454.	12	455.	13
456.	11	457.	12	458.	12	459.	11	460.	12
461.	11	462.	12	463.	9	464.	12	465.	14
466.	10	467.	11	468.	11	469.	11	470.	11
471.	12	472.	10	473.	14	474.	14	475.	12
476.	12	477.	12	478.	13	479.	15	480.	11
481.	11	482.	11	483.	11	484.	12	485.	12
486.	12	487.	12	488.	13	489.	12	490.	12
491.	10	492.	12	493.	12	494.	12	495.	10
496.	12	497.	11	498.	13	499.	10	500.	13

Premature Infants

Subject	Time	Subject	Time	Subject	Time	Subject	Time	Subject	Time
1.	12	2.	14	3.	16	4.	14	5.	14
6.	14	7.	15	8.	13	9.	12	10.	15
11.	14	12.	15	13.	15	14.	14	15.	13
16.	13	17.	14	18.	17	19.	12	20.	15
21.	13	22.	14	23.	15	24.	14	25.	14
26.	13	27.	16	28.	12	29.	12	30.	13
31.	14	32.	14	33.	14	34.	16	35.	13
36.	15	37.	16	38.	16	39.	17	40.	15
41.	16	42.	14	43.	13	44.	14	45.	13
46.	15	47.	15	48.	15	49.	15	50.	14
51.	15	52.	13	53.	13	54.	14	55.	15
56.	14	57.	14	58.	13	59.	16	60.	13
61.	13	62.	14	63.	13	64.	14	65.	13
66.	15	67.	14	68.	15	69.	14	70.	13
71.	15	72.	14	73.	14	74.	16	75.	15
76.	12	77.	12	78.	14	79.	14	80.	14
81.	13	82.	15	83.	18	84.	14	85.	16
86.	14	87.	15	88.	14	89.	16	90.	13
91.	12	92.	16	93.	14	94.	15	95.	14
96.	16	97.	13	98.	17	99.	15	100.	15
101.	14	102.	18	103.	16	104.	14	105.	14
106.	13	107.	12	108.	14	109.	12	110.	16
111.	15	112.	14	113.	12	114.	15	115.	16
116.	14	117.	13	118.	15	119.	9	120.	16
121.	16	122.	14	123.	14	124.	15	125.	14
126.	13	127.	13	128.	16	129.	12	130.	14
131.	13	132.	11	133.	15	134.	16	135.	14
136.	13	137.	15	138.	15	139.	14	140.	12
141.	13	142.	13	143.	12	144.	14	145.	13
146.	11	147.	16	148.	14	149.	14	150.	12
151.	12	152.	13	153.	15	154.	17	155.	15
156.	14	157.	13	158.	14	159.	13	160.	15
161.	13	162.	15	163.	14	164.	16	165.	13
166.	13	167.	12	168.	13	169.	12	170.	14
171.	12	172.	15	173.	13	174.	11	175.	13
176.	13	177.	15	178.	13	179.	13	180.	15
181.	14	182.	12	183.	12	184.	16	185.	13
186.	13	187.	14	188.	12	189.	14	190.	16
191.	14	192.	14	193.	14	194.	12	195.	15
196.	12	197.	14	198.	16	199.	16	200.	14
201.	13	202.	12	203.	12	204.	13	205.	15
206.	13	207.	14	208.	12	209.	15	210.	14
211.	12	212.	15	213.	14	214.	14	215.	16
216.	17	217.	13	218.	14	219.	15	220.	16
221.	14	222.	15	223.	14	224.	13	225.	13
226.	14	227.	15	228.	13	229.	15	230.	13

PREMATURE INFANTS

Subject	Time	Subject	Time	Subject	Time	Subject	Time	Subject	Time
231.	14	232.	11	233.	15	234.	13	235.	14
236.	15	237.	13	238.	13	239.	14	240.	11
241.	13	242.	14	243.	15	244.	14	245.	13
246.	14	247.	13	248.	14	249.	15	250.	15
251.	14	252.	13	253.	15	254.	17	255.	14
256.	15	257.	13	258.	12	259.	15	260.	15
261.	14	262.	14	263.	14	264.	14	265.	15
266.	14	267.	15	268.	15	269.	15	270.	11
271.	12	272.	14	273.	11	274.	16	275.	12
276.	12	277.	13	278.	14	279.	15	280.	14
281.	12	282.	16	283.	14	284.	12	285.	15
286.	14	287.	14	288.	14	289.	17	290.	19
291.	13	292.	13	293.	14	294.	13	295.	15
296.	14	297.	12	298.	16	299.	14	300.	15
301.	16	302.	11	303.	14	304.	12	305.	15
306.	12	307.	13	308.	13	309.	16	310.	15
311.	13	312.	15	313.	14	314.	14	315.	15
316.	12	317.	11	318.	15	319.	12	320.	16
321.	13	322.	13	323.	13	324.	16	325.	13
326.	15	327.	14	328.	16	329.	15	330.	15
331.	14	332.	15	333.	12	334.	15	335.	15
336.	13	337.	13	338.	14	339.	14	340.	16
341.	16	342.	13	343.	14	344.	17	345.	15
346.	13	347.	9	348.	13	349.	14	350.	15
351.	12	352.	14	353.	14	354.	13	355.	15
356.	15	357.	14	358.	14	359.	15	360.	12
361.	14	362.	16	363.	14	364.	17	365.	14
366.	13	367.	15	368.	15	369.	16	370.	11
371.	13	372.	14	373.	15	374.	15	375.	12
376.	15	377.	16	378.	13	379.	15	380.	15
381.	15	382.	15	383.	15	384.	16	385.	15
386.	15	387.	15	388.	16	389.	13	390.	13
391.	17	392.	12	393.	11	394.	12	395.	17
396.	13	397.	13	398.	16	399.	12	400.	16
401.	13	402.	15	403.	14	404.	15	405.	14
406.	13	407.	16	408.	12	409.	14	410.	13
411.	13	412.	13	413.	14	414.	17	415.	15
416.	14	417.	14	418.	16	419.	12	420.	14
421.	14	422.	14	423.	14	424.	13	425.	15
426.	13	427.	14	428.	11	429.	12	430.	15
431.	15	432.	15	433.	13	434.	13	435.	15
436.	18	437.	14	438.	14	439.	14	440.	12
441.	15	442.	15	443.	14	444.	14	445.	16
446.	15	447.	16	448.	15	449.	15	450.	13
451.	15	452.	15	453.	14	454.	14	455.	13
456.	15	457.	12	458.	13	459.	14	460.	14

PREMATURE INFANTS

Subject	Time	Subject	Time	Subject	Time	Subject	Time	Subject	Time
461.	15	462.	16	463.	16	464.	14	465.	13
466.	16	467.	15	468.	15	469.	16	470.	15
471.	12	472.	14	473.	14	474.	12	475.	13
476.	14	477.	15	478.	13	479.	14	480.	16
481.	12	482.	15	483.	14	484.	13	485.	14
486.	13	487.	16	488.	14	489.	12	490.	14
491.	13	492.	13	493.	15	494.	15	495.	15
496.	11	497.	13	498.	13	499.	12	500.	16

28. The following data on head circumference (in centimeters) were collected on a population of adolescent males with a sex chromosome abnormality (SCA) and a population of normal controls (NC). Subjects were matched on the basis of age, ethnic group, and socioeconomic status. Select a simple random sample of size 20 from the population and perform an appropriate hypothesis test to determine if one can conclude that subjects with the sex chromosome abnormality tend to have smaller heads than normal subjects. Let $\alpha = .05$. Construct a 95 percent confidence interval for the population mean difference. What assumptions are necessary? Compare your results with those of your classmates.

Matched Pair	SCA	NC	Matched Pair	SCA	NC
1.	50.3	53.2	2.	55.2	58.1
3.	54.5	56.0	4.	49.7	53.7
5.	51.5	55.5	6.	58.2	58.2
7.	52.2	54.7	8.	56.5	58.6
9.	58.9	58.8	10.	53.6	57.8
11.	54.7	57.8	12.	53.7	55.7
13.	55.9	60.9	14.	55.9	58.7
15.	59.3	60.2	16.	56.8	58.4
17.	55.7	58.5	18.	55.3	57.6
19.	54.9	58.7	20.	57.9	58.3
21.	52.5	57.7	22.	57.6	59.4
23.	54.8	59.3	24.	54.6	57.7
25.	57.4	58.2	26.	54.6	56.6
27.	56.9	58.8	28.	55.6	58.9
29.	54.3	58.1	30.	53.1	56.5
31.	55.7	57.6	32.	50.2	52.8
33.	54.4	56.6	34.	57.2	58.2
35.	53.7	58.3	36.	56.0	57.6
37.	54.9	56.4	38.	58.1	58.1

Matched Pair	SCA	NC	Matched Pair	SCA	NC
39.	52.5	53.3	40.	50.4	52.9
41.	52.4	53.0	42.	56.1	57.4
43.	55.7	57.8	44.	56.6	57.1
45.	53.3	57.1	46.	56.5	58.8
47.	54.8	56.3	48.	58.1	58.2
49.	52.6	54.4	50.	51.8	54.1
51.	52.4	56.8	52.	57.4	57.2
53.	55.1	58.1	54.	54.1	57.5
55.	54.6	58.6	56.	57.7	58.5
57.	53.9	58.5	58.	51.2	50.3
59.	55.8	56.6	60.	58.3	58.1
61.	57.3	57.9	62.	50.8	55.8
63.	55.5	58.4	64.	58.3	58.1
65.	55.0	58.7	66.	57.2	58.4
67.	51.1	52.8	68.	53.8	58.2
69.	50.7	50.8	70.	56.7	58.8
71.	56.6	58.7	72.	52.1	55.5
73.	50.7	54.1	74.	56.4	58.2
75.	55.3	57.7	76.	53.1	54.6
77.	57.8	58.0	78.	50.5	50.0
79.	53.1	56.5	80.	55.3	57.1
81.	57.0	58.7	82.	57.0	58.3
83.	57.9	58.1	84.	54.8	57.2
85.	56.0	58.1	86.	50.2	50.3
87.	56.8	58.3	88.	52.7	54.5
89.	52.7	55.6	90.	54.7	57.2
91.	56.7	58.7	92.	52.8	55.3
93.	55.1	58.2	94.	52.4	57.6
95.	56.0	58.3	96.	56.7	58.3
97.	50.3	57.1	98.	53.2	56.0
99.	47.8	50.7	100.	50.4	50.9
101.	54.5	57.7	102.	54.4	57.9
103.	54.4	56.8	104.	57.8	58.4
105.	55.3	58.1	106.	53.9	57.5
107.	57.2	58.3	108.	55.7	57.3
109.	53.2	54.8	110.	56.2	58.7
111.	53.1	55.2	112.	55.3	58.3
113.	56.3	58.1	114.	54.6	58.7
115.	56.3	58.2	116.	57.0	58.1
117.	53.6	55.8	118.	57.5	58.3
119.	50.8	54.7	120.	53.1	56.0
121.	56.0	57.3	122.	54.0	56.7
123.	55.4	56.3	124.	50.8	53.2
125.	55.0	57.4	126.	53.7	55.3
127.	55.3	58.3	128.	57.3	57.2
129.	55.5	57.5	130.	53.8	57.0

Matched Pair	SCA	NC	Matched Pair	SCA	NC
131.	53.4	58.5	132.	57.1	58.8
133.	57.0	58.4	134.	49.9	54.6
135.	52.2	55.8	136.	55.8	58.5
137.	50.9	53.3	138.	54.0	55.4
139.	55.3	58.9	140.	54.7	56.1
141.	53.2	56.1	142.	56.8	58.5
143.	54.8	58.5	144.	56.0	58.7
145.	54.0	56.2	146.	53.0	54.9
147.	52.5	54.8	148.	48.5	58.2
149.	56.7	58.6	150.	58.3	58.4
151.	50.4	33.1	152.	57.4	57.4
153.	52.6	56.1	154.	55.3	58.2
155.	53.6	58.6	156.	54.8	57.5
157.	51.3	57.1	158.	53.7	58.2
159.	51.8	55.2	160.	55.2	58.1
161.	57.6	58.3	162.	55.9	58.8
163.	58.3	58.5	164.	55.6	58.8
165.	53.8	56.5	166.	55.7	58.5
167.	55.3	56.7	168.	56.0	58.9
169.	56.0	58.1	170.	55.7	58.2
171.	54.3	58.9	172.	54.6	57.2
173.	57.9	58.7	174.	54.9	58.7
175.	55.3	58.1	176.	52.8	56.6
177.	53.2	55.8	178.	56.7	58.4
179.	58.5	58.2	180.	54.8	57.6
181.	54.2	56.1	182.	57.1	58.1
183.	52.8	55.6	184.	57.6	58.4
185.	56.3	58.0	186.	53.7	56.5
187.	56.9	58.2	188.	56.3	58.4
189.	54.7	56.5	190.	55.1	57.7
191.	56.0	58.3	192.	54.5	57.6
193.	55.2	57.0	194.	54.7	56.8
195.	54.9	58.1	196.	56.6	58.1
197.	56.2	58.2	198.	54.1	57.8
199.	57.2	58.3	200.	55.5	57.7
201.	54.4	58.6	202.	52.8	56.4
203.	53.0	57.1	204.	52.2	54.4
205.	58.2	58.5	206.	52.4	55.6
207.	48.0	58.0	208.	53.6	57.6
209.	52.7	55.2	210.	55.8	58.9
211.	53.7	56.5	212.	51.5	55.0
213.	55.1	56.5	214.	57.7	58.3
215.	54.4	56.7	216.	53.8	56.9
217.	52.9	55.1	218.	54.9	58.1
219.	56.1	58.4	220.	57.8	58.6
221.	52.3	53.8	222.	51.9	55.2

Matched Pair	SCA	NC	Matched Pair	SCA	NC
223.	54.4	57.3	224.	57.6	58.3
225.	55.0	58.5	226.	55.6	57.1
227.	55.7	57.0	228.	53.6	56.7
229.	53.5	55.4	230.	54.3	57.8
231.	52.8	54.7	232.	56.2	58.0
233.	57.5	58.8	234.	57.9	58.1
235.	54.9	58.5	236.	57.1	58.5
237.	56.1	57.8	238.	53.2	57.3
239.	52.9	56.4	240.	56.1	57.5
241.	54.9	55.5	242.	56.0	58.2
243.	52.4	56.0	244.	52.4	56.5
245.	50.4	54.0	246.	51.2	55.1
247.	53.9	57.9	248.	54.2	58.5
249.	54.8	57.1	250.	51.1	55.1
251.	51.1	53.0	252.	57.1	58.5
253.	51.8	56.9	254.	53.8	58.7
255.	54.7	57.0	256.	56.5	58.4
257.	52.9	56.3	258.	54.7	56.9
259.	55.8	58.5	260.	57.6	58.9
261.	53.8	57.7	262.	57.3	58.9
263.	53.4	56.7	264.	53.9	56.9
265.	48.1	53.3	266.	57.6	58.1
267.	51.4	54.9	268.	53.8	58.0
269.	57.7	58.5	270.	50.0	54.4
271.	53.9	56.3	272.	54.5	58.9
273.	54.2	57.7	274.	52.3	55.7
275.	54.9	58.1	276.	52.0	56.4
277.	55.7	58.0	278.	50.9	52.1
279.	55.3	57.1	280.	55.2	56.8
281.	47.0	51.7	282.	50.0	53.6
283.	50.1	51.4	284.	55.4	57.0
285.	55.5	57.4	286.	46.8	50.5
287.	54.0	54.8	288.	53.7	58.7
289.	48.1	50.5	290.	45.7	50.8
291.	52.5	53.8	292.	52.9	55.8
293.	53.9	55.5	294.	54.1	56.7
295.	55.2	58.2	296.	57.4	50.5
297.	56.3	58.9	298.	54.8	58.1
299.	52.6	54.6	300.	56.6	57.4
301.	56.1	56.8	302.	50.2	53.5
303.	55.6	58.7	304.	50.5	53.4
305.	47.8	55.8	306.	52.4	54.9
307.	55.2	58.5	308.	54.7	57.6
309.	49.2	53.1	310.	55.9	58.4
311.	53.7	55.7	312.	57.1	58.5
313.	54.9	55.1	314.	51.9	55.2

Matched Pair	SCA	NC	Matched Pair	SCA	NC
315.	53.9	56.8	316.	54.3	57.5
317.	55.9	56.4	318.	54.0	56.9
319.	50.8	55.3	320.	57.9	58.8
321.	57.4	58.6	322.	56.6	58.3
323.	52.2	55.7	324.	57.1	58.9
325.	50.2	53.1	326.	48.1	50.4
327.	54.7	56.1	328.	47.1	50.5
329.	56.6	58.8	330.	51.4	53.9
331.	49.6	52.3	332.	54.2	57.2
333.	55.5	56.2	334.	54.5	56.0
335.	54.3	58.7	336.	51.0	51.9
337.	52.2	56.8	338.	51.4	54.9
339.	55.7	58.5	340.	55.4	57.4
341.	55.8	58.4	342.	48.3	52.1
343.	56.3	58.4	344.	54.1	56.2
345.	56.1	56.9	346.	54.3	56.9
347.	48.7	50.5	348.	54.3	57.4
349.	47.5	51.9	350.	54.4	57.9
351.	54.1	57.5	352.	55.6	56.2
353.	54.8	58.1	354.	47.8	50.0
355.	53.2	55.8	356.	47.2	50.3
357.	49.3	53.4	358.	47.0	50.3
359.	54.1	58.1	360.	49.4	51.7
361.	52.8	56.2	362.	53.8	56.2
363.	48.3	53.0	364.	52.5	56.9
365.	54.2	54.8	366.	56.0	58.6
367.	56.8	57.5	368.	47.2	51.0
369.	48.5	51.1	370.	54.5	57.5
371.	55.1	55.6	372.	55.2	58.7
373.	55.9	58.5	374.	55.3	58.0
375.	52.4	55.2	376.	54.3	57.9
377.	52.8	54.9	378.	57.5	58.1
379.	51.3	53.5	380.	54.3	56.5
381.	52.2	55.6	382.	48.5	49.7
383.	48.5	50.4	384.	47.4	50.3
385.	54.9	58.0	386.	56.5	58.9
387.	57.4	58.7	388.	52.5	55.9
389.	55.2	56.8	390.	47.0	50.2
391.	54.9	56.4	392.	53.1	56.5
393.	48.0	50.5	394.	53.4	58.6
395.	49.7	53.4	396.	56.0	58.3
397.	53.9	55.7	398.	47.9	51.2
399.	52.2	53.8	400.	55.4	58.0
401.	56.5	58.7	402.	54.2	56.7
403.	56.0	58.2	404.	55.2	58.1
405.	47.7	50.6	406.	54.1	56.2

Matched Pair	SCA	NC	Matched Pair	SCA	NC
407.	47.4	50.8	408.	52.1	54.9
409.	47.4	50.1	410.	54.7	57.9
411.	53.5	57.7	412.	55.4	56.9
413.	56.4	58.1	414.	54.4	57.8
415.	53.6	57.2	416.	56.9	58.0
417.	53.9	57.1	418.	52.3	55.9
419.	54.8	58.2	420.	53.4	56.7
421.	55.5	58.9	422.	54.9	58.4
423.	55.0	56.9	424.	47.8	50.2
425.	52.1	55.5	426.	57.0	58.7
427.	55.5	57.9	428.	51.4	53.6
429.	48.8	51.5	430.	55.4	57.5
431.	56.1	58.4	432.	53.1	56.4
433.	54.0	57.2	434.	55.5	57.5
435.	55.4	56.6	436.	55.6	58.7
437.	56.6	58.6	438.	54.6	56.3
439.	56.4	58.0	440.	55.3	56.2
441.	56.0	57.4	442.	52.5	54.9
443.	55.8	56.6	444.	49.2	51.6
445.	55.3	56.8	446.	54.5	56.5
447.	56.2	58.3	448.	54.3	55.9
449.	52.7	57.7	450.	48.1	51.7
451.	54.0	57.1	452.	54.8	58.7
453.	48.5	53.7	454.	55.1	58.5
455.	54.9	56.0	456.	53.6	56.1
457.	53.8	58.0	458.	53.1	54.8
459.	55.9	58.0	460.	54.4	57.6
461.	47.1	51.9	462.	47.9	50.3
463.	51.1	53.2	464.	56.4	57.2
465.	51.7	53.0	466.	55.8	58.6
467.	52.6	55.7	468.	56.9	57.9
469.	52.4	56.6	470.	53.5	57.1
471.	55.2	57.8	472.	52.8	57.4
473.	54.1	57.1	474.	54.4	58.5
475.	56.1	57.4	476.	53.9	56.2
477.	53.6	57.3	478.	54.0	57.6
479.	53.8	55.5	480.	56.3	58.4
481.	55.3	57.0	482.	48.1	51.1
483.	52.6	54.8	484.	53.8	56.0
485.	49.0	52.1	486.	55.8	58.8
487.	56.2	57.1	488.	54.4	56.1
489.	56.3	57.6	490.	54.7	57.2
491.	53.0	54.2	492.	57.3	58.9
493.	49.1	54.0	494.	47.1	50.7
495.	55.7	58.9	496.	49.1	52.1
497.	56.5	58.7	498.	53.8	57.0

Matched Pair	SCA	NC	Matched Pair	SCA	NC
499.	54.9	58.0	500.	55.1	58.1
501.	54.3	58.1	502.	57.1	57.7
503.	53.5	57.0	504.	54.7	58.1
505.	52.9	54.3	506.	47.0	50.1
507.	54.3	56.4	508.	54.4	55.9
509.	54.0	58.5	510.	55.1	57.7
511.	55.0	57.2	512.	56.8	57.5
513.	55.8	56.5	514.	56.6	57.6
515.	50.2	54.3	516.	53.4	56.7
517.	53.6	56.9	518.	55.9	56.2
519.	53.0	56.1	520.	52.1	54.2
521.	54.9	58.3	522.	47.9	50.6
523.	50.3	52.4	524.	53.0	56.4
525.	54.0	56.0	526.	56.3	57.9
527.	48.9	51.9	528.	47.5	54.0
529.	54.0	58.3	530.	52.8	56.1
531.	52.9	56.3	532.	54.9	57.9
533.	54.1	56.5	534.	56.3	57.3
535.	54.0	58.5	536.	48.6	51.8
537.	54.6	58.1	538.	56.3	57.7
539.	50.5	51.4	540.	55.8	57.2
541.	51.5	54.3	542.	57.1	58.6
543.	54.8	55.2	544.	56.3	56.5
545.	56.9	57.3	546.	57.5	58.9
547.	55.6	58.2	548.	55.4	56.1
549.	49.9	53.2	550.	53.8	58.3
551.	55.8	57.6	552.	48.3	50.5
553.	47.8	50.7	554.	56.8	57.3
555.	54.7	57.2	556.	52.9	55.7
557.	53.6	57.2	558.	56.3	57.4
559.	53.4	57.1	560.	47.0	51.3
561.	54.0	57.3	562.	53.9	57.4
563.	51.7	55.9	564.	55.9	57.9
565.	53.5	54.5	566.	55.9	57.4
567.	55.8	58.6	568.	56.1	57.0
569.	50.0	53.0	570.	52.0	55.0
571.	52.6	56.3	572.	55.9	57.9
573.	53.6	55.5	574.	47.3	50.1
575.	54.9	56.0	576.	55.8	57.9
577.	52.3	53.3	578.	56.0	56.6
579.	51.6	54.3	580.	48.7	52.1
581.	55.4	58.5	582.	48.1	50.0
583.	53.9	56.6	584.	55.0	57.2
585.	55.2	57.3	586.	54.6	58.4
587.	54.5	56.2	588.	48.9	52.2
589.	47.5	49.5	590.	55.6	58.8

Matched Pair	SCA	NC	Matched Pair	SCA	NC
591.	55.0	57.3	592.	55.5	57.9
593.	53.2	56.4	594.	53.7	56.5
595.	55.8	57.2	596.	56.2	57.7
597.	55.5	57.0	598.	54.4	56.9
599.	49.9	50.7	600.	55.9	57.0
601.	48.7	52.0	602.	53.3	56.8
603.	52.6	55.2	604.	56.7	57.3
605.	48.4	52.3	606.	52.5	54.9
607.	57.1	58.5	608.	53.3	57.1
609.	48.1	51.9	610.	48.3	51.4
611.	53.3	58.0	612.	47.7	50.0
613.	56.5	57.8	614.	50.2	53.7
615.	55.1	58.1	616.	56.3	57.0
617.	47.3	50.8	618.	54.1	56.2
619.	55.5	58.0	620.	49.5	52.5
621.	55.6	56.9	622.	57.1	57.5
623.	53.9	55.5	624.	54.8	57.6
625.	56.3	57.4	626.	55.4	56.9
627.	52.0	56.2	628.	52.4	56.0
629.	54.0	57.0	630.	52.7	55.9
631.	52.9	57.3	632.	58.6	58.7
633.	48.1	50.8	634.	53.0	56.0
635.	54.2	57.2	636.	48.9	53.3
637.	54.0	55.7	638.	54.4	54.7
639.	53.9	57.6	640.	49.2	53.1
641.	53.2	56.9	642.	55.7	57.4
643.	51.8	54.1	644.	55.2	58.4
645.	54.5	58.0	646.	55.8	56.4
647.	54.1	57.3	648.	49.7	53.5
649.	54.6	56.7	650.	53.5	58.1
651.	48.8	52.8	652.	49.6	52.1
653.	56.8	57.5	654.	53.2	55.3
655.	54.1	57.3	656.	50.8	55.2
657.	50.4	54.1	658.	52.9	56.6
659.	55.5	56.0	660.	53.4	58.5
661.	54.1	57.4	662.	51.0	54.1
663.	56.9	57.7	664.	57.7	58.5
665.	52.7	55.4	666.	52.6	57.2
667.	53.9	57.2	668.	51.7	55.5
669.	53.5	55.8	670.	55.5	58.2
671.	52.8	56.6	672.	53.2	56.2
673.	56.0	56.3	674.	53.8	57.3
675.	48.4	50.9	676.	54.6	57.7
677.	49.5	54.7	678.	54.0	56.8
679.	48.0	51.4	680.	56.2	58.3
681.	51.9	53.3	682.	54.6	56.6

Matched Pair	SCA	NC	Matched Pair	SCA	NC
683.	55.6	57.4	684.	50.9	54.1
685.	55.5	57.4	686.	56.2	56.7
687.	50.5	53.6	688.	56.7	58.8
689.	52.6	54.4	690.	55.2	58.0
691.	55.8	58.4	692.	53.7	57.4
693.	48.3	52.1	694.	54.3	57.3
695.	47.4	51.5	696.	52.4	55.1
697.	54.4	57.4	698.	54.0	57.0
699.	52.2	56.2	700.	56.2	57.7
701.	55.1	57.2	702.	54.9	58.4
703.	52.8	55.7	704.	52.4	56.4
705.	55.7	57.1	706.	56.8	57.4
707.	53.3	57.9	708.	55.0	58.2
709.	52.7	56.4	710.	54.6	57.5
711.	53.8	57.3	712.	51.5	52.5
713.	52.6	54.5	714.	56.4	57.8
715.	54.8	56.1	716.	55.7	57.1
717.	52.9	56.4	718.	54.9	57.9
719.	56.4	58.7	720.	54.9	57.2
721.	54.0	56.9	722.	53.1	54.7
723.	48.4	50.2	724.	49.8	53.3
725.	47.2	50.0	726.	57.3	58.8
727.	55.0	57.9	728.	55.8	56.9
729.	55.4	57.4	730.	53.3	56.9
731.	55.5	57.5	732.	56.6	58.8
733.	51.7	53.8	734.	54.5	57.0
735.	55.1	58.7	736.	55.8	58.1
737.	48.0	50.4	738.	55.9	56.4
739.	47.1	50.3	740.	50.4	50.6
741.	52.2	55.5	742.	47.5	50.2
743.	51.6	53.8	744.	47.4	50.8
745.	48.6	52.4	746.	51.9	54.6
747.	50.5	50.9	748.	52.6	54.5
749.	56.5	57.1	750.	56.5	57.0
751.	53.5	58.3	752.	57.4	57.9
753.	47.0	50.4	754.	53.8	57.9
755.	54.0	57.3	756.	49.2	52.3
757.	55.3	57.6	758.	54.1	58.9
759.	51.6	53.4	760.	56.9	57.7
761.	55.3	58.6	762.	52.9	54.9
763.	56.8	57.3	764.	56.3	57.3
765.	50.7	50.6	766.	55.4	56.7
767.	54.0	56.9	768.	52.6	56.9
769.	55.2	58.0	770.	56.0	56.5
771.	55.0	57.1	772.	53.2	57.7
773.	54.7	57.7	774.	51.2	54.4

Matched Pair	SCA	NC	Matched Pair	SCA	NC
775.	52.5	55.2	776.	47.8	50.4
777.	55.7	57.5	778.	52.2	55.6
779.	54.2	57.6	780.	50.2	50.0
781.	56.9	57.2	782.	48.0	50.9
783.	55.0	58.9	784.	57.4	58.9
785.	50.9	51.7	786.	50.3	51.9
787.	56.7	58.7	788.	49.2	52.2
789.	55.4	57.9	790.	53.7	57.6
791.	52.1	54.3	792.	56.9	57.3
793.	47.0	50.5	794.	54.0	57.2
795.	47.2	50.6	796.	53.8	57.3
797.	51.2	52.7	798.	55.0	57.7
799.	55.0	56.1	800.	55.6	58.1
801.	56.0	56.7	802.	55.3	57.2
803.	55.1	58.0	804.	54.4	56.4
805.	54.0	55.0	806.	49.7	52.4
807.	55.2	56.4	808.	55.9	58.5
809.	55.1	57.5	810.	49.3	52.4
811.	56.1	56.4	812.	49.4	53.2
813.	54.6	58.9	814.	48.0	51.9
815.	57.7	58.5	816.	55.2	58.6
817.	54.6	55.7	818.	56.8	57.3
819.	53.5	56.6	820.	54.0	57.9
821.	54.6	57.0	822.	55.1	57.8
823.	52.6	55.6	824.	56.7	57.4
825.	56.5	57.3	826.	52.7	55.5
827.	56.6	57.7	828.	48.0	50.9
829.	54.2	58.7	830.	54.3	58.0
831.	51.6	55.6	832.	51.7	54.7
833.	53.5	55.4	834.	54.5	56.9
835.	56.3	58.2	836.	48.0	51.6
837.	51.3	54.5	838.	56.8	57.4
839.	47.6	51.1	840.	56.2	57.0
841.	47.0	50.4	842.	56.3	56.5
843.	55.7	58.7	844.	54.2	57.0
845.	47.1	50.5	846.	48.5	51.1
847.	47.1	50.4	848.	53.8	58.4
849.	52.6	56.3	850.	55.0	57.3
851.	56.3	57.0	852.	55.4	56.5
853.	52.1	55.4	854.	52.7	57.5
855.	49.5	52.4	856.	52.1	55.4
857.	53.4	56.9	858.	53.5	56.2
859.	52.4	56.2	860.	54.0	56.7
861.	53.7	57.3	862.	55.7	57.5
863.	57.1	58.8	864.	47.8	52.5
865.	54.5	56.9	866.	50.9	53.6

Matched Pair	SCA	NC	Matched Pair	SCA	NC
867.	56.3	58.9	868.	54.0	56.6
869.	47.6	50.5	870.	50.6	53.6
871.	47.6	50.1	872.	49.3	52.2
873.	54.3	57.9	874.	54.8	58.6
875.	56.9	57.3	876.	56.9	57.6
877.	53.6	57.9	878.	47.1	50.5
879.	55.6	58.6	880.	48.4	51.2
881.	56.3	57.2	882.	53.0	54.3
883.	56.1	58.1	884.	54.8	55.6
885.	49.2	50.7	886.	54.7	56.9
887.	53.5	55.8	888.	47.9	50.0
889.	55.6	58.9	890.	52.4	55.6
891.	54.1	56.7	892.	54.4	57.3
893.	54.4	57.7	894.	51.9	56.0
895.	49.8	53.2	896.	52.1	53.8
897.	54.7	55.7	898.	55.5	58.2
899.	52.0	56.9	900.	54.3	56.8
901.	54.2	55.1	902.	56.7	57.0
903.	57.2	58.5	904.	56.9	58.5
905.	55.0	58.0	906.	53.9	56.7
907.	56.8	56.8	908.	56.1	56.2
909.	57.2	57.0	910.	55.5	57.9
911.	53.3	56.8	912.	47.2	50.7
913.	47.7	50.1	914.	47.8	49.1
915.	55.3	57.3	916.	52.8	57.0
917.	57.2	58.6	918.	54.3	57.1
919.	55.1	56.2	920.	53.6	56.3
921.	53.9	56.4	922.	54.0	57.7
923.	47.5	50.8	924.	53.8	55.2
925.	47.7	51.3	926.	56.2	57.9
927.	55.2	58.7	928.	55.2	58.8
929.	52.1	55.9	930.	55.3	57.6
931.	48.8	51.2	932.	51.3	55.2
933.	48.9	54.1	934.	54.5	57.1
935.	56.0	56.0	936.	53.2	56.9
937.	52.4	56.6	938.	52.7	55.1
939.	57.0	57.0	940.	48.1	50.5
941.	53.8	56.8	942.	54.6	58.5
943.	54.5	56.6	944.	56.2	58.7
945.	55.1	56.1	946.	53.7	57.4
947.	53.7	55.6	948.	53.6	54.7
949.	55.3	56.9	950.	56.8	57.1
951.	51.4	54.5	952.	52.6	55.0
953.	53.8	55.7	954.	52.7	55.7
955.	47.5	52.0	956.	54.1	55.0
957.	48.7	52.4	958.	57.4	57.9

Matched Pair	SCA	NC	Matched Pair	SCA	NC
959.	55.3	55.5	960.	55.9	56.6
961.	52.3	54.6	962.	47.3	51.2
963.	49.2	52.6	964.	47.4	50.2
965.	55.2	56.7	966.	57.3	58.7
967.	54.5	58.0	968.	58.0	58.3
969.	55.6	56.0	970.	58.2	58.8
971.	53.1	55.6	972.	55.8	58.0
973.	55.2	58.9	974.	48.7	51.8
975.	52.7	55.2	976.	51.4	54.4
977.	52.7	55.1	978.	56.9	58.4
979.	54.8	57.6	980.	53.4	57.0
981.	53.0	55.7	982.	48.0	53.3
983.	52.5	55.0	984.	56.3	56.5
985.	55.1	57.2	986.	55.7	57.1
987.	55.6	57.9	988.	50.8	55.6
989.	56.0	56.7	990.	55.0	57.9
991.	47.6	50.6	992.	55.8	56.6
993.	55.0	56.6	994.	56.6	57.1
995.	54.4	57.8	996.	54.2	56.5
997.	47.2	51.5	998.	47.5	50.5
999.	55.3	57.7	1000.	48.2	51.2

29. The following hemoglobin (Hb) measurements were collected on two populations of children. Subjects in population A had iron deficiency anemia, while those in population B were apparently healthy children in the same age group as population A. Select a simple random sample of size 16 from population A and an independent simple random sample of size 16 from population B. Do your sample data provide sufficient evidence to indicate that the two populations differ with respect to mean Hb value? Let $\alpha = .05$. What assumptions are necessary for your procedure to be valid? Compare your results with those of your classmates.

POPULATION A—CHILDREN WITH IRON DEFICIENCY ANEMIA									
Subject	Hb	Subject	Hb	Subject	Hb	Subject	Hb	Subject	Hb
1.	5.7	2.	6.5	3.	6.1	4.	6.5	5.	6.9
6.	6.7	7.	7.9	8.	8.0	9.	4.3	10.	4.8
11.	7.1	12.	8.1	13.	9.0	14.	3.7	15.	2.8
16.	8.0	17.	4.6	18.	5.5	19.	8.6	20.	5.0
21.	7.1	22.	8.1	23.	4.5	24.	6.9	25.	8.0
26.	10.0	27.	6.7	28.	6.8	29.	7.6	30.	9.4

Subject	Hb	Subject	Hb	Subject	Hb	Subject	Hb	Subject	Hb
31.	7.1	32.	3.9	33.	8.2	34.	5.6	35.	8.6
36.	8.2	37.	5.7	38.	8.0	39.	3.7	40.	6.9
41.	5.0	42.	5.7	43.	5.1	44.	8.3	45.	8.3
46.	7.1	47.	7.1	48.	5.4	49.	4.8	50.	4.8
51.	7.1	52.	2.7	53.	7.2	54.	3.6	55.	5.6
56.	8.1	57.	6.2	58.	9.1	59.	6.2	60.	5.9
61.	4.9	62.	4.7	63.	9.8	64.	8.5	65.	7.5
66.	9.1	67.	8.9	68.	5.1	69.	6.1	70.	7.4
71.	7.6	72.	4.4	73.	7.1	74.	3.8	75.	5.5
76.	5.1	77.	9.7	78.	8.4	79.	4.7	80.	4.9
81.	6.5	82.	3.1	83.	7.2	84.	5.5	85.	6.4
86.	7.2	87.	4.5	88.	8.3	89.	9.3	90.	12.4
91.	7.5	92.	5.7	93.	7.7	94.	6.0	95.	7.9
96.	3.9	97.	8.9	98.	9.5	99.	3.8	100.	10.6
101.	9.0	102.	3.8	103.	3.5	104.	7.6	105.	5.9
106.	8.1	107.	5.6	108.	5.3	109.	8.8	110.	7.2
111.	8.0	112.	2.2	113.	7.3	114.	3.2	115.	7.1
116.	5.2	117.	5.3	118.	8.0	119.	9.0	120.	3.3
121.	7.9	122.	6.2	123.	6.7	124.	9.4	125.	4.6
126.	6.4	127.	6.8	128.	7.0	129.	9.2	130.	5.3
131.	2.4	132.	9.3	133.	3.8	134.	8.1	135.	7.6
136.	7.4	137.	8.1	138.	5.0	139.	8.6	140.	8.3
141.	8.7	142.	8.9	143.	6.7	144.	6.3	145.	6.9
146.	8.7	147.	3.3	148.	7.1	149.	6.8	150.	7.6
151.	10.2	152.	6.0	153.	5.3	154.	6.0	155.	6.9
156.	5.5	157.	7.8	158.	4.7	159.	4.2	160.	9.7
161.	4.7	162.	7.8	163.	6.9	164.	4.0	165.	10.9
166.	7.8	167.	8.9	168.	10.0	169.	5.6	170.	8.5
171.	2.7	172.	6.6	173.	10.6	174.	7.9	175.	6.2
176.	6.2	177.	8.0	178.	4.5	179.	5.5	180.	7.3
181.	7.2	182.	7.4	183.	10.6	184.	4.8	185.	6.6
186.	7.1	187.	5.0	188.	7.2	189.	8.6	190.	5.5
191.	7.4	192.	6.8	193.	10.4	194.	6.0	195.	6.7
196.	9.9	197.	4.7	198.	4.7	199.	5.6	200.	6.8
201.	11.0	202.	6.2	203.	7.0	204.	8.5	205.	9.6
206.	7.5	207.	5.6	208.	8.1	209.	8.2	210.	8.1
211.	8.7	212.	7.8	213.	8.0	214.	8.3	215.	7.3
216.	4.7	217.	4.5	218.	5.5	219.	5.1	220.	8.5
221.	5.5	222.	6.1	223.	5.7	224.	8.2	225.	6.9
226.	6.0	227.	9.5	228.	8.8	229.	9.4	230.	6.2
231.	6.1	232.	8.8	233.	6.7	234.	6.4	235.	8.4
236.	9.6	237.	7.1	238.	7.3	239.	4.3	240.	9.2
241.	4.7	242.	5.7	243.	6.7	244.	5.0	245.	8.4
246.	5.8	247.	5.9	248.	8.2	249.	5.9	250.	7.2
251.	7.8	252.	5.2	253.	7.8	254.	5.9	255.	6.1
256.	6.9	257.	8.0	258.	7.8	259.	4.0	260.	6.8
261.	5.9	262.	3.7	263.	7.7	264.	5.5	265.	5.8

Subject	Hb	Subject	Hb	Subject	Hb	Subject	Hb	Subject	Hb
266.	5.7	267.	4.3	268.	3.2	269.	4.8	270.	5.8
271.	5.5	272.	7.7	273.	4.6	274.	6.6	275.	10.5
276.	6.9	277.	6.6	278.	4.6	279.	6.7	280.	4.8
281.	6.3	282.	5.5	283.	7.9	284.	8.3	285.	5.7
286.	6.9	287.	4.6	288.	8.4	289.	4.9	290.	5.5
291.	2.1	292.	5.3	293.	6.3	294.	7.9	295.	7.2
296.	5.9	297.	6.0	298.	6.1	299.	7.2	300.	5.2
301.	10.8	302.	6.9	303.	5.0	304.	6.6	305.	6.0
306.	8.7	307.	6.9	308.	5.5	309.	3.7	310.	8.2
311.	7.2	312.	4.7	313.	8.7	314.	3.3	315.	4.3
316.	7.3	317.	7.2	318.	6.3	319.	7.8	320.	10.8
321.	5.4	322.	8.2	323.	8.5	324.	3.4	325.	4.7
326.	8.4	327.	5.5	328.	10.0	329.	5.6	330.	7.4
331.	6.9	332.	8.8	333.	4.3	334.	5.4	335.	5.6
336.	10.7	337.	7.3	338.	5.5	339.	9.1	340.	8.4
341.	8.7	342.	6.8	343.	11.2	344.	7.7	345.	8.4
346.	11.6	347.	9.4	348.	6.2	349.	9.1	350.	7.2
351.	5.3	532.	7.9	353.	8.1	354.	8.6	355.	6.5
356.	9.6	357.	9.1	358.	6.7	359.	7.6	360.	5.2
361.	3.3	362.	9.4	363.	7.4	364.	6.8	365.	8.5
366.	5.6	367.	6.3	368.	3.3	369.	10.7	370.	9.9
371.	6.1	372.	5.4	373.	5.8	374.	7.2	375.	8.0
376.	4.0	377.	6.6	378.	6.2	379.	6.3	380.	4.1
381.	8.3	382.	9.0	383.	5.5	384.	7.7	385.	11.5
386.	6.1	387.	8.4	388.	3.0	389.	8.6	390.	7.9
391.	6.9	392.	8.4	393.	8.7	394.	6.3	395.	2.8
396.	7.9	397.	8.9	398.	8.0	399.	8.7	400.	7.0
401.	8.3	402.	7.4	403.	10.1	404.	8.1	405.	7.5
406.	6.7	407.	5.2	408.	7.1	409.	7.6	410.	7.0
411.	8.6	412.	6.2	413.	5.5	414.	5.0	415.	2.9
416.	5.5	417.	4.2	418.	11.9	419.	8.0	420.	10.4
421.	6.2	422.	7.6	423.	5.2	424.	6.9	425.	9.7
426.	3.7	427.	4.9	428.	4.4	429.	9.0	430.	5.4
431.	6.3	432.	5.9	433.	6.8	434.	8.6	435.	9.7
436.	8.6	437.	6.0	438.	5.3	439.	7.6	440.	9.5
441.	7.8	442.	6.2	443.	6.5	444.	5.6	445.	4.3
446.	9.3	447.	4.4	448.	4.0	449.	10.9	450.	11.3
451.	7.0	452.	5.8	453.	8.8	454.	4.3	455.	4.8
456.	5.7	457.	4.6	458.	7.5	459.	8.4	460.	6.9
461.	7.4	462.	5.7	463.	2.6	464.	4.9	465.	5.5
466.	3.9	467.	8.0	468.	5.6	469.	7.0	470.	4.2
471.	4.6	472.	7.0	473.	10.2	474.	4.0	475.	6.0
476.	6.9	477.	8.0	478.	8.2	479.	6.9	480.	8.2
481.	4.2	482.	10.1	483.	10.1	484.	3.3	485.	6.7
486.	9.9	487.	10.1	488.	7.4	489.	5.0	490.	8.4
491.	3.5	492.	9.7	493.	5.4	494.	6.0	495.	6.7
496.	7.5	497.	5.2	498.	10.3	499.	7.9	500.	8.4

POPULATION B—HEALTHY CHILDREN

Subject	Hb	Subject	Hb	Subject	Hb	Subject	Hb	Subject	Hb
1.	13.0	2.	13.6	3.	15.0	4.	11.9	5.	10.9
6.	12.3	7.	11.7	8.	11.4	9.	13.2	10.	14.3
11.	10.1	12.	9.9	13.	14.5	14.	10.0	15.	12.5
16.	10.2	17.	13.5	18.	13.8	19.	10.6	20.	12.7
21.	12.2	22.	14.5	23.	14.2	24.	13.6	25.	11.6
26.	15.3	27.	12.7	28.	13.2	29.	13.0	30.	16.8
31.	11.3	32.	12.5	33.	12.9	34.	13.0	35.	14.1
36.	10.5	37.	11.8	38.	12.2	39.	10.2	40.	13.0
41.	17.0	42.	16.1	43.	15.4	44.	12.9	45.	15.2
46.	15.4	47.	11.3	48.	13.0	49.	13.6	50.	10.8
51.	15.2	52.	13.5	53.	15.2	54.	11.2	55.	12.1
56.	12.3	57.	13.7	58.	11.6	59.	13.9	60.	11.8
61.	14.8	62.	9.7	63.	15.8	64.	14.2	65.	12.5
66.	12.3	67.	14.5	68.	12.2	69.	12.4	70.	12.5
71.	9.1	72.	12.8	73.	16.6	74.	14.8	75.	11.7
76.	11.6	77.	13.0	78.	13.3	79.	10.8	80.	12.2
81.	13.6	82.	10.7	83.	14.6	84.	15.7	85.	16.2
86.	15.0	87.	12.6	88.	12.2	89.	13.0	90.	13.5
91.	12.9	92.	15.8	93.	15.2	94.	11.3	95.	13.5
96.	14.7	97.	12.0	98.	13.0	99.	13.0	100.	15.9
101.	15.3	102.	11.7	103.	11.6	104.	10.2	105.	12.9
106.	15.6	107.	11.9	108.	10.8	109.	14.6	110.	10.4
111.	14.7	112.	13.5	113.	15.6	114.	12.5	115.	17.5
116.	10.3	117.	13.3	118.	11.5	119.	11.2	120.	10.5
121.	9.8	122.	11.8	123.	14.7	124.	14.8	125.	12.7
126.	11.1	127.	14.6	128.	14.1	129.	14.1	130.	12.3
131.	12.7	132.	14.3	133.	10.8	134.	12.1	135.	13.3
136.	10.6	137.	12.5	138.	11.0	139.	13.9	140.	9.4
141.	12.7	142.	12.8	143.	15.6	144.	12.4	145.	16.2
146.	13.8	147.	12.9	148.	10.4	149.	13.6	150.	15.2
151.	16.5	152.	11.3	153.	14.3	154.	12.3	155.	12.3
156.	10.6	157.	15.1	158.	10.8	159.	11.8	160.	11.9
161.	15.7	162.	9.8	163.	15.4	164.	12.4	165.	13.1
166.	11.4	167.	9.2	168.	15.7	169.	14.7	170.	13.6
171.	10.8	172.	10.2	173.	13.1	174.	13.3	175.	13.2
176.	12.0	177.	11.0	178.	16.2	179.	11.8	180.	11.1
181.	11.8	182.	14.6	183.	16.0	184.	11.3	185.	12.2
186.	11.4	187.	12.9	188.	12.8	189.	11.0	190.	13.8
191.	15.3	192.	13.2	193.	11.8	194.	9.3	195.	13.0
196.	13.6	197.	9.8	198.	13.8	199.	16.4	200.	11.1
201.	13.0	202.	12.4	203.	13.8	204.	15.7	205.	13.6
206.	12.7	207.	2.4	208.	13.3	209.	12.9	210.	14.5
211.	14.2	212.	11.1	213.	14.6	214.	11.0	215.	13.0
216.	9.3	217.	10.4	218.	11.8	219.	12.7	220.	10.0
221.	13.3	222.	12.7	223.	14.4	224.	12.0	225.	9.9
226.	13.1	227.	13.0	228.	15.7	229.	14.4	230.	13.9

Subject	Hb	Subject	Hb	Subject	Hb	Subject	Hb	Subject	Hb
231.	14.4	232.	14.7	233.	12.5	234.	11.8	235.	10.7
236.	15.1	237.	14.6	238.	9.1	239.	12.2	240.	14.5
241.	10.0	242.	10.8	243.	16.1	244.	12.1	245.	11.3
246.	10.6	247.	14.0	248.	14.9	249.	13.7	250.	12.7
251.	10.4	252	12.8	253.	14.0	254.	10.9	255.	16.2
256.	9.1	257.	13.5	258	12.4	259.	9.4	260.	10.6
261.	14.3	262.	13.0	263.	11.1	264.	13.8	265.	14.3
266.	14.3	267.	15.2	268.	14.2	269.	13.5	270.	10.5
271.	11.7	272.	10.9	273.	10.6	274.	13.6	275.	13.1
276.	9.3	277.	11.1	278.	12.1	279.	12.1	280.	13.5
281.	14.0	282.	12.9	283.	15.5	284.	12.3	285.	15.9
286.	12.9	287.	14.6	288.	15.6	289.	13.5	290.	13.1
291.	11.9	292.	14.9	293.	13.4	294.	12.1	295.	14.1
296.	11.5	297.	14.4	298.	10.1	299.	14.1	300.	9.9
301.	11.6	302.	14.5	303.	10.4	304.	11.9	305.	11.8
306.	11.9	307.	10.2	308.	12.6	309.	14.7	310.	11.4
311.	11.9	312.	14.4	313.	12.4	314.	14.3	315.	14.9
316.	14.4	317.	9.6	318.	11.7	319.	12.6	320.	12.5
321.	12.0	322.	16.1	323.	13.5	324.	13.4	325.	13.9
326.	13.1	327.	14.0	328.	10.0	329.	14.3	330.	13.7
331.	13.9	332.	14.8	333.	11.5	334.	12.5	335.	14.1
336.	14.0	337.	15.7	338.	10.1	339.	10.2	340.	13.4
341.	14.3	342.	10.4	343.	14.6	344.	12.8	345.	13.2
346.	11.6	347.	14.4	348.	9.1	349.	12.4	350.	13.6
351.	11.7	352.	12.9	353.	13.8	354.	12.9	355.	13.6
356.	12.7	357.	12.1	358.	14.3	359.	13.9	360.	10.3
361.	14.5	362.	14.3	363.	13.6	364.	11.8	365.	10.1
366.	16.0	367.	13.4	368.	11.9	369.	16.0	370.	13.7
371.	12.7	372.	14.3	373.	16.9	374.	13.4	375.	13.6
376.	13.6	377.	13.6	378.	12.8	379.	13.1	380.	9.5
381.	9.7	382.	13.9	383.	13.0	384.	15.9	385.	13.8
386.	11.4	387.	14.3	388.	13.2	389.	12.3	390.	10.1
391.	15.7	392.	10.8	393.	10.4	394.	13.3	395.	12.6
396.	11.1	397.	15.3	398.	13.5	399.	15.4	400.	12.1
401.	12.8	402.	14.5	403.	14.1	404.	12.2	405.	12.8
406.	12.7	407.	12.6	408.	13.9	409.	16.0	410.	14.0
411.	14.1	412.	12.8	413.	12.4	414.	12.3	415.	12.7
416.	13.1	417.	12.9	418.	14.0	419.	12.3	420.	9.6
421.	12.6	422.	11.9	423.	15.0	424.	10.6	425.	10.1
426.	10.7	427.	15.0	428.	15.2	429.	15.9	430.	11.5
431.	14.1	432.	15.5	433.	13.4	434.	11.6	435.	10.8
436.	12.2	437.	11.3	438.	11.9	439.	14.1	440.	12.2
441.	13.7	442.	12.7	443.	13.8	444.	12.7	445.	12.3
446.	14.6	447.	15.4	448.	10.4	449.	11.8	450.	16.3
451.	13.0	452.	13.0	453.	14.4	454.	11.6	455.	9.6
456.	14.1	457.	9.8	458.	14.3	459.	14.6	460.	10.1
461.	12.8	462.	15.5	463.	11.9	464.	14.0	465.	11.0

Subject	Hb	Subject	Hb	Subject	Hb	Subject	Hb	Subject	Hb
466.	11.3	467.	13.6	468.	16.2	469.	14.6	470.	15.2
471.	10.4	472.	13.0	473.	14.8	474.	12.8	475.	13.7
476.	10.9	477.	14.1	478.	13.5	479.	12.4	480.	14.6
481.	16.1	482.	16.1	483.	14.5	484.	12.4	485.	9.0
486.	12.1	487.	12.2	488.	14.3	489.	13.4	490.	9.9
491.	10.3	492.	14.6	493.	13.5	494.	10.6	495.	14.1
496.	11.3	497.	13.4	498.	13.6	499.	11.6	500.	14.3

30. The following manual dexterity scores are available on two populations. Population A consists of learning-disabled children, and population B consists of children with no known learning disability. Select a simple random sample of size 10 from population A and an independent simple random sample of size 15 from population B. Do your samples provide sufficient evidence for you to conclude that learning-disabled children, on the average, have lower manual dexterity scores than children without a learning disability? Let $\alpha = .05$. What assumptions are necessary in order for your procedure to be valid? Compare your results with those of your classmates.

LEARNING - DISABLED CHILDREN

Subject	Score	Subject	Score	Subject	Score	Subject	Score	Subject	Score
1.	33.0	2.	20.0	3.	27.5	4.	22.5	5.	25.8
6.	19.8	7.	29.9	8.	28.4	9.	20.8	10.	25.6
11.	21.4	12.	19.1	13.	25.3	14.	10.7	15.	23.5
16.	17.7	17.	20.6	18.	25.2	19.	20.9	20.	24.6
21.	21.2	22.	20.5	23.	24.3	24.	28.8	25.	17.7
26.	25.2	27.	22.4	28.	19.7	29.	29.4	30.	27.3
31.	28.9	32.	22.4	33.	24.7	34.	30.4	35.	20.9
36.	28.8	37.	19.5	38.	30.3	39.	25.3	40.	23.3
41.	20.1	42.	19.1	43.	18.5	44.	19.5	45.	27.3
46.	25.8	47.	26.7	48.	21.7	49.	20.0	50.	24.2
51.	21.9	52.	24.2	53.	30.9	54.	30.4	55.	28.0
56.	27.1	57.	22.4	58.	28.6	59.	27.0	60.	21.1
61.	18.4	62.	23.1	63.	20.9	64.	25.6	65.	32.6
66.	21.9	67.	21.5	68.	22.6	69.	25.2	70.	22.2
71.	22.7	72.	19.7	73.	27.6	74.	31.6	75.	23.5
76.	30.6	77.	25.9	78.	15.9	79.	26.0	80.	17.1
81.	25.2	82.	22.2	83.	22.7	84.	31.2	85.	26.2
86.	25.3	87.	22.4	88.	22.3	89.	21.6	90.	24.8
91.	26.2	92.	25.9	93.	23.7	94.	23.1	95.	26.4

Subject	Score	Subject	Score	Subject	Score	Subject	Score	Subject	Score
96.	26.0	97.	29.2	98.	20.8	99.	17.1	100.	21.1
101.	30.7	102.	31.1	103.	22.2	104.	14.7	105.	24.2
106.	23.4	107.	27.9	108.	34.4	109.	29.1	110.	23.1
111.	24.1	112.	25.2	113.	23.8	114.	20.8	115.	23.3
116.	29.8	117.	35.4	118.	23.4	119.	21.9	120.	27.3
121.	25.7	122.	24.4	123.	25.2	124.	24.2	125.	8.9
126.	23.3	127.	24.6	128.	29.9	129.	27.8	130.	31.3
131.	23.5	132.	28.0	133.	22.9	134.	17.0	135.	27.7
136.	27.7	137.	27.1	138.	20.4	139.	20.7	140.	23.6
141.	21.6	142.	29.1	143.	31.0	144.	19.5	145.	21.8
146.	27.9	147.	24.0	148.	21.5	149.	18.4	150.	23.6
151.	24.0	152.	24.0	153.	26.0	154.	20.7	155.	20.4
156.	22.7	157.	26.8	158.	22.0	159.	30.4	160.	24.7
161.	22.1	162.	22.2	163.	27.9	164.	31.9	165.	13.4
166.	18.6	167.	29.2	168.	17.6	169.	31.8	170.	28.3
171.	10.8	172.	20.4	173.	27.4	174.	23.6	175.	17.9
176.	18.5	177.	19.6	178.	30.9	179.	23.8	180.	21.9
181.	18.1	182.	25.4	183.	28.6	184.	29.5	185.	20.3
186.	25.6	187.	22.9	188.	27.2	189.	19.0	190.	18.9
191.	24.5	192.	22.3	193.	22.1	194.	18.5	195.	21.9
196.	28.2	197.	37.6	198.	16.4	199.	23.5	200.	25.0
201.	26.1	202.	20.1	203.	24.2	204.	29.4	205.	15.4
206.	25.0	207.	19.3	208.	21.9	209.	30.5	210.	28.3
211.	24.6	212.	16.1	213.	21.2	214.	25.7	215.	33.0
216.	28.8	217.	26.4	218.	23.3	219.	27.1	220.	18.4
221.	20.0	222.	22.7	223.	19.6	224.	24.1	225.	14.0
226.	20.6	227.	12.5	228.	23.5	229.	19.7	230.	31.7
231.	24.4	232.	23.1	233.	22.0	234.	26.6	235.	33.6
236.	14.6	237.	26.2	238.	24.9	239.	24.8	240.	30.9
241.	25.7	242.	31.3	243.	23.1	244.	17.5	245.	22.7
246.	23.7	247.	24.0	248.	20.7	249.	23.8	250.	26.9
251.	23.6	252.	20.9	253.	29.7	254.	18.1	255.	27.9
256.	13.0	257.	23.2	258.	21.5	259.	26.4	260.	22.7
261.	26.6	262.	29.1	263.	28.6	264.	24.2	265.	34.9
266.	22.6	267.	26.2	268.	11.3	269.	30.0	270.	22.7
271.	22.1	272.	18.4	273.	18.2	274.	18.1	275.	26.9
276.	27.0	277.	23.8	278.	21.9	279.	17.4	280.	29.0
281.	19.0	282.	30.4	283.	26.4	284.	26.3	285.	28.1
286.	27.9	287.	19.9	288.	24.5	289.	22.2	290.	24.4
291.	29.2	292.	23.7	293.	24.9	294.	17.6	295.	24.5
296.	31.5	297.	15.1	298.	23.5	299.	25.2	300.	26.6
301.	21.5	302.	31.4	303.	27.3	304.	24.2	305.	27.1
306.	27.2	307.	27.4	308.	19.5	309.	21.6	310.	21.3
311.	26.1	312.	30.1	313.	27.8	314.	31.9	315.	22.3
316.	24.6	317.	25.5	318.	25.9	319.	21.8	320.	21.9
321.	26.6	322.	30.1	323.	26.4	324.	16.4	325.	29.5

Subject	Score	Subject	Score	Subject	Score	Subject	Score	Subject	Score
326.	20.5	327.	21.8	328.	22.9	329.	14.7	330.	25.7
331.	24.9	332.	23.2	333.	20.5	334.	24.0	335.	22.9
336.	23.6	337.	17.5	338.	24.9	339.	10.1	340.	25.3
341.	23.8	342.	24.8	343.	27.3	344.	25.1	345.	22.9
346.	22.2	347.	26.2	348.	29.3	349.	20.7	350.	25.1
351.	17.4	352.	26.7	353.	22.2	354.	23.1	355.	27.5
356.	25.5	357.	11.0	358.	22.6	359.	39.8	360.	14.8
361.	15.9	362.	26.3	363.	24.5	364.	16.3	365.	16.1
366.	21.7	367.	21.2	368.	20.1	369.	18.2	370.	23.2
371.	20.2	372.	23.3	373.	16.2	374.	24.4	375.	25.3
376.	30.4	377.	22.7	378.	23.2	379.	19.8	380.	31.3
381.	33.9	382.	33.1	383.	19.8	384.	22.6	385.	27.5
386.	26.5	387.	25.6	388.	23.7	389.	21.2	390.	19.7
391.	20.1	392.	25.8	393.	20.3	394.	29.9	395.	17.5
396.	20.6	397.	29.4	398.	16.0	399.	21.8	400.	23.3
401.	30.3	402.	33.4	403.	27.0	404.	25.8	405.	28.9
406.	16.5	407.	24.5	408.	24.7	409.	31.4	410.	24.6
411.	26.0	412.	26.9	413.	22.9	414.	21.8	415.	22.9
416.	23.1	417.	23.1	418.	26.5	419.	11.4	420.	25.3
421.	28.5	422.	23.8	423.	28.1	424.	29.5	425.	27.2
426.	23.4	427.	15.8	428.	23.2	429.	25.3	430.	30.6
431.	23.2	432.	30.8	433.	22.9	434.	27.5	435.	20.2
436.	26.6	437.	28.9	438.	23.3	439.	21.8	440.	18.5
441.	29.7	442.	30.1	443.	24.1	444.	19.8	445.	18.5
446.	26.1	447.	13.2	448.	23.2	449.	37.7	450.	30.0
451.	30.9	452.	23.3	453.	23.7	454.	30.9	455.	16.2
456.	19.9	457.	24.8	458.	28.3	459.	29.0	460.	35.0
461.	23.6	462.	28.5	463.	33.2	464.	29.4	465.	25.8
466.	25.5	467.	17.8	468.	23.8	469.	31.6	470.	25.3
471.	24.7	472.	21.5	473.	26.0	474.	18.2	475.	19.8
476.	25.6	477.	20.1	478.	23.7	479.	25.2	480.	19.8
481.	29.3	482.	29.0	483.	15.1	484.	12.8	485.	27.2
486.	25.9	487.	26.1	488.	25.0	489.	28.8	490.	22.9
491.	28.9	492.	17.2	493.	19.2	494.	21.2	495.	28.3
496.	31.1	497.	25.0	498.	21.9	499.	25.2	500.	20.8

CHILDREN WITHOUT A LEARNING DISABILITY

Subject	Score	Subject	Score	Subject	Score	Subject	Score	Subject	Score
1.	32.0	2.	30.6	3.	29.1	4.	31.8	5.	25.3
6.	26.0	7.	24.2	8.	27.5	9.	20.8	10.	28.2
11.	28.3	12.	23.9	13.	28.4	14.	22.3	15.	30.4
16.	28.4	17.	30.0	18.	30.0	19.	31.2	20.	33.5
21.	28.5	22.	28.1	23.	25.7	24.	32.8	25.	24.5

Subject	Score	Subject	Score	Subject	Score	Subject	Score	Subject	Score
26.	24.5	27.	24.9	28.	30.6	29.	30.4	30.	24.0
31.	22.7	32.	26.4	33.	27.0	34.	25.8	35.	28.5
36.	26.4	37.	29.5	38.	26.4	39.	28.4	40.	25.1
41.	26.2	42.	28.0	43.	31.4	44.	29.5	45.	30.1
46.	24.7	47.	29.0	48.	28.0	49.	25.4	50.	25.8
51.	23.8	52.	27.2	53.	27.0	54.	28.8	55.	29.9
56.	22.4	57.	24.1	58.	25.0	59.	30.2	60.	28.3
61.	29.1	62.	27.0	63.	30.7	64.	32.6	65.	23.2
66.	29.9	67.	25.8	68.	20.5	69.	30.5	70.	28.1
71.	20.6	72.	29.1	73.	22.8	74.	21.3	75.	24.5
76.	29.1	77.	22.8	78.	23.3	79.	22.1	80.	32.4
81.	31.4	82.	25.6	83.	32.2	84.	30.6	85.	25.1
86.	30.4	87.	29.3	88.	28.8	89.	25.1	90.	26.4
91.	28.1	92.	32.1	93.	26.3	94.	26.5	95.	26.4
96.	32.8	97.	26.7	98.	31.7	99.	27.0	100.	28.3
101.	28.5	102.	29.4	103.	27.2	104.	26.3	105.	27.6
106.	25.7	107.	28.1	108.	23.0	109.	27.5	110.	26.5
111.	28.4	112.	23.5	113.	30.6	114.	23.4	115.	27.2
116.	22.1	117.	25.4	118.	27.3	119.	23.2	120.	27.6
121.	32.6	122.	28.0	123.	27.1	124.	27.8	125.	26.9
126.	25.4	127.	26.1	128.	23.6	129.	29.7	130.	26.2
131.	30.4	132.	27.6	133.	26.2	134.	26.7	135.	28.1
136.	30.1	137.	23.2	138.	24.3	139.	23.8	140.	23.2
141.	24.0	142.	29.0	143.	20.0	144.	29.8	145.	23.5
146.	22.8	147.	25.4	148.	27.7	149.	26.3	150.	27.3
151.	28.0	152.	28.5	153.	31.8	154.	26.6	155.	20.4
156.	25.7	157.	28.0	158.	27.4	159.	25.0	160.	32.2
161.	29.4	162.	30.9	163.	24.2	164.	23.7	165.	27.0
166.	27.0	167.	28.0	168.	29.2	169.	29.2	170.	27.2
171.	24.1	172.	27.0	173.	23.8	174.	24.4	175.	27.6
176.	22.7	177.	28.9	178.	26.7	179.	27.5	180.	23.5
181.	30.3	182.	27.7	183.	24.9	184.	21.8	185.	30.7
186.	27.2	187.	29.0	188.	31.2	189.	24.7	190.	28.9
191.	26.3	192.	32.0	193.	25.3	194.	28.7	195.	27.0
196.	28.5	197.	24.9	198.	25.1	199.	30.4	200.	24.9
201.	27.3	202.	25.8	203.	28.6	204.	24.8	205.	26.9
206.	28.9	207.	21.6	208.	21.6	209.	22.3	210.	30.2
211.	24.6	212.	30.1	213.	23.3	214.	29.7	215.	32.9
216.	31.8	217.	30.7	218.	27.2	219.	32.0	220.	19.9
221.	24.2	222.	30.2	223.	32.2	224.	34.5	225.	31.6
226.	23.0	227.	32.6	228.	31.1	229.	26.6	230.	28.9
231.	22.2	232.	30.6	233.	26.6	234.	29.7	235.	28.7
236.	30.9	237.	27.0	238.	26.0	239.	23.7	240.	26.6
241.	28.7	242.	29.9	243.	32.3	244.	26.3	245.	27.2
246.	26.2	247.	23.6	248.	20.9	249.	31.5	250.	27.2
251.	27.8	252.	25.3	253.	27.7	254.	31.6	255.	23.2
256.	27.8	257.	23.7	258.	22.2	259.	28.6	260.	29.6

Subject	Score	Subject	Score	Subject	Score	Subject	Score	Subject	Score
261.	25.2	262.	28.4	263.	25.6	264.	27.8	265.	27.9
266.	25.7	267.	29.1	268.	32.0	269.	26.0	270.	26.8
271.	23.5	272.	25.5	273.	22.5	274.	23.6	275.	24.1
276.	25.1	277.	28.3	278.	28.2	279.	26.8	280.	31.0
281.	21.3	282.	26.7	283.	24.9	284.	24.3	285.	25.1
286.	27.3	287.	28.5	288.	25.8	289.	21.3	290.	26.7
291.	29.4	292.	28.2	293.	28.5	294.	24.5	295.	26.2
296.	24.5	297.	24.4	298.	35.5	299.	28.0	300.	29.9
301.	27.0	302.	27.1	303.	23.0	304.	29.7	305.	32.8
306.	29.5	307.	23.3	308.	33.4	309.	27.5	310.	23.8
311.	26.1	312.	27.0	313.	28.4	314.	27.2	315.	26.9
316.	26.2	317.	29.5	318.	28.8	319.	24.6	320.	26.3
321.	23.8	322.	28.8	323.	22.8	324.	31.5	325.	25.5
326.	27.2	327.	28.9	328.	24.5	329.	28.7	330.	33.5
331.	30.0	332.	27.8	333.	25.0	334.	25.9	335.	26.1
336.	28.1	337.	27.2	338.	28.9	339.	25.5	340.	23.0
341.	33.8	342.	23.5	343.	23.4	344.	25.6	345.	27.5
346.	30.1	347.	27.1	348.	33.5	349.	26.1	350.	26.2
351.	29.4	352.	26.9	353.	24.8	354.	28.6	355.	30.3
356.	22.6	357.	25.9	358.	26.8	359.	32.9	360.	27.6
361.	25.4	362.	22.3	363.	25.8	364.	27.5	365.	29.1
366.	31.2	367.	31.6	368.	18.8	369.	29.8	370.	25.4
371.	25.6	372.	30.4	373.	28.0	374.	22.7	375.	27.1
376.	31.6	377.	26.0	378.	27.7	379.	28.0	380.	31.3
381.	26.0	382.	27.5	383.	25.2	384.	26.2	385.	28.8
386.	29.8	387.	25.0	388.	23.2	389.	26.0	390.	30.4
391.	28.2	392.	27.1	393.	34.3	394.	29.3	395.	25.8
396.	27.8	397.	28.5	398.	27.3	399.	26.9	400.	31.0
401.	27.6	402.	24.2	403.	33.2	404.	23.6	405.	21.3
406.	25.3	407.	25.9	408.	28.1	409.	24.8	410.	25.6
411.	26.3	412.	24.7	413.	26.1	414.	30.8	415.	27.4
416.	31.9	417.	27.5	418.	30.2	419.	25.6	420.	26.9
421.	30.3	422.	33.1	423.	31.6	424.	31.7	425.	25.9
426.	27.7	427.	26.8	428.	31.5	429.	22.1	430.	22.2
431.	30.3	432.	26.1	433.	32.7	434.	22.9	435.	28.2
436.	24.9	437.	28.3	438.	34.3	439.	23.6	440.	28.9
441.	28.9	442.	26.0	443.	23.4	444.	26.1	445.	28.4
446.	25.3	447.	29.5	448.	27.9	449.	24.7	450.	26.2
451.	29.7	452.	28.5	453.	31.7	454.	18.3	455.	28.0
456.	24.2	457.	30.1	458.	24.1	459.	27.2	460.	26.8
461.	25.2	462.	31.1	463.	24.0	464.	27.8	465.	28.0
466.	30.8	467.	24.8	468.	26.1	469.	26.0	470.	30.4
471.	25.1	472.	27.5	473.	23.5	474.	28.7	475.	27.5
476.	22.6	477.	21.6	478.	23.8	479.	28.1	480.	30.8
481.	28.3	482.	27.8	483.	25.2	484.	22.9	485.	26.8
486.	29.2	487.	27.7	488.	28.3	489.	20.6	490.	29.0
491.	25.4	492.	26.5	493.	23.3	494.	27.9	495.	28.0
496.	26.4	497.	32.3	498.	28.7	499.	21.9	500.	27.7

REFERENCES

References Cited

1. Jean D. Gibbons and John W. Pratt, "*P*-values: Interpretation and Methodology," *The American Statistician*, *29* (1975), 20–25.

2. Anita K. Bahn, "*P* and the Null Hypothesis," *Annals of Internal Medicine*, *76* (1972), 674.

3. Wayne W. Daniel, "What Are *p*-Values? How Are They Calculated? How Are They Related to Levels of Significance?" *Nursing Research*, *26* (1977), 304–306.

4. Wayne W. Daniel, *Introductory Statistics with Applications*, Houghton Mifflin, Boston, 1977.

Other References, Journal Articles

1. Arnold Binder, "Further Considerations of Testing the Null Hypothesis and the Strategy and Tactics of Investigating Theoretical Models," *Psychological Review*, *70* (1963), 107–115.

2. C. Allen Boneau, "The Effects of Violations of Assumptions Underlying the *t* Test," *Psychological Bulletin*, *57* (1960), 49–64.

3. Donna R. Brogan, "Choosing an Appropriate Statistical Test of Significance for a Nursing Research Hypothesis or Question," *Western Journal of Nursing*, *3* (1981), 337–363.

4. C. J. Burke, "A Brief Note on One-tailed Tests," *Psychological Bulletin*, *50* (1953), 384–387.

5. C. J. Burke, "Further Remarks on One-tailed Tests," *Psychological Bulletin*, *51* (1954), 587–590.

6. Wayne W. Daniel, "Statistical Significance versus Practical Significance," *Science Education*, *61* (1977) 423–427.

7. Eugene S. Edgington, "Statistical Inference and Nonrandom Samples," *Psychological Bulletin*, *66* (1966), 485–487.

8. Ward Edwards, "Tactical Note on the Relation Between Scientific and Statistical Hypotheses," *Psychological Bulletin*, *63* (1965), 400–402.

9. William E. Feinberg, "Teaching the Type I and II Errors: The Judical Process," *The American Statistician*, *25* (June 1971), 30–32.

10. I. J. Good, "What Are Degrees of Freedom?" *The American Statistician*, *27* (1973), 227–228.

11. David A. Grant, "Testing the Null Hypothesis and the Strategy and Tactics of Investigating Theoretical Models," *Psychological Review*, *69* (1962), 54–61.

12. W. E. Hick, "A Note on One-tailed and Two-tailed Tests," *Psychological Review*, *59* (1952), 316–318.

13. L. V Jones, "Tests of Hypotheses: One-sided vs. Two-sided Alternatives," *Psychological Bulletin*, *46* (1949), 43–46.

14. L. V. Jones, "A Rejoinder on One-tailed Tests," *Psychological Bulletin*, *51* (1954), 585–586.

15. Sanford Labovitz, "Criteria for Selecting a Significance Level: A Note on the Sacredness of .05," *The American Sociologist*, *3* (1968), 220–222.

16. William Lurie, "The Impertinent Questioner: The Scientist's Guide to the Statistician's Mind," *American Scientist*, *46* (1958), 57–61.

17. M. R. Marks, "One- and Two-tailed Tests," *Psychological Review*, *60* (1953), 207–208.

18. C. A. McGilchrist and J. Y. Harrison, "Testing of Means with Different Alternatives," *Technometrics*, *10* (1968), 195–198.

19. Mary G. Natrella, "The Relation Between Confidence Intervals and Tests of Significance," *American Statistician*, *14* (1960), 20–22, 33.

20. O. B. Ross, Jr., "Use of Controls in Medical Research," *Journal of the American Medical Association*, *145* (1951), 72–75.

21. William W. Rozeboom, "The Fallacy of the Null-Hypothesis Significance Test," *Psychological Bulletin*, *57* (1960), 416–428.

22. H. C. Selvin, "A Critique of Tests of Significance in Survey Research," *American Sociological Review*, *22* (1957), 519–527.

23. James K. Skipper, Anthony L. Guenther, and Gilbert Nass, "The Sacredness of .05: A Note Concerning the Uses of Significance in Social Science," *The American Sociologist*, *2* (1967), 16–18.

24. Warren R. Wilson and Howard Miller, "A Note on the Inconclusiveness of Accepting the Null Hypothesis," *Psychological Review*, *71* (1964), 238–242.

25. Warren Wilson, Howard L. Miller, and Jerold S. Lower, "Much Ado About the Null Hypothesis," *Psychological Bulletin*, *67* (1967), 188–196.

26. R. F. Winch and D. T. Campbell, "Proof? No. Evidence? Yes. The Significance of Tests of Significance," *American Sociologist*, *4* (1969), 140–143.

Other References, Books

1. W. I. B. Beveridge, *The Art of Scientific Investigation*, Third Edition, W. W. Norton, New York, 1957.

2. Dwight J. Ingle, *Principles of Research in Biology and Medicine*, J. B. Lippincott, Philadelphia, 1958.

3. E. L. Lehman, *Testing Statistical Hypotheses*, Wiley, New York, 1959.

Other References, Other Publications

1. David Bakan, "The Test of Significance in Psychological Research," in *One Method: Toward a Reconstruction of Psychological Investigation*, Jossey-Bass, San Francisco, 1967.

2. William H. Kruskal, "Tests of Significance," in *International Encyclopedia of the Social Sciences*, *14* (1968), 238–250.

CHAPTER 7

Analysis of Variance

7.1 INTRODUCTION

In the preceding chapters the basic concepts of statistics have been examined, and they provide a foundation for the present and succeeding chapters.

The present chapter is concerned with *analysis of variance*, which may be defined as *a technique whereby the total variation present in a set of data is partitioned into several components. Associated with each of these components is a specific source of variation, so that in the analysis it is possible to ascertain the magnitude of the contributions of each of these sources to the total variation.*

The development of this subject is mainly due to the work of the late R. A. Fisher (1), whose contributions to statistics, spanning the years 1912 to 1962, have had a tremendous influence on modern statistical thought (2, 3).

Analysis of variance finds its widest application in the analysis of data derived from experiments. The principles of the design of experiments are well covered in several books, including those of Cochran and Cox (4), Cox (5), Davies (6), Federer (7), Finney (8), Fisher (1), John (9), Kempthorne (10), Li (11), and Mendenhall (12). We do not study this topic in detail, since to do it justice would require a minimum of an additional chapter. Some of the important concepts in experimental design, however, will become apparent as we discuss analysis of variance.

The relationship between the two topics may be summarized by pointing out that when experiments are designed with the analysis in mind, researchers can, before conducting experiments, identify those sources of variation that they consider important and choose a design that will allow them to measure the extent of the contribution of these sources to the total variation. Analysis of variance is used for two different purposes: (1) to estimate and test hypotheses about population variances and (2) to estimate and test hypotheses about population means. We are concerned here with the latter use. However, as we shall see, our conclusions regarding the means will depend on the magnitudes of the observed variances.

Underlying the valid use of analysis of variance as a tool of statistical inference are a set of fundamental assumptions. For a detailed discussion of these assumptions, the reader is referred to the paper by Eisenhart (13). Although an experimenter must not expect to find all the assumptions met to perfection, it is important that the user of analysis of variance techniques be aware of the underlying assumptions and be able to recognize when they are substantially unsatisfied. The consequences of the failure to meet the assumptions are discussed by Cochran (14) in a companion paper to that of Eisenhart. Because experiments in which all the assumptions are perfectly met are rare, Cochran suggests that analysis of variance results be considered as approximate rather than exact. These assumptions are pointed out at appropriate points in the following sections.

We discuss analysis of variance as it is used to analyze the results of two different experimental designs, the completely randomized and the randomized complete block designs. In addition to these, the concept of a factorial experiment is given through its use in a completely randomized design. These do not exhaust the possibilities. A discussion of additional designs will be found in the references (4–12).

In our presentation of the analysis of variance for the different designs, we follow the nine-step procedure presented in Chapter 6. The following is a restatement of the steps of the procedure, including some new concepts necessary for its adaptation to analysis of variance.

1. *Description of Data*   In addition to describing the data in the usual way, we display the sample data in tabular form.

2. *Assumptions*   Along with the assumptions underlying the analysis we present the model for each design we discuss. The model consists of a symbolic representation of a typical value from the data being analyzed.

3. *Hypotheses*

4. *Test Statistic*

5. *Distribution of the Test Statistic*

6. *Decision Rule*

7. *Calculation of the Test Statistic*   The results of the arithmetic calculations will be summarized in a table called the analysis of variance (ANOVA) table. The entries in the table make it easy to evaluate the results of the analysis.

8. *Statistical Decision*

9. *Conclusion*

The Use of Computers   The calculations required by analysis of variance are lengthier and more complicated than those we have encountered in preceding chapters. For this reason the computer assumes an important role in analysis of variance. All of the exercises appearing in this chapter are suitable for computer analysis and may be used with the statistical packages mentioned in Chapter 1.

The output of the statistical packages may vary slightly from that presented in this chapter, but this should pose no major problem to those who use a computer to analyze the data of the exercises. The basic concepts of analysis of variance that we present here should provide the necessary background for understanding the description of the programs and their output in any of the statistical packages.

7.2 THE COMPLETELY RANDOMIZED DESIGN

We saw in Chapter 6 how it is possible to test the null hypothesis of no difference between two population means. It is not unusual for the investigator to be interested in testing the null hypothesis of no difference among several population means. The student first encountering this problem might be inclined to suggest that all possible pairs of sample means be tested separately by means of the Student t test. Suppose there are five populations involved. The number of possible pairs of sample mean is $\binom{5}{2} = 10$. As the amount of work involved in carrying out this many t tests is substantial, it would be worthwhile if a more efficient alternative for analysis were available. A more important consequence of performing all possible t tests, however, is that it is very likely to lead to a false conclusion.

Suppose we draw five samples from populations having equal means. As we have seen, there would be 10 tests if we were to do each of the possible tests separately. If we select a significance level of $\alpha = .05$ for each test, the probability of failing to reject a hypothesis of no difference in each case would be .95. By the multiplication rule of probability, if the tests were independent of one another, the probability of failing to reject a hypothesis of no difference in all 10 cases would be $(.95)^{10} = .5987$. The probability of rejecting, at least, one hypothesis of no difference, then, would be $1 - .5987 = .4013$. Since we know that the null hypothesis is true in every case in this illustrative example, rejecting the null hypothesis constitutes the committing of a type I error. In the long run, then, in testing all possible pairs of means from five samples, we would commit a type I error 40 percent of the time. The problem becomes even more complicated in practice, since three or more t tests based on the same data would not be independent of one another.

It becomes clear, then, that some other method for testing for a significant difference among several means is needed. Analysis of variance provides such a method.

The simplest type of analysis of variance is that known as *one-way analysis of variance*, in which only one source of variation, or *factor*, is investigated. It is an extension to three or more samples of the *t*-test procedure (discussed in Chapter 6) for use with two independent samples. Stated another way, we can say that the t test for use with two independent samples is a special case of one-way analysis of variance.

TABLE 7.2.1

Table of Sample Values for the Completely Randomized Design

	TREATMENT					
	1	*2*	*3*	$\cdots$	*k*	
	x_{11}	x_{12}	x_{13}	$\cdots$	x_{1k}	
	x_{21}	x_{22}	x_{23}	$\cdots$	x_{2k}	
	x_{31}	x_{32}	x_{33}	$\cdots$	x_{3k}	
	$\vdots$	$\vdots$	$\vdots$	$\vdots$	$\vdots$	
	$x_{n_1 1}$	$x_{n_2 2}$	$x_{n_3 3}$	$\cdots$	$x_{n_k k}$	
Total	$T_{.1}$	$T_{.2}$	$T_{.3}$		$T_{.k}$	$T_{..}$
Mean	$\bar{x}_{.1}$	$\bar{x}_{.2}$	$\bar{x}_{.3}$		$\bar{x}_{.k}$	$\bar{x}_{..}$

In a typical situation we want to use one-way analysis of variance to test the null hypothesis that three or more treatments are equally effective. The necessary experiment is designed in such a way that the treatments of interest are assigned completely at random to the subjects or objects on which the measurements to determine treatment effectiveness are to be made. For this reason the design is called the *completely randomized experimental design*. The measurements (or observations) resulting from a completely randomized experimental design, along with the means and totals that can be computed from them, may be displayed for convenience as in Table 7.2.1. The symbols used in Table 7.2.1 are defined as follows:

$$x_{ij} = \text{the } i\text{th observation resulting from the } j\text{th treatment}$$

$$i = 1, 2, \ldots, n_j, \; j = 1, 2, \ldots, k$$

$$T_{.j} = \sum_{i=1}^{n_j} x_{ij} = \text{total of the } j\text{th column}$$

$$\bar{x}_{.j} = \frac{T_{.j}}{n_j} = \text{mean of the } j\text{th column}$$

$$T_{..} = \sum_{j=1}^{k} T_{.j} = \sum_{j=1}^{k} \sum_{i=1}^{n_j} x_{ij} = \text{total of all observations}$$

$$\bar{x}_{..} = \frac{T_{..}}{N}, \qquad N = \sum_{j=1}^{k} n_j$$

Example 7.2.1 In a study of the effect of glucose on insulin release, specimens of pancreatic tissue from experimental animals were randomly assigned to be treated with one of five different stimulants. Later, a determination was made

TABLE 7.2.2

Insulin Released

	STIMULANT					
	1	*2*	*3*	*4*	*5*	
	1.53	3.15	3.89	8.18	5.86	
	1.61	3.96	3.68	5.64	5.46	
	3.75	3.59	5.70	7.36	5.69	
	2.89	1.89	5.62	5.33	6.49	
	3.26	1.45	5.79	8.82	7.81	
		1.56	5.33	5.26	9.03	
				7.10	7.49	
					8.98	
Total	13.04	15.60	30.01	47.69	56.81	163.15
Mean	2.61	2.60	5.00	6.81	7.10	5.10

on the amount of insulin released. The experimenters wished to know if they could conclude that there is a difference among the five treatments with respect to the mean amount of insulin released.

1. *Description of Data*   Treatment 1 was applied to five specimens, treatments 2 and 3 were each used on six specimens, seven specimens received treatment 4, and treatment 5 was applied to eight specimens. The resulting measurements of amount of insulin released following treatment are displayed in Table 7.2.2.

2. *Assumptions*   Before stating the assumptions, let us specify the model for the experiment described here.

The Model.   To write down the model for this example, let us begin by identifying a typical value from the set of data represented by the sample displayed in Table 7.2.2. We use the symbol x_{ij} to represent this typical value.

Within each treatment group (or population) represented by our data, any particular value bears this relationship to the true mean, μ_j, of that group: it is equal to the true group mean plus some amount that is either zero, a positive amount, or a negative amount. This means that a particular value in a given group may be equal to the group mean, larger than the group mean, or smaller than the group mean. Let us call the amount by which any value differs from its group mean the *error* and represent this error by the symbol e_{ij}. By the term error, we do not mean a mistake. The term is used to refer to the uncontrolled variation that exists among members of any population. Given a population of adult males, for example, we know that the heights of some individuals are above the true mean height of the population, while some heights are below the population mean. This variation is caused by a myriad of hereditary and environmental factors. If to any group mean, μ_j, we add a given error, e_{ij}, the result will be x_{ij}, the observation that deviates from the group mean by the amount e_{ij}.

We may write this relationship symbolically as

$$x_{ij} = \mu_j + e_{ij} \tag{7.2.1}$$

Solving for e_{ij} we have

$$e_{ij} = x_{ij} - \mu_j \tag{7.2.2}$$

If we have in mind k populations (in the present example $k = 5$), we may refer to the grand mean of all the observations in all the populations as μ. Given k populations, we could compute μ by taking the average of the k population means, that is,

$$\mu = \frac{\sum \mu_j}{k} \tag{7.2.3}$$

Just as a particular observation within a group, in general, differs from its group mean by some amount, a particular group mean differs from the grand mean by some amount. The amount by which a group mean differs from the grand mean we refer to as the *treatment effect*. We may write the jth treatment effect as

$$\tau_j = \mu_j - \mu \tag{7.2.4}$$

τ_j is a measure of the effect on μ_j of having been computed from observations receiving the jth treatment.

We may solve Equation 7.2.4 for μ_j to obtain

$$\mu_j = \mu + \tau_j \tag{7.2.5}$$

If we substitute the right-hand portion of Equation 7.2.5 for μ_j in Equation 7.2.1 we have

$$x_{ij} = \mu + \tau_j + e_{ij}, \qquad i = 1, 2, \ldots, n_j; \quad j = 1, 2, \ldots, k \tag{7.2.6}$$

and our model is specified.

By looking at our model we can see that a typical observation from the total set of data under study is composed of the grand mean, a treatment effect, and an error term representing the deviation of the observation from its group mean.

In the present example we assume that we are interested only in the k treatments represented in our experiment. Any inferences that we make apply only to these treatments. We do not wish to extend our inferences to any larger collection of treatments. When we place such a restriction on our inference goals, we refer to our model as the *fixed-effects model*, or *model I*. The discussion in this book is limited to this model.

Information on other models will be found in the articles by Eisenhart (13), Wilk and Kempthorne (15), Crump (16), Cunningham and Henderson (17), Henderson (18), Rutherford (19), Schultz (20), and Searle (21).

Assumptions of the Model. The assumptions for the fixed effects model are as follows.

(a) The k sets of observed data constitute k independent random samples from the respective populations.

(b) Each of the populations from which the samples come is normally distributed with mean, μ_j, and variance σ_j^2.

(c) Each of the populations has the same variance. That is, $\sigma_1^2 = \sigma_2^2 = \cdots = \sigma_k^2 = \sigma^2$, the common variance.

(d) The τ_j's are unknown constants and $\sum \tau_j = 0$, since the sum of all deviations of the μ_j from their mean, μ, is zero.

Three consequences of the relationship

$$e_{ij} = x_{ij} - \mu_j$$

specified in Equation 7.2.2 are

(i) The e_{ij} have a mean of 0, since the mean of x_{ij} is μ_j.

(ii) The e_{ij} have a variance equal to the variance of the x_{ij}, since the e_{ij} and x_{ij} differ only by a constant; that is, the error variance is equal to σ^2, the common variance specified in assumption (c) above.

(iii) The e_{ij} are normally (and independently) distributed.

3. *Hypotheses* We may test the null hypothesis that all population or treatment means are equal against the alternative that the members of at least one pair are not equal. We may state the hypotheses formally as follows:

$$H_0: \mu_1 = \mu_2 = \cdots = \mu_k$$
$$H_A: \text{not all } \mu_j \text{ are equal}$$

If the population means are equal, each treatment effect is equal to zero, so that, alternatively, the hypotheses may be stated as

$$H_0: \tau_j = 0, \qquad j = 1, 2, \ldots, k$$
$$H_A: \text{not all } \tau_j = 0$$

For our present example we have

$$H_0: \mu_1 = \mu_2 = \mu_3 = \mu_4 = \mu_5$$
$$H_A: \text{not all } \mu_j \text{ are equal.}$$

At this point a level of significance, α, is chosen. Let us use .05.

If H_0 is true and the assumption of equal variances and normally distributed populations are met, a picture of the populations will look like Figure 7.2.1. When H_0 is true the population means are all equal, and the populations are centered at the same point (the common mean) on the horizontal axis. If the

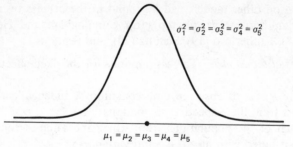

FIGURE 7.2.1

Picture of the Populations Represented by Example 7.2.1 When H_0 Is True and the Assumptions Are Met

populations are all normally distributed with equal variances the distributions will be identical, so that in drawing their pictures each is superimposed on each of the others, and a single picture sufficiently represents them all.

When H_0 is false it may be false because one of the population means is different from the others, which are all equal. Or, perhaps, all of the population means are different. These are only two of the possibilities when H_0 is false. There are many other possible combinations of equal and unequal means. Figure 7.2.2 shows a picture of the populations when the assumptions are met, but H_0 is false because no two population means are equal.

4. *Test Statistic*   The test statistic for one-way analysis of variance is a computed variance ratio, which we designate by V.R. as we did in Chapter 6. The two variances from which V.R. is calculated are themselves computed from the sample data. The methods by which they are calculated will be given in the discussion that follows.

5. *Distribution of the Test Statistic*   As discussed in Section 6.8, V.R. is distributed as the F distribution when H_0 is true and the assumptions are met.

6. *Decision Rule*   In general, the decision rule is: Reject the null hypothesis if the computed value of V.R. is equal to or greater than the critical value of F for the chosen α level.

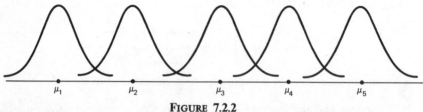

FIGURE 7.2.2

Picture of the Populations Represented in Example 7.2.1 When the Assumptions of Equal Variances and Normally Distributed Populations Are Met, But H_0 Is False Because None of the Population Means Are Equal

7. *Calculation of the Test Statistic* We have defined analysis of variance as a process whereby the total variation present in a set of data is partitioned into components that are attributable to different sources. The term *variation* used in this context refers to the *sum of squared deviations of observations from the their mean,* or *sum of squares* for short.

The data for Exercise 7.2.1 are graphed in Figure 7.2.3. Here we see the relationship between $\bar{x}_{..}$, which estimates μ of the model; the $\bar{x}_{.j}$'s, the

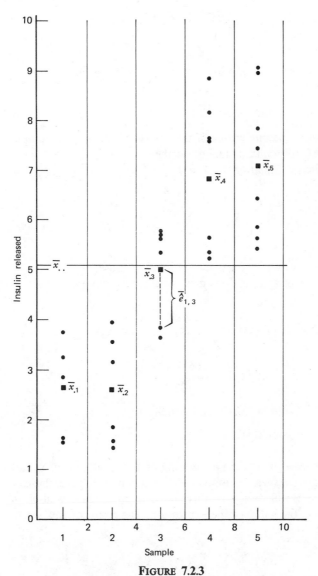

FIGURE 7.2.3

Graph of Data Given in Table 7.2.1 for Example 7.2.1

sample means; the x_{ij}'s, the actual observations; and the $\hat{e}_{ij}$'s, the estimates of the e_{ij}'s of the model.

The Total Sum of Squares   Before we can do any partitioning, we must first obtain the total sum of squares. The total sum of squares is the sum of the squares of the deviations of individual observations from the mean of all the observations taken together. This *total sum of squares* is defined as

$$SST = \sum_{j=1}^{k} \sum_{i=1}^{n_j} (x_{ij} - \bar{x}_{..})^2 \tag{7.2.7}$$

where $\displaystyle\sum_{i=1}^{n_j}$ tells us to sum the squared deviations for each group, and $\displaystyle\sum_{j=1}^{k}$ tells us to add the k group totals obtained by applying $\displaystyle\sum_{i=1}^{n_j}$. The reader will recognize Equation 7.2.7 as the numerator of the variance that may be computed from the complete set of observations taken together.

We may rewrite Equation 7.2.7 as

$$SST = \sum_{j=1}^{k} \sum_{i=1}^{n_j} x_{ij}^2 - \frac{T_{..}^2}{N} \tag{7.2.8}$$

which is more convenient for computational purposes.

The total sum of squares for our illustrative example is

$$SST = (1.53)^2 + (1.61)^2 + \cdots + (8.98)^2 - \frac{(163.15)^2}{32}$$

$$= 994.3529 - \frac{26617.923}{32}$$

$$= 994.3529 - 831.81008$$

$$= 162.54282$$

We now proceed with the partitioning of the total sum of squares. We may, without changing its value, insert $-\bar{x}_{.j} + \bar{x}_{.j}$ in the parentheses of Equation 7.2.7. The reader will recognize this added quantity as a well-chosen zero that does not change the value of the expression. The result of this addition is

$$SST = \sum_{j=1}^{k} \sum_{i=1}^{n_j} (x_{ij} - \bar{x}_{.j} + \bar{x}_{.j} - \bar{x}_{..})^2 \tag{7.2.9}$$

If we group terms and expand, we have

$$SST = \sum_{j=1}^{k} \sum_{i=1}^{n_j} \left[(x_{ij} - \bar{x}_{.j}) + (\bar{x}_{.j} + \bar{x}_{..})\right]^2$$

$$= \sum_{j=1}^{k} \sum_{i=1}^{n_j} (x_{ij} - \bar{x}_{.j})^2 + 2 \sum_{j=1}^{k} \sum_{i=1}^{n_j} (x_{ij} - \bar{x}_{.j})(\bar{x}_{.j} - \bar{x}_{..})$$

$$+ \sum_{j=1}^{k} \sum_{i=1}^{n_j} (\bar{x}_{.j} - \bar{x}_{..})^2 \qquad (7.2.10)$$

The middle term above may be written as

$$2 \sum_{j=1}^{k} (\bar{x}_{.j} - \bar{x}_{..}) \sum_{i=1}^{n_j} (x_{ij} - \bar{x}_{.j}) \qquad (7.2.11)$$

Examination of Equation 7.2.11 reveals that this term is equal to zero, since the sum of the deviations of a set of values from their mean as in $\sum_{i=1}^{n_j} (x_{ij} - \bar{x}_{.j})$ is equal to zero.

We now may write Equation 7.2.10 as

$$SST = \sum_{j=1}^{k} \sum_{i=1}^{n_j} (x_{ij} - \bar{x}_{.j})^2 + \sum_{j=1}^{k} \sum_{i=1}^{n_j} (\bar{x}_{.j} + \bar{x}_{..})^2$$

$$= \sum_{j=1}^{k} \sum_{i=1}^{n_j} (x_{ij} - \bar{x}_{.j})^2 + \sum_{j=1}^{k} n_j (\bar{x}_{.j} - \bar{x}_{..})^2 \qquad (7.2.12)$$

When the number of observations is the same in each group the last term on the right may be rewritten to give

$$SST = \sum_{j=1}^{k} \sum_{i=1}^{n_j} (x_{ij} - \bar{x}_{.j})^2 + n \sum_{j=1}^{k} (\bar{x}_{.j} - \bar{x}_{..})^2 \qquad (7.2.13)$$

where $n = n_1 = n_2 = \cdots = n_k$.

The Within Groups Sum of Squares   The partitioning of the total sum of squares is now complete, and we see that in the present case there are two components. Let us now investigate the nature and source of these two components of variation.

If we look at the first term on the right of the Equation 7.2.12, we see that the first step in the indicated computation calls for performing certain calculations *within* each group. These calculations involve computing within each group the sum of the squared deviations of the individual observations from their mean.

When these calculations have been performed within each group, the symbol $\sum\limits_{j=1}^{k}$ tells us to obtain the sum of the individual group results. This component of variation is called the *within groups sum of squares* and may be designated *SSW*. This quantity is sometimes referred to as the *residual* or *error* sum of squares. The expression may be written in a computationally more convenient form as follows:

$$SSW = \sum_{j=1}^{k} \sum_{i=1}^{n_j} (x_{ij} - \bar{x}_{.j})^2 = \sum_{j=1}^{k} \sum_{i=1}^{n_j} x_{ij}^2 - \sum_{j=1}^{k} \frac{(T_{.j})^2}{n_j} \qquad (7.2.14)$$

By using the data from Table 7.2.1, we find the within sum of squares for Example 7.2.1 to be

$$SSW = (1.53)^2 + (1.61)^2 + \cdots + (8.98)^2$$
$$- \left[\frac{(13.04)^2}{5} + \frac{(15.60)^2}{6} + \frac{(30.01)^2}{6} + \frac{(47.69)^2}{7} + \frac{(56.81)^2}{8} \right]$$
$$= 994.3529 - (34.00832 + 40.56$$
$$+ 150.10002 + 324.90516 + 403.42201)$$
$$= 994.3529 - 952.99551$$
$$= 41.35739$$

The Among Groups Sum of Squares   Now let us examine the second term on the right in Equation 7.2.12. The operation called for by this term is to obtain for each group the squared deviation of the group mean from the grand mean and to multiply the result by the size of the group. Finally, we must add these results over all groups. This quantity is a measure of the variation among groups and is referred to as the *sum of squares among groups* or *SSA*. The computing formula is as follows:

$$SSA = \sum_{j=1}^{k} n_j (\bar{x}_{.j} - \bar{x}_{..})^2 = \sum_{j=1}^{k} \frac{T_{.j}^2}{n_j} - T_{..}^2 / N \qquad (7.2.15)$$

The *SSA* for our illustrative example is computed as follows:

$$SSA = \frac{(13.04)^2}{5} + \frac{(15.60)^2}{6} + \frac{(30.01)^2}{6} + \frac{(47.69)^2}{7} + \frac{(56.81)^2}{8} - \frac{(163.15)^2}{32}$$
$$= 952.99551 - 831.81008$$
$$= 121.18543$$

In summary, then, we have found that the total sum of squares is equal to the sum of the among sum of squares and the within sum of squares. We express this relationship as follows:

$$SST = SSA + SSW$$

For our illustrative example we have

$$162.54282 = 121.18543 + 41.35739$$
$$162.54282 = 162.54282$$

From the sums of squares that we have just computed, it is possible to obtain two estimates of the common population variance, σ^2. It can be shown that when the assumptions are met and the population means are all equal, both the among sum of squares and the within sum of squares, when divided by their respective degrees of freedom, yield independent and unbiased estimates of σ^2.

The First Estimate of σ^2   Within any sample

$$\frac{\sum\limits_{i=1}^{n_j} (x_{ij} - \bar{x}_{.j})^2}{n_j - 1}$$

provides an unbiased estimate of the true variance of the population from which the sample came. Under the assumption that the population variances are all equal, we may pool the k estimates to obtain

$$\frac{\sum\limits_{j=1}^{k} \sum\limits_{i=1}^{n_j} (x_{ij} - \bar{x}_{.j})^2}{\sum\limits_{j=1}^{k} (n_j - 1)} \tag{7.2.16}$$

This is our first estimate of σ^2 and may be called the *within groups variance*, since it is the within groups sum of squares of Equation 7.2.14 divided by the appropriate degrees of freedom. The student will recognize this as an extension to k samples of the pooling of variances procedure encountered in Chapters 5 and 6 when the variances from two samples were pooled in order to use the t distribution. The quantity in Equation 7.2.16 is customarily referred to as the within groups *mean square* rather than the within groups variance. The within groups mean square for our illustrative example is

$$MSW = \frac{SSW}{27} = \frac{41.35739}{27} = 1.5317552$$

The within groups mean square is a valid estimate of σ^2 only if the population variances are equal. It is not necessary, however, for H_0 to be true in order for the within groups mean square to be a valid estimate of σ^2. That is, the within groups mean square estimates σ^2 regardless of whether H_0 is true or false, as long as the population variances are equal.

The Second Estimation of σ^2 The second estimate of σ^2 may be obtained from the familiar formula for the variance of sample means, $\sigma_{\bar{x}}^2 = \sigma^2/n$. If we solve this equation for σ^2, the variance of the population from which the samples were drawn, we have

$$\sigma^2 = n\sigma_{\bar{x}}^2 \tag{7.2.17}$$

An unbiased estimate of $\sigma_{\bar{x}}^2$, computed from sample data, is provided by

$$\frac{\sum_{j=1}^{k} (\bar{x}_{.j} - \bar{x}_{..})^2}{k-1}$$

If we substitute this quantity into Equation 7.2.17, we obtain the desired estimate of σ^2,

$$\frac{n\sum_{j=1}^{k} (\bar{x}_{.j} - \bar{x}_{..})^2}{k-1} \tag{7.2.18}$$

The reader will recognize the numerator of Equation 7.2.18 as the among groups sum of squares for the special case when all sample sizes are equal. This sum of squares when divided by the associated degrees of freedom $k-1$ is referred to as the *among groups mean square*.

When the sample sizes are not all equal, an estimate of σ^2 based on the variability among sample means is provided by

$$\frac{\sum_{j=1}^{k} n_j(\bar{x}_{.j} - \bar{x}_{..})^2}{k-1} \tag{7.2.19}$$

For our illustrative example we have the following among groups mean square:

$$MSA = \frac{SSA}{(5-1)} = \frac{121.18543}{4} = 30.296358$$

If, indeed, the null hypothesis is true we would expect these two estimates of σ^2 to be fairly close in magnitude. If the null hypothesis is false, that is, if all population means are not equal, we would expect the among groups mean square, which is computed by using the squared deviations of the sample means from the overall mean, to be larger than the within groups mean square.

In order to understand analysis of variance we must realize that the among groups mean square provides a valid estimate of σ^2 when the assumption of equal population variances is met *and when H_0 is true.* Both conditions, a true null hypothesis and equal population variances, must be met in order for the among groups mean square to be a valid estimate of σ^2.

The Variance Ratio   What we need to do now is to compare these two estimates of σ^2, and we do this by computing the following variance ratio, which is the desired test statistic:

$$\text{V.R.} = \frac{\text{among groups mean square}}{\text{within groups mean square}}$$

If the two estimates are about equal, V.R. will be close to 1. A ratio close to 1 tends to support the hypothesis of equal population means. If, on the other hand, the among groups mean square is considerably larger than the within groups mean square, V.R. will be considerably greater than 1. A value of V.R. sufficiently greater than 1 will cast doubt on the hypothesis of equal population means.

We know that because of the vagaries of sampling, even when the null hypothesis is true, it is unlikely that the among and within groups mean squares will be equal. We must decide, then, how big the observed difference has to be before we can conclude that the difference is due to something other than sampling fluctuation. In other words, how large a value of V.R. is required for us to be willing to conclude that the observed difference between our two estimates of σ^2 is not the result of chance alone?

The F Test   To answer the question posed above, we must consider the sampling distribution of the ratio of two sample variances. In Chapter 5 we learned that the quantity $(s_1^2/\sigma_1^2)/(s_2^2/\sigma_2^2)$ follows a distribution known as the F distribution when the sample variances are computed from random and independently drawn samples from normal populations. The F distribution, introduced by R. A. Fisher in the early 1920s, has become one of the most widely used distributions in modern statistics. We have already become acquainted with its use in constructing confidence intervals for, and testing hypotheses about, population variances. In this chapter we shall see that it is the distribution fundamental to analysis of variance.

In Chapter 6 we learned that when the population variances are the same, they cancel in the expression $(s_1^2/\sigma_1^2)/(s_2^2/\sigma_2^2)$, leaving s_1^2/s_2^2, which is itself distributed as F. The F distribution is really a family of distributions, and the particular F distribution we use in a given situation depends on the number of degrees of freedom associated with the sample variance in the numerator (*numerator degrees of freedom*) and the number of degrees of freedom associated with the sample variance in the denominator (*denominator degrees of freedom*).

TABLE 7.2.3
Analysis of Variance Table for the Completely Randomized Design

Source of variation	Sum of squares	Degrees of freedom	Mean square	Variance ratio
Among samples	$SSA = \sum\limits_{j=1}^{k} n_j(\bar{x}_{.j} - \bar{x}_{..})^2$ $= \sum\limits_{j=1}^{k} \dfrac{T_{.j}^2}{n_j} - \dfrac{T_{..}^2}{N}$	$k - 1$	MSA $= SSA/(k-1)$	$V.R. = \dfrac{MSA}{MSW}$
Within samples	$SSW = \sum\limits_{j=1}^{k} \sum\limits_{i=1}^{n_j} (x_{ij} - \bar{x}_{.j})^2$ $= \sum\limits_{j=1}^{k} \sum\limits_{i=1}^{n_j} x_{ij}^2 - \sum\limits_{i=1}^{k} \dfrac{(T_{.j})^2}{n_j}$	$N - k$	MSW $= SSW/(N-k)$	
Total	$SST = \sum\limits_{j=1}^{k} \sum\limits_{i=1}^{n_j} (x_{ij} - \bar{x}_{..})^2$ $= \sum\limits_{j=1}^{k} \sum\limits_{i=1}^{n_j} x_{ij}^2 - \dfrac{T_{..}^2}{N}$	$N - 1$		

Once the appropriate F distribution has been determined, the size of the observed V.R. that will cause rejection of the hypothesis of equal population variances depends on the significance level chosen. The significance level chosen determines the critical value of F, the value that separates the acceptance region from the rejection region.

As we have seen, we compute V.R. in situations of this type by placing the among groups mean square in the numerator and the within groups mean square in the denominator, so that the numerator degrees of freedom is equal to the number of groups minus 1, $(k - 1)$, and the denominator degrees of freedom value is equal to $\sum\limits_{j=1}^{k} (n_j - 1) = \sum\limits_{j=1}^{k} n_j - k = N - k$. For our present example we have

$$\text{numerator degrees of freedom} = 5 - 1 = 4$$
$$\text{denominator degrees of freedom} = 32 - 5 = 27$$

The ANOVA Table. The calculations that we have performed may be summarized and displayed in a table such as Table 7.2.3, which is called the ANOVA table.

Table 7.2.4 is the ANOVA table for our illustrative example.

TABLE 7.2.4

ANOVA Table, Example 7.2.1

Source	SS	d.f.	MS	V.R.
Among samples	121.18543	4	30.296358	19.78
Within samples	41.35739	27	1.5317552	
Total	162.54282	31		

Notice that in Table 7.2.3 the term $\sum_{j=1}^{k} \sum_{i=1}^{n_j} x_{ij}^2 / N = T_{..}^2 / N$ occurs in the expression for both SSA and SST. A savings in computational time and labor may be realized if we take advantage of this fact. We need only to compute this quantity, which is called the correction term and designated by the letter C, once and use it as needed.

The computational burden may be lightened in still another way. Since SST is equal to the sum of SSA and SSW, and since SSA is easier to compute than SSW, we may compute SST and SSA and subtract the latter from the former to obtain SSW.

8. *Statistical Decision*   To reach a decision we must compare our computed V.R. with the critical value of F, which we obtain by entering Table G with 4 and 27 degrees of freedom. If we choose a .05 level of significance, the value of F we seek is 2.73. The resulting acceptance and rejection regions are shown in Figure 7.2.4.

Since our computed V.R., 19.78, is greater than our critical F, 2.73, we must try to explain the difference. There are two possible explanations.

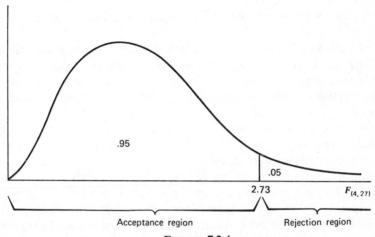

FIGURE 7.2.4

Acceptance and Rejection Regions, Example 7.2.1

If the null hypothesis is true, that is, if the two sample variances are estimates of a common variance, we know that the probability of getting a value as large or larger than 2.73 is .05. We obtained a value much greater than 2.73 which, if the null hypothesis is true, we would consider to be a very rare occurrence. We may, if we wish, conclude that the null hypothesis is true and assume that because of chance we got a set of data that gave rise to a rare event. On the other hand, we may prefer to take the position that our large computed V.R. value does not represent a rare event brought about by chance but, instead, reflects the fact that something other than chance is operative. This other something we conclude to be a false null hypothesis.

It is this latter explanation that we usually give for computed values of V.R. that exceed the critical value of F. In other words, if the computed value of V.R. is greater than the critical value of F, we reject the null hypothesis.

It will be recalled that the original hypothesis we set out to test was

$$H_0: \mu_1 = \mu_2 = \mu_3 = \mu_4 = \mu_5$$

Does rejection of the hypothesis about variances imply a rejection of the hypothesis of equal population means? The answer is, yes. A large value of V.R. resulted from the fact that the among groups mean square was considerably larger than the within groups mean square. Since the among groups mean square is based on the dispersion of the sample means about their mean, this quantity will be large when there is a large discrepancy among the sizes of the sample means. Because of this, then, a significant value of V.R. tells us to reject the null hypothesis that all population means are equal. Since, in our present example, 19.78 is greater than 2.73, we reject the hypothesis that the observed sample variances are estimates of a common variance. For this test, $p < .005$, since $19.78 > F_{.995} = 4.74$.

9. *Conclusion* We conclude that not all population means are equal.

A Word of Caution The completely randomized design is simple and, therefore, widely used. It should be used, however, only when the units receiving the treatments are homogeneous. If the experimental units are not homogeneous, the researcher should consider an alternative design such as one of those to be discussed later in this chapter.

Although in our illustrative example the sample sizes were not equal, this is not a requirement, as the completely randomized design and its analysis may be used when the sample sizes are equal.

In our illustrative example the five treatments are treatments in the usual sense of the word. This is not always the case, however, as the term "treatment" as used in experimental design is quite general. We might, for example, wish to study the response to the same treatment (in the usual sense of the word) of several breeds of animal. We would, however, refer to the breed of animal as the "treatment."

We must also point out that, although the techniques of analysis of variance are more often applied to data resulting from controlled experiments, the techniques also may be used to analyze data collected by a survey, provided that the underlying assumptions are reasonably well met.

Computer Analysis   Figure 7.2.5 shows part of the computer output for Example 7.2.1 provided by a one-way analysis of variance program found in the MINITAB package. When you compare the ANOVA table on the printout with the one given in Table 7.2.4, you see that the printout uses the label "factor" instead of "among samples." The different treatments are referred to on the printout as levels. Thus level 1 = treatment 1, level 2 = treatment 2, and so on. The printout gives the five sample means and standard deviations as well as the pooled standard deviation. This last quantity is equal to the square root of the error mean square shown in the ANOVA table. Finally, the computer output gives graphic representations of the 95 percent confidence intervals for the mean of each of the five populations represented by the sample data. The MINITAB package can provide additional analyses through the use of appropriate commands. The package also contains programs for two-way analysis of variance, to be discussed in the following sections.

```
ANALYSIS OF VARIANCE

DUE TO           DF           SS        MS=SS/DF       F-RATIO
FACTOR            4        121.19        30.30          19.78
ERROR            27         41.36         1.53
TOTAL            31        162.54


LEVEL          N          MEAN         ST. DEV.
C1             5          2.61           1.00
C2             6          2.60           1.10
C3             6          5.00            .96
C4             7          6.81           1.43
C5             8          7.10           1.44

POOLED ST. DEV. =          1.24
```

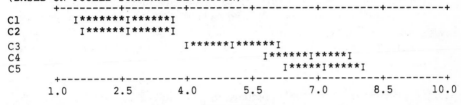

```
INDIVIDUAL 95 PERCENT C. I. FOR LEVEL MEANS
(BASED ON POOLED STANDARD DEVIATION)
      +---------+---------+---------+---------+---------+---------+
C1      I*******I******I
C2      I******I******I
C3                      I******I******I
C4                            I******I*****I
C5                               I*****I*****I
      +---------+---------+---------+---------+---------+---------+
    1.0       2.5       4.0       5.5       7.0       8.5      10.0
```

FIGURE 7.2.5

Computer Output for Example 7.2.1

If the data available for analysis do not meet the assumptions for one-way analysis of variance as discussed here, one may wish to consider the use of the Kruskal–Wallis procedure, a nonparametric technique discussed in Chapter 11.

Testing for Significant Differences Between Individual Pairs of Means Whenever the analysis of variance leads to a rejection of the null hypothesis of no difference among population means, the question naturally arises regarding just which pairs of means are different. In fact, the desire, more often than not, is to carry out a significance test on each and every pair of treatment means. For instance, in Example 7.2.1, where there are five treatments, we may wish to know, after rejecting H_0: $\mu_1 = \mu_2 = \mu_3 = \mu_4 = \mu_5$, which of the 10 possible individual hypotheses should be rejected. The experimenter, however, must exercise caution in testing for significant differences between individual means and must always make certain that his procedure is valid. The critical issue in the procedure is the level of significance. Although the probability, α, of rejecting a true null hypothesis for the test as a whole is made small, the probability of rejecting at least one true hypothesis when several pairs of means are tested is, as we have seen, greater than α.

Over the years several procedures for making individual comparisons have been suggested. The oldest procedure, and perhaps the one most widely used in the past, is the *least significant difference* (LSD) procedure due to Fisher, who first discussed it in the 1935 edition of his book, *The Design of Experiments* (1). The LSD procedure, which is a Student's t test using a pooled error variance, is valid only when making independent comparisons or comparisons planned before the data are analyzed. A difference between any two means that exceeds a least significant difference is considered significant at the level of significance used in computing the LSD. The LSD procedure usually is used only when the overall analysis of variance leads to a significant V.R. For an example of the use of the LSD, see Steel and Torrie (22).

Duncan (23–26) has contributed a considerable amount of research to the subject of multiple comparisons with the result that at the present time a widely used procedure is *Duncan's new multiple range test*. The extension of the test to the case of unequal sample sizes is discussed by Kramer (27).

When the objective of an experiment is to compare several treatments with a control, and not with each other, a procedure due to Dunnett (28, 29) for comparing the control against each of the other treatments is usually followed.

Other multiple comparison procedures in use are those proposed by Tukey (30, 31), Newman (32), Keuls (33), and Scheffé (34, 35). The advantages and disadvantages of the various procedures are discussed by Bancroft (36), Daniel and Coogler (37), and Winer (38). Daniel (39) has prepared a bibliography on multiple comparison procedures.

Tukey's HSD Test A multiple comparison procedure developed by Tukey (31) is frequently used for testing the null hypotheses that all possible pairs of treatment means are equal when the samples are all of the same size. When this test is employed we select an overall significance level of α. The probability is α, then, that one or more of the null hypotheses is false.

Tukey's test, which is usually referred to as the HSD (*honestly significant difference*) test, makes use of a single value against which all differences are compared. This value, called the HSD, is given by

$$\text{HSD} = q_{\alpha, k, N-k} \sqrt{\frac{\text{MSE}}{n}} \qquad (7.2.20)$$

where α is the chosen level of significance, k is the number of means in the experiment, N is the total number of observations in the experiment, n is the number of observations in a treatment, MSE is the error mean square from the ANOVA table, and q is obtained by entering Appendix Table H with α, k, and $N - k$.

All possible differences between pairs of means are computed and any difference which yields an absolute value that exceeds HSD is declared to be significant.

When the samples are not all of the same size, as is the case in Example 7.2.1, Tukey's HSD test given by Equation 7.2.20 is not applicable. Spjøtvoll and Stoline (40), however, have extended the Tukey procedure to the case where the sample sizes are different. Their procedure, which is applicable for experiments involving three or more treatments and significance levels of .05 or less, consists of replacing n in Equation 7.2.20 with n_j^*, the smallest of the two sample sizes associated with the two sample means that are to be compared. If we designate the new quantity by HSD*, we have as the new test criterion

$$\text{HSD}^* = q_{\alpha, k, N-k} \sqrt{\frac{\text{MSE}}{n_j^*}} \qquad (7.2.21)$$

Any absolute value of the difference between two sample means, one of which is computed from a sample of size n_j^* (which is smaller than the sample from which the other mean is computed), that exceeds HSD* is declared to be significant.

Example 7.2.2 Let us illustrate the use of the HSD test with the data from Example 7.2.1. The first step is to prepare a table of all possible differences between means. The results of this step for the present example are displayed in Table 7.2.5.

Suppose we let $\alpha = .05$. Entering Table H with $\alpha = .05$, $k = 5$, and $N - k = 27$, we find that q is about 4.14, if we interpolate. In Table 7.2.4 we have MSE = 1.5317552.

The hypotheses that can be tested, the value of HSD*, and the statistical decision for each test are shown in Table 7.2.6.

The results of the hypotheses tests displayed in Table 7.2.6 may be summarized by a technique suggested by Duncan (26). The sample means are

TABLE 7.2.5

Differences Between Means (Absolute Value) for Example 7.2.2 (Means Are Taken from Table 7.2.2)

	$\bar{x}_{.2}$	$\bar{x}_{.1}$	$\bar{x}_{.3}$	$\bar{x}_{.4}$	$\bar{x}_{.5}$
$\bar{x}_{.2} = 2.60$	—	.01	2.40	4.21	4.50
$\bar{x}_{.1} = 2.61$		—	2.39	4.20	4.49
$\bar{x}_{.3} = 5.00$			—	1.81	2.10
$\bar{x}_{.4} = 6.81$				—	.29
$\bar{x}_{.5} = 7.10$					—

TABLE 7.2.6

Multiple Comparison Tests Using Data of Example 7.2.1 and HSD*

Hypotheses	HSD*	Statistical decision
$H_0: \mu_1 = \mu_2$	$\text{HSD*} = 4.14\sqrt{\dfrac{1.5317552}{5}} = 2.29$	Do not reject H_0 since .01 < 2.29
$H_0: \mu_1 = \mu_3$	$\text{HSD*} = 4.14\sqrt{\dfrac{1.5317552}{5}} = 2.29$	Reject H_0 since 2.39 > 2.29
$H_0: \mu_1 = \mu_4$	$\text{HSD*} = 4.14\sqrt{\dfrac{1.5317552}{5}} = 2.29$	Reject H_0 since 4.20 > 2.29
$H_0: \mu_1 = \mu_5$	$\text{HSD*} = 4.14\sqrt{\dfrac{1.5317552}{5}} = 2.29$	Reject H_0 since 4.49 > 2.29
$H_0: \mu_2 = \mu_3$	$\text{HSD*} = 4.14\sqrt{\dfrac{1.5317552}{6}} = 2.09$	Reject H_0 since 2.40 > 2.09
$H_0: \mu_2 = \mu_4$	$\text{HSD*} = 4.14\sqrt{\dfrac{1.5317552}{6}} = 2.09$	Reject H_0 since 4.21 > 2.09
$H_0: \mu_2 = \mu_5$	$\text{HSD*} = 4.14\sqrt{\dfrac{1.5317552}{6}} = 2.09$	Reject H_0 since 4.50 > 2.09
$H_0: \mu_3 = \mu_4$	$\text{HSD*} = 4.14\sqrt{\dfrac{1.5317552}{6}} = 2.09$	Do not reject H_0 since 1.81 < 2.09
$H_0: \mu_3 = \mu_5$	$\text{HSD*} = 4.14\sqrt{\dfrac{1.5317552}{6}} = 2.09$	Reject H_0 since 2.10 > 2.09
$H_0: \mu_4 = \mu_5$	$\text{HSD*} = 4.14\sqrt{\dfrac{1.5317552}{7}} = 1.94$	Do not reject H_0 since .29 < 1.94

displayed in a line approximately to scale. Any two that are not significantly different are underscored by the same line. Any two sample means that are not underscored by the same line are significantly different. Thus, for the present example, we may write

2.60   2.61   5.00   6.81   7.10

EXERCISES

In each of the following exercises go through the nine steps of analysis of variance hypothesis testing. Determine the p value for each test.

7.2.1   We wish to know if four treatments for joint swelling differ in effectiveness. Four groups of patients were subjected to the different treatments. At the end of a specified time period each was given a test to measure treatment effectiveness. The following scores were obtained.

TREATMENT

1	2	3	4
64	76	58	95
88	70	74	90
72	90	66	80
80	80	60	87
79	75	82	88
71	82	75	85

Let $\alpha = .05$.

7.2.2   An experiment was conducted to compare three methods of packaging a certain frozen food. The criterion was the ascorbic acid content (mg/100 gm) after a specified period of time. The following data were obtained.

PACKAGING METHOD

A	B	C
14.29	20.06	20.04
19.10	20.64	26.23
19.09	18.00	22.74
16.25	19.56	24.04
15.09	19.47	23.37
16.61	19.07	25.02
19.63	18.38	23.27

Do these data provide sufficient evidence at the .01 level of significance to indicate a difference in packaging methods?

7.2.3 Three groups of animals were used in an experiment to compare response times, in seconds, to three different stimuli. The following results were obtained.

	STIMULUS	
I	*II*	*III*
16	6	8
14	7	10
14	7	9
13	8	10
13	4	6
12	8	7
12	9	10
17	6	9
17	8	11
17	6	11
19	4	9
14	9	10
15	5	9
20	5	5

Do these data provide sufficient evidence to indicate a true difference among population means? Let $\alpha = .05$.

7.2.4 Blood sugar determinations (mg/100 ml) were made on 10 specimens from each of five breeds of a certain experimental animal with the following results.

		BREED		
A	*B*	*C*	*D*	*E*
124	111	117	104	142
116	101	142	128	139
101	130	121	130	133
118	108	123	103	120
118	127	121	121	127
120	129	148	119	149
110	122	141	106	150
127	103	122	107	149
106	122	139	107	120
130	127	125	115	116

Do these data provide sufficient evidence to indicate a difference in mean blood sugar level among breeds? Let $\alpha = .05$.

7.2.5 It was desired to compare three physicians with regard to the length of stay of their patients in the hospital following a certain minor surgical procedure without complications. A sample of eight records were selected from the records of each physician, and the following lengths of stay were observed.

PHYSICIAN

A	B	C
4	4	5
5	5	3
5	4	3
4	3	3
6	4	3
6	5	3
4	3	4
5	3	5

Do these data suggest a difference in average length of stay among the three physicians? Let $\alpha = .01$.

7.2.6 Refer to Exercise 7.2.1 and apply Tukey's HSD test. Let $\alpha = .01$.

7.2.7 Refer to Exercise 7.2.2 and apply Tukey's HSD test. Let $\alpha = .05$.

7.2.8 Refer to Exercise 7.2.3 and apply Tukey's HSD test. Let $\alpha = .05$.

7.2.9 Refer to Exercise 7.2.4 and apply Tukey's HSD test. Let $\alpha = .01$.

7.2.10 Refer to Exercise 7.2.5 and apply Tukey's HSD test. Let $\alpha = .01$.

7.2.11 The following are the weights of a certain organ as a percentage of body weight for 30 laboratory animals. The four treatments are the different diets the animals were fed. Test the null hypothesis of no difference in diet effects and test for significant differences between all possible pairs of means. Let $\alpha = .05$.

DIET

A	B	C	D
4.34	4.47	4.72	4.48
4.73	4.65	4.99	5.02
4.84	4.62	5.24	4.58
4.57	4.41	5.00	4.89
4.72	4.43	4.82	4.90
4.55	4.23	4.95	4.81
	4.54	5.28	5.26
	4.45	4.90	
		4.98	

7.2.12 Researchers wish to compare four physical fitness programs designed for business executives. Thirty executives are randomly assigned to one of the four programs. The following table shows the differences between the executives' physical fitness scores before and after participating in the program.

PROGRAM			
A	B	C	D
13	11	12	22
24	13	19	26
19	20	9	22
18	14	14	22
9	11	21	26
21	21	7	19
17	14	6	
22	8		
24			

Can we conclude from these data that the four programs differ in effectiveness? Let $\alpha = .05$.

7.3   THE RANDOMIZED COMPLETE BLOCK DESIGN

Of all the experimental designs that are in use, the *randomized complete block* design appears to be the most widely used by far. This design was developed about 1925 by R. A. Fisher (3, 41), who was seeking methods of improving agricultural field experiments. The name of the design reflects its origin in agricultural experiments, where the land was divided into *blocks* and the blocks were divided into *plots* that received the treatments under investigation.

The randomized complete block design is a design in which the units (called *experimental units*) to which the treatments are applied are subdivided into homogeneous groups called *blocks*, so that the number of experimental units in a block is equal to the number (or some multiple of the number) of treatments being studied. The treatments are then assigned at random to the experimental units within each block. It should be emphasized that each treatment appears in every block, and each block receives every treatment.

The objective in using the randomized complete block design is to isolate and remove from the error term the variation attributable to the blocks, while assuring that treatment means will be free of block effects. The effectiveness of the design depends on the ability to achieve homogeneous blocks of experimental units. The ability to form homogeneous blocks depends on the researcher's knowledge of the experimental material. When blocking is used effectively, the error mean square in the ANOVA table will be reduced, the V.R. will be increased, and the chance of rejecting the null hypothesis will be improved.

In animal experiments, if it is felt that different breeds of animal will respond differently to the same treatment, the breed of animal may be used as a blocking factor. Litters may also be used as blocks, in which case an animal from each litter receives a treatment. In experiments involving human beings, if it is desired

TABLE 7.3.1

Table of Sample Values for the Randomized Complete Block Design

| Blocks | TREATMENTS | | | | | Total | Mean |
	1	2	3	$\cdots$	k		
1	x_{11}	x_{12}	x_{13}	$\cdots$	x_{1k}	$T_{1.}$	$\bar{x}_{1.}$
2	x_{21}	x_{22}	x_{23}	$\cdots$	x_{2k}	$T_{2.}$	$\bar{x}_{2.}$
3	x_{31}	x_{32}	x_{33}	$\cdots$	x_{3k}	$T_{3.}$	$\bar{x}_{3.}$
$\vdots$	$\vdots$	$\vdots$	$\vdots$	$\vdots$	$\vdots$	$\vdots$	$\vdots$
n	x_{n1}	x_{n2}	x_{n3}	$\cdots$	x_{nk}	$T_{n.}$	$\bar{x}_{n.}$
Total	$T_{.1}$	$T_{.2}$	$T_{.3}$	$\cdots$	$T_{.k}$	$T_{..}$	
Mean	$\bar{x}_{.1}$	$\bar{x}_{.2}$	$\bar{x}_{.3}$	$\cdots$	$\bar{x}_{.k}$		$\bar{x}_{..}$

that differences resulting from age be eliminated, then subjects may be grouped according to age so that one person of each age receives each treatment. The randomized complete block design also may be employed effectively when an experiment must be carried out in more than one laboratory (block) or when several days (blocks) are required for completion.

Some of the advantages of the randomized complete block design include the fact that it is both easily understood and computationally simple. Furthermore, certain complications that may arise in the course of an experiment are easily handled when this design is employed.

It is instructive here to point out that the paired comparisons analysis presented in Chapter 6 is a special case of the randomized complete block design. Example 6.4.1, for example, may be treated as a randomized complete block design in which the two cholesterol levels are the treatment effects and the 12 individuals on whom the measurements were taken are the blocks.

In general, the data from an experiment utilizing the randomized complete block design may be displayed in a table such as Table 7.3.1. The following new notation in this table should be observed:

$$\text{the total of the } i\text{th block} = T_{i.} = \sum_{j=1}^{k} x_{ij}$$

$$\text{the mean of the } i\text{th block} = \bar{x}_{i.} = \frac{\sum_{j=1}^{k} x_{ij}}{k}$$

$$\text{and the grand total} = T_{..} = \sum_{j=1}^{k} T_{.j} = \sum_{i=1}^{n} T_{i.}$$

indicating that the grand total may be obtained either by adding row totals or by adding column totals.

The technique for analyzing the data from a randomized complete block design is called *two-way analysis of variance* since an observation is categorized on the basis of two criteria—the block to which it belongs as well as the treatment group to which it belongs.

Example 7.3.1 A physical therapist wished to compare three methods for teaching patients to use a certain prosthetic device. He felt that the rate of learning would be different for patients of different ages and wished to design an experiment in which the influence of age could be taken into account. The randomized complete block design is the appropriate design for achieving this goal.

1. *Data* Three patients in each of five age groups were selected to participate in the experiment, and one patient in each age group was randomly assigned to each of the teaching methods. The methods of instruction constitute our three treatments, and the five age groups are the blocks. The data shown in Table 7.3.2 were obtained.

2. *Assumptions* The model for the randomized complete block design and its underlying assumptions are as follows.

The Model

$$x_{ij} = \mu + \beta_i + \tau_j + e_{ij} \qquad (7.3.1)$$

$$i = 1, 2, \ldots, n; \qquad j = 1, 2, \ldots, k$$

In this model

x_{ij} is a typical value from the overall population.
μ is an unknown constant.

TABLE 7.3.2

Time (in Days) Required to Learn the Use of a Certain Prosthetic Device

Age group	TEACHING METHOD			Total	Mean
	A	B	C		
Under 20	7	9	10	26	8.67
20 to 29	8	9	10	27	9.00
30 to 39	9	9	12	30	10.00
40 to 49	10	9	12	31	10.33
50 and over	11	12	14	37	12.33
Total	45	48	58	151	
Mean	9.0	9.6	11.6		10.07

β_i represents a block effect reflecting the fact that the experimental unit fell in the ith block.

τ_j represents a treatment effect, reflecting the fact that the experimental unit received the jth treatment.

e_{ij} is a residual component representing all sources of variation other than treatments and blocks.

For our illustrative example we may say that the value of a particular observation is a result of a patient's being in the study, plus the effect of the method of treatment, plus the effect of his age, plus other factors that are not accounted for.

Assumptions of the Model

(a)   Each x_{ij} that is observed constitutes a random independent sample of size 1 from one of the kn populations represented.

(b)   Each of these kn populations is normally distributed with mean μ_{ij} and the same variance σ^2. This implies that the e_{ij} are independently and normally distributed with mean 0 and variance σ^2.

(c)   The block and treatment effects are additive. This assumption may be interpreted to mean that there is no *interaction* between treatments and blocks. In other words, a particular block–treatment combination does not produce an effect that is greater or less than the sum of their individual effects. It can be shown that when this assumption is met

$$\sum_{j=1}^{k} \tau_j = \sum_{i=1}^{n} \beta_i = 0$$

The consequences of a violation of this assumption are misleading results. Anderson and Bancroft (42) suggest that one need not become concerned with the violation of the additivity assumption unless the largest mean is more than 50 percent greater than the smallest. The nonadditivity problem is also dealt with by Tukey (43) and Mandel (44).

When these assumptions hold true, the τ_j and β_i are a set of fixed constants, and we have a situation that fits the fixed-effects model.

3.   *Hypotheses*   We may test

$$H_0: \tau_j = 0, \qquad j = 1, 2, \ldots, k$$

against the alternative

$$H_A: \text{not all } \tau_j = 0$$

A hypothesis test regarding block effects is not usually carried out under the assumptions of the fixed-effects model for two reasons. First, the primary interest is in treatment effects, the usual purpose of the blocks being to provide a means of eliminating an extraneous source of variation. Second, although the experimental units are randomly assigned to the treatments, the blocks are obtained in a nonrandom manner.

For our illustrative example we have the following hypotheses:

$$H_0: \tau_j = 0, \qquad j = 1, 2, 3$$
$$H_A: \text{not all } \tau_j = 0$$
$$\text{Let } \alpha = .05$$

4. *Test Statistic*   The test statistic is V.R.

5. *Distribution of the Test Statistic*   When H_0 is true and the assumptions are met, V.R. follows an F distribution.

6. *Decision Rule*   Reject the null hypothesis if the computed value of the test statistic is equal to or greater than the critical value of F.

7. *Calculation of Test Statistic*   It can be shown that the total sum of squares for the randomized complete block design can be partitioned into three components, one each attributable to treatments ($SSTr$), blocks ($SSBl$), and error (SSE). The algebra is somewhat tedious and will be omitted. The partitioned sum of squares may be expressed by the following equation:

$$\sum_{j=1}^{k} \sum_{i=1}^{n} (x_{ij} - \bar{x}_{..})^2 = \sum_{j=1}^{k} \sum_{i=1}^{n} (\bar{x}_{i.} - \bar{x}_{..})^2 + \sum_{j=1}^{k} \sum_{i=1}^{n} (\bar{x}_{.j} - \bar{x}_{..})^2$$
$$+ \sum_{j=1}^{k} \sum_{i=1}^{n} (x_{ij} - \bar{x}_{i.} - \bar{x}_{.j} + \bar{x}_{..})^2 \qquad (7.3.2)$$

that is,

$$SST = SSBl + SSTr + SSE \qquad (7.3.3)$$

The computing formulas for the quantities in Equations 7.3.2 and 7.3.3 are as follows:

$$SST = \sum_{j=1}^{k} \sum_{i=1}^{n} x_{ij}^2 - C \qquad (7.3.4)$$

$$SSBl = \sum_{i=1}^{n} \frac{T_{i.}^2}{k} - C \qquad (7.3.5)$$

$$SSTr = \sum_{j=1}^{k} \frac{T_{.j}^2}{n} - C \qquad (7.3.6)$$

$$SSE = SST - SSBl - SSTr \qquad (7.3.7)$$

It will be recalled that C is a correction term, and in the present situation it is computed as follows:

$$C = \left(\sum_{j=1}^{k} \sum_{i=1}^{n} x_{ij} \right)^2 \Big/ kn$$

$$= T_{..}^2/kn \tag{7.3.8}$$

The appropriate degrees of freedom for each component of Equation 7.3.3 are

$$
\begin{array}{cccc}
\text{total} & \text{blocks} & \text{treatments} & \text{residual (error)} \\
kn - 1 = (n - 1) + & (k - 1) & +(n - 1)(k - 1)
\end{array}
$$

The residual degrees of freedom, like the residual sum of squares, may be obtained by subtraction as follows:

$$(kn - 1) - (n - 1) - (k - 1) = kn - 1 - n + 1 - k + 1$$
$$= n(k - 1) - 1(k - 1) = (n - 1)(k - 1)$$

For our illustrative example we compute the following sums of squares:

$$C = \frac{(151)^2}{(3)(5)} = \frac{22801}{15} = 1520.0667$$

$$SST = 7^2 + 9^2 + \cdots + 14^2 - 1520.0667 = 46.9333$$

$$SSBl = \frac{26^2 + 27^2 + \cdots + 37^2}{3} - 1520.0667 = 24.9333$$

$$SSTr = \frac{45^2 + 48^2 + 58^2}{5} - 1520.0667 = 18.5333$$

$$SSE = 46.9333 - 24.9333 - 18.5333 = 3.4667$$

The degrees of freedom are total $= (3)(5) - 1 = 14$, blocks $= 5 - 1 = 4$, treatments $= 3 - 1 = 2$, and residual $= (5 - 1)(3 - 1) = 8$.

The ANOVA Table. The results of the calculations for the randomized complete block design may be displayed in an ANOVA table such as Table 7.3.3.

For our illustrative example we have the ANOVA table shown in Table 7.3.4.

8. *Statistical Decision* It can be shown that when the fixed-effects model applies and the null hypothesis of no treatment effects (all $\tau_i = 0$) is true, both the error, or residual, mean square and the treatments mean square are

TABLE 7.3.3

ANOVA Table for the Randomized Complete Block Design

Source	SS	d.f.	MS	V.R.
Treatments	SSTr	$(k-1)$	$MSTr = $ $SSTr/(k-1)$	$MSTr/$ MSE
Blocks	SSBl	$(n-1)$	$MSBl = $ $SSBl/(n-1)$	
Residual	SSE	$(n-1)(k-1)$	$MSE = $ $SSE/(n-1)(k-1)$	
Total	SST	$kn - 1$		

estimates of the common variance σ^2. When the null hypothesis is true, therefore, the quantity

$$MSTr/MSE$$

is distributed as F with $k-1$ numerator degrees of freedom and $(n-1) \times (k-1)$ denominator degrees of freedom. The computed variance ratio, therefore, is compared with the critical value of F.

For our illustrative example the critical value of F with 2 and 8 degrees of freedom and $\alpha = .05$ is 4.46. Since our computed variance ratio, 21.38, is greater than 4.46, we reject the null hypothesis of no treatment effects on the assumption that such a large V.R. reflects the fact that the two sample mean squares are not estimating the same quantity. The only other explanation for this large V.R. would be that the null hypothesis is really true, and we have just observed an unusual set of results. We rule out the second explanation in favor of the first.

9. *Conclusion* We conclude that not all treatment effects are equal to zero, or equivalently, that not all treatment means are equal. For this test $p < .005$.

When the data available for analysis do not meet the assumptions of the randomized complete block design as discussed here, the Friedman procedure discussed in Chapter 11 may prove to be a suitable nonparametric alternative.

TABLE 7.3.4

ANOVA Table for Example 7.3.1

Source	SS	d.f.	MS	V.R.
Treatments	18.5333	2	9.26665	21.38
Blocks	24.9333	4	6.233325	
Residual	3.4667	8	.4333375	
Total	46.9333	14		

EXERCISES

For each of the following exercises perform the nine-step hypothesis testing procedure for analysis of variance. Determine the p value for each exercise.

7.3.1 Three food service systems were studied in five hospitals. The variable of interest was time (in minutes) utilized per mean served. The noon meal was served by each hospital by each method with the following results.

	METHOD		
Hospital	A	B	C
1	7.56	9.68	11.65
2	9.98	9.69	10.69
3	7.23	10.49	11.77
4	8.22	8.55	10.72
5	7.59	8.30	12.36

After eliminating hospital effects, do these data suggest a difference among the methods in mean time utilized per meal served? Let $\alpha = .05$.

7.3.2 Sixteen overweight subjects participated in a study to compare four weight reducing regimens. Subjects were grouped according to initial weight, and each of four subjects from each initial weight group was randomly assigned to one of the four reducing regimens. At the end of the experimental period the following weight losses, in pounds, were recorded.

Initial	REGIMEN			
weight (pounds)	A	B	C	D
150 to 174	12	26	24	23
175 to 199	15	29	23	25
200 to 225	15	27	25	24
Over 225	18	38	33	31

After eliminating differences due to initial weight, do these data provide sufficient evidence to indicate a difference in regimen effects? Let $\alpha = .01$.

7.3.3 A remotivation team in a psychiatric hospital conducted an experiment to compare five methods for remotivating patients. Patients were grouped according to level of initial motivation. Patients in each group were randomly assigned to the five methods. At the end of the experimental period the patients were evaluated by a team composed of a psychiatrist, a psychologist, a nurse, and a social worker, none of whom was aware of the method to which patients had been assigned. The team assigned each patient a composite score as a measure of his or her level of motivation. The results were as follows.

Level of initial motivation	REMOTIVATION METHOD				
	A	B	C	D	E
Nil	58	68	60	68	64
Very low	62	70	65	80	69
Low	67	78	68	81	70
Average	70	81	70	89	74

Do these data provide sufficient evidence to indicate a difference in mean scores among methods? Let $\alpha = .05$.

7.3.4 The nursing supervisor in a local health department wished to study the influence of time of day on length of home visits by the nursing staff. It was felt that individual differences among nurses might be large, so the nurse was used as a blocking factor. The nursing supervisor collected the following data.

	LENGTH OF HOME VISIT BY TIME OF DAY			
Nurse	Early morning	Late morning	Early afternoon	Late afternoon
A	27	28	30	23
B	31	30	27	20
C	35	38	34	30
D	20	18	20	14

Do these data provide sufficient evidence to indicate a difference in length of home visit among the different times of day? Let $\alpha = .05$.

7.3.5 Analyze the data of Exercise 6.4.1 as a randomized block design.

7.3.6 Four subjects participated in an experiment to compare three methods of relieving stress. Each subject was placed in a stressful situation on three different occasions. Each time a different method for reducing stress was used with the subject. The response variable is the amount of decrease in stress level as measured before and after treatment application. The results were as follows.

	TREATMENT		
Subject	A	B	C
1	16	26	22
2	16	20	23
3	17	21	22
4	28	29	36

Can we conclude from these data that the three methods differ in effectiveness? Let $\alpha = .05$.

7.4 THE FACTORIAL EXPERIMENT

In the experimental designs that we have considered up to this point we have been interested in the effects of only one variable, the treatments. Frequently, however, we may be interested in studying, simultaneously, the effects of two or more variables. We refer to the variables in which we are interested as *factors*. The experiment in which two or more factors are investigated simultaneously is called a *factorial experiment*.

The different designated categories of the factors are called *levels*. Suppose, for example, that we are studying the effect on reaction time of three dosages of some drug. The drug factor, then, is said to occur at three levels. Suppose the second factor of interest in the study is age, and it is felt that two age groups, under 65 years and 65 years and over, should be included. We then have two levels of the age factor. In general, we say that factor A occurs at a levels and factor B occurs at b levels.

In a factorial experiment we may study not only the effects of individual factors but also, if the experiment is properly conducted, the *interaction* between factors. To illustrate the concept of interaction let us consider the following example.

Example 7.4.1 Suppose, in terms of effect on reaction time, that the true relationship between three dosage levels of some drug and the age of human subjects taking the drug is known. Suppose further that age occurs at two levels —"young" (under 65) and "old" (65 and over). If the true relationship between the two factors is known, we will know, for the three dosage levels, the mean effect on reaction time of subjects in the two age groups. Let us assume that effect is measured in terms of reduction in reaction time to some stimulus. Suppose these means are as shown in Table 7.4.1.

The following important features of the data in Table 7.4.1 should be noted.

1. For both levels of factor A the difference between the means for any two levels of factor B is the same. That is, for both levels of factor A, the difference between means for levels 1 and 2 is 5, for levels 2 and 3 the difference is 10, and for levels 1 and 3 the difference is 15.

TABLE 7.4.1

Mean Reduction in Reaction Time (Milliseconds) of Subjects in Two Age Groups at Three Drug Dosage Levels

Factor A—age	FACTOR B—DRUG DOSAGE		
	$j = 1$	$j = 2$	$j = 3$
Young ($i = 1$)	$\mu_{11} = 5$	$\mu_{12} = 10$	$\mu_{13} = 20$
Old ($i = 2$)	$\mu_{21} = 10$	$\mu_{22} = 15$	$\mu_{23} = 25$

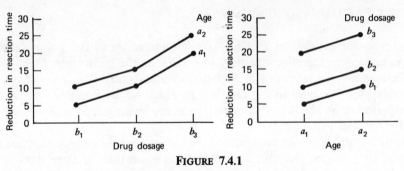

FIGURE 7.4.1

Age and Drug Effects, No Interaction Present

2. For all levels of factor B the difference between means for the two levels of factor A is the same. In the present case the difference is 5 at all three levels of factor B.

3. A third characteristic is revealed when the data are plotted as in Figure 7.4.1. We note that the curves corresponding to the different levels of a factor are all parallel.

When population data possess the three characteristics listed above, we say that there is no interaction present.

The presence of interaction between two factors can affect the characteristics of the data in a variety of ways depending on the nature of the interaction. We illustrate the effect of one type of interaction by altering the data of Table 7.4.1 as shown in Table 7.4.2.

The important characteristics of the data in Table 7.4.2 are as follows.

1. The difference between means for any two levels of factor B is not the same for both levels of factor A. We note in Table 7.4.2, for example, that the difference between levels 1 and 2 of factor B is -5 for the young age group and $+5$ for the old age group.

2. The difference between means for both levels of factor A is not the same at all levels of factor B. The differences between factor A means are -10, 0, and 15 for levels 1, 2, and 3, respectively, of factor B.

TABLE 7.4.2

Data of Table 7.4.1 Altered to Show the Effect of One Type of Interaction

Factor A—age	FACTOR B—DRUG DOSAGE		
	$j = 1$	$j = 2$	$j = 3$
Young ($i = 1$)	$\mu_{11} = 5$	$\mu_{12} = 10$	$\mu_{13} = 20$
Old ($i = 2$)	$\mu_{21} = 15$	$\mu_{22} = 10$	$\mu_{23} = 5$

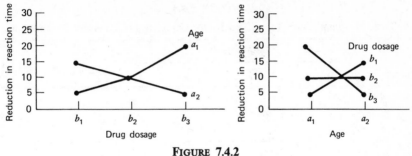

FIGURE 7.4.2

Age and Drug Effects, Interaction Present

3. The factor level curves are not parallel as shown in Figure 7.4.2.

When population data exhibit the characteristics illustrated by Table 7.4.2 and Figure 7.4.2 we say that there is interaction between the two factors. We emphasize that the kind of interaction illustrated by the present example is only one of many types of interaction that may occur between two factors.

In summary, then, we can say that *there is interaction between two factors if a change in one of the factors produces a change in response at one level of the other factor different from that produced at other levels of this factor.*
The advantages of the factorial experiment include the following.

1. The interaction of the factors may be studied.
2. There is a saving of time and effort.

In the factorial experiment all the observations may be used to study the effects of each of the factors under investigation. The alternative, when two factors are being investigated, would be to conduct two different experiments, one to study each of the two factors. If this were done, some of the observations would yield information only on one of the factors, and the remainder would yield information only on the other factor. To achieve the level of accuracy of the factorial experiment, more experimental units would be needed if the factors were studied through two experiments. It is seen, then, that 1 two-factor experiment is more economical than 2 one-factor experiments.

3. Since the various factors are combined in one experiment, the results have a wider range of application.

A factorial arrangement may be studied with either of the designs that have been discussed. We illustrate the analysis of a factorial experiment by means of a two-factor completely randomized design.

TABLE 7.4.3

Table of Sample Data from a Two-Factor Completely Randomized Experiment

Factor A	**FACTOR B**				Totals	Means
	_1_	_2_	$\cdots$	_b_		
1	x_{111} $\vdots$ x_{11n}	x_{121} $\vdots$ x_{12n}	$\cdots$ $\vdots$ $\cdots$	x_{1b1} $\vdots$ x_{1bn}	$T_{1..}$	$\bar{x}_{1..}$
2	x_{211} $\vdots$ x_{21n}	x_{221} $\vdots$ x_{22n}	$\cdots$ $\vdots$ $\cdots$	x_{2b1} $\vdots$ x_{2bn}	$T_{2..}$	$\bar{x}_{2..}$
$\vdots$	$\vdots$	$\vdots$	$\vdots$	$\vdots$	$\vdots$	$\vdots$
a	x_{a11} $\vdots$ x_{a1n}	x_{a21} $\vdots$ x_{a2n}	$\cdots$ $\vdots$ $\cdots$	x_{ab1} $\vdots$ x_{abn}	$T_{a..}$	$\bar{x}_{a..}$
Totals	$T_{.1.}$	$T_{.2.}$	$\cdots$	$T_{.b.}$	$T_{...}$	
Means	$\bar{x}_{.1.}$	$\bar{x}_{.2.}$	$\cdots$	$\bar{x}_{1b1}$		$\bar{x}_{...}$

The results from a two-factor completely randomized design may be presented in tabular form as shown in Table 7.4.3.

Here we have a levels of factor A, b levels of factor B, and n observations for each combination of levels. Each of the ab combinations of levels of factor A with levels of factor B is a treatment. In addition to the totals and means shown in Table 7.4.3, we note that the total and mean of the ijth cell are $T_{ij.} = \sum_{k=1}^{n} x_{ijk}$ and $\bar{x}_{ij.} = T_{ij.}/n$, respectively. The subscript i runs from 1 to a and j runs from 1 to b. The total number of observations is nab.

To show that Table 7.4.3 represents data from a completely randomized design, we consider that each combination of factor levels is a treatment and that we have n observations for each treatment. An alternative arrangement of the data would be obtained by listing the observations of each treatment in a separate column. Table 7.4.3 may also be used to display data from a two-factor randomized block design if we consider the first observation in each cell as belonging to block 1, the second observation in each cell as belonging to block 2, and so on to the nth observation in each cell, which may be considered as belonging to block n.

Note the similarity of the data display for the factorial experiment as shown in Table 7.4.3 to the randomized complete block data display of Table 7.3.1. The factorial experiment, in order that the experimenter may test for interaction, requires at least two observations per cell, whereas the randomized complete block design only requires one observation per cell. We use two-way analysis of variance to analyze the data from a factorial experiment of the type presented here.

Example 7.4.2 In a study of length of time spent on individual home visits by public health nurses, data were reported on length of home visit, in minutes, by a sample of 80 nurses. A record was made also of each nurses's age and the type of illness of each patient visited. The researchers wished to obtain from their investigation answers to the following questions:

1. Does the mean length of home visit differ among the different age groups of nurses?
2. Does type of patient affect the mean length of home visit?
3. Is there interaction between nurse's age and type of patient?

1. *Data* The data on length of home visit that were obtained during the study, are shown in Table 7.4.4.

2. *Assumptions* To analyze these data we assume a fixed-effects model and a two-factor completely randomized design.

The Model. The fixed-effects model for the two-factor completely randomized design may be written as

$$x_{ijk} = \mu + \alpha_i + \beta_j + (\alpha\beta)_{ij} + e_{ijk}$$

$$i = 1, 2, \ldots, a; \qquad j = 1, 2, \ldots, b; \qquad k = 1, 2, \ldots, n$$

(7.4.1)

where x_{ijk} is a typical observation, μ is a constant, α represents an effect due to factor A, β represents an effect due to factor B, $(\alpha\beta)$ represents an effect due to the interaction of factors A and B, and e_{ijk} represents the experimental error.

Assumptions of the model.

(a) The observations in each of the *ab* cells constitute a random independent sample of size *n* drawn from the population defined by the particular combination of the levels of the two factors. In our illustrative example we have a sample from 16 populations.

(b) Each of the *ab* populations is normally distributed.

(c) The populations all have the same variance.

3. *Hypotheses*
   The following hypotheses may be tested.

   (1) H_0: $\alpha_i = 0$ $i = 1, 2, \ldots, a$
        H_A: not all $\alpha_i = 0$

   (2) H_0: $\beta_j = 0$ $j = 1, 2, \ldots, b$
        H_A: not all $\beta_j = 0$

   (3) H_0: $(\alpha\beta)_{ij} = 0$ $i = 1, 2, \ldots, a; \qquad j = 1, 2, \ldots, b$
        H_A: not all $(\alpha\beta)_{ij} = 0$

TABLE 7.4.4

Length of Home Visit in Minutes by Public Health Nurses by Nurse's Age Group and Type of Patient

Factor A (type of patient) levels	FACTOR B (NURSE'S AGE GROUP) LEVELS				Totals	Means
	1 (20 to 29)	2 (30 to 39)	3 (40 to 49)	4 (50 and over)		
1 (Cardiac)	20	25	24	28		
	25	30	28	31		
	22	29	24	26	534	26.70
	27	28	25	29		
	21	30	30	32		
2 (Cancer)	30	30	39	40		
	45	29	42	45		
	30	31	36	50	765	38.25
	35	30	42	45		
	36	30	40	60		
3 (C.V.A.)	31	32	41	42		
	30	35	45	50		
	40	30	40	40	766	38.30
	35	40	40	55		
	30	30	35	45		
4 (Tuberculosis)	20	23	24	29		
	21	25	25	30		
	20	28	30	28	509	25.45
	20	30	26	27		
	19	31	23	30		
Totals	557	596	659	762	2574	
Means	27.85	29.8	32.95	38.10		32.18

Cell	$a_1 b_1$	$a_1 b_2$	$a_1 b_3$	$a_1 b_4$	$a_2 b_1$	$a_2 b_2$	$a_2 b_3$	$a_2 b_4$
Totals	115	142	131	146	176	150	199	240
Means	23.0	28.4	26.2	29.2	35.2	30.0	39.8	48.0

Cell	$a_3 b_1$	$a_3 b_2$	$a_3 b_3$	$a_3 b_4$	$a_4 b_1$	$a_4 b_2$	$a_4 b_3$	$a_4 b_4$
Totals	166	167	201	232	100	137	128	144
Means	33.2	33.4	40.2	46.4	20.0	27.4	25.6	28.8

Before collecting data, the researchers may decide to test only one of the possible hypotheses. In this case they select the hypothesis they wish to test, choose a significance level α, and proceed in the familiar, straightforward fashion. This procedure is free of the complications that arise if the researchers wish to test all three hypotheses.

When all three hypotheses are tested, the situation is complicated by the fact that the three tests are not independent in the probability sense. If we let α be the significance level associated with the test as a whole, and α', α'', and α''' the significance levels associated with hypotheses 1, 2, and 3, respectively, Kimball (45) has shown that

$$\alpha < 1 - (1 - \alpha')(1 - \alpha'')(1 - \alpha''') \qquad (7.4.2)$$

If $\alpha' = \alpha'' = \alpha''' = .05$, then $\alpha < 1 - (.95)^3$, or $\alpha < .143$. This means that the probability of rejecting one or more of the three hypotheses is something less than .143 when a significance level of .05 has been chosen for the hypotheses and all are true. To demonstrate the hypothesis testing procedure for each case, we perform all three tests. The reader, however, should be aware of the problem involved in interpreting the results. The problem is discussed by Dixon and Massey (46) and Guenther (47).

For our illustrative example we may test the following hypotheses subject to the conditions mentioned above.

(1) H_0: $\alpha_1 = \alpha_2 = \alpha_3 = \alpha_4 = 0$
      H_A: not all $\alpha_i = 0$

(2) H_0: $\beta_1 = \beta_2 = \beta_3 = \beta_4 = 0$
      H_A: not all $\beta_j = 0$

(3) H_0: $(\alpha\beta)_{ij} = 0$
      H_A: not all $(\alpha\beta)_{ij} = 0$
      Let $\alpha = .05$

4. *Test Statistic*   The test statistic for each hypothesis set is V.R.

5. *Distribution of Test Statistic*   When H_0 is true and the assumptions are met each of the test statistics is distributed as F.

6. *Decision Rule*   Reject H_0 if the computed value of the test statistic is equal to or greater than the critical value of F.

7. *Calculation of Test Statistic*   By an adaptation of the procedure used in partitioning the total sum of squares for the completely randomized design, it can be shown that the total sum of squares under the present model can be partitioned into two parts as follows:

$$\sum_{i=1}^{a} \sum_{j=1}^{b} \sum_{k=1}^{n} (x_{ijk} - \bar{x}_{...})^2 = \sum_{i=1}^{a} \sum_{j=1}^{b} \sum_{k=1}^{n} (\bar{x}_{ij.} - \bar{x}_{...})^2$$
$$+ \sum_{i=1}^{a} \sum_{j=1}^{b} \sum_{k=1}^{n} (x_{ijk} - \bar{x}_{ij.})^2 \qquad (7.4.3)$$

or

$$SST = SSTr + SSE \qquad (7.4.4)$$

The sum of squares for treatments can be partitioned into three parts as follows:

$$\sum_{i=1}^{a} \sum_{j=1}^{b} \sum_{k=1}^{n} (\bar{x}_{ij.} - \bar{x}_{...})^2 = \sum_{i=1}^{a} \sum_{j=1}^{b} \sum_{k=1}^{n} (\bar{x}_{i..} - \bar{x}_{...})^2$$

$$+ \sum_{i=1}^{a} \sum_{j=1}^{b} \sum_{k=1}^{n} (\bar{x}_{.j.} - \bar{x}_{...})^2$$

$$+ \sum_{i=1}^{a} \sum_{j=1}^{b} \sum_{k=1}^{n} (\bar{x}_{ij.} - \bar{x}_{i..} - \bar{x}_{.j.} + \bar{x}_{...})^2 \quad (7.4.5)$$

or

$$SSTr = SSA + SSB + SSAB$$

The computing formulas for the various components are as follows:

$$SST = \sum_{i=1}^{a} \sum_{j=1}^{b} \sum_{k=1}^{n} x_{ijk}^2 - C \qquad (7.4.6)$$

$$SSTr = \frac{\sum_{i=1}^{a} \sum_{j=1}^{b} T_{ij.}^2}{n} - C \qquad (7.4.7)$$

$$SSE = SST - SSTr \qquad (7.4.8)$$

$$SSA = \frac{\sum_{i=1}^{a} T_{i..}^2}{bn} - C \qquad (7.4.9)$$

$$SSB = \frac{\sum_{j=1}^{b} T_{.j.}^2}{an} - C \qquad (7.4.10)$$

and

$$SSAB - SSTr - SSA - SSB \qquad (7.4.11)$$

In the above equations

$$C = \left(\sum_{i=1}^{a} \sum_{j=1}^{b} \sum_{k=1}^{n} x_{ijk} \right)^2 \bigg/ abn \qquad (7.4.12)$$

TABLE 7.4.5

Analysis of Variance Table for a Two-Factor Completely Randomized Experiment (Fixed-Effects Model)

Source	SS	d.f.	MS	V.R.
A	SSA	$a - 1$	$MSA = SSA/(a-1)$	MSA/MSE
B	SSB	$b - 1$	$MSB = SSB/(b-1)$	MSB/MSE
AB	SSAB	$(a-1)(b-1)$	$MSAB = SSAB/(a-1)(b-1)$	$MSAB/MSE$
Treatments	SSTr	$ab - 1$		
Residual	SSE	$ab(n-1)$	$MSE = \dfrac{SSE}{ab(n-1)}$	
Total	SST	$abn - 1$		

For our illustrative example we have

$$C = (2574)^2/80 = 82818.45$$

$$SST = (20^2 + 25^2 + \cdots + 30^2) - 82818.45 = 5741.55$$

$$SSTr = \frac{115^2 + 142^2 + \cdots + 144^2}{5} - 82818.45 = 4801.95$$

$$SSA = \frac{534^2 + 765^2 + 766^2 + 509^2}{20} - 82818.45 = 2992.45$$

$$SSB = \frac{557^2 + 596^2 + 659^2 + 762^2}{20} - 82818.45 = 1201.05$$

$$SSAB = 4801.95 - 2992.45 - 1201.05 = 608.45$$

$$SSE = 5741.55 - 4801.95 = 939.60$$

The ANOVA Table. The results of the calculations for the fixed-effects model for a two-factor completely randomized experiment may, in general, be displayed as shown in Table 7.4.5.

The results for our illustrative example are displayed in Table 7.4.6.

8. *Statistical Decision* If the assumptions stated earlier hold true, and if each hypothesis is true, it can be shown that each of the variance ratios shown in Table 7.4.5 follows an *F* distribution with the indicated degrees of freedom. The critical values of *F* for testing the three hypotheses of our illustrative example are 2.76, 2.76, and 2.04, respectively. Since denominator degrees of freedom equal to 64 are not shown in Table G, 60 was used as the

TABLE 7.4.6

ANOVA Table for Example 7.4.1

Source	SS	d.f.	MS	V.R.
A	2992.45	3	997.48	67.95
B	1201.05	3	400.35	27.27
AB	608.45	9	67.61	4.61
Treatments	4801.95	15		
Residual	939.60	64	14.68	
Total	5741.55	79		

denominator degrees of freedom. We see that each of the three hypotheses would be rejected at the .05 level of significance. The reader is again cautioned that if all three tests are made simultaneously, the probability of rejecting at least one hypothesis when all are true is something greater than .05. Corresponding to the hypotheses stated in step 3 above we have $p_1 < .005$, $p_2 < .005$, and $p_3 < .005$.

9. *Conclusion* When H_0: $\alpha_1 = \alpha_2 = \alpha_3 = \alpha_4$ is rejected, we conclude that there are differences among the levels of A, that is, differences in the average amount of time spent in home visits with different types of patients. Similarly, when H_0: $\beta_1 = \beta_2 = \beta_3 = \beta_4$ is rejected, we conclude that there are differences among the levels of B, or differences in the average amount of time spent on home visits among the different nurses when grouped by age. When H_0: $(\alpha\beta)_{ij} = 0$ is rejected, we conclude that factors A and B interact; that is, different combinations of levels of the two factors produce different effects. When the hypothesis of no interaction is rejected interest in the levels of factors A and B usually become subordinate to interest in the interaction effects. In other words, we are more interested in learning what combinations of levels are significantly different.

We haved treated only the case where the number of observations in each cell is the same. When the number of observations per cell is not the same for every cell the analysis becomes more complex. Many of the references that have been cited cover the analysis appropriate to such a situation, and the reader is referred to them for further information.

EXERCISES

For each of the following exercises perform the analysis of variance, test appropriate hypotheses at the indicated level of significance, and determine the p value associated with each test.

7.4.1 In a treatment facility for the mentally retarded an experiment was conducted to compare three different methods for teaching basic self-care skills and three different reward systems. At the end of the experimental period the subjects were given a test designed to measure the degree to which the skills had been learned. The results are shown in the following table.

	METHODS		
Rewards	A	B	C
None	52	58	58
	76	56	24
	60	68	32
	58	74	39
Praise	60	60	56
	78	70	66
	75	74	54
	72	77	49
Tangible	98	76	72
	94	80	74
	96	84	76
	98	80	70

(a) Perform an analysis of variance of these data and test the hypotheses that reward effects are zero, method effects are zero, and interaction effects are zero. Let $\alpha' = \alpha'' = \alpha''' = .01$.

(b) What is the magnitude of α, the significance level for the test as a whole?

7.4.2 A mental health team in a local health department conducted a study of mothers' understanding of communicated child-care instructions regarding their children who had been seen in the mental health diagnostic and evaluation clinic. Three methods of consultation were compared, the routine consultation procedure and two experimental procedures. Mothers from three socioeconomic groups were identified. Following the consultations the mothers were interviewed by a second team of professionals, who assigned each mother a score designed to measure her level of understanding of the child-care instructions provided by the first team. The following results were obtained.

Socioeconomic group	METHOD OF CONSULTATION					
	Routine		A		B	
I	56	46	47	28	28	42
	29	33	28	30	53	42
II	51	21	59	47	68	75
	36	27	45	53	77	84
III	45	41	55	71	44	68
	31	37	76	85	77	76

(a)   Perform an analysis of variance of these data and test the three possible hypotheses. Let $\alpha' = .01$, $\alpha'' = .05$, and $\alpha''' = .01$.

(b)   What is the magnitude of α, the level of significance for the test as a whole?

7.4.3   In a study of the effects of different diets and group therapy in helping overweight persons lose weight, a researcher used 40 overweight women who were carefully matched on as many relevant variables as possible such as age, initial weight, and physical condition. The subjects were randomly assigned to 4 groups of 10 each, and each group of 10 was placed on a different diet. Each group of 10 was further divided at random into two groups. One of these 2 groups participated in a group therapy session twice a week, while the other group did not. At the end of the experimental period the amount of weight lost per person was recorded as shown in the table. Do an analysis of variance of these data and test the three possible hypotheses. Let $\alpha' = \alpha'' = \alpha''' = .05$.

		DIET		
Group therapy	I	II	III	IV
	15	25	19	22
	12	19	24	22
Yes	18	21	18	18
	16	22	16	19
	13	19	21	15
	9	13	13	33
	9	15	13	30
No	13	12	15	31
	7	15	18	27
	9	12	15	28

7.4.4   A study was conducted to compare the abilities of three drugs to retard reaction time of experimental animals to a certain stimulus. The same strain of animals was trained to respond by three different methods and three animals were used for each training method/drug combination. The following table shows the response time in seconds following administration of the drug.

		DRUG	
Training method	A	B	C
	5	8	10
I	8	8	12
	7	10	10
	6	10	15
II	8	12	14
	6	11	14
	7	12	16
III	8	12	16
	10	14	18

Perform an analysis of variance of these data and test the three possible hypotheses. Let $\alpha' = \alpha'' = \alpha''' = .05$.

7.4.5 Researchers at a trauma center wished to develop a program to help brain-damaged trauma victims regain an acceptable level of independence. An experiment involving 72 subjects with the same degree of brain damage was conducted. The objective was to compare different combinations of psychiatric treatment and physical therapy. Each subject was assigned to one of 24 different combinations of four types of psychiatric treatment and six physical therapy programs. There were three subjects in each combination. The response variable is the number of months elapsing between initiation of therapy and time at which the patient was able to function independently. The results were as follows.

Physical therapy program	PSYCHIATRIC TREATMENT			
	A	B	C	D
I	11.0	9.4	12.5	13.2
	9.6	9.6	11.5	13.2
	10.8	9.6	10.5	13.5
II	10.5	10.8	10.5	15.0
	11.5	10.5	11.8	14.6
	12.0	10.5	11.5	14.0
III	12.0	11.5	11.8	12.8
	11.5	11.5	11.8	13.7
	11.8	12.3	12.3	13.1
IV	11.5	9.4	13.7	14.0
	11.8	9.1	13.5	15.0
	10.5	10.8	12.5	14.0
V	11.0	11.2	14.4	13.0
	11.2	11.8	14.2	14.2
	10.0	10.2	13.5	13.7
VI	11.2	10.8	11.5	11.8
	10.8	11.5	10.2	12.8
	11.8	10.2	11.5	12.0

Can one conclude on the basis of these data that the different psychiatric treatment programs have different effects? Can one conclude that the physical therapy programs differ in effectiveness? Can one conclude that there is interaction between psychiatric treatment programs and physical therapy programs? Let $\alpha = .05$ for each test.

7.5 MISCELLANY

A general introductory text cannot begin to cover every aspect of every included topic. Although this chapter, for example, has covered the major concepts of analysis of variance, many topics have not been discussed, in the interest of brevity. To provide an awareness of some of the problems that may be encountered and of some of the additional analyses that are possible in analysis of variance, brief comments on additional topics of importance are given here. For further information refer to the experimental design and analysis of variance texts listed in the references. Most contain a more complete coverage of one or more of these topics.

Missing Data   Even though an experiment may be carried out with the greatest of care, accidents may occur that result in some of the observations being missing when it comes time to do the analysis. Missing observations may result in animal experiments if an animal dies. In experiments involving human beings a subject may be unable to participate in part of the experiment because of illness. In laboratory experiments, cultures and solutions may be destroyed as a result of breakage. An experiment may be completed without mishap, but some of the recorded results may be lost or destroyed.

   Generally, one of three procedures may be followed when data are found to be missing:

a. In some instances the analysis may proceed in the usual manner. If in a randomized complete block design, for example, a complete block or a complete treatment is missing, the usual analysis may be performed on the remaining data provided that there are at least two blocks and two treatments left.

b. The remaining data may be analysed by methods appropriate for the case of unequal subclass numbers. These procedures, which are more complex than the methods presented in this chapter, are discussed in some of the references.

c. It is usually preferable, however, to estimate the missing values. One of two methods is generally employed for this purpose.

   One method involves estimating the missing values in such a way that the experimental error sum of squares will be minimized when the usual analysis is performed. A disadvantage of this method is that it results in a biased treatment sum of squares that has to be corrected. Ostle (48) gives a numerical example of this method. Missing values are also frequently estimated by a technique known as *analysis of covariance*. An advantage of the use of covariance analysis is that it gives an unbiased treatment sum of squares as well as a minimum residual sum of squares. This method is illustrated by Steel and Torrie (22). Federer (7) gives an extensive list of references on missing values. References not cited by Federer include the articles by Glenn and Kramer (49), Kramer and Glass (50), and Baird and Kramer (51).

Transformations Occasionally the assumptions underlying the analysis of variance are not met in the data. An alternative procedure when this happens is to perform a transformation on the data in such a way that the assumptions are more nearly met. By transformation is meant a change in the scale of measurement.

Three frequent objectives in the employment of transformations are to make the variance independent of the mean, to achieve normality, and to achieve additivity of effects (eliminate interaction). Frequently a single transformation will achieve two or more objectives simultaneously. Some of the more commonly used transformations include the following.

1. *The Logarithmic Transformation* When this transformation is used, the logarithms (usually common logarithms) of the measurements are taken. This transformation is used when the mean is positively correlated with the variance.

2. *The Square Root Transformation* With this transformation the square roots of the measurements are taken. It is used when the data consist of counts, such as the number of deaths occurring in various groups of experimental animals. If zero counts are encountered, .5 is added to each count before the square root is taken.

3. *The Arcsine Transformation* This transformation is used when the data are proportions or percentages. The transformation yields $\theta = \arcsin \sqrt{p}$, where θ is the transformed measurement and p is the original measurement.

4. *The Reciprocal Transformation* When the variance increases as the fourth power of the mean, the reciprocal transformation may be used. Each original observation, θ, is replaced by its reciprocal, $1/\theta$.

In most of the general statistics and experimental design textbooks already mentioned, transformations are discussed. In addition, the subject is covered in more detail in the books by Quenouille (52), Sokal and Rohlf (53), and Pearce (54) and in an article by Bartlett (55). Federer (7) gives additional references.

Nonparametric Alternatives An alternative procedure that may be followed when the assumptions for analysis of variance are not met is to make use of what is known as a *nonparametric* method of analysis. These methods, in general, are discussed more fully in Chapter 11, where appropriate nonparametric alternatives are given for some of the experimental designs we have discussed.

Efficiency It is usually of interest to know how much improvement in the experiment as a whole may be expected if one type of design rather than some other type of design is employed. For example, the investigator may wonder if it is worth the effort to use a randomized complete block design instead of a completely randomized design. Even if it is not possible or desirable to choose among designs, the investigator may wish to investigate the merits of increasing sample size or the number of samples. Comparisons like these are judged on the basis of *efficiency* or *relative efficiency*. Basically, efficiency is a ratio of the variances resulting from the use of the designs being compared and is usually expressed as a percentage. Further discussion of this topic may be found in Cochran and Cox (4), Steel and Torrie (22), Scheffé (34), Ostle (48), and Sokal and Rohlf (53).

Other Designs   The two experimental designs considered in this chapter are by no means the only ones employed. The condensation necessary in a general introductory textbook obviates a complete discussion of those that are more frequently encountered. The reader who wishes to investigate the possibilities should refer to the experimental design texts listed in the references. Additional designs discussed in these references include the following.

1. *The Latin Square Design*   When it is possible and desirable to identify and isolate two extraneous sources of variation in an experiment, an extension of the randomized complete block design known as the *Latin square design* may be employed. The term, Latin square, was first used in an analysis of variance context by R. A. Fisher (41), who borrowed it from the Swiss mathematician Leonhard Euler (1707–1783).

   In the Latin square design one source of extraneous variation is assigned to the columns of the square, the second source of extraneous variation is assigned to the rows of the square, and the treatments, usually designated by Latin letters, are assigned in such a way that each treatment occurs once and only once in each row and each column. The number of rows, the number of columns, and the number of treatments are all equal.

2. *The Graeco–Latin Square Design*   This is an extension of the Latin square design which allows for the identification and isolation of three extraneous sources of variation. This design gets its name from the fact that another variable, represented by Greek letters, is superimposed on the Latin letters (treatments) of a Latin square in such a way that each Greek letter occurs once in each row, once in each column, and once with each Latin letter.

3. *The Incomplete Block Design*   If a large number of treatments must be included in an experiment, the use of the randomized complete block design may not be feasible. An alternative is to include only a part of the treatments in a block. The blocks that result from this procedure are called *incomplete blocks*.

4. *The Split Plot Design*   The *split plot design* is a special kind of incomplete block design that is frequently used for factorial experiments. Levels of one or more factors are applied to *whole plots* that are divided into *subplots* to which levels of one or more additional factors are applied.

7.6 SUMMARY

The purpose of this chapter is to introduce the student to the basic ideas and techniques of analysis of variance. Two experimental designs, the completely randomized and the randomized complete block, are discussed in considerable detail. In addition, the concept of a factorial experiment as used with the completely randomized design is introduced. Some additional topics are examined briefly. Individuals who wish to pursue further any aspect of analysis of variance will find the references at the end of the chapter most helpful. The extensive bibliography by Herzberg and Cox (56) indicates further readings.

REVIEW QUESTIONS AND EXERCISES

1. Define analysis of variance.

2. For each of the following designs describe a situation in your particular field of interest where the design would be an appropriate experimental design. Use real or realistic data and do the appropriate analysis of variance for each one:
   (a) Completely randomized design.
   (b) Randomized complete block design.
   (c) Completely randomized design with a factorial experiment.

3. Heart rate (beats per minute) was measured in four groups of adult subjects: normal controls (A), angina patients (B), hypertensive subjects (C), and patients with healed myocardial infarction (D). The results were as follows.

A	B	C	D
83	81	75	61
61	65	68	75
80	77	80	78
63	87	80	80
67	95	74	68
89	89	78	65
71	103	69	68
73	89	72	69
70	78	76	70
66	83	75	79
57	91	69	61

Do these data provide sufficient evidence to indicate a difference in the mean heart rate among these four types of subject? Let $\alpha = .05$. Determine the p value.

4. Random samples were drawn from each of three types of subject: high exposure level employees of a pesticide manufacturing firm, farm laborers exposed to the pesticide for several weeks each year, and persons with no known exposure to the pesticide. Acetylcholinesterase determinations were made on blood samples from each subject with the following results.

Employees	Farm laborers	Unexposed
6.4	6.5	7.3
6.6	6.8	7.5
6.8	7.0	7.8
6.9	7.1	7.9
9.5	9.7	10.8
6.1	6.2	6.9
7.5	7.7	8.5
8.2	8.4	9.4
4.1	4.2	4.6
5.5	5.6	6.3

Do these data provide sufficient evidence to indicate a difference, on the average, among the three groups? Let $\alpha = .05$. Determine the p value.

5. Respiratory rate (breaths per minute) was measured in eight experimental animals under three levels of exposure to carbon monoxide. The results were as follows.

	EXPOSURE LEVEL		
Animal	Low	Moderate	High
1	36	43	45
2	33	38	39
3	35	41	33
4	39	34	39
5	41	28	33
6	41	44	26
7	44	30	39
8	45	31	29

Can one conclude on the basis of these data that the three exposure levels, on the average, have a different effect on respiratory rate? Let $\alpha = .05$. Determine the p value.

6. An experiment was designed to study the effects of three different drugs and three types of stressful situation in producing anxiety in adolescent subjects.

The table below shows the difference between the pre- and posttreatment scores of 18 subjects who participated in the experiment.

Stressful situation (factor A)	DRUG (FACTOR B)		
	A	B	C
I	4	1	1
	5	3	0
II	6	6	6
	6	6	3
III	5	7	4
	4	4	5

Perform an analysis of variance of these data and test the three possible hypotheses. Let $\alpha' = \alpha'' = \alpha''' = .05$. Determine the p values.

7. The following table shows the emotional maturity scores of 27 young adult males cross-classified by age and the extent to which they use marijuana.

Age (factor A)	MARIJUANA USAGE (FACTOR B)		
	Never	Occasionally	Daily
15–19	25	18	17
	28	23	24
	22	19	19
20–24	28	16	18
	32	24	22
	30	20	20
25–29	25	14	10
	35	16	8
	30	15	12

Perform an analysis of variance of these data. Let $\alpha' = \alpha'' = \alpha''' = .05$. Compute the p values.

8. The following are the amounts of a certain chemical present in a uniform amount of tissue from 25 laboratory animals representing four different species. Test the null hypothesis that, on the average, the four species contain the same amount of the chemical, and test for a significant difference between all possible pairs of sample means. Let $\alpha = .05$.

SPECIES

1	2	3	4
65.7	186.7	86.8	139.1
70.3	176.0	102.6	147.5
76.1	188.5	84.4	130.0
78.3	178.9	90.3	150.1
68.7	180.2	98.0	142.1
72.1		88.5	144.4
76.1			138.3
80.2			

9. An experiment was conducted to test the effect of four different drugs on blood coagulation time (in minutes). Specimens of blood drawn from 10 subjects were divided equally into four parts that were randomly assigned to one of the four drugs. The results were as follows.

DRUG

Subject	W	X	Y	Z
A	1.5	1.8	1.7	1.9
B	1.4	1.4	1.3	1.5
C	1.8	1.6	1.5	1.9
D	1.3	1.2	1.2	1.4
E	2.0	2.1	2.2	2.3
F	1.1	1.0	1.0	1.2
G	1.5	1.6	1.5	1.7
H	1.5	1.5	1.5	1.7
I	1.2	1.0	1.3	1.5
J	1.5	1.6	1.6	1.9

Can we conclude on the basis of these data that the drugs have different effects? Let $\alpha = .05$.

10. Determinations of serum aldolase in the serum of three samples of children gave the following results.

Controls	Subjects with progressive muscular dystropy	Subjects with poliomyelitis
.30	14.00	1.50
.30	5.50	1.10
.30	12.00	.80
.40	7.10	1.30
.40	8.00	.90
.40	4.20	.60
.50	6.30	.50
.50	3.00	
.50	6.00	
.50		

Do these data provide sufficient evidence to indicate that the three represented populations differ with respect to mean serum aldolase levels? Let $\alpha = .01$. Test for a significant difference between all possible pairs of sample means.

11. The following table shows the arterial plasma epinephrine concentrations (nanograms per milliiter) found in 10 laboratory animals during three types of anesthesia.

	ANIMAL									
Anesthesia	*1*	*2*	*3*	*4*	*5*	*6*	*7*	*8*	*9*	*10*
A	.28	.50	.68	.27	.31	.99	.26	.35	.38	.34
B	.20	.38	.50	.29	.38	.62	.42	.87	.37	.43
C	1.23	1.34	.55	1.06	.48	.68	1.12	1.52	.27	.35

Can we conclude from these data that the three types of anesthesia, on the average, have different effects? Let $\alpha = .05$.

12. The nutritive value of a certain edible fruit was measured in a total of 72 specimens representing 6 specimens of each of four varieties grown in each of three geographic regions. The results were as follows.

Geographic region	**VARIETY**			
	W	X	Y	Z
	6.9	11.0	13.1	13.4
	11.8	7.8	12.1	14.1
	6.2	7.3	9.9	13.5
A	9.2	9.1	12.4	13.0
	9.2	7.9	11.3	12.3
	6.2	6.9	11.0	13.7
	8.9	5.8	12.1	9.1
	9.2	5.1	7.1	13.1
	5.2	5.0	13.0	13.2
B	7.7	9.4	13.7	8.6
	7.8	8.3	12.9	9.8
	5.7	5.7	7.5	9.9
	6.8	7.8	8.7	11.8
	5.2	6.5	10.5	13.5
	5.0	7.0	10.0	14.0
C	5.2	9.3	8.1	10.8
	5.5	6.6	10.6	12.3
	7.3	10.8	10.5	14.0

Test for a difference among varieties, a difference among regions, and interaction. Let $\alpha = .05$ for all tests.

13. A random sample of the records of single births was selected from each of four populations. The weights (grams) of the babies at birth were as follows.

SAMPLE

A	B	C	D
2946	3186	2300	2286
2913	2857	2903	2938
2280	3099	2572	2952
3685	2761	2584	2348
2310	3290	2675	2691
2582	2937	2571	2858
3002	3347		2414
2408			2008
			2850
			2762

Do these data provide sufficient evidence to indicate, at the .05 level of significance, that the four populations differ with respect to mean birth weight? Test for a significant difference between all possible pairs of means.

14. The following table shows the aggression scores of 30 laboratory animals reared under three different conditions. One animal from each of 10 litters was randomly assigned to each of the three rearing conditions.

REARING CONDITION

Litter	Extremely crowded	Moderately crowded	Not crowded
1	30	20	10
2	30	10	20
3	30	20	10
4	25	15	10
5	35	25	20
6	30	20	10
7	20	20	10
8	30	30	10
9	25	25	10
10	30	20	20

Do these data provide sufficient evidence to indicate that level of crowding has an effect on aggression? Let $\alpha = .05$.

15. The following table shows the vital capacity measurements of 60 adult males classified by occupation and age group.

	OCCUPATION			
Age group	A	B	C	D
	4.31	4.68	4.17	5.75
	4.89	6.18	3.77	5.70
1	4.05	4.48	5.20	5.53
	4.44	4.23	5.28	5.97
	4.59	5.92	4.44	5.52
	4.13	3.41	3.89	4.58
	4.61	3.64	3.64	5.21
2	3.91	3.32	4.18	5.50
	4.52	3.51	4.48	5.18
	4.43	3.75	4.27	4.15
	3.79	4.63	5.81	6.89
	4.17	4.59	5.20	6.18
3	4.47	4.90	5.34	6.21
	4.35	5.31	5.94	7.56
	3.59	4.81	5.56	6.73

Test for differences among occupations, for differences among age groups, and for interaction. Let $\alpha = .05$ for all tests.

16. Complete the following ANOVA table and state which design was used.

Source	SS	d.f.	MS	VR	p
Treatments	154.9199	4			
Error					
Total	200.4773	39			

17. Complete the following ANOVA table and state which design was used.

Source	SS	d.f.	MS	VR	p
Treatments		3			
Blocks	183.5	3			
Error	26.0				
Total	709.0	15			

18. Consider the following ANOVA table.

Source	SS	d.f.	MS	VR	p
A	12.3152	2	6.15759	29.4021	< .005
B	19.7844	3	6.59481	31.4898	< .005
AB	8.94165	6	1.49027	7.11596	< .005
Treatments	41.0413	11			
Error	10.0525	48	0.209427		
Total	51.0938	59			

(a) What sort of analysis was employed?
(b) What can one conclude from the analysis? Let $\alpha = .05$.

19. Consider the following ANOVA table.

Source	SS	d.f.	MS	VR
Treatments	5.05835	2	2.52917	1.0438
Error	65.42090	27	2.4230	

(a) What design was employed?
(b) How many treatments were compared?
(c) How many observations were analyzed?
(d) At the .05 level of significance, can one conclude that there is a difference among treatments? Why?

20. Consider the following ANOVA table.

Source	SS	d.f.	MS	VR
Treatments	231.5054	2	115.7527	2.824
Blocks	98.5000	7	14.0714	
Error	573.7500	14	40.9821	

(a) What design was employed?
(b) How many treatments were compared?
(c) How many observations were analyzed?
(d) At the .05 level of significance, can one conclude that the treatments have different effects? Why?

21. We wish to conduct a study to determine if the measurement of serum lipid-bound sialic acid (LSA) might be of use in the detection of breast cancer. The following are the LSA measurements (mg/dl) in four populations of subjects: normal controls, A; patients with benign breast disease, B; patients with primary breast cancer, C; and patients with recurrent metastatic breast cancer, D. Select a simple random sample of size 10 from each population and perform an appropriate analysis to determine if we may conclude that the four population means are different. Let $\alpha = .05$ and determine the p value. Test all possible pairs of sample means for significance. What conclusions can one draw from the analysis? Prepare a verbal report of the findings. Compare your results with those of your classmates.

Observation number	POPULATION			
	A	B	C	D
1.	18.8	24.3	19.0	22.3
2.	17.3	22.2	12.5	17.3
3.	16.4	19.0	21.0	20.8
4.	18.9	21.5	21.5	26.1

Observation number	POPULATION			
	A	B	C	D
5.	14.7	24.6	18.5	25.1
6.	13.3	21.2	22.1	24.8
7.	16.9	18.8	26.2	23.0
8.	17.9	21.2	19.1	23.8
9.	18.0	20.3	22.2	25.1
10.	20.9	22.6	18.8	25.3
11.	18.8	18.6	16.4	22.9
12.	11.1	15.9	20.9	23.4
13.	18.3	19.2	17.8	19.7
14.	13.3	22.3	26.5	21.9
15.	17.9	21.6	20.5	22.6
16.	18.6	20.3	23.1	20.2
17.	15.1	15.2	20.0	27.6
18.	18.8	20.3	19.4	21.5
19.	12.1	17.3	25.2	14.5
20.	16.4	23.0	17.0	19.4
21.	20.1	24.7	22.5	22.7
22.	16.7	20.2	22.6	23.3
23.	18.3	12.6	25.8	22.2
24.	16.3	16.8	21.4	20.3
25.	12.1	18.8	20.4	23.7
26.	17.9	23.9	21.4	13.1
27.	18.6	15.1	20.5	18.7
28.	18.5	21.3	17.4	22.1
29.	13.7	21.5	20.8	23.0
30.	18.3	16.6	17.5	22.4
31.	14.5	22.5	18.2	22.4
32.	18.5	23.4	22.2	17.9
33.	17.1	24.1	18.4	24.7
34.	14.8	21.0	22.0	19.1
35.	12.3	19.9	18.1	25.1
36.	12.3	15.2	18.6	23.5
37.	12.3	22.1	16.8	23.6
38.	19.4	19.5	21.5	26.6
39.	15.6	25.9	22.5	23.3
40.	14.3	24.6	18.5	19.5
41.	15.8	23.0	17.5	25.2
42.	18.0	16.6	19.7	22.4
43.	18.8	23.0	19.0	23.6
44.	13.4	25.5	19.4	22.8
45.	19.7	17.5	18.9	22.4
46.	17.5	16.5	19.2	28.0
47.	15.6	22.6	16.0	18.2
48.	10.8	23.8	22.4	20.8
49.	20.2	19.9	20.2	26.6

| Observation | POPULATION | | | |
number	A	B	C	D
50.	15.7	23.5	20.7	20.6
51.	18.2	14.9	21.0	18.7
52.	13.0	21.6	20.2	25.7
53.	17.5	13.4	18.6	24.7
54.	17.5	18.9	19.9	19.1
55.	11.2	24.1	19.8	22.9
56.	17.2	26.6	26.9	29.5
57.	17.8	23.3	22.3	21.1
58.	17.8	10.4	27.5	29.9
59.	15.7	20.6	20.8	20.2
60.	20.3	14.0	16.9	22.9
61.	15.7	22.7	23.2	22.2
62.	19.6	24.0	19.1	26.1
63.	11.4	17.8	19.9	22.7
64.	16.1	21.8	22.8	20.6
65.	14.5	20.5	20.0	25.0
66.	14.7	21.5	21.9	18.3
67.	20.2	18.2	18.2	20.4
68.	12.1	18.0	22.3	23.0
69.	18.4	22.2	17.8	26.7
70.	15.0	17.1	21.0	26.9
71.	20.9	18.6	19.9	23.0
72.	13.4	18.6	16.7	24.1
73.	12.7	18.9	21.4	22.3
74.	15.9	19.8	18.4	19.7
75.	20.8	16.2	17.5	25.1
76.	15.4	19.5	21.8	26.1
77.	16.2	22.0	24.5	27.5
78.	16.9	18.3	15.4	22.0
79.	17.0	20.1	23.4	24.1
80.	19.0	19.5	27.8	19.8
81.	20.4	20.6	19.8	25.8
82.	16.9	20.3	19.8	25.4
83.	14.1	20.7	26.6	21.8
84.	20.8	18.3	23.5	27.2
85.	15.4	16.7	16.2	22.9
86.	16.3	16.3	20.5	24.4
87.	13.7	20.9	22.9	28.3
88.	18.4	21.5	18.6	24.9
89.	16.6	23.6	25.6	25.0
90.	13.7	25.3	16.9	23.2
91.	16.9	24.6	16.2	19.7
92.	19.4	19.2	17.9	20.6
93.	18.7	21.3	19.1	24.7
94.	15.8	16.3	15.1	29.3

Observation number	POPULATION			
	A	B	C	D
95.	14.3	19.9	20.0	23.6
96.	12.2	13.3	19.9	23.0
97.	13.7	21.9	23.4	22.4
98.	13.0	17.4	14.8	23.2
99.	13.6	22.2	24.7	24.9
100.	14.5	26.3	19.9	26.7
101.	13.6	19.0	23.8	25.8
102.	19.1	27.1	19.8	21.6
103.	11.1	19.5	13.6	24.2
104.	13.5	25.4	19.6	22.2
105.	13.7	24.3	22.6	24.3
106.	18.2	20.8	16.6	21.2
107.	17.1	19.3	16.0	24.0
108.	16.3	22.2	16.9	21.4
109.	16.6	19.7	21.5	23.0
110.	16.6	19.5	21.5	26.4
111.	19.3	18.0	13.5	25.0
112.	13.2	22.2	17.2	23.7
113.	18.9	20.5	17.2	25.4
114.	12.2	15.9	14.8	25.9
115.	19.6	21.7	21.2	21.6
116.	14.1	13.4	17.7	13.9
117.	10.6	20.1	26.5	27.3
118.	14.7	20.8	18.3	22.2
119.	15.0	17.7	15.5	26.8
120.	13.3	19.4	19.9	24.4
121.	18.1	23.5	22.3	21.9
122.	17.0	16.6	19.5	24.5
123.	15.7	16.7	20.0	23.8
124.	16.4	16.7	20.3	21.4
125.	19.1	15.2	22.4	23.5
126.	14.1	20.7	22.1	20.6
127.	17.7	17.8	23.8	21.7
128.	15.6	19.0	23.4	17.8
129.	12.4	16.1	19.8	24.1
130.	17.4	17.0	16.7	23.0
131.	14.1	24.5	20.8	22.5
132.	20.6	21.1	25.9	27.7
133.	16.2	20.7	20.3	20.3
134.	19.1	23.7	15.7	21.2
135.	12.6	18.9	21.4	25.3
136.	20.6	19.3	22.8	18.2
137.	15.3	25.5	18.5	23.8
138.	16.3	19.2	17.5	22.3
139.	13.4	19.9	22.8	25.8

Observation	POPULATION			
number	A	B	C	D
140.	20.6	20.6	24.0	28.4
141.	16.8	15.5	18.0	22.8
142.	15.2	16.9	17.0	19.7
143.	15.5	21.9	23.3	23.4
144.	15.4	18.1	15.5	24.7
145.	12.7	22.8	22.6	21.9
146.	18.5	19.9	20.6	30.5
147.	19.6	22.6	23.9	21.8
148.	18.9	20.8	20.7	22.8
149.	15.5	23.9	13.1	22.1
150.	16.9	16.5	18.2	22.0
151.	15.7	21.4	21.0	21.9
152.	19.3	16.7	18.5	23.9
153.	20.5	21.8	21.0	22.8
154.	11.1	19.5	19.7	24.5
155.	17.8	20.0	19.1	25.3
156.	14.4	22.7	28.0	23.0
157.	16.8	19.3	18.6	23.1
158.	16.6	23.2	20.4	21.1
159.	11.4	11.7	15.5	26.9
160.	16.7	19.2	18.1	21.7
161.	20.9	20.2	22.0	19.3
162.	15.0	24.4	19.6	17.8
163.	14.0	26.1	21.1	20.1
164.	17.5	18.7	17.8	18.1
165.	14.1	25.3	22.9	22.0
166.	20.6	19.8	21.7	26.9
167.	15.5	14.8	23.3	24.2
168.	14.7	21.1	22.3	23.8
169.	20.2	19.1	20.8	21.0
170.	17.2	24.5	19.8	24.5
171.	12.8	14.6	21.6	19.6
172.	17.6	23.3	21.8	18.6
173.	17.2	21.5	22.8	21.3
174.	11.2	21.0	19.6	20.1
175.	15.1	23.6	17.3	27.2
176.	12.4	21.1	20.8	25.2
177.	11.8	23.7	20.8	21.9
178.	13.2	20.2	13.8	25.3
179.	16.4	21.6	19.1	17.8
180.	16.3	18.0	19.0	22.6
181.	17.5	17.1	14.2	27.6
182.	11.0	16.8	24.1	17.7
183.	15.9	22.0	17.7	21.9
184.	20.8	17.9	17.9	24.0
185.	16.4	19.7	13.3	22.2

Observation number	POPULATION			
	A	B	C	D
186.	20.9	22.9	18.4	24.5
187.	15.4	13.0	20.1	24.0
188.	16.2	17.2	18.1	27.6
189.	17.1	22.4	18.8	22.9
190.	11.3	20.8	16.2	18.3
191.	11.4	17.9	18.0	20.2
192.	15.5	20.7	22.4	18.7
193.	18.1	13.3	16.7	22.7
194.	18.9	18.4	19.8	24.7
195.	20.7	18.1	19.5	24.1
196.	15.3	22.6	17.6	28.9
197.	15.2	21.2	22.8	28.0
198.	19.3	22.4	19.1	17.7
199.	20.8	22.2	21.0	24.0
200.	18.8	25.0	21.7	25.3
201.	15.0	18.1	18.1	25.1
202.	11.5	16.1	20.2	25.9
203.	13.9	19.8	23.4	23.4
204.	14.3	27.0	19.8	19.8
205.	20.1	21.8	22.2	17.6
206.	14.3	21.3	19.3	21.7
207.	19.5	16.5	21.6	20.5
208.	13.3	17.9	18.0	17.8
209.	17.0	22.0	15.0	25.7
210.	16.0	19.3	19.6	24.6
211.	12.5	18.7	18.0	26.7
212.	14.8	16.3	19.8	21.4
213.	16.6	18.6	15.2	28.0
214.	11.5	22.4	17.5	24.5
215.	18.6	16.5	22.7	24.6
216.	15.4	17.4	19.3	20.7
217.	20.4	15.3	23.2	24.7
218.	17.7	20.7	19.2	29.0
219.	20.3	20.4	21.7	22.7
220.	20.3	20.8	18.6	27.9
221.	17.7	25.5	23.0	18.3
222.	15.1	23.1	20.6	25.0
223.	18.5	21.0	15.5	20.9
224.	17.5	23.3	21.9	27.6
225.	16.4	16.5	17.4	20.2
226.	11.5	17.7	21.1	25.6
227.	14.0	19.4	25.4	20.2
228.	13.3	14.1	18.2	16.4
229.	16.5	16.9	22.3	25.7
230.	15.9	19.3	21.5	21.1
231.	16.8	20.2	21.7	22.5

Observation number	POPULATION			
	A	B	C	D
232.	17.8	16.4	20.6	24.7
233.	10.7	24.2	21.9	26.9
234.	15.2	18.6	20.2	22.7
235.	20.7	21.5	20.0	26.7
236.	15.0	20.4	23.2	24.5
237.	14.1	24.5	20.7	21.7
238.	17.5	22.4	18.6	24.4
239.	20.1	22.3	19.0	21.9
240.	16.2	20.3	21.8	22.0
241.	18.4	16.3	21.7	21.2
242.	17.8	19.6	18.2	26.1
243.	13.1	20.3	15.9	23.1
244.	15.2	15.5	20.8	21.7
245.	19.5	22.0	17.4	26.6
246.	15.5	20.9	22.6	22.0
247.	19.9	17.9	21.6	21.0
248.	15.3	16.7	22.8	21.9
249.	18.9	25.2	20.0	25.0
250.	20.9	23.2	18.2	24.2
251.	18.3	18.8	18.0	28.0
252.	14.0	20.0	26.7	24.5
253.	13.0	21.9	21.7	26.7
254.	16.3	17.0	21.1	23.0
255.	17.0	20.6	22.0	29.9
256.	16.4	20.7	22.4	24.4
257.	15.7	22.5	17.0	22.3
258.	15.5	23.7	18.1	27.3
259.	17.0	16.2	23.4	21.7
260.	18.4	16.6	17.1	18.7
261.	12.7	22.6	22.7	24.6
262.	16.6	17.9	23.0	25.1
263.	15.6	17.4	16.3	24.0
264.	14.7	20.1	25.6	19.0
265.	11.3	16.1	22.4	29.0
266.	16.4	23.6	21.7	20.0
267.	11.9	18.5	23.0	20.2
268.	15.6	16.8	24.2	24.1
269.	18.9	26.0	21.7	19.5
270.	11.2	23.3	20.9	19.9
271.	16.9	16.5	13.2	22.8
272.	13.9	19.1	20.6	24.3
273.	20.7	17.9	17.0	26.9
274.	15.1	26.1	19.1	26.5
275.	14.3	24.4	20.6	26.4
276.	13.5	23.3	24.7	27.5
277.	17.1	15.1	20.8	15.2

Observation number	POPULATION			
	A	B	C	D
278.	18.4	17.9	14.7	20.2
279.	14.2	19.4	16.2	24.4
280.	16.3	13.3	25.1	17.6
281.	14.7	20.4	19.5	22.6
282.	19.3	21.6	20.1	29.9
283.	13.7	19.9	19.4	33.7
284.	20.7	19.6	22.7	26.0
285.	11.5	17.5	23.4	23.1
286.	19.0	15.3	19.2	25.7
287.	12.3	18.5	18.9	20.2
288.	20.0	19.3	19.2	20.9
289.	15.8	21.7	22.5	27.9
290.	15.5	21.9	19.4	26.7
291.	20.8	14.6	21.0	22.6
292.	20.6	25.1	22.7	18.8
293.	13.4	17.7	21.5	24.4
294.	11.2	23.8	19.0	22.1
295.	18.0	19.3	19.6	17.3
296.	19.0	23.4	25.4	21.0
297.	19.5	22.8	20.4	19.8
298.	20.0	23.8	21.6	25.6
299.	12.4	20.5	19.4	25.6
300.	16.7	18.0	25.3	20.2
301.	14.5	19.1	20.7	28.2
302.	20.3	17.1	19.8	20.8
303.	11.3	19.7	15.5	22.1
304.	19.2	14.3	14.8	22.1
305.	20.4	22.5	20.5	22.1
306.	14.8	20.6	17.2	26.8
307.	18.4	20.3	23.4	24.4
308.	11.8	19.3	19.7	23.3
309.	18.9	22.5	22.3	21.8
310.	13.4	22.0	18.0	19.8
311.	16.4	22.8	13.3	23.4
312.	11.3	21.1	15.2	23.6
313.	16.3	19.7	15.7	22.0
314.	12.7	16.8	19.9	22.3
315.	17.0	22.5	19.9	25.9
316.	16.5	23.0	16.7	25.0
317.	16.7	13.3	18.7	20.9
318.	17.9	18.4	21.7	20.6
319.	13.6	17.4	26.6	20.4
320.	13.7	25.7	22.6	21.3
321.	18.0	19.1	23.1	21.5
322.	17.6	18.6	18.6	25.7
323.	13.5	18.4	23.6	21.8

| Observation | POPULATION | | | |
number	A	B	C	D
324.	13.6	15.2	23.6	25.7
325.	11.6	17.3	27.1	25.4
326.	15.8	22.2	21.9	22.3
327.	15.9	21.0	18.2	24.1
328.	13.0	22.5	22.0	24.1
329.	14.9	18.7	19.1	24.4
330.	13.0	25.2	21.5	23.6
331.	16.1	20.4	18.8	23.5
332.	14.3	22.9	16.5	14.9
333.	18.1	24.1	16.7	26.8
334.	16.0	24.2	24.4	19.9
335.	16.0	23.0	21.5	22.7
336.	16.9	19.3	24.8	20.1
337.	20.7	22.1	19.3	24.9
338.	19.0	16.5	17.8	24.0
339.	11.6	19.4	19.4	30.5
340.	12.3	19.5	17.5	23.4
341.	12.3	16.9	21.2	23.0
342.	14.4	20.5	15.5	25.2
343.	14.3	19.0	17.5	24.2
344.	20.9	18.1	15.8	22.8
345.	13.2	26.1	23.6	27.5
346.	19.2	13.2	19.8	25.1
347.	19.1	25.1	16.9	25.7
348.	19.3	21.8	16.1	16.6
349.	12.9	16.4	22.0	24.1
350.	14.2	26.7	18.3	21.6

22. Sarcoidosis, found throughout the world, is a systemic granulomatous disease of unknown cause. The assay of serum angiotensin-converting enzyme (SACE) is helpful in the diagnosis of active sarcoidosis. The activity of SACE is usually increased in patients with the disease, while normal levels occur in subjects who have not had the disease, those who have recovered, and patients with other granulomatous disorders. The following data are the SACE values for four populations of subjects classified according to status regarding sarcoidosis: never had, A; active, B; stable, C; recovered, D. Select a simple random sample of 15 subjects from each population and perform an analysis to determine if you can conclude that the population means are different. Let $\alpha = .05$. Use Tukey's test to test for significant differences among individual pairs of means. Prepare a written report on your findings. Compare your results with those of your classmates.

Observation number	POPULATION			
	A	B	C	D
1.	108.4	198.3	89.1	64.8
2.	73.8	174.1	117.0	70.4
3.	65.9	168.2	113.3	65.0
4.	86.7	189.9	108.4	85.6
5.	102.7	169.9	132.1	98.5
6.	101.2	183.6	115.1	81.2
7.	98.9	163.4	102.3	80.5
8.	94.2	193.8	130.2	69.3
9.	118.4	183.7	114.2	76.4
10.	87.4	192.9	112.9	88.3
11.	101.8	174.1	120.2	70.4
12.	88.2	168.8	126.0	72.9
13.	87.5	187.6	110.8	72.7
14.	73.9	191.6	103.8	75.5
15.	90.5	183.8	105.4	87.8
16.	87.6	190.6	103.8	78.2
17.	87.5	188.4	139.4	74.4
18.	88.5	187.5	107.9	61.6
19.	96.3	171.7	121.7	89.3
20.	81.0	169.0	109.9	89.5
21.	63.3	171.2	108.9	61.4
22.	102.2	184.3	108.7	81.4
23.	74.6	174.6	135.8	90.1
24.	77.3	157.2	107.5	71.0
25.	91.7	171.5	109.0	86.2
26.	97.0	186.9	114.4	97.7
27.	72.4	160.7	107.5	89.5
28.	94.6	164.3	113.8	70.4
29.	88.7	166.3	115.0	84.4
30.	73.2	150.0	112.1	70.3
31.	84.3	161.4	91.8	81.6
32.	90.3	163.7	97.3	57.7
33.	71.3	166.3	137.0	87.5
34.	98.8	170.9	113.3	89.1
35.	86.8	173.9	137.0	80.8
36.	86.2	178.0	115.6	83.9
37.	81.7	162.2	94.4	55.1
38.	105.9	173.1	99.2	75.7
39.	79.8	175.8	120.2	71.2
40.	85.9	157.7	106.2	86.8
41.	92.7	159.0	108.0	83.4
42.	117.3	156.4	107.8	107.1
43.	103.8	196.2	101.5	113.7
44.	86.1	180.2	126.7	88.2
45.	114.0	166.9	113.4	74.1
46.	78.3	153.8	110.9	102.1

Observation number	POPULATION			
	A	B	C	D
47.	87.8	158.5	127.3	75.9
48.	92.4	179.7	109.1	81.3
49.	78.4	209.5	79.9	114.6
50.	78.7	172.2	113.7	94.2
51.	80.7	180.3	109.2	77.6
52.	84.1	185.4	117.8	69.1
53.	107.5	162.0	108.6	60.1
54.	104.5	165.5	115.8	54.5
55.	107.8	163.1	124.4	72.5
56.	79.6	186.3	130.6	79.6
57.	74.0	167.4	108.2	72.9
58.	100.2	174.1	82.1	66.0
59.	73.6	209.4	101.4	71.3
60.	93.1	176.3	114.2	60.0
61.	109.2	178.2	110.6	81.4
62.	97.5	157.4	95.5	55.1
63.	90.8	195.0	111.3	67.8
64.	82.5	161.5	126.7	54.5
65.	94.1	194.2	111.1	83.4
66.	82.9	169.8	121.5	70.7
67.	91.9	161.6	112.4	83.9
68.	80.7	182.3	91.0	84.4
69.	112.6	167.3	105.8	90.8
70.	98.0	156.8	114.6	81.0
71.	90.1	161.0	137.4	69.0
72.	69.3	171.1	93.7	87.4
73.	100.9	159.4	117.5	67.0
74.	81.9	173.3	117.8	89.9
75.	78.6	171.7	118.9	82.1
76.	100.2	173.7	99.8	76.4
77.	92.6	171.2	90.1	59.2
78.	93.3	178.1	86.5	56.9
79.	109.2	178.8	111.6	95.8
80.	91.6	176.7	115.7	86.6
81.	76.3	165.5	109.7	62.4
82.	95.8	194.6	104.9	89.5
83.	97.3	183.6	102.0	70.1
84.	90.3	185.9	126.8	90.7
85.	108.4	180.0	96.7	83.0
86.	75.5	197.0	100.5	83.4
87.	104.6	175.1	115.2	73.8
88.	101.5	178.8	95.8	63.1
89.	100.4	152.1	101.2	76.6
90.	85.3	158.0	110.4	92.5
91.	88.4	158.8	118.3	68.2
92.	91.5	158.7	127.0	90.4

Observation number	POPULATION			
	A	B	C	D
93.	82.5	164.4	104.7	69.3
94.	102.9	187.9	129.5	76.8
95.	95.5	190.2	124.7	69.3
96.	98.7	188.6	120.7	71.3
97.	108.1	197.3	108.8	87.4
98.	68.7	177.2	116.1	64.0
99.	72.9	182.0	84.0	79.4
100.	87.4	170.4	112.1	77.1
101.	59.8	188.6	133.1	93.8
102.	83.5	160.3	110.9	75.2
103.	91.7	185.7	98.9	74.9
104.	90.4	172.5	104.4	75.3
105.	77.8	182.1	94.1	77.3
106.	67.0	184.4	108.7	68.1
107.	79.4	188.3	112.3	82.8
108.	85.4	168.8	127.8	72.8
109.	89.1	174.1	123.1	73.7
110.	87.2	190.8	110.4	74.3
111.	81.9	183.3	119.9	80.2
112.	111.1	157.6	100.1	85.3
113.	108.9	174.1	109.1	96.9
114.	105.9	183.7	142.2	66.4
115.	96.8	178.1	103.1	65.2
116.	81.3	173.4	110.6	67.1
117.	93.4	185.0	103.6	84.8
118.	104.5	184.8	120.2	101.7
119.	72.7	193.7	104.3	69.5
120.	79.2	193.1	86.9	110.3
121.	61.6	193.0	106.4	73.0
122.	87.2	177.2	102.2	62.8
123.	103.1	175.4	92.3	73.2
124.	97.6	180.9	75.2	68.8
125.	76.4	175.8	103.6	82.0
126.	63.8	148.6	109.1	86.3
127.	88.1	181.8	120.5	60.5
128.	96.8	192.3	118.2	80.2
129.	94.4	173.6	121.3	70.3
130.	85.2	179.7	103.2	68.4
131.	115.9	185.9	110.9	84.7
132.	116.2	184.3	127.3	73.5
133.	101.1	173.1	92.7	82.3
134.	96.1	177.6	117.5	70.7
135.	111.3	141.9	122.3	82.5
136.	95.7	168.4	93.0	87.4
137.	84.1	176.0	103.0	66.5
138.	76.9	203.5	89.2	88.2

Observation number	POPULATION			
	A	B	C	D
139.	77.1	183.8	92.8	95.2
140.	101.8	174.1	131.0	77.4
141.	81.6	186.3	125.4	70.5
142.	83.7	165.7	103.6	67.6
143.	88.9	181.0	116.2	78.7
144.	88.0	167.9	97.4	104.9
145.	85.4	190.1	124.8	58.3
146.	77.6	173.8	101.1	79.6
147.	98.5	176.4	114.9	61.4
148.	98.2	181.2	122.3	78.5
149.	88.0	199.0	124.0	75.1
150.	85.8	188.9	98.7	85.7
151.	97.2	179.1	133.3	85.2
152.	76.1	175.5	125.0	69.2
153.	92.2	185.2	83.7	57.1
154.	101.8	182.9	120.9	79.4
155.	63.2	190.9	110.3	86.2
156.	72.5	182.2	109.5	61.9
157.	97.0	179.6	93.7	61.5
158.	83.0	172.2	110.2	61.1
159.	86.8	171.0	102.6	84.0
160.	72.8	183.1	121.8	61.6
161.	108.0	169.4	100.5	80.6
162.	104.9	186.2	108.3	91.6
163.	113.0	177.2	107.1	55.7
164.	102.4	165.1	117.5	83.6
165.	89.1	167.0	111.6	71.3
166.	87.8	167.7	115.5	64.9
167.	99.2	181.5	109.8	55.1
168.	102.9	175.9	106.4	62.7
169.	77.1	167.4	108.6	75.8
170.	77.3	169.0	105.1	97.9
171.	99.0	188.6	107.1	77.3
172.	91.5	170.6	89.4	75.5
173.	75.8	183.2	105.5	64.8
174.	67.5	180.4	124.4	67.4
175.	108.9	170.1	111.8	90.9
176.	96.4	184.5	119.6	73.5
177.	88.2	161.6	124.9	69.2
178.	101.2	175.3	136.4	78.2
179.	125.2	169.4	138.2	67.0
180.	87.6	162.5	101.6	92.4
181.	95.1	179.9	95.9	53.1
182.	110.5	186.8	108.8	65.6
183.	105.3	188.7	106.8	58.4
184.	85.6	173.4	102.2	75.2

Observation number	POPULATION			
	A	B	C	D
185.	94.2	188.0	114.0	89.0
186.	94.5	169.3	107.9	72.8
187.	87.8	187.3	98.4	87.8
188.	84.4	185.1	112.8	73.3
189.	86.1	178.5	102.0	69.7
190.	85.5	172.5	107.7	79.8
191.	79.7	213.0	133.9	57.8
192.	96.8	196.1	129.9	48.1
193.	81.9	181.5	98.9	74.6
194.	68.9	175.0	92.1	73.6
195.	105.5	189.2	131.5	75.7
196.	95.7	168.6	97.0	83.8
197.	91.5	185.3	132.9	73.0
198.	86.9	183.8	106.1	70.2
199.	100.4	193.8	90.5	93.6
200.	105.5	157.9	103.3	97.7
201.	85.6	173.7	105.6	82.9
202.	80.0	158.3	124.4	77.0
203.	91.9	196.4	98.6	84.9
204.	90.4	186.9	129.4	62.7
205.	105.0	171.9	109.1	87.4
206.	114.2	189.4	131.3	72.3
207.	91.6	195.4	107.6	79.0
208.	107.7	185.8	82.5	87.8
209.	107.7	185.1	111.8	84.6
210.	88.4	188.7	94.7	86.1
211.	100.3	185.3	109.2	47.3
212.	78.3	175.0	93.9	60.2
213.	95.1	197.9	113.7	95.3
214.	96.6	201.9	122.5	78.0
215.	85.7	160.5	113.7	67.3
216.	73.6	181.7	126.9	97.6
217.	82.3	195.8	115.5	66.2
218.	94.7	185.4	105.2	82.2
219.	69.6	169.3	116.2	68.0
220.	99.1	164.9	96.4	55.7
221.	91.4	182.3	114.9	64.3
222.	96.7	165.2	118.0	91.4
223.	73.3	165.7	93.3	96.0
224.	84.5	139.3	121.9	73.9
225.	105.4	177.0	118.4	76.7
226.	78.9	204.3	136.1	78.2
227.	92.1	184.4	126.8	62.3
228.	79.1	184.1	142.6	87.5
229.	108.2	185.5	92.5	94.0
230.	73.0	175.8	100.9	74.2

Observation number	POPULATION			
	A	B	C	D
231.	103.2	171.5	105.0	76.4
232.	96.1	185.1	107.0	72.2
233.	101.3	174.0	78.3	88.2
234.	78.7	183.5	94.1	57.2
235.	94.0	200.2	103.0	85.7
236.	86.6	181.6	93.9	85.7
237.	81.2	162.1	112.9	70.7
238.	83.4	182.7	109.9	78.3
239.	90.9	153.1	89.9	99.4
240.	83.1	191.9	116.2	82.9
241.	71.8	197.6	99.1	86.4
242.	100.4	170.9	121.9	62.5
243.	104.5	187.4	101.5	74.8
244.	102.2	180.1	106.2	85.1
245.	79.1	177.6	114.8	85.5
246.	90.1	175.8	115.2	73.1
247.	81.7	174.0	84.1	72.1
248.	95.2	173.6	120.8	74.7
249.	107.1	186.0	104.1	65.9
250.	89.3	167.5	126.7	66.6
251.	89.9	177.6	124.5	91.4
252.	90.6	170.7	121.6	88.9
253.	63.8	196.3	126.0	62.1
254.	92.1	179.9	104.0	96.5
255.	82.5	184.7	120.6	105.7
256.	101.8	199.0	120.4	72.9
257.	97.4	177.9	112.2	68.9
258.	77.7	156.3	109.1	82.5
259.	84.5	177.7	115.7	88.9
260.	99.0	160.6	108.7	63.9
261.	103.7	176.1	89.8	76.9
262.	94.9	170.6	106.9	77.8
263.	60.9	186.0	124.2	75.6
264.	96.4	171.3	98.1	83.3
265.	89.7	173.3	116.7	86.1
266.	94.7	168.6	102.4	91.9
267.	82.4	162.7	102.4	47.4
268.	104.3	175.8	95.4	79.7
269.	93.7	169.3	95.2	73.6
270.	105.1	180.9	83.7	82.9
271.	100.7	162.3	105.8	69.2
272.	90.4	184.5	104.1	76.7
273.	97.2	180.1	117.3	86.8
274.	86.5	174.0	114.0	58.7
275.	84.2	162.5	86.8	81.0

Observation number	POPULATION			
	A	B	C	D
276.	91.1	166.8	98.4	72.5
277.	89.0	175.2	120.0	60.4
278.	70.9	170.1	101.7	78.7
279.	90.5	200.9	120.8	88.6
280.	95.6	193.9	124.3	75.6
281.	87.1	148.4	106.4	76.2
282.	84.4	171.0	99.8	95.3
283.	86.4	204.5	90.0	78.7
284.	95.5	166.1	102.5	84.9
285.	104.8	161.2	110.5	73.9
286.	90.8	178.7	99.9	73.0
287.	100.7	174.7	112.1	72.0
288.	112.6	153.4	114.5	71.2
289.	68.7	189.4	111.3	68.8
290.	90.8	180.2	109.1	63.1
291.	93.4	194.2	122.6	62.1
292.	86.4	181.8	126.0	78.8
293.	87.9	174.5	94.8	90.8
294.	107.0	188.4	117.0	89.8
295.	76.7	178.5	105.4	74.5
296.	74.4	175.2	108.1	80.7
297.	87.1	177.1	106.5	78.1
298.	78.6	173.3	120.8	80.5
299.	88.8	189.6	94.1	72.2
300.	89.6	201.9	109.5	75.6
301.	94.6	180.8	106.0	73.0
302.	113.1	196.8	106.2	93.8
303.	104.1	183.4	98.7	82.1
304.	78.3	173.7	122.8	78.8
305.	99.2	171.1	128.8	54.1
306.	99.3	171.3	120.5	87.8
307.	87.9	180.5	129.9	78.0
308.	61.7	197.0	133.8	81.3
309.	91.4	179.8	97.6	86.7
310.	93.2	184.9	108.4	76.4
311.	104.3	189.0	101.2	85.3
312.	91.1	173.8	119.4	60.2
313.	74.7	188.1	110.5	76.7
314.	73.1	180.7	104.1	73.0
315.	99.5	173.6	114.9	85.4
316.	73.1	168.1	130.0	70.9
317.	104.4	171.5	129.5	87.5
318.	70.3	185.5	116.4	63.0
319.	73.7	183.1	121.3	73.3
320.	84.2	177.8	81.2	81.4

Observation number	POPULATION			
	A	B	C	D
321.	93.2	162.7	112.6	81.6
322.	99.8	172.8	121.7	102.3
323.	76.3	170.5	114.8	76.6
324.	79.7	187.2	116.8	74.2
325.	95.1	180.7	103.1	72.6
326.	83.1	174.2	111.7	71.2
327.	77.9	185.4	108.9	81.2
328.	89.4	175.8	97.7	71.3
329.	72.6	170.2	95.3	65.4
330.	82.0	184.9	101.3	79.2
331.	104.7	199.4	107.1	76.8
332.	97.4	154.4	105.1	81.9
333.	93.3	191.0	106.5	82.6
334.	91.2	191.0	102.9	70.1
335.	89.5	177.4	103.4	87.7
336.	86.5	179.9	116.9	87.9
337.	92.6	171.9	133.8	85.4
338.	102.2	147.2	92.9	82.9
339.	96.9	199.9	99.1	94.0
340.	87.9	186.6	109.1	89.8
341.	84.5	166.6	118.3	83.2
342.	103.7	172.9	110.4	89.9
343.	80.5	168.5	129.3	60.7
344.	83.7	188.4	121.3	71.5
345.	57.2	174.4	121.2	75.2
346.	103.9	174.3	135.3	71.8
347.	92.1	183.8	92.0	81.4
348.	104.6	183.0	104.9	85.0
349.	94.9	185.0	94.5	66.1
350.	115.5	186.5	131.0	65.9
351.	79.6	172.7	104.3	82.4
352.	96.1	182.0	121.9	83.9
353.	83.7	181.7	85.8	88.2
354.	100.9	166.4	106.2	84.5
355.	94.0	154.1	128.6	74.8
356.	93.4	162.1	110.6	81.0
357.	110.0	192.2	115.5	72.2
358.	82.8	182.8	114.9	75.5
359.	82.4	172.1	97.2	84.0
360.	96.2	191.0	115.8	71.6
361.	108.4	179.7	105.2	88.6
362.	105.2	175.1	95.5	97.6
363.	87.1	181.1	123.4	99.2
364.	103.4	184.5	103.0	84.1
365.	83.7	196.0	105.1	91.1
366.	92.8	170.3	106.3	42.9

Observation number	POPULATION			
	A	B	C	D
367.	101.5	182.7	98.4	92.6
368.	98.2	193.4	139.7	86.8
369.	112.7	154.1	116.7	67.8
370.	107.7	166.1	96.9	69.4
371.	94.2	166.4	116.7	91.7
372.	97.6	166.7	94.2	77.2
373.	81.9	166.6	109.8	80.4
374.	75.6	191.1	112.6	84.9
375.	105.2	166.0	134.8	63.2
376.	92.5	182.4	107.1	79.7
377.	79.1	180.8	88.5	110.3
378.	98.8	189.6	110.0	82.3
379.	88.0	174.6	110.9	72.5
380.	92.1	181.4	127.3	80.2
381.	74.0	199.5	105.0	89.2
382.	59.8	171.6	130.3	88.3
383.	83.2	189.1	99.6	71.9
384.	75.2	178.5	114.5	47.2
385.	87.7	156.1	113.0	81.6
386.	76.6	164.8	115.9	78.9
387.	99.1	174.1	109.8	90.6
388.	77.8	168.9	96.4	68.7
389.	79.7	182.6	127.0	90.0
390.	93.6	171.2	116.4	77.7
391.	93.7	181.7	106.6	68.8
392.	93.5	188.0	117.9	70.1
393.	88.2	180.8	125.4	83.4
394.	104.5	164.2	111.5	59.8
395.	99.0	158.0	101.8	75.9
396.	88.5	154.8	98.8	75.7
397.	98.7	184.7	106.8	97.9
398.	90.1	177.8	103.8	68.1
399.	97.0	204.1	124.1	65.0
400.	76.3	173.5	98.0	82.7

23. The following are the urinary colony-stimulating factor (CSF) levels in five populations: normal subjects and subjects with four different diseases. Each observation represents the mean colony count of four plates from a single urine specimen from a given subject. Select a simple random sample of size 15 from each of the five populations and perform an analysis of variance to determine if one may conclude that the population means are different. Let $\alpha = .05$. Use Tukey's HSD statistic to test for significant differences among all possible pairs of sample means. Prepare a narrative report on the results of your analysis. Compare your results with those of your classmates.

Observation number	SUBJECTS WITH DISEASE				
	None	A	B	C	D
1.	102	64	130	82	123
2.	74	56	136	51	113
3.	63	42	137	72	138
4.	67	39	107	77	126
5.	68	29	155	45	135
6.	58	42	137	85	138
7.	77	61	138	80	124
8.	55	67	120	51	102
9.	80	40	138	76	125
10.	78	89	165	95	103
11.	87	47	138	82	124
12.	89	44	163	92	128
13.	60	42	144	67	133
14.	75	69	129	55	96
15.	64	37	131	69	117
16.	53	55	138	45	120
17.	54	38	148	92	125
18.	59	46	127	38	131
19.	73	71	134	66	117
20.	55	38	170	53	128
21.	54	59	131	63	128
22.	52	67	144	62	110
23.	84	49	172	56	115
24.	64	58	115	57	123
25.	85	72	154	90	132
26.	68	42	140	82	118
27.	74	66	128	88	126
28.	83	64	157	73	108
29.	53	50	154	53	112
30.	56	39	151	36	113
31.	54	50	151	88	108
32.	61	69	154	81	105
33.	59	63	137	81	97
34.	66	66	157	68	95
35.	95	38	133	93	142
36.	75	48	173	58	99
37.	77	76	172	61	119
38.	69	37	130	86	109
39.	73	58	145	52	135
40.	71	35	136	69	105
41.	60	55	140	70	145
42.	79	56	140	63	135
43.	80	61	141	82	104
44.	69	30	142	72	125
45.	85	67	149	97	104
46.	51	51	145	64	105

Observation number	SUBJECTS WITH DISEASE				
	None	A	B	C	D
47.	54	42	135	114	107
48.	55	67	164	83	125
49.	62	54	147	94	124
50.	71	69	155	94	124
51.	72	58	160	80	139
52.	77	57	138	67	125
53.	74	81	155	101	96
54.	76	54	127	83	122
55.	68	75	175	93	113
56.	87	38	133	61	137
57.	94	75	148	75	143
58.	43	13	124	96	115
59.	69	24	150	62	102
60.	68	52	131	88	113
61.	51	44	130	70	117
62.	63	62	143	66	111
63.	99	56	143	71	82
64.	77	62	148	82	105
65.	83	60	161	87	97
66.	61	23	166	93	134
67.	93	62	149	63	98
68.	42	73	134	52	120
69.	80	75	145	67	103
70.	82	47	125	66	133
71.	92	55	154	80	143
72.	88	74	141	94	133
73.	60	59	156	89	104
74.	78	63	143	76	117
75.	54	20	134	83	121
76.	50	67	145	69	127
77.	91	72	135	67	130
78.	88	74	141	56	120
79.	90	36	148	100	118
80.	78	63	161	80	107
81.	63	53	148	63	96
82.	98	44	155	60	103
83.	59	30	154	93	114
84.	94	51	171	110	115
85.	85	59	119	57	113
86.	74	57	125	103	109
87.	51	43	115	64	110
88.	72	72	146	71	122
89.	90	69	119	91	118
90.	92	36	175	64	107
91.	83	57	144	77	109
92.	53	52	143	87	149

Observation number	SUBJECTS WITH DISEASE				
	None	A	B	C	D
93.	78	44	140	82	142
94.	103	48	166	69	112
95.	64	35	157	79	107
96.	52	40	153	63	96
97.	99	39	142	67	109
98.	50	59	146	78	128
99.	85	59	137	61	108
100.	64	32	128	62	136
101.	52	50	167	75	123
102.	84	40	132	64	108
103.	44	59	151	88	119
104.	74	51	158	93	119
105.	75	61	120	95	110
106.	78	61	127	52	122
107.	77	56	149	74	126
108.	67	35	159	88	119
109.	59	47	134	87	116
110.	53	34	124	51	110
111.	48	56	129	79	109
112.	66	44	139	54	118
113.	42	59	130	81	119
114.	70	64	138	79	100
115.	34	43	119	70	116
116.	92	51	164	78	123
117.	69	55	161	85	135
118.	72	44	117	64	111
119.	76	74	136	78	125
120.	72	40	137	77	95
121.	76	55	151	48	97
122.	64	38	144	50	122
123.	64	52	140	74	126
124.	62	49	131	91	119
125.	62	47	141	78	115
126.	56	48	133	67	109
127.	74	49	125	86	118
128.	88	32	157	92	131
129.	75	60	156	50	131
130.	82	30	168	57	108
131.	89	55	154	64	104
132.	58	71	140	76	112
133.	93	45	128	73	120
134.	76	36	131	68	89
135.	91	51	166	75	88
136.	88	50	144	68	126
137.	67	78	147	74	92
138.	85	65	148	50	125

Observation number	SUBJECTS WITH DISEASE				
	None	*A*	*B*	*C*	*D*
139.	83	63	135	81	122
140.	71	59	136	66	122
141.	55	27	130	69	110
142.	63	55	136	85	114
143.	65	68	165	49	132
144.	48	70	138	58	101
145.	86	44	140	65	125
146.	77	28	140	97	129
147.	59	42	128	79	119
148.	67	81	161	94	144
149.	79	44	139	55	109
150.	54	39	136	73	141
151.	46	44	144	66	111
152.	75	45	145	83	122
153.	57	64	128	42	108
154.	59	26	157	83	94
155.	107	26	189	81	110
156.	86	32	130	60	111
157.	50	50	156	67	114
158.	57	42	141	95	107
159.	56	60	148	41	117
160.	68	53	148	112	138
161.	58	44	158	67	117
162.	72	46	165	68	125
163.	84	35	154	56	129
164.	46	34	164	50	113
165.	77	52	161	91	94
166.	66	36	125	55	127
167.	95	53	129	69	121
168.	31	43	161	88	117
169.	65	69	153	114	101
170.	45	31	133	85	122
171.	60	71	145	49	112
172.	103	79	168	48	120
173.	80	62	149	102	109
174.	68	63	153	54	107
175.	79	75	140	73	100
176.	78	40	128	54	102
177.	70	48	121	65	81
178.	90	20	134	77	124
179.	66	39	166	80	134
180.	88	73	186	84	118
181.	74	41	138	39	138
182.	78	79	164	92	150
183.	74	58	130	68	177
184.	42	40	165	105	142

Observation number	SUBJECTS WITH DISEASE				
	None	*A*	*B*	*C*	*D*
185.	61	66	111	79	112
186.	66	42	141	84	112
187.	69	54	124	66	93
188.	66	61	134	49	116
189.	61	50	135	91	124
190.	65	89	129	65	129
191.	44	62	145	54	103
192.	69	61	165	49	104
193.	77	39	168	76	158
194.	69	38	129	78	118
195.	51	86	144	91	123
196.	73	44	133	42	140
197.	88	46	129	97	147
198.	78	63	138	57	130
199.	81	44	133	65	133
200.	73	47	134	56	125
201.	60	60	123	68	114
202.	82	75	135	60	136
203.	81	36	160	114	121
204.	103	57	152	53	130
205.	91	34	144	75	127
206.	56	61	151	66	126
207.	57	77	115	77	109
208.	31	34	137	60	113
209.	62	39	157	78	111
210.	58	52	159	84	130
211.	60	58	138	74	87
212.	65	36	160	63	115
213.	59	49	154	72	137
214.	76	48	178	77	115
215.	47	53	128	89	94
216.	60	33	114	86	114
217.	58	71	137	86	122
218.	49	50	129	71	130
219.	57	64	143	72	138
220.	71	37	142	82	119
221.	48	60	145	64	86
222.	114	62	153	60	98
223.	60	43	152	64	113
224.	63	51	172	81	128
225.	72	41	151	70	142
226.	89	66	139	40	136
227.	34	54	109	54	137
228.	88	50	146	65	110
229.	73	63	135	44	128
230.	82	35	121	62	115

Observation number	SUBJECTS WITH DISEASE				
	None	A	B	C	D
231.	68	72	160	58	109
232.	80	54	149	71	113
233.	59	72	136	48	147
234.	65	39	168	65	127
235.	64	45	170	76	107
236.	98	64	132	73	130
237.	56	60	162	76	97
238.	57	76	160	51	132
239.	48	37	168	62	124
240.	67	55	161	84	137
241.	83	15	170	71	115
242.	58	58	152	54	119
243.	71	50	161	97	100
244.	60	47	128	40	103
245.	59	76	128	89	113
246.	89	71	159	69	121
247.	64	41	170	68	122
248.	70	57	159	70	154
249.	71	65	159	84	112
250.	104	64	119	76	101
251.	86	46	147	71	131
252.	83	72	130	47	99
253.	81	37	135	74	118
254.	59	31	158	47	117
255.	88	66	139	64	133
256.	45	56	141	80	112
257.	48	47	181	83	106
258.	47	59	157	48	110
259.	106	69	141	78	120
260.	101	64	153	68	102
261.	73	50	136	58	109
262.	53	75	151	73	135
263.	69	41	151	74	105
264.	88	60	141	97	121
265.	52	48	156	84	111
266.	60	32	128	92	103
267.	76	47	148	86	128
268.	83	51	126	81	72
269.	46	68	117	78	109
270.	61	63	148	88	125
271.	91	38	168	68	109
272.	85	63	161	57	91
273.	85	62	141	85	113
274.	64	55	165	72	126
275.	78	38	161	54	116
276.	67	41	132	66	107

Observation number	SUBJECTS WITH DISEASE				
	None	A	B	C	D
277.	71	68	119	95	109
278.	87	79	102	71	118
279.	79	41	134	104	130
280.	82	69	124	60	125
281.	42	72	139	97	138
282.	73	36	136	62	117
283.	97	68	162	71	138
284.	74	63	148	92	128
285.	54	80	140	68	115
286.	61	39	122	57	122
287.	110	49	126	87	84
288.	79	47	165	68	140
289.	45	61	153	66	130
290.	77	77	128	63	114
291.	89	64	151	81	117
292.	80	33	158	90	89
293.	90	48	113	59	130
294.	77	56	138	41	88
295.	62	30	125	62	128
296.	54	42	142	71	129
297.	92	61	125	78	115
298.	58	37	147	90	115
299.	63	59	134	79	128
300.	75	40	167	70	103

24. Suppose that you are a statistical consultant to a medical researcher who is interested in learning something about the relationship between blood folate concentrations in adult females and the quality of their diet. The researcher has available three populations of subjects: those whose diet quality is rated as good, those whose diets are fair, and those with poor diets. For each subject there is also available her red blood cell (RBC) folate value (in μg/liter of red cells). Draw a simple random sample of size 10 from each population and determine whether the researcher can conclude that the three populations differ with respect to mean RBC folate value. Use Tukey's test to make all possible comparisons. Let $\alpha = .05$ and find the p value for each test. Compare your results with those of your classmates.

Observation number	QUALITY OF DIET			Observation number	QUALITY OF DIET		
	Good	Fair	Poor		Good	Fair	Poor
1.	285	233	174	2.	283	238	190
3.	314	211	197	4.	311	248	175
5.	319	199	203	6.	311	236	165

Observation number	QUALITY OF DIET			Observation number	QUALITY OF DIET		
	Good	Fair	Poor		Good	Fair	Poor
7.	307	214	171	8.	299	244	195
9.	290	254	196	10.	281	248	199
11.	285	241	176	12.	289	240	180
13.	289	255	197	14.	277	218	178
15.	282	218	181	16.	301	250	180
17.	283	254	202	18.	322	241	188
19.	274	233	174	20.	292	221	190
21.	276	245	200	22.	280	231	194
23.	312	220	167	24.	289	259	188
25.	300	219	197	26.	308	227	213
27.	266	218	198	28.	291	223	208
29.	294	269	202	30.	285	243	205
31.	289	232	179	32.	302	231	184
33.	298	236	217	34.	287	251	211
35.	252	223	212	36.	275	205	192
37.	266	228	195	38.	281	240	195
39.	293	230	182	40.	303	239	197
41.	281	235	218	42.	246	232	185
43.	302	236	208	44.	312	212	236
45.	289	233	173	46.	287	229	187
47.	297	235	194	48.	276	263	170
49.	301	229	195	50.	280	233	215
51.	304	260	204	52.	285	270	226
53.	282	215	192	54.	328	233	197
55.	306	220	205	56.	291	239	190
57.	278	226	183	58.	316	220	222
59.	311	212	195	60.	274	226	192
61.	299	225	202	62.	311	233	197
63.	291	240	185	64.	307	246	205
65.	282	254	189	66.	305	214	199
67.	324	239	203	68.	299	241	207
69.	293	210	217	70.	286	235	180
71.	278	260	218	72.	295	245	246
73.	296	257	186	74.	307	239	207
75.	318	254	213	76.	313	229	178
77.	291	229	207	78.	285	254	197
79.	303	229	179	80.	277	238	208
81.	277	240	190	82.	287	266	221
83.	299	235	200	84.	272	239	194
85.	296	206	200	86.	288	220	176
87.	294	221	190	88.	298	243	215
89.	287	232	208	90.	298	240	187
91.	289	253	176	92.	282	236	201
93.	307	225	226	94.	290	218	198
95.	265	223	233	96.	297	222	199
97.	296	229	192	98.	289	251	181

Observation number	QUALITY OF DIET			Observation number	QUALITY OF DIET		
	Good	Fair	Poor		Good	Fair	Poor
99.	287	223	210	100.	288	232	203
101.	287	243	196	102.	282	252	202
103.	269	235	194	104.	272	250	178
105.	306	204	187	106.	302	230	200
107.	303	228	203	108.	302	222	192
109.	305	239	246	110.	300	232	197
111.	271	240	201	112.	274	230	203
113.	278	220	195	114.	316	237	197
115.	271	206	191	116.	281	250	196
117.	296	239	198	118.	281	262	212
119.	323	227	183	120.	301	249	174
121.	297	232	190	122.	288	222	231
123.	299	223	214	124.	296	212	167
125.	289	239	195	126.	309	227	200
127.	279	231	203	128.	319	212	182
129.	300	263	180	130.	304	220	187
131.	289	245	174	132.	321	231	164
133.	288	254	188	134.	274	221	204
135.	282	231	203	136.	291	254	182
137.	300	224	221	138.	287	262	190
139.	321	227	172	140.	297	236	199
141.	281	231	182	142.	286	252	213
143.	317	269	197	144.	300	232	217
145.	297	229	181	146.	324	245	198
147.	305	224	187	148.	312	262	211
149.	313	241	179	150.	270	261	182
151.	290	242	200	152.	283	220	191
153.	254	244	233	154.	269	217	203
155.	303	207	197	156.	279	237	207
157.	272	234	193	158.	293	250	177
159.	271	242	175	160.	295	249	216
161.	278	254	216	162.	295	228	213
163.	311	234	179	164.	260	235	176
165.	273	232	164	166.	322	228	189
167.	319	216	196	168.	314	240	177
169.	306	230	207	170.	283	233	191
171.	278	227	193	172.	300	248	204
173.	319	246	217	174.	302	271	204
175.	265	269	205	176.	284	223	200
177.	305	225	194	178.	306	237	189
179.	273	247	209	180.	299	214	173
181.	286	234	205	182.	269	236	221
183.	259	230	226	184.	282	266	235
185.	273	246	207	186.	297	232	189
187.	314	214	177	188.	273	271	183
189.	322	261	208	190.	286	254	192

Observation number	Quality of Diet			Observation number	Quality of Diet		
	Good	Fair	Poor		Good	Fair	Poor
191.	321	228	196	192.	292	229	203
193.	298	230	181	194.	299	240	183
195.	294	225	208	196.	300	246	182
197.	313	255	208	198.	279	252	196
199.	282	258	174	200.	276	229	196
201.	325	238	190	202.	296	233	178
203.	289	236	174	204.	311	232	216
205.	314	230	213	206.	303	210	212
207.	302	242	241	208.	328	237	192
209.	302	228	182	210.	301	243	205
211.	309	222	186	212.	300	269	187
213.	277	252	185	214.	309	247	209
215.	307	234	188	216.	331	220	191
217.	288	265	173	218.	285	208	190
219.	277	223	211	220.	308	224	180
221.	291	226	193	222.	274	223	212
223.	306	238	192	224.	324	230	194
225.	271	226	182	226.	290	226	233
227.	298	232	191	228.	271	248	182
229.	295	243	214	230.	325	246	227
231.	311	250	193	232.	322	256	207
233.	328	258	190	234.	302	204	222
235.	281	251	218	236.	295	240	199
237.	299	244	204	238.	302	258	186
239.	289	218	212	240.	312	238	203
241.	273	222	189	242.	298	250	177
243.	303	234	193	244.	324	261	191
245.	291	252	181	246.	316	227	176
247.	309	262	206	248.	258	227	194
249.	331	236	208	250.	306	229	217
251.	289	234	203	252.	302	260	222
253.	282	226	204	254.	310	218	213
255.	320	231	209	256.	297	226	199
257.	316	260	209	258.	289	224	192
259.	307	234	191	260.	307	249	191
261.	297	229	197	262.	283	229	187
263.	284	258	199	264.	285	232	204
265.	300	220	203	266.	306	224	190
267.	288	223	189	268.	292	249	228
269.	294	223	200	270.	283	261	199
271.	305	257	186	272.	271	245	215
273.	290	230	190	274.	279	247	192
275.	305	227	188	276.	298	251	199
277.	292	215	208	278.	301	227	203
279.	279	235	187	280.	286	224	208
281.	294	237	210	282.	298	255	184

Observation number	QUALITY OF DIET			Observation number	QUALITY OF DIET		
	Good	Fair	Poor		Good	Fair	Poor
283.	315	240	191	284.	300	258	209
285.	273	245	199	286.	303	224	197
287.	284	225	224	288.	286	236	213
289.	317	253	191	290.	321	240	188
291.	292	217	204	292.	290	219	184
293.	283	216	176	294.	301	240	187
295.	305	226	185	296.	294	226	192
297.	292	242	202	298.	284	219	167
299.	295	234	198	300.	293	253	208
301.	268	253	201	302.	288	220	193
303.	291	258	167	304.	276	245	187
305.	244	233	192	306.	285	236	206
307.	302	243	183	308.	273	232	196
309.	328	227	200	310.	313	256	193
311.	308	220	186	312.	272	225	188
313.	313	228	190	314.	301	251	175
315.	274	225	206	316.	294	232	214
317.	301	250	202	318.	283	226	210
319.	291	242	193	320.	285	243	173
321.	277	242	168	322.	308	249	213
323.	301	259	192	324.	309	234	210
325.	276	226	195	326.	276	250	222
327.	280	229	217	328.	302	262	214
329.	286	246	191	330.	305	258	198
331.	278	234	192	332.	292	210	236
333.	266	248	188	334.	295	233	174
335.	302	237	199	336.	285	238	178
337.	286	255	194	338.	284	222	216
339.	296	229	214	340.	285	261	184
341.	297	209	191	342.	295	244	192
343.	299	240	208	344.	299	209	210
345.	331	221	202	346.	288	202	189
347.	314	213	191	348.	273	237	218
349.	295	248	200	350.	284	245	196

25. Three-hundred-fifty adult males between the ages of 30 and 65 participated in a study to investigate the relationship between the consumption of meat and serum cholesterol levels. Each subject ate beef as his only meat for a period of 20 weeks, pork as his only meat for another period of 20 weeks, and chicken or fish as his only meat for another 20-week period. At the end of each period serum cholesterol determinations (mg/100 ml) were made on each subject. The results are shown in the following table.

Select a simple random sample of 10 subjects from the population of 350. Use two-way analysis of variance to determine whether one should conclude that there is a difference in population mean serum cholesterol levels among the three diets. Let $\alpha = .05$. Compare your results with those of your classmates.

	DIET				DIET		
Subject	Beef	Pork	Chicken or fish	Subject	Beef	Pork	Chicken or fish
1.	241	245	249	2.	218	197	222
3.	261	199	221	4.	190	162	215
5.	238	191	207	6.	256	182	193
7.	248	160	205	8.	224	180	227
9.	225	208	203	10.	238	227	180
11.	178	174	200	12.	185	209	154
13.	194	225	211	14.	224	271	204
15.	221	187	169	16.	211	169	271
17.	241	197	204	18.	164	141	188
19.	169	208	202	20.	215	226	166
21.	268	225	176	22.	181	209	147
23.	203	212	192	24.	180	194	179
25.	208	213	203	26.	187	227	176
27.	216	175	251	28.	182	163	193
29.	192	222	146	30.	271	166	209
31.	235	202	212	32.	166	222	193
33.	218	207	179	34.	193	238	233
35.	168	220	223	36.	184	214	176
37.	212	202	216	38.	268	179	163
39.	252	217	216	40.	233	206	228
41.	175	186	225	42.	202	208	188
43.	259	191	227	44.	195	172	205
45.	198	212	198	46.	152	195	181
47.	212	217	217	48.	245	230	227
49.	243	251	272	50.	176	246	203
51.	217	190	177	52.	160	222	204
53.	222	227	158	54.	190	212	259
55.	250	237	217	56.	269	186	215
57.	187	198	188	58.	136	283	209
59.	194	236	233	60.	210	242	178
61.	213	175	273	62.	227	216	190
63.	201	133	239	64.	229	211	205
65.	239	244	205	66.	209	263	171
67.	196	214	253	68.	206	213	194
69.	173	187	154	70.	188	173	174
71.	214	216	188	72.	225	202	182
73.	251	168	220	74.	191	198	170
75.	203	203	174	76.	218	229	205
77.	260	181	201	78.	191	198	249

| Subject | | Diet | | Subject | | Diet | |
	Beef	Pork	Chicken or fish		Beef	Pork	Chicken or fish
79.	172	264	180	80.	194	166	199
81.	213	206	218	82.	239	226	199
83.	185	218	176	84.	193	254	229
85.	175	231	234	86.	171	151	155
87.	196	206	247	88.	206	239	188
89.	176	233	201	90.	193	236	266
91.	255	202	255	92.	180	207	243
93.	203	263	181	94.	162	221	249
95.	229	125	242	96.	198	206	233
97.	179	206	238	98.	234	234	227
99.	230	213	220	100.	227	219	202
101.	239	217	250	102.	162	204	183
103.	220	203	191	104.	232	240	224
105.	209	212	218	106.	201	182	208
107.	219	202	198	108.	162	160	140
109.	251	225	255	110.	214	189	198
111.	222	189	229	112.	201	267	213
113.	149	151	203	114.	147	178	192
115.	258	184	227	116.	217	235	211
117.	213	226	229	118.	181	211	207
119.	154	206	198	120.	155	231	195
121.	214	183	230	122.	196	251	198
123.	177	190	182	124.	246	164	255
125.	199	232	231	126.	242	197	198
127.	240	201	265	128.	196	216	235
129.	264	171	218	130.	216	193	183
131.	194	206	238	132.	179	197	170
133.	219	202	222	134.	190	231	245
135.	243	216	150	136.	234	213	236
137.	159	253	177	138.	216	267	176
139.	181	238	161	140.	166	210	245
141.	193	296	213	142.	260	201	294
143.	184	237	169	144.	198	134	185
145.	156	233	195	146.	177	217	207
147.	210	208	197	148.	190	218	224
149.	186	234	268	150.	194	211	215
151.	204	242	210	152.	193	198	281
153.	191	276	194	154.	200	199	240
155.	275	220	226	156.	193	252	186
157.	204	254	171	158.	198	253	208
159.	198	222	244	160.	197	186	217
161.	193	214	198	162.	192	227	217
163.	228	211	165	164.	207	173	199
165.	210	175	179	166.	246	228	222
167.	205	189	168	168.	176	180	246
169.	215	229	228	170.	129	194	199

		DIET				**DIET**	
Subject	Beef	Pork	Chicken or fish	Subject	Beef	Pork	Chicken or fish
171.	177	185	193	172.	217	221	217
173.	222	194	192	174.	181	196	175
175.	225	154	251	176.	125	229	171
177.	157	189	202	178.	221	239	177
179.	170	200	215	180.	278	251	133
181.	249	211	211	182.	220	201	187
183.	185	208	224	184.	239	201	166
185.	222	191	247	186.	205	175	209
187.	237	199	255	188.	255	213	225
189.	218	219	220	190.	209	173	182
191.	200	228	238	192.	170	203	233
193.	200	175	233	194.	227	270	178
195.	182	213	196	196.	210	188	159
197.	239	222	193	198.	176	187	207
199.	237	185	203	200.	229	188	209
201.	172	171	190	202.	188	200	215
203.	245	241	254	204.	172	240	239
205.	232	238	193	206.	234	180	206
207.	245	184	219	208.	165	221	285
209.	225	168	264	210.	234	228	192
211.	241	213	175	212.	178	200	261
213.	197	271	216	214.	215	191	131
215.	240	140	227	216.	203	197	200
217.	157	187	231	218.	200	234	137
219.	247	219	200	220.	205	178	187
221.	179	198	163	222.	215	210	164
223.	208	202	208	224.	225	244	217
225.	200	237	163	226.	165	213	210
227.	207	228	206	228.	178	249	177
229.	173	132	210	230.	160	195	266
231.	224	203	199	232.	246	230	203
233.	134	224	247	234.	228	190	239
235.	209	176	157	236.	193	231	203
237.	192	223	197	238.	193	211	229
239.	238	167	199	240.	244	193	185
241.	228	276	187	242.	258	239	269
243.	224	185	199	244.	231	214	218
245.	207	220	162	246.	172	190	186
247.	234	187	190	248.	245	156	258
249.	206	240	209	250.	196	179	186
251.	199	176	236	252.	219	205	200
253.	186	272	220	254.	168	221	211
255.	264	194	203	256.	176	196	217
257.	202	212	143	258.	169	216	225
259.	229	168	227	260.	252	181	239
261.	154	198	255	262.	231	206	191

| Subject | DIET | | | Subject | DIET | | |
	Beef	Pork	Chicken or fish		Beef	Pork	Chicken or fish
263.	165	226	183	264.	211	229	210
265.	243	210	177	266.	228	298	243
267.	188	210	176	268.	179	166	213
269.	229	162	178	270.	158	248	209
271.	180	221	162	272.	233	233	173
273.	206	188	230	274.	211	152	232
275.	264	242	203	276.	208	246	193
277.	250	220	221	278.	214	187	204
279.	239	219	182	280.	212	255	251
281.	214	207	175	282.	234	261	172
283.	215	173	173	284.	229	224	217
285.	198	213	155	286.	201	238	203
287.	256	235	215	288.	170	178	218
289.	207	204	165	290.	190	188	204
291.	225	226	254	292.	213	213	204
293.	208	204	201	294.	237	198	244
295.	177	232	203	296.	170	203	211
297.	193	243	208	298.	231	209	190
299.	189	200	196	300.	185	181	200
301.	169	194	223	302.	237	188	177
303.	237	182	173	304.	160	213	195
305.	163	224	220	306.	165	157	260
307.	241	223	216	308.	200	234	193
309.	211	177	224	310.	170	203	233
311.	178	201	153	312.	280	241	190
313.	254	207	237	314.	194	187	258
315.	162	258	163	316.	232	211	250
317.	152	219	223	318.	190	219	131
319.	228	110	210	320.	228	207	159
321.	270	233	225	322.	170	178	202
323.	216	245	189	324.	216	199	253
325.	212	165	185	326.	237	224	282
327.	227	185	166	328.	193	229	159
329.	285	229	219	330.	182	222	159
331.	236	171	174	332.	217	240	225
333.	260	218	213	334.	180	160	182
335.	224	132	186	336.	200	156	216
337.	216	194	215	338.	178	211	221
339.	177	186	224	340.	171	214	189
341.	230	175	214	342.	240	167	216
343.	210	202	223	344.	260	181	178
345.	220	233	192	346.	202	196	140
347.	157	201	243	348.	194	189	178
349.	150	195	212	350.	205	227	141

REFERENCES

References Cited

1. R. A. Fisher, *The Design of Experiments*, Eighth Edition, Oliver and Boyd, Edinburgh, 1966.

2. R. A. Fisher, *Contributions to Mathematical Statistics*, Wiley, New York, 1950.

3. Ronald A. Fisher, *Statistical Methods for Research Workers*, Thirteenth Edition, Hafner, New York, 1958.

4. William G. Cochran and Gertrude M. Cox, *Experimental Designs*, Wiley, New York, 1957.

5. D. R. Cox, *Planning of Experiments*, Wiley, New York, 1958.

6. Owen L. Davies (ed.), *The Design and Analysis of Experiments*, Hafner, New York, 1960.

7. Walter T. Federer, *Experimental Design*, Macmillan, New York, 1955.

8. D. J. Finney, *Experimental Design and Its Statistical Basis*, The University of Chicago Press, Chicago, 1955.

9. Peter W. M. John, *Statistical Design and Analysis of Experiments*, Macmillan, New York, 1971.

10. Oscar Kempthorne, *The Design and Analysis of Experiments*, Wiley, New York, 1952.

11. C. C. Li, *Introduction to Experimental Statistics*, McGraw-Hill, New York, 1964.

12. William Mendenhall, *Introduction to Linear Models and the Design and analysis of Experiments*, Wadsworth, Belmont, Cal., 1968.

13. Churchill Eisenhart, "The Assumptions Underlying the Analysis of Variance," *Biometrics*, *21* (1947), 1–21.

14. W. G. Cochran, "Some Consequences When the Assumptions for the Analysis of Variance Are Not Satisfied," *Biometrics*, *3* (1947), 22–38.

15. M. B. Wilk and O. Kempthorne, "Fixed, Mixed and Random Models," *Journal of the American Statistical Association*, *50* (1955), 1144–1167.

16. S. L. Crump, "The Estimation of Variance Components in Analysis of Variance," *Biometrics*, *2* (1946), 7–11.

17. E. P. Cunningham and C. R. Henderson, "An Iterative Procedure for Estimating Fixed Effects and Variance Components in Mixed Model Situations," *Biometrics*, *24* (1968), 13–25.

18. C. R. Henderson, "Estimation of Variance and Covariance Components," *Biometrics*, *9* (1953), 226–252.

19. J. R. Rutherford, "A Note on Variances in the Components of Variance Model," *The American Statistician*, *25* (June 1971), 1, 2.

20. E. F. Schultz, Jr., "Rules of Thumb for Determining Expectations of Mean Squares in Analysis of Variance," *Biometrics*, *11* (1955), 123–135.

21. S. R. Searle, "Topics in Variance Component Estimation," *Biometrics*, *27* (1971), 1–76.

22. Robert G. D. Steel and James H. Torrie, *Principles and Procedures of Statistics*, McGraw-Hill, New York, 1960.

23. David B. Duncan, *Significance Tests for Differences Between Ranked Variates Drawn from Normal Populations*, Ph.D. Thesis (1949), Iowa State College, 117 pp.

24. David B. Duncan, "A Significance Test for Differences Between Ranked Treatments in an Analysis of Variance," *Virginia Journal of Science*, *2* (1951), 171–189.

25. David B. Duncan, "On the Properties of the Multiple Comparisons Test," *Virginia Journal of Science*, *3* (1952), 50–67.

26. David B. Duncan, "Multiple Range and Multiple-*F* Tests," *Biometrics*, *11* (1955), 1–42.

27. C. Y. Kramer, "Extension of Multiple Range Tests to Group Means with Unequal Numbers of Replications," *Biometrics*, *12* (1956) 307–310.

28. C. W. Dunnett, "A Multiple Comparisons Procedure for Comparing Several Treatments with a Control," *Journal of the American Statistical Association*, *50* (1955), 1096–1121.

29. C. W. Dunnett, "New Tables for Multiple Comparisons with a Control," *Biometrics*, *20* (1964), 482–491.

30. J. W. Tukey, "Comparing Individual Means in the Analysis of Variance," *Biometrics*, *5* (1949), 99–114.

31. J. W. Tukey, "The Problem of Multiple Comparisons," Ditto, Princeton University, 1953; cited in Roger E. Kirk, *Experimental Design: Procedures for the Behaviorial Sciences*, Brooks/Cole, Belmont, California, 1968.

32. D. Newman, "The Distribution of the Range in Samples from a Normal Population in Terms of an Independent Estimate of Standard Deviation," *Biometrika*, *31* (1939), 20–30.

33. M. Keuls, "The Use of the Studentized Range in Connection with the Analysis of Variance," *Euphytica*, *1* (1952), 112–122.

34. Henry Scheffé, "A Method for Judging All Contrasts in the Analysis of Variance," *Biometrika*, *40* (1953), 87–104.

35. Henry Scheffé, *Analysis of Variance*, Wiley, New York (1959).

36. T. A. Bancroft, *Topics in Intermediate Statistical Methods*, Volume I, The Iowa State University Press, Ames, 1968.

37. Wayne W. Daniel and Carol E. Coogler, "Beyond Analysis of Variance: A Comparison of Some Multiple Comparison Procedures," *Physical Therapy*, *55* (1975), 144–150.

38. B. J. Winer, *Statistical Principles in Experimental Design*, Second Edition, McGraw-Hill, New York, 1971.

39. Wayne W. Daniel, *Multiple Comparison Procedures: A Selected Bibliography*, Vance Bibliographies, Monticello, Illinois, June 1980.

40. Emil Spjøtvoll and Michael R. Stoline, "An Extension of the T-Method of Multiple Comparison to Include the Cases with Unequal Sample Sizes," *Journal of the American Statistical Association, 68* (1973), 975–978.

41. R. A. Fisher, "The Arrangement of Field Experiments," *Journal of Ministry of Agriculture, 33* (1926), 503–513.

42. R. L. Anderson and T. A. Bancroft, *Statistical Theory in Research*, McGraw-Hill, New York, 1952.

43. J. W. Tukey, "One Degree of Freedom for Non-Additivity," *Biometrics, 5* (1949), 232–242.

44. John Mandel, "A New Analysis of Variance Model for Non-Additive Data," *Technometrics, 13* (1971), 1–18.

45. A. W. Kimball, "On Dependent Tests of Significance in the Analysis of Variance," *Annals of Mathematical Statistics, 22* (1951), 600–602.

46. Wilfred J. Dixon and Frank J. Massey, *Introduction to Statistical Analysis*, Third Edition, McGraw-Hill, New York, 1969.

47. William C. Guenther, *Analysis of Variance*, Prentice-Hall, Englewood Cliffs, N.J., 1964.

48. Bernard Ostle, *Statistics in Research*, Second Edition, The Iowa State University Press, Ames, Iowa, 1963.

49. William Alexander Glenn and Clyde Young Kramer, "Analysis of Variance of a Randomized Block Design with Missing Observations," *Applied Statistics, 7* (1958), 173–185.

50. Clyde Young Kramer and Suzanne Glass, "Analysis of Variance of a Latin Square Design with Missing Observations," *Applied Statistics, 9* (1960), 43–50.

51. Hugh Robert Baird and Clyde Young Kramer," Analysis of Variance of a Balanced Incomplete Block Design with Missing Observations," *Applied Statistics, 9* (1960), 189–198.

52. M. H. Quenoille, *Introductory Statistics*, Pergamon, London 1950.

53. Robert R. Sokal and F. James Rohlf, *Biometry*, W. H. Freeman, San Francisco, 1969.

54. S. C. Pearce, *Biological Statistics: An Introduction*, McGraw-Hill, New York, 1965.

55. M. S. Bartlett, "The Use of Transformations," *Biometrics, 3* (1947), 39–52.

56. Agnes M. Herzberg and D. R. Cox, "Recent Work on the Design of Experiments: A Bibliography and a Review," *Journal of the Royal Statistical Society* (Series A), *132* (1969), 29–67.

Other References Books
1. Geoffrey M. Clarke, *Statistics and Experimental Design*, American Elsevier, New York, 1969.
2. D. J. Finney, *An Introduction to the Theory of Experimental Design*, The University of Chicago Press, Chicago, 1960.
3. E. G. Olds, T. B. Mattson, and R. E. Odeh: *Notes on the Use of Transformations in the Analysis of Variance*, WADC Tech Rep 56-308, 1956, Wright Air Development Center.

Other References, Journal Articles
1. Benjamin A. Barnes, Elinor Pearson, and Eric Reiss, "The Analysis of Variance: A Graphical Representation of a Statistical Concept," *Journal of the American Statistical Association*, *50* (1955), 1064–1072.
2. David B. Duncan, "Bayes Rules for a Common Multiple Comparisons Problem and Related Student-*t* Problems," *Annals of Mathematical Statistics*, *32* (1961), 1013–1033.
3. David B. Duncan, "A Bayesian Approach to Multiple Comparisons," *Technometrics*, *7* (1965), 171–222.
4. Ray A. Waller and David B. Duncan, "A Bayes Rule for the Symmetric Multiple Comparisons Problem," *Journal of the American Statistical Association*, *64* (1969), 1484–1503.
5. Alva R. Feinstein, "Clinical Biostatistics. II. Statistics Versus Science in the Design of Experiments," *Clinical Pharmacology and Therapeutics*, *11* (1970), 282–292.
6. B. G. Greenberg, "Why Randomize?" *Biometrics*, *7* (1951), 309–322.
7. M. Harris, D. G. Horvitz, and A. M. Mood, "On the Determination of Sample Sizes in Designing Experiments," *Journal of the American Statistical Association*, *43* (1948), 391–402.
8. H. Leon Harter, "Multiple Comparison Procedures for Interactions," *The American Statistician*, *24* (December 1970), 30–32.
9. Carl E. Hopkins and Alan J. Gross, "Significance Levels in Multiple Comparison Tests," *Health Services Research*, *5* (Summer 1970), 132–140.
10. Richard J. Light and Barry H. Margolin, "An Analysis of Variance for Categorical Data," *Journal of the American Statistical Association*, *66* (1971), 534–544.
11. Ken Sirotnik, "On the Meaning of the Mean in ANOVA (or the Case of the Missing Degree of Freedom)," *The American Statistician*, *25* (October 1971), 36–37.

CHAPTER 8

Simple Linear Regression and Correlation

8.1 INTRODUCTION

In analyzing data for the health sciences disciplines, we find that it is frequently desirable to learn something about the relationship between two variables. We may, for example, be interested in studying the relationship between blood pressure and age, height and weight, the concentration of an injected drug and heart rate, the consumption level of some nutrient and weight gain, the intensity of a stimulus and reaction time, or total family income and medical care expenditures. The nature and strength of the relationships between variables such as these may be examined by *regression* and *correlation* analysis, two statistical techniques that, although related, serve different purposes.

Regression analysis is helpful in ascertaining the probable form of the relationship between variables, and the ultimate objective when this method of analysis is employed usually is to *predict* or *estimate* the value of one variable corresponding to a given value of another variable. The ideas of regression were first elucidated by the English scientist Sir Francis Galton (1822–1911) in reports of his research on heredity—first in sweet peas and later in human stature (1–3). He described a tendency of adult offspring, having either short or tall parents, to revert back toward the average height of the general population. He first used the word *reversion*, and later *regression*, to refer to this phenomenon.

Correlation analysis, on the other hand, is concerned with measuring the strength of the relationship between variables. When we compute measures of correlation from a set of data, we are interested in the degree of the *correlation* between variables. Again, the concepts and terminology of correlation analysis originated with Galton, who first used the word *correlation* in 1888 (4).

In this chapter our discussion is limited to the exploration of the relationship between two variables. The concepts and methods of regression are covered first, beginning in the next section. In Section 8.6 the ideas and techniques of correlation are introduced. In the next chapter we consider the case where there is an interest in the relationships among three or more variables.

Regression and correlation analysis are areas in which the speed and accuracy of a computer are most appreciated. The data for the exercises of this chapter, therefore, are presented in a way that makes them suitable for computer processing. As is always the case, the input requirements and output features of the particular programs to be used should be studied carefully.

8.2 THE REGRESSION MODEL

In the typical regression problem, as in most problems in applied statistics, researchers have available for analysis a sample of observations from some real or hypothetical population. Based on the results of their analysis of the sample data, they are interested in reaching decisions about the population from which the sample is presumed to have been drawn. It is important, therefore, that the researchers understand the nature of the population in which they are interested. They should know enough about the population to be able either to construct a mathematical model for its representation or to determine if it reasonably fits some established model. A researcher about to analyze a set of data by the methods of simple linear regression, for example, should be secure in the knowledge that the simple linear regression model is, at least, an approximate representation of the population. It is unlikely that the model will be a perfect portrait of the real situation, since this characteristic is seldom found in models of practical value. A model constructed so that it corresponds precisely with the details of the situation is usually too complicated to yield any information of value. On the other hand, the results obtained from the analysis of data which have been forced into a model that does not fit are also worthless. Fortunately, however, a perfectly fitting model is not a requirement for obtaining useful results. Researchers, then, should be able to distinguish between the occasion when their chosen models and the data are sufficiently compatible for them to proceed and the case where their chosen model must be abandoned.

Assumptions Underlying Simple Linear Regression   In the simple linear regression model two variables, X and Y, are of interest. The variable X is usually referred to as the *independent variable*, since frequently it is controlled by the investigator; that is, values of X may be selected by the investigator and, corresponding to each preselected value of X, one or more values of Y are obtained. The other variable, Y, accordingly, is called the *dependent variable*, and we speak of the regression of Y on X. The following are the assumptions underlying the simple linear regression model.

1. Values of the independent variable X are said to be "fixed." This means that the values of X are preselected by the investigator so that in the collection of the data they are not allowed to vary from these preselected values. In this

model, X is referred to by some writers as a *nonrandom* variable and by others as a *mathematical* variable. It should be pointed out at this time that the statement of this assumption classifies our model as the *classical regression model*. Regression analysis also can be carried out on data in which X is a random variable.

2. The variable X is measured without error. Since no measuring procedure is perfect, this means that the magnitude of the measurement error in X is negligible.

3. For each value of X there is a subpopulation of Y values. For the usual inferential procedures of estimation and hypothesis testing to be valid, these subpopulations must be normally distributed. In order that these procedures may be presented it will be assumed that the Y values are normally distributed in the examples and exercises that follow.

4. The variances of the subpopulations of Y are all equal.

5. The means of the subpopulations of Y all lie on the same straight line. This is known as the *assumption of linearity*. This assumption may be expressed symbolically as

$$\mu_{y|x} = \alpha + \beta x \qquad (8.2.1)$$

where $\mu_{y|x}$ is the mean of the subpopulation of Y values for a particular value of X, and α and β are called population regression coefficients. Geometrically, α and β represent the y-intercept and slope, respectively, of the line on which all the means are assumed to lie.

6. The Y values are statistically independent. In other words, in drawing the sample, it is assumed that the values of Y chosen at one value of X in no way depend on the values of Y chosen at another value of X.

These assumptions may be summarized by means of the following equation, which is called the regression model:

$$y = \alpha + \beta x + e \qquad (8.2.2)$$

where y is a typical value from one of the subpopulations of Y, α and β are as defined for Equation 8.2.1, and e is called the error term. If we solve 8.2.2 for e, we have

$$\begin{aligned} e &= y - (\alpha + \beta x) \\ &= y - \mu_{y|x} \end{aligned} \qquad (8.2.3)$$

and we see that e shows the amount by which y deviates from the mean of the subpopulation of Y values from which it is drawn. As a consequence of the assumption that the subpopulations of Y values are normally distributed with equal variances, the e's for each subpopulation are normally distributed with a variance equal to the common variance of the subpopulations of Y values.

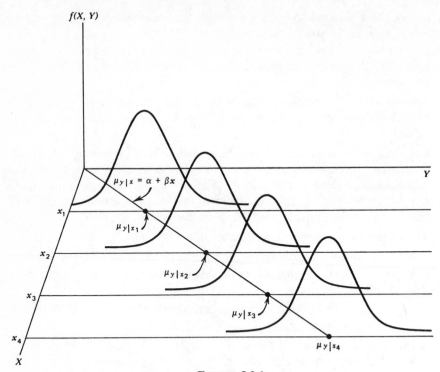

FIGURE 8.2.1

Representation of the Simple Linear Regression Model

The following acronym will help the reader remember most of the assumptions necessary for inference in linear regression analysis:

LINE   [Linear (assumption 5), Independent (assumption 6),

Normal (assumption 3), Equal variances (assumption 4)]

A graphical representation of the regression model is given in Figure 8.2.1.

8.3   THE SAMPLE REGRESSION EQUATION

In simple linear regression the object of the researcher's interest is the population regression equation—the equation that describes the true relationship between the dependent variable Y and the independent variable X. In an effort to reach a decision regarding the likely form of this relationship, the researcher draws a sample from the population of interest and, using the resulting data, computes a sample regression equation that forms the basis for reaching conclusions regarding the unknown population regression equation.

In the absence of extensive information regarding the nature of the variables of interest, a frequently employed strategy is to assume initially that they are linearly related. Subsequent analysis, then, involves the following steps:

1. Determine whether or not the assumptions underlying a linear relationship are met in the data available for analysis.
2. Obtain the equation for the line that best fits the sample data.
3. Evaluate the equation to obtain some idea of the strength of the relationship and the usefulness of the equation for predicting and estimating.
4. If the data appear to conform satisfactorily to the linear model, use the equation obtained from the sample data to predict and to estimate.

When we use the regression equation to *predict*, we will be predicting the value Y is likely to have when X has a given value. When we use the equation to *estimate*, we will be estimating the mean of the subpopulation of Y values assumed to exist at a given value of X.

Example 8.3.1 A team of professional mental health workers in a long-stay psychiatric hospital wished to measure the level of response of withdrawn patients to a program of remotivation therapy. A standardized test was available for this purpose, but it was expensive and time-consuming to administer. To overcome this obstacle the team developed a test that was much easier to administer. To test the usefulness of the new instrument for measuring the level of patient response, the team decided to examine the relationship between scores made on the new test and scores made on the standardized test. The objective was to use the new test if it could be shown that it was a good predictor of a patient's score on the standardized test. The team was interested only in carrying out the analysis for standardized scores between 50 and 100, since a score below 50 did not represent a significant level of response, and scores above 100, although possible, were seldom made by the type of patient under consideration. The team also felt that the use of scores in increments of 5 would give a good coverage of the range of scores between 50 and 100. Accordingly, 11 patients who had made scores on the new test of 50, 55, 60, 65, 70, 75, 80, 85, 90, 95, and 100, respectively, were selected to take the standardized test. The independent and dependent variables, respectively, are scores made on the new test and scores made on the standardized test. The results are shown in Table 8.3.1.

The reader will recognize that the assumption of fixed values of X is satisfied in this example. The values of X were selected in advance and were not allowed to vary as would have been the case had the patients been selected at random before taking the tests.

A first step that is usually useful in studying the relationship between two variables is to prepare a *scatter diagram* of the data such as is shown in Figure 8.3.1. The points are plotted by assigning values of the independent variable X to the horizontal axis and values of the dependent variable Y to the vertical axis.

TABLE 8.3.1

Patients' Scores on Standardized Test and New Test, Example 8.3.1

Patient number	Score on new test (X)	Score on standardized test (Y)
1	50	61
2	55	61
3	60	59
4	65	71
5	70	80
6	75	76
7	80	90
8	85	106
9	90	98
10	95	100
11	100	114

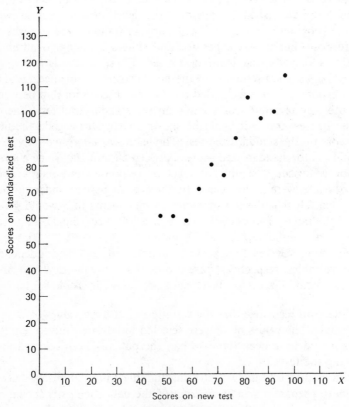

FIGURE 8.3.1

Scatter Diagram of Data Shown in Table 8.3.1

The pattern made by the points plotted on the scatter diagram usually suggests the basic nature of the relationship between two variables. As we look at Figure 8.3.1, for example, the points seem to be scattered around an invisible straight line. The scatter diagram also shows that, in general, patients who score high on the new test also make high scores on the standardized test. These impressions suggest that the relationship between scores on the two tests may be described by a straight line crossing the Y-axis near the origin and making approximately a 45-degree angle with the X-axis. It looks as if it would be simple to draw, freehand, through the data points the line that describes the relationship between X and Y. It is highly unlikely, however, that the lines drawn by any two people would be exactly the same. In other words, for every person drawing such a line by eye, or freehand, we would expect a slightly different line. The question then arises as to which line best describes the relationship between the variables. We cannot obtain an answer to this question by inspecting the lines. In fact, it is not likely that any freehand line drawn through the data will be the line that best describes the relationship between X and Y, since freehand lines will reflect any defects of vision or judgment of the person drawing the line. Similarly, when judging which of two lines best describes the relationship, subjective evaluation is liable to the same deficiencies.

What is needed for obtaining the desired line is some method that is not fraught with these difficulties.

The Least-Squares Line   The method usually employed for obtaining the desired line is known as the *method of least squares*, and the resulting line is called the *least-squares line*. The reason for calling the method by this name will be explained in the discussion that follows.

We recall from algebra that the general equation for a straight line is given by

$$y = a + bx \qquad (8.3.1)$$

where y is a value on the vertical axis, x is a value on the horizontal axis, a is the point where the line crosses the vertical axis, and b shows the amount by which y changes for each unit change in x. We refer to a as the y-intercept and b as the *slope* of the line. To draw a line based on Equation 8.3.1, we need the numerical values of the constants a and b. Given these constants, we may substitute various values of x into the equation to obtain corresponding values of y. The resulting points may be plotted. Since any two such coordinates determine a straight line, we may select any two, locate them on a graph, and connect them to obtain the line corresponding to the equation. It can be shown by mathematics beyond the scope of this book that a and b may be obtained by the simultaneous solution of the following two equations, which are known as the *normal equations* for a set of data:

$$\sum y_i = na + b\sum x_i \qquad (8.3.2)$$

$$\sum x_i y_i = a\sum x_i + b\sum x_i^2 \qquad (8.3.3)$$

TABLE 8.3.2

Intermediate Computations for Normal Equations, Example 8.3.1

Score on New Test x	Score on Standardized Test y	x^2	y^2	xy
50	61	2500	3721	3050
55	61	3025	3721	3355
60	59	3600	3481	3540
65	71	4225	5041	4615
70	80	4900	6400	5600
75	76	5625	5776	5700
80	90	6400	8100	7200
85	106	7225	11236	9010
90	98	8100	9604	8820
95	100	9025	10000	9500
100	114	10000	12996	11400
Total 825	916	64625	80076	71790

In Table 8.3.2 we have the necessary values for substituting into the normal equations.

Substituting appropriate values from Table 8.3.2 into Equations 8.3.2 and 8.3.3 gives

$$916 = 11a + 825b$$
$$71790 = 825a + 64625b$$

We may solve these equations by any familiar method to obtain

$$a = -.9973 \quad \text{and} \quad b = 1.1236$$

The linear equation for the least-squares line that describes the relationship between scores on the standardized test and scores on the new test may be written, then, as

$$\hat{y} = -.9973 + 1.1236x \qquad (8.3.4)$$

This equation tells us that since a is negative, the line crosses the Y-axis below the origin, and that since b, the slope, is positive, the line extends from the lower left-hand corner of the graph to the upper right-hand corner. We see further that for each unit increase in x, y increases by an amount equal to 1.1236. The symbol $\hat{y}$ denotes that it is a value of y computed from the equation, rather than an observed value of Y.

By substituting two convenient values of X into Equation 8.3.4, we may obtain the necessary coordinates for drawing the line. Suppose, first, we let

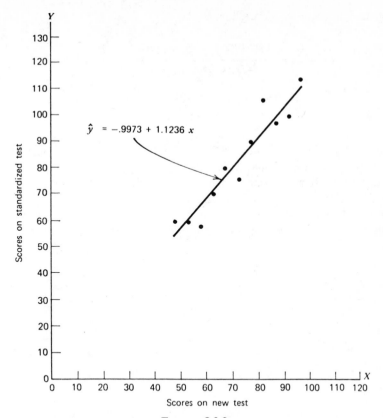

$\hat{y} = -.9973 + 1.1236\,x$

FIGURE 8.3.2

Original Data and Least-Squares Line for Example 8.3.1

$X = 50$ and obtain

$$\hat{y} = -.9973 + 1.1236(50) = 55.1827$$

If we let $X = 100$ we obtain

$$\hat{y} = -.9973 + 1.1236(100) = 111.3627$$

The line, along with the original data, is shown in Figure 8.3.2.

Numerical values for a and b may be obtained by alternative formulas that do not directly involve the normal equations. The formulas are as follows:

$$b = \frac{n\sum xy - \left(\sum x\right)\left(\sum y\right)}{n\sum x^2 - \left(\sum x\right)^2} \tag{8.3.5}$$

$$a = \frac{\sum y - b\sum x}{n} \tag{8.3.6}$$

For the present example we have

$$b = \frac{11(71790) - (825)(916)}{11(64625) - (825)^2} = 1.1236$$

$$a = \frac{916 - 1.1236(825)}{11} = -.9973$$

Thus, we see that Equations 8.3.5 and 8.3.6 yield the same results as one obtains by solving the normal equations.

Now that we have obtained what we call the "best" line for describing the relationship between our two variables, we need to determine by what criterion it is considered best. Before the criterion is stated, let us examine Figure 8.3.3. We note that the least-squares line does not pass through any of the observed points

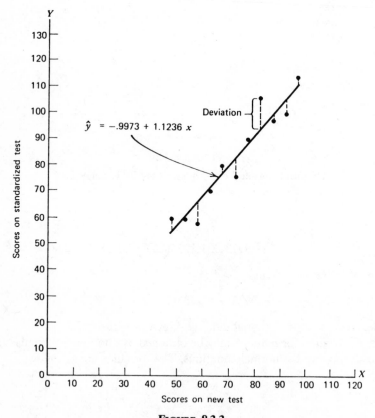

FIGURE 8.3.3

Original Data, Least-Squares Line, and Vertical Deviations for Example 8.3.1

that are plotted on the scatter diagram. In other words, the observed points *deviate* from the line by varying amounts.

The line that we have drawn through the points is best in this sense:

The sum of the squared vertical deviations of the observed data points (y_i) from the least-squares line is smaller than the sum of the squared vertical deviations of the data points from any other line.

In other words, if we square the vertical distance from each observed point (y_i) to the least-squares line and add these squared values for all points, the resulting total will be smaller than the similarly computed total for any other line that can be drawn through the points. For this reason the line we have drawn is called the least-squares line.

EXERCISES

For each of the following exercises (a) draw a scatter diagram and (b) obtain the regression equation and plot it on the scatter diagram.

8.3.1 An experiment was conducted to study the effect of a certain drug in lowering heart rate in adults. The independent variable is dosage in milligrams of the drug, and the dependent variable is the difference between lowest rate following administration of the drug and a predrug control. The following data were collected.

Dose (mg) X	Reduction in heart rate (beats / min) Y
.5	10
.75	8
1.00	12
1.25	12
1.50	14
1.75	12
2.00	16
2.25	18
2.50	17
2.75	20
3.00	18
3.25	20
3.50	21

8.3.2 The following data show the optical densities of a certain substance at different concentration levels.

Level of concentration (X)	Optical density (Y)
80	.08
120	.12
160	.18
200	.21
240	.28
280	.28
320	.38
360	.40
400	.42
440	.50
480	.52
520	.60

8.3.3 A hospital administrator collected the following data on the cost per meal of a standard meal at different volumes of preparation.

Number of meals served (X)	Cost per meal (Y)
30	$1.15
35	1.10
40	.98
45	1.01
50	.97
55	.90
60	.89
70	.85
75	.78
80	.70
65	.80

8.3.4 An experiment was conducted to study the relationship between an objective measurement of anxiety and heart rate in adults. The following data were obtained on 12 normal adults.

Heart rate per minute (X)	Objective measurement of anxiety (Y)
50	48
55	41
60	45
65	41
70	42
75	36
80	38
85	36
90	30
95	32
100	34
105	25

8.3.5 The following data were collected in a study of the relationship between intelligence and family size.

Number of children in family	Average measured intelligence score of all children in the family
1	105
2	102
3	104
4	100
5	97
6	101
7	95
8	93
9	97
10	88

8.3.6 The measurements in the following table were made on 20 geographic areas. The variable X is a combined measure of air and water pollution in the area, and Y is a measure of the health status of residents of the area. The higher the value of X the greater the pollution, and the higher the value of Y the poorer the quality of health.

X	Y	X	Y
73	90	67	76
52	74	73	82
68	91	71	93
47	62	57	73
60	63	86	82
71	78	76	88
67	60	91	97
80	89	69	80
86	82	87	87
91	105	77	95

8.4 EVALUATING THE REGRESSION EQUATION

Once the regression equation has been obtained it must be evaluated to determine whether it adequately describes the relationship between the two variables and whether it can be used effectively for prediction and estimation purposes.

If in the population the relationship between X and Y is linear, β, the slope of the line that describes this relationship, will be either positive, negative, or zero. If β is zero, sample data drawn from the population will, in the long run, yield regression equations that are of little or no value for prediction and estimation purposes. Furthermore, even though we assume that the relationship between X and Y is linear, it may be that the relationship could be described better by some nonlinear model. When this is the case, sample data when fitted to a linear model will tend to yield results compatible with a population slope of zero. Thus, following a test in which the null hypothesis that β equals zero is not rejected, we may conclude (assuming that we have not made a type II error by accepting a false null hypothesis) either (1) that although the relationship between X and Y may be linear it is not strong enough for X to be of much value in predicting and estimating Y, or (2) that the relationship between X and Y is not linear; that is, some curvilinear model provides a better fit to the data, and again X is of limited or no value in predicting and estimating Y. Figure 8.4.1 shows the kinds of relationships between X and Y in a population that may prevent rejection of the null hypothesis that $\beta = 0$.

Now let us consider the situations in a population that may lead to rejection of the null hypothesis that $\beta = 0$. Assuming that we do not commit a type I error, rejection of the null hypothesis that $\beta = 0$ may be attributed to one of the following conditions in the population: (1) The relationship is linear and of sufficient strength to justify the use of sample regression equations to predict and

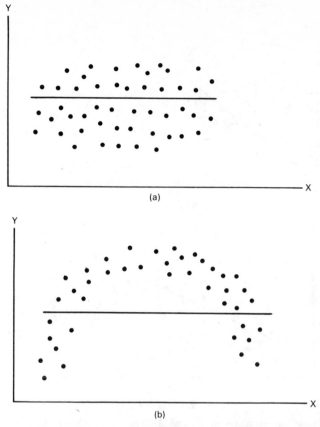

FIGURE 8.4.1

Conditions in a Population That May Prevent Rejection of the Null Hypothesis That β = zero. (*a*) The Relationship Between X and Y Is Linear, but β Is So Close to Zero That Sample Data Are Not Likely to Yield Equations That Are Useful for Predicting Y when X Is Given. (*b*) The Relationship between X and and Y Is not Linear. A Curvilinear Model Provides a Better Fit to the Data. Sample Data Are not Likely to Yield Equations That Are Useful for Predicting Y when X Is Given.

estimate Y for given values of X. (2) There is a good fit of the data to a linear model, but some curvilinear model might provide an even better fit. Figure 8.4.2 illustrates the two population conditions that may lead to rejection of H_0: $\beta = 0$.

Thus we see that before using a sample regression equation to predict and estimate, it is desirable to test H_0: $\beta = 0$. We may do this either by using analysis of variance and the F statistic or by using the t statistic. We shall illustrate both methods. Before we do this, however, let us see how we may investigate the strength of the relationship between X and Y.

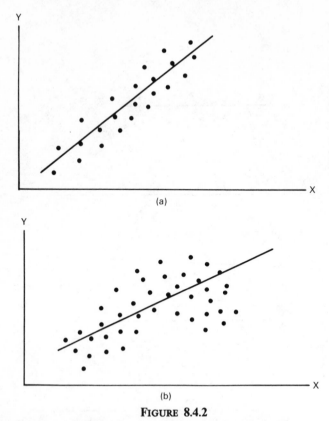

FIGURE 8.4.2

Population Conditions Relative to X and Y That May Cause Rejection of the Null Hypothesis that $\beta = 0$. (a) The Relationship between X and Y Is Linear and of Sufficient Strength to Justify the Use of a Sample Regression Equation to Predict and Estimate Y for Given Values of X. (b) A Linear Model Provides a Good Fit to the Data, but Some Curvilinear Model Would Provide an Even Better Fit.

The Coefficient of Determination   One way to evaluate the strength of the regression equation is to compare the scatter of the points about the regression line with the scatter about $\bar{y}$, the mean of the sample values of Y. If we take the scatter diagram for Example 8.3.1 and draw through the points a line that intersects the Y-axis at $\bar{y}$ and is parallel to the X-axis, we may obtain a visual impression of the relative magnitudes of the scatter of the points about this line and the regression line. This has been done in Figure 8.4.3.

It appears rather obvious from Figure 8.4.3 that the scatter of the points about the regression line is much less than the scatter about the $\bar{y}$ line. We would not wish, however, to decide on this basis alone that the equation is a useful one. The situation may not be always this clear-cut, so that an objective measure of some

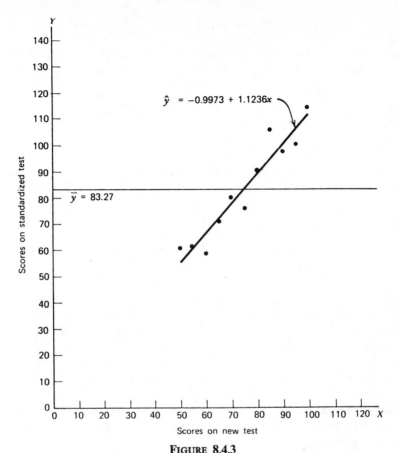

FIGURE 8.4.3

Scatter Diagram, Sample Regression Line, and $\bar{y}$ Line for Example 8.3.1

sort would be much more desirable. Such an objective measure, called the *coefficient of determination*, is available.

Before defining the coefficient of determination, let us justify its use by examining the logic behind its computation. We begin by considering the point corresponding to any observed value, y_i, and by measuring its vertical distance from the $\bar{y}$ line. We call this the *total deviation* and designate it by $(y_i - \bar{y})$.

If we measure the vertical distance from the regression line to the $\bar{y}$ line, we obtain $(\hat{y} - \bar{y})$, which is called the *explained deviation*, since it shows by how much the total deviation is reduced when the regression line is fitted to the points.

Finally, we measure the vertical distance of the observed point from the regression line to obtain $(y_i - \hat{y})$, which is called the *unexplained deviation*, since it represents the portion of the total deviation not "explained" or accounted for by the introduction of the regression line. These three quantities are shown for a typical value of Y in Figure 8.4.4.

It is seen, then, that the total deviation for a particular y_i is equal to the sum of the explained and unexplained deviations. We may write this symbolically as

$$(y_i - \bar{y}) = (\hat{y} - \bar{y}) + (y_i - \hat{y}) \tag{8.4.1}$$

$$\begin{array}{ccc} \text{total} & \text{explained} & \text{unexplained} \\ \text{deviation} & \text{deviation} & \text{deviation} \end{array}$$

If we measure these deviations for each value of y_i and $\hat{y}$, square each deviation, and add up the squared deviations, we have

$$\sum(y_i - \bar{y})^2 = \sum(\hat{y} - \bar{y})^2 + \sum(y_i - \hat{y})^2 \tag{8.4.2}$$

$$\begin{array}{ccc} \text{total} & \text{explained} & \text{unexplained} \\ \text{sum} & \text{sum} & \text{sum} \\ \text{of squares} & \text{of squares} & \text{of squares} \end{array}$$

These quantities may be considered measures of dispersion or variability. The *total sum of squares* (*SST*), for example, is a measure of the dispersion of the observed values of Y about their mean $\bar{y}$; that is, this term is a measure of the total variation in the observed values of Y. The reader will recognize this term as the numerator of the familiar formula for the sample variance.

The *explained sum of squares* measures the amount of the total variability in the observed values of Y that is accounted for by the linear relationship between the observed values of X and Y. This quantity is referred to also as the *sum of squares due to linear regression* (*SSR*).

The *unexplained sum of squares* is a measure of the dispersion of the observed Y values about the regression line and is sometimes called the *error sum of squares*, or the *residual sum of squares* (*SSE*). It is this quantity that is minimized when the least-squares line is obtained.

We may express the relationship among the three sums of squares values as

$$SST = SSR + SSE$$

The necessary calculations for obtaining the total, the regression, and The error sums of squares for our illustrative example are displayed in Table 8.4.1.

For our illustrative example we have

$$SST = SSR + SSE$$
$$3798.1822 \approx 3471.8116 + 326.1455$$
$$3798.1822 \approx 3797.9571$$

The failure of the two components on the right of the equality to add to the total on the left is due to rounding.

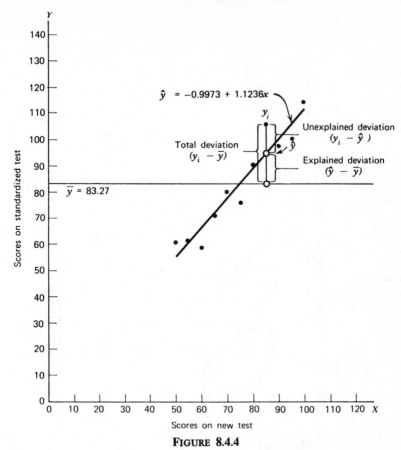

FIGURE 8.4.4

Scatter Diagram Showing the Total, Explained, and Unexplained Deviations for a Selected Value of Y, Example 8.3.1

We may calculate the total sum of squares by the more convenient formula

$$\sum (y_i - \bar{y})^2 = \sum y_i^2 - \frac{(\sum y_i)^2}{n} \qquad (8.4.3)$$

and the regression sum of squares may be computed by

$$\sum (\hat{y} - \bar{y})^2 + b^2 \sum (x_i - \bar{x})^2 = b^2 \left[\sum x_i^2 - (\sum x_i)^2/n \right] \qquad (8.4.4)$$

The error sum of squares is more conveniently obtained by subtraction. For our

TABLE 8.4.1

Calculation of Total, Explained, and Unexplained Sums of Squares

Patient number	Scores on standardized test (y_i)	$\hat{y} = -.9973 + 1.1236x$	$(y_i - \bar{y})$	$(y_i - \bar{y})^2$	$(y_i - \hat{y})$	$(y_i - \hat{y})^2$	$(\hat{y} - \bar{y})$	$(\hat{y} - \bar{y})^2$
1	61	55.1827	−22.2727	496.0732	5.8173	33.8410	−28.0900	789.0481
2	61	60.8007	−22.2727	496.0732	.1993	.0397	−22.4720	504.9908
3	59	66.4187	−24.2727	589.1640	−7.4187	55.0371	−16.8540	284.0573
4	71	72.0367	−12.2727	150.6192	−1.0367	1.0747	−11.2360	126.2477
5	80	77.6547	−3.2727	10.7106	2.3453	5.5004	−5.6180	31.5619
6	76	83.2727	−7.2727	52.8922	−7.2727	52.8922	0.0000	0.0000
7	90	88.8907	6.7273	45.2566	1.1093	1.2305	5.6180	31.5619
8	106	94.5087	22.7273	516.5302	11.4913	132.0500	11.2360	126.2477
9	98	100.1267	14.7273	216.8934	−2.1267	4.5229	16.8540	284.0573
10	100	105.7447	16.7273	279.8026	−5.7447	33.0016	22.4720	504.9908
11	114	111.3627	30.7273	944.1670	2.6373	6.9554	28.0900	789.0481
Total	916		$SST = 3798.1822$		$SSE = 326.1455$		$SSR = 3471.8116$	
$\bar{y} = 83.2727$								

illustrative example we have

$$SST = 61^2 + 61^2 + \cdots + 114^2 - \frac{(916)^2}{11} = 3798.1818$$

$$SSR = (1.1236)^2 \left[50^2 + 55^2 + \cdots + 100^2 - \frac{(825)^2}{11} \right]$$

$$= 3471.8116$$

and

$$SSE = SST - SSR$$
$$= 3798.1818 - 3471.8116$$
$$= 326.3702$$

The results using the computationally more convenient formulas are the same results as those shown in Table 8.4.1 except for rounding.

It is intuitively appealing to speculate that if a regression equation does a good job of describing the relationship between two variables, the explained or regression sum of squares should constitute a large proportion of the total sum of squares. It would be of interest, then, to determine the magnitude of this proportion by computing the ratio of the explained sum of squares to the total sum of squares. This is exactly what is done in evaluating a regression equation based on sample data, and the result is called the sample *coefficient of determination*, r^2. In other words,

$$r^2 = \frac{\sum(\hat{y} - \bar{y})^2}{\sum(y_i - \bar{y})^2} = \frac{b^2 \left[\sum x_i^2 - (\sum x_i)^2/n \right]}{\sum y_i^2 - \frac{(\sum y_i)^2}{n}} = SSR/SST$$

In our present example we have, using the sums of squares values computed by Equations 8.4.3 and 8.4.4,

$$r^2 = \frac{3471.8116}{3798.1818} = .91$$

The sample coefficient of determination measures the closeness of fit of the sample regression equation to the observed values of Y. If we look at Table 8.4.1, we see that when the quantities $(y_i - \hat{y})$, the vertical distances of the observed values of Y from the equation, are small, the unexplained sum of squares is small. This leads to a large explained sum of squares that leads, in turn, to a large value of r^2. This is illustrated in Figure 8.4.5.

In Figure 8.4.5a we see that the observations all lie close to the regression line, and we would expect r^2 to be large. In fact, the computed r^2 for these data is

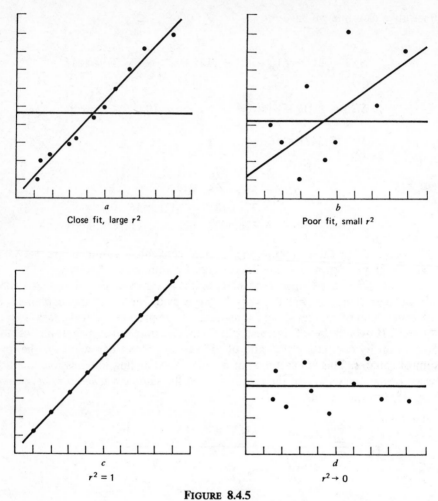

a
Close fit, large r^2

b
Poor fit, small r^2

c
$r^2 = 1$

d
$r^2 \rightarrow 0$

FIGURE 8.4.5

r^2 as a Measure of Closeness-of-Fit of the Sample Regression Line to the Sample Observations

.986, indicating that about 99 percent of the total variation in the y_i is explained by the regression.

In Figure 8.4.5b we illustrate a case where the y_i are widely scattered about the regression line, and there we suspect that r^2 is small. The computed r^2 for the data is .403, that is, less than 50 percent of the total variation in the y_i is explained by the regression.

The largest value that r^2 can assume is 1, a result that occurs when all of the variation in the y_i is explained by the regression. When $r^2 = 1$ all the observations fall on the regression line. This situation is shown in Figure 8.4.5c.

The lower limit of r^2 is 0. This result is obtained when the regression line and the line drawn through $\bar{y}$ coincide. In this situation none of the variation in the y_i is explained by the regression. Figure 8.4.5d illustrates a situation in which r^2 is close to zero.

When r^2 is large, then, the regression has accounted for a large proportion of the total variability in the observed values of Y, and we look with favor on the regression equation. On the other hand, a small r^2, which indicates a failure of the regression to account for a large proportion of the total variation in the observed values of Y, tends to cast doubt on the usefulness of the regression equation for predicting and estimating purposes. We do not, however, pass final judgment on the equation until it has been subjected to an objective statistical test.

Testing H_0: $\beta = 0$ with the F Statistic The steps in the hypothesis testing procedure are as follows:

1. *Data* The data were described in the opening statement of Example 8.3.1.
2. *Assumptions* We presume that the simple linear regression model and its underlying assumptions as given in Section 8.2 are applicable.
3. *Hypotheses*

$$H_0: \beta = 0$$
$$H_A: \beta \neq 0$$
$$\alpha = .05$$

4. *Test Statistic* The test statistic is V.R.

From the three sum-of-squares terms and their associated degrees of freedom the analysis of variance table of Table 8.4.2 may be constructed.

In general, the degrees of freedom associated with the sum of squares due to regression is equal to the number of constants in the regression equation minus 1. In the simple linear case we have two constants, a and b, hence the degrees of freedom for regression are $2 - 1 = 1$.

TABLE 8.4.2

ANOVA Table for Simple Linear Regression

Source of variation	SS	d.f.	MS	V.R.
Linear regression	SSR	1	SSR/1	MSR/MSE
Residual	SSE	$n - 2$	$SSE/(n - 2)$	
Total	SST	$n - 1$		

TABLE 8.4.3

ANOVA Table for Example 8.3.1

Source of variation	SS	d.f.	MS	V.R.
Linear regression	3471.8116	1	3471.8116	95.74
Residual	326.3702	9	36.2634	
Total	3798.1818	10		

5. *Distribution of Test Statistic* It can be shown that when the hypothesis of no linear relationship between X and Y is true, and when the assumptions underlying regression are met, the ratio obtained by dividing the regression mean square by the residual mean square is distributed as F with 1 and $n - 2$ degrees of freedom.

6. *Decision Rule* Reject H_0 if the computed value of V.R. is equal to or greater than the critical value of F.

7. *Calculation of the Test Statistic* Substituting appropriate numerical values into Table 8.4.2 gives Table 8.4.3.

8. *Statistical Decision* Since 95.74 is greater than 5.12, the critical value of F for 1 and 9 degrees of freedom, the null hypothesis is rejected.

9. *Conclusion* We conclude that the linear model provides a good fit to the data. For this test, since $95.74 > 13.61$, we have $\hat{p} < .005$.

The sample coefficient of determination provides a point estimate of ρ^2, the *population coefficient of determination*. The population coefficient of determination, ρ^2, has the same function relative to the population as r^2 has to the sample. It shows what proportion of the total population variation in Y is explained by the regression of Y on X. When the number of degrees of freedom are small, r^2 is positively biased. That is, r^2 tends to be large. An unbiased estimator of ρ^2 is provided by

$$\tilde{r}^2 = 1 - \frac{\sum(y_i - \hat{y})^2/(n - 2)}{\sum(y_i - \bar{y})^2/(n - 1)} \tag{8.4.5}$$

Observe that the numerator of the fraction in Equation 8.4.5 is the unexplained mean square and the denominator is the total mean square. These quantities appear in the analysis of variance table. For our illustrative example we have, using the data from Table 8.4.3,

$$\tilde{r}^2 = 1 - \frac{326.3702/9}{3798.1818/10} = 1 - \frac{36.2634}{379.81818} = .9045$$

We see that this value is slightly less than

$$r^2 = 1 - \frac{326.3702}{3798.1818} = 1 - .08593 = .91407$$

We see that the difference in r^2 and $\tilde{r}^2$ is due to the factor $(n - 1)/(n - 2)$. When n is large, this factor will approach 1 and the difference between r^2 and $\tilde{r}^2$ will approach zero.

Testing H_0: $\beta = 0$ with the t Statistic When the assumptions stated in Section 8.2 are met, a and b are unbiased point estimators of the corresponding parameters α and β, and since, under these assumptions, the subpopulations of Y values are normally distributed. We may also construct confidence intervals for and test hypotheses about α and β.

When the assumptions of Section 8.2 hold true, the sampling distributions of a and b are each normally distributed with means and variances as follows:

$$\mu_a = \alpha \tag{8.4.6}$$

$$\sigma_a^2 = \frac{\sigma_{y|x}^2 \sum x_i^2}{n \sum (x_i - \bar{x})^2} \tag{8.4.7}$$

$$\mu_b = \beta \tag{8.4.8}$$

and

$$\sigma_b^2 = \frac{\sigma_{y|x}^2}{\sum (x_i - \bar{x})^2} \tag{8.4.9}$$

In Equations 8.4.7 and 8.4.9 $\sigma_{y|x}^2$ is the unexplained variance of the subpopulations of Y values.

With knowledge of the sampling distributions of a and b we may construct confidence intervals and test hypotheses relative to α and β in the usual manner. Inferences regarding α are usually not of interest. On the other hand, as we have seen, a great deal of interest centers on inferential procedures with respect to β. The reason for this is the fact that β tells us so much about the form of the relationship between X and Y. When X and Y are linearly related a positive β indicates that, in general, Y increases as X increases, and we say that there is a *direct linear relationship* between X and Y. A negative β indicates that values of Y tend to decrease as values of X increase, and we say that there is an *inverse linear relationship* between X and Y. When there is no linear relationship between X and Y, β is equal to zero. These three situations are illustrated in Figure 8.4.6.

For testing hypotheses about β the test statistic when $\sigma_{y|x}^2$ is known is

$$z = \frac{b - \beta_0}{\sigma_b} \tag{8.4.10}$$

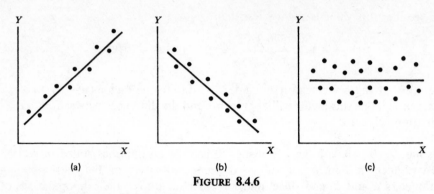

FIGURE 8.4.6

Scatter Diagrams Showing Direct, Inverse, and no Linear Relationships between X and Y. (a) Direct Linear Relationship. (b) Inverse Linear Relationship. (c) No Linear Relationship.

where β_0 is the hypothesized value of β. The hypothesized value of β does not have to be zero, but in practice, more often than not, the null hypothesis of interest is that $\beta = 0$.

As a rule $\sigma_{y|x}^2$ is unknown. When this is the case, the test statistic is

$$t = \frac{b - \beta_0}{s_b} \qquad (8.4.11)$$

where s_b is an estimate of σ_b, and t is distributed as Student's t with $n - 2$ degrees of freedom. To obtain s_b, we must first estimate $\sigma_{y|x}^2$. An unbiased estimator of this parameter is provided by the unexplained variance computed from the sample data. That is,

$$s_{y|x}^2 = \frac{\sum (y_i - \hat{y})^2}{n - 2} \qquad (8.4.12)$$

is an unbiased estimator of $\sigma_{y|x}^2$. This is the unexplained mean square which appears in the analysis of variance table.

The terms $(y_i - \hat{y})$ in Equation 8.4.12 are called the *residuals*. Some computer programs for regression analysis routinely give the residuals as part of the output. When this is the case one may obtain $s_{y|x}^2$ by squaring the residuals, adding the squared terms, and dividing the result by $n - 2$. An alternative formula for $s_{y|x}^2$ is

$$s_{y|x}^2 = \frac{n - 1}{n - 2} \left(s_y^2 - b^2 s_x^2 \right) \qquad (8.4.13)$$

where s_y^2 and s_x^2 are the variances of the y and x observations, respectively. For our illustrative example we have

$$s_y^2 = 379.8182 \quad \text{and} \quad s_x^2 = 275.0000$$

so that

$$s_{y|x}^2 = \frac{10}{9}\left[379.8182 - (1.1236)^2(275.0000)\right] = 36.2634$$

a result that agrees with the residual mean square in Table 8.4.3.

The square root of $s_{y|x}^2$, $s_{y|x}$, is the standard deviation of the observations about the fitted regression line and measures the dispersion of these points about the line. The greater $s_{y|x}$, the poorer the fit of the line to the observed data.

When $s_{y|x}^2$ is used to estimate $\sigma_{y|x}^2$ we may obtain the desired and unbiased estimator of σ_b^2 by

$$s_b^2 = \frac{s_{y|x}^2}{\sum(x_i - \bar{x})^2} \tag{8.4.14}$$

We may write Equation 8.4.14 in the following computationally more convenient form:

$$s_b^2 = \frac{s_{y|x}^2}{\sum x_i^2 - \left(\sum x_i\right)^2/n} \tag{8.4.15}$$

If the probability of observing a value as extreme as the value of the test statistic computed by Equation 8.4.11 when the null hypothesis is true is less than $\alpha/2$ (since we have a two-sided test), the null hypothesis is rejected.

Let us use our now familiar example to illustrate the procedure of testing the null hypothesis that $\beta = 0$ with the t statistic.

1. *Data*   See Example 8.3.1.
2. *Assumptions*   We presume that the simple linear regression model and its underlying assumptions are applicable.
3. *Hypotheses*

$$H_0: \beta = 0$$
$$H_A: \beta \neq 0$$
$$\alpha = .05$$

4. *Test Statistic*   The test statistic is given by Equation 8.4.11.
5. *Distribution of the Test Statistic*   When the assumptions are met and H_0 is true, the test statistic is distributed as Student's t with $n - 2$ degrees of freedom.

6. *Decision Rule*   Reject H_0 if the computed value of t is either greater than or equal to 2.2622 or less than or equal to -2.2622.

7. *Calculation of the Statistic*   We first compute $s_b{}^2$. From Table 8.4.3 we have $s_{y|x}^2 = 36.2634$, so that we may compute

$$s_b{}^2 = \frac{36.2634}{[50^2 + 55^2 + \cdots + 100^2] - \dfrac{(825)^2}{11}} = .013$$

We may compute our test statistic

$$t = \frac{1.1236 - 0}{\sqrt{.013}} = 9.85$$

8. *Statistical Decision*   Reject H_0 because $9.85 > 2.2622$. The p value for this test is less than .01, since, when H_0 is true, the probability of getting a value of t as large as or larger than 3.2498 is .005 and the probability of getting a value of t as small as or smaller than -3.2498 is also .005. Since 9.85 is greater than 3.2498, the probability of observing a value of t as large as or larger than 9.85, when the null hypothesis is true, is less than .005. We double this value to obtain $2(.005) = .01$.

9. *Conclusion*   We conclude that the slope of the true regression line is not zero. The practical implication is that we can expect to get better predictions and estimates of Y if we use the sample regression equation than we would get if we ignore the relationship between X and Y. The fact that b is positive leads us to believe that β is positive and that the relationship between X and Y is a direct linear relationship. As has already been pointed out, Equation 8.4.11 may be used to test the null hypothesis that β is equal to some value other than 0. The hypothesized value for β, β_0, is substituted into Equation 8.4.11 rather than 0. All other quantities, as well as the computations, are the same as in the illustrative example. The degrees of freedom and the method of determining significance are also the same.

A Confidence Interval for β   Once it has been determined that it is unlikely, in light of sample evidence, that β is zero, the researcher may be interested in obtaining an interval estimate of β. The general formula for a confidence interval,

$$\text{estimator} \pm (\text{reliability factor})(\text{standard error of the estimate})$$

may be used. When obtaining a confidence interval for β, the estimator is b, the reliability factor is some value of z or t (depending on whether or not $\sigma_{y|x}^2$ is known), and the standard error of the estimator is

$$\sigma_b = \sqrt{\frac{\sigma_{y|x}^2}{\sum (x_i - \bar{x})^2}}$$

When $\sigma_{y|x}^2$ is unknown, σ_b is estimated by

$$s_b = \sqrt{\frac{s_{y|x}^2}{\sum(x_i - \bar{x})^2}}$$

so that in most practical situations our $100(1 - \alpha)$ percent confidence interval for β is

$$b \pm t_{(1-\alpha/2)}\sqrt{\frac{s_{y|x}^2}{\sum(x_i - \bar{x})^2}} \qquad (8.4.16)$$

For our illustrative example we construct the following 95 percent confidence interval for β:

$$1.1236 \pm 2.2622\sqrt{.013}$$
$$1.1236 \pm .2579$$
$$.8657, 1.3815$$

We interpret this interval in the usual manner. From the probabilistic point of view we say that in repeated sampling 95 percent of the intervals constructed in this way will include β. The practical interpretation is that we are 95 percent confident that the single interval constructed includes β.

It is instructive to note that the confidence interval we constructed does not include zero, so that zero is not a candidate for the parameter being estimated. We feel, then that it is unlikely that $\beta = 0$. This is compatible with the results of our hypothesis test in which we rejected the null hypothesis that $\beta = 0$. Actually, we can always test H_0: $\beta = 0$ at the α significance level by constructing the $100(1 - \alpha)$ percent confidence interval for β, and we can reject or fail to reject the hypothesis on the basis of whether or not the interval includes zero. If the interval contains zero, the null hypothesis is not rejected; and if zero is not contained in the interval, we reject the null hypothesis.

It must be emphasized that failure to reject the null hypothesis that $\beta = 0$ does not mean that X and Y are not related. Not only is it possible that a type II error may have been committed but it may be true that X and Y are related in some nonlinear manner. On the other hand, when we reject the null hypothesis that $\beta = 0$, we cannot conclude that the *true* relationship between X and Y is linear. Again, it may be that although the data fit the linear regression model fairly well (as evidenced by the fact that the null hypothesis that $\beta = 0$ is rejected), some nonlinear model would provide an even better fit. Consequently, when we reject H_0 that $\beta = 0$, the best we can say is that more useful results (discussed below) may be obtained by taking into account the regression of Y on X than in ignoring it.

EXERCISES

8.4.1 to 8.4.6 refer to Exercises 8.3.1 to 8.3.6 and for each one do the following:

(a) Compute the coefficient of determination.
(b) Prepare an ANOVA table and use the F statistic to test the null hypothesis that $\beta = 0$.
(c) Use the t statistic to test the null hypothesis that $\beta = 0$ at the .05 level of significance.
(d) Determine the p value for each hypothesis test.
(e) State your conclusions in terms of the problem.
(f) Construct the 95 percent confidence interval for β.

8.5 USING THE REGRESSION EQUATION

If the results of the evaluation of the sample regression equation indicate that there is a relationship between the two variables of interest, we can put the regression equation to practical use. There are two ways in which the equation can be used. It can be used to *predict* what value Y is likely to assume given a particular value of X. When the assumptions of Section 8.2 are met, a *prediction interval* for this predicted value of Y may be constructed.

We may also use the regression equation to *estimate* the mean of the subpopulation of Y values assumed to exist at any particular value of X. Again, if the previously stated assumptions hold, a confidence interval for this parameter may be constructed. The predicted value of Y and the point estimate of the mean of the subpopulation of Y will be numerically equivalent for any particular value of X but, as we shall see, the prediction interval will be wider than the confidence interval.

Predicting Y for a Given X Suppose, in our illustrative example, we have a patient who makes a score of 70 on the new test and we want to predict his score on the standardized test. To obtain the predicted value, we substitute 70 for x in the sample regression equation to obtain

$$\hat{y} = -.9973 + 1.1236(70)$$

$$= 78$$

rounded to the nearest whole number.

If we decide to give the patient the standardized test, we would predict his score to be 78. Since we have no confidence in this prediction, we would prefer an interval with an associated level of confidence. If it is known, or if we are willing to assume that the assumptions of Section 8.2 are met, and when $\sigma_{y|x}^2$ is unknown, the $100(1 - \alpha)$ percent prediction interval for Y is given by

$$\hat{y} \pm t_{(1-\alpha/2)} s_{y|x} \sqrt{1 + \frac{1}{n} + \frac{(x_p - \bar{x})^2}{\sum(x_i - \bar{x})^2}} \qquad (8.5.1)$$

where x_p is the particular value of x at which we wish to obtain a prediction interval for Y and the degrees of freedom used in selecting t are $n - 2$. For our illustrative example we may construct the following 95 percent prediction interval:

$$78 \pm 2.2622\sqrt{36.2634} \sqrt{1 + \frac{1}{11} + \frac{(70 - 75)^2}{2750}}$$

$$78 \pm 2.2622(6.02)(1.0488)$$

$$78 \pm 14$$

$$64, 92$$

Our interpretation of a prediction interval is similar to the interpretation of a confidence interval. If we repeatedly draw samples, do a regression analysis, and construct prediction intervals for patients who score 70 on the new test, about 95 percent of them will include the patient's score on the standardized test. This is the probabilistic interpretation. The practical interpretation is that we are 95 percent confident that the single prediction interval constructed includes the patient's score on the standardized test.

Estimating the Mean of Y for a Given X   If, in our illustrative example, we are interested in estimating the mean Y score for a subpopulation of patients all of whom make a score of 70 on the new test, we would again calculate

$$\hat{y} = -.9973 + 1.1236(70)$$

$$= 78$$

The $100(1 - \alpha)$ percent confidence interval for $\mu_{y|x}$, when $\sigma_{y|x}^2$ is unknown, is given by

$$\hat{y} \pm t_{(1-\alpha/2)} s_{y|x} \sqrt{\frac{1}{n} + \frac{(x_p - \bar{x})^2}{\sum(x_i - \bar{x})^2}} \qquad (8.5.2)$$

TABLE 8.5.1

95 Percent Confidence Limits for $\mu_{y|x}$, for Each Value of X, Example 8.3.1

x	$\hat{y}$	Lower limit	Upper limit
50	55.1827	47.5009	62.8645
55	60.8007	54.1798	67.4216
60	66.4187	60.7588	72.0786
65	72.0367	67.1783	76.8951
70	77.6547	73.3482	81.9612
75	83.2727	79.1666	87.3788
80	88.8907	84.5842	93.1972
85	94.5087	89.6503	99.3671
90	100.1267	94.4668	105.7866
95	105.7447	99.1238	112.3656
100	111.3627	103.6809	119.0445

We can obtain the following 95 percent confidence interval for $\mu_{y|x=70}$ of our present example by making proper substitutions:

$$78 \pm 2.2622\sqrt{36.2634}\ \sqrt{\frac{1}{11} + \frac{(70 - 75)^2}{2750}}$$

$$78 \pm 4$$

$$74, 82$$

If we repeatedly drew samples from our population of patients, performed a regression analysis, and estimated $\mu_{y|x=70}$ with a similarly constructed confidence interval, about 95 percent of such intervals would include the true mean. For this reason we are 95 percent confident that the single interval constructed contains the population mean.

If confidence intervals are constructed for several subpopulation means and the upper and lower limits plotted on the same scatter diagram with the regression line, we can obtain a *confidence band* by connecting all the upper limits with one curve and all the lower limits with another curve. Table 8.5.1 contains the 95 percent upper and lower confidence limits for $\mu_{y|x}$ for each value of X in our illustrative example, and Figure 8.5.1 shows the 95 percent confidence band that results when these values are plotted.

Notice that the confidence band is wider at the ends than in the middle. In fact, the confidence band is narrowest for $x_p = \bar{x} = 75$. The reason for this is that when $\bar{x} = 75$ is substituted for x_p in Equation 8.5.2, the quantity under the radical is a minimum that results in the narrowest confidence interval. As x_p increases or decreases, the quantity under the radical becomes larger, as do the associated confidence intervals.

Graybill and Bowden (5) present a method for constructing confidence bands that are straight lines. They suggest that bands of this type may be preferable to

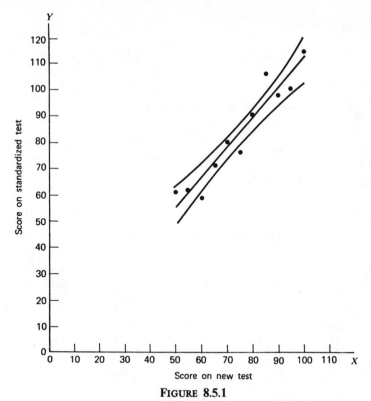

FIGURE 8.5.1

The 95 Percent Confidence Band for $\mu_{y|x}$, Example 8.3.1

the curvilinear type given above because they are easier to compute and graph and have a smaller average width.

Dunn (6), who has constructed a confidence band which she states is preferable to that of Graybill and Bowden, suggests that the usual curvilinear confidence band is to be preferred over any of the straight-line variety. This topic is also discussed by Hoel (7), Gafarian (8), Halperin and Gurian (9), and Elston and Grizzle (10).

In a similar manner, prediction bands may be constructed by plotting the prediction intervals at several values of X.

Computer Analysis Now let us look at a computer analysis of the data of Example 8.3.1. Figure 8.5.2 shows part of the computer printout from the MINITAB simple linear regression program. Note that the column labeled ST. DEV. PRED. Y contains, for each observed value of X, numerical values of

$$s_{y|x}\sqrt{\frac{1}{n}+\frac{(x_p-\bar{x})^2}{\sum(x-\bar{x})^2}}$$

```
THE REGRESSION EQUATION IS
Y = -  1.00 +   1.12 X1
```

	COLUMN	COEFFICIENT	ST. DEV. OF COEF.	T-RATIO = COEF/S.D.
	--	-1.000	8.799	-0.11
X1	C1	1.1236	0.1148	9.79

```
THE ST. DEV. OF Y ABOUT REGRESSION LINE IS
S = 6.020
WITH (  11- 2) =   9 DEGREES OF FREEDOM

R-SQUARED = 91.4 PERCENT
R-SQUARED = 90.5 PERCENT, ADJUSTED FOR D.F.
```

ANALYSIS OF VARIANCE

DUE TO	DF	SS	MS=SS/DF
REGRESSION	1	3472.04	3472.04
RESIDUAL	9	326.15	36.24
TOTAL	10	3798.18	

ROW	X1 C1	Y C2	PRED. Y VALUE	ST.DEV. PRED. Y	RESIDUAL	ST.RES.
1	50	61.00	55.18	3.40	5.82	1.17
2	55	61.00	60.80	2.93	0.20	0.04
3	60	59.00	66.42	2.50	-7.42	-1.35
4	65	71.00	72.04	2.15	-1.04	-0.18
5	70	80.00	77.65	1.90	2.35	0.41
6	75	76.00	83.27	1.82	-7.27	-1.27
7	80	90.00	88.89	1.90	1.11	0.19
8	85	106.00	94.51	2.15	11.49	2.04
9	90	98.00	100.13	2.50	-2.13	-0.39
10	95	100.00	105.75	2.93	-5.75	-1.09
11	100	114.00	111.36	3.40	2.64	0.53

FIGURE 8.5.2

Partial Printout of the Computer Analysis of the Data Given in Example 8.3.1, Using the MINITAB Software Package

EXERCISES

In each of Exercises 8.5.1 to 8.5.5 refer to the appropriate exercise and, for the value of X indicated, (a) construct the 95 percent confidence interval for $\mu_{y|x}$ and (b) construct the 95 percent prediction interval for Y.

8.5.1 Refer to Exercise 8.3.1 and let $X = 2$.

8.5.2 Refer to Exercise 8.3.2 and let $X = 400$.

8.5.3 Refer to Exercise 8.3.3 and let $X = 50$.

8.5.4 Refer to Exercise 8.3.4 and let $X = 100$.

8.5.5 Refer to Exercise 8.3.5 and let $X = 5$.

8.5.6 Construct the 95 percent confidence band for $\mu_{y|x}$ for Exercise 8.3.5.

8.5.7 Refer to Exercise 8.3.6 and let $X = 90$.

8.6   THE CORRELATION MODEL

In the classic regression model, which has been the underlying model in our discussion up to this point, only Y, which has been called the dependent variable, is random. The variable X is defined as a fixed (nonrandom or mathematical) variable and is referred to as the independent variable. Recall, also, that observations have been described as being obtained by preselecting values of X and determining corresponding values of Y.

When both Y and X are random variables, we have what is called the *correlation model*. Typically, under the correlation model, sample observations are obtained by selecting a random sample of the *units of association* (which may be persons, places, animals, points in time, or any other element on which the two measurements are taken) and by taking on each a measurement of X and a measurement of Y. In this procedure, values of X are not preselected, but occur at random, depending on the unit of association selected in the sample.

Although correlation analysis cannot be carried out meaningfully under the classic regression model, regression analysis can be carried out under the correlation model. Correlation involving two variables implies a co-relationship between variables that puts both on an equal footing and does not distinguish between them by referring to one as the dependent and the other as the independent variable. In fact, in the basic computational procedures, which are the same as for the regression model, we may fit a straight line to the data either by minimizing $\sum (y_i - \hat{y})^2$ or by minimizing $\sum (x_i - \hat{x})^2$. In other words, we may do a regression of X on Y as well as a regression of Y on X. The fitted lines in the two cases in general will be different, and a logical question arises as to which line to fit.

If the objective is solely to obtain a measure of the strength of the relationship between the two variables, it does not matter which line is fitted, since the measure usually computed will be the same in either case. If, however, it is desired to use the equation describing the relationship between the two variables for the purposes discussed in the preceding sections, it does matter which line is fitted. The variable for which we wish to estimate means or to make predictions should be treated as the dependent variable; that is, this variable should be regressed on the other variable.

Under the correlation model, X and Y are assumed to vary together in what is called a *joint distribution*. If the form of this joint distribution is normally distributed, it is referred to as a *bivariate normal distribution*. Inferences regarding this population may be made based on the results of samples properly drawn from it. If, on the other hand, the form of the joint distribution is known to be nonnormal, or if the form is unknown and there is no justification for assuming normality, inferential procedures are invalid, although descriptive measures may be computed.

The following assumptions must hold for inferences about the population to be valid when sampling is from a bivariate distribution.

1. For each value of X there is a normally distributed subpopulation of Y values.

2. For each value of Y there is a normally distributed subpopulation of X values.

3. The joint distribution of X and Y is a normal distribution called the *bivariate normal distribution*.

4. The subpopulations of Y values all have the same variance.

5. The subpopulations of X values all have the same variance.

The bivariate normal distribution is represented graphically in Figure 8.6.1. In this illustration we see that if we slice the mound parallel to Y at some value of

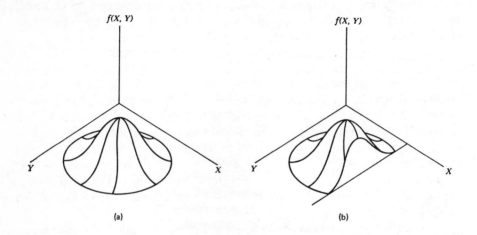

(a)   (b)

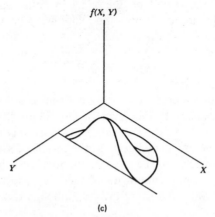

(c)

FIGURE 8.6.1

A Bivariate Normal Distribution. (*a*) A Bivariate Normal Distribution. (*b*) A Cutaway Showing Normally Distributed Subpopulation of *Y* for Given *X*. (*c*) A Cutaway Showing Normally Distributed Subpopulation of *X* for Given *Y*

X, the cutaway reveals the corresponding normal distribution of Y. Similarly, a slice through the mound parallel to X at some value of Y reveals the corresponding normally distributed subpopulation of X.

8.7 THE CORRELATION COEFFICIENT

The bivariate normal distribution discussed in Section 8.6 has five parameters, σ_x, σ_y, μ_x, μ_y, and ρ. The first four are, respectively, the standard deviations and means associated with the individual distributions. The other parameter, ρ, is called the population *correlation coefficient* and measures the strength of the linear relationship between X and Y.

The population correlation coefficient is the square root of ρ^2, the population coefficient of determination previously discussed, and since the coefficient of determination takes on values between 0 and 1 inclusive, ρ may assume any value between -1 and $+1$. If $\rho = 1$ there is a perfect direct linear correlation between the two variables, while $\rho = -1$ indicates perfect inverse linear correlation. If $\rho = 0$ the two variables are not correlated. The sign of ρ will always be the same as the sign of β, the slope of the population regression line for X and Y.

The sample correlation coefficient, r, describes the relationship between the sample observations on two variables in the same way that ρ describes the relationship in a population.

Figure 8.4.5d, and 8.4.5c, respectively, show typical scatter diagrams where $r \to 0$ $(r^2 \to 0)$ and $r = +1$ $(r^2 = 1)$. Figure 8.7.1 shows a typical scatter diagram where $r = -1$.

We are usually interested in knowing if we may conclude that $\rho \neq 0$, that is, that X and Y are correlated. Since ρ is usually unknown, we draw a random sample from the population of interest, compute r, the estimator of ρ, and test

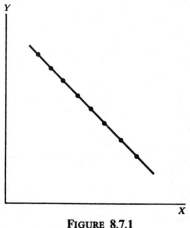

FIGURE 8.7.1

Scatter Diagram for $r = -1$

H_0: $\rho = 0$ against the alternative $\rho \neq 0$. The procedure will be illustrated in the following example.

Example 8.7.1   Blood pressure readings by two different methods were made on 25 patients with essential hypertension. The systolic readings by the two methods are shown in Table 8.7.1.

The clinician wished to investigate the strength of the relationship between the two measurements. The scatter diagram and least-squares regression line are shown in Figure 8.7.2.

The preliminary calculations necessary for obtaining the least-squares regression line are shown in Table 8.7.2. Let us assume that the investigator wishes to obtain a regression equation to use for estimating and predicting purposes. In that case the sample correlation coefficient will be obtained by the methods discussed under the regression model.

TABLE 8.7.1

Systolic Blood Pressure Readings (mm Hg) by Two Methods in 25 Patients with Essential Hypertension

Patient number	Method I	Method II
1	132	130
2	138	134
3	144	132
4	146	140
5	148	150
6	152	144
7	158	150
8	130	122
9	162	160
10	168	150
11	172	160
12	174	178
13	180	168
14	180	174
15	188	186
16	194	172
17	194	182
18	200	178
19	200	196
20	204	188
21	210	180
22	210	196
23	216	210
24	220	190
25	220	202

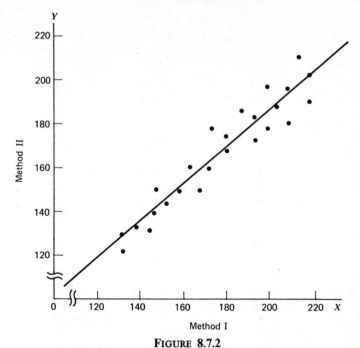

FIGURE 8.7.2

Systolic Blood Pressure Readings (mm Hg), 25 Patients with Essential Hypertension

Let us also assume that the clinician anticipated that he might want to predict a method II reading given a method I reading. We treat method I readings, then, as the independent variable.

By substituting from Table 8.7.2 into Equations 8.3.2 and 8.3.3, the following normal equations are obtained:

$$4172 = 25a + 440b$$
$$757{,}276 = 4440a + 808408b$$

When these equations are solved we have

$$a = 20.8928$$
$$b = .8220$$

The least-squares equation is

$$\hat{y} = 20.8928 + .8220x$$

The coefficient of determination, which is equal to the explained sum of squares divided by the total sum of squares, is (using Equations 8.4.3 and 8.4.4)

$$r^2 = \frac{b^2\left[\sum x_i^2 - \left(\sum x_i\right)^2/n\right]}{\sum y_i^2 - \left(\sum y_i\right)^2/n} \tag{8.7.1}$$

TABLE 8.7.2
Preliminary Calculations for Obtaining Least-Squares Regression Line, Example 8.7.1

Method I (x)	Method II (y)	x^2	y^2	xy
132	130	17424	16900	17160
138	134	19044	17956	18492
144	132	20736	17424	19008
146	140	21316	19600	20440
148	150	21904	22500	22200
152	144	23104	20736	21888
158	150	24964	22500	23700
130	122	16900	14884	15860
162	160	26244	25600	25920
168	150	28224	22500	25200
172	160	29584	25600	27520
174	178	30276	31684	30972
180	168	32400	28224	30240
180	174	32400	30276	31320
188	186	35344	34596	34968
194	172	37636	29584	33368
194	182	37636	33124	35308
200	178	40000	31684	35600
200	196	40000	38416	39200
204	188	41616	35344	38352
210	180	44100	32400	37800
210	196	44100	38416	41160
216	210	46656	44100	45360
220	190	48400	36100	41800
220	202	48400	40804	44440
Total 4440	4172	808408	710952	757276

Substituting values from Table 8.7.2 and the regression equation into Equation 8.7.1 gives

$$r^2 = \frac{(.8220)^2 \left[808{,}408 - (4440)^2/25 \right]}{710{,}952 - (4172)^2/25}$$
$$= .9112713$$

Finally, the correlation coefficient is

$$r = \sqrt{r^2} = \sqrt{.9112713} = .954605 \approx .95$$

An alternative formula for computing r is given by

$$r = \frac{n\sum x_i y_i - \left(\sum x_i\right)\left(\sum y_i\right)}{\sqrt{n\sum x_i^2 - \left(\sum x_i\right)^2} \sqrt{n\sum y_i^2 - \left(\sum y_i\right)^2}} \tag{8.7.2}$$

An advantage of this formula is that r may be computed without first computing *b.* This is the desirable procedure when it is not anticipated that the regression equation will be used. Substituting from Table 8.7.2 in Equation 8.7.2 gives

$$r = \frac{25(757{,}276) - (4440)(4172)}{\sqrt{25(808{,}408) - (4440)^2}\sqrt{25(710{,}952) - (4172)^2}}$$

$$= .95$$

To see if this value of r is of sufficient magnitude to indicate that in the population the two variables of interest are correlated, we conduct a hypothesis test.

1. *Data* See the initial discussion of Example 8.7.1.

2. *Assumptions* We presume that the assumptions given in Section 8.6 are applicable.

3. *Hypotheses*

$$H_0: \rho = 0$$
$$H_A: \rho \neq 0$$

4. *Test Statistic* When $\rho = 0$, it can be shown that the appropriate test statistic is

$$t = r\sqrt{\frac{n-2}{1-r^2}} \qquad (8.7.3)$$

5. *Distribution of Test Statistic* When H_0 is true and the assumptions are met, the test statistic is distributed as Student's t distribution with $n - 2$ degrees of freedom.

6. *Decision Rule* If we let $\alpha = .05$, the critical values of t in the present example are ± 2.0687. If, from our data, we compute a value of t that is either greater than or equal to $+2.0687$ or less than or equal to -2.0687, we will reject the null hypothesis.

7. *Calculation of Test Statistic* Our calculated value of t is

$$t = .954605\sqrt{\frac{23}{1 - .9112713}}$$

$$= 15.37$$

8. *Statistical Decision* Since the computed value of the test statistic does exceed the critical value of t, we reject the null hypothesis.

9. *Conclusion* We conclude that the two variables are correlated. Since 15.37 > 2.8073, we have for this test, $p < .01$.

A Test for Use When the Hypothesized ρ is a Nonzero Value The use of the t statistic computed in the above test is appropriate only for testing $H_0: \rho = 0$. If it is desired to test $H_0: \rho = \rho_0$, where ρ_0 is some value other than zero, we must use another approach. Fisher (11) suggests that r be transformed to z_r as follows

$$z_r = \frac{1}{2}\ln\frac{1 + r}{1 - r} \tag{8.7.4}$$

where ln is a natural logarithm. It can be shown that Z_r is approximately normally distributed with a mean of $z_\rho = \frac{1}{2}\ln\{(1 + \rho)/(1 - \rho)\}$ and estimated standard deviation of

$$\frac{1}{\sqrt{n - 3}} \tag{8.7.5}$$

To test the null hypothesis that ρ is equal to some value other than zero the test statistic is

$$Z = \frac{z_r - z_\rho}{1/\sqrt{n - 3}} \tag{8.7.6}$$

which follows approximately the unit normal distribution.

To determine z_r for an observed r and z_ρ for a hypothesized ρ, we consult Table I, thereby avoiding the direct use of natural logarithms.

Suppose in our present example we wish to test

$$H_0: \rho = .98$$

against the alternative

$$H_A: \rho \neq .98$$

at the .05 level of significance. By consulting Table I we find that for

$$r = .95, \qquad z_r = 1.83178$$

and for

$$\rho = .98, \qquad z_\rho = 2.29756$$

Our test statistic, then, is

$$Z = \frac{1.83178 - 2.29756}{1/\sqrt{25 - 3}}$$
$$= -2.18$$

Since -2.18 is less than the critical value of $z = -1.96$, we must reject H_0 and conclude that the population correlation coefficient is not .98.

For sample sizes less than 25, Fisher's Z transformation should be used with caution, if at all. An alternative procedure due to Hotelling (12) may be used for sample sizes equal to or greater than 10. In this procedure the following transformation of r is employed

$$z^* = z_r - \frac{3z_r + r}{4n} \tag{8.7.7}$$

The standard deviation of z^* is

$$\sigma_{z^*} = \frac{1}{\sqrt{n-1}} \qquad (8.7.8)$$

The test statistic is

$$Z^* = \frac{z^* - \zeta^*}{1/\sqrt{n-1}} = (z^* - \zeta^*)\sqrt{n-1} \qquad (8.7.9)$$

where

$$\zeta^*(\text{pronounced zeta}) = z_\rho - \frac{(3z_\rho + \rho)}{4n}$$

Critical values for comparison purposes are obtained from the unit normal distribution.

In our present example, to test H_0: $\rho = .98$ against H_A: $\rho \neq .98$ ($\alpha = .05$), using the Hotelling transformation we have

$$z^* = 1.83178 - \frac{3(1.83178) + .95}{4(25)} = 1.76733$$

$$\zeta^* = 2.29756 - \frac{3(2.29756) + .98}{4(25)} = 2.21883$$

$$Z^* = (1.76733 - 2.21883)\sqrt{25 - 1}$$
$$= -2.21$$

Since -2.21 is smaller than -1.96, the null hypothesis is rejected and the same conclusion is reached as when the Fisher transformation is used.

In some situations the data available for analysis do not meet the assumptions necessary for the valid use of the procedures discussed here for testing hypotheses about a population correlation coefficient. In such cases it may be more appropriate to use the Spearman rank correlation technique discussed in Chapter 11.

Confidence Interval for ρ Fisher's transformation may be used to construct $100(1 - \alpha)$ percent confidence intervals for ρ. The general formula for a confidence interval

$$\text{estimator} \pm (\text{reliability factor})(\text{standard error})$$

is employed. We first convert our estimator, r, to z_r, construct a confidence interval about z_ρ, and then reconvert the limits to obtain a $100(1 - \alpha)$ percent confidence interval about ρ. The general formula then becomes

$$z_r \pm z(1/\sqrt{n-3}) \qquad (8.7.10)$$

For our present example the 95 percent confidence interval for z_ρ is given by

$$1.83178 \pm 1.96(1/\sqrt{25 - 3})$$
$$1.83178 \pm .41787$$
$$1.41391, 2.24965$$

Converting these limits, which are values of z_r, into values of r gives

z_r	r
1.41391	.890
2.24965	.975

We are 95 percent confident, then, that ρ is contained in the interval .890 to .975. Because of the limited entries in the table, these limits must be considered as only approximate.

An alternative method of constructing confidence intervals for ρ is to use the special charts prepared by David (13).

EXERCISES

In each of the following exercises:

(a) Prepare a scatter diagram.
(b) Compute the sample correlation coefficient.
(c) Test H_0: $\rho = 0$ at the .05 level of significance and state your conclusions.
(d) Determine the p value for the test.
(e) Construct the 95 percent confidence interval for ρ.

8.7.1   A random sample of the records of a certain hospital yielded the following information on the length of hospital stay in days and the annual family income (rounded to the nearest $500) of 15 discharged patients.

Annual family income (X)	Length of stay (Y)
$2000	11
2500	12
3000	9
3500	8
4000	9
4500	10
5000	7
5500	8
6000	4
6500	7
7000	5
7500	6
8000	3
8500	4
9000	4

$$\sum x_i = 82{,}500 \qquad \sum y_i^2 = 871$$

$$\sum x_i^2 = 523{,}750{,}000 \qquad \sum x_i y_i = 510{,}500$$

$$\sum y_i = 107$$

8.7.2 The following table shows the systolic blood pressures of each of 25 pairs of identical twins.

First twin, (X)	Second twin, (Y)
118	115
116	119
118	116
120	119
122	118
122	138
122	124
120	128
124	126
125	130
138	130
140	125
142	164
144	160
145	158
162	145
180	184
180	190
182	188
185	180
170	174
172	170
150	160
152	155
155	160

$$\sum x_i = 3{,}604 \qquad \sum x_i^2 = 532{,}832$$
$$\sum y_i = 3{,}676 \qquad \sum y_i^2 = 555{,}618$$
$$\sum x_i y_i = 543{,}120$$

8.7.3 Two methods of measuring cardiac output were compared in 10 experimental animals with the following results

CARDIAC OUTPUT (1/min)

Method I (X)	Method II (Y)
.8	.5
1.0	1.2
1.3	1.1
1.4	1.3
1.5	1.1
1.4	1.8
2.0	1.6
2.4	2.0
2.7	2.4
3.0	2.8

8.7.4 The following are 15 readings on traffic volume and carbon monoxide concentration taken at a metropolitan air quality sampling site.

Traffic volume (cars per hour) (X)	CO (ppm) (Y)
100	8.8
110	9.0
125	9.5
150	10.0
175	10.5
190	10.5
200	10.5
225	10.6
250	11.0
275	12.1
300	12.1
325	12.5
350	13.0
375	13.2
400	14.5

$$\sum x_i = 3{,}550 \qquad \sum y_i = 167.8$$

$$\sum x_i^2 = 974{,}450 \qquad \sum y_i^2 = 1{,}915.36$$

$$\sum x_i y_i = 41{,}945$$

8.7.5 A random sample of 25 nurses selected from a state registry of nurses yielded the following information on each nurse's score on the state board examination and his or her final score in school. Both scores relate to the nurse's area of affiliation.

Final score (X)	State board score (Y)
87	440
87	480
87	535
88	460
88	525
89	480
89	510
89	530
89	545
89	600
90	495
90	545
90	575
91	525

(Table continued)

Final score (X)	State board score (Y)
91	575
91	600
92	490
92	510
92	575
93	540
93	595
94	525
94	545
94	600
94	625

$$\sum x_i = 2{,}263 \qquad \sum y_i = 13{,}425$$

$$\sum x_i^2 = 204{,}977 \qquad \sum y_i^2 = 7{,}264{,}525$$

$$\sum x_i y_i = 1{,}216{,}685$$

8.7.6 A simple random sample of 15 apparently healthy children between the ages of 6 months and 15 years yielded the following data on age, X, and liver volume per unit of body weight (ml/kg), Y.

X	Y	X	Y
.5	41	10.0	26
.7	55	10.1	35
2.5	41	10.9	25
4.1	39	11.5	31
5.9	50	12.1	31
6.1	32	14.1	29
7.0	41	15.0	23
8.2	42		

8.8 SOME PRECAUTIONS

Regression and correlation analysis are very powerful statistical tools when properly employed. Their inappropriate use, however, can lead only to meaningless results. To aid in the proper use of these techniques, we make the following suggestions,

1. The assumptions underlying regression and correlation analysis should be carefully reviewed before the data are collected. Although it is rare to find

that assumptions are met to perfection, practitioners should have some idea about the magnitude of the gap that exists between the data to be analyzed and the assumptions of the proposed model, so that they may decide whether they should choose another model; proceed with the analysis, but use caution in the interpretation of the results; or use the chosen model with confidence.

2. In simple linear regression and correlation analysis, the two variables of interest are measured on the same entity, called the unit of association. If we are interested in the relationship between height and weight, for example, these two measurements are taken on the same individual. It usually does not make sense to speak of the correlation, say, between the heights of one group of individuals and the weights of another group.

3. No matter how strong is the indication of a relationship between two variables, it should not be interpreted as one of cause and effect. If, for example, a significant sample correlation coefficient between two variables X and Y is observed, it can mean one of several things:
   (a) X causes Y.
   (b) Y causes X.
   (c)   Some third factor, either directly or indirectly, causes both X and Y.

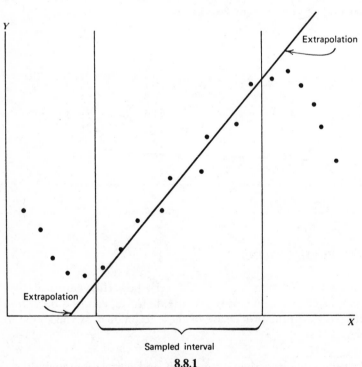

8.8.1

Example of Extrapolation

(d) An unlikely event has occurred and a large sample correlation coefficient has been generated by chance from a population in which X and Y are, in fact, not correlated.

(e) The correlation is purely nonsensical, a situation that may arise when measurements of X and Y are not taken on a common unit of association.

4. The sample regression equation should not be used to predict or estimate outside the range of values of the independent variable represented in the sample. This practice, called *extrapolation*, is risky. The true relationship between two variables, although linear over an interval of the independent variable, sometimes may be described best as a curve outside this interval. If our sample by chance is drawn only from the interval where the relationship is linear, we have only a limited representation of the population, and to project the sample results beyond the interval represented by the sample may lead to false conclusions. Figure 8.8.1 illustrates the possible pitfalls of extrapolation.

8.9   SUMMARY

In this chapter two important tools of statistical analysis, simple linear regression and correlation, are examined. The following outline for the application of these techniques has been suggested.

1. *Identify the Model*   The practitioner must know whether the regression model or the correlation model is the appropriate one for answering his questions.

2. *Review Assumptions*   It has been pointed out several times that the validity of the conclusions depends on how well the data analyzed fit the chosen model.

3. *Obtain the Regression Equation*   We have seen how the regression equation is obtained by the method of least squares. Although the computations, when done by hand, are rather lengthy, involved, and subject to error, this is not the problem today that it has been in the past. Electronic computers are now in such widespread use that the researcher or statistician without access to one is the exception rather than the rule. No apology for lengthy computations is necessary to the researcher who has a computer available.

4. *Evaluate the Equation*   We have seen that the usefulness of the regression equation for estimating and predicting purposes is determined by means of the analysis of variance, which tests the significance of the regression mean square. The strength of the relationship between two variables under the correlation model is assessed by testing the null hypothesis that there is no correlation in the population. If this hypothesis can be rejected we may

conclude, at the chosen level of significance, that the two variables are correlated.

5. *Use the Equation* Once it has been determined that it is likely that the regression equation provides a good description of the relationship between two variables, X and Y, it may be used for one of two purposes:
   (a) to predict what value Y is likely to assume, given a particular value of X, or
   (b) to estimate the mean of the subpopulation of Y values for a particular value of X.

This necessarily abridged treatment of simple linear regression and correlation may have raised more questions than it has answered. It may have occurred to the reader, for example, that a dependent variable can be more precisely predicted using two or more independent variables rather than one. Or, perhaps, he or she may feel that knowledge of the strength of the relationship among several variables might be of more interest than knowledge of the relationship between only two variables. The exploration of these possibilities is the subject of the next chapter, and the reader's curiosity along these lines should be, at least, partially relieved.

For those who would like to pursue the topic further, a number of excellent references, in addition to those already cited, follow at the end of this chapter.

REVIEW QUESTIONS AND EXERCISES

1. What are the assumptions underlying simple linear regression analysis when one of the objectives is to make inferences about the population from which the sample data were drawn?

2. Why is the regression equation called the least-squares equation?

3. Explain the meaning of a in the sample regression equation.

4. Explain the meaning of b in the sample regression equation.

5. Explain the following terms:
   (a) Total sum of squares
   (b) Explained sum of squares
   (c) Unexplained sum of squares

6. Explain the meaning of and the method of computing the coefficient of determination.

7. What is the function of the analysis of variance in regression analysis?

8. Describe two ways in which one may test the null hypothesis that $\beta = 0$.

9. For what two purposes can a regression equation be used?

10. What are the assumptions underlying simple correlation analysis when inference is an objective?

11. What is meant by the unit of association in regression and correlation analysis?

12. What are the possible explanations for a significant sample correlation coefficient?

13. Explain why it is risky to use a sample regression equation to predict or to estimate outside the range of values of the independent variable represented in the sample.

14. Describe a situation in your particular area of interest where simple regression analysis would be useful. Use real or realistic data and do a complete regression analysis.

15. Describe a situation in your particular area of interest where simple correlation analysis would be useful. Use real or realistic data and do a complete correlation analysis.

In each of the following exercises carry out the required analysis and test hypotheses at the indicated significance levels. Compute the p value for each test.

16. The following are the plasma testosterone levels and age at first conviction for violent and aggressive crimes collected on a sample of young male prisoners.

Testosterone level	Age at 1st conviction	Testosterone level	Age at 1st conviction
1305	11	710	17
1000	12	1150	18
1175	13	605	20
1495	14	690	21
1060	15	700	23
800	16	625	24
1005	16	610	27
		450	30

Do these data provide sufficient evidence, at the .05 level of significance, to indicate that the two variables are correlated?

17. Some health researchers have reported an inverse relationship between central nervous system malformations and the hardness of the related water supplies. Suppose data were collected on a sample of 20 geographic areas with the following results.

C.N.S. malformation rate (per 1000 births)	Water hardness (ppm)	C.N.S. malformation rate (per 1000 births)	Water hardness (ppm)
7.2	50	6.3	160
8.1	25	12.5	50
11.2	15	15.0	45
9.3	75	6.5	60
9.4	100	8.0	100
5.0	150	10.0	155
5.8	180	5.3	200
3.3	250	4.9	240
3.6	275	7.2	40
4.8	220	11.9	65

Construct a scatter diagram and compute r. What are your conclusions? ($\alpha = .05$)

18. The mean systolic blood pressure during surgery and amount of blood loss were recorded for 16 patients.

Mean systolic BP during surgery (mm Hg)	Blood loss (ml)	Mean systolic BP during surgery (mm Hg)	Blood loss (ml)
95	274	80	190
90	170	110	288
125	352	90	205
105	317	105	150
110	171	90	175
105	150	110	64
90	245	110	276
90	120	140	318

Prepare a scatter diagram, compute r, and test for significance at the .05 level.

19. In a study of the effect of a dietary component on plasma lipid composition, the following data were obtained on a sample of 15 experimental animals.

Measure of dietary component (X)	Measure of plasma lipid level (Y)
18	38
21	40
28	47
35	54
47	66
33	52
40	59
41	60
·28	47
21	40
30	49
46	65
44	63
38	57
19	38

Obtain the regression equation for these data, compute r^2, and test H_0: $\beta = 0$ by both the F test and the t test.

20. The following are the pulmonary blood flow (PBF) and pulmonary blood volume (PBV) values recorded for 16 infants and children with congenital heart disease.

Y PBV (ml/sqM)	X PBF (L/min/sqM)
168	4.31
280	3.40
391	6.20
420	17.30
303	12.30
429	13.99
605	8.73
522	8.90
224	5.87
291	5.00
233	3.51
370	4.24
531	19.41
516	16.61
211	7.21
439	11.60

Find the regression equation describing the linear relationship between the two variables, compute r^2, and test H_0: $\beta = 0$ by both the F test and the t test. Let $\alpha = .05$.

21. Fifteen specimens of human sera were tested comparatively for tuberculin antibody by two methods. The logarithms of the titers obtained by the two methods were as follows.

METHOD

A(X)	B(Y)
3.31	4.09
2.41	3.84
2.72	3.65
2.41	3.20
2.11	2.97
2.11	3.22
3.01	3.96
2.13	2.76
2.41	3.42
2.10	3.38
2.41	3.28
2.09	2.93
3.00	3.54
2.08	3.14
2.11	2.76

Find the regression equation describing the relationship between the two variables, compute r^2, and test H_0: $\beta = 0$ by both the F test and the t test.

22. The following table shows the methyl mercury intake and whole blood mercury values in 12 subjects exposed to methyl mercury through consumption of contaminated fish.

X Methyl mercury intake ($\mu g\ Hg/day$)	Y Mercury in whole blood (ng/g)
180	90
200	120
230	125
410	290
600	310
550	290
275	170
580	375
105	70
250	105
460	205
650	480

Find the regression equation describing the linear relationship between the two variables, compute r^2, and test H_0: $\beta = 0$ by both the F and t tests. Let $\alpha = .05$. Construct a 95 percent confidence interval for β.

23. The following are the weights (kg) and blood glucose levels (mg/100 ml) of 16 apparently healthy adult males.

Weight (X)	Glucose (Y)
64.0	108
75.3	109
73.0	104
82.1	102
76.2	105
95.7	121
59.4	79
93.4	107
82.1	101
78.9	85
76.7	99
82.1	100
83.9	108
73.0	104
64.4	102
77.6	87

Find the simple linear regression equation and test H_0: $\beta = 0$ using both ANOVA and the t test. Test H_0: $\rho = 0$ and construct a 95 percent confidence interval for ρ. What is the predicted glucose level for a man who weighs 95 kg? Construct the 95 percent prediction interval for his weight. Let $\alpha = .05$ for all tests.

24. The following are the ages (years) and systolic blood pressures of 20 apparently healthy adults.

Age (X)	B.P. (Y)	Age (X)	B.P. (Y)
20	120	46	128
43	128	53	136
63	141	70	146
26	126	20	124
53	134	63	143
31	128	43	130
58	136	26	124
46	132	19	121
58	140	31	126
70	144	23	123

Find the simple linear regression equation and test H_0: $\beta = 0$ using both

ANOVA and the t test. Test H_0: $\rho = 0$, and construct a 95 percent confidence interval for ρ. Find the 95 percent prediction interval for the systolic blood pressure of a person who is 25 years old. Let $\alpha = .05$ for all tests.

25. The following data were collected during an experiment in which laboratory animals were inoculated with a pathogen. The variables are time in hours after inoculation and temperature in degrees Celsius.

Time	Temperature	Time	Temperature
24	38.8	44	41.1
28	39.5	48	41.4
32	40.3	52	41.6
36	40.7	56	41.8
40	41.0	60	41.9

Find the simple linear regression equation and test H_0: $\beta = 0$ using both ANOVA and the t test. Test H_0: $\rho = 0$ and construct a 95 percent confidence interval for ρ. Construct the 95 percent prediction interval for the temperature at 50 hours after inoculation. Let $\alpha = .05$ for all tests.

26. Cerebral edema with consequent increased intracranial pressure frequently accompanies lesions resulting from head injury and other conditions that adversely affect the integrity of the brain. Available treatments for cerebral edema vary in effectiveness and undesirable side effects. One such treatment is glycerol, administered either orally or intravenously. Of interest to clinicians is the relationship between intracranial pressure and glycerol plasma concentration. The following data are available for a population of subjects hospitalized with a diagnosis of cerebral edema who were treated with glycerol. Intracranial pressure (Y) is measured in torr units and plasma glycerol concentration (X) is measured in milligrams per milliliter. Suppose you are a statistical consultant with a research team investigating the relationship between these two variables. Select a simple random sample from the population and perform the analysis that you think would be useful to the researchers. Present your findings and conclusions in narrative form and illustrate with graphs where appropriate. Compare your results with those of your classmates.

Subject	X	Y	Subject	X	Y	Subject	X	Y
1.	1.8	19.1	2.	3.3	7.6	3.	2.5	10.2
4.	2.7	14.7	5.	1.8	15.9	6.	1.9	21.0
7.	2.7	10.2	8.	2.2	15.7	9.	1.8	19.3
10.	2.0	16.1	11.	2.1	12.9	12.	2.4	15.5
13.	2.5	13.9	14.	2.4	14.0	15.	1.8	19.9
16.	2.5	12.0	17.	2.7	9.3	18.	2.3	14.4
19.	1.6	20.7	20.	2.5	17.3	21.	2.3	13.2

Subject	X	Y	Subject	X	Y	Subject	X	Y
22.	2.9	8.6	23.	3.3	3.6	24.	2.4	15.4
25.	1.2	24.2	26.	2.3	13.0	27.	1.8	16.9
28.	2.0	17.3	29.	2.5	15.0	30.	1.5	23.0
31.	2.0	11.6	32.	2.4	5.9	33.	2.4	14.6
34.	2.0	18.5	35.	1.8	15.6	36.	1.8	16.8
37.	0.7	28.7	38.	2.6	10.9	39.	2.8	11.3
40.	2.1	15.3	41.	2.8	4.3	42.	1.5	22.7
43.	2.3	13.5	44.	2.1	18.4	45.	2.5	9.6
46.	2.3	7.7	47.	2.6	12.8	48.	2.8	13.7
49.	2.2	17.2	50.	2.4	16.8	51.	2.0	17.0
52.	2.2	19.0	53.	2.0	21.0	54.	1.8	18.5
55.	1.7	20.9	56.	2.2	16.2	57.	2.4	14.4
58.	1.4	21.3	59.	3.1	9.3	60.	2.7	11.4
61.	2.4	11.2	62.	1.5	21.2	63.	2.8	13.0
64.	2.2	16.9	65.	2.1	17.3	66.	1.5	22.6
67.	2.1	14.6	68.	2.0	18.2	69.	1.9	16.2
70.	2.1	14.4	71.	1.5	18.9	72.	1.3	22.3
73.	1.3	24.1	74.	0.8	29.7	75.	1.8	20.2
76.	2.2	15.5	77.	2.0	12.7	78.	1.5	18.2
79.	1.5	21.8	80.	2.4	11.9	81.	1.4	23.5
82.	1.5	16.9	83.	1.7	18.4	84.	2.2	17.7
85.	2.0	18.9	86.	1.4	17.2	87.	2.3	10.1
88.	2.3	12.9	89.	1.4	22.8	90.	2.5	13.0
91.	2.1	15.7	92.	2.0	20.7	93.	1.6	18.5
94.	2.2	13.3	95.	1.4	22.1	96.	2.0	12.0
97.	2.2	15.0	98.	2.4	13.0	99.	1.4	23.0
100.	1.9	17.9	101.	1.7	22.4	102.	2.2	13.4
103.	1.8	15.4	104.	2.1	13.1	105.	2.0	13.8
106.	2.4	13.3	107.	2.6	7.9	108.	2.2	14.6
109.	2.0	16.7	110.	1.6	14.9	111.	3.1	5.0
112.	3.0	9.9	113.	2.3	10.5	114.	2.7	10.1
115.	1.6	15.8	116.	2.8	7.7	117.	2.5	14.5
118.	1.2	29.0	119.	1.9	16.3	120.	2.0	18.6
121.	2.5	14.2	122.	2.6	11.4	123.	1.9	17.3
124.	2.3	10.8	125.	2.7	8.2	126.	2.0	14.8
127.	2.9	10.9	128.	1.5	19.3	129.	1.7	15.1
130.	2.1	15.5	131.	2.3	14.8	132.	1.6	17.3
133.	2.2	15.5	134.	3.1	7.6	135.	1.8	19.7
136.	1.7	18.8	137.	2.1	17.2	138.	2.2	10.3
139.	1.9	16.1	140.	1.8	17.3	141.	1.9	17.8
142.	2.0	19.0	143.	2.2	11.8	144.	2.5	13.4
145.	2.0	18.3	146.	1.8	19.0	147.	2.0	12.7
148.	2.7	10.1	149.	2.4	9.1	150.	2.3	15.7
151.	1.9	17.6	152.	2.5	13.4	153.	2.1	20.6
154.	3.0	9.8	155.	2.3	10.9	156.	1.6	17.6
157.	2.2	9.8	158.	1.9	15.5	159.	1.3	21.9
160.	1.9	16.3	161.	2.5	9.1	162.	3.0	5.6
163.	2.8	12.7	164.	2.4	14.8	165.	2.1	19.9

Subject	X	Y	Subject	X	Y	Subject	X	Y
166.	2.3	14.8	167.	2.4	12.9	168.	2.1	10.1
169.	2.5	11.5	170.	1.7	18.1	171.	2.5	15.1
172.	2.7	11.6	173.	1.3	23.6	174.	2.8	12.1
175.	2.6	13.5	176.	1.4	21.0	177.	2.5	13.7
178.	2.6	11.7	179.	2.8	8.2	180.	1.2	24.0
181.	2.3	14.3	182.	1.3	21.4	183.	3.5	7.9
184.	2.3	7.9	185.	2.2	14.5	186.	2.2	12.7
187.	1.8	15.9	188.	2.6	7.7	189.	1.6	19.1
190.	1.4	24.0	191.	2.2	13.6	192.	2.1	13.0
193.	1.7	16.6	194.	2.1	14.7	195.	2.3	12.1
196.	1.6	22.8	197.	1.7	17.9	198.	2.2	9.5
199.	3.2	8.1	200.	2.5	16.8	201.	2.3	9.4
202.	2.4	12.7	203.	2.3	13.6	204.	2.5	10.2
205.	1.9	11.9	206.	2.2	17.0	207.	2.0	13.5
208.	2.3	12.6	209.	1.1	21.1	210.	1.9	16.2
211.	2.4	16.4	212.	1.8	19.4	213.	2.4	12.8
214.	2.5	11.4	215.	2.2	14.8	216.	1.6	25.1
217.	1.7	22.0	218.	2.2	12.0	219.	2.6	8.6
220.	2.4	15.7	221.	2.3	7.5	222.	2.1	14.6
223.	2.3	12.4	224.	2.1	17.3	225.	2.9	7.1
226.	2.8	9.4	227.	2.3	12.8	228.	2.0	17.9
229.	2.7	9.8	230.	1.1	26.3	231.	2.2	13.8
232.	2.1	15.0	233.	2.5	8.4	234.	2.1	15.8
235.	1.6	18.7	236.	2.3	15.5	237.	2.6	10.8
238.	2.6	11.2	239.	1.6	16.1	240.	2.6	10.0
241.	2.1	11.8	242.	2.0	17.5	243.	1.7	16.2
244.	1.4	23.9	245.	2.5	12.3	246.	2.3	13.0
247.	1.4	20.7	248.	2.3	14.6	249.	2.2	14.0
250.	2.2	12.6	251.	2.3	15.2	252.	3.2	4.2
253.	1.6	19.2	254.	1.9	18.7	255.	3.2	6.5
256.	2.7	9.6	257.	1.5	20.2	258.	2.5	13.8
259.	1.9	21.0	260.	2.1	15.1	261.	3.2	9.2
262.	2.3	13.4	263.	2.1	14.5	264.	2.5	10.5
265.	2.1	13.8	266.	2.6	6.2	267.	2.6	11.7
268.	1.8	15.0	269.	2.1	7.4	270.	1.5	19.7
271.	1.4	23.0	272.	1.7	20.2	273.	2.1	18.1
274.	2.0	20.9	275.	2.9	3.3	276.	1.7	17.3
277.	2.1	16.9	278.	2.2	20.5	279.	1.8	21.5
280.	1.4	19.8	281.	2.1	13.9	282.	2.6	13.7
283.	2.1	15.5	284.	2.7	13.2	285.	2.4	18.2
286.	1.7	22.3	287.	2.0	22.1	288.	2.5	10.2
289.	2.7	10.1	290.	1.5	25.0	291.	1.3	21.3
292.	1.5	24.6	293.	1.8	13.4	294.	2.4	15.7
295.	2.5	9.9	296.	2.3	8.2	297.	2.7	6.2
298.	2.4	14.1	299.	2.1	12.8	300.	2.8	8.9
301.	1.4	26.0	302.	2.3	17.7	303.	2.5	10.5
304.	2.8	11.8	305.	1.9	17.1	306.	2.4	13.3
307.	1.3	20.7	308.	1.6	22.9	309.	1.8	21.5

Subject	X	Y	Subject	X	Y	Subject	X	Y
310.	2.4	14.8	311.	2.3	7.2	312.	2.9	8.9
313.	2.2	11.6	314.	2.1	16.1	315.	2.5	15.8
316.	1.8	19.5	317.	2.3	13.8	318.	1.9	22.6
319.	1.5	16.6	320.	2.4	16.3	321.	2.9	10.1
322.	2.0	11.0	323.	1.5	16.3	324.	2.6	11.5
325.	1.5	16.9	326.	1.8	16.3	327.	1.5	19.9
328.	2.5	12.7	329.	1.7	21.1	330.	2.5	16.6
331.	1.9	19.4	332.	1.8	15.3	333.	2.5	14.0
334.	1.8	16.5	335.	0.9	24.6	336.	1.6	16.7
337.	2.1	20.7	338.	2.3	15.4	339.	2.2	12.8
340.	1.2	21.7	341.	2.0	18.3	342.	3.7	5.7
343.	2.0	19.9	344.	1.6	20.3	345.	2.4	10.5
346.	1.8	9.8	347.	1.8	18.3	348.	1.8	15.0
349.	2.2	17.9	350.	1.7	19.2	351.	2.2	14.1
352.	2.3	16.0	353.	3.4	0.3	354.	2.6	11.5
355.	2.2	12.0	356.	2.1	16.2	357.	2.6	11.7
358.	2.7	12.2	359.	2.3	16.8	360.	1.8	16.1
361.	2.2	12.2	362.	1.8	17.6	363.	2.5	10.4
364.	2.5	10.6	365.	1.9	15.3	366.	2.7	10.8
367.	2.1	17.2	368.	1.6	15.9	369.	2.1	18.3
370.	2.1	16.7	371.	2.9	7.6	372.	2.2	12.8
373.	1.0	23.8	374.	0.9	28.1	375.	2.0	15.1
376.	1.3	21.2	377.	2.3	11.8	378.	1.5	20.7
379.	2.2	12.7	380.	1.4	22.4	381.	1.7	19.2
382.	2.3	13.3	383.	2.7	9.7	384.	1.4	19.3
385.	2.4	13.7	386.	2.4	12.4	387.	2.0	16.9
388.	2.4	9.2	389.	2.0	17.4	390.	2.7	8.9
391.	2.9	3.1	392.	2.6	9.1	393.	1.4	21.6
394.	2.0	16.4	395.	1.4	22.9	396.	1.5	22.9
397.	1.7	17.4	398.	2.2	9.8	399.	1.7	14.0
400.	2.0	14.7	401.	2.0	17.2	402.	2.7	9.3
403.	2.6	12.1	404.	2.0	23.3	405.	2.3	10.7
406.	2.1	17.1	407.	1.5	21.5	408.	2.4	13.2
409.	2.9	4.8	410.	1.8	18.6	411.	2.3	14.6
412.	1.9	14.2	413.	2.5	9.1	414.	2.7	9.6
415.	1.8	21.4	416.	2.3	15.2	417.	1.2	20.5
418.	2.0	13.5	419.	2.1	17.6	420.	2.1	15.7
421.	2.4	11.2	422.	1.5	19.1	423.	2.4	9.3
424.	1.9	18.3	425.	2.5	11.8	426.	2.7	6.6
427.	2.4	16.1	428.	1.4	15.7	429.	2.0	15.3
430.	2.8	9.6	431.	3.1	1.7	432.	2.3	14.3
433.	2.2	12.9	434.	2.2	11.9	435.	1.8	19.4
436.	2.2	13.8	437.	2.7	7.7	438.	2.1	13.7
439.	2.5	10.9	440.	2.2	11.3	441.	2.1	16.3
442.	2.0	14.3	443.	1.2	24.8	444.	1.8	17.4
445.	2.0	14.5	446.	2.9	7.9	447.	3.2	6.0
448.	1.9	13.8	449.	2.4	8.9	450.	2.2	12.8
451.	2.5	15.1	452.	2.6	10.7	453.	1.9	12.8

Subject	X	Y	Subject	X	Y	Subject	X	Y
454.	1.9	14.3	455.	2.4	15.2	456.	2.2	12.0
457.	2.6	18.3	458.	2.3	7.5	459.	2.6	10.3
460.	2.3	13.9	461.	2.0	17.0	462.	2.9	9.0
463.	2.2	12.5	464.	1.9	18.5	465.	2.5	9.0
466.	1.2	23.7	467.	2.9	5.6	468.	1.6	24.7
469.	2.2	19.2	470.	2.1	15.9	471.	1.8	22.0
472.	2.2	14.2	473.	1.6	21.7	474.	2.4	13.6
475.	2.3	15.3	476.	1.2	23.7	477.	2.0	12.6
478.	1.1	28.0	479.	2.6	9.3	480.	2.6	13.2
481.	1.6	21.2	482.	2.1	18.6	483.	3.6	2.1
484.	2.4	10.2	485.	2.1	15.4	486.	2.8	11.4
487.	2.3	15.7	488.	2.8	7.6	489.	1.8	20.4
490.	2.2	14.2	491.	1.4	24.2	492.	1.7	18.0
493.	2.4	16.7	494.	2.9	9.0	495.	1.7	16.1
496.	1.9	15.2	497.	1.4	24.3	498.	1.3	19.6
499.	1.7	20.2	500.	1.8	17.0	501.	2.4	11.3
502.	2.2	17.6	503.	2.4	11.6	504.	2.1	15.5
505.	2.5	16.7	506.	2.3	17.2	507.	1.9	18.2
508.	2.5	13.9	509.	1.5	14.6	510.	3.2	4.1
511.	2.3	11.9	512.	2.0	16.0	513.	1.6	18.2
514.	1.2	23.1	515.	2.5	10.3	516.	1.6	17.1
517.	1.8	16.0	518.	2.0	22.7	519.	2.3	17.6
520.	2.0	16.1	521.	2.3	18.5	522.	2.0	20.9
523.	2.0	17.2	524.	1.4	20.6	525.	3.0	8.8
526.	2.6	9.5	527.	2.2	16.3	528.	2.5	15.4
529.	2.6	15.1	530.	2.0	15.2	531.	1.5	26.3
532.	1.5	23.3	533.	2.5	10.6	534.	2.0	16.5
535.	2.3	8.5	536.	2.4	10.1	537.	2.1	14.9
538.	2.6	8.7	539.	2.7	9.4	540.	2.3	16.0
541.	2.2	16.2	542.	2.0	15.2	543.	2.5	14.4
544.	2.9	9.7	545.	1.9	18.1	546.	2.0	12.9
547.	2.2	16.1	548.	2.7	10.7	549.	1.5	21.2
550.	2.1	16.3	551.	1.8	19.1	552.	2.8	13.9
553.	2.1	13.3	554.	2.4	18.5	555.	1.7	23.4
556.	2.0	12.6	557.	2.1	19.1	558.	2.2	13.4
559.	2.2	14.2	560.	1.5	24.0	561.	2.2	17.6
562.	2.9	10.1	563.	2.2	17.2	564.	1.9	15.4
565.	2.4	12.1	566.	2.2	14.4	567.	1.8	19.1
568.	2.3	15.3	569.	3.2	4.7	570.	2.0	19.9
571.	1.4	21.5	572.	2.1	17.0	573.	2.4	13.8
574.	3.0	2.1	575.	2.1	14.8	576.	2.2	17.1
577.	2.3	14.7	578.	3.0	8.4	579.	2.6	8.8
580.	2.6	9.7	581.	2.0	13.4	582.	2.5	17.0
583.	2.2	19.4	584.	1.9	15.3	585.	2.4	14.0
586.	2.6	9.0	587.	3.0	6.6	588.	2.5	10.4
589.	2.5	7.6	590.	2.3	12.5	591.	2.9	10.9
592.	2.3	17.8	593.	2.8	4.8	594.	1.6	18.1
595.	2.1	13.5	596.	2.3	12.6	597.	1.7	18.7

Subject	X	Y	Subject	X	Y	Subject	X	Y
598.	2.5	13.8	599.	1.9	18.4	600.	3.1	6.0
601.	2.1	18.2	602.	1.3	22.3	603.	1.1	19.1
604.	1.6	18.8	605.	2.2	11.8	606.	2.6	14.6
607.	2.0	14.3	608.	2.3	19.1	609.	2.1	15.3
610.	1.4	22.4	611.	2.5	9.4	612.	1.2	23.6
613.	2.3	14.5	614.	2.1	15.8	615.	1.9	14.5
616.	2.3	11.5	617.	3.1	9.3	618.	2.6	9.1
619.	1.8	22.7	620.	2.1	13.8	621.	1.3	18.6
622.	1.6	21.5	623.	1.3	22.2	624.	1.7	18.1
625.	1.2	24.0	626.	3.4	4.7	627.	3.4	0.6
628.	2.0	14.6	629.	2.6	12.8	630.	2.2	12.3
631.	2.6	9.9	632.	2.2	13.5	633.	2.5	15.9
634.	2.2	13.1	635.	1.9	19.1	636.	2.0	19.0
637.	3.0	8.3	638.	2.5	11.5	639.	2.2	15.1
640.	2.0	10.7	641.	1.4	17.8	642.	2.2	16.1
643.	2.6	9.1	644.	1.7	15.0	645.	2.5	12.4
646.	1.9	17.8	647.	1.7	18.5	648.	2.5	10.6
649.	2.6	11.0	650.	1.6	18.9	651.	1.5	16.3
652.	1.8	20.7	653.	2.9	5.9	654.	1.3	23.2
655.	3.1	3.7	656.	1.6	20.2	657.	2.3	12.4
658.	2.1	18.1	659.	2.1	13.7	660.	2.4	7.7
661.	3.1	5.3	662.	2.0	9.5	663.	2.0	11.5
664.	1.3	21.7	665.	1.8	18.1	666.	2.0	20.4
667.	1.9	16.7	668.	2.0	13.0	669.	2.3	13.0
670.	2.7	15.0	671.	1.7	15.4	672.	2.4	10.2
673.	2.1	18.6	674.	1.8	14.1	675.	1.3	19.2
676.	2.3	14.8	677.	2.0	15.5	678.	1.9	19.6
679.	2.9	9.1	680.	2.8	4.9	681.	1.9	19.2
682.	2.1	18.9	683.	2.0	17.2	684.	1.6	19.1
685.	2.1	17.7	686.	1.9	18.3	687.	1.3	21.9
688.	2.4	15.4	689.	2.5	11.5	690.	2.7	14.0
691.	2.7	8.3	692.	2.5	7.8	693.	2.9	7.3
694.	1.0	26.7	695.	2.7	13.1	696.	2.2	14.6
697.	2.4	15.4	698.	1.2	23.5	699.	1.6	22.1
700.	2.4	11.4	701.	2.3	10.2	702.	2.0	15.3
703.	2.5	12.3	704.	2.5	13.0	705.	1.1	24.7
706.	2.2	16.1	707.	1.8	12.1	708.	2.2	19.3
709.	1.8	15.7	710.	2.4	15.2	711.	2.1	18.1
712.	2.5	11.5	713.	1.8	15.2	714.	1.9	17.1
715.	2.0	18.3	716.	2.5	15.5	717.	2.2	10.3
718.	1.9	19.3	719.	1.4	19.9	720.	1.6	19.6
721.	2.0	18.8	722.	2.1	17.9	723.	2.5	16.3
724.	2.4	17.9	725.	2.0	16.7	726.	2.8	7.5
727.	2.2	12.4	728.	3.5	3.4	729.	2.4	7.4
730.	1.8	12.9	731.	2.4	8.7	732.	2.5	12.7
733.	2.2	14.4	734.	1.6	20.1	735.	1.9	19.5
736.	2.5	9.6	737.	2.1	16.4	738.	2.1	17.7
739.	2.1	13.2	740.	2.5	16.5	741.	2.7	11.0

Subject	X	Y	Subject	X	Y	Subject	X	Y
742.	2.7	7.5	743.	2.0	17.3	744.	1.0	21.2
745.	1.7	19.4	746.	1.9	11.8	747.	1.3	21.7
748.	2.5	9.9	749.	2.4	15.2	750.	3.0	10.2
751.	2.2	11.8	752.	2.6	10.4	753.	2.5	17.1
754.	1.8	16.5	755.	1.7	17.7	756.	2.2	15.0
757.	2.8	12.2	758.	2.8	7.8	759.	2.5	8.1
760.	2.4	17.1	761.	2.0	16.5	762.	2.5	13.7
763.	2.4	13.0	764.	2.2	16.8	765.	1.1	26.5
766.	3.3	9.2	767.	1.7	17.5	768.	2.9	4.5
769.	1.7	17.7	770.	2.0	15.0	771.	2.0	16.9
772.	1.9	22.5	773.	1.2	20.9	774.	3.0	3.8
775.	2.2	14.8	776.	2.7	7.1	777.	3.7	1.5
778.	2.5	16.4	779.	1.8	19.5	780.	2.2	18.9
781.	1.2	22.5	782.	1.3	20.5	783.	2.1	17.6
784.	1.7	17.7	785.	2.0	14.5	786.	2.3	16.4
787.	1.9	19.1	788.	2.1	18.6	789.	2.4	12.7
790.	2.5	13.5	791.	2.9	8.5	792.	1.1	19.2
793.	2.9	8.9	794.	1.9	13.2	795.	2.8	10.7
796.	2.7	11.9	797.	1.7	21.2	798.	1.7	18.6
799.	2.1	17.7	800.	2.2	14.5	801.	1.7	16.1
802.	2.0	17.6	803.	3.0	10.4	804.	2.2	16.4
805.	2.8	7.2	806.	1.7	20.6	807.	1.7	18.3
808.	2.2	14.1	809.	1.4	18.7	810.	2.5	14.3
811.	2.2	15.3	812.	1.9	16.4	813.	1.3	24.8
814.	2.4	7.3	815.	2.6	10.6	816.	1.8	18.4
817.	1.9	19.2	818.	2.6	12.3	819.	3.3	3.0
820.	1.5	18.1	821.	2.8	9.0	822.	2.7	6.3
823.	2.7	10.2	824.	1.7	17.8	825.	1.9	20.1
826.	2.1	6.9	827.	2.4	11.0	828.	1.9	18.1
829.	1.9	8.4	830.	2.4	9.7	831.	3.0	6.7
832.	1.6	23.7	833.	2.5	7.6	834.	3.0	3.2
835.	1.8	16.4	836.	2.1	17.2	837.	2.2	17.9
838.	2.0	14.7	839.	2.2	15.9	840.	2.1	14.4
841.	2.0	19.2	842.	2.7	8.7	843.	1.7	17.1
844.	1.9	16.4	845.	2.4	16.6	846.	1.6	23.3
847.	3.0	5.8	848.	2.4	16.1	849.	0.9	21.1
850.	3.0	8.6	851.	1.8	17.4	852.	1.5	25.8
853.	1.5	17.3	854.	1.6	19.7	855.	2.1	15.3
856.	1.4	28.0	857.	2.4	15.7	858.	2.2	13.4
859.	2.2	12.0	860.	2.6	9.0	861.	1.4	19.5
862.	2.4	12.0	863.	1.9	17.6	864.	1.7	20.4
865.	1.8	19.2	866.	1.7	18.2	867.	1.7	15.6
868.	2.1	18.5	869.	2.2	12.8	870.	1.8	18.9
871.	0.9	22.0	872.	2.8	11.2	873.	2.9	14.7
874.	3.2	12.2	875.	1.7	21.5	876.	1.9	17.9
877.	3.0	4.9	878.	2.4	11.6	879.	2.3	12.3
880.	2.3	13.6	881.	1.0	25.9	882.	2.0	15.1
883.	2.3	16.7	884.	2.4	16.6	885.	2.5	11.5

Subject	X	Y	Subject	X	Y	Subject	X	Y
886.	1.7	17.8	887.	2.8	8.9	888.	1.7	13.4
889.	2.7	13.1	890.	1.5	23.3	891.	1.6	21.7
892.	2.4	16.3	893.	1.9	17.5	894.	2.3	9.0
985.	1.4	21.7	896.	2.0	19.5	897.	1.6	21.4
898.	1.8	17.6	899.	2.0	15.8	900.	2.5	9.8
901.	2.7	6.7	902.	2.6	6.4	903.	1.5	20.4
904.	2.1	14.5	905.	3.1	6.5	906.	1.7	20.3
907.	2.5	11.1	908.	1.8	14.5	909.	2.5	14.5
910.	2.7	12.3	911.	2.2	13.3	912.	1.5	21.5
913.	2.0	13.3	914.	3.0	7.7	915.	1.6	17.4
916.	2.6	11.2	917.	2.6	10.7	918.	2.3	12.9
919.	2.9	7.0	920.	1.4	23.0	921.	2.6	12.5
922.	1.8	20.4	923.	1.0	29.1	924.	1.5	21.6
925.	2.4	11.5	926.	2.8	12.3	927.	1.4	19.0
928.	2.8	10.7	929.	2.7	7.2	930.	1.5	23.9
931.	1.6	16.3	932.	1.6	18.0	933.	1.7	17.5
934.	2.3	16.2	935.	2.2	14.5	936.	2.1	15.4
937.	1.9	17.0	938.	2.0	12.9	939.	2.3	15.4
940.	2.1	19.6	941.	2.6	7.8	942.	2.5	12.5
943.	1.6	20.7	944.	2.9	14.3	945.	1.5	19.0
946.	3.2	6.6	947.	2.0	20.4	948.	1.7	18.8
949.	2.7	9.5	950.	1.9	14.1	951.	1.9	15.7
952.	2.4	9.8	953.	1.0	24.3	954.	2.9	8.3
955.	1.8	18.2	956.	1.7	17.0	957.	1.8	14.6
958.	1.8	20.2	959.	2.4	18.8	960.	2.0	15.8
961.	2.0	16.1	962.	2.0	13.7	963.	2.1	14.7
964.	1.5	23.3	965.	2.4	12.0	966.	1.8	17.0
967.	1.7	18.6	968.	2.1	12.7	969.	2.4	8.8
970.	2.2	13.1	971.	2.7	9.7	972.	2.1	17.0
973.	2.6	8.8	974.	1.9	19.4	975.	1.7	17.2
976.	1.9	18.3	977.	3.1	10.3	978.	2.9	7.1
979.	2.2	16.8	980.	2.6	9.6	981.	1.7	15.7
982.	2.3	15.1	983.	1.8	13.7	984.	2.0	18.5
985.	2.1	11.3	986.	2.2	14.9	987.	2.5	16.9
988.	1.9	20.2	989.	2.6	9.4	990.	1.9	19.3
991.	2.0	16.1	992.	1.8	17.6	993.	2.4	12.0
994.	2.2	14.3	995.	2.6	9.2	996.	1.9	18.7
997.	2.8	12.7	998.	2.4	12.9	999.	2.3	15.4
1000.	2.1	16.8	1001.	2.1	18.3	1002.	1.5	16.2
1003.	1.4	19.1	1004.	2.9	12.6	1005.	3.1	2.9
1006.	1.7	18.1	1007.	2.2	15.9	1008.	2.7	7.5
1009.	2.7	6.6	1010.	1.6	20.0	1011.	1.8	23.1
1012.	2.5	13.9	1013.	2.1	18.0	1014.	1.7	25.5
1015.	2.0	18.1	1016.	1.8	18.7	1017.	2.0	15.2
1018.	1.8	15.6	1019.	1.8	19.8	1020.	1.5	18.5
1021.	2.3	11.1	1022.	2.2	15.5	1023.	2.0	16.7
1024.	2.8	11.6	1025.	1.7	20.5	1026.	2.7	9.5
1027.	1.9	19.3	1028.	1.9	19.2	1029.	1.5	21.2

Subject	X	Y	Subject	X	Y	Subject	X	Y
1030.	2.2	17.9	1031.	2.1	16.8	1032.	2.2	11.1
1033.	2.2	12.3	1034.	3.1	5.0	1035.	1.9	18.6
1036.	2.6	13.6	1037.	2.6	17.3	1038.	2.2	12.7
1039.	2.2	20.5	1040.	1.7	18.2	1041.	2.4	12.7
1042.	1.7	20.7	1043.	2.1	14.5	1044.	1.6	18.7
1045.	1.7	22.5	1046.	1.7	14.4	1047.	3.0	8.6
1048.	1.9	16.0	1049.	3.2	3.6	1050.	1.7	19.9

27. The following data are available on a population of patients with essential hypertension. The dependent variable Y is systolic blood pressure (mm Hg) and the independent variable X is a measure of body sodium. Suppose you are a statistical consultant to a medical research team interested in essential hypertension. Select a simple random sample from the population and perform the analyses that you think would be useful to the researchers. Present your findings and conclusions in narrative form and illustrate with graphs where appropriate. Compare your results with those of your class-mates. Consult with your instructor regarding the size of sample you should select.

Subject	X	Y	Subject	X	Y	Subject	X	Y
1.	98	147	2.	99	149	3.	96	175
4.	109	145	5.	91	135	6.	107	149
7.	87	121	8.	110	170	9.	102	163
10.	103	141	11.	117	149	12.	92	135
13.	90	127	14.	93	132	15.	113	181
16.	99	152	17.	114	164	18.	103	144
19.	96	148	20.	111	180	21.	128	183
22.	92	132	23.	84	135	24.	102	141
25.	103	147	26.	117	167	27.	89	129
28.	107	168	29.	104	137	30.	84	120
31.	89	138	32.	102	160	33.	93	131
34.	95	144	35.	104	162	36.	90	138
37.	106	173	38.	101	174	39.	110	152
40.	96	157	41.	98	151	42.	108	153
43.	90	143	44.	94	129	45.	117	170
46.	92	133	47.	110	177	48.	98	138
49.	98	147	50.	102	162	51.	87	131
52.	105	167	53.	92	140	54.	114	179
55.	102	148	56.	97	158	57.	115	161
58.	97	150	59.	99	162	60.	100	128
61.	101	145	62.	96	140	63.	107	167
64.	84	120	65.	95	120	66.	95	144

Subject	X	Y	Subject	X	Y	Subject	X	Y
67.	96	132	68.	101	166	69.	110	170
70.	120	170	71.	99	144	72.	104	137
73.	93	149	74.	116	174	75.	93	131
76.	88	146	77.	94	143	78.	99	146
79.	86	120	80.	93	145	81.	96	141
82.	113	185	83.	85	132	84.	87	127
85.	117	168	86.	107	152	87.	106	151
88.	107	164	89.	94	130	90.	96	170
91.	103	161	92.	104	178	93.	94	132
94.	105	158	95.	120	190	96.	118	163
97.	87	137	98.	101	161	99.	92	154
100.	110	156	101.	85	128	102.	94	141
103.	100	146	104.	104	164	105.	99	138
106.	96	156	107.	89	137	108.	96	138
109.	116	178	110.	102	145	111.	93	129
112.	106	162	113.	104	153	114.	107	159
115.	103	167	116.	109	156	117.	106	155
118.	88	127	119.	109	180	120.	108	188
121.	102	150	122.	101	160	123.	93	126
124.	100	143	125.	103	153	126.	88	133
127.	90	139	128.	93	135	129.	91	133
130.	100	121	131.	105	179	132.	118	168
133.	87	138	134.	96	157	135.	101	168
136.	103	156	137.	114	164	138.	101	136
139.	107	173	140.	95	130	141.	90	125
142.	97	141	143.	112	173	144.	98	145
145.	97	146	146.	113	178	147.	101	150
148.	107	154	149.	88	127	150.	84	139
151.	103	148	152.	94	139	153.	85	130
154.	107	153	155.	107	135	156.	85	120
157.	99	138	158.	101	148	159.	102	140
160.	99	139	161.	87	153	162.	110	168
163.	110	169	164.	102	159	165.	93	153
166.	113	151	167.	102	170	168.	108	149
169.	86	120	170.	102	165	171.	124	187
172.	96	155	173.	108	160	174.	80	125
175.	96	147	176.	94	153	177.	99	155
178.	118	174	179.	119	173	180.	106	161
181.	96	147	182.	100	152	183.	102	150
184.	112	171	185.	100	155	186.	121	165
187.	87	137	188.	110	180	189.	104	144
190.	95	134	191.	95	147	192.	96	138
193.	80	124	194.	103	144	195.	106	175
196.	105	149	197.	111	144	198.	95	135
199.	104	152	200.	80	130	201.	95	147
202.	80	125	203.	103	165	204.	104	159
205.	93	142	206.	107	151	207.	103	161
208.	96	141	209.	92	154	210.	108	160

Subject	X	Y	Subject	X	Y	Subject	X	Y
211.	108	172	212.	80	121	213.	103	170
214.	104	145	215.	105	141	216.	111	173
217.	100	136	218.	111	164	219.	102	160
220.	96	138	221.	87	129	222.	112	164
223.	105	163	224.	87	124	225.	100	148
226.	112	178	227.	114	168	228.	98	145
229.	105	138	230.	106	150	231.	106	153
232.	108	161	233.	103	153	234.	96	142
235.	106	163	236.	84	120	237.	95	146
238.	102	152	239.	99	158	240.	92	145
241.	111	158	242.	96	154	243.	96	139
244.	100	145	245.	102	130	246.	103	152
247.	90	133	248.	100	162	249.	114	158
250.	115	162	251.	106	181	252.	80	124
253.	95	144	254.	107	170	255.	85	124
256.	93	129	257.	80	138	258.	99	151
259.	107	151	260.	115	168	261.	102	191
262.	99	152	263.	112	158	264.	93	152
265.	117	170	266.	102	130	267.	120	159
268.	104	163	269.	104	151	270.	87	148
271.	96	129	272.	104	153	273.	118	183
274.	94	130	275.	98	167	276.	105	154
277.	101	155	278.	110	158	279.	99	150
280.	104	158	281.	120	179	282.	92	131
283.	103	144	284.	108	156	285.	101	131
286.	101	145	287.	97	156	288.	117	156
289.	94	138	290.	91	121	291.	100	144
292.	104	156	293.	112	142	294.	111	195
295.	92	161	296.	99	149	297.	113	165
298.	91	147	299.	103	161	300.	106	154
301.	102	149	302.	107	176	303.	100	120
304.	120	196	305.	100	149	306.	96	123
307.	106	175	308.	99	161	309.	107	160
310.	104	139	311.	96	148	312.	89	163
313.	105	161	314.	108	159	315.	84	135
316.	99	139	317.	80	120	318.	89	141
319.	108	161	320.	89	154	321.	106	162
322.	90	134	323.	106	167	324.	102	153
325.	97	152	326.	90	147	327.	84	147
328.	91	149	329.	96	155	330.	108	158
331.	111	163	332.	80	121	333.	101	158
334.	94	137	335.	84	124	336.	92	120
337.	101	139	338.	104	153	339.	99	130
340.	120	179	341.	105	172	342.	88	128
343.	91	148	344.	121	185	345.	85	127
346.	89	121	347.	80	133	348.	117	171
349.	110	156	350.	95	139	351.	109	163
352.	117	197	353.	91	127	354.	103	163

Subject	X	Y	Subject	X	Y	Subject	X	Y
355.	102	171	356.	110	164	357.	108	176
358.	105	162	359.	90	132	360.	80	120
361.	105	164	362.	107	166	363.	89	142
364.	100	178	365.	113	168	366.	109	144
367.	98	164	368.	112	175	369.	110	154
370.	110	159	371.	103	137	372.	89	121
373.	109	155	374.	102	176	375.	129	187
376.	94	133	377.	93	147	378.	77	123
379.	98	156	380.	115	166	381.	90	152
382.	108	138	383.	90	161	384.	95	149
385.	88	122	386.	96	162	387.	90	134
388.	96	146	389.	101	139	390.	84	120
391.	103	142	392.	105	155	393.	102	152
394.	99	155	395.	122	180	396.	99	142
397.	100	154	398.	98	142	399.	120	169
400.	118	183	401.	97	147	402.	100	159
403.	94	138	404.	89	126	405.	96	165
406.	95	162	407.	117	191	408.	103	163
409.	95	138	410.	95	158	411.	98	151
412.	110	174	413.	120	170	414.	95	160
415.	107	155	416.	101	152	417.	81	121
418.	86	129	419.	95	150	420.	94	123
421.	107	161	422.	97	163	423.	105	160
424.	93	128	425.	120	180	426.	91	137
427.	109	165	428.	112	161	429.	103	134
430.	94	128	431.	102	152	432.	111	169
433.	80	123	434.	102	156	435.	117	178
436.	113	159	437.	93	146	438.	101	139
439.	95	160	440.	100	147	441.	90	143
442.	102	153	443.	108	167	444.	94	138
445.	87	131	446.	99	156	447.	121	179
448.	84	127	449.	107	158	450.	99	154
451.	93	153	452.	95	133	453.	96	140
454.	96	144	455.	103	147	456.	86	145
457.	106	150	458.	104	161	459.	92	144
460.	113	174	461.	92	138	462.	99	151
463.	101	174	464.	94	120	465.	98	144
466.	100	145	467.	119	171	468.	101	166
469.	102	165	470.	101	150	471.	93	145
472.	100	168	473.	100	137	474.	97	157
475.	92	135	476.	100	152	477.	98	163
478.	109	172	479.	85	141	480.	110	167
481.	99	142	482.	91	140	483.	103	155
484.	96	150	485.	87	145	486.	81	143
487.	120	200	488.	95	144	489.	91	137
490.	105	157	491.	107	138	492.	86	143
493.	81	129	494.	113	187	495.	99	163
496.	100	158	497.	83	126	498.	97	145

Subject	X	Y	Subject	X	Y	Subject	X	Y
499.	108	146	500.	113	197	501.	107	150
502.	92	128	503.	100	137	504.	105	158
505.	104	163	506.	90	156	507.	97	144
508.	98	134	509.	113	186	510.	87	126
511.	107	167	512.	104	171	513.	89	143
514.	90	126	515.	98	146	516.	113	156
517.	96	142	518.	96	154	519.	82	123
520.	89	147	521.	103	157	522.	118	173
523.	102	162	524.	100	152	525.	115	187
526.	104	171	527.	113	178	528.	98	163
529.	108	191	530.	94	164	531.	101	160
532.	100	166	533.	80	143	534.	120	191
535.	101	143	536.	105	159	537.	101	146
538.	120	175	539.	99	145	540.	108	149
541.	86	132	542.	83	139	543.	94	147
544.	90	134	545.	105	182	546.	92	154
547.	96	159	548.	119	157	549.	108	157
550.	103	157	551.	88	139	552.	110	155
553.	103	160	554.	107	182	555.	105	156
556.	117	165	557.	100	154	558.	96	145
559.	108	159	560.	97	162	561.	100	152
562.	104	164	563.	110	146	564.	104	163
565.	106	161	566.	102	133	567.	101	169
568.	87	157	569.	117	169	570.	96	150
571.	102	155	572.	108	140	573.	97	152
574.	99	157	575.	92	149	576.	84	124
577.	108	152	578.	93	162	579.	92	154
580.	96	137	581.	95	131	582.	96	155
583.	100	166	584.	94	132	585.	100	150
586.	97	143	587.	103	174	588.	88	145
589.	99	139	590.	105	139	591.	108	158
592.	90	147	593.	115	194	594.	96	138
595.	105	150	596.	103	150	597.	102	149
598.	103	165	599.	80	125	600.	96	146
601.	114	173	602.	124	182	603.	110	166
604.	80	126	605.	98	147	606.	97	137
607.	107	144	608.	108	161	609.	106	168
610.	93	137	611.	114	180	612.	84	126
613.	111	158	614.	91	126	615.	97	145
616.	98	127	617.	101	151	618.	113	164
619.	120	177	620.	90	129	621.	95	148
622.	114	159	623.	102	159	624.	109	151
625.	94	143	626.	93	132	627.	120	165
628.	101	141	629.	90	146	630.	92	137
631.	107	156	632.	113	169	633.	87	127
634.	115	183	635.	94	156	636.	103	156
637.	101	156	638.	110	173	639.	98	152
640.	80	135	641.	84	136	642.	105	153

Subject	X	Y	Subject	X	Y	Subject	X	Y
643.	83	144	644.	108	156	645.	109	153
646.	112	172	647.	107	150	648.	97	129
649.	115	160	650.	95	134	651.	99	164
652.	103	135	653.	103	150	654.	101	154
655.	106	126	656.	114	161	657.	107	174
658.	99	151	659.	88	144	660.	108	160
661.	113	179	662.	97	130	663.	115	150
664.	101	160	665.	92	122	666.	104	157
667.	102	130	668.	84	146	669.	119	171
670.	95	135	671.	115	173	672.	116	154
673.	80	126	674.	107	168	675.	97	144
676.	84	136	677.	93	135	678.	99	148
679.	97	155	680.	101	164	681.	119	171
682.	92	147	683.	95	139	684.	88	122
685.	92	137	686.	96	151	687.	88	131
688.	102	164	689.	93	144	690.	97	137
691.	90	148	692.	93	155	693.	97	156
694.	121	183	695.	90	136	696.	111	147
697.	87	134	698.	108	189	699.	104	172
700.	120	171	701.	94	148	702.	114	158
703.	111	166	704.	96	143	705.	101	143
706.	102	153	707.	102	148	708.	109	155
709.	114	166	710.	101	140	711.	90	133
712.	84	140	713.	101	177	714.	89	155
715.	97	174	716.	87	159	717.	120	167
718.	88	149	719.	99	136	720.	100	171
721.	115	159	722.	98	148	723.	96	167
724.	85	136	725.	103	160	726.	108	169
727.	104	147	728.	92	136	729.	95	142.
730.	94	133	731.	100	147	732.	92	128
733.	90	140	734.	86	142	735.	91	123
736.	117	159	737.	114	151	738.	113	160
739.	92	149	740.	106	160	741.	112	178
742.	104	168	743.	103	151	744.	80	124
745.	92	141	746.	98	165	747.	106	159
748.	90	125	749.	97	148	750.	106	158
751.	103	151	752.	101	129	753.	81	125
754.	97	150	755.	89	128	756.	97	164
757.	90	123	758.	108	148	759.	107	172
760.	87	121	761.	89	125	762.	97	149
763.	108	151	764.	109	158	765.	98	143
766.	111	145	767.	88	141	768.	116	173
769.	111	160	770.	94	136	771.	95	165
772.	103	166	773.	93	134	774.	92	120
775.	119	155	776.	113	186	777.	99	152
778.	88	124	779.	104	137	780.	99	141
781.	111	170	782.	94	150	783.	115	140
784.	93	151	785.	99	143	786.	102	158

Subject	X	Y	Subject	X	Y	Subject	X	Y
787.	81	120	788.	108	173	789.	98	139
790.	103	165	791.	91	131	792.	119	189
793.	88	139	794.	101	162	795.	85	127
796.	111	146	797.	105	154	798.	83	125
799.	96	144	800.	119	195	801.	117	174
802.	119	165	803.	104	149	804.	105	158
805.	109	171	806.	91	145	807.	93	123
808.	117	155	809.	114	150	810.	101	157
811.	93	150	812.	104	178	813.	87	132
814.	87	142	815.	94	147	816.	103	141
817.	103	151	818.	84	123	819.	123	143
820.	99	132	821.	83	125	822.	101	153
823.	92	144	824.	89	124	825.	113	177
826.	90	151	827.	107	155	828.	96	143
829.	98	157	830.	120	179	831.	87	154
832.	114	157	833.	110	172	834.	97	135
835.	101	153	836.	91	143	837.	101	164
838.	110	164	839.	112	172	840.	97	144
841.	100	146	842.	96	158	843.	106	155
844.	98	149	845.	120	159	846.	120	175
847.	101	143	848.	97	139	849.	95	125
850.	106	170	851.	99	150	852.	98	129
853.	97	152	854.	120	172	855.	81	148
856.	108	147	857.	89	157	858.	89	138
859.	99	146	860.	86	150	861.	109	153
862.	90	141	863.	91	139	864.	103	151
865.	104	166	866.	96	127	867.	115	177
868.	88	132	869.	93	148	870.	95	139
871.	111	179	872.	97	144	873.	99	141
874.	107	171	875.	94	134	876.	117	191
877.	108	161	878.	86	145	879.	111	165
880.	97	153	881.	99	171	882.	111	163
883.	113	171	884.	100	146	885.	90	141
886.	99	146	887.	94	131	888.	74	132
889.	91	148	890.	99	141	891.	97	142
892.	99	154	893.	107	161	894.	114	159
895.	94	134	896.	88	157	897.	81	120
898.	80	128	899.	102	156	900.	118	191
901.	111	181	902.	91	130	903.	84	130
904.	84	126	905.	98	154	906.	111	157
907.	100	148	908.	120	175	909.	112	179
910.	80	126	911.	82	142	912.	106	143
913.	80	129	914.	107	177	915.	91	131
916.	86	137	917.	108	150	918.	98	150
919.	113	168	920.	99	131	921.	109	161
922.	109	153	923.	104	155	924.	107	170
925.	105	172	926.	110	127	927.	103	171
928.	100	144	929.	94	120	930.	100	177

Subject	X	Y	Subject	X	Y	Subject	X	Y
931.	99	146	932.	104	180	933.	108	163
934.	107	162	935.	90	144	936.	99	178
937.	118	181	938.	83	137	939.	105	152
940.	94	151	941.	85	135	942.	93	137
943.	86	156	944.	104	144	945.	89	144
946.	120	166	947.	99	163	948.	95	140
949.	91	134	950.	97	145	951.	110	145
952.	104	187	953.	102	145	954.	97	144
955.	116	152	956.	99	160	957.	80	124
958.	111	168	959.	102	156	960.	91	135
961.	115	179	962.	97	154	963.	108	152
964.	93	155	965.	101	140	966.	86	125
967.	112	179	968.	98	139	969.	99	149
970.	108	154	971.	109	169	972.	104	171
973.	93	162	974.	103	145	975.	102	152
976.	101	160	977.	93	142	978.	99	146
979.	112	152	980.	102	154	981.	115	153
982.	105	152	983.	102	148	984.	108	162
985.	106	163	986.	111	175	987.	93	127
988.	110	174	989.	101	141	990.	117	194
991.	105	145	992.	92	125	993.	97	147
994.	111	157	995.	105	142	996.	106	167
997.	103	146	998.	80	126	999.	104	161
1000.	101	171	1001.	113	172	1002.	105	145
1003.	85	120	1004.	93	155	1005.	85	137
1006.	94	143	1007.	106	149	1008.	92	127
1009.	104	164	1010.	103	167	1011.	106	154
1012.	103	158	1013.	116	165	1014.	107	142
1015.	101	139	1016.	112	161	1017.	101	146
1018.	106	171	1019.	112	162	1020.	107	157
1021.	82	140	1022.	95	128	1023.	108	148
1024.	92	126	1025.	82	122	1026.	94	151
1027.	106	156	1028.	102	164	1029.	113	170
1030.	103	160	1031.	107	163	1032.	95	134
1033.	101	159	1034.	116	159	1035.	90	141
1036.	92	132	1037.	114	172	1038.	100	153
1039.	85	129	1040.	105	164	1041.	89	122
1042.	95	128	1043.	108	162	1044.	103	145
1045.	95	147	1046.	97	142	1047.	93	130
1048.	93	135	1049.	101	170	1050.	118	189

28. The following data on total body calcium (expressed as a percentage of the mean normal value) are available for a population of rheumatoid arthritis patients receiving prednisolone treatment. There are a total of 1200 patients. One hundred patients received the medicine at each dose level. Suppose you are a medical researcher wishing to gain insight into the nature of the relationship between dose level of prednisolone and total body calcium. Select a simple random sample of three patients from each dose level group and do the following.

(a) Use the total number of pairs of observations to obtain the least-squares equation describing the relationship between dose level (the independent variable) and total body calcium.

(b) Draw a scatter diagram of the data and plot the equation.

(c) Compute r and test for significance at the .05 level. Find the p value.

(d) Compare your results with those of your classmates.

Patient number	DOSE LEVEL											
	1	2	3	4	5	6	7	8	9	10	11	12
1.	99	98	77	91	87	89	85	86	81	76	77	66
2.	99	98	91	85	85	82	77	79	70	71	65	67
3.	95	99	97	89	90	86	86	81	71	81	71	61
4.	98	99	99	93	93	88	81	80	78	75	75	71
5.	101	96	89	85	89	93	79	78	77	75	74	63
6.	98	96	91	96	78	79	83	86	77	81	78	71
7.	99	94	94	85	86	80	80	83	83	72	71	65
8.	99	104	88	77	90	83	79	79	76	78	81	71
9.	95	97	94	90	86	86	77	73	74	74	74	69
10.	102	100	96	97	82	83	74	80	73	76	70	61
11.	95	97	91	94	90	78	80	79	76	70	73	63
12.	100	105	87	92	84	81	85	87	70	74	80	67
13.	99	91	88	90	87	85	82	80	73	76	67	70
14.	95	101	84	89	84	87	90	82	82	79	74	72
15.	99	96	84	88	88	79	74	78	75	74	66	59
16.	95	93	87	93	93	79	86	75	75	72	75	66
17.	100	102	89	94	86	83	78	82	81	66	67	64
18.	103	96	87	96	86	89	84	89	77	65	65	65
19.	102	98	94	81	89	85	86	75	75	79	70	68
20.	98	98	88	90	90	82	74	77	74	71	72	63
21.	95	101	92	88	84	77	90	78	82	74	62	69
22.	103	99	90	85	88	84	81	78	75	66	71	60
23.	106	98	99	86	82	85	78	77	72	74	73	64
24.	99	91	92	89	86	82	82	82	77	80	69	66
25.	110	98	91	94	83	84	80	78	71	77	69	74
26.	95	95	84	92	91	85	85	79	78	72	78	66
27.	89	101	90	92	89	86	73	83	81	75	70	75
28.	97	93	90	87	85	78	83	83	79	75	71	67
29.	94	97	97	92	90	86	80	82	71	74	69	67
30.	97	90	85	89	96	79	85	81	79	70	67	59
31.	94	103	87	92	92	86	81	81	73	77	76	67
32.	103	106	85	89	88	80	83	81	75	77	66	66
33.	98	96	99	86	89	79	82	78	76	67	68	63
34.	104	100	89	94	91	87	80	82	76	71	72	68

Patient number	DOSE LEVEL											
	1	2	3	4	5	6	7	8	9	10	11	12
35.	105	98	89	88	96	84	80	77	74	71	75	72
36.	96	99	93	86	88	85	81	83	76	72	74	70
37.	103	101	97	82	89	89	83	81	74	76	69	69
38.	97	103	89	86	84	78	82	77	82	70	63	70
39.	104	91	86	94	90	81	79	78	74	66	71	71
40.	100	103	87	95	87	84	88	82	74	72	63	68
41.	100	96	89	90	89	83	76	77	68	76	74	68
42.	101	103	92	83	88	73	75	77	84	73	70	66
43.	98	90	87	86	90	81	76	79	83	75	66	66
44.	100	103	90	86	84	80	81	79	72	76	69	61
45.	94	103	89	85	93	83	80	82	69	75	67	64
46.	89	92	81	89	86	84	83	76	71	74	78	70
47.	93	108	80	88	86	79	83	84	71	73	77	70
48.	102	101	92	100	84	75	78	80	76	78	61	72
49.	91	99	88	85	84	86	80	85	75	75	67	74
50.	90	99	90	89	91	87	87	76	79	73	70	66
51.	104	99	92	84	91	85	80	80	72	68	73	72
52.	95	104	89	89	87	84	86	77	77	70	70	76
53.	94	103	89	92	93	88	89	79	74	67	73	66
54.	101	102	86	92	92	83	83	81	78	75	78	66
55.	95	100	88	95	81	80	83	76	72	67	70	70
56.	94	100	88	87	89	81	80	88	75	71	72	75
57.	100	94	85	93	87	77	81	81	76	73	71	65
58.	92	97	89	98	88	83	91	83	79	69	76	67
59.	109	93	96	86	86	82	75	78	80	69	63	70
60.	93	93	91	88	87	87	82	77	70	79	77	60
61.	96	96	86	92	94	81	84	83	78	77	75	67
62.	94	95	88	94	84	87	83	81	75	74	65	73
63.	99	97	94	96	87	80	81	83	71	70	73	70
64.	96	98	97	86	82	74	88	74	81	70	67	70
65.	93	103	90	88	95	88	87	87	84	68	69	74
66.	90	98	93	86	89	84	82	75	70	77	66	63
67.	99	95	96	92	90	77	81	78	78	73	74	72
68.	100	97	88	92	89	82	77	77	74	73	71	66
69.	103	106	88	93	87	92	80	91	74	74	67	65
70.	103	104	86	90	82	84	78	82	77	65	72	68
71.	92	94	92	94	90	77	86	86	73	76	69	72
72.	96	100	94	86	91	77	84	76	76	79	74	64
73.	94	97	88	91	93	73	79	88	66	71	70	60
74.	99	98	89	86	92	74	74	75	84	72	71	62
75.	97	93	98	89	80	89	85	85	78	67	74	65
76.	98	91	86	85	93	80	85	81	76	76	64	70
77.	97	102	83	95	92	83	73	84	79	74	72	63
78.	96	97	93	93	91	81	84	82	78	78	70	72
79.	98	93	92	89	84	77	82	84	79	77	63	77
80.	95	86	88	88	95	81	84	82	78	68	74	72

Patient number	DOSE LEVEL											
	1	2	3	4	5	6	7	8	9	10	11	12
81.	94	92	85	84	89	85	90	83	74	73	72	67
82.	96	92	87	90	97	82	81	86	67	72	67	66
83.	100	95	87	89	87	88	79	84	78	72	67	66
84.	89	96	95	85	86	84	79	84	78	74	73	67
85.	106	106	87	85	91	85	80	81	76	74	76	67
86.	97	92	87	89	96	83	83	78	76	70	74	66
87.	102	91	85	89	91	86	80	80	79	72	72	68
88.	95	96	96	85	92	86	80	79	83	78	73	68
89.	95	95	93	83	83	85	74	81	78	66	68	68
90.	103	94	86	90	94	74	79	80	79	72	71	68
91.	96	93	92	91	87	88	83	80	72	67	75	69
92.	100	97	83	86	90	86	83	87	73	80	66	73
93.	97	95	86	92	79	87	83	88	74	79	74	74
94.	93	98	84	87	82	84	84	79	67	69	74	72
95.	98	93	87	91	88	76	84	77	70	79	70	71
96.	90	99	87	92	91	83	77	81	72	67	67	67
97.	97	96	90	90	88	83	79	81	78	68	65	69
98.	95	89	83	96	85	78	83	83	71	81	69	62
99.	99	98	82	92	90	79	75	82	79	69	70	65
100.	104	96	85	85	94	78	84	76	75	73	64	75

REFERENCES

References Cited
1. Francis Galton, *Natural Inheritance*, Macmillan, London (1899).
2. Francis Galton, *Memories of My Life*, Methuen, London, 1908.
3. Karl Pearson, *The Life, Letters and Labours of Francis Galton*, Volume III A, Cambridge at the University Press, 1930.
4. Francis Galton, "Co-relations and Their Measurement, Chiefly from Anthropometric Data," *Proceedings of the Royal Society*, XLV (1888), 135–145.
5. F. A. Graybill and D. C. Bowden, "Linear Segment Confidence Bands for Simple Linear Models," *Journal of the American Statistical Association*, 62 (1967), 403–408.
6. Olive Jean Dunn, "A Note on Confidence Bands for a Regression Line Over a Finite Range," *Journal of the American Statistical Association*, 63 (1968), 1028–1033.
7. P. G. Hoel, "Confidence Regions for Linear Regression," *Proceedings of the Second Berkeley Symposium on Mathematical Statistics and Probability*, University of California Press, Berkeley, 1951, pp. 75–81.

8. A. V. Gafarian, "Confidence Bands in Straight Line Regression," *Journal of the American Statistical Association*, *59* (1964), 182–213.

9. M. Halperin and J. Gurian, "Confidence Bands in Linear Regression with Constraints in the Independent Variables," *Journal of the American Statistical Association*, *63* (1968), 1020–1027.

10. R. C. Elston and J. E. Grizzle, "Estimation of Time Response Curves and Their Confidence Bands," *Biometrics*, *18* (1962), 148–159.

11. R. A. Fisher, "On the Probable Error of a Coefficient of Correlation Deduced from a Small Sample," *Metron*, *1* (1921), 3–21.

12. H. Hotelling, "New Light on the Correlation Coefficient and Its Transforms," *Journal of the Royal Statistical Society*, Ser B, *15* (1953), 193–232.

13. F. N. David, *Tables of the Ordinates and Probability Integral of the Distribution of the Correlation Coefficient in Small Samples*, Cambridge University Press, Cambridge, 1938.

Other References, Books

1. F. S. Acton, *Analysis of Straight Line Data*, Wiley, New York, 1959.

2. Andrew R. Baggaley, *Intermediate Correlational Methods*, Wiley, New York, 1964.

3. Cuthbert Daniel and Fred S. Wood, *Fitting Equations to Data*, Wiley-Interscience, New York, 1971.

4. N. R. Draper and H. Smith, *Applied Regression Analysis*, Second Edition, Wiley, New York, 1981.

5. Mordecai Ezekiel and Karl A. Fox, *Methods of Correlation and Regression Analysis*, Third Edition, Wiley, New York, 1959.

6. Arthur S. Goldberger, *Topics in Regression Analysis*, Macmillan, Toronto, 1968.

7. David G. Kleinbaum and Lawrence L. Kupper, *Applied Regression Analysis and Other Multivariable Methods*, Duxbury Press, North Scituate, Mass., 1978.

8. R. L. Plackett, *Principles of Regression Analysis*, Oxford University Press, London, 1960.

9. K. W. Smillie, *An Introduction to Regression and Correlation*, Academic Press, New York, 1966.

10. Peter Sprent, *Models in Regression*, Methuen, London, 1969.

11. E. J. Williams, *Regression Analysis*, Wiley, New York, 1959.

12. Stephen Wiseman, *Correlation Methods*, Manchester University Press, Manchester, 1966.

13. Mary Sue Younger, *Handbook for Linear Regression*, Duxbury Press, North Scituate, Mass., 1979.

Other References, Journal Articles

1. R. G. D. Allen, "The Assumptions of Linear Regression," *Economica*, *6 N.S.* (1939), 191–204.

2. M. S. Bartlett, "The Fitting of Straight Lines if Both Variables Are Subject to Error," *Biometrics*, *5* (1949), 207–212.

3. J. Berkson, "Are There Two Regressions?" *Journal of the American Statistical Association*, *45* (1950), 164–180.

4. Dudley J. Cowden, "A Procedure for Computing Regression Coefficients," *Journal of the American Statistical Association*, *53* (1958), 144–150.

5. A. S. C. Ehrenberg, "Bivariate Regression Is Useless," *Applied Statistics*, *12* (1963), 161–179.

6. M. G. Kendall, "Regression, Structure, and Functional Relationship, Part I," *Biometrika*, *28* (1951), 11–25.

7. M. G. Kendall, "Regression, Structure, and Functional Relationship II," *Biometrika*, *39* (1952), 96–108.

8. D. V. Lindley, "Regression Lines and the Linear Functional Relationship," *Journal of the Royal Statistical Society* (Supplement), *IX* (1947), 218–244.

9. A. Madansky, "The Fitting of Straight Lines When Both Variables Are Subject to Error," *Journal of the American Statistical Association*, *54* (1959), 173–205.

10. A. Wald, "The Fitting of Straight Lines if Both Variables Are Subject to Error," *Annals of Mathematical Statistics*, *11* (1940), 284–300.

11. W. G. Warren, "Correlation or Regression: Bias or Precision," *Applied Statistics*, *20* (1971), 148–164.

12. Charles P. Winsor, "Which Regression?" *Biometrics*, *2* (1946), 101–109.

CHAPTER 9

Multiple Regression and Correlation

9.1 INTRODUCTION

In Chapter 8 we explored the concepts and techniques for analyzing and making use of the relationship between two variables. We saw that this analysis may lead to an equation that can be used to predict the value of some dependent variable given the value of an associated independent variable.

Intuition tells us that, in general, we ought to be able to improve our predicting ability by including more independent variables in such an equation. For example, a researcher may find that intelligence scores of individuals may be predicted from physical factors such as birth order, birth weight, and length of gestation along with certain hereditary and external environmental factors. Length of stay in a chronic disease hospital may be related to the patient's age, marital status, sex, and income, not to mention the obvious factor of diagnosis. The response of an experimental animal to some drug may depend on the size of the dose and the age and weight of the animal. A nursing supervisor may be interested in the strength of the relationship between a nurse's performance on the job, score on the state board examination, scholastic record, and score on some achievement or aptitude test. Or a hospital administrator studying admissions from various communities served by the hospital may be interested in determining what factors seem to be responsible for differences in admission rates.

The concepts and techniques for analyzing the association among several variables are natural extensions of those explored in the previous chapters. The computations, as one would expect, are more complex and tedious. However, as is pointed out in Chapter 8, this presents no real problem when a computer is available. It is not unusual to find researchers investigating the relationships among a dozen or more variables. For those who have access to a computer, the decision as to how many variables to include in an analysis is based not on the complexity and length of the computations but on such considerations as their meaningfulness, the cost of their inclusion, and the importance of their contribution.

In this chapter we follow closely the sequence of the previous chapter. The regression model is considered first, followed by a discussion of the correlation model. In considering the regression model, the following points are covered: a description of the model, methods for obtaining the regression equation, evaluation of the equation, and the uses that may be made of the equation. In both models the possible inferential procedures and ther underlying assumptions are discussed.

9.2 THE MULTIPLE REGRESSION MODEL

In the multiple regression model we assume that a linear relationship exists between some variable Y, which we call the dependent variable, and k independent variables, $X_1, X_2, \ldots, X_k$. The independent variables are sometimes referred to as *explanatory variables*, because of their use in explaining the variation in Y, and *predictor variables*, because of their use in predicting Y.

The accompanying assumptions are as follows.

1. The X_i are nonrandom (fixed) variables. This assumption distinguishes the multiple regression model from the multiple correlation model, which will be presented in Section 9.7. This condition indicates that any inferences which are drawn from sample data apply only to the set of X values observed and not to some larger collection of X's. Under the regression model, correlation analysis is not meaningful. Under the correlation model to be presented later, the regression techniques that follow may be applied.

2. For each set of X_i values there is a subpopulation of Y values. To construct certain confidence intervals and test hypotheses it must be known, or the researcher must be willing to assume, that these subpopulations of Y values are normally distributed. Since we will want to demonstrate these inferential procedures, the assumption of normality will be made in the examples and exercises in this chapter.

3. The variances of the subpopulations of Y are all equal.

4. The Y values are independent. That is, the values of Y selected for one set of X values do not depend on the values of Y selected at another set of X values.

These assumptions may be stated in more compact fashion as

$$y_j = \beta_0 + \beta_1 x_{1j} + \beta_2 x_{2j} + \cdots + \beta_k x_{kj} + e_j \qquad (9.2.1)$$

where y_j is a typical value from one of the subpopulations of Y values, the β_i are called the regression coefficients, $x_{1j}, x_{2j}, \ldots, x_{kj}$ are, respectively, particular values of the independent variables $X_1, X_2, \ldots, X_k$, and e_j is a random variable with mean 0 and variance σ^2, the common variance of the subpopulations of Y values. To construct confidence intervals for and test hypotheses about the regression coefficients, we assume that the e_j are normally and independently distributed. The statements regarding e_j are a consequence of the assumptions regarding the distributions of Y values. We will refer to Equation 9.2.1 as the *multiple regression model*.

When Equation 9.2.1 consists of one dependent variable and two independent variables, that is, when the model is written

$$y_j = \beta_0 + \beta_1 x_{1j} + \beta_2 x_{2j} + e_j \qquad (9.2.2)$$

a *plane* in three-dimensional space may be fitted to the data points as illustrated in Figure 9.2.1. When the model contains more than two independent variables, it is described geometrically as a *hyperplane*.

In Figure 9.2.1 the observer should visualize some of the points as being located above the plane and some as being located below the plane. The deviation of a point from the plane is represented by

$$e_j = y_j - \beta_0 - \beta_1 x_{1j} - \beta_2 x_{2j} \qquad (9.2.3)$$

In Equation 9.2.2, β_0 represents the point where the plane cuts the Y-axis, that is, it represents the Y-intercept of the plane. β_1 measures the average change in Y

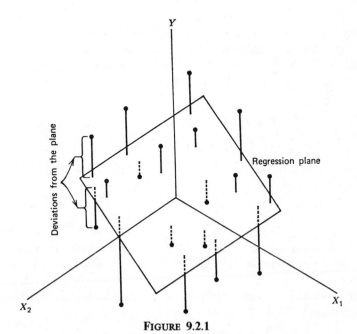

FIGURE 9.2.1

Multiple Regression Plane and Scatter of Points

for a unit change in X_1 when X_2 remains unchanged, and β_2 measures the average change in Y for a unit change in X_2 when X_1 remains unchanged. For this reason β_1 and β_2 are referred to as *partial regression coefficients*.

9.3   OBTAINING THE MULTIPLE REGRESSION EQUATION

Unbiased estimates of the parameters, $\beta_0, \beta_1, \ldots, \beta_k$ of the model specified in Equation 9.2.1 are obtained by the method of least squares. This means that the sum of the squared deviations of the observed values of Y from the resulting regression surface is minimized. In the three-variable case, as illustrated in Figure 9.2.1, the sum of the squared deviations of the observations from the plane are a minimum when β_0, β_1, and β_2 are estimated by the method of least squares. In other words, by the method of least squares, sample estimates of $\beta_0, \beta_1, \ldots, \beta_k$ are selected in such a way that the quantity

$$\sum e_j^2 = \sum \left(y_j - \beta_0 x_{1j} - \beta_1 x_{2j} - \cdots - \beta x_{kj} \right)$$

is minimized. This quantity, referred to as the sum of squares of the residuals may also be written as

$$\sum (y_j - \hat{y})^2 \qquad (9.3.1)$$

indicating the fact that the sum of squares of deviations of the observed values of Y from the values of Y calculated from the estimated equation is minimized.

Estimates, $b_0, b_1, b_2, \ldots, b_k$, of the regression coefficients are obtained by solving the following set of normal equations:

$$\left.\begin{aligned}
nb_0 + b_1 \sum x_{1j} + b_2 \sum x_{2j} + \cdots + b_k \sum x_{kj} &= \sum y_j \\
b_0 \sum x_{1j} + b_1 \sum x_{1j}^2 + b_2 \sum x_{1j} x_{2j} + \cdots + b_k \sum x_{1j} x_{kj} &= \sum x_{1j} y_j \\
b_0 \sum x_{2j} + b_1 \sum x_{2j} x_{1j} + b_2 \sum x_{2j}^2 + \cdots + b_k \sum x_{2j} x_{kj} &= \sum x_{2j} y_j \\
\cdots \quad \cdots \quad \cdots \quad \cdots \quad \cdots \quad \cdots \quad \cdots \quad \cdots \quad \cdots \\
b_0 \sum x_{kj} + b_1 \sum x_{kj} x_{1j} + b_2 \sum x_{kj} x_{2j} + \cdots + b_k \sum x_{kj}^2 &= \sum x_{kj} y_j
\end{aligned}\right\} \quad (9.3.2)$$

Let us illustrate the calculation of the estimated regression equation and its uses for the case where we have one dependent variable and two independent variables.

When the model contains only two independent variables, the sample regression equation is

$$\hat{y}_j = b_0 + b_1 x_{1j} + b_2 x_{2j} \qquad (9.3.3)$$

TABLE 9.3.1

Length of Stay in Days, Age in Years, and Number of Previous Admissions, 15 Patients Admitted to a Chronic Disease Hospital

Length of stay (Y)	Number of previous admissions (X_1)	Age (X_2)
15	0	21
15	0	18
21	0	22
28	1	24
30	1	25
35	1	25
40	1	26
35	2	34
30	2	25
45	2	38
50	3	44
60	3	51
45	4	39
60	4	54
50	5	55

Example 9.3.1 In a study of length of stay of patients in a chronic disease hospital researchers wished to see how well they could predict length of stay (Y) given the independent variables number of previous admissions (X_1) and age (X_2). The data shown in Table 9.3.1 were obtained for a sample of $n = 15$ patients.

To obtain the least-squares equation, the following normal equations must be solved for the sample regression coefficients:

$$\left. \begin{array}{l} nb_0 + b_1\sum x_{1j} + b_2\sum x_{2j} = \sum y_j \\ b_0\sum x_{1j} + b_1\sum x_{1j}^2 + b_2\sum x_{1j}x_{2j} = \sum x_{1j}y_j \\ b_0\sum x_{2j} + b_1\sum x_{1j}x_{2j} + b_2\sum x_{2j}^2 = \sum x_{2j}y_j \end{array} \right\} \qquad \textbf{(9.3.4)}$$

When a computer is available the solution of these equations poses no particular computational burden since the computer can be programmed to do the work or commercial software packages may be used. If, however, the calculations must be done with the aid of the desk calculator or a hand-held calculator the arithmetic can be formidable, as the discussion that follows well demonstrates. Those who have access to a computer, therefore, may, if they wish, skip most of the following explanation of computations.

Note that the number of equations is equal to the number of parameters to be estimated. We may solve the three equations as they stand or we may reduce them to a set of two equations by transforming each value into a deviation from its mean. If we designate these deviations by y'_j, x'_{1j}, and x'_{2j}, we have

$$\left. \begin{aligned} y'_j &= y_j - \bar{y} \\ x'_{1j} &= x_{1j} - \bar{x}_1 \\ x'_{2j} &= x_{2j} - \bar{x}_2 \end{aligned} \right\} \tag{9.3.5}$$

If we restate the original sample regression equation (Equation 9.3.3) in terms of these transformations, we have

$$\hat{y}'_j = b'_0 + b'_1 x'_{1j} + b'_2 x'_{2j} \tag{9.3.6}$$

where b'_0, b'_1, and b'_2 are appropriate coefficients for the transformed variables. The relationship between the two sets of coefficients can be determined by substituting the deviations from the means of the original variables into Equation 9.3.6 and then simplifying. Thus,

$$\hat{y}_j - \bar{y} = b'_0 + b'_1(x_{1j} - \bar{x}_1) + b'_2(x_{2j} - \bar{x}_2)$$

$$\hat{y}_j = \bar{y} + b'_0 + b'_1 x_{1j} - b'_1 \bar{x}_1 + b'_2 x_{2j} - b'_2 \bar{x}_2 \tag{9.3.7}$$

$$\hat{y}_j = b'_0 + \bar{y} - b'_1 \bar{x}_1 - b'_2 \bar{x}_2 + b'_1 x_{1j} + b'_2 x_{2j}$$

When we set the coefficients of like terms in Equations 9.3.3 and 9.3.7, equal to each other, we obtain

$$b_1 = b'_1$$

$$b_2 = b'_2$$

and, therefore,

$$b_0 = b'_0 + \bar{y} - b'_1 \bar{x}_1 - b'_2 \bar{x}_2 = b'_0 + \bar{y} - b_1 \bar{x}_1 - b_2 \bar{x}_2 \tag{9.3.8}$$

A new set of normal equations based on Equation 9.3.6 is

$$nb'_0 + b'_1 \sum x'_{1j} + b'_2 \sum x'_{2j} = \sum y'_j$$

$$b'_0 \sum x'_{1j} + b'_1 \sum x'^2_{1j} + b'_2 \sum x'_{1j} x'_{2j} = \sum x'_{1j} y'_j$$

$$b'_0 \sum x'_{2j} + b'_1 \sum x'_{1j} x'_{2j} + b'_2 \sum x'^2_{2j} = \sum x'_{2j} y'_j$$

Using the transformations from Equation 9.3.5 and the property that $\sum(x_i - \bar{x}) = 0$, we obtain

$$nb_0' + b_1'(0) + b_2'(0) = 0$$

$$b_0'(0) + b_1'\sum(x_{1j} - \bar{x}_1)^2 + b_2'\sum(x_{1j} - \bar{x}_1)(x_{2j} - \bar{x}_2) = \sum(x_{1j} - \bar{x}_1)(y_j - \bar{y})$$

$$b_0'(0) + b_1'\sum(x_{1j} - \bar{x}_1)(x_{2j} - \bar{x}_2) + b_2'\sum(x_{2j} - \bar{x}_2)^2 = \sum(x_{2j} - \bar{x}_2)(y_j - \bar{y})$$

Note that $b_0' = 0$. Thus by Equation 9.3.8

$$b_0 = \bar{y} - b_1\bar{x}_1 - b_2\bar{x}_2$$

and when we substitute b_1 and b_2 for b_1' and b_2', respectively, our normal equations collapse to the following.

$$\left. \begin{array}{l} b_1\sum x_{1j}'^2 + b_2\sum x_{1j}'x_{2j}' = \sum x_{1j}'y_j' \\ b_1\sum x_{1j}'x_{2j}' + b_2\sum x_{2j}'^2 = \sum x_{2j}'y_j' \end{array} \right\} \tag{9.3.9}$$

Table 9.3.2 contains the sums of squares and cross products of the original values necessary for computing the sums of squares and cross products of y_j', x_{1j}', and x_{2j}'.

TABLE 9.3.2

Sums of Squares and Sums of Cross Products for Example 9.3.1

y_j	x_{1j}	x_{2j}	$x_{1j}x_{2j}$	$x_{1j}y_j$	$x_{2j}y_j$	x_{1j}^2	x_{2j}^2	y_j^2
15	0	21	0	0	315	0	441	225
15	0	18	0	0	270	0	324	225
21	0	22	0	0	462	0	484	441
28	1	24	24	28	672	1	576	784
30	1	25	25	30	750	1	625	900
35	1	25	25	35	875	1	625	1225
40	1	26	26	40	1040	1	676	1600
35	2	34	68	70	1190	4	1156	1225
30	2	25	50	60	750	4	625	900
45	2	38	76	90	1710	4	1444	2025
50	3	44	132	150	2200	9	1936	2500
60	3	51	153	180	3060	9	2601	3600
45	4	39	156	180	1755	16	1521	2025
60	4	54	216	240	3240	16	2916	3600
50	5	55	275	250	2750	25	3025	2500
Totals 559	29	501	1,226	1,353	21,039	91	18,975	23,775
Means 37.27	1.93	33.40						

Using the data in Table 9.3.2, we compute

$$\sum x_{1j}'^2 = \sum (x_{1j} - \bar{x}_1)^2 = \sum x_{1j}^2 - \left(\sum x_{1j}\right)^2/n$$
$$= 91 - (29)^2/15 = 34.93$$

$$\sum x_{2j}'^2 = \sum (x_{2j} - \bar{x}_2)^2 = \sum x_{2j}^2 - \left(\sum x_{2j}\right)^2/n$$
$$= 18{,}975 - (501)^2/15 = 2241.60$$

$$\sum x_{1j}' x_{2j}' = \sum (x_{1j} - \bar{x}_1)(x_{2j} - \bar{x}_2) = \sum x_{1j}x_{2j} - \sum x_{1j}\sum x_{2j}/n$$
$$= 1226 - (29)(501)/15 = 257.40$$

$$\sum x_{1j}' y_j' = \sum (x_{1j} - \bar{x}_1)(y_j - \bar{y}) = \sum x_{1j}y_j - \sum x_{1j}\sum y_j/n$$
$$= 1353 - (29)(559)/15 = 272.27$$

$$\sum x_{2j}' y_j' = \sum (x_{2j} - \bar{x}_2)(y_j - \bar{y}) = \sum x_{2j}y_j - \sum x_{2j}\sum y_j/n$$
$$= 21{,}039 - (501)(559)/15 = 2368.40$$

When we substitute these values into Equations 9.3.9, we have

$$\left.\begin{array}{c} 34.93b_1 + 257.40b_2 = 272.27 \\ 257.40b_1 + 2241.60b_2 = 2368.40 \end{array}\right\} \qquad \textbf{(9.3.10)}$$

These equations may be solved by any standard method to obtain

$$b_1 = .06$$
$$b_2 = 1.05$$

We obtain b_0 from the relationship

$$b_0 = \bar{y} - b_1\bar{x}_1 - b_2\bar{x}_2 \qquad \textbf{(9.3.11)}$$

For our example, we have

$$b_0 = 37.27 - (.06)(1.93) - (1.05)(33.40)$$
$$= 2.08$$

Our sample multiple regression equation, then, is

$$\hat{y} = 2.08 + .06x_{1j} + 1.05x_{2j} \qquad \textbf{(9.3.12)}$$

We have used an example containing only three variables for simplicity. As the number of variables increases, the algebraic manipulations and arithmetic calculations become more tedious and subject to error, although they are natural extensions of the procedures given in the present example.

Snedecor and Cochran (1) and Steel and Torrie (2) give numerical examples for four variables, and Anderson and Bancroft (3) illustrate the calculations involved when there are five variables. The techniques used by these authors are applicable for any number of variables.

After the multiple regression equation has been obtained, the next step involves its evaluation and interpretation. It is this facet of the analysis that is covered in the next section.

EXERCISES

Obtain the regression equation for each of the following sets of data.

9.3.1 In a study designed to discover what factors might be related to birth weight, the following data were obtained on 10 newborn babies.

Birth weight in grams (Y)	Socioeconomic status score (X₁)	Birth order (X₂)
1361	8	4
1588	7	3
1815	4	4
2087	5	3
2268	5	2
2404	4	2
3402	3	2
3629	3	1
3765	2	1
4083	1	1
26402	42	23

$$\sum x_{1j}^2 = 218 \quad \sum x_{2j}^2 = 65 \quad \sum y_j^2 = 78{,}536{,}258$$

$$\sum x_{1j}x_{2j} = 114 \quad \sum x_{1j}y_j = 93{,}361 \quad \sum x_{2j}y_j = 51{,}354$$

9.3.2 A researcher collected the following data on 15 children.

Intelligence score (Y)	Birth order (X₁)	Mother's age at birth of child (X₂)
110	1	25
115	1	24
120	1	22
118	1	24
110	2	20
108	2	20
105	2	20
104	3	24
98	3	25
99	4	30
98	4	24
100	5	29
90	5	30
93	5	30
90	6	28
Totals 1558	45	375

$$\sum x_{1j}^2 = 177 \quad \sum x_{2j}^2 = 9563 \quad \sum y_j^2 = 163{,}112$$

$$\sum x_{1j}x_{2j} = 1191 \quad \sum x_{1j}y_j = 4458 \quad \sum x_{2j}y_j = 38{,}620$$

9.3.3 In a study of factors thought to be related to admission patterns to a large general hospital, a hospital administrator obtained these data on 10 communities in the hospital's catchment area.

Community	Persons per 1000 population admitted during study period (Y)	Index of availability of other health services (X_1)	Index of indigency (X_2)
1	61.6	6.0	6.3
2	53.2	4.4	5.5
3	65.5	9.1	3.6
4	64.9	8.1	5.8
5	72.7	9.7	6.8
6	52.2	4.8	7.9
7	50.2	7.6	4.2
8	44.0	4.4	6.0
9	53.8	9.1	2.8
10	53.5	6.7	6.7
Total	571.6	69.9	55.6

$$\sum x_{1j}^2 = 525.73 \qquad \sum x_{2j}^2 = 331.56 \qquad \sum y_j^2 = 33,349.92$$

$$\sum x_{1j} x_{2j} = 374.31 \qquad \sum x_{1j} y_j = 4104.32 \qquad \sum x_{2j} y_j = 3183.57$$

9.3.4 The administrator of a general hospital obtained the following data on 20 surgery patients during a study to determine what factors appear to be related to length of stay.

Postoperative length of stay in days (Y)	Number of current medical problems (X_1)	Preoperative length of stay in days (X_2)
6	1	1
6	2	1
11	2	2
9	1	3
16	3	3
16	1	5
4	1	1
8	3	1
11	2	2
13	3	2
13	1	4
9	1	2
17	3	3
17	2	4
12	4	1
6	1	1
5	1	1

(Table continued)

	Postoperative length of stay in days (Y)	Number of current medical problems (X₁)	Preoperative length of stay in days (X₂)
	12	3	2
	8	1	2
	9	2	2
Total	208	38	43

$$\sum x_{1j}^2 = 90 \qquad \sum x_{2j}^2 = 119 \qquad \sum y_j^2 = 2478$$

$$\sum x_{1j}x_{2j} = 79 \qquad \sum x_{1j}y_j = 430 \qquad \sum x_{2j}y_j = 519$$

9.3.5 Refer to Exercise 8.7.5. Additional information on the score made by each nurse on an aptitude test, taken at the time of entering nursing school, was made available to the researcher. The complete data are as follows.

	State board score (Y)	Final score (X₁)	Aptitude test score (X₂)
	440	87	92
	480	87	79
	535	87	99
	460	88	91
	525	88	84
	480	89	71
	510	89	78
	530	89	78
	545	89	71
	600	89	76
	495	90	89
	545	90	90
	575	90	73
	525	91	71
	575	91	81
	600	91	84
	490	92	70
	510	92	85
	575	92	71
	540	93	76
	595	93	90
	525	94	94
	545	94	94
	600	94	93
	625	94	73
Total	13,425	2263	2053

$$\sum x_{1j}^2 = 204{,}977 \qquad \sum x_{2j}^2 = 170{,}569 \qquad \sum y_j^2 = 7{,}264{,}525$$

$$\sum x_{1j}x_{2j} = 185{,}838 \qquad \sum x_{1j}y_j = 1{,}216{,}685 \qquad \sum x_{2j}y_j = 1{,}101{,}220$$

9.3.6 The following data were collected on a simple random sample of 20 patients with hypertension. The variables are

$$Y = \text{mean arterial blood pressure (mm Hg)}$$

$$X_1 = \text{age (years)}$$

$$X_2 = \text{weight (kg)}$$

$$X_3 = \text{body surface area (sq m)}$$

$$X_4 = \text{duration of hypertension (years)}$$

$$X_5 = \text{basal pulse (beats/min)}$$

$$X_6 = \text{measure of stress}$$

Patient	Y	X_1	X_2	X_3	X_4	X_5	X_6
1.	105	47	85.4	1.75	5.1	63	33
2.	115	49	94.2	2.10	3.8	70	14
3.	116	49	95.3	1.98	8.2	72	10
4.	117	50	94.7	2.01	5.8	73	99
5.	112	51	89.4	1.89	7.0	72	95
6.	121	48	99.5	2.25	9.3	71	10
7.	121	49	99.8	2.25	2.5	69	42
8.	110	47	90.9	1.90	6.2	66	8
9.	110	49	89.2	1.83	7.1	69	62
10.	114	48	92.7	2.07	5.6	64	35
11.	114	47	94.4	2.07	5.3	74	90
12.	115	49	94.1	1.98	5.6	71	21
13.	114	50	91.6	2.05	10.2	68	47
14.	106	45	87.1	1.92	5.6	67	80
15.	125	52	101.3	2.19	10.0	76	98
16.	114	46	94.5	1.98	7.4	69	95
17.	106	46	87.0	1.87	3.6	62	18
18.	113	46	94.5	1.90	4.3	70	12
19.	110	48	90.5	1.88	9.0	71	99
20.	122	56	95.7	2.09	7.0	75	99

9.4 EVALUATING THE MULTIPLE REGRESSION EQUATION

Before one makes use of a multiple regression equation, it is desirable to determine first whether it is, in fact, worth using. In our study of simple linear regression we have learned that the usefulness of a regression equation may be evaluated by a consideration of the sample coefficient of determination and estimated slope. In evaluating our multiple regression equation we focus our attention on the *coefficient of multiple determination* and the partial regression coefficients.

The Coefficient of Multiple Determination   In Chapter 8 the coefficient of determination is discussed in considerable detail. The concept extends logically to the multiple regression case. The total variation present in the Y values may be partitioned into two components—the explained variation, which measures the amount of the total variation that is explained by the fitted regression surface, and the unexplained variation, which is that part of the total variation not explained by fitting the regression surface. The measure of variation in each case is a sum of squared deviations. The total variation is the sum of squared deviations of each observation of Y from the mean of the observations and is designated by $\sum(y_j - \bar{y})^2$. The explained variation, designated by $\sum(\hat{y} - \bar{y})^2$, is the sum of squared deviations of the calculated values from the mean of the observed Y values. This sum of squared deviations is called the *sum of squares due to regression* (*SSR*). The unexplained variation, written as $\sum(y_j - \hat{y})^2$, is the sum of squared deviations of the original observations from the calculated values. This quantity is referred to as the *sum of squares about regression* or the *error sum of squares* (*SSE*). We may summarize the relationship among the three sums of squares with the following equation:

$$\sum(y_j - \bar{y})^2 = \sum(\hat{y} - \bar{y})^2 + \sum(y_j - \hat{y})^2 \qquad (9.4.1)$$

$$SST = SSR + SSE$$

total sum of squares = explained (regression) sum of squares

+ unexplained (error) sum of squares

The total, explained, and unexplained sum of squares are computed as follows:

$$SST = \sum(y_j - \bar{y})^2 = \sum y_j^2 - \left(\sum y_j\right)^2 / n \qquad (9.4.2)$$

$$SSR = \sum(\hat{y} - \bar{y})^2$$

$$= b_1 \sum x'_{1j} y'_j + b_2 \sum x'_{2j} y'_j + \cdots + b_k \sum x'_{kj} y'_j \qquad (9.4.3)$$

$$SSE = SST - SSR \qquad (9.4.4)$$

For our illustrative example we have (using data from Table 9.3.2 and some previous calculations)

$$SST = 23{,}775 - (559)^2 / 15 = 2942.93$$

$$SSR = (.06)(272.27) + (1.05)(2368.40) = 2503.16$$

$$SSE = 2942.93 - 2503.16 = 439.77$$

The coefficient of multiple determination, $R^2_{y.12\ldots k}$, is obtained by dividing the explained sum of squares by the total sum of squares. That is,

$$R^2_{y.12\ldots k} = \frac{\sum(\hat{y} - \bar{y})^2}{\sum(y_j - \bar{y})^2} \qquad (9.4.5)$$

The subscript $y.12\dots k$ indicates that in the analysis Y is treated as the dependent variable and the X variables from X_1 through X_k are treated as the independent variables. The value of $R^2_{y.12\dots k}$ indicates what proportion of the total variation in the observed Y values is explained by the regression of Y on $X_1, X_2, \dots, X_k$. In other words, we may say that $R^2_{y.12\dots k}$ is a measure of the goodness of fit of the regression surface. This quantity is analogous to r^2, which was computed in Chapter 8.

For our present example we have

$$R^2_{y.12} = \frac{2503.16}{2942.93} = .85$$

We say that 85 percent of the total variation in the Y values is explained by the fitted regression plane.

To determine whether the overall regression is significant (that is, to determine whether $R^2_{y.12}$ is significant), we may perform a hypothesis test as follows.

1. *Data* See the description of the data given in the initial statement of the problem.

2. *Assumptions* We presume that the multiple regression model and its underlying assumptions as presented in Section 9.2 are applicable.

3. *Hypotheses* In general, the null hypothesis is H_0: $\beta_1 = \beta_2 = \beta_3 = \cdots = \beta_k = 0$ and the alternative is H_A: not all $\beta_j = 0$. For our example we have

$$H_0: \beta_1 = \beta_2 = 0$$
$$H_A: \text{not all } \beta_j = 0$$

   In words, the null hypothesis states that all the independent variables are of no value in explaining the variation in the Y values.

4. *Test Statistic* The appropriate test statistic is V.R., which is computed as part of an analysis of variance. The general ANOVA table is shown as Table 9.4.1. In Table 9.4.1, MSR stands for mean square due to regression and MSE stands for mean square about regression or, as it is sometimes called, the error mean square.

For our illustrative example the analysis of variance is shown in Table 9.4.2.

5. *Distribution of the Test Statistic* When H_0 is true and the assumptions are met V.R. is distributed as F with k and $n - k - 1$ degrees of freedom.

6. *Decision Rule* Reject H_0 if the computed value of V.R. is equal to or greater than the critical value of F.

7. *Calculation of the Test Statistic* See Table 9.4.2.

8. *Statistical Decision* When we consult Table G with 2 and 12 degrees of freedom we find that our computed V.R. of 34.15 is significant at the .005 level (that is, $p < .005$).

TABLE 9.4.1

ANOVA Table for Multiple Regression

Source	SS	d.f.	MS	V.R.
Due to regression	SSR	k	$MSR = SSR/k$	MSR/MSE
About regression	SSE	$n - k - 1$	$MSE = SSE/(n - k - 1)$	
Total	SST	$n - 1$		

TABLE 9.4.2

ANOVA Table, Example 9.3.1

Source	SS	d.f.	MS	V.R.
Due to regression	2503.16	2	1251.58	34.15
About regression	439.77	12	36.65	
Total	2942.93	14		

9. *Conclusion* We conclude that, in the sample, the regression explains a significant proportion of the total variation in Y. In other words, we conclude that the fitted plane provides a good fit to the data.

Inferences Regarding Individual β's   An alternative procedure for evaluating the strength of the linear relationship between Y and the independent variables is to test the null hypothesis that $\beta_i = 0$ against the alternative $\beta_i \neq 0$ $(i = 1, 2, \ldots, k)$. The validity of this procedure rests on the assumptions stated earlier: that for each combination of X_i values there is a normally distributed subpopulation of Y values with variance σ^2. When these assumptions hold true, it can be shown that each of the sample estimates, b_i, is normally distributed with mean β_i and variance $c_{ii}\sigma^2$. Since σ^2 is unknown, it will have to be estimated. An estimate is provided by the mean square about regression which is shown in the ANOVA table. We may designate this quantity generally as $s_{y.12\ldots k}^2$. For our particular example we have, from Table 9.4.2, $s_{y.12}^2 = 36.65$, and $s_{y.12} = 6.05$. We must digress briefly at this point, however, to explain the computation of c_{ii}.

Computation of the c_{ii}   The c_{ii} values are called *Gauss multipliers*. For those familiar with matrix algebra it may be enlightening to point out that they may be obtained by inverting the matrix of sums of squares and cross products that can be constructed by using the left-hand terms of the normal equations given in Equation 9.3.5. The c's may be found without the use of matrix algebra by solving the following two sets of equations:

$$c_{11}\sum x_{1j}'^2 + c_{12}\sum x_{1j}'x_{2j}' = 1 \atop c_{11}\sum x_{1j}'x_{2j}' + c_{12}\sum x_{2j}'^2 = 0 \Big\} \qquad (9.4.6)$$

$$c_{21}\sum x_{1j}'^2 + c_{22}\sum x_{1j}'x_{2j}' = 0 \atop c_{21}\sum x_{1j}'x_{2j}' + c_{22}\sum x_{2j}'^2 = 1 \Big\} \qquad (9.4.7)$$

In the above equations $c_{12} = c_{21}$. Note also that the sums of squares and cross products are the same as those in the normal equations (Equation 9.3.5).

When the analysis involves more than two independent variables, the c's are obtained by expanding Equations 9.4.6 and 9.4.7 so that there is one set of equations for each independent variable. Each set of equations also contains as many individual equations as there are independent variables. The 1 is placed to the right of the equal sign in all equations containing a term of the form $c_{ii}\sum x_i'^2$. For example, in Equation 9.4.6, a 1 is to the right of the equal sign in the equation containing $c_{11}\sum x_{1j}'^2$. Ezekiel and Fox (4) have written out the equations for the case of three independent variables, and they, as well as Snedecor and Cochran (1), Steel and Torrie (2), and Anderson and Bancroft (3), demonstrate the use of the abbreviated Doolittle method (5) of obtaining the c's as well as the regression coefficients. Anderson and Bancroft (3) give a numerical example for four independent variables.

When we substitute data from our illustrative example into Equations 9.4.6 and 9.4.7, we have

$$34.93c_{11} + 257.40c_{12} = 1 \atop 257.40c_{11} + 2241.60c_{12} = 0 \Big\}$$

$$34.93c_{21} + 257.40c_{22} = 0 \atop 257.40c_{21} + 2241.60c_{22} = 1 \Big\}$$

The solution of these equations yields

$$c_{11} = .1861$$
$$c_{12} = c_{21} = -.0213711$$
$$c_{22} = .0029001$$

We may now return to the problem of inference procedures regarding the individual partial regression coefficients. To test the null hypothesis that β_i is equal to some particular value, say β_{i0}, the following t statistic may be computed:

$$t = \frac{b_i - \beta_{i0}}{s_{bi}} \qquad (9.4.8)$$

where the degrees of freedom are equal to $n - k - 1$, and

$$s_{bi} = s_{y.12...k} \sqrt{c_{ii}} \qquad (9.4.9)$$

The standard error of the difference between two partial regression coefficients is given by

$$s_{(bi-bj)} = \sqrt{s_{y.12...k}^2 \left(c_{ii} + c_{jj} - 2c_{ij}\right)} \qquad (9.4.10)$$

so that we may test H_0: $\beta_i = \beta_j$ by computing

$$t = \frac{b_i - b_j}{s_{(bi-bj)}} \qquad (9.4.11)$$

which has $n - k - 1$ degrees of freedom. Confidence intervals for β_i and $\beta_i - \beta_j$ may be constructed in the usual way by using a value from the t distribution for the reliability factor and standard errors given above.

For our hospital length of stay example let us test the null hypothesis that number of previous admissions is irrelevant in predicting length of hospital stay. We proceed as follows:

1. *Data*   See Example 9.3.1.
2. *Assumptions*   See Section 9.2.
3. *Hypotheses*

$$H_0: \beta_1 = 0$$
$$H_A: \beta_1 \neq 0$$
$$\text{Let } \alpha = .05$$

4. *Test Statistic*   See Equation 9.4.8.
5. *Distribution of the Test Statistic*   When H_0 is true and the assumptions are met, the test statistic is distributed as Student's t with 12 degrees of freedom.
6. *Decision Rule*   Reject H_0 if the computed t is either greater than or equal to 2.1788 or less than or equal to -2.1788.
7. *Calculation of the Test Statistic*   By Equation 9.4.8 we compute

$$t = \frac{b_1 - 0}{s_{b_1}} = \frac{.06}{6.05\sqrt{.1861}} = \frac{.06}{2.609928} = .023$$

8. *Statistical Decision*   The null hypothesis is not rejected, since the computed value of t, .023, is between -2.1788 and $+2.1788$, the critical values of t for a two-sided test when $\alpha = .05$ and the degrees of freedom are 12.

9. *Conclusion*   We conclude, then, that there may not be a significant linear relationship between Y and X_1 when X_2 is held constant. At least these data do not provide evidence for such a relationship. In other words, the data of the present sample do not provide sufficient evidence to indicate that number of previous admissions, when used in a regression equation along with age, is a useful variable in predicting length of hospital stay. [For this test, $p > 2(.10) = .20$.]

Now, let us perform a similar test for the second partial regression coefficient, β_2.

$$H_0: \beta_2 = 0$$
$$H_A: \beta_2 \neq 0$$
$$\alpha = .05$$
$$t = \frac{b_2 - 0}{s_{b_2}} = \frac{1.05}{6.05\sqrt{.0029001}} = \frac{1.05}{.325808} = 3.223$$

In this case the null hypothesis is rejected, since 3.223 is greater than 2.1788. We conclude that there is a linear relationship between X_2 and Y when X_1 is held constant, and that age, used in this manner, is a useful variable for predicting length of hospital stay. [For this test, $p < 2(.005) = .01$.]

When the researcher has been led to conclude that a partial regression coefficient is not 0, he or she may be interested in obtaining a confidence interval for this β_i. A $100(1 - \alpha)$ percent confidence interval for a β_i is given by

$$b_i \pm t_{(1-\alpha/2, \, n-k-1)} s_{y.12\ldots k} \sqrt{c_{ii}}$$

For our illustrative example we may compute the following 95 percent confidence interval for β_2:

$$1.05 \pm (2.1788)(6.05)\sqrt{.0029001}$$
$$1.05 \pm .71$$
$$.34, 1.76$$

We may give this interval the usual probabilistic and practical interpretations. We are 95 percent confident that β_2 is contained in the interval from .34 to 1.76 since, in repeated sampling, 95 percent of the intervals that may be constructed in this manner will include the true parameter.

One should be aware of the problems involved in carrying out multiple hypothesis tests and constructing multiple confidence intervals from the same sample data. The effect on α of performing multiple hypothesis tests from the same data is discussed in Section 7.6. A similar problem arises when one wishes to construct confidence intervals for two or more partial regression coefficients. The intervals will not be independent, so that the tabulated confidence coefficient does not, in general apply. In other words, all such intervals would not be $100(1 - \alpha)$ percent confidence intervals. Durand (6) gives a procedure that may be followed when confidence intervals for more than one partial regression coefficient are desired. See also the book by Neter, Wasserman, and Kutner (7).

Another problem sometimes encountered in the application of multiple regression is an apparent incompatibility in the results of the various tests of significance that one may perform. In a given problem for a given level of significance, one or the other of the following situations may be observed.

1. R^2 and all b_i significant.

2. R^2 and some but not all b_i significant.

3. R^2 but none of the b_i significant.

4. All b_i significant but not R^2.

5. Some b_i significant, but not all nor R^2.

6. Neither R^2 nor any b_i significant.

Geary and Leser (8) identify these six situations and, after pointing out that situations 1 and 6 present no problem (since they both imply compatible results), discuss each of the other situations in some detail.

Notice that situation 2 exists in our illustrative example, where we have a significant R^2 but only one out of two significant regression coefficients. Geary and Leser (8) point out that this situation is very common, especially when a large number of independent variables have been included in the regression equation, and that the only problem is to decide whether or not to eliminate from the analysis one or more of the variables associated with nonsignificant coefficients.

EXERCISES

9.4.1 Refer to Exercise 9.3.1. (a) Calculate the coefficient of multiple determination; (b) perform an analysis of variance; (c) test the significance of each b_i $(i > 0)$. Let $\alpha = .05$ for all tests of significance. Determine the p value for all tests.

9.4.2 Refer to Exercise 9.3.2. Do the analysis suggested in 9.4.1.

9.4.3 Refer to Exercise 9.3.3. Do the analysis suggested in 9.4.1.

9.4.4 Refer to Exercise 9.3.4. Do the analysis suggested in 9.4.1.

9.4.5 Refer to Exercise 9.3.5. Do the analysis suggested in 9.4.1.

9.4.6 Refer to Exercise 9.3.6. Do the analysis suggested in 9.4.1.

9.5 USING THE MULTIPLE REGRESSION EQUATION

As we learned in the previous chapter, a regression equation may be used to obtain a computed value of Y, $\hat{y}$, when a particular value of X is given. Similarly we may use our multiple regression equation to obtain a $\hat{y}$ value when we are given particular values of the two or more X variables present in the equation.

Just as was the case in simple linear regression, we may, in multiple regression, interpret a $\hat{y}$ value in one of two ways. First we may interpret $\hat{y}$ as an estimate of the mean of the subpopulation of Y values assumed to exist for particular combinations of X_i values. Under this interpretation $\hat{y}$ is called an *estimate*, and when it is used for this purpose, the equation is thought of as an *estimating equation*. The second interpretation of $\hat{y}$ is that it is the value Y is most likely to assume for given values of the X_i. In this case $\hat{y}$ is called the *predicted value* of Y, and the equation is called a *prediction equation*. In both cases, intervals may be constructed about the $\hat{y}$ value when the assumptions of Section 9.2 hold true. When $\hat{y}$ is interpreted as an estimate of a population mean, the interval is called a *confidence interval*, and when $\hat{y}$ is interpreted as a predicted value of Y, the interval is called a *prediction interval*. Now let us see how each of these intervals is constructed.

The Confidence Interval for the Mean of a Subpopulation of Y Values Given Particular Values of the X_i We have seen that a $100(1 - \alpha)$ percent confidence interval for a parameter may be constructed by the general procedure of adding to and subtracting from the estimator a quantity equal to the reliability factor corresponding to $1 - \alpha$ multiplied by the standard error of the estimator. We have also seen that in the present situation our estimator is

$$\hat{y} = b_0 + b_1 x_{1j} + b_2 x_{2j} + \cdots + b_k x_{kj} \tag{9.5.1}$$

The standard error of this estimator for the case of two independent variables is given by

$$s_{y.12}\sqrt{\frac{1}{n} + c_{11}x_{1j}'^2 + c_{22}x_{2j}'^2 + 2c_{12}x_{1j}'x_{2j}'} \tag{9.5.2}$$

where x_{ij}' values are particular values of the X_i expressed as deviations from their mean. Expression 9.5.2, is easily generalized to any number of independent variables. See for example, Anderson and Bancroft (3). The $100(1 - \alpha)$ percent confidence interval for the three-variable case, then, is as follows:

$$\hat{y} \pm t_{(1-\alpha/2,\, n-k-1)} s_{y.12}\sqrt{\frac{1}{n} + c_{11}x_{1j}'^2 + c_{22}x_{2j}'^2 + 2c_{12}x_{1j}'x_{2j}'} \tag{9.5.3}$$

For our illustrative example, suppose we wish to estimate the mean length of stay for all 25-year-old patients with two previous admissions. In other words, given $x_{1j} = 2$ and $x_{2j} = 25$, what is our estimate of the mean of the corresponding subpopulation of Y values? To obtain a point estimate, we substitute $x_{1j} = 2$ and $x_{2j} = 25$ into Equation 9.3.8:

$$\hat{y} = 2.08 + .06(2) + 1.05(25)$$
$$= 28.45$$

To compute the standard error we first obtain $x'_{1j} = (x_{1j} - \bar{x}_1) = (2 - 1.93) = .07$ and $x'_{2j} = (x_{2j} - \bar{x}_2) = (25 - 33.40) = -8.40$. Substituting these and other appropriate values into Expression 9.5.3 gives

$$28.45 \pm (2.1788)(6.05)$$
$$\times \sqrt{\tfrac{1}{15} + (.1861)(.07)^2 + (.0029001)(-8.40)^2 + 2(-.0213711)(.07)(-8.40)}$$
$$28.45 \pm 7.19$$
$$21.26, 35.64$$

We interpret this interval in the usual ways. We are 95 percent confident that the interval from 21.26 to 35.64 includes the mean of the subpopulation of Y values for the specified combination of X_i values, since this parameter would be included in about 95 percent of the intervals that can be constructed in the manner shown.

The Prediction Interval for a Particular Value of Y Given Particular Values of the X_i When we interpret $\hat{y}$ as the value Y is most likely to assume when particular values of the X_i are observed, we may construct a prediction interval in the same way in which the confidence interval was constructed. The only difference in the two is the standard error. The standard error of the prediction is slightly larger than the standard error of the estimate, which causes the prediction interval to be wider than the confidence interval.

The standard error of the prediction for the three-variable case is given by

$$s_{y.12}\sqrt{1 + \frac{1}{n} + c_{11}x'^2_{1j} + c_{22}x'^2_{2j} + 2c_{12}x'_{1j}x'_{2j}} \qquad (9.5.4)$$

so that the $100(1 - \alpha)$ percent prediction interval is

$$\hat{y} \pm t_{(1-\alpha/2,\, n-k-1)}s_{y.12}\sqrt{1 + \frac{1}{n} + c_{11}x'^2_{1j} + c_{22}x'^2_{2j} + 2c_{12}x'_{1j}x'_{2j}} \qquad (9.5.5)$$

Suppose, for our illustrative example, a 25-year-old patient with two previous admissions is admitted. How long is that patient likely to stay? Our best point prediction is found by computing $\hat{y}$. That is,

$$\hat{y} = 2.08 + .06(2) + 1.05(25)$$
$$= 28.45$$

the same value obtained previously when we interpreted $\hat{y}$ as a point estimate of the subpopulation mean.

```
THE REGRESSION EQUATION IS
Y =     2.09 + .0570 X1 +  1.05 X2

                                     ST. DEV.      T-RATIO =
           COLUMN     COEFFICIENT    OF COEF.      COEF/S.D.
           --           2.086         6.739          0.31
X1         C2           0.057         2.613          0.02
X2         C3           1.0500        0.3262         3.22

THE ST. DEV. OF Y ABOUT REGRESSION LINE IS
S = 6.059
WITH (  15- 3) =  12 DEGREES OF FREEDOM

R-SQUARED = 85.0 PERCENT
R-SQUARED = 82.5 PERCENT, ADJUSTED FOR D.F.

ANALYSIS OF VARIANCE

  DUE TO      DF           SS       MS=SS/DF
REGRESSION     2       2502.39      1251.20
RESIDUAL      12        440.54        36.71
TOTAL         14       2942.93

FURTHER ANALYSIS OF VARIANCE
SS EXPLAINED BY EACH VARIABLE WHEN ENTERED IN THE ORDER GIVEN

  DUE TO      DF           SS
REGRESSION     2       2502.39
C2             1       2122.02
C3             1        380.37

            X1           Y      PRED. Y    ST.DEV.
ROW         C2          C1       VALUE     PRED. Y    RESIDUAL    ST.RES.
 1         0.00       15.00      24.14       2.60       -9.14      -1.67
 2         0.00       15.00      20.99       2.55       -5.99      -1.09
 3         0.00       21.00      25.19       2.69       -4.19      -0.77
 4         1.00       28.00      27.34       2.01        0.66       0.11
 5         1.00       30.00      28.39       1.90        1.61       0.28
 6         1.00       35.00      28.39       1.90        6.61       1.15
 7         1.00       40.00      29.44       1.84       10.56       1.83
 8         2.00       35.00      37.90       1.57       -2.90      -0.50
 9         2.00       30.00      28.45       3.30        1.55       0.30
10         2.00       45.00      42.10       2.06        2.90       0.51
11         3.00       50.00      48.46       2.11        1.54       0.27
12         3.00       60.00      55.81       3.71        4.19       0.87
13         4.00       45.00      43.26       4.10        1.74       0.39
14         4.00       60.00      59.01       3.16        0.99       0.19
15         5.00       50.00      60.12       3.53      -10.12      -2.05
```

FIGURE 9.4.1

Partial Printout of Analysis of Data of Example 9.3.1, Using MINITAB
Regression Analysis Program

We may obtain the following 95 percent prediction interval by using Expression 9.5.5:

$$28.45 \pm (2.1788)(6.05)$$

$$\times \sqrt{1 + \tfrac{1}{15} + (.1861)(.07)^2 + (.0029001)(-8.40)^2 + 2(-.0213711)(.07)(-8.40)}$$

$$28.45 \pm 15.01$$

$$13.44, 43.46$$

We are 95 percent confident that this patient would remain in the hospital somewhere between 13.44 and 43.46 days.

Computer Analysis Figure 9.4.1 shows part of the results of a computer analysis of the data of Example 9.3.1 using the MINITAB multiple regression program. Note that the column labeled ST. DEV. PRED. Y contains, for each set of observed X's, numerical values of the standard error obtained by Expression 9.5.2.

EXERCISES

For each of the following exercises compute the $\hat{y}$ value and construct (a) 95 percent confidence and (b) 95 percent prediction intervals for the specified values of X_i.

9.5.1 Refer to Exercise 9.3.1 and let $x_{1j} = 5$ and $x_{2j} = 3$.
9.5.2 Refer to Exercise 9.3.2 and let $x_{1j} = 2$ and $x_{2j} = 20$.
9.5.3 Refer to Exercise 9.3.3 and let $x_{1j} = 5$ and $x_{2j} = 6$.
9.5.4 Refer to Exercise 9.3.4 and let $x_{1j} = 1$ and $x_{2j} = 2$.
9.5.5 Refer to Exercise 9.3.5 and let $x_{1j} = 90$ and $x_{2j} = 80$.
9.5.6 Refer to Exercise 9.3.6 and let $x_{1j} = 50$, $x_{2j} = 95.0$, $x_{3j} = 2.00$, $x_{4j} = 6.00$, $x_{5j} = 75$, and $x_{6j} = 70$.

9.6 QUALITATIVE INDEPENDENT VARIABLES

The independent variables considered in the preceding discussion were all quantitative; that is, they yielded numerical values that either were counts or were measurements in the usual sense of the word. For example, some of the independent variables used in our examples and exercises were age, aptitude test scores, and number of current medical problems. Frequently, however, it is desirable to use one or more qualitative variables as independent variables in the regression model. Qualitative variables, it will be recalled, are those variables whose "values" are categories and convey the concept of attribute rather than amount or quantity. The variable marital status, for example, is a qualitative variable whose categories are "single," "married," "widowed," and "divorced." Other examples of qualitative variables include sex (male or female), diagnosis, race, occupation, and immunity status to some disease. In certain situations an investigator may suspect that including one or more variables such as these in the regression equation would contribute significantly to the reduction of the error sum of squares and thereby provide more precise estimates of the parameters of interest.

Suppose, for example, that we are studying the relationship between the dependent variable systolic blood pressure and the independent variables weight and age. We might also want to include the qualitative variable sex as one of the independent variables. Or suppose we wish to gain insight into the nature of the relationship between lung capacity and other relevant variables. Candidates for inclusion in the model might consist of such quantitative variables as height, weight, and age, as well as qualitative variables like sex, area of residence (urban, suburban, rural), and smoking status (current smoker, ex-smoker, never smoked).

In order to incorporate a qualitative independent variable in the multiple regression model it must be quantified in some manner. This may be accomplished through the use of what are known as *dummy variables*.

A dummy variable is a variable that assumes only a finite number of values (such as 0 or 1) for the purpose of identifying the different categories of a qualitative variable.

The term "dummy" is used to indicate the fact that the numerical values (such as 0 and 1) assumed by the variable have no quantitative meaning but are used merely to identify different categories of the qualitative variable under consideration.

The following are some examples of qualitative variables and the dummy variables used to quantify them.

Qualitative variable	*Dummy variable*
Sex (male, female):	$x_1 = \begin{cases} 1 \text{ for male} \\ 0 \text{ for female} \end{cases}$
Place of residence (urban, rural, suburban):	$x_1 = \begin{cases} 1 \text{ for urban} \\ 0 \text{ for rural and suburban} \end{cases}$
	$x_2 = \begin{cases} 1 \text{ for rural} \\ 0 \text{ for urban and suburban} \end{cases}$
Smoking status [current smoker, ex-smoker (has not smoked for 5 years or less), ex-smoker (has not smoked for more than 5 years), never smoked]:	$x_1 = \begin{cases} 1 \text{ for current smoker} \\ 0 \text{ otherwise} \end{cases}$
	$x_2 = \begin{cases} 1 \text{ for ex-smoker } (\leq 5 \text{ years}) \\ 0 \text{ otherwise} \end{cases}$
	$x_3 = \begin{cases} 1 \text{ for ex-smoker } (> 5 \text{ years}) \\ 0 \text{ otherwise} \end{cases}$

Note in these examples that when the qualitative variable has k categories, $k - 1$ dummy variables must be defined for all the categories to be properly coded. This rule is applicable for any multiple regression containing an intercept constant. The variable sex, with two categories, can be quantified by the use of only one dummy variable, while three dummy variables are required to quantify the variable smoking status, which has four categories.

The following examples illustrate some of the uses of qualitative variables in multiple regression. In the first example we assume that there is no interaction between the independent variables. Since the assumption of no interaction is not realistic in many instances, we illustrate, in the second example, the analysis that is appropriate when interaction between variables is accounted for.

Example 9.6.1 In a study of factors thought to be associated with birth weight, data from a simple random sample of 32 birth records were examined. Table 9.6.1 shows part of the data that were extracted from each record. There we see that we have two independent variables: length of gestation in weeks, which is quantitative; and smoking status of mother, a qualitative variable. For the analysis of the data we will quantify smoking status by means of a dummy variable that is coded 1 if the mother is a smoker and 0 if she is a nonsmoker. The data in Table 9.6.1 are plotted as a scatter diagram in Figure 9.6.1. The scatter diagram suggests that, in general, longer periods of gestation are associated with larger birth weights.

To obtain additional insight into the nature of these data we may enter them into a computer and employ an appropriate program to perform further analyses. For example, we enter the observations $y_1 = 2940$, $x_{11} = 38$, $x_{21} = 1$ for the first case, $y_2 = 3130$, $x_{12} = 38$, $x_{22} = 0$ for the second case, and so on. Figure 9.6.2 shows the computer output obtained with the use of the MINITAB multiple regression program.

We see in the printout that the multiple regression equation is

$$\hat{y}_j = b_0 + b_1 x_{1j} + b_2 x_{2j}$$
$$\hat{y}_j = -2390 + 143 x_{1j} - 245 x_{2j} \tag{9.6.1}$$

To observe the effect on this equation when we wish to consider only the births to smoking mothers, we let $x_{2j} = 1$. The equation then becomes

$$\hat{y}_j = -2390 + 143 x_{1j} - 245(1)$$
$$= -2635 + 143 x_{1j} \tag{9.6.2}$$

which has a y-intercept of -2635 and a slope of 143. Note that the y-intercept for the new equation is equal to $(b_0 + b_1) = [-2390 + (-245)] = -2635$.

Now let us consider only births to nonsmoking mothers. When we let $x_2 = 0$, our regression equation reduces to

$$\hat{y}_j = -2390 + 143 x_{1j} - 245(0)$$
$$= -2390 + 143 x_{1j} \tag{9.6.3}$$

TABLE 9.6.1

Data Collected on a Simple Random Sample of 32 Births, Example 9.6.1

Case	Y Birth weight (grams)	X_1 Gestation (weeks)	X_2 Smoking status of mother
1	2940	38	Smoker (S)
2	3130	38	Nonsmoker (N)
3	2420	36	S
4	2450	34	N
5	2760	39	S
6	2440	35	S
7	3226	40	N
8	3301	42	S
9	2729	37	N
10	3410	40	N
11	2715	36	S
12	3095	39	N
13	3130	39	S
14	3244	39	N
15	2520	35	N
16	2928	39	S
17	3523	41	N
18	3446	42	S
19	2920	38	N
20	2957	39	S
21	3530	42	N
22	2580	38	S
23	3040	37	N
24	3500	42	S
25	3200	41	S
26	3322	39	N
27	3459	40	N
28	3346	42	S
29	2619	35	N
30	3175	41	S
31	2740	38	S
32	2841	36	N

The slope of this equation is the same as the slope of the equation for smoking mothers, but the y-intercepts are different. The y-intercept for the equation associated with nonsmoking mothers is larger than the one for the smoking mothers. These results show that for this sample babies born to mothers who do not smoke weighed, on the average, more than babies born to mothers who do smoke, when length of gestation is taken into account. The amount of the difference, on the average, is 245 grams. Stated another way, we can say that for this sample babies born to mothers who smoke weighed, on the average, 245

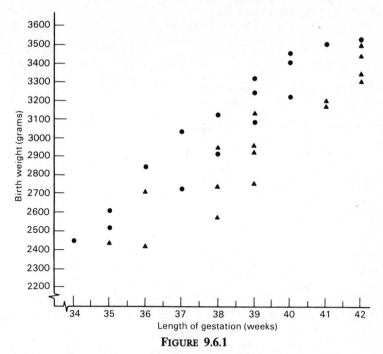

FIGURE 9.6.1

Birth Weights and Length of Gestation for 32 Births: (▲) Smoking and (●) Nonsmoking Mothers

grams less than the babies born to mothers who do not smoke, when length of gestation is taken into account. Figure 9.6.3 shows the scatter diagram of the original data along with a plot of the two regression lines (Equations 9.6.2 and 9.6.3).

At this point a question arises regarding what inferences we can make about the sampled population on the basis of these sample results. First of all, we may ask if the sample difference of 245 grams is significant. In other words, does smoking have an effect on birth weight? We may answer this question through the following hypothesis testing procedure.

1. *Data* The data are as given in Example 9.6.1.

2. *Assumptions* We presume that the assumptions underlying multiple regression analysis are met.

3. *Hypotheses* $H_0: \beta_2 = 0$; $H_A: \beta_2 \neq 0$. Suppose we let $\alpha = .05$.

4. *Test Statistic* The test statistic is $t = (b_2 - 0)/s_{b_2}$.

5. *Distribution of Test Statistic* When the assumptions are met and H_0 is true the test statistic is distributed as Student's t with 29 degrees of freedom.

```
THE REGRESSION EQUATION IS
Y = - 2390. + 143. X1 - 245. X2
```

	COLUMN	COEFFICIENT	ST. DEV. OF COEF.	T-RATIO = COEF/S.D.
	--	-2389.6	349.2	-6.84
K1	C2	143.100	9.128	15.68
K2	C3	-244.54	41.98	-5.83

```
THE ST. DEV. OF Y ABOUT REGRESSION LINE IS
S = 115.5
WITH ( 32- 3) = 29 DEGREES OF FREEDOM

R-SQUARED = 89.6 PERCENT
R-SQUARED = 88.9 PERCENT, ADJUSTED FOR D.F.

ANALYSIS OF VARIANCE
```

DUE TO	DF	SS	MS=SS/DF
REGRESSION	2	3348719	1674360
RESIDUAL	29	387070	13347
TOTAL	31	3735789	

FIGURE 9.6.2

Partial Computer Printout, MINITAB Multiple Regression Analysis, Example 9.6.1

6. *Decision Rule* We reject H_0 if the computed t is either greater than or equal to 2.0452 or less than or equal to -2.0452.

7. *Calculation of Test Statistic* The calculated value of the test statistic appears in Figure 9.6.2 as the t ratio for the coefficient associated with the variable appearing in column 3 of Table 9.6.1. This coefficient, of course, is b_2. We see that the computed t is -5.83.

8. *Statistical Decision* Since $-5.83 < -2.0452$, we reject H_0.

9. *Conclusion* We conclude that, in the sampled population, whether or not the mothers smoke does have an effect on the birth weights of their babies.

For this test we have $p < 2(.005)$ since -5.83 is less than -2.7564.

Given that we are able to conclude that in the sampled population the smoking status of the mothers does have an effect on the birth weights of their babies, we may now inquire as to the magnitude of the effect. Our best point estimate of the average difference in birth weights, when length of gestation is taken into account, is 245 grams in favor of babies born to mothers who do not smoke. We may obtain an interval estimate of the mean amount of the difference by using information from the computer printout by means of the following expression:

$$b_2 \pm ts_{b_2}$$

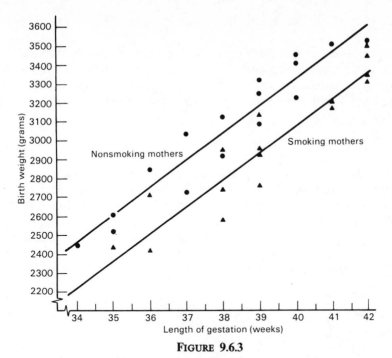

FIGURE 9.6.3

Birth Weights and Length of Gestation for 32 Births and the Fitted Regression Lines: (▲) Smoking and (●) Nonsmoking Mothers

For a 95 percent confidence interval we have

$$-244.54 \pm 2.0452(41.98)$$

$$-330.3975, -158.6825$$

Thus we are 95 percent confident that the difference is somewhere between about 159 grams and 331 grams.

   The reader may have correctly surmised that an alternative analysis of the data of Example 9.6.1 would consist of fitting two separate regression equations: one to the subsample of mothers who smoke and another to the subsample of those who do not. Such an approach, however, lacks some of the advantages of the dummy variable technique and is a less desirable procedure when the latter procedure is valid. If we can justify the assumption that the two separate regression lines have the same slope, we can get a better estimate of this common slope through the use of dummy variables, which entails pooling of the data from the two subsamples. In Example 9.6.1 the estimate using a dummy variable is based on a total sample size of 32 observations, whereas separate estimates would each be based on a sample of only 16 observations. The dummy variable approach also yields more precise inferences regarding other parameters since more degrees of freedom are available for the calculation of the error mean square.

Now let us consider the situation in which interaction between the variables is assumed to be present. Suppose, for example, that we have two independent variables: one quantitative variable x_1 and one qualitative variable with three response levels yielding the two dummy variables X_2 and X_3. The model, then, would be

$$y_j = \beta_0 + \beta_1 X_{1j} + \beta_2 X_{2j} + \beta_3 X_{3j} + \beta_4 X_{1j} X_{2j} + \beta_5 X_{1j} X_{3j} + e_j \quad (9.6.4)$$

in which $\beta_4 X_{1j} X_{2j}$ and $\beta_5 X_{1j} X_{3j}$ are called *interaction terms* and represent the interaction between the quantitative and the qualitative independent variables. Note that there is no need to include in the model the term containing $X_{2j} X_{3j}$; it will always be zero because when $X_2 = 1$, $X_3 = 0$ and when $X_3 = 1$, $X_2 = 0$. The model of Equation 9.6.4 allows for a different slope and Y-intercept for each level of the qualitative variable.

Suppose we use dummy variable coding to quantify the qualitative variable as follows:

$$X_2 = \begin{cases} 1 \text{ for level 1} \\ 0 \text{ otherwise} \end{cases}$$

$$X_3 = \begin{cases} 1 \text{ for level 2} \\ 0 \text{ otherwise} \end{cases}$$

The three sample regression equations for the three levels of the qualitative variable, then, are as follows:

Level 1 ($X_2 = 1$, $X_3 = 0$):

$$\hat{y}_j = b_0 + b_1 x_{1j} + b_2(1) + b_3(0) + b_4 x_{1j}(1) + b_5 x_{1j}(0)$$
$$= b_0 + b_1 x_{1j} + b_2 + b_4 x_{1j}$$
$$= (b_0 + b_2) + (b_1 + b_4) x_{1j} \quad (9.6.5)$$

Level 2 ($X_2 = 0$, $X_3 = 1$):

$$\hat{y}_j = b_0 + b_1 x_{1j} + b_2(0) + b_3(1) + b_4 x_{1j}(0) + b_5 x_{1j}(1)$$
$$= b_0 + b_1 x_{1j} + b_3 + b_5 x_{1j}$$
$$= (b_0 + b_3) + (b_1 + b_5) x_{1j} \quad (9.6.6)$$

Level 3 ($X_2 = 0$, $X_3 = 0$):

$$\hat{y} = b_0 + b_1 x_{1j} + b_2(0) + b_3(0) + b_4 x_{1j}(0) + b_5 x_{1j}(0)$$
$$\hat{y}_j = b_0 + b_1 x_{1j} \quad (9.6.7)$$

Let us illustrate these results by means of an example.

Example 9.6.2 A team of mental health researchers wish to compare three methods (A, B, and C) of treating severe depression. They would also like to study the relationship between age and treatment effectiveness as well as the interaction (if any) between age and treatment. Each member of a simple random sample of 36 patients, comparable with respect to diagnosis and severity of depression, was randomly assigned to receive either treatment A, B, or C. The results are shown in Table 9.6.2. The dependent variable Y is treatment effectiveness, the quantitative independent variable X_1 is patient's age at nearest birth-

TABLE 9.6.2

Data for Example 9.6.2

Measure of effectiveness	Age	Method of treatment
56	21	A
41	23	B
40	30	B
28	19	C
55	28	A
25	23	C
46	33	B
71	67	C
48	42	B
63	33	A
52	33	A
62	56	C
50	45	C
45	43	B
58	38	A
46	37	C
58	43	B
34	27	C
65	43	A
55	45	B
57	48	B
59	47	C
64	48	A
61	53	A
62	58	B
36	29	C
69	53	A
47	29	B
73	58	A
64	66	B
60	67	B
62	63	A
71	59	C
62	51	C
70	67	A
71	63	C

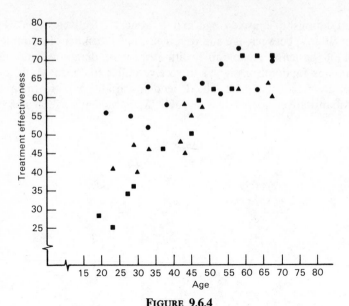

FIGURE 9.6.4

Scatter diagram of Data for Example 9.6.2. (●) Treatment A; (▲) Treatment B; (■) Treatment C

day, and the independent variable type of treatment is a qualitative variable that occurs at three levels. The following dummy variable coding is used to quantify the qualitative variable:

$$X_2 = \begin{cases} 1 \text{ if treatment A} \\ 0 \text{ otherwise} \end{cases}$$

$$X_3 = \begin{cases} 1 \text{ if treatment B} \\ 0 \text{ otherwise} \end{cases}$$

The scatter diagram for these data is shown in Figure 9.6.4. Table 9.6.3 shows the data as they were entered into a computer for analysis, and Figure 9.6.5 contains the printout of the analysis using the MINITAB multiple regression program.

Now let us examine the printout to see what it provides in the way of insight into the nature of the relationships among the variables. The least-squares equation is

$$\hat{y}_j = 6.21 + 1.03x_{1j} + 41.3x_{2j} + 22.7x_{3j} - .703x_{1j}x_{2j} - .510x_{1j}x_{3j}$$

TABLE 9.6.3

Data for Example 9.6.2 Coded for Computer Analysis

Y	X_1	X_2	X_3	$X_1 X_2$	$X_1 X_3$
56	21	1	0	21	0
55	28	1	0	28	0
63	33	1	0	33	0
52	33	1	0	33	0
58	38	1	0	38	0
65	43	1	0	43	0
64	48	1	0	48	0
61	53	1	0	53	0
69	53	1	0	53	0
73	58	1	0	58	0
62	63	1	0	63	0
70	67	1	0	67	0
41	23	0	1	0	23
40	30	0	1	0	30
46	33	0	1	0	33
48	42	0	1	0	42
45	43	0	1	0	43
58	43	0	1	0	43
55	45	0	1	0	45
57	48	0	1	0	48
62	58	0	1	0	58
47	29	0	1	0	29
64	66	0	1	0	66
60	67	0	1	0	67
28	19	0	0	0	0
25	23	0	0	0	0
71	67	0	0	0	0
62	56	0	0	0	0
50	45	0	0	0	0
46	37	0	0	0	0
34	27	0	0	0	0
59	47	0	0	0	0
36	29	0	0	0	0
71	59	0	0	0	0
62	51	0	0	0	0
71	63	0	0	0	0

The three regression equations for the three treatments are as follows:
Treatment A (Equation 9.6.5):

$$\hat{y}_j = (6.21 + 41.3) + (1.03 - .703) x_{1j}$$
$$= 47.51 + .327 x_{1j}$$

```
THE REGRESSION EQUATION IS
Y =    6.21 +  1.03 X1 +  41.3 X2
    +  22.7 X3 -  .703 X4 -  .510 K5
```

	COLUMN	COEFFICIENT	ST. DEV. OF COEF.	T-RATIO = COEF/S.D.
	--	6.211	3.350	1.85
X1	C2	1.03339	0.07233	14.29
K2	C3	41.304	5.085	8.12
K3	C4	22.707	5.091	4.46
K4	C5	-0.7029	0.1090	-6.45
K5	C6	-0.5097	0.1134	-4.62

```
THE ST. DEV. OF Y ABOUT REGRESSION LINE IS
S = 3.925
WITH ( 36- 6) =  30 DEGREES OF FREEDOM


R-SQUARED = 91.4 PERCENT
R-SQUARED = 90.0 PERCENT, ADJUSTED FOR D.F.


ANALYSIS OF VARIANCE
```

DUE TO	DF	SS	MS=SS/DF
REGRESSION	5	4932.85	986.57
RESIDUAL	30	462.15	15.40
TOTAL	35	5395.00	

```
FURTHER ANALYSIS OF VARIANCE
SS EXPLAINED BY EACH VARIABLE WHEN ENTERED IN THE ORDER GIVEN
```

DUE TO	DF	SS
REGRESSION	5	4932.85
C2	1	3424.43
C3	1	803.80
C4	1	1.19
C5	1	375.00
C6	1	328.42

FIGURE 9.6.5

Computer Printout, MINITAB Multiple Regression Analysis, Example 9.6.2

Treatment B (Equation 9.6.6):

$$\hat{y}_j = (6.21 + 22.7) + (1.03 - .510)x_{1j}$$
$$= 28.91 + .520x_{1j}$$

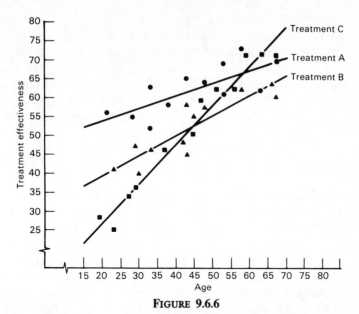

FIGURE 9.6.6

**Scatter Diagram of Data for Example 9.6.2 with the Fitted Regression Lines. (●)
Treatment A; (▲) Treatment B; (■) Treatment C**

Treatment C (Equation 9.6.7):

$$\hat{y}_j = 6.21 + 1.03x_{1j}$$

Figure 9.6.6 contains the scatter diagram of the original data along with the regression equations for the three treatments. Visual inspection of Figure 9.6.6 suggests that treatments A and B do not differ greatly with respect to their slopes, but their y-intercepts are considerably different. The graph suggests that treatment A is better than treatment B for younger patients, but the difference is less dramatic with older patients. Treatment C appears to be decidedly less desirable than both treatments A and B for younger patients, but is about as effective as treatment B for older patients. These subjective impressions are compatible with the contention that there is interaction between treatments and age.

What we see in Figure 9.6.6, however, are sample results. What can we conclude about the population from which the sample was drawn?

For an answer let us look at the t ratios on the computer printout in Figure 9.6.5. Each of these is the test statistic

$$t = \frac{b_j - 0}{s_{b_j}}$$

for testing H_0: $\beta_i = 0$. We see by Equation 9.6.5 that the y-intercept of the regression line for treatment A is equal to $b_0 + b_2$. Since the t ratio of 8.12 for testing H_0: $\beta_2 = 0$ is greater than the critical t of 2.0423 (for $\alpha = .05$) , we can reject H_0 that $\beta_2 = 0$ and conclude that the y-intercept of the population regression line for treatment A is different from the y-intercept of the population regression line for treatment C, which has a y-intercept of β_0. Similarly, since the t ratio of 4.45 for testing H_0: $\beta_3 = 0$ is also greater than the critical t of 2.0423, we can conclude (at the .05 level of significance) that the y-intercept of the population regression line for treatment B is also different from the y-intercept of the population regression line for treatment C. (See the y-intercept of Equation 9.6.6.)

Now let us consider the slopes. We see by Equation 9.6.5 that the slope of the regression line for treatment A is equal to b_1 (the slope of the line for treatment C) + b_4. Since the t ratio of -6.45 for testing H_0: $\beta_4 = 0$ is less than the critical t of -2.0423, we can conclude (for $\alpha = .05$) that the slopes of the population regression lines for treatments A and C are different. Similarly, since the computed t ratio for testing H_0: $\beta_5 = 0$ is also less than -2.0423, we conclude (for $\alpha = .05$) that the population regression lines for treatments B and C have different slopes (see the slope of Equation 9.6.6). Thus we conclude that there is interaction between age and type of treatment. This is reflected by a lack of parallelism among the regression lines in Figure 9.6.6.

Another question of interest is this: Is the slope of the population regression line for treatment A different from the slope of the population regression line for treatment B? To answer this question requires computational techniques beyond the scope of this text. The interested reader is referred to the books by Neter, Wasserman, and Kutner (7) and Kleinbaum and Kupper (9) for help with this problem.

In Section 9.4 the reader was warned that there are problems involved in making multiple inferences from the same sample data. The references cited in that section may be consulted for procedures to be followed when multiple inferences, such as those discussed in this section, are desired.

We have discussed only two situations in which the use of dummy variables is appropriate. More complex models involving the use of one or more qualitative independent variables in the presence of two or more quantitative variables may be appropriate in certain circumstances. More complex models are discussed by Mendenhall and McClave (10), Kleinbaum and Kupper (9), Draper and Smith (11), and Neter, Wasserman, and Kutner (7).

EXERCISES

9.6.1 The following data were collected on a simple random sample of 22 subjects seen at a health fair conducted at a suburban shopping mall.

Health status (Y)	Age	Hypertensive?
45.8	37	Yes
60.0	62	Yes
52.9	49	Yes
43.9	61	No
30.1	32	Yes
51.8	57	Yes
46.1	57	No
32.5	38	Yes
16.0	41	No
27.0	36	No
30.3	46	No
35.0	54	No
64.1	57	Yes
45.3	45	Yes
48.6	39	Yes
51.0	72	No
22.1	35	No
59.9	73	No
29.8	48	No
38.2	32	Yes
51.8	67	No
29.1	42	No

A larger value of the dependent variable indicates a poorer quality of health. Use dummy variable coding to quantify the qualitative variable and use multiple regression to obtain a least-squares equation describing the relationship between health status and age for the hypertensive group and the group with normal blood pressures. Can we conclude that blood pressure status has an effect on level of health? Let $\alpha = .05$ and find the p value. Construct the 95 percent confidence interval for the difference between the mean level of health for the two groups when age is taken into account. Draw a scatter diagram of the data and plot the fitted regression equations.

9.6.2 Researchers wished to study the effect of biofeedback and manual dexterity on the ability of patients to accurately perform a complicated task. Twenty-eight patients were randomly selected from those referred for physical therapy. The 28 were then randomly assigned to either receive or not receive biofeedback. The dependent variable is the number of consecutive repetitions of the task completed before an error was made. The following are the results.

Biofeedback?	Manual dexterity score	Number of repetitions (Y)
Yes	225	88
Yes	88	102
No	162	73

(Table continued)

Biofeedback?	Manual dexterity score	Number of repetitions (Y)
Yes	90	105
No	245	51
Yes	150	52
Yes	87	106
Yes	212	76
Yes	112	100
Yes	77	112
No	137	89
No	171	52
No	199	49
Yes	137	75
No	149	50
Yes	251	75
No	102	75
Yes	90	112
No	180	55
Yes	25	115
No	142	50
No	88	87
No	87	106
No	101	91
Yes	211	75
Yes	136	70
No	100	100
Yes	100	100

Use dummy variable coding and regression techniques to analyze these data. Draw a scatter diagram of the original data and plot the fitted regression equations.

9.6.3 A simple random sample of persons responding to a newspaper advertisement for volunteers to participate in a medical research project was given a test to measure level of knowledge of good health practices. The scores on the test, along with the subjects' ages and areas of residence, are given in the following table.

Score (Y)	Age in years	Area of residence	Score (Y)	Age in years	Area of residence
112	65	C	240	37	A
230	30	B	88	56	B
280	23	A	120	67	A
205	59	A	220	22	B
236	48	A	260	30	A
260	25	B	45	69	B
176	39	C	210	30	C
245	20	A	225	40	A

(Table continued)

Score (Y)	Age in years	Area of residence	Score (Y)	Age in years	Area of residence
123	66	A	126	45	B
176	40	B	200	26	C
216	25	B	240	21	B
176	63	A	170	56	A
90	75	C	116	58	C
176	69	A	120	50	B
200	52	A	194	33	B
45	75	B	156	48	C
110	75	C	100	75	A
150	62	C	115	71	A
148	70	C	65	69	B
265	20	B	240	25	A

Use dummy variable coding to quantify the qualitative variable and obtain the regression equation for each area of residence. What can one conclude from the results of the analysis? Is there interaction between age and area of residence? Plot the original data and the fitted regression equations. If there appears to be interaction between the two independent variables, how do you explain it?

9.7 THE MULTIPLE CORRELATION MODEL

We pointed out in the preceding chapter that while regression analysis is concerned with the form of the relationship between variables, the objective of correlation analysis is to gain insight into the strength of the relationship. This is also true in the multivariable case, and in this section we investigate methods for measuring the strength of the relationship among several variables. First, however, let us define the model and assumptions on which our analysis rests.

We may write the correlation model as

$$y_j = \beta_0 + \beta_1 x_{1j} + \beta_2 x_{2j} + \cdots + \beta_k x_{kj} + e_j \qquad (9.7.1)$$

Where y_j is a typical value from the population of values of the variable, Y, the β's are the regression coefficients defined in Section 9.2, the x_{ij} are particular (known) values of the random variables X_i. This model is similar to the multiple regression model, but there is one important distinction. In the multiple regression model, given in Equation 9.2.1, the X_i are nonrandom variables, but in the multiple correlation model the X_i are random variables. In other words, in the correlation model there is a joint distribution of Y and the X_i that we call a *multivariate distribution*. Under this model, the variables are no longer thought of as being dependent or independent, since logically they are interchangeable and either of the X_i may play the role of Y.

Typically random samples of units of association are drawn from a population of interest, and measurements of Y and the X_i are made.

A least-squares plane or hyperplane is fitted to the sample data by methods described in Section 9.3, and the same uses may be made of the resulting

TABLE 9.7.1

**Measurements Taken on 11 Apparently Normal
Males Between the Ages of 14 and 24 Years**

Serum cholesterol, mg/100 cc (Y)	Weight in kilograms (X_1)	Systolic blood pressure (X_2)
162.2	51.0	108
158.0	52.9	111
157.0	56.0	115
155.0	56.5	116
156.0	58.0	117
154.1	60.1	120
169.1	58.0	124
181.0	61.0	127
174.9	59.4	122
180.2	56.1	121
174.0	61.2	125

equation. Inferences may be made about the population from which the sample was drawn if it can be assumed that the underlying distribution is normal, that is, if it can be assumed that the joint distribution of Y and the X_i is a *multivariate normal distribution*. In addition, sample measures of the degree of the relationship among the variables may be computed and, under the assumption that sampling is from a multivariate normal distribution, the corresponding parameters may be estimated by means of confidence intervals, and hypothesis tests may be carried out. Specifically, we may compute an estimate of the *multiple correlation coefficient* that measures the dependence between Y and the X_i. This is a straightforward extension of the concept of correlation between two variables that we discuss in Chapter 8. We may also compute *partial correlation coefficients* that measure the intensity of the relationship between any two variables when the influence of all other variables has been removed.

To illustrate the concepts and techniques of correlation analysis, let us consider an example.

Example 9.7.1 The measurements shown in Table 9.7.1 were taken on 11 apparently normal males between the ages of 14 and 24 years. The researcher who gathered the data wished to investigate the nature and strength of the relationships among the three variables.

The Multiple Correlation Coefficient As a first step in analyzing the relationships among the variables, we look at the multiple correlation coefficient.

The multiple correlation coefficient is the square root of the coefficient of determination and, consequently, the sample value may be computed by taking the square root of Equation 9.4.5. That is,

$$R_{y.12\ldots k} = \sqrt{R_{y.12\ldots k}^2} = \sqrt{\frac{\sum(\hat{y} - \bar{y})^2}{\sum(y_j - \bar{y})^2}} \qquad (9.7.2)$$

The numerator of the term under the radical in Equation 9.7.2, which is the explained sum of squares, is given by Equation 9.4.3, which we recall contains b_1 and b_2, the sample partial regression coefficients. We compute these by the methods of Section 9.3.

First, the various sums, sums of squares, and sums of cross products must be computed. They are as follows:

$$\sum x_{1j} = 630.20 \qquad \sum x_{2j} = 1306.00 \qquad \sum y_j = 1821.50$$
$$\sum x_{1j}^2 = 36{,}209.68 \qquad \sum x_{2j}^2 = 155{,}410.00 \qquad \sum y_j^2 = 302{,}723.51$$
$$\sum x_{1j}x_{2j} = 74{,}995.80 \qquad \sum x_{1j}y_j = 104{,}485.19 \qquad \sum x_{2j}y_j = 216{,}682.00$$

When we compute the sums of squares and cross products of

$$y_j' = (y_j - \bar{y})$$
$$x_{1j}' = (x_{1j} - \bar{x}_1) \quad \text{and} \quad x_{2j}' = (x_{2j} - \bar{x}_2)$$

we have

$$\sum x_{1j}'^2 = 36{,}209.68 - (630.20)^2/11 = 104.95$$
$$\sum x_{2j}'^2 = 155{,}410.00 - (1306.00)^2/11 = 352.18$$
$$\sum y_j'^2 = 302{,}723.51 - (1821.5)^2/11 = 1099.67$$
$$\sum x_{1j}'x_{2j}' = 74{,}995.8 - (630.20)(1306.00)/11 = 173.87$$
$$\sum x_{1j}'y_j' = 104{,}485.19 - (630.20)(1821.50)/11 = 129.80$$
$$\sum x_{2j}'y_j' = 216{,}682.00 - (1306.00)(1821.50)/11 = 420.27$$

The normal equations, by Equations 9.3.9, are

$$104.95b_1 + 173.87b_2 = 129.80$$
$$173.87b_1 + 352.18b_2 = 420.27$$

The simultaneous solution of these equations gives $b_1 = -4.06$, $b_2 = 3.20$. We obtain b_0 by substituting appropriate values into Equation 9.3.11:

$$b_0 = 165.59 - (-4.06)(57.29) - (3.20)(118.73) = 18.25$$

The least-squares equation, then, is

$$\hat{y} = 18.25 - 4.06x_{1j} + 3.20x_{2j}$$

This equation may be used for estimation and prediction purposes and may be evaluated by the methods discussed in Section 9.4.

We now have the necessary quantities for computing the multiple correlation coefficient. We first compute the explained sum of squares by Equation 9.4.3:

$$SSR = (-4.06)(129.80) + (3.20)(420.27)$$
$$= 817.876$$

The total sum of squares by Equation 9.4.2 is

$$SST = 302{,}723.51 - (1821.5)^2/11$$
$$= 1{,}099.669$$

The coefficient of multiple determination, then, is

$$R_{y.12}^2 = \frac{817.876}{1099.669} = .7437$$

and the multiple correlation coefficient is

$$R_{y.12} = \sqrt{.7437} = .86$$

We interpret $R_{y.12}$ as a measure of the correlation between serum cholesterol level, weight, and systolic blood pressure in our sample of apparently normal males between 14 and 24 years of age. If our data constitute a random sample from the population of such persons we may use $R_{y.12}$ as an estimate of $\rho_{y.12}$, the true population multiple correlation coefficient. We may also interpret $R_{y.12}$ as the simple correlation coefficient between y_i and $\hat{y}$, respectively, the observed and calculated values of the "dependent" variable. Perfect correspondence between the observed and calculated values of Y will result in a correlation coefficient of 1, while a complete lack of a linear relationship between observed and calculated values yields a correlation coefficient of 0. The multiple correlation coefficient is always given a positive sign.

We may test the null hypothesis that $\rho_{y.12\ldots k} = 0$ by computing

$$F = \frac{R_{y.12\ldots k}^2}{1 - R_{y.12\ldots k}^2} \cdot \frac{n - k - 1}{k} \tag{9.7.3}$$

The numerical value obtained from Equation 9.7.3 is compared with the tabulated value of F with k and $n - k - 1$ degrees of freedom. The reader will recall that this is identical to the test of H_0: $\beta_1 = \beta_2 = \cdots = \beta_k = 0$ described in Section 9.4.

For our present example let us test the null hypothesis that $\rho_{y.12} = 0$ against the alternative that $\rho_{y.12} \neq 0$. We compute

$$F = \frac{.7437}{(1 - .7437)} \cdot \frac{(11 - 2 - 1)}{2} = 11.61$$

Since 11.61 is greater than 11.04, $p < .005$, so that we may reject the null hypothesis at the .005 level of significance and conclude that serum cholesterol level is linearly correlated with weight and systolic blood pressure in the sampled population.

Further comments on the significance of observed multiple correlation coefficients may be found in Ezekiel and Fox (4), who discuss a paper on the subject by R. A. Fisher (12) and present graphs for constructing confidence intervals when the number of variables is eight or less. Kramer (13) presents tables for constructing confidence limits when the number of variables is greater than eight.

Partial Correlation   The researcher may wish to have a measure of the strength of the linear relationship between two variables when the effect of the remaining variables has been removed. Such a measure is provided by the *partial correlation* coefficient. For example, the partial correlation coefficient $\rho_{y.12}$ is a measure of the correlation between Y and X_1 when X_2 is held constant.

The partial correlation coefficients may be computed from the *simple correlation coefficients*. The simple correlation coefficients measure the correlation between two variables when no effort has been made to control other variables. In other words, they are the coefficients for any pair of variables that would be obtained by the methods of simple correlation discussed in Chapter 8.

Suppose we have three variables, Y, X_1, and X_2. The partial correlation coefficient measuring the correlation between Y and X_1 with X_2 held constant, for example, is written $r_{y1.2}$. In the subscript, the symbol to the right of the decimal point indicates which variable is held constant, while the two symbols to the left of the decimal point indicate which variables are being correlated. For the three-variable case, there are two other partial correlation coefficients that we may compute. They are $r_{y2.1}$ and $r_{12.y}$.

The square of the partial correlation coefficient is called the coefficient of partial determination. It provides useful information about the interrelationships among variables. Consider $r_{y1.2}$, for example. Its square, $r_{y1.2}^2$, tells us what proportion of the remaining variability in Y is explained by X_1 after X_2 has explained as much of the total variability in Y as it can.

For three variables these simple correlation coefficients may be obtained:

r_{y1}, the simple correlation between Y and X_1

r_{y2}, the simple correlation between Y and X_2

r_{12}, the simple correlation between X_1 and X_2

These may be computed as follows:

$$r_{y1} = \sum x'_{1j} y'_j \Big/ \sqrt{\sum x'^2_{1j} \sum y'^2_j} \tag{9.7.4}$$

$$r_{y2} = \sum x'_{2j} y'_j \Big/ \sqrt{\sum x'^2_{2j} \sum y'^2_j} \tag{9.7.5}$$

$$r_{12} = \sum x'_{1j} x'_{2j} \Big/ \sqrt{\sum x'^2_{1j} \sum x'^2_{2j}} \tag{9.7.6}$$

For our illustrative example we may use previously computed quantities to obtain

$$r_{y1} = 129.80 \Big/ \sqrt{(104.95)(1099.67)} = .382$$

$$r_{y2} = 420.27 \Big/ \sqrt{(352.18)(1099.67)} = .675$$

$$r_{12} = 173.87 \Big/ \sqrt{(104.95)(352.18)} = .904$$

The sample partial correlation coefficients that may be computed in the three-variable case are

1. The partial correlation between Y and X_1 when X_2 is held constant:

$$r_{y1.2} = (r_{y1} - r_{y2} r_{12}) \Big/ \sqrt{(1 - r_{y2}^2)(1 - r_{12}^2)} \tag{9.7.7}$$

2. The partial correlation between Y and X_2 when X_1 is held constant:

$$r_{y2.1} = (r_{y2} - r_{y1} r_{12}) \Big/ \sqrt{(1 - r_{y1}^2)(1 - r_{12}^2)} \tag{9.7.8}$$

3. The partial correlation between X_1 and X_2 when Y is held constant:

$$r_{12.y} = (r_{12} - r_{y1} r_{y2}) \Big/ \sqrt{(1 - r_{y1}^2)(1 - r_{y2}^2)} \tag{9.7.9}$$

If we use the simple correlation coefficients already computed, we obtain the following partial correlation coefficients for our illustrative example:

$$r_{y1.2} = [.382 - (.675)(.904)] \Big/ \sqrt{(1 - .675^2)(1 - .904^2)} = -.723$$

$$r_{y2.1} = [.675 - (.382)(.904)] \Big/ \sqrt{(1 - .382^2)(1 - .904^2)} = .834$$

$$r_{12.y} = [.904 - (.382)(.675)] \Big/ \sqrt{(1 - .382^2)(1 - .675^2)} = .948$$

We may test the null hypothesis that the population partial correlation coefficient corresponding to any one of the above is 0 by means of the t test. For example, to test H_0: $\rho_{y1.2\ldots k} = 0$, we compute

$$t = r_{y1.2\ldots k}\sqrt{\frac{n - k - 1}{1 - r^2_{y1.2\ldots k}}} \qquad (9.7.10)$$

which is distributed as Student's t with $n - k - 1$ degrees of freedom.

Let us illustrate the procedure for our current example by testing H_0: $\rho_{12.y} = 0$ against the alternative, H_A: $\rho_{12.y} \neq 0$. The computed t is

$$t = .948\sqrt{\frac{11 - 2 - 1}{1 - .948^2}}$$
$$= 8.425$$

Since the computed t of 8.425 is larger than the tabulated t of 3.5554 for 8 degrees of freedom and $\alpha = .01$, (two-sided test), we may reject H_0 at the .01 level of significance and conclude that there is a significant correlation between systolic blood pressure and weight when serum cholesterol level is held constant. Significance tests for the other two partial correlation coefficients will be left as an exercise for the reader.

Although our illustration of correlation analysis is limited to the three-variable case, the concepts and techniques extend logically to the case of four or more variables. The number and complexity of the calculations increases rapidly as the number of variables increases.

EXERCISES

9.7.1 The following data were obtained on 12 males between the ages of 12 and 18 years (all measurements are in centimeters).

Height (Y)	Radius length (X₁)	Femur length (X₂)
149.0	21.00	42.50
152.0	21.79	43.70
155.7	22.40	44.75
159.0	23.00	46.00
163.3	23.70	47.00
166.0	24.30	47.90
169.0	24.92	48.95
172.0	25.50	49.90
174.5	25.80	50.30
176.1	26.01	50.90
176.5	26.15	50.85
179.0	26.30	51.10
Total 1992.1	290.87	573.85

$\sum x_{1j}^2 = 7087.6731$ $\sum x_{2j}^2 = 27541.8575$ $\sum y_j^2 = 331851.09$

$\sum x_{1j}x_{2j} = 13970.5835$ $\sum x_{1j}y_j = 48492.886$ $\sum x_{2j}y_j = 95601.09$

(a) Find the sample multiple correlation coefficient and test the null hypothesis that $\rho_{y.12} = 0$
(b) Find each of the partial correlation coefficients and test each for signifi-cance. Let $\alpha = .05$ for all tests.
(c) Determine the p value for each test.
(d) State your conclusions.

9.7.2 The following data were collected on 15 obese girls.

Weight in kilograms (Y)	Lean body weight (X₁)	Mean daily caloric intake (X₂)
79.2	54.3	2670
64.0	44.3	820
67.0	47.8	1210
78.4	53.9	2678
66.0	47.5	1205
63.0	43.0	815
65.9	47.1	1200
63.1	44.0	1180
73.2	44.1	1850
66.5	48.3	1260
61.9	43.5	1170
72.5	43.3	1852
101.1	66.4	1790
66.2	47.5	1250
99.9	66.1	1789
Total 1087.9	741.1	22739

$$\sum x_{1j}^2 = 37,439.95 \qquad \sum x_{2j}^2 = 39,161,759 \qquad \sum y_j^2 = 81,105.63$$

$$\sum x_{1j} x_{2j} = 1,154,225.2 \qquad \sum x_{1j} y_j = 55,021.31 \qquad \sum x_{2j} y_j = 1,707,725.3$$

(a) Find the multiple correlation coefficient and test it for significance.
(b) Find each of the partial correlation coefficients and test each for signifi-cance. Let $\alpha = .05$ for all tests.
(c) Determine the p value for each test.
(d) State your conclusions.

9.7.3 Refer to Exercise 9.3.5. Assume that the data fit the correlation model.

(a) Compute the multiple correlation coefficient and test it for significance.
(b) Compute each of the partial correlation coefficients and test each for significance. Let $\alpha = .05$ for all tests.

(c) Determine the p value for each test.
(d) What are your conclusions?

9.7.4 Refer to Example 9.7.1. Test $r_{y1.2}$ and $r_{y2.1}$ for significance at the .05 level. Determine the p value for each test.

9.7.5 A research project was conducted to study the relationships among intelligence, aphasia, and apraxia. The subjects were patients with focal left hemisphere damage. Scores on the following variables were obtained through application of standard tests.

$$Y = \text{intelligence}$$
$$X_1 = \text{ideomotor apraxia}$$
$$X_2 = \text{constructive apraxia}$$
$$X_3 = \text{lesion volume (pixels)}$$
$$X_4 = \text{severity of aphasia}$$

The results are shown in the following table. Find the multiple correlation coefficient and test for significance. Let $\alpha = .05$ and find the p value.

Subject	Y	X_1	X_2	X_3	X_4
1.	66	7.6	7.4	2296.87	2
2.	78	13.2	11.9	2975.82	8
3.	79	13.0	12.4	2839.38	11
4.	84	14.2	13.3	3136.58	15
5.	77	11.4	11.2	2470.50	5
6.	82	14.4	13.1	3136.58	9
7.	82	13.3	12.8	2799.55	8
8.	75	12.4	11.9	2565.50	6
9.	81	10.7	11.5	2429.49	11
10.	71	7.6	7.8	2369.37	6
11.	77	11.2	10.8	2644.62	7
12.	74	9.7	9.7	2647.45	9
13.	77	10.2	10.0	2672.92	7
14.	74	10.1	9.7	2640.25	8
15.	68	6.1	7.2	1926.60	5

9.8 CHOOSING THE INDEPENDENT VARIABLES FOR THE MULTIPLE REGRESSION EQUATION

One of the more troublesome problems in the use of multiple regression analysis is deciding what independent variables should be included in the equation. A final decision in most cases is the result of both statistical and nonstatistical considerations.

In the initial stages of determining the regression equation, the investigator will be guided by his knowledge of relevant variables and which ones are likely to be useful for prediction and estimation purposes. He or she will also take into account the relative costs and ease of obtaining measurements.

In this computer era the volume and complexity of calculations associated with a large number of variables do not pose the problem they once did. Even so, it is usually desirable (particularly from the point of view of cost and convenience of obtaining measurements) to include in the final equation fewer independent variables than are available. It is at this point that the investigator may employ statistical analysis as an aid to arriving at a decision. There are several procedures that can be followed. The most thorough, but also the most tedious and time-consuming method, even with the aid of a computer, is to regress Y on every subset of the X_i; that is, Y is regressed on each X, then on each pair of X's, then on each group of three X's, and so on. The best, according to some statistical criterion, of these regression equations will be the one selected unless some nonstatistical consideration such as cost necessitates a compromise. Because of the magnitude of the computations this method is not widely used.

A second approach, called the step-up or forward method, consists of introducing independent variables, one at a time, into the regression and at each stage statistically evaluating the "goodness" of the equation. The procedure continues until, according to some statistical criterion, a satisfactory equation is obtained.

A third method, called the step-down or backward method, is the reverse of the forward method. By this method the regression of Y on all X_i is carried out, and then independent variables are eliminated one at a time until a satisfactory equation has been obtained.

For a discussion of these and other methods for determining what independent variables to include in the analysis, refer to books and articles by Allen (14), Beale et al. (15), Draper and Smith (11), Garside (16), Gorman and Toman (17), Hocking and Leslie (18), Larson and Bancroft (19), Lindley (20), Schultz and Goggans (21), Smillie (22), Sprent (23), and Summerfield and Lubin (24).

9.9 SUMMARY

In this chapter we examine how the concepts and techniques of simple linear regression and correlation analysis are extended to the multiple-variable case. The least-squares method of obtaining the regression equation is presented and illustrated. This chapter also is concerned with the calculation of descriptive measures, tests of significance, and the uses to be made of the multiple regression equation. In addition, the methods and concepts of correlation analysis, including partial correlation, are discussed. For those who wish to extend their knowledge of multiple regression and correlation analysis, the references at the end of the chapter provide a good beginning.

When the assumptions underlying the methods of regression and correlation presented in this and the previous chapter are not met, the researcher must resort to alternative techniques. One alternative is to use a nonparametric procedure such as the ones discussed by Daniel (25, 26).

REVIEW QUESTIONS AND EXERCISES

1. What are the assumptions underlying multiple regression analysis when one wishes to infer about the population from which the sample data have been drawn?

2. What are the assumptions underlying the correlation model when inference is an objective?

3. Explain fully the following terms:
   (a) Coefficient of multiple determination
   (b) Multiple correlation coefficient
   (c) Simple correlation coefficient
   (d) Partial correlation coefficient

4. Describe a situation in your particular area of interest where multiple regression analysis would be useful. Use real or realistic data and do a complete regression analysis.

5. Describe a situation in your particular area of interest where multiple correlation analysis would be useful. Use real or realistic data and do a complete correlation analysis.

   In the exercises that follow carry out the indicated analysis and test hypotheses at the indicated significance levels. Compute the *p*-value for each test.

6. The following table shows certain pulmonary function values observed in 10 hospitalized patients.

X_1 Vital capacity (liters)	X_2 Total lung capacity (liters)	Y Forced expiratory volume (liters) per second
2.2	2.5	1.6
1.5	3.2	1.0
1.6	5.0	1.4
3.4	4.4	2.6
2.0	4.4	1.2
1.9	3.3	1.5
2.2	3.2	1.6
3.3	3.3	2.3
2.4	3.7	2.1
.9	3.6	.7

   Compute the multiple correlation coefficient and test for significance at the .05 level.

7. The following table shows the weight and total cholesterol and triglyceride levels in 15 patients with primary type II hyperlipoproteinemia just prior to initiation of treatment.

Y Weight (kg)	X_1 Total cholesterol (mg/100 ml)	X_2 Triglyceride (mg/100 ml)
76	302	139
97	336	101
83	220	57
52	300	56
70	382	113
67	379	42
75	331	84
78	332	186
70	426	164
99	399	205
75	279	230
78	332	186
70	410	160
77	389	153
76	302	139

Compute the multiple correlation coefficient and test for significance at the .05 level.

8. In a study of the relationship between creatinine excretion, height, and weight the data shown in the following table were collected on 20 infant males.

Infant	Creatinine excretion (mg/day) Y	Weight (kg) X_1	Height (cm) X_2
1	100	9	72
2	115	10	76
3	52	6	59
4	85	8	68
5	135	10	60
6	58	5	58
7	90	8	70
8	60	7	65
9	45	4	54
10	125	11	83
11	86	7	64
12	80	7	66
13	65	6	61
14	95	8	66
15	25	5	57
16	125	11	81
17	40	5	59
18	95	9	71
19	70	6	62
20	120	10	75

(a) Find the multiple regression equation describing the relationship among these variables.

(b) Compute R^2 and do an analysis of variance.

(c) Let $X_1 = 10$ and $X_2 = 60$ and find the predicted value of Y.

9. A study was conducted to examine those variables thought to be related to the job satisfaction of nonprofessional hospital employees. A random sample of 15 employees gave the following results.

Score on job satisfaction test (Y)	Coded intelligence score (X₁)	Index of personal adjustment (X₂)
54	15	8
37	13	1
30	15	1
48	15	7
37	10	4
37	14	2
31	8	3
49	12	7
43	1	9
12	3	1
30	15	1
37	14	2
61	14	10
31	9	1
31	4	5

(a) Find the multiple regression equation describing the relationship among these variables.

(b) Compute the coefficient of multiple determination and do an analysis of variance.

(c) Let $X_1 = 10$ and $X_2 = 5$ and find the predicted value of Y.

10. A medical research team obtained the index of adiposity, basal insulin, and basal glucose values on 21 normal subjects. The results are shown in the following table. The researchers wished to investigate the strength of the association among these variables.

Index of adiposity Y	Basal insulin (µU/ml) X₁	Basal glucose (mg/100 ml) X₂
90	12	98
112	10	103
127	14	101
137	11	102
103	10	90
140	38	108
105	9	100
92	6	101
92	8	92

(Table continued)

Index of adiposity Y	Basal insulin (μU / ml) X_1	Basal glucose (mg / 100ml) X_2
96	6	91
114	9	95
108	9	95
160	41	117
91	7	101
115	9	86
167	40	106
108	9	84
156	43	117
167	17	99
165	40	104
168	22	85

Compute the multiple correlation coefficient and test for significance at the .05 level.

11. As part of a study to investigate the relationship between stress and certain other variables the following data were collected on a simple random sample of 15 corporate executives.

(a) Find the least-squares regression equation for these data.

(b) Construct the analysis of variance table and test the null hypothesis of no relationship among the five variables.

(c) Test the null hypothesis that each slope in the regression model is equal to zero.

(d) Find the multiple coefficient of determination and the multiple correlation coefficient. Let $\alpha = .05$ and find the p value for each test.

Measure of stress (Y)	Measure of firm size (X_1)	Number of years in present position (X_2)	Annual salary ($\times 1,000$) (X_3)	Age (X_4)
101	812	15	$30	38
60	334	8	20	52
10	377	5	20	27
27	303	10	54	36
89	505	13	52	34
60	401	4	27	45
16	177	6	26	50
184	598	9	52	60
34	412	16	34	44
17	127	2	28	39
78	601	8	42	41
141	297	11	84	58
11	205	4	31	51
104	603	5	38	63
76	484	8	41	30

12. In the following table are the cardiac output (l/min) and oxygen consumption ($\dot{V}O_2$) values for a sample of adults (A) and children (C) who participated in a study designed to investigate the relationships among these variables.

   Measurements were taken both at rest and during exercise. Treat cardiac output as the dependent variable and use dummy variable coding and analyze the data by regression techniques. Explain the results. Plot the original data and the fitted regression equations.

Cardiac output (l/min)	$\dot{V}O_2$ (l/min)	Age group
4.0	.21	A
7.5	.91	C
3.0	.22	C
8.9	.60	A
5.1	.59	C
5.8	.50	A
9.1	.99	A
3.5	.23	C
7.2	.51	A
5.1	.48	C
6.0	.74	C
5.7	.70	C
14.2	1.60	A
4.1	.30	C
4.0	.25	C
6.1	.22	A
6.2	.61	C
4.9	.45	C
14.0	1.55	A
12.9	1.11	A
11.3	1.45	A
5.7	.50	C
15.0	1.61	A
7.1	.83	C
8.0	.61	A
8.1	.82	A
9.0	1.15	C
6.1	.39	A

13. A simple random sample of normal subjects between the ages of 6 and 18 yielded the data on total body potassium (mEq) and total body water (liters) shown in the following table.

   Let total potassium be the dependent variable and use dummy variable coding to quantify the qualitative variable. Analyze the data using regression techniques. Explain the results. Plot the original data and the fitted regression equations.

Total body potassium	Total body water	Sex
795	13	M
1590	16	F
1250	15	M
1680	21	M
800	10	F
2100	26	M
1700	25	F
1260	16	M
1370	18	F
1000	11	F
1100	14	M
1500	20	F
1450	19	M
1100	14	M
950	12	F
2400	26	M
1600	24	F
2400	30	M
1695	26	F
1510	21	F
2000	27	F
3200	33	M
1050	14	F
2600	31	M
3000	37	M
1900	25	F
2200	30	F

14. The data shown in the following table were collected as a part of a study in which the subjects were preterm infants with low birth weights born in three different hospitals.

   Use dummy variable coding and multiple regression techniques to analyze these data. May we conclude that the three sample hospital populations differ with respect to mean birth weight when gestational age is taken into account? May we conclude that there is interaction between hospital of birth and gestational age? Plot the original data and the fitted regression equations.

Birth weight (kg)	Gestation age (weeks)	Hospital of birth
1.4	30	A
.9	27	B
1.2	33	A
1.1	29	C
1.3	35	A

(Table continued)

Birth weight (kg)	Gestation age (weeks)	Hospital of birth
.8	27	B
1.0	32	A
.7	26	A
1.2	30	C
.8	28	A
1.5	32	B
1.3	31	A
1.4	32	C
1.5	33	B
1.0	27	A
1.8	35	B
1.4	36	C
1.2	34	A
1.1	28	B
1.2	30	B
1.0	29	C
1.4	33	C
.9	28	A
1.0	28	C
1.9	36	B
1.3	29	B
1.7	35	C
1.0	30	A
.9	28	A
1.0	31	A
1.6	31	B
1.6	33	B
1.7	34	B
1.6	35	C
1.2	28	A
1.5	30	B
1.8	34	B
1.5	34	C
1.2	30	A
1.2	32	C

15. Weight loss among cancer patients is a well-known phenomenon. Of interest to clinicians is the role played by metabolic abnormalities in the process. One investigation into the relationships among these variables yielded the following data on whole-body protein turnover (Y) and percentage of ideal body weight for height (X). Subjects were lung cancer patients and healthy controls of the same age. Select a simple random sample of size 15 from each group and do the following.

(a) Draw a scatter diagram of the sample data using different symbols for each of the two groups.

(b) Use dummy variable coding to analyze these data.

(c) Plot the two regression lines on the scatter diagram.

May one conclude that the two sampled populations differ with respect to mean protein turnover when percentage of ideal weight is taken into account?

May one conclude that there is interaction between health status and percentage of ideal body weight?

Prepare a verbal interpretation of the results of your analysis and compare your results with those of your classmates.

CANCER PATIENTS

Subject	X	Y	Subject	X	Y
1.	96.2	2.52	2.	85.0	4.28
3.	89.5	3.67	4.	98.1	3.33
5.	89.7	3.34	6.	84.3	3.55
7.	90.0	3.38	8.	90.9	2.54
9.	86.1	3.01	10.	103.9	1.69
11.	91.8	2.83	12.	86.1	2.44
13.	93.0	3.37	14.	98.2	2.35
15.	93.7	2.05	16.	87.4	3.52
17.	88.8	3.28	18.	96.3	2.46
19.	89.8	3.12	20.	100.0	2.32
21.	94.4	2.99	22.	92.9	2.86
23.	88.9	2.48	24.	78.8	3.88
25.	82.9	3.66	26.	86.4	3.79
27.	78.8	5.07	28.	90.1	3.19
29.	94.8	2.91	30.	93.2	3.24
31.	93.7	2.75	32.	88.7	3.40
33.	83.5	3.10	34.	96.9	2.46
35.	85.6	4.05	36.	93.1	3.29
37.	98.4	2.44	38.	92.3	3.11
39.	91.3	3.29	40.	88.5	3.50
41.	90.0	2.28	42.	88.5	3.91
43.	98.8	2.68	44.	92.7	2.94
45.	90.1	3.32	46.	88.7	3.62
47.	94.0	2.78	48.	84.7	3.08
49.	99.2	1.70	50.	96.2	3.32
51.	86.2	3.50	52.	87.6	2.74
53.	89.4	3.65	54.	92.4	3.23
55.	96.8	3.34	56.	80.6	3.68
57.	79.0	3.61	58.	83.2	4.17
59.	89.7	3.35	60.	82.6	4.17
61.	91.3	4.00	62.	81.7	3.67
63.	90.4	2.80	64.	94.9	3.37
65.	97.9	2.64	66.	88.5	2.97

Subject	X	Y	Subject	X	Y
67.	94.1	2.44	68.	90.2	2.90
69.	92.2	3.37	70.	80.6	4.61
71.	91.9	2.99	72.	80.9	3.58
73.	96.4	2.84	74.	86.6	3.81
75.	82.1	4.56	76.	85.6	3.56
77.	92.6	2.35	78.	96.9	3.32
79.	80.8	3.99	80.	82.8	3.71
81.	96.4	2.91	82.	89.0	3.46
83.	87.5	3.41	84.	94.1	2.95
85.	94.7	2.06	86.	77.7	4.61
87.	90.7	2.62	88.	95.0	3.06
89.	95.3	3.20	90.	92.9	2.56
91.	86.2	3.28	92.	94.3	2.51
93.	89.2	3.31	94.	82.4	4.52
95.	85.4	4.46	96.	92.8	3.36
97.	88.0	2.33	98.	88.4	3.13
99.	85.7	4.33	100.	89.3	2.97
101.	83.8	3.98	102.	75.9	4.16
103.	85.8	3.58	104.	93.8	3.39
105.	96.3	3.05	106.	95.0	3.48
107.	93.9	3.52	108.	91.0	3.22
109.	84.3	3.63	110.	86.7	3.87
111.	94.1	3.44	112.	92.7	2.78
113.	95.1	3.18	114.	86.8	3.67
115.	81.3	3.29	116.	93.1	2.82
117.	88.0	3.26	118.	100.1	3.04
119.	95.5	2.51	120.	94.7	3.44
121.	90.7	3.70	122.	97.8	2.83
123.	92.8	3.27	124.	88.3	3.92
125.	89.7	3.34	126.	78.3	3.98
127.	99.4	2.81	128.	86.6	2.77
129.	90.6	3.01	130.	94.4	2.58
131.	102.1	1.69	132.	89.7	3.07
133.	94.5	3.13	134.	98.4	2.57
135.	83.4	3.66	136.	93.8	3.52
137.	95.5	2.88	138.	81.3	3.89
139.	78.7	4.41	140.	93.6	2.60
141.	83.9	4.08	142.	88.5	2.98
143.	88.6	2.97	144.	88.0	3.07
145.	88.7	3.29	146.	90.8	3.74
147.	87.2	3.49	148.	89.8	3.33
149.	81.6	4.40	150.	92.5	3.16
151.	89.9	3.52	152.	91.7	2.97
153.	90.6	3.26	154.	96.3	3.00
155.	98.8	2.14	156.	90.7	2.76
157.	95.8	3.02	158.	87.4	3.46
159.	88.0	4.51	160.	101.3	1.75

Subject	X	Y	Subject	X	Y
161.	93.0	3.26	162.	89.3	3.47
163.	85.0	4.17	164.	100.8	1.66
165.	87.0	3.99	166.	90.2	3.00
167.	93.0	2.98	168.	95.2	2.98
169.	91.7	2.94	170.	91.4	3.47
171.	94.4	1.80	172.	86.4	4.34
173.	80.7	4.62	174.	89.1	3.38
175.	91.3	3.33	176.	106.2	1.06
177.	84.6	3.72	178.	87.2	3.14
179.	85.7	4.06	180.	86.1	4.19
181.	98.7	2.05	182.	92.9	2.38
183.	91.7	2.99	184.	95.8	2.73
185.	86.5	3.93	186.	91.7	3.09
187.	94.9	2.78	188.	96.9	2.59
189.	90.0	2.25	190.	95.3	2.20
191.	93.0	2.74	192.	88.2	3.20
193.	90.3	3.72	194.	97.0	2.43
195.	97.2	2.69	196.	93.5	3.29
197.	101.6	1.73	198.	85.2	3.40
199.	92.5	3.34	200.	95.1	2.80
201.	86.1	3.44	202.	96.1	3.01
203.	88.9	3.31	204.	92.6	3.20
205.	85.4	3.66	206.	91.7	3.17
207.	98.0	2.53	208.	85.8	4.19
209.	84.2	4.48	210.	97.1	2.00
211.	91.3	3.76	212.	92.2	3.24
213.	86.3	3.61	214.	91.1	2.85
215.	79.5	4.77	216.	87.4	3.57
217.	94.5	2.97	218.	97.2	2.67
219.	90.5	2.82	220.	94.8	3.08
221.	89.4	3.50	222.	99.0	2.77
223.	87.6	2.93	224.	98.3	1.90
225.	92.6	2.66	226.	94.1	2.79
227.	92.1	3.22	228.	92.9	3.29
229.	96.4	1.92	230.	84.2	4.14
231.	91.6	3.46	232.	86.5	3.81
233.	97.2	2.16	234.	79.0	4.41
235.	90.2	3.25	236.	87.4	3.34
237.	95.6	2.28	238.	99.5	2.34
239.	90.7	2.89	240.	83.3	3.83
241.	85.7	3.47	242.	95.3	2.16
243.	96.1	2.79	244.	87.4	3.53
245.	93.8	3.08	246.	81.2	3.79
247.	92.4	2.68	248.	94.3	2.86
249.	82.7	3.61	250.	87.4	4.01
251.	85.0	3.94	252.	93.4	2.29
253.	81.2	3.84	254.	90.4	2.73
255.	92.1	2.68	256.	80.4	3.61

Subject	X	Y	Subject	X	Y
257.	90.5	3.08	258.	83.5	4.39
259.	94.7	2.49	260.	88.8	2.83
261.	89.6	3.55	262.	95.2	2.61
263.	93.9	2.57	264.	95.7	2.80
265.	94.9	2.69	266.	91.9	3.04
267.	88.0	3.47	268.	90.3	4.04
269.	91.6	2.55	270.	102.8	2.52
271.	96.5	2.93	272.	89.2	3.07
273.	98.7	2.33	274.	86.8	2.98
275.	84.2	3.57	276.	87.4	3.66
277.	87.2	3.24	278.	87.1	3.20
279.	85.6	3.77	280.	94.3	2.47
281.	92.5	2.75	282.	103.7	2.30
283.	91.0	3.71	284.	92.1	2.93
285.	81.3	3.46	286.	93.2	2.48
287.	95.0	2.76	288.	84.4	3.27
289.	95.7	3.26	290.	91.3	2.72
291.	96.5	2.43	292.	84.0	3.83
293.	99.2	2.11	294.	93.0	2.65
295.	94.3	2.51	296.	89.8	3.45
297.	89.0	2.76	298.	88.0	2.81
299.	89.3	2.73	300.	89.8	2.85
301.	86.5	3.03	302.	96.4	2.14
303.	95.2	2.54	304.	92.3	4.09
305.	86.3	2.69	306.	96.0	2.55
307.	87.6	3.52	308.	102.5	2.03
309.	85.5	2.76	310.	96.0	1.77
311.	101.9	1.41	312.	91.0	2.82
313.	91.1	2.78	314.	85.4	3.91
315.	91.7	3.11	316.	92.7	3.03
317.	97.0	2.27	318.	92.4	3.02
319.	82.0	3.57	320.	93.7	3.15
321.	93.0	2.72	322.	95.5	2.40
323.	86.6	3.77	324.	83.0	4.17
325.	87.9	2.42	326.	99.4	3.02
327.	101.3	2.29	328.	94.0	2.69
329.	96.1	2.50	330.	96.0	3.34
331.	88.1	3.84	332.	89.5	3.08
333.	96.2	2.74	334.	85.8	3.55
335.	79.3	4.12	336.	76.4	4.87
337.	80.8	4.13	338.	100.9	2.75
339.	78.1	4.41	340.	96.8	3.29
341.	81.9	3.16	342.	98.7	2.91
343.	100.1	3.33	344.	99.8	2.08
345.	88.8	3.25	346.	77.5	5.19
347.	88.5	3.85	348.	94.5	3.26
349.	97.5	2.76	350.	78.2	5.19
351.	83.1	3.53	352.	98.1	2.22

Subject	X	Y	Subject	X	Y
353.	82.0	4.01	354.	89.8	3.74
355.	98.8	3.09	356.	97.9	2.30
357.	81.7	3.80	358.	86.9	3.37
359.	91.2	2.25	360.	77.8	3.72
361.	90.0	3.67	362.	86.2	2.00
363.	94.9	2.35	364.	80.7	4.25
365.	86.9	3.42	366.	83.3	3.50
367.	95.7	2.69	368.	97.6	2.51
369.	86.9	3.06	370.	92.9	2.44
371.	86.1	4.48	372.	101.3	2.25
373.	100.0	2.90	374.	91.4	2.43
375.	97.7	3.12	376.	88.7	3.85
377.	91.0	2.49	378.	92.5	2.85
379.	102.0	2.49	380.	92.6	3.19
381.	97.7	3.17	382.	93.9	3.50
383.	85.6	3.29	384.	82.4	3.86
385.	88.5	3.87	386.	81.6	4.31
387.	83.5	4.55	388.	98.2	2.93
389.	95.3	2.91	390.	98.6	2.31
391.	91.4	3.16	392.	90.3	2.07
393.	102.4	2.00	394.	86.6	3.53
395.	86.7	3.94	396.	89.1	2.66
397.	82.7	4.33	398.	85.1	3.22
399.	100.9	2.71	400.	107.7	1.44
401.	87.3	3.63	402.	95.0	2.99
403.	96.6	3.45	404.	101.4	1.81
405.	97.9	2.99	406.	90.6	4.09
407.	93.3	3.09 ·	408.	92.5	3.64
409.	86.5	3.73	410.	96.5	2.12
411.	88.9	3.28	412.	98.2	2.43
413.	92.2	2.67	414.	86.5	3.36
415.	91.0	4.15	416.	84.1	3.21
417.	96.2	2.11	418.	90.7	3.63
419.	90.4	3.03	420.	82.0	3.14
421.	88.6	3.57	422.	89.6	3.90
423.	81.0	3.66	424.	107.7	1.12
425.	102.2	2.29	426.	86.6	3.78
427.	87.8	4.10	428.	80.8	3.89
429.	97.8	2.17	430.	98.2	2.57
431.	86.9	2.93	432.	95.9	1.96
433.	91.1	2.47	434.	78.2	4.97
435.	80.7	4.27	436.	83.8	3.47
437.	96.3	4.08	438.	94.8	2.59
439.	82.7	4.31	440.	90.2	3.82
441.	94.9	2.84	442.	94.0	2.44
443.	100.2	1.97	444.	97.5	2.37
445.	90.1	3.97	446.	88.9	3.08

Subject	X	Y	Subject	X	Y
447.	90.6	3.53	448.	88.7	3.52
449.	86.3	3.58	450.	97.3	2.31
451.	88.8	3.62	452.	87.9	3.61
453.	89.5	3.60	454.	96.3	2.30
455.	88.1	3.80	456.	92.7	2.92
457.	90.6	3.29	458.	90.8	3.64
459.	100.9	2.69	460.	94.2	3.08
461.	95.4	3.64	462.	89.4	3.31
463.	87.2	3.77	464.	86.6	3.24
465.	92.5	2.55	466.	93.5	3.43
467.	92.1	3.06	468.	94.2	2.92
469.	96.7	2.52	470.	90.5	2.79
471.	90.2	3.72	472.	93.6	2.86
473.	88.8	3.25	474.	90.6	3.20
475.	93.9	3.41	476.	92.7	3.47
477.	84.2	3.40	478.	94.9	2.97
479.	95.0	2.63	480.	89.6	2.80
481.	96.3	2.93	482.	87.8	3.67
483.	89.7	3.29	484.	90.1	2.98
485.	86.8	3.52	486.	90.7	3.06
487.	91.0	3.25	488.	100.0	2.33
489.	83.0	3.71	490.	90.1	3.59
491.	84.1	3.93	492.	95.4	2.70
493.	93.0	3.38	494.	89.1	3.31
495.	91.3	3.18	496.	94.6	2.82
497.	92.2	3.03	498.	95.3	2.28
499.	86.2	3.90	500.	91.9	3.27
501.	95.1	2.94	502.	87.2	3.48
503.	90.2	2.97	504.	88.2	3.41
505.	89.8	3.18	506.	91.1	3.61
507.	94.3	2.36	508.	83.3	3.71
509.	91.1	3.06	510.	95.3	2.51
511.	94.6	3.22	512.	84.0	4.15
513.	92.1	2.75	514.	87.8	3.52
515.	86.4	3.70	516.	89.8	3.48
517.	91.9	2.92	518.	89.8	3.22
519.	90.4	2.71	520.	89.3	3.08
521.	89.4	3.46	522.	91.4	3.42
523.	90.9	3.13	524.	95.4	3.00
525.	86.3	3.81	526.	85.1	3.86
527.	90.4	3.39	528.	98.4	2.25
529.	83.5	3.89	530.	83.3	3.27
531.	92.9	2.76	532.	85.7	3.33
553.	87.8	3.31	534.	88.0	3.74
535.	90.5	2.92	536.	100.5	2.07
537.	92.4	3.28	538.	82.4	3.80
539.	92.0	3.17	540.	93.2	3.14

Subject	X	Y	Subject	X	Y
541.	96.7	2.76	542.	86.7	3.72
543.	96.6	1.96	544.	95.6	2.65
545.	92.4	3.60	546.	94.6	2.97
547.	97.6	2.00	548.	93.2	2.90
549.	84.7	3.83	550.	99.5	2.75
551.	86.9	3.26	552.	93.4	2.61
553.	86.5	3.72	554.	92.6	3.42
555.	93.5	3.28	556.	86.3	3.86
557.	87.0	3.45	558.	93.6	2.78
559.	91.8	2.37	560.	91.0	3.59
561.	93.5	2.79	562.	89.9	3.31
563.	84.0	3.71	564.	85.2	3.58
565.	93.1	2.29	566.	88.8	3.63
567.	86.6	3.85	568.	90.1	3.41
569.	93.3	2.90	570.	98.5	2.30
571.	88.2	3.42	572.	89.4	3.18
573.	91.4	3.45	574.	99.0	2.38
575.	87.8	3.81	576.	79.5	4.66
577.	90.5	3.33	578.	85.4	3.98
579.	102.3	2.03	580.	93.6	2.48
581.	92.0	2.53	582.	90.9	2.94
583.	84.0	3.68	584.	90.6	2.64
585.	86.0	3.90	586.	91.8	3.07
587.	89.4	3.08	588.	86.1	3.06
589.	95.3	2.49			

HEALTHY SUBJECTS

Subject	X	Y	Subject	X	Y
1.	105.5	1.74	2.	106.0	2.30
3.	103.8	2.38	4.	101.7	2.37
5.	104.0	1.94	6.	110.9	1.70
7.	102.3	2.18	8.	105.5	2.25
9.	100.3	2.35	10.	104.8	1.93
11.	106.0	2.63	12.	109.9	1.75
13.	103.1	2.18	14.	106.6	2.28
15.	105.0	1.89	16.	107.2	1.41
17.	104.6	2.10	18.	109.0	1.81
19.	107.2	1.93	20.	106.9	2.35
21.	101.8	2.65	22.	107.8	2.39
23.	103.4	3.05	24.	104.2	2.60
25.	103.5	2.51	26.	107.8	1.91
27.	106.7	2.03	28.	106.9	1.92
29.	107.4	2.28	30.	111.3	1.81
31.	109.6	1.97	32.	109.5	1.55
33.	103.5	2.70	34.	106.7	1.91

Subject	X	Y	Subject	X	Y
35.	110.8	1.62	36.	105.8	2.06
37.	103.8	2.20	38.	106.5	2.69
39.	105.5	1.96	40.	104.2	2.24
41.	104.5	1.99	42.	102.3	2.21
43.	104.8	2.41	44.	108.8	2.26
45.	105.8	2.63	46.	110.5	1.98
47.	103.8	1.52	48.	104.7	2.27
49.	107.3	1.99	50.	108.9	2.14
51.	112.4	1.36	52.	104.1	2.49
53.	111.1	1.86	54.	99.6	2.57
55.	105.3	1.99	56.	107.7	2.10
57.	103.9	2.25	58.	104.5	2.56
59.	109.0	2.20	60.	104.4	2.01
61.	105.9	2.12	62.	103.6	2.46
63.	103.9	2.13	64.	98.5	2.75
65.	107.7	1.99	66.	108.8	1.67
67.	107.0	1.95	68.	106.0	2.28
69.	107.1	1.95	70.	110.8	1.76
71.	102.1	2.23	72.	103.2	2.23
73.	103.3	2.23	74.	103.5	2.87
75.	107.8	1.79	76.	106.8	2.10
77.	106.6	1.74	78.	106.4	2.39
79.	103.0	2.31	80.	106.7	1.99
81.	105.3	2.31	82.	103.5	2.53
83.	109.6	1.29	84.	107.3	1.78
85.	110.8	1.80	86.	104.2	2.43
87.	100.4	2.64	88.	105.9	2.44
89.	109.0	1.79	90.	110.1	1.73
91.	108.5	1.70	92.	106.0	1.62
93.	104.2	2.31	94.	107.2	2.50
95.	109.1	1.87	96.	106.7	1.86
97.	108.6	1.71	98.	106.9	1.87
99.	106.5	2.18	100.	105.5	2.14
101.	105.4	2.04	102.	102.1	2.52
103.	102.2	2.44	104.	104.1	2.49
105.	100.4	2.91	106.	102.3	2.42
107.	110.2	2.13	108.	110.7	1.53
109.	105.4	1.95	110.	104.3	2.22
111.	106.1	1.93	112.	106.7	2.11
113.	105.5	2.30	114.	108.5	1.73
115.	102.4	2.49	116.	105.7	2.29
117.	101.1	2.67	118.	102.6	2.11
119.	110.0	2.29	120.	104.9	2.47
121.	108.0	1.71	122.	104.4	2.21
123.	109.0	2.18	124.	104.0	2.02
125.	106.1	1.79	126.	106.3	1.89
127.	106.8	2.21	128.	106.1	1.90

Subject	X	Y	Subject	X	Y
129.	109.0	1.60	130.	109.1	2.03
131.	107.9	1.72	132.	109.1	1.78
133.	107.2	2.15	134.	105.7	2.18
135.	103.5	2.32	136.	105.5	2.01
137.	108.6	2.02	138.	102.1	2.37
139.	97.3	2.83	140.	107.0	1.90
141.	105.2	2.56	142.	106.4	2.00
143.	105.3	2.46	144.	103.3	2.52
145.	110.2	1.41	146.	103.0	2.05
147.	103.4	2.28	148.	104.1	2.44
149.	110.5	1.65	150.	110.5	1.55
151.	101.7	2.86	152.	101.3	2.05
153.	104.8	2.05	154.	108.8	2.04
155.	106.0	2.07	156.	105.2	1.67
157.	100.0	2.64	158.	109.4	2.09
159.	107.7	1.88	160.	104.5	2.86
161.	108.0	1.54	162.	104.9	2.02
163.	106.4	1.92	164.	105.0	1.73
165.	105.5	2.09	166.	103.5	2.32
167.	107.7	2.20	168.	109.3	2.08
169.	107.7	1.79	170.	109.3	1.73
171.	105.9	1.92	172.	106.7	1.77
173.	109.2	2.20	174.	103.5	2.42
175.	102.6	2.31	176.	108.4	1.83
177.	110.7	1.84	178.	107.9	1.71
179.	105.9	2.05	180.	104.5	2.28
181.	101.7	2.42	182.	108.5	2.10
183.	113.2	1.07	184.	104.9	2.55
185.	107.0	2.01	186.	103.3	2.25
187.	109.9	2.01	188.	109.3	1.98
189.	102.6	2.01	190.	100.4	2.66
191.	109.8	1.66	192.	106.7	1.98
193.	108.9	1.82	194.	109.0	1.72
195.	106.8	1.92	196.	99.7	2.88
187.	109.5	1.43	198.	101.4	2.42
199.	110.1	1.76	200.	103.9	2.27
201.	102.3	2.41	202.	107.0	2.07
203.	108.5	2.40	204.	102.1	2.71
205.	103.5	2.23	206.	103.5	2.57
207.	103.2	2.58	208.	104.5	2.13
209.	108.0	1.82	210.	108.2	1.98
211.	107.5	2.23	212.	102.4	2.43
213.	103.4	2.29	214.	104.4	2.43
215.	106.4	1.84	216.	100.2	2.55
217.	103.1	2.31	218.	103.6	2.23
219.	101.9	2.53	220.	98.4	3.07
221.	103.5	2.77	222.	103.3	2.35

Subject	X	Y	Subject	X	Y
223.	101.7	2.48	224.	101.7	2.61
225.	104.2	2.48	226.	103.2	2.46
227.	100.7	2.77	228.	107.5	2.07
229.	107.3	1.83	230.	109.1	1.36
231.	105.8	2.31	232.	105.9	2.20
233.	107.9	1.84	234.	103.4	2.16
235.	107.5	2.09	236.	103.4	2.46
237.	108.7	1.91	238.	103.0	2.11
239.	106.1	1.92	240.	103.9	2.40
241.	104.5	1.90	242.	108.3	1.88
243.	107.8	2.31	244.	104.5	2.34
245.	107.7	2.04	246.	103.1	2.44
247.	106.7	1.89	248.	106.7	2.00
249.	98.8	3.02	250.	107.0	1.58
251.	106.6	2.07	252.	104.2	2.54
253.	105.8	2.10	254.	109.9	1.82
255.	110.9	1.36	256.	108.3	1.58
257.	107.5	1.84	258.	100.8	2.45
259.	110.5	1.66	260.	105.1	2.47
261.	104.7	2.62	262.	103.8	2.50
263.	103.1	1.98	264.	106.1	2.31
265.	104.5	2.53	266.	108.0	1.72
267.	111.2	1.33	268.	104.0	2.18
269.	102.7	2.45	270.	109.0	1.76
271.	106.7	1.87	272.	107.5	1.81
273.	104.3	2.43	274.	101.4	2.66
275.	107.8	2.02	276.	100.8	2.26
277.	110.4	1.89	278.	107.0	1.54
279.	104.0	2.11	280.	109.7	1.65
281.	108.8	2.03	282.	114.1	1.49
283.	108.3	1.93	284.	105.0	2.00
285.	105.7	2.93	286.	106.3	2.10
287.	106.4	1.52	288.	109.4	2.05
289.	109.2	1.66	290.	104.3	2.10
291.	104.3	1.96	292.	107.8	2.42
293.	101.5	2.02	294.	99.7	2.77
295.	100.4	2.34	296.	105.2	1.76
297.	107.1	2.58	298.	104.4	2.52
299.	105.9	2.32	300.	110.9	1.63
301.	106.8	1.87	302.	105.4	2.07
303.	109.1	1.44	304.	105.9	1.80
305.	105.9	1.95	306.	104.6	2.07
307.	104.1	2.08	308.	105.5	1.83
309.	102.6	2.37	310.	103.0	2.18
311.	104.7	2.39	312.	109.6	1.44
313.	109.4	1.65	314.	104.2	1.98
315.	109.6	1.62	316.	108.3	1.92

Subject	X	Y	Subject	X	Y
317.	102.7	2.22	318.	109.7	1.71
319.	101.9	2.32	320.	107.5	2.24
321.	101.9	2.46	322.	108.4	1.39
323.	107.7	1.71	324.	106.3	1.63
325.	104.8	2.35	326.	101.4	2.36
327.	105.9	2.24	328.	102.0	2.39
329.	105.2	2.16	330.	103.5	2.29
331.	105.0	1.96	332.	102.6	2.18
333.	106.1	1.72	334.	108.0	1.76
335.	105.2	1.76	336.	106.7	2.17
337.	105.7	1.55	338.	106.3	1.88
339.	108.0	1.43	340.	104.4	2.18
341.	106.1	2.02	342.	107.8	2.00
343.	104.1	2.48	344.	101.1	2.20
345.	105.4	2.19	346.	102.5	2.32
347.	102.1	2.57	348.	106.5	2.37
349.	109.5	1.49	350.	103.0	2.18
351.	103.7	2.36	352.	99.3	2.67
353.	106.2	1.64	354.	105.0	2.39
355.	109.3	1.85	356.	105.8	1.91
357.	106.6	1.85	358.	105.5	2.02
359.	107.6	1.84	360.	104.1	1.87
361.	112.6	1.28	362.	109.2	1.43
363.	105.1	2.07	364.	105.1	1.45
365.	105.6	2.46	366.	110.7	1.41
367.	107.6	2.12	368.	104.9	1.97
369.	104.7	2.25	370.	103.8	2.10
371.	108.9	1.92	372.	104.4	2.41
373.	108.8	1.93	374.	104.3	2.17
375.	107.8	1.83	376.	112.2	1.19
377.	102.2	2.54	378.	105.7	1.84
379.	104.5	2.17	380.	107.3	1.32
381.	105.0	1.98	382.	110.7	1.42
383.	97.8	3.06	384.	110.3	1.59
385.	105.7	2.02	386.	106.3	1.82
387.	104.9	2.08	388.	103.1	2.45
389.	105.8	2.12	390.	102.5	2.40
391.	104.2	2.06	392.	105.2	2.01
393.	107.1	1.80	394.	106.4	2.14
395.	104.2	2.22	396.	108.3	1.69
397.	107.2	2.04	398.	106.4	1.94
399.	106.0	1.90	400.	107.8	1.93
401.	102.3	2.32	402.	106.3	1.97
403.	105.8	2.27	404.	104.0	2.19
405.	103.2	2.06	406.	106.7	1.68
407.	102.3	2.23	408.	105.8	1.81
409.	103.2	2.14	410.	106.3	2.10
411.	110.3	1.13	412.	108.8	1.55

Subject	X	Y	Subject	X	Y
413.	101.6	2.57	414.	110.5	1.48
415.	107.1	2.03	416.	104.6	2.18
417.	108.1	1.59	418.	104.1	2.02
419.	103.7	2.16	420.	106.7	1.91
421.	109.9	1.44	422.	107.7	1.76
423.	108.8	1.75	424.	104.7	1.95
425.	110.0	1.30	426.	109.6	1.62
427.	103.8	2.45	428.	108.4	1.66
429.	101.0	2.83	430.	108.1	1.83
431.	106.3	2.09	432.	107.7	1.83
433.	105.3	1.75	434.	107.5	2.12
435.	109.6	1.94	436.	103.8	2.48
437.	106.1	1.78	438.	103.8	2.17
439.	104.0	2.12	440.	104.2	2.21
441.	100.7	2.47	442.	103.6	2.19
443.	101.7	2.18	444.	109.1	1.87
445.	104.8	2.36	446.	109.8	1.64
447.	106.3	1.85	448.	105.4	2.00
449.	110.7	1.84	450.	105.8	1.80
451.	105.8	2.32	452.	98.9	2.93
453.	109.9	2.02	454.	109.2	1.73
455.	111.5	1.80	456.	102.3	3.08
457.	106.8	1.89	458.	105.5	2.00
459.	105.9	2.45	460.	106.4	2.33
461.	109.5	1.40	462.	107.8	2.30
463.	101.6	3.02	464.	106.7	2.15
465.	106.3	2.47	466.	111.4	1.78
467.	111.0	1.71	468.	106.3	2.16
469.	108.0	2.29	470.	103.9	2.64
471.	107.8	2.33	472.	105.8	1.86
473.	111.3	1.54	474.	107.2	1.78
475.	109.5	1.68	476.	99.7	3.11
477.	106.1	2.31	478.	100.7	2.70
479.	100.6	3.04	480.	108.5	1.85
481.	106.3	1.99	482.	110.0	1.96
483.	107.0	2.10	484.	104.8	2.21
485.	110.2	1.89	486.	103.7	2.46
487.	108.1	1.97	488.	107.3	2.39
489.	107.7	1.77	490.	109.1	1.76
491.	104.2	2.32	492.	103.7	2.57
493.	102.1	2.74	494.	103.9	2.47
495.	95.9	3.15	496.	111.8	1.33
497.	107.0	2.10	498.	107.5	2.09
499.	106.0	1.99	500.	106.9	1.74
501.	103.5	1.98	502.	106.7	1.69
503.	106.1	2.14	504.	106.3	1.61
505.	104.9	2.30	506.	112.7	1.20
507.	109.6	2.15	508.	105.1	2.44

Subject	X	Y	Subject	X	Y
509.	110.0	2.08	510.	110.8	1.84
511.	111.4	1.67	512.	107.2	1.93
513.	103.9	2.25	514.	107.6	1.76
515.	103.8	2.02	516.	102.9	2.35
517.	107.3	1.69	518.	104.7	2.16
519.	105.1	2.15	520.	108.0	2.14
521.	108.3	1.93	522.	106.1	1.81
523.	103.6	2.61	524.	111.5	1.41
525.	104.7	2.20	526.	106.6	2.47
527.	108.4	2.24	528.	107.3	1.78
529.	105.7	2.29	530.	108.4	2.10
531.	106.9	2.15	532.	111.2	1.81
533.	110.7	1.64	534.	104.5	2.59
535.	105.9	2.48	536.	108.0	2.19
537.	102.6	2.76	538.	103.5	2.31
539.	102.7	2.32	540.	105.9	2.14
541.	105.2	2.33	542.	101.2	2.53
543.	105.3	2.45	544.	105.9	2.34
545.	101.7	2.79	546.	105.3	2.11
547.	109.9	1.96	548.	104.8	2.66
549.	96.8	3.19	550.	100.4	2.41
551.	107.5	2.07	552.	105.1	2.22
553.	104.0	2.10	554.	103.9	1.91
555.	106.6	2.18	556.	105.6	2.11
557.	105.9	2.48	558.	102.2	2.50
559.	100.1	2.75	560.	107.3	2.23
561.	104.6	2.33	562.	106.4	1.93
563.	107.0	1.90	564.	107.8	1.98
565.	104.4	2.87	566.	102.5	2.49
567.	111.8	1.77	568.	104.6	2.00
569.	104.6	2.56	570.	108.5	2.15
571.	104.3	2.32	572.	105.1	2.32
573.	101.1	2.90	574.	109.4	1.49
575.	111.5	1.62	576.	104.1	2.44
577.	107.2	2.08	578.	108.1	1.94
579.	106.8	1.92	580.	99.6	2.80
581.	105.9	2.19	582.	109.0	2.15
583.	103.5	2.21	584.	104.2	2.27
585.	106.8	2.11	586.	110.0	2.17
587.	99.0	2.62	588.	106.1	2.76
589.	109.8	2.22	590.	104.3	2.35
591.	103.8	2.66	592.	107.7	2.07
593.	107.2	2.07	594.	108.9	2.13
595.	110.8	2.05	596.	103.6	2.03
597.	106.9	2.22	598.	111.2	1.59
599.	104.0	2.44	600.	109.9	1.76

16. A medical research team is conducting a study to determine what factors may be related to respiratory disease. The following data are available on a population of patients who have sought treatment for the relief of respiratory disease symptoms. The dependent variable Y is a measure of the severity of the disease. A larger value indicates a more serious condition. The independent variables are as follows.

X_1 = education (highest grade completed)

X_2 = measure of crowding of living quarters

X_3 = measure of air quality at place of residence

(a larger number indicates Poorer quality)

X_4 = nutritional status (a large number indicates

a higher level of nutrition)

X_5 = smoking status (0 = smoker, 1 = nonsmoker)

Select a simple random sample of subjects from this population and conduct a statistical analysis that you think would be of value to the research team. prepare a narrative report of your results and conclusions. Use graphic illustrations where appropriate. Compare your results with those of your classmates. Consult your instructor regarding the size of sample you should select.

Subject	Y	X_1	X_2	X_3	X_4	X_5	Subject	Y	X_1	X_2	X_3	X_4	X_5
1.	40	7	25	22	94	0	2.	67	7	33	61	18	1
3.	30	6	19	30	103	0	4.	71	15	29	50	17	1
5.	47	11	21	43	109	0	6.	53	10	24	54	0	1
7.	39	8	21	28	33	0	8.	55	14	22	35	21	1
9.	47	10	26	22	76	0	10.	56	9	32	43	97	1
11.	43	8	22	48	104	0	12.	41	8	19	27	−37	0
13.	51	9	28	32	87	1	14.	48	8	22	62	131	0
15.	36	8	19	37	53	0	16.	39	7	16	53	−4	0
17.	55	11	26	52	183	1	18.	40	7	25	22	11	0
19.	61	11	27	48	19	1	20.	41	9	22	21	59	0
21.	64	10	28	76	79	1	22.	62	11	25	79	11	1
23.	37	8	20	29	85	0	24.	58	10	26	50	−48	1
25.	52	14	22	35	75	1	26.	57	12	26	54	156	1
27.	62	12	28	45	35	1	28.	48	11	22	36	41	0
29.	64	12	29	35	28	1	30.	52	8	29	30	46	1
31.	51	10	27	33	112	1	32.	57	10	24	56	56	1
33.	47	11	24	32	124	0	34.	43	11	22	37	148	0
35.	48	8	25	44	70	0	36.	32	8	20	13	63	0
37.	48	11	25	38	118	0	38.	34	6	19	35	29	0
39.	48	9	26	50	167	0	40.	52	10	27	56	118	1
41.	43	7	23	30	13	0	42.	21	4	14	33	50	0
43.	52	11	21	61	90	1	44.	46	8	19	55	27	0

Subject	Y	X_1	X_2	X_3	X_4	X_5	Subject	Y	X_1	X_2	X_3	X_4	X_5
45.	34	5	19	36	48	0	46.	34	10	18	15	60	0
47.	35	8	18	25	61	0	48.	58	12	24	47	36	1
49.	54	9	29	49	153	1	50.	54	12	28	39	102	1
51.	27	5	18	16	59	0	52.	45	9	19	44	−10	0
53.	51	9	26	31	51	1	54.	47	8	25	39	79	0
55.	63	11	34	37	142	1	56.	39	7	22	31	9	0
57.	62	11	33	39	133	1	58.	64	13	27	70	106	1
59.	40	11	26	8	250	0	60.	50	8	25	56	73	1
61.	51	9	26	37	42	1	62.	53	14	26	25	109	1
63.	57	10	32	37	140	1	64.	45	10	22	41	137	0
65.	42	9	18	59	64	0	66.	45	9	18	59	89	0
67.	51	9	26	37	27	1	68.	31	5	22	14	63	0
69.	57	10	29	50	113	1	70.	41	11	20	23	85	0
71.	42	7	25	42	150	0	72.	58	9	26	44	10	1
73.	28	7	15	20	0	0	74.	37	7	23	33	132	0
75.	53	6	31	65	64	1	76.	47	10	24	21	17	0
77.	39	8	20	39	122	0	78.	56	12	28	48	115	1
79.	42	8	21	35	75	0	80.	48	6	27	37	46	0
81.	49	10	28	31	86	0	82.	50	10	23	46	78	1
83.	31	8	21	16	103	0	84.	39	8	21	32	51	0
85.	43	11	19	31	80	0	86.	52	12	25	45	156	1
87.	60	15	28	44	175	1	88.	60	10	28	52	44	1
89.	43	7	25	31	80	0	90.	64	13	25	65	50	1
91.	60	11	28	46	29	1	92.	34	9	12	51	27	0
93.	59	12	23	63	83	1	94.	46	9	28	47	140	0
95.	51	10	27	35	42	1	96.	52	10	26	39	104	1
97.	49	11	22	47	87	0	98.	57	11	30	43	77	1
99.	51	8	26	41	44	1	100.	61	9	32	51	58	1
101.	45	8	23	37	46	0	102.	26	7	13	39	56	0
103.	37	8	22	19	97	0	104.	56	9	27	33	43	1
105.	40	9	18	46	119	0	106.	47	11	23	52	171	0
107.	51	8	30	42	88	1	108.	44	11	17	60	104	0
109.	55	9	26	38	5	1	110.	70	15	26	74	85	1
111.	54	10	28	46	81	1	112.	46	9	23	51	122	0
113.	48	6	25	46	30	0	114.	82	15	35	70	34	1
115.	67	11	35	27	−17	1	116.	59	13	25	35	5	1
117.	41	9	25	16	29	0	118.	36	8	21	11	44	0
119.	58	10	31	37	75	1	120.	38	7	18	50	83	0
121.	38	10	18	39	103	0	122.	50	7	25	56	74	0
123.	62	10	34	32	117	1	124.	53	13	23	53	114	1
125.	50	11	24	48	88	1	126.	46	8	23	46	63	0
127.	78	13	38	50	57	1	128.	45	7	24	41	79	0
129.	78	12	34	61	0	1	130.	36	8	19	25	88	0
131.	46	10	23	28	46	0	132.	62	11	33	40	112	1
133.	45	7	21	54	56	0	134.	43	10	20	44	123	0
135.	54	12	28	18	59	1	136.	40	6	21	26	30	0
137.	61	10	30	61	98	1	138.	45	9	20	41	84	0
139.	53	7	27	61	85	1	140.	45	10	21	27	44	0

Subject	Y	X_1	X_2	X_3	X_4	X_5	Subject	Y	X_1	X_2	X_3	X_4	X_5
141.	38	6	19	33	111	0	142.	29	6	16	29	66	0
143.	57	8	30	36	6	1	144.	42	8	15	56	63	0
145.	59	11	28	58	61	1	146.	51	9	33	17	136	1
147.	41	9	19	42	52	0	148.	50	12	22	45	74	0
149.	50	9	24	23	36	0	150.	45	8	18	58	77	0
151.	50	10	25	54	123	0	152.	41	8	23	10	25	0
153.	52	10	26	52	88	1	154.	44	12	28	19	144	0
155.	36	9	12	44	5	0	156.	46	10	21	36	42	0
157.	43	9	21	41	124	0	158.	48	12	19	43	−4	0
159.	55	8	27	53	85	1	160.	42	10	20	20	56	0
161.	52	10	27	28	123	1	162.	63	10	29	63	88	1
163.	48	10	22	42	105	0	164.	35	10	14	36	34	0
165.	62	10	35	21	70	1	166.	44	7	18	52	58	0
167.	61	11	28	41	37	1	168.	53	12	25	23	63	1
169.	48	10	27	30	88	0	170.	42	13	23	5	115	0
171.	43	7	24	43	159	0	172.	53	12	27	39	113	1
173.	48	9	26	35	145	0	174.	35	7	20	37	110	0
175.	58	9	33	13	44	1	176.	53	10	23	59	65	1
177.	68	11	33	31	−5	1	178.	53	11	22	48	28	1
179.	62	10	30	56	98	1	180.	67	10	28	78	59	1
181.	56	7	30	47	44	1	182.	48	10	24	33	94	0
183.	41	6	22	29	17	0	184.	51	8	26	38	71	1
185.	47	10	28	42	179	0	186.	52	9	22	55	128	1
187.	41	9	18	53	77	0	188.	42	8	22	47	150	0
189.	49	10	22	52	75	0	190.	54	10	25	51	36	1
191.	59	13	27	42	106	1	192.	56	11	29	52	171	1
193.	59	13	24	53	51	1	194.	37	8	22	5	29	0
195.	57	8	29	48	26	1	196.	50	9	20	49	4	0
197.	55	11	31	27	83	1	198.	60	10	30	42	118	1
199.	59	11	25	65	49	1	200.	49	9	23	47	96	0
201.	49	12	28	22	152	0	202.	47	11	20	54	82	0
203.	62	9	28	49	24	1	204.	63	11	26	56	2	1
205.	38	7	21	36	45	0	206.	55	9	34	26	117	1
207.	42	12	20	20	155	0	208.	44	10	20	24	39	0
209.	45	11	25	30	144	0	210.	63	9	30	50	64	1
211.	57	11	28	53	105	1	212.	41	10	27	22	124	0
213.	56	12	25	31	62	1	214.	55	11	27	54	89	1
215.	44	12	22	22	104	0	216.	55	9	27	52	43	1
217.	59	10	29	57	100	1	218.	42	9	23	40	77	0
219.	57	6	33	42	80	1	220.	62	9	32	33	56	1
221.	49	10	24	19	27	0	222.	53	11	24	43	56	1
223.	43	7	25	40	102	0	224.	46	10	19	28	−5	0
225.	29	10	16	14	106	0	226.	56	9	24	53	26	1
227.	47	4	25	39	−92	0	228.	48	9	21	52	70	0
229.	50	9	27	32	64	0	230.	65	12	32	52	41	1
231.	70	14	35	49	128	1	232.	56	9	26	70	106	1
233.	44	7	29	39	99	0	234.	53	10	27	55	155	1
235.	37	8	19	32	87	0	236.	43	9	23	26	114	0

Subject	Y	X_1	X_2	X_3	X_4	X_5	Subject	Y	X_1	X_2	X_3	X_4	X_5
237.	51	11	22	51	67	1	238.	40	10	19	51	2	0
239.	58	11	24	58	47	1	240.	70	13	36	36	124	1
241.	36	9	17	27	69	0	242.	50	9	29	18	47	1
243.	58	11	28	32	−21	1	244.	45	9	24	10	−40	0
245.	41	9	23	11	39	0	246.	55	10	28	35	95	1
247.	52	11	28	22	19	1	248.	42	10	21	49	115	0
249.	69	14	30	48	104	1	250.	62	11	27	68	55	1
251.	42	8	23	36	61	0	252.	54	11	25	43	131	1
253.	45	7	24	31	25	0	254.	40	9	21	5	48	0
255.	63	12	30	25	−7	1	256.	48	9	29	23	156	0
257.	52	10	22	52	36	1	258.	38	8	20	37	116	0
259.	46	10	29	4	34	0	260.	38	8	22	31	50	0
261.	57	11	26	38	108	1	262.	62	11	31	35	16	1
263.	63	13	30	53	106	1	264.	50	7	30	16	14	0
265.	60	11	26	63	68	1	266.	41	10	24	15	91	0
267.	65	12	32	38	24	1	268.	64	10	26	72	57	1
269.	55	8	27	47	38	1	270.	53	11	25	31	20	1
271.	55	9	32	29	56	1	272.	42	8	22	33	95	0
273.	50	11	21	43	94	1	274.	30	6	18	23	48	0
275.	54	13	23	35	60	1	276.	37	3	26	28	55	0
277.	47	10	23	41	74	0	278.	38	10	21	34	188	0
279.	33	9	15	11	19	0	280.	42	10	18	35	17	0
281.	50	9	25	18	2	0	282.	42	8	24	35	65	0
283.	57	9	25	53	5	1	284.	31	7	17	31	26	0
285.	29	7	19	30	111	0	286.	59	10	28	47	50	1
287.	55	9	24	67	53	1	288.	45	9	25	25	120	0
289.	67	11	29	50	74	1	290.	51	11	21	42	40	1
291.	57	12	25	50	0	1	292.	59	13	24	57	153	1
293.	51	7	25	66	105	1	294.	42	10	19	29	35	0
295.	44	6	20	58	87	0	296.	44	10	24	25	70	0
297.	49	11	25	30	88	0	298.	49	7	31	26	130	0
299.	24	8	17	−1	137	0	300.	38	9	17	38	−7	0
301.	48	9	20	34	−44	0	302.	61	10	30	46	41	1
303.	71	13	34	51	16	1	304.	50	13	23	25	39	1
305.	51	9	26	60	117	1	306.	52	13	25	30	106	1
307.	43	9	22	25	68	0	308.	72	11	31	61	−11	1
309.	62	10	30	56	60	1	310.	58	8	32	55	163	1
311.	49	7	27	41	79	0	312.	28	9	14	23	135	0
313.	35	8	17	39	69	0	314.	68	14	26	60	67	1
315.	56	11	23	54	24	1	316.	59	10	27	62	89	1
317.	49	10	23	45	83	0	318.	40	10	19	39	71	0
319.	54	11	27	49	123	1	320.	53	10	28	34	7	1
321.	47	10	20	57	143	0	322.	57	10	28	54	83	1
323.	73	13	31	69	60	1	324.	44	9	28	11	103	0
325.	60	11	28	54	61	1	326.	44	12	21	28	88	0
327.	52	10	27	51	59	1	328.	56	11	25	50	116	1
329.	44	9	23	22	29	0	330.	39	8	20	54	256	0
331.	70	12	42	19	81	1	332.	40	7	28	10	82	0

Subject	Y	X_1	X_2	X_3	X_4	X_5	Subject	Y	X_1	X_2	X_3	X_4	X_5
333.	45	9	25	30	125	0	334.	49	12	25	21	115	0
335.	50	6	23	75	45	1	336.	51	6	24	68	42	1
337.	56	10	31	41	97	1	338.	47	7	22	52	63	0
339.	47	10	23	40	135	0	340.	43	9	22	20	65	0
341.	45	6	26	39	89	0	342.	47	10	17	48	−6	0
343.	55	8	28	31	−10	1	344.	64	9	33	61	47	1
345.	39	12	15	49	132	0	346.	64	9	32	65	65	1
347.	34	7	17	32	20	0	348.	52	9	24	32	60	1
349.	38	9	10	78	117	0	350.	69	12	34	42	107	1
351.	43	8	25	41	100	0	352.	62	8	34	56	114	1
353.	75	12	36	50	95	1	354.	55	11	28	23	29	1
355.	58	9	28	75	72	1	356.	40	9	21	26	64	0
357.	56	7	29	62	114	1	358.	29	5	19	30	120	0
359.	67	11	32	45	28	1	360.	68	11	40	16	76	1
361.	56	9	27	41	20	1	362.	45	9	25	26	38	0
363.	67	10	29	68	34	1	364.	52	12	24	32	118	1
365.	49	10	20	38	13	0	366.	42	7	22	47	102	0
367.	35	10	16	25	97	0	368.	57	11	28	42	58	1
369.	57	12	31	34	126	1	370.	46	9	22	42	87	0
371.	64	7	37	49	99	1	372.	29	9	15	31	8	0
373.	25	7	15	3	27	0	374.	46	8	24	43	61	0
375.	35	9	20	9	28	0	376.	52	12	24	19	50	1
377.	51	8	23	57	76	1	378.	36	8	20	25	102	0
379.	51	8	28	42	71	1	380.	52	10	29	31	97	1
381.	45	7	24	22	11	0	382.	65	12	32	43	57	1
383.	35	5	16	30	−2	0	384.	36	6	16	42	−33	0
385.	37	8	21	19	64	0	386.	46	9	25	33	98	0
387.	26	10	12	22	94	0	388.	77	14	34	61	48	1
389.	55	11	24	59	54	1	390.	50	9	29	32	139	0
391.	66	13	34	25	13	1	392.	64	15	28	42	109	1
393.	38	7	20	45	65	0	394.	50	12	17	65	130	0
395.	50	8	25	44	80	0	396.	54	11	25	40	81	1
397.	51	8	24	33	18	1	398.	33	9	17	36	84	0
399.	49	10	23	44	71	0	400.	42	6	20	45	34	0
401.	46	8	25	39	87	0	402.	44	8	24	30	71	0
403.	37	6	20	43	33	0	404.	55	8	25	58	37	1
405.	65	10	32	64	51	1	406.	47	9	22	48	60	0
407.	45	7	31	25	150	0	408.	40	9	27	5	51	0
409.	38	8	16	50	76	0	410.	45	10	18	53	45	0
411.	53	13	22	42	23	1	412.	53	10	24	45	−8	1
413.	53	10	26	51	108	1	414.	54	9	30	29	73	1
415.	46	9	21	48	11	0	416.	57	10	30	30	79	1
417.	58	8	30	38	17	1	418.	40	10	24	2	100	0
419.	39	5	27	39	120	0	420.	46	11	22	33	65	0
421.	46	10	21	22	17	0	422.	60	9	27	47	−3	1
423.	60	11	28	60	99	1	424.	43	4	27	52	43	0
425.	44	9	25	25	78	0	426.	42	11	16	29	43	0
427.	37	7	22	18	60	0	428.	64	8	31	60	78	1

Subject	Y	X_1	X_2	X_3	X_4	X_5	Subject	Y	X_1	X_2	X_3	X_4	X_5
429.	33	5	25	12	49	0	430.	43	8	24	37	77	0
431.	56	9	33	29	88	1	432.	57	10	32	34	73	1
433.	42	9	22	42	84	0	434.	52	7	30	42	26	1
435.	54	12	28	17	49	1	436.	60	13	23	70	24	1
437.	53	9	24	59	102	1	438.	59	10	23	62	21	1
439.	49	14	21	29	52	0	440.	57	12	27	45	96	1
441.	44	9	22	33	94	0	442.	52	12	22	54	101	1
443.	71	10	36	39	−8	1	444.	50	11	24	42	54	0
445.	55	8	30	17	49	1	446.	39	9	20	21	64	0
447.	49	12	23	28	104	0	448.	42	8	24	26	57	0
449.	55	13	24	38	77	1	450.	39	7	21	35	33	0
451.	49	11	20	65	104	0	452.	57	8	29	46	100	1
453.	42	9	23	15	65	0	454.	40	7	23	42	97	0
455.	53	9	26	58	44	1	456.	40	10	15	52	54	0
457.	56	13	28	38	114	1	458.	40	8	21	31	−28	0
459.	55	10	29	56	171	1	460.	54	11	24	52	89	1
461.	55	9	29	42	58	1	462.	53	8	27	45	83	1
463.	60	8	32	51	24	1	464.	51	10	24	38	45	1
465.	29	7	16	36	86	0	466.	59	12	29	41	78	1
467.	49	9	26	30	45	0	468.	41	11	20	32	134	0
469.	42	9	23	40	102	0	470.	41	7	21	27	64	0
471.	51	9	24	44	14	1	472.	55	15	28	17	157	1
473.	47	8	21	42	−58	0	474.	54	12	23	53	117	1
475.	27	7	12	19	59	0	476.	53	6	31	43	76	1
477.	57	9	24	74	88	1	478.	30	7	14	48	47	0
479.	43	10	20	38	66	0	480.	51	11	27	14	52	1
481.	42	10	20	47	141	0	482.	51	10	24	24	15	1
483.	47	7	26	50	74	0	484.	55	11	22	50	8	1
485.	37	7	18	42	24	0	486.	42	10	18	58	139	0
487.	37	6	19	53	97	0	488.	65	14	28	45	123	1
489.	53	10	30	28	146	1	490.	42	9	17	50	15	0
491.	51	8	24	52	0	1	492.	53	9	31	36	109	1
493.	37	6	17	50	28	0	494.	46	11	18	65	72	0
495.	36	8	19	19	59	0	496.	36	8	17	38	54	0
497.	48	14	24	38	192	0	498.	50	9	24	41	58	1
499.	55	9	25	36	4	1	500.	49	8	24	44	33	0

17. Suppose you wish to study cardiovascular risk factors among adult males engaged in sedentary occupations. For a population of these subjects data on the following variables are available as shown in the table that follows. The variables are

$$Y = \text{oxygen consumption}$$

$$X_1 = \text{systolic blood pressure (mm Hg)}$$

$$X_2 = \text{total cholesterol (mg/DL)}$$

$$X_3 = \text{HDL cholesterol (mg/DL)}$$

$$X_4 = \text{triglycerides (mg/DL)}$$

Select a simple random sample from this population and carry out an appropriate statistical analysis. Prepare a narrative report of your findings and compare them with those of your classmates. Consult with your instructor regarding the size of the sample.

Subject	Y	X_1	X_2	X_3	X_4	Subject	Y	X_1	X_2	X_3	X_4
1.	44	130	199	55	205	2.	22	128	260	47	137
3.	25	131	272	42	211	4.	31	132	254	53	157
5.	37	133	212	51	154	6.	39	142	236	45	197
7.	26	127	228	37	172	8.	31	132	259	60	168
9.	27	127	181	34	128	10.	31	133	208	36	172
11.	36	136	197	43	131	12.	35	133	198	45	126
13.	33	121	179	39	146	14.	29	117	200	37	190
15.	28	125	234	45	163	16.	26	148	261	41	160
17.	35	137	240	51	188	18.	40	132	221	48	198
19.	36	134	209	44	209	20.	32	122	210	49	170
21.	45	126	154	61	115	22.	32	132	240	39	232
23.	36	113	193	46	190	24.	28	123	199	31	189
25.	19	112	227	37	154	26.	43	136	186	48	162
27.	40	136	196	47	159	28.	37	139	211	38	201
29.	36	137	234	48	180	30.	35	141	240	47	168
31.	38	116	149	51	111	32.	32	130	240	50	166
33.	28	162	296	41	194	34.	38	113	173	49	138
35.	48	125	109	44	185	36.	31	135	225	38	158
37.	32	135	261	51	182	38.	29	120	211	44	178
39.	44	145	236	55	215	40.	37	131	188	46	146
41.	24	138	273	37	179	42.	29	147	251	37	165
43.	32	125	197	41	168	44.	20	134	248	38	133
45.	22	133	245	28	185	46.	35	115	170	49	122
47.	34	136	238	52	211	48.	38	141	239	53	170
49.	41	129	172	60	118	50.	32	145	239	35	190
51.	46	131	217	46	239	52.	33	129	245	43	198
53.	36	114	162	54	108	54.	27	141	234	28	159
55.	34	118	246	61	205	56.	36	114	183	51	132
57.	35	138	186	28	180	58.	22	119	245	49	110
59.	47	116	141	60	163	60.	38	126	207	57	178
61.	28	124	153	30	144	62.	26	142	227	23	179
63.	28	133	224	42	160	64.	24	129	247	43	175
65.	37	147	276	49	223	66.	37	113	199	57	138
67.	31	135	254	41	179	68.	19	127	238	40	92
69.	34	108	154	44	170	70.	21	136	262	33	176
71.	32	133	207	36	207	72.	30	120	200	44	149
73.	33	133	231	46	213	74.	28	139	257	40	193
75.	38	131	169	40	155	76.	31	145	266	46	197
77.	43	117	165	40	221	78.	35	129	195	38	181

Subject	Y	X_1	X_2	X_3	X_4	Subject	Y	X_1	X_2	X_3	X_4
79.	31	118	228	61	169	80.	34	114	181	52	145
81.	25	155	260	29	165	82.	37	128	216	51	180
83.	28	136	212	29	134	84.	30	128	222	42	176
85.	29	124	206	37	156	86.	41	111	193	66	186
87.	24	124	256	49	111	88.	26	126	186	32	129
89.	34	124	195	41	167	90.	35	123	165	47	99
91.	27	135	213	37	123	92.	33	119	137	30	106
93.	35	123	235	60	184	94.	34	135	224	45	171
95.	32	128	199	42	154	96.	45	115	155	48	217
97.	40	133	182	41	203	98.	33	125	196	51	123
99.	32	134	215	44	143	100.	38	107	173	53	170
101.	29	120	217	44	162	102.	24	146	270	35	165
103.	27	143	283	54	155	104.	40	127	217	58	158
105.	42	142	209	48	176	106.	21	139	223	24	160
107.	40	114	189	55	190	108.	51	117	206	81	179
109.	31	121	168	36	134	110.	30	134	215	42	139
111.	38	135	200	45	169	112.	34	131	284	61	185
113.	30	138	212	38	170	114.	37	123	201	50	181
115.	26	130	259	47	158	116.	34	116	219	62	152
117.	41	120	179	51	148	118.	32	143	269	45	165
119.	43	126	175	58	164	120.	28	127	264	64	130
121.	29	147	252	38	213	122.	33	132	248	42	177
123.	27	134	259	45	165	124.	23	125	262	47	160
125.	24	123	225	40	171	126.	41	127	206	57	172
127.	29	123	227	38	203	128.	28	134	225	42	201
129.	26	125	232	55	133	130.	32	114	231	55	196
131.	24	123	232	43	141	132.	38	130	222	43	222
133.	16	127	220	23	135	134.	33	143	204	38	152
135.	31	137	196	31	144	136.	32	108	177	47	149
137.	35	146	223	42	178	138.	27	132	263	46	198
139.	33	145	212	35	158	140.	34	135	204	44	130
141.	25	145	230	29	175	142.	39	119	162	41	164
143.	43	108	139	54	184	144.	27	138	269	48	147
145.	27	123	217	46	166	146.	40	117	145	52	123
147.	22	137	258	38	149	148.	30	128	232	39	168
149.	41	133	229	49	226	150.	30	107	144	36	137
151.	34	111	210	60	165	152.	27	117	234	58	131
153.	34	148	236	31	210	154.	35	129	146	26	134
155.	29	120	225	50	154	156.	32	133	211	42	156
157.	30	114	228	52	163	158.	25	123	212	25	159
159.	41	113	170	47	170	160.	21	129	244	40	138
161.	30	150	230	35	166	162.	29	137	231	41	151
163.	39	136	228	56	158	164.	40	133	167	41	182
165.	34	138	211	41	148	166.	39	157	234	38	209
167.	33	127	201	47	155	168.	32	122	227	49	125
169.	30	140	219	34	146	170.	26	127	239	45	160
171.	45	124	173	57	167	172.	37	144	250	48	191

Subject	Y	X_1	X_2	X_3	X_4	Subject	Y	X_1	X_2	X_3	X_4
173.	31	138	229	40	166	174.	37	123	205	61	161
175.	32	134	192	29	185	176.	40	119	162	50	163
177.	27	133	237	42	159	178.	39	133	221	56	155
179.	30	128	132	29	90	180.	40	133	177	45	199
181.	27	142	287	41	196	182.	28	117	158	25	159
183.	38	129	161	31	196	184.	36	135	176	30	179
185.	41	117	216	63	177	186.	27	131	199	26	163
187.	27	124	233	42	174	188.	35	126	150	38	123
189.	31	120	162	39	152	190.	34	135	205	50	95
191.	18	133	215	27	96	192.	27	121	186	34	125
193.	37	138	178	38	165	194.	38	133	214	55	144
195.	27	124	164	31	112	196.	40	114	175	54	159
197.	37	128	188	41	170	198.	35	136	199	38	192
199.	32	118	236	63	99	200.	32	130	246	49	192
201.	34	142	281	48	217	202.	24	121	205	40	114
203.	24	136	288	36	210	204.	42	130	184	52	170
205.	33	123	200	54	103	206.	39	122	212	63	172
207.	32	124	205	43	174	208.	37	120	135	38	126
209.	29	140	207	24	186	210.	25	110	202	44	121
211.	23	146	268	34	189	212.	32	109	190	54	133
213.	33	134	255	55	167	214.	27	146	230	29	173
215.	31	117	180	45	124	216.	33	120	183	38	172
217.	34	129	205	45	144	218.	28	131	220	39	147
219.	37	141	239	55	164	220.	29	105	115	31	138
221.	30	123	158	31	146	222.	44	103	145	54	195
223.	40	132	230	48	198	224.	44	130	188	52	183
225.	35	117	212	48	183	226.	27	111	222	44	162
227.	38	133	196	45	191	228.	36	131	172	31	211
229.	41	143	243	45	224	230.	30	129	246	50	180
231.	32	133	210	38	171	232.	36	121	177	42	176
233.	37	128	177	49	144	234.	36	150	200	30	221
235.	33	119	151	35	151	236.	30	146	248	30	207
237.	30	123	180	32	178	238.	37	125	225	48	236
239.	20	137	212	27	141	240.	34	118	211	54	199
241.	21	118	253	48	151	242.	28	124	190	47	97
243.	40	122	182	58	157	244.	34	106	213	55	184
245.	43	116	159	49	207	246.	33	130	230	43	203
247.	49	127	182	61	183	248.	36	127	196	50	181
249.	33	142	200	32	143	250.	29	149	265	55	132
251.	33	117	97	22	111	252.	37	128	215	48	195
253.	40	129	171	34	188	254.	29	129	191	37	136
255.	34	112	225	55	192	256.	27	121	157	22	84
257.	31	138	196	28	162	258.	33	129	193	43	142
259.	35	140	212	48	132	260.	22	137	254	36	177
261.	24	143	271	31	188	262.	33	126	194	41	183
263.	32	124	207	42	186	264.	29	140	184	23	163
265.	35	132	172	32	121	266.	43	143	216	57	159

Subject	Y	X_1	X_2	X_3	X_4	Subject	Y	X_1	X_2	X_3	X_4
267.	40	128	230	62	173	268.	35	131	238	48	195
269.	31	155	281	36	236	270.	33	132	239	51	156
271.	24	123	180	24	134	272.	31	113	188	46	157
273.	34	125	219	60	164	274.	33	127	143	30	116
275.	28	120	169	36	136	276.	40	120	178	53	160
277.	30	139	230	29	201	278.	30	111	205	50	134
279.	32	148	251	37	207	280.	31	115	215	54	132
281.	27	127	230	50	141	282.	39	127	169	44	158
283.	37	115	133	35	124	284.	27	135	247	42	149
285.	30	130	225	35	221	286.	32	141	224	31	227
287.	30	133	232	40	193	288.	38	119	149	47	144
289.	38	129	185	39	184	290.	32	143	278	40	213
291.	37	134	181	45	150	292.	34	148	226	39	160
293.	33	127	211	49	142	294.	34	121	176	43	135
295.	33	123	233	62	120	296.	37	131	203	47	160
297.	33	131	204	47	114	298.	41	137	210	51	155
299.	40	137	212	49	178	300.	26	134	201	31	129
301.	35	135	217	51	154	302.	22	131	186	26	99
303.	17	144	277	27	144	304.	35	138	222	39	169
305.	26	119	195	35	161	306.	23	129	232	34	176
307.	36	125	203	45	153	308.	19	129	247	39	162
309.	27	123	248	49	157	310.	18	136	266	40	118
311.	32	136	210	36	158	312.	18	139	216	31	72
313.	34	119	211	45	170	314.	41	112	163	50	172
315.	43	105	104	33	170	316.	33	129	222	48	165
317.	35	119	159	52	115	318.	26	113	157	30	108
319.	29	134	253	41	233	320.	23	149	257	31	165
321.	34	123	198	51	117	322.	27	130	192	26	187
323.	39	121	200	52	178	324.	31	145	222	40	157
325.	32	127	269	55	143	326.	36	156	255	46	189
327.	40	139	195	51	140	328.	28	117	253	50	158
329.	36	116	158	39	144	330.	42	121	212	63	209
331.	32	134	233	40	185	332.	34	112	161	35	189
333.	29	142	198	24	160	334.	25	160	274	25	177
335.	26	135	173	23	102	336.	20	142	247	23	171
337.	37	112	140	41	161	338.	45	119	162	49	197
339.	38	123	216	52	176	340.	38	146	257	51	201
341.	26	146	263	28	219	342.	31	116	186	46	143
343.	31	131	213	43	171	344.	41	114	228	67	209
345.	33	132	244	53	199	346.	39	116	194	56	193
347.	43	120	139	50	166	348.	21	120	229	40	129
349.	37	138	278	56	223	350.	40	128	160	40	135
351.	39	112	201	54	191	352.	27	128	232	42	136
353.	40	136	222	47	182	354.	34	122	165	36	159
355.	38	127	139	30	153	356.	41	131	178	46	157
357.	42	142	190	39	140	358.	34	124	209	50	215
359.	39	156	243	49	189	360.	28	132	257	51	175

Subject	Y	X_1	X_2	X_3	X_4	Subject	Y	X_1	X_2	X_3	X_4
361.	33	129	201	42	171	362.	31	136	212	37	162
363.	27	123	184	29	173	364.	42	134	184	61	99
365.	36	116	189	44	181	366.	46	128	171	57	179
367.	39	116	121	39	126	368.	34	122	202	49	154
369.	43	133	197	57	143	370.	37	117	213	63	169
371.	40	125	168	52	150	372.	37	117	186	52	182
372.	35	125	201	51	132	374.	31	131	229	43	163
375.	36	126	180	48	122	376.	37	124	216	49	193
377.	32	132	232	46	160	378.	35	126	154	39	120
379.	27	146	229	27	170	380.	29	131	201	32	167
381.	30	138	246	45	184	382.	31	127	203	45	110
383.	34	131	151	27	179	384.	33	131	208	46	163
385.	42	109	113	47	108	386.	33	123	237	58	151
387.	44	142	199	48	181	388.	28	146	219	32	153
389.	35	136	268	52	238	390.	36	119	203	51	178
391.	38	133	192	51	174	392.	38	114	168	45	189
393.	28	140	228	35	144	394.	45	142	250	57	260
395.	35	128	173	31	145	396.	28	132	230	38	205
397.	39	131	182	41	188	398.	35	135	166	36	133
399.	22	121	197	41	130	400.	38	128	214	47	203
401.	18	129	205	30	142	402.	21	121	268	47	194
403.	21	136	260	44	165	404.	16	144	232	26	222
405.	26	125	233	54	174	406.	18	133	268	48	175
407.	28	150	289	47	204	408.	47	149	285	63	220
409.	50	120	158	48	169	410.	18	127	212	43	114
411.	44	130	199	51	159	412.	29	122	207	48	158
413.	44	125	201	57	172	414.	27	105	151	50	84
415.	38	131	212	58	151	416.	39	125	208	57	139
417.	50	127	162	51	189	418.	31	136	205	37	142
419.	21	137	282	37	222	420.	27	126	242	55	148
421.	43	136	216	46	195	422.	33	122	176	47	124
423.	44	115	181	53	194	424.	13	134	230	29	157
425.	24	128	249	48	118	426.	26	124	193	35	179
427.	43	129	174	38	171	428.	15	128	269	41	197
429.	31	139	196	40	119	430.	38	120	190	41	195
431.	44	127	220	55	193	432.	53	132	172	53	179
433.	39	122	141	31	156	434.	24	131	209	31	184
435.	48	128	169	45	149	436.	25	124	200	43	121
437.	18	114	163	33	125	438.	38	104	177	58	105
439.	29	119	159	35	101	440.	42	113	202	58	209
441.	35	134	192	35	169	442.	53	113	117	49	127
443.	31	117	231	48	181	444.	18	114	186	33	149
445.	18	134	257	38	179	446.	31	151	242	36	170
447.	37	130	174	41	159	448.	44	115	178	55	154
449.	26	134	206	39	155	450.	45	123	155	42	126
451.	40	133	149	37	180	432.	26	137	221	28	194
453.	34	127	144	42	126	454.	22	122	201	27	160

Subject	Y	X_1	X_2	X_3	X_4	Subject	Y	X_1	X_2	X_3	X_4
455.	34	139	226	50	154	456.	30	160	252	36	169
457.	37	143	228	54	157	458.	33	134	205	39	182
459.	29	118	192	42	159	460.	25	153	291	32	193
461.	30	115	205	41	185	462.	36	136	205	43	177
463.	20	128	237	28	160	464.	41	117	131	46	111
465.	39	140	223	55	202	466.	36	143	229	52	136
467.	43	148	226	56	172	468.	25	125	221	42	171
469.	35	118	182	48	153	470.	36	144	209	42	166
471.	36	118	195	45	180	472.	41	119	171	63	98
473.	34	135	191	39	184	474.	26	140	210	34	111
475.	30	117	245	46	171	476.	39	121	144	48	108
477.	39	115	197	64	130	478.	27	144	284	46	187
479.	26	113	168	41	106	480.	22	150	288	27	240
481.	25	141	217	29	168	482.	26	136	177	20	167
483.	45	117	210	73	149	484.	26	141	257	26	228
485.	38	130	201	49	188	486.	36	135	248	50	183
487.	28	137	207	39	165	488.	35	129	177	48	111
489.	27	116	186	49	98	490.	35	132	207	44	189
491.	33	135	262	43	203	492.	24	113	175	33	129
493.	30	130	238	44	158	494.	30	138	217	29	165
495.	26	127	235	44	158	496.	34	146	244	38	206
497.	32	112	163	33	135	498.	38	108	125	52	136
499.	43	147	218	43	193	500.	42	119	208	60	179
501.	25	131	188	20	147	502.	34	127	215	49	156
503.	32	125	169	48	83	504.	32	135	180	31	166
505.	32	125	227	53	121	506.	32	132	208	42	190
507.	31	128	270	49	207	508.	36	133	216	50	167
509.	33	120	225	54	177	510.	35	127	197	49	130
511.	37	135	241	40	223	512.	31	124	159	39	96
513.	39	128	175	42	176	514.	31	119	242	48	209
515.	35	125	227	46	213	516.	29	110	198	45	150
517.	31	122	226	47	184	518.	30	124	217	47	144
519.	36	117	209	44	181	520.	44	111	121	51	104
521.	34	122	231	57	195	522.	32	145	268	35	230
523.	35	125	214	49	157	524.	29	126	175	23	152
525.	34	128	188	42	129	526.	19	136	257	25	153
527.	25	133	242	40	149	528.	37	133	181	36	201
529.	44	108	134	64	83	530.	34	130	217	41	204
531.	37	129	166	31	190	532.	34	108	176	52	144
533.	21	151	295	30	220	534.	23	140	286	37	212
535.	38	125	203	57	152	536.	26	132	270	54	145
537.	39	130	210	52	183	538.	39	128	201	50	218
539.	41	135	176	44	182	540.	37	121	148	40	178
541.	26	122	229	42	158	542.	41	112	166	54	125
543.	40	109	160	48	159	544.	31	128	249	54	189
545.	31	143	263	51	176	546.	33	137	254	52	158
547.	41	135	194	39	179	548.	26	142	261	33	206

Subject	Y	X_1	X_2	X_3	X_4	Subject	Y	X_1	X_2	X_3	X_4
549.	29	134	269	53	205	550.	30	119	166	41	122
551.	43	134	199	51	179	552.	33	117	186	53	146
553.	20	134	265	33	149	554.	43	124	222	69	148
555.	29	112	198	51	127	556.	38	138	211	52	139
557.	32	136	228	35	197	558.	30	123	173	31	111
559.	32	124	215	57	63	560.	34	119	201	42	186
561.	38	127	175	45	198	562.	31	130	184	33	153
563.	41	112	156	50	128	564.	42	106	130	47	208
565.	38	145	249	44	183	566.	31	141	239	45	195
567.	29	131	206	45	106	568.	35	127	176	39	102
569.	29	133	256	47	154	570.	44	132	168	46	134
571.	28	129	201	41	63	572.	39	139	211	49	246
573.	38	137	238	45	241	574.	41	130	201	45	246
575.	32	136	247	40	205	576.	35	143	222	37	180
577.	37	122	220	56	207	578.	33	124	198	42	222
579.	26	126	225	38	108	580.	34	112	200	60	139
581.	36	134	264	60	234	582.	34	131	244	54	217
583.	33	121	181	40	129	584.	27	107	226	48	164
585.	40	130	188	48	211	586.	32	137	242	49	156
587.	21	145	287	29	188	588.	31	128	183	31	119
589.	36	134	189	34	253	590.	34	126	188	51	95
591.	40	138	180	41	178	592.	33	131	248	48	180
593.	30	128	203	41	100	594.	43	118	159	50	185
595.	40	134	214	55	192	596.	36	125	194	51	113
597.	34	123	198	46	218	598.	31	126	191	43	73
599.	29	127	238	55	133	600.	32	126	223	49	121
601.	30	126	238	48	145	602.	31	118	203	48	125
603.	23	130	203	35	100	604.	24	120	233	44	153
605.	29	115	214	44	147	606.	20	120	224	30	164
607.	27	110	181	45	138	608.	34	147	194	30	207
609.	42	137	215	51	224	610.	39	130	230	53	192
611.	28	102	176	51	79	612.	41	119	201	58	158
613.	32	120	164	29	190	614.	34	121	266	59	289
615.	48	128	137	38	205	616.	27	124	206	42	65
617.	32	143	252	49	124	618.	39	127	206	53	169
619.	42	130	165	50	162	620.	40	112	182	57	201
621.	37	126	196	55	177	622.	26	148	262	39	178
623.	30	119	175	32	168	624.	27	116	230	51	175
625.	31	135	251	43	210	626.	31	120	191	45	161
627.	36	116	155	56	95	628.	39	114	170	58	115
629.	20	138	277	38	172	630.	27	139	273	53	153
631.	31	121	186	39	131	632.	37	125	192	49	194
633.	32	139	253	48	217	634.	34	136	177	45	89
635.	35	135	273	57	234	636.	22	135	278	49	132
637.	39	128	176	47	165	638.	45	103	109	58	82
639.	30	127	231	51	151	640.	35	132	180	37	135
641.	34	120	195	55	119	642.	35	128	238	55	194

Subject	Y	X_1	X_2	X_3	X_4	Subject	Y	X_1	X_2	X_3	X_4
643.	35	132	184	35	230	644.	21	139	237	28	134
645.	36	145	198	29	270	646.	36	160	243	32	213
647.	36	122	166	46	121	648.	36	142	225	41	255
649.	33	122	173	33	215	650.	31	137	200	38	94
651.	31	119	178	36	172	652.	40	119	166	43	162
653.	27	129	272	42	155	654.	35	119	187	43	189
655.	35	103	109	38	71	656.	35	123	201	42	208
657.	37	143	231	51	190	658.	32	137	215	45	171
659.	36	148	224	31	240	660.	35	140	227	43	182
661.	30	108	180	50	130	662.	30	151	191	29	91
663.	37	112	137	47	114	664.	29	140	218	33	125
665.	40	137	206	43	212	666.	29	135	201	43	78
667.	28	127	206	47	120	668.	22	143	316	46	207
669.	31	149	217	32	137	670.	30	140	246	36	157
671.	39	118	220	72	218	672.	22	146	269	34	192
673.	29	114	191	40	119	674.	31	118	196	42	189
675.	42	144	222	55	146	676.	23	134	286	44	192
677.	33	115	194	42	190	678.	37	131	180	37	145
679.	20	149	273	35	161	680.	40	121	143	50	68
681.	23	147	284	37	212	682.	34	118	190	51	159
683.	27	126	254	46	187	684.	21	132	228	21	161
685.	35	110	126	38	92	686.	33	137	192	33	184
687.	28	142	246	35	232	688.	30	139	203	29	193
689.	21	119	201	30	90	690.	34	124	224	44	265
691.	33	124	210	45	207	692.	36	142	262	51	238
693.	27	120	217	40	101	694.	30	125	204	46	158
695.	28	116	250	62	143	696.	35	120	205	47	168
697.	35	119	200	55	91	698.	36	108	159	41	218
699.	26	131	237	43	110	700.	17	122	286	35	161
701.	20	109	212	39	115	702.	22	138	249	24	216
703.	39	122	173	42	167	704.	25	142	224	27	193
705.	43	121	144	44	197	706.	35	128	217	37	216
707.	40	137	255	54	278	708.	31	128	163	26	156
709.	33	119	202	46	134	710.	29	123	228	46	184
711.	34	135	235	50	207	712.	30	122	198	37	140
713.	37	146	239	60	126	714.	25	113	202	41	103
715.	32	130	269	53	205	716.	38	111	165	61	57
717.	39	132	206	55	192	718.	28	151	246	34	137
719.	32	136	225	43	175	720.	38	121	144	45	129
721.	25	121	245	48	181	722.	27	127	178	33	133
723.	36	123	153	35	134	724.	31	122	213	41	227
725.	34	121	122	28	100	726.	33	130	212	46	157
727.	41	118	184	55	183	728.	34	135	240	54	120
729.	31	145	284	43	249	730.	38	142	193	44	182
731.	30	132	282	64	158	732.	34	133	207	48	167
733.	33	121	202	51	187	734.	36	129	202	45	171
735.	26	123	258	51	186	736.	37	133	184	50	105
737.	37	125	178	40	205	738.	25	126	240	42	158

Subject	Y	X_1	X_2	X_3	X_4	Subject	Y	X_1	X_2	X_3	X_4
739.	26	135	221	37	162	740.	37	132	197	47	158
741.	40	124	161	50	90	742.	35	144	203	31	215
743.	43	123	205	56	226	744.	31	127	242	51	173
745.	27	141	259	46	146	746.	32	115	220	55	174
747.	33	128	231	50	153	748.	27	135	239	45	122
749.	33	129	197	41	151	750.	31	108	161	51	74
751.	22	142	204	30	49	752.	32	123	179	42	164
753.	37	130	182	43	147	754.	46	121	168	58	155
755.	32	115	211	48	204	756.	26	132	203	39	55
757.	23	129	231	40	130	758.	38	148	229	39	224
759.	26	129	276	54	123	760.	48	120	140	57	154
761.	31	125	208	45	189	762.	37	128	225	52	207
763.	44	125	152	41	167	764.	29	133	211	36	88
765.	35	103	199	54	219	766.	40	129	159	43	118
767.	35	130	259	57	221	768.	23	134	258	47	76
769.	36	122	185	44	142	770.	29	125	182	32	121
771.	30	118	185	45	152	772.	25	132	245	51	123
773.	29	129	195	27	208	774.	45	142	182	34	265
775.	22	152	288	27	238	776.	30	121	242	55	160
777.	36	139	212	47	130	778.	41	125	206	51	244
779.	30	124	181	45	82	780.	27	131	188	35	43
781.	34	151	251	40	211	782.	37	128	182	54	49
783.	25	130	260	40	154	784.	44	99	116	57	153
785.	33	123	196	46	112	786.	35	103	168	57	105
787.	41	133	195	56	123	788.	19	150	265	23	128
789.	36	134	193	40	159	790.	32	123	194	44	163
791.	34	133	210	43	126	792.	41	121	114	36	161
793.	20	136	205	29	51	794.	25	138	268	30	232
795.	41	108	143	54	135	796.	34	130	213	41	238
797.	34	148	277	58	176	798.	25	122	208	34	158
799.	35	134	223	52	186	800.	30	130	228	44	183
801.	29	135	230	47	90	802.	31	140	223	41	197
803.	35	129	266	56	284	804.	31	135	204	31	197
805.	22	145	274	34	156	806.	30	141	261	46	209
807.	29	130	216	44	145	808.	34	125	212	49	185
809.	42	128	196	52	192	810.	20	136	248	39	113
811.	35	120	153	41	165	812.	43	122	157	52	177
813.	28	144	259	48	162	814.	21	142	308	44	180
815.	46	125	150	54	171	816.	24	134	196	27	130
817.	32	124	187	45	138	818.	28	136	214	39	147
819.	34	137	205	44	164	820.	37	135	228	53	197
821.	40	139	205	42	187	822.	24	132	250	44	155
823.	40	143	226	50	198	824.	34	121	199	42	192
825.	26	134	211	34	160	826.	30	122	240	45	196
827.	31	129	234	47	187	828.	32	111	187	43	119
829.	29	118	211	41	185	830.	39	115	167	56	157
831.	28	112	199	44	147	832.	40	130	173	48	161
833.	30	133	217	49	109	834.	31	128	202	32	195

Subject	Y	X_1	X_2	X_3	X_4	Subject	Y	X_1	X_2	X_3	X_4
835.	30	137	279	52	200	836.	34	140	238	40	185
837.	37	129	254	57	198	838.	31	91	146	47	146
839.	38	136	204	40	187	840.	34	116	180	33	165
841.	27	130	250	38	193	842.	32	132	223	46	170
843.	46	124	163	56	152	844.	29	121	178	39	129
845.	32	131	270	50	223	846.	29	130	230	42	199
847.	35	134	202	39	151	848.	34	132	261	53	184
849.	37	139	183	39	134	850.	27	125	206	36	160
851.	34	125	206	45	209	852.	26	142	247	33	146
853.	34	124	186	43	139	854.	44	125	170	54	165
855.	29	127	212	48	118	856.	43	127	179	49	169
857.	34	116	174	48	128	858.	44	120	181	67	147
859.	28	112	213	49	149	860.	25	134	218	29	149
861.	22	122	233	40	133	862.	42	128	164	51	162
863.	19	142	254	23	177	864.	26	134	244	51	123
865.	27	140	267	46	183	866.	41	126	201	56	167
867.	23	110	208	43	138	868.	41	132	201	58	136
869.	27	141	259	38	158	870.	37	114	165	44	144
871.	39	131	244	67	180	872.	30	128	274	49	195
873.	42	111	111	43	137	874.	32	131	271	58	203
875.	39	130	216	52	214	876.	25	138	272	36	234
877.	34	135	185	43	102	878.	42	127	166	43	169
879.	24	146	261	39	142	880.	24	121	275	53	178
881.	28	125	225	41	172	882.	39	111	206	64	175
883.	28	121	227	50	111	884.	38	110	118	41	94
885.	31	122	246	55	174	886.	21	132	197	26	164
887.	20	136	302	46	169	888.	32	137	274	49	219
889.	31	125	196	31	186	890.	31	129	183	40	146
891.	26	123	197	31	183	892.	39	116	127	36	137
893.	49	134	203	60	211	894.	37	110	183	51	167
895.	33	137	255	49	200	896.	46	126	140	44	163
897.	38	143	203	41	189	898.	31	143	243	49	170
899.	38	124	223	51	236	900.	32	100	147	51	95
901.	35	121	164	42	133	902.	38	117	253	70	201
903.	33	121	154	35	157	904.	28	118	231	47	166
905.	35	127	235	58	188	906.	37	134	224	53	179
907.	33	114	160	41	132	908.	37	120	204	54	142
909.	31	134	237	44	163	910.	25	129	248	38	166
911.	34	142	238	40	206	912.	27	126	189	37	178
913.	36	128	199	44	158	914.	20	149	257	27	168
915.	34	132	192	44	144	916.	28	135	229	33	202
917.	36	118	187	49	170	918.	28	126	226	45	149
919.	28	128	239	40	206	920.	33	115	160	47	165
921.	39	121	171	53	152	922.	25	133	195	27	87
923.	37	115	205	59	125	924.	16	144	300	33	193
925.	32	121	213	43	165	926.	41	133	178	50	115
927.	34	139	180	30	150	928.	29	120	191	43	110
929.	40	111	168	54	155	930.	46	130	160	42	190

Subject	Y	X_1	X_2	X_3	X_4	Subject	Y	X_1	X_2	X_3	X_4
931.	34	132	188	41	174	932.	34	130	255	51	195
933.	27	138	241	34	182	934.	28	111	159	36	120
935.	16	151	272	34	111	936.	31	120	172	38	160
937.	33	123	243	57	207	938.	26	120	229	47	135
939.	35	112	104	28	108	940.	35	136	240	57	182
941.	33	127	168	41	107	942.	38	130	173	44	128
943.	18	117	233	32	161	944.	27	127	197	33	161
945.	41	122	203	55	180	946.	45	124	148	51	167
947.	36	127	237	54	185	948.	36	120	169	56	90
949.	29	124	215	41	192	950.	30	127	198	35	198
951.	45	122	141	46	194	952.	30	127	235	44	175
953.	27	129	216	35	129	954.	28	131	254	44	208
955.	30	127	232	44	188	956.	24	149	235	26	133
957.	38	119	160	56	127	958.	40	133	212	46	188
959.	28	107	165	42	144	960.	38	127	197	52	167
961.	34	121	176	37	146	962.	36	127	199	41	166
963.	38	110	167	45	196	964.	30	110	181	52	100
965.	44	118	167	53	179	966.	29	136	247	48	159
967.	33	137	272	51	193	968.	44	120	193	54	171
969.	15	134	248	33	92	970.	34	159	308	47	230
971.	28	114	184	43	143	972.	23	125	211	39	158
973.	16	128	192	20	138	974.	41	128	185	51	156
975.	32	131	201	42	176	976.	30	142	246	50	161
977.	38	141	214	45	186	978.	32	137	198	39	173
979.	33	117	188	45	156	980.	38	118	246	59	224
981.	38	132	159	41	132	982.	31	135	219	34	194
983.	24	128	235	40	160	984.	42	134	262	77	161
985.	35	122	195	48	168	986.	33	133	185	40	117
987.	26	141	224	26	185	988.	35	135	191	39	146
989.	28	128	155	29	97	990.	24	134	217	37	93
991.	32	137	237	44	183	992.	23	136	213	31	113
993.	30	111	225	55	182	994.	22	137	201	27	145
995.	26	121	200	41	131	996.	30	155	247	34	149
997.	38	121	199	52	218	998.	38	146	246	51	190
999.	36	133	174	39	129	1000.	36	129	237	54	217

REFERENCES

References Cited
1. George W. Snedecor and William G. Cochran, *Statistical Methods*, Sixth Edition, The Iowa State University Press, Ames, Iowa, 1967.
2. Robert G. D. Steel and James H. Torrie, *Principles and Procedures of Statistics*, McGraw-Hill, New York, 1960.

3. R. L. Anderson and T. A. Bancroft, *Statistical Theory in Research*, McGraw-Hill, New York, 1952.

4. Mordecai Ezekiel and Karl A. Fox, *Methods of Correlation and Regression Analysis*, Third Edition, Wiley, New York, 1959.

5. M. H. Doolittle, "Method Employed in the Solution of Normal Equation and the Adjustment of a Triangulation," *U.S. Coast and Geodetic Survey Report*, 1878.

6. David Durand, "Joint Confidence Region for Multiple Regression Coefficients," *Journal of the American Statistical Association*, *49* (1954), 130–146.

7. John Neter, William Wasserman, and Michael H. Kutner, *Applied Linear Regression Models*, Irwin, Homewood, Illinois, 1983.

8. R. C. Geary and C. E. V. Leser, "Significance Tests in Multiple Regression," *The American Statistician*, *22* (February 1968), 20–21.

9. David G. Kleinbaum and Lawrence L. Kupper, *Applied Regression Analysis and Other Multivariable Methods*, Duxbury Press, North Scituate, Massachusetts, 1978.

10. William Mendenhall and James T. McClave, *A Second Course in Business Statistics: Regression Analysis*, Dellen Publishing Company, San Francisco, 1981.

11. N. R. Draper and H. Smith, *Applied Regression Analysis*, Second Edition, Wiley, New York, 1981.

12. R. A. Fisher, "The General Sampling Distribution of the Multiple Correlation Coefficient," *Proceedings of the Royal Society A*, *121* (1928), 654–673.

13. K. H. Kramer, "Tables for Constructing Confidence Limits on the Multiple Correlation Coefficient," *Journal of the American Statistical Association*, *58* (1963), 1082–1085.

14. David M. Allen, "Mean Square Error of Prediction as a Criterion for Selecting Variables," *Technometrics*, *13* (1971), 469–475.

15. E. M. L. Beale, M. G. Kendall, and D. W. Mann, "The Discarding of Variables in Multivariate Analysis," *Biometrika*, *54* (1967), 357–366.

16. M. J. Garside, "The Best Sub-Set in Multiple Regression Analysis," *Applied Statistics*, *14* (1965), 196–200.

17. J. W. Gorman and R. J. Toman, "Selection of Variables for Fitting Equations to Data," *Technometrics*, *8* (1966), 27–52.

18. R. R. Hocking and R. N. Leslie, "Selection of the Best Sub-Set in Regression Analysis," *Technometrics*, *9* (1967), 531–540.

19. H. J. Larson and T. A. Bancroft, "Sequential Model Building for Prediction in Regression Analysis I," *Annals of Mathematical Statistics*, *34* (1963), 462–479.

20. D. V. Lindley, "The Choice of Variables in Multiple Regression," *Journal of the Royal Statistical Society B*, *30* (1968), 31–66.

21. E. F. Schultz, Jr. and J. F. Goggans, "A Systematic Procedure for Determining Potent Independent Variables in Multiple Regression and Discriminant Analysis," Agri. Exp. Sta. Bull. 336, Auburn University, 1961.

22. K. W. Smillie, *An Introduction to Regression and Correlation*, Academic Press, New York, 1966.

23. Peter Sprent, *Models in Regression*, Methuen, London, 1969.

24. A. Summerfield and A. Lubin, "A Square Root Method of Selecting a Minimum Set of Variables in Multiple Regression: I, The Method," *Psychometrika*, *16* (3):271 (1951).

25. Wayne W. Daniel, *Applied Nonparametric Statistics*, Houghton Mifflin, Boston, 1978.

26. Wayne W. Daniel, *Nonparametric, Distribution-Free, and Robust Procedures in Regression Analysis: A Selected Bibliography*, Vance Bibliographies, Monticello, Ill., June 1980.

Other References, Books
1. Frank Andrews, James Morgan, and John Sonquist, *Multiple Classification Analysis*, Ann Arbor, Survey Research Center, 1967.

2. T. W. Anderson, *Introduction to Multivariate Statistical Analysis*, Wiley, New York, 1958.

3. Cuthbert Daniel and Fred S. Wood, *Fitting Equations to Data*, Second Edition, Wiley-Interscience, New York, 1979.

4. Alan E. Treloar, *Correlation Analysis*, Burgess, Minneapolis, 1949.

Other References, Journal Articles
1. R. G. Newton and D. J. Spurrell, "A Development of Multiple Regression for the Analysis of Routine Data," *Applied Statistics*, *16* (1967), 51–64.

2. Potluri Rao, "Some Notes on Misspecification in Multiple Regression," *The American Statistician*, *25* (December 1971), 37–39.

3. Neil S. Weiss, "A Graphical Representation of the Relationships Between Multiple Regression and Multiple Correlation," *The American Statistician*, *24* (April 1970), 25–29.

4. E. J. Williams, "The Analysis of Association Among Many Variates," *Journal of the Royal Statistical Society*, *29* (1967), 199–242.

Other References, Other Publications
1. R. L. Bottenbery and Y. H. Ward, Jr., *Applied Multiple Linear Regression*, U.S. Department of Commerce, Office of Technical Services, AD 413128, 1963.

2. Wayne W. Daniel, *Ridge Regression: A Selected Bibliography*, Vance Bibliographies, Monticello, Ill., October 1980.

3. Wayne W. Daniel, *The Use of Dummy Variables in Regression Analysis: A Selected Bibliography With Annotations*, Vance Bibliographies, Monticello, Ill., November 1979.

4. Wayne W. Daniel, *Outliers in Research Data: A Selected Bibliography*, Vance Bibliographies, Monticello, Ill., July 1980.

5. Jean Draper, *Interpretation of Multiple Regression Analysis Part I, Problems of Interpreting Large Samples of Data*, The University of Arizona, College of Business and Public Administration, Division of Economics and Business Research, 1968.

CHAPTER 10

The Chi-Square Distribution and the Analysis of Frequencies

10.1 INTRODUCTION

In the chapters on estimation and hypothesis testing brief mention is made of the chi-square distribution in the construction of confidence intervals for and the testing of hypotheses concerning a population variance. This distribution, which is one of the most widely used distributions in statistical applications, has many other uses. Some of the more common ones are presented in this chapter along with a more complete description of the distribution itself, which follows in the next section.

10.2 THE MATHEMATICAL PROPERTIES OF THE CHI-SQUARE DISTRIBUTION

The chi-square distribution may be derived from normal distributions. Suppose that from a normally distributed random variable Y with mean μ and variance σ^2 we randomly and independently select samples of size $n = 1$. Each value selected may be transformed to the unit normal variable z by the familiar formula

$$z = \frac{y_i - \mu}{\sigma} \tag{10.2.1}$$

Each value of z may be squared to obtain z^2. When we investigate the sampling distribution of z^2, we find that it follows a chi-square distribution with 1 degree of freedom. That is,

$$\chi^2_{(1)} = \left(\frac{y - \mu}{\sigma}\right)^2 = z^2$$

Now suppose that we randomly and independently select samples of size $n = 2$ from the normally distributed population of Y values. Within each sample we may transform each value of y to the unit normal variable z and square as before. If the resulting values of z^2 for each sample are added, we may designate this sum by

$$\chi^2_{(2)} = \left(\frac{y_1 - \mu}{\sigma}\right)^2 + \left(\frac{y_2 - \mu}{\sigma}\right)^2 = z_1^2 + z_2^2$$

since it follows the chi-square distribution with 2 degrees of freedom, the number of independent squared terms that are added together.

The procedure may be repeated for any sample size n. The sum of the resulting z^2 values in each case will be distributed as chi-square with n degrees of freedom. In general, then,

$$\chi^2_{(n)} = z_1^2 + z_2^2 + \cdots + z_n^2 \tag{10.2.2}$$

follows the χ^2 distribution with n degrees of freedom. The mathematical form of the chi-square distribution is as follows:

$$f(u) = \frac{1}{\left(\frac{k}{2} - 1\right)!} \frac{1}{2^{k/2}} u^{(k/2)-1} e^{-(u/2)}, \qquad u > 0 \tag{10.2.3}$$

where e is the irrational number $2.71828\cdots$ and k is the number of degrees of freedom. The variate u is usually designated by the Greek letter chi (χ) and, hence, the distribution is called the χ^2 distribution. As we pointed out in Chapter 5, the χ^2 distribution has been tabulated in Table F. Further use of the table is demonstrated as the need arises in succeeding sections.

The mean and variance of the χ^2 distribution are k and $2k$, respectively. The modal value of the distribution is $k - 2$ for values of k greater than or equal to 2 and is zero for $k = 1$.

The shapes of the χ^2 distributions for several values of k are shown in Figure 5.9.1. We observe in this figure that the shapes for $k = 1$ and $k = 2$ are quite different from the general shape of the distribution for $k > 2$. We also see from this figure that χ^2 assumes values between 0 and infinity. It cannot take on negative values, since it is the sum of values that have been squared. A final characteristic of the χ^2 distribution worth noting is that the sum of two or more independent χ^2 variables also follows a χ^2 distribution.

We make use of the χ^2 distribution in this chapter in testing hypotheses where the data available for analysis are in the form of frequencies. These hypothesis testing procedures are discussed under the topics of *tests of goodness-of-fit*, *tests of independence*, and *tests of homogeneity*. We shall discover that, in a sense, all of the χ^2 tests that we employ may be thought of as goodness-of-fit tests, in that they test the goodness-of-fit of observed frequencies to frequencies that one would expect if the data were generated under some particular theory or hypothesis. We, however, reserve the phrase "goodness-of-fit" for use in a more restricted sense. We use the phrase "goodness-of-fit" to refer to a comparison of a sample distribution to some theoretical distribution that it is assumed describes the population from which the sample came. The justification of our use of the distribution in these situations is due to Karl Pearson (1), who showed that the χ^2 distribution may be used as a test of the agreement between observation and hypothesis whenever the data are in the form of frequencies.

An extensive treatment of the χ^2 distribution is to be found in the book by Lancaster (2).

10.3 TESTS OF GOODNESS-OF-FIT

As we have pointed out, a goodness-of-fit test is appropriate when one wishes to decide if an observed distribution of frequencies is incompatible with some preconceived or hypothesized distribution.

We may, for example, wish to determine whether or not a sample of observed values of some random variable is compatible with the hypothesis that it was drawn from a population of values that is normally distributed. The procedure for reaching a decision consists of placing the values into mutually exclusive categories or class intervals and noting the frequency of occurrence of values in each category. We then make use of our knowledge of normal distributions to determine the frequencies for each category that one could expect if the sample had come from a normal distribution. If the discrepancy between what was observed and what one would expect, given that sampling was from a normal distribution, is too great to be attributed to chance, we conclude that the sample did not come from a normal distribution. If the discrepancy is of such magnitude that it could have come about due to chance, we conclude that the sample may have come from a normal distribution. In a similar manner, tests of goodness-of-fit may be carried out in cases where the hypothesized distribution is the binomial, the Poisson, or any other distribution. Let us illustrate in more detail with some examples of tests of hypotheses of goodness-of-fit.

Example 10.3.1 The Normal Distribution

1. *Data* Suppose that a research team making a study of hospitals in the United States collects data on a sample of 250 hospitals which enables the team to compute for each the inpatient occupancy ratio, a variable that shows, for a 12-month period, the ratio of average daily census to the average number of beds maintained. Suppose the sample yielded the distribution of ratios (expressed as percents), shown in Table 10.3.1.

TABLE 10.3.1

Results of Study Described in Example 10.3.1

Inpatient occupancy ratio	Number of hospitals
0.0 to 39.9	16
40.0 to 49.9	18
50.0 to 59.9	22
60.0 to 69.9	51
70.0 to 79.9	62
80.0 to 89.9	55
90.0 to 99.9	22
100.0 to 109.9	4
Total	250

We wish to know whether these data provide sufficient evidence to indicate that the sample did not come from a normally distributed population.

2. *Assumptions*   We assume that the sample available for analysis is a simple random sample.

3. *Hypotheses*

   H_0:   In the population from which the sample was drawn, inpatient occupancy ratios are normally distributed.

   H_A: The sampled population is not normally distributed.

4. *Test Statistic*   The test statistic is

$$X^2 = \sum_{i=1}^{k} \frac{(O_i - E_i)^2}{E_i} \qquad (10.3.1)$$

In Equation 10.3.1 O_i refers to the ith observed frequency and E_i refers to the ith expected frequency.

5. *Distribution of the Test Statistic*   When the null hypothesis is true X^2 is distributed approximately as χ^2 with $k - r$ degrees of freedom. In determining the degrees of freedom, k is equal to the number of groups for which observed and expected frequencies are available, and r is the number of restrictions or constraints imposed on the given comparison. A restriction is imposed when we force the sum of the expected frequencies to equal the sum of the observed frequencies, and an additional restriction is imposed for each parameter that is estimated from the sample. For a full discussion of the theoretical justification for subtracting one degree of freedom for each estimated parameter, see Cramer (3).

6. *Decision Rule*   The quantity $\sum[(O_i - E_i)^2/E_i]$ will be small if the observed and expected frequencies are close together, and will be large if the differences are large.

   The computed value of X^2 is compared with the tabulated value of χ^2 with $k - r$ degrees of freedom. The decision rule, then, is: Reject H_0 if X^2 is greater than or equal to the tabulated χ^2 for the chosen value of α.

7. *Calculation of Test Statistic*   Since the mean and variance of the hypothesized distribution are not specified, the sample data must be used to estimate them. These parameters, or their estimates, will be needed to compute the frequency that would be expected in each class interval when the null hypothesis is true. The mean and standard deviation computed from the grouped data of Table 10.3.1 by the methods of Sections 1.7 and 1.8 are

$$\bar{x} = 69.91$$
$$s = 19.02$$

   As the next step in the analysis we must obtain for each class interval the frequency of occurrence of values that we would expect when the null hypothesis is true, that is, if the sample were, in fact, drawn from a normally distributed population of values. To do this, we first determine the expected relative frequency of occurrence of values for each class interval and then multiply these expected relative frequencies by the total number of values to obtain the expected number of values for each interval.

   It will be recalled from our study of the normal distribution that the relative frequency of occurrence of values equal to or less than some specified value, say x_0, of the normally distributed random variable X is equivalent to the area under the curve and to the left of x_0 as represented by the shaded area in Figure 10.3.1. We obtain the numerical value of this area by converting x_0 to a standard normal deviate by the formula $z_0 = (x_0 - \mu)/\sigma$ and finding the appropriate value in Table C. We use this procedure to obtain the expected relative frequencies corresponding to each of the class

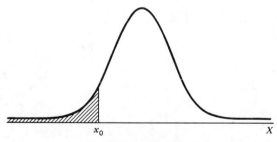

FIGURE 10.3.1

A Normal Distribution Showing the Relative Frequency of Occurrence of Values Less Than or Equal to x_0. The Shaded Area Represents the Relative Frequency of Occurrence of Values Equal to or Less Than x_0.

intervals in Table 10.3.1. We estimate μ and σ with $\bar{x}$ and s as computed from the grouped sample data. The first step consists of obtaining z values corresponding to the lower limit of each class interval. The area between two successive z values will give the expected relative frequency of occurrence of values for the corresponding class interval.

For example, to obtain the expected relative frequency of occurrence of values in the interval 40.0 to 49.9 we proceed as follows:

$$\text{The } z \text{ value corresponding to } X = 40.0 \text{ is } z = \frac{40.0 - 69.91}{19.02} = -1.57$$

$$\text{The } z \text{ value corresponding to } X = 50.0 \text{ is } z = \frac{50.0 - 69.91}{19.02} = -1.05$$

In Table C we find that the area to the left of -1.05 is .1469, and the area to the left of -1.57 is .0582. The area between -1.05 and -1.57 is equal to $.1469 - .0582 = .0887$, which is equal to the expected relative frequency of occurrence of values of occupancy ratios within the interval 40.0 to 49.9. This tells us that if the null hypothesis is true, that is, if the occupancy ratio values are normally distributed, we should expect 8.9 percent of the values in our sample to be between 40.0 and 49.9. When we multiply our total sample size, 250, by .0887 we find the expected frequency for the interval to be 22.18. Similar calculations will give the expected frequencies for the other intervals as shown in Table 10.3.2.

We are now interested in examining the magnitudes of the discrepancies between the observed frequencies and the expected frequencies, since we note that the two sets of frequencies do not agree. We know that even if our

TABLE 10.3.2

Class Intervals and Expected Frequencies for Example 10.3.1

Class interval	$z = (x_i - \bar{x})/s$ at lower limit of interval	Expected relative frequency	Expected frequency
< 40.0		.0582	14.55
40.0 to 49.9	−1.57	.0887	22.18
50.0 to 59.9	−1.05	.1546	38.65
60.0 to 69.9	−.52	.1985	49.62
70.0 to 79.9	.00	.2019	50.48
80.0 to 89.9	.53	.1535	38.38
90.0 to 99.9	1.06	.0875	21.88
100.0 to 109.9	1.58	.0397	9.92
110.0 and greater	2.11	.0174	4.35
Total		1.0000	250.00

TABLE 10.3.3

Observed and Expected Frequencies and $(O_i - E_i)^2 / E_i$ for Example 10.3.1

Class interval	Observed frequency (O_i)	Expected frequency (E_i)	$(O_i - E_i)^2 / E_i$
< 40.0	16	14.55	.145
40.0 to 49.9	18	22.18	.788
50.0 to 59.9	22	38.65	7.173
60.0 to 69.9	51	49.62	.038
70.0 to 79.9	62	50.48	2.629
80.0 to 89.9	55	38.38	7.197
90.0 to 99.9	22	21.88	.001
100.0 to 109.9	4	9.92	3.533
110.0 and greater	0	4.35	4.350
Total	250	250.00	25.854

sample were drawn from a normal distribution of values, sampling variability alone would make it highly unlikely that the observed and expected frequencies would agree perfectly. We wonder, then, if the discrepancies between the observed and expected frequencies are small enough that we feel it reasonable that they could have occurred by chance alone, when the null hypothesis is true. If they are of this magnitude, we will be unwilling to reject the null hypothesis that the sample came from a normally distributed population.

If the discrepancies are so large that it does not seem reasonable that they could have occurred by chance alone when the null hypothesis is true, we will want to reject the null hypothesis. The criterion against which we judge whether the discrepancies are "large" or "small" is provided by the χ^2 distribution.

The observed and expected frequencies along with each value of $(O_i - E_i)^2/E_i$ are shown in Table 10.3.3. The first entry in the last column, for example, is computed from $(16 - 14.55)^2/14.55 = .145$. The other values of $(O_i - E_i)^2/E_i$ are computed in a similar manner.

From Table 10.3.3 we see that $X^2 = \sum[(O_i - E_i)^2/E_i] = 25.854$. The appropriate degrees of freedom are 9 (the number of groups or class intervals) $- 3$ (for the three restrictions: making $\sum E_i = \sum O_i$, and estimating μ and σ from the sample data) $= 6$.

8. *Statistical Decision*   When we compare $X^2 = 25.854$ with values of χ^2 in Table F, we see that it is larger than $\chi^2_{.995} = 18.548$, so that we can reject the null hypothesis that the sample came from a normally distributed population at the .005 level of significance. In other words, the probability of obtaining a value of X^2 as large as 25.854, when the null hypothesis is true, is less than 5 in 1000 ($p < .005$). We say that such a rare event did not occur due to

chance alone (when H_0 is true), so we look for another explanation. The other explanation is that the null hypothesis is false.

9. *Conclusion*   We conclude that in the sampled population, inpatient occupancy ratios are not normally distributed.

Sometimes the parameters are specified in the null hypothesis. It should be noted that had the mean and variance of the population been specified as part of the null hypothesis in Example 10.3.1, we would not have had to estimate them from the sample and our degrees of freedom would have been $9 - 1 = 8$.

If parameters are estimated from ungrouped sample data rather than from grouped data as in our example, the distribution of X^2 may not be sufficiently approximated by the chi-square distribution to give satisfactory results. The problem is discussed by Dahiya and Gurland (4) and Watson (5, 6, 7). The same problem is encountered when parameters are estimated independently of the sample as discussed by Chase (8).

Small Expected Frequencies   Frequently in applications of the chi-square test the expected frequency for one or more categories will be small, perhaps much less than 1. In the literature the point is frequently made that the approximation of X^2 to χ^2 is not strictly valid when some of the expected frequencies are small. There is disagreement among writers, however, over what size expected frequencies are allowable before making some adjustment or abandoning χ^2 in favor of some alternative test. Some writers, especially the earlier ones, suggest lower limits of 10 while others suggest that all expected frequencies should be no less than 5. Cochran (9, 10), writing in the early 1950s, suggests that for goodness-of-fit tests of unimodal distributions (such as the normal) the minimum expected frequency can be as low as 1. If, in practice, one encounters one or more expected frequencies less than 1, adjacent categories may be combined to achieve the suggested minimum. Combining reduces the number of categories and, therefore, the number of degrees of freedom. Cochran's suggestions appear to have been followed extensively by practitioners in recent years. More recent research on the subject of small expected frequencies includes that of Roscoe and Byars (11), Yarnold (12), Tate and Hyer (13), Slakter (14, 15), and Lewontin and Felsentein (16).

Although one frequently encounters in the literature the use of chi-square to test for normality, it is not the most appropriate test to use when the hypothesized distribution is continuous. The Kolmogorov–Smirnov test, described in Chapter 11, was especially designed for goodness-of-fit tests involving continuous distributions.

Example 10.3.2   The Binomial Distribution   In a study designed to determine patient acceptance of a new pain reliever, 100 physicians each selected a sample of 25 patients to participate in the study. Each patient, after trying the new pain reliever for a specified period of time, was asked whether it was preferable to the pain reliever used regularly in the past.

TABLE 10.3.4
Results of Study Described in Example 10.3.2

Number of patients out of 25 preferring new pain reliever	Number of doctors reporting this number	Total number of patients preferring new pain reliever by doctor
0	5	0
1	6	6
2	8	16
3	10	30
4	10	40
5	15	75
6	17	102
7	10	70
8	10	80
9	9	81
10 or more	0	0
Total	100	500

The results of the study are shown in Table 10.3.4.

We are interested in determining whether or not these data are compatible with the hypothesis that they were drawn from a population that follows a binomial distribution. Again, we employ a chi-square goodness-of-fit test.

Since the binomial parameter, p, is not specified, it must be estimated from the sample data. A total of 500 patients out of the 2500 patients participating in the study said they preferred the new pain reliever, so that our point estimate of p is $\hat{p} = 500/2500 = .20$. The expected relative frequencies can be obtained by evaluating the binomial function

$$f(x) = \binom{25}{x}.2^x.8^{25-x}$$

for $x = 0, 1, \ldots, 25$. For example, to find the probability that out of a sample of 25 patients none would prefer the new pain reliever, when in the total population the true proportion preferring the new pain reliever is .2, we would evaluate

$$f(0) = \binom{25}{0}.2^0.8^{25-0}$$

This can be done most easily by consulting Table A, where we see that $P(X = 0) = .0038$. The relative frequency of occurrence of samples of size 25 in which no patients prefer the new pain reliever is .0038. To obtain the corresponding expected frequency we multiply .0038 by 100 to get .38. Similar calculations

TABLE 10.3.5

Calculations for Example 10.3.2

Number of patients out of 25 preferring new pain reliever	Number of doctors reporting this number (observed frequency, O_i)	Expected relative frequency	Expected frequency E_i
0	5 ⎫ 11	.0038	.38 ⎫ 2.74
1	6 ⎭	.0236	2.36 ⎭
2	8	.0708	7.08
3	10	.1358	13.58
4	10	.1867	18.67
5	15	.1960	19.60
6	17	.1633	16.33
7	10	.1109	11.09
8	10	.0623	6.23
9	9	.0295	2.95
10 or more	0	.0173	1.73
Total	100	1.0000	100.00

yield the remaining expected frequencies which, along with the observed frequencies, are shown in Table 10.3.5. We see in this table that the first expected frequency is less than 1, so that we follow Cochran's suggestion and combine this group with the second group. When we do this, all of the expected frequencies are greater than 1.

From the data we compute

$$X^2 = \frac{(11 - 2.74)^2}{2.74} + \frac{(8 - 7.08)^2}{7.08} + \cdots + \frac{(0 - 1.73)^2}{1.73} = 47.624$$

The appropriate degrees of freedom are 10 (the number of groups left after combining the first two) less 2, or 8. One degree of freedom is lost because we force the total of the expected frequencies to equal the total observed frequencies, and one degree of freedom is sacrificed because we estimate p from the sample data.

We compare our computed X^2 with the tabulated χ^2 with 8 degrees of freedom and find that it is significant at the .005 level of significance. We reject the null hypothesis that the data came from a binomial distribution.

Example 10.3.3. The Poisson Distribution   A hospital administrator wishes to test the null hypothesis that emergency admissions follow a Poisson distribution with $\lambda = 3$. Suppose that over a period of 90 days the numbers of emergency admissions were as shown in Table 10.3.6.

TABLE 10.3.6
Number of Emergency Admissions to a Hospital During a 90-Day Period

Day	Emergency admissions	Day	Emergency admissions	Day	Emergency admissions	Day	Emergency admissions
1.	2	24.	5	47.	4	70.	3
2.	3	25.	3	48.	2	71.	5
3.	4	26.	2	49.	2	72.	4
4.	5	27.	4	50.	3	73.	1
5.	3	28.	4	51.	4	74.	1
6.	2	29.	3	52.	2	75.	6
7.	3	30.	5	53.	3	76.	3
8.	0	31.	1	54.	1	77.	3
9.	1	32.	3	55.	2	78.	5
10.	0	33.	2	56.	3	79.	2
11.	1	34.	4	57.	2	80.	1
12.	0	35.	2	58.	5	81.	7
13.	6	36.	5	59.	2	82.	7
14.	4	37.	0	60.	7	83.	1
15.	4	38.	6	61.	8	84.	5
16.	4	39.	4	62.	3	85.	1
17.	3	40.	4	63.	1	86.	4
18.	4	41.	5	64.	3	87.	4
19.	3	42.	1	65.	1	88.	9
20.	3	43.	3	66.	0	89.	2
21.	3	44.	1	67.	3	90.	3
22.	4	45.	2	68.	2		
23.	3	46.	3	69.	1		

The data of Table 10.3.6 are summarized in Table 10.3.7.

To obtain the expected frequencies we first obtain the expected relative frequencies by evaluating the Poisson function given by Equation 3.4.1 for each entry in the left-hand column of Table 10.3.7. For example, the first expected relative frequency is obtained by evaluating

$$f(0) = \frac{e^{-3}3^0}{0!}$$

We may use Table B of the Appendix to find this and all the other expected relative frequencies that we need. Each of the expected relative frequencies is multiplied by 90 to obtain the corresponding expected frequencies. These values along with the observed and expected frequencies and the components of X^2, $(O_i - E_i)^2/E_i$, are displayed in Table 10.3.8.

TABLE 10.3.7

Summary of Data Presented in Table 10.3.6

Number of emergency admissions in a day	Number of days this number of emergency admissions occurred
0	5
1	14
2	15
3	23
4	16
5	9
6	3
7	3
8	1
9	1
10 or more	0
Total	90

TABLE 10.3.8

Observed and Expected Frequencies and Components of X^2 for Example 10.3.3

Number of emergency admissions	Number of days this number occurred, O_i	Expected relative frequency	Expected frequency	$\dfrac{(O_i - E_i)^2}{E_i}$
0	5	.050	4.50	.056
1	14	.149	13.41	.026
2	15	.224	20.16	1.321
3	23	.224	20.16	.400
4	16	.168	15.12	.051
5	9	.101	9.09	.001
6	3	.050	4.50	.500
7	3	.022	1.98	.525
8	1 ⎱	.008	.72 ⎱	
9	1 ⎰ 2	.003	.27 ⎰ 1.08	.784
10 or more	0 ⎰	.001	.09 ⎰	
Total	90	1.000	90.00	3.664

In Table 10.3.8 we see that

$$X^2 = \sum \frac{(O_i - E_i)^2}{E_i} = \frac{(5 - 4.50)^2}{4.50} + \cdots + \frac{(2 - 1.08)^2}{1.08} = 3.664$$

We also note that the last three expected frequencies are less than 1, so that they must be combined to avoid having any expected frequencies less than 1. This means that we have only nine effective categories for computing degrees of freedom. Since the parameter, λ, was specified in the null hypothesis, we do not lose a degree of freedom for reasons of estimation, so that the appropriate degrees of freedom are $9 - 1 = 8$. By consulting Table F of the Appendix, we find that the critical value of χ^2 for 8 degrees of freedom and $\alpha = .05$ is 15.507, so that we cannot reject the null hypothesis at the .05, or for that matter any reasonable level of significance ($p > .10$). We conclude, therefore, that emergency admissions at this hospital may follow a Poisson distribution with $\lambda = 3$. At least the observed data do not cast any doubt on that hypothesis.

EXERCISES

10.3.1 The following table shows the distribution of uric acid determinations taken on 250 patients. Test the goodness-of-fit of these data to a normal distribution with $\mu = 5.74$ and $\sigma = 2.01$. Let $\alpha = .01$.

Uric acid determination	Observed frequency
< 1	1
1 to 1.99	5
2 to 2.99	15
3 to 3.99	24
4 to 4.99	43
5 to 5.99	50
6 to 6.99	45
7 to 7.99	30
8 to 8.99	22
9 to 9.99	10
10 or higher	5
Total	250

10.3.2 The following data were collected on 300 eight-year-old girls. Test, at the .05 level of significance, the null hypothesis that the data are drawn from a normally distributed population. Use the methods of Chapter 1 to compute the sample mean and standard deviation from grouped data.

Height in centimeters	Observed frequency
114 to 115.9	5
116 to 117.9	10
118 to 119.9	14
120 to 121.9	21
122 to 123.9	30
124 to 125.9	40
126 to 127.9	45
128 to 129.9	43
130 to 131.9	42
132 to 133.9	30
134 to 135.9	11
136 to 137.9	5
138 to 139.9	4
	300

10.3.3 The face sheet of patients' records maintained in a local health department contains 10 entries. A sample of 100 records revealed the following distribution of erroneous entries.

Number of erroneous entries out of 10	Number of records
0	8
1	25
2	32
3	24
4	10
5 or more	1
	100

Test the goodness-of-fit of these data to the binomial distribution with $p = .20$. Find the p value for this test.

10.3.4 Over the past 10 years the following distribution of accidents was observed by the physician in charge of the employee health clinic of a large industrial firm.

Number of accidents in a month	Number of months this number occurred
0	2
1	10
2	15
3	30
4	28

(Table continued)

Number of accidents in a month	Number of months this number occurred
5	15
6	10
7	6
8	2
9	1
10 or more	1
	120

Are these data compatible with the hypothesis that accidents in this firm are Poisson distributed with $\lambda = 4$? What is the p value for this test?

10.3.5 The following are the numbers of a particular organism found in 100 samples of water from a pond.

Number of organisms per sample	Frequency
0	15
1	30
2	25
3	20
4	5
5	4
6	1
7	0
	100

Test the null hypothesis that these data were drawn from a Poisson distribution. Determine the p value for this test.

10.3.6 A research team conducted a survey in which the subjects were adult smokers. Each subject in a sample of 200 was asked to indicate the extent to which he/she agreed with the statement: "I would like to quit smoking." The results were as follows:

Response:	Strongly agree	Agree	Disagree	Strongly disagree
Number responding:	102	30	60	8

Can one conclude on the basis of these data that, in the sampled population, opinions are not equally distributed over the four levels of agreement. Let the probability of committing a type I error be .05 and find the p value.

10.4   TESTS OF INDEPENDENCE

Another, and perhaps the most frequent, use of the chi-square distribution is to test the null hypothesis that two criteria of classification, when applied to the same set of entities, are independent. We say that two criteria of classification are independent if the distribution of one criterion is the same no matter what the distribution of the other criterion. For example, if socioeconomic status and area of residence of the inhabitants of a certain city are independent, we would expect to find the same proportion of families in the low, medium, and high socioeconomic groups in all areas of the city.

The classification, according to two criteria, of a set of entities, say people, can be shown by a table in which the r rows represent the various levels of one criterion of classification and the c columns represent the various levels of the second criterion. Such a table is generally called a *contingency table*. The classification according to two criteria of a finite population of entities is shown in Table 10.4.1.

We will be interested in testing the null hypothesis that in the population the two criteria of classification are independent. If the hypothesis is rejected, we will conclude that the two criteria of classification are not independent. A sample of size n will be drawn from the population of entities, and the frequency of occurrence of entities in the sample corresponding to the cells formed by the intersections of the rows and columns of Table 10.4.1 along with the marginal totals will be displayed in a table such as Table 10.4.2.

The expected frequencies, under the null hypothesis that the two criteria of classification are independent, are calculated for each cell. The expected frequencies and observed frequencies are compared. If the discrepancy is "small," the null hypothesis is tenable. If the discrepancy is "large," the null hypothesis is rejected, and we conclude that the two criteria of classification are not independent. The decision as to whether the discrepancy between observed and expected

TABLE 10.4.1

Two-Way Classification of a Finite Population of Entities

Second criterion of classification level	First Criterion of Classification Level					
	1	*2*	*3*	$\cdots$	*c*	*Total*
1	N_{11}	N_{12}	N_{13}	$\cdots$	N_{1c}	$N_{1.}$
2	N_{21}	N_{22}	N_{23}	$\cdots$	N_{2c}	$N_{2.}$
3	N_{31}	N_{32}	N_{33}	$\cdots$	N_{3c}	$N_{3.}$
$\vdots$	$\vdots$	$\vdots$	$\vdots$	$\vdots$	$\vdots$	$\vdots$
r	N_{r1}	N_{r2}	N_{r3}	$\cdots$	N_{rc}	$N_{r.}$
Total	$N_{.1}$	$N_{.2}$	$N_{.3}$	$\cdots$	$N_{.c}$	N

TABLE 10.4.2

Two-Way Classification of a Sample of Entities

Second criterion of classification level	FIRST CRITERION OF CLASSIFICATION LEVEL					
	1	*2*	*3*	$\cdots$	*c*	*Total*
1	n_{11}	n_{12}	n_{13}	$\cdots$	n_{1c}	$n_{1.}$
2	n_{21}	n_{22}	n_{23}	$\cdots$	n_{2c}	$n_{2.}$
3	n_{31}	n_{32}	n_{33}	$\cdots$	n_{3c}	$n_{3.}$
$\vdots$	$\vdots$	$\vdots$	$\vdots$		$\vdots$	$\vdots$
r	n_{r1}	n_{r2}	n_{r3}	$\cdots$	n_{rc}	$n_{r.}$
Total	$n_{.1}$	$n_{.2}$	$n_{.3}$	$\cdots$	$n_{.c}$	n

frequencies is "large" or "small" will be made on the basis of the size of the quantity computed when we use Equation 10.3.1, where O_i and E_i refer, respectively, to the observed and expected frequencies in the cells of Table 10.4.2. It would be more logical to designate the observed and expected frequencies in these cells by O_{ij} and E_{ij}, but to keep the notation simple and to avoid the introduction of another formula, we have elected to use the simpler notation. It will be helpful to think of the cells as being numbered from 1 to k, where 1 refers to cell 11 and k refers to cell rc. It can be shown that X^2 as defined in this manner is distributed approximately as χ^2 with $(r - 1)(c - 1)$ degrees of freedom when the null hypothesis is true. If the computed value of X^2 is equal to or larger than the tabulated value of χ^2 for some α, the null hypothesis is rejected at the α level of significance. The hypothesis testing procedure is illustrated with the following example.

Example 10.4.1

1. *Data* A research team studying the relationship between blood type and severity of a certain condition in a population collected data on 1500 subjects as displayed in the contingency table shown as Table 10.4.3.
The researchers wished to know if these data were compatible with the hypothesis that severity of condition and blood type are independent.

2. *Assumptions* We assume that the sample available for analysis is a simple random sample drawn from the population of interest.

3. *Hypotheses*

> H_0: Blood type and severity of the condition are independent.
> H_A: The two variables are not independent.

Let $\alpha = .05$.

TABLE 10.4.3

Fifteen-Hundred Subjects Classified by Severity of Condition and Blood Type

Severity of condition	BLOOD TYPE				
	A	B	AB	O	Total
Absent	543	211	90	476	1320
Mild	44	22	8	31	105
Severe	28	9	7	31	75
Total	615	242	105	538	1500

4. *Test Statistic*   The test statistic is

$$X^2 = \sum_{i=1}^{k} \left[\frac{(O_i - E_i)^2}{E_i} \right]$$

5. *Distribution of the Test Statistic*   When H_0 is true X^2 is distributed approximately as χ^2 with $(r - 1)(c - 1) = (3 - 1)(4 - 1) = (2)(3) = 6$ degrees of freedom.

6. *Decision Rule*   Reject H_0 if the computed value of X^2 is equal to or greater than 12.592.

7. *Calculation of Test Statistic*   The first step in the analysis is to obtain the frequency for each cell that one would expect if, in fact, the two criteria of classification are independent. We may begin by computing estimates of the various marginal probabilities from the marginal totals shown in Table 10.4.3. Our estimate of the probability that a subject picked at random from the population from which the sample was drawn does not have the condition is $1320/1500 = .88$; the probability that a subject picked at random has a mild form of the condition is estimated to be $105/1500 = .07$; and the probability that a subject picked at random has a severe form of the condition is estimated to be $75/1500 = .05$. Similarly, we estimate the probabilities that a subject picked at random from the population under study will have type A, B, AB, and O blood to be $615/1500 = .41$, $242/1500 = .16$, $105/1500 = .07$, and $538/1500 = .36$, respectively.

We learned in Chapter 2 (see Equation 2.6.4) that if two events are independent, the probability of their joint occurrence is equal to the product of their individual probabilities. Under the null hypothesis that blood type and severity of condition are independent, then, we would compute the estimated probability that a subject picked at random is, say, both free of the condition and has blood type A as follows:

$P(\text{free of condition} \cap \text{type A blood}) = P(\text{free of condition}) P(\text{type A blood})$

$$= \tfrac{1320}{1500} \cdot \tfrac{615}{1500}$$

$$= .3608$$

The probability just computed is our estimate of the probability that a subject picked at random would fall in the first cell of Table 10.4.3 when the null hypothesis is true. Our estimate of the probability that a subject picked at random would be both free of the condition and have blood type B is

$$\frac{1320}{1500} \cdot \frac{242}{1500} = .141973$$

The estimated probabilities associated with each of the remaining cells are computed in a similar manner.

After computing the estimated probabilities associated with each cell, we may obtain the corresponding expected frequencies by multiplying each cell probability by 1500, the total sample size. For example, the expected frequency for the first cell is $(.3608)(1500) = 541.2$, and the expected number of subjects out of 1500 who are both free of the condition and have blood type B is $(.141973)(1500) = 212.96$.

A shortcut that saves considerable time may be employed in computing the expected frequencies. Note that the expected frequency for the first cell may be expressed as

$$P(\text{free of condition} \cap \text{type A blood})(1500) = \frac{1320}{1500} \cdot \frac{615}{1500} \cdot 1500$$

One of the denominators of 1500 cancels into the 1500 in the numerator, leaving

$$\frac{(1320)(615)}{1500} = 541.20$$

This leads to the shortcut procedure of obtaining expected cell frequencies by dividing the product of corresponding marginal totals by the total sample size. As another example, note that the expected number of subjects out of 1500 who are both free of the condition and have type B blood is $(242)(1320)/1500 = 212.96$. The observed and expected frequencies for our illustrative example are displayed together in Table 10.4.4, where the expected frequencies are enclosed in parentheses.

TABLE 10.4.4

Observed and Expected Frequencies, Example 10.4.1

Severity of condition	BLOOD TYPE				
	A	B	AB	O	Total
Absent	543(541.2)	211(212.96)	90(92.40)	476(473.44)	1320
Mild	44(43.05)	22(16.94)	8(7.35)	31(37.66)	105
Severe	28(30.75)	9(12.10)	7(5.25)	31(26.90)	75
Total	615	242	105	538	1500

From the observed and expected frequencies we may compute

$$X^2 = \sum \frac{(O_i - E_i)^2}{E_i}$$

$$= \frac{(543 - 541.2)^2}{541.2} + \frac{(211 - 212.96)^2}{212.96} + \cdots + \frac{(31 - 26.90)^2}{26.90}$$

$$= .005987 + .018039 + \cdots + .624907$$

$$= 5.12$$

8. *Statistical Decision*   The computed value of $X^2 = 5.12$ is compared with tabulated values of χ^2 with 6 degrees of freedom. It is seen that the computed X^2 is not significant at any reasonable level. For example, the critical value of χ^2 for $\alpha = .10$ and 6 degrees of freedom is 10.645.

9. *Conclusion*   We conclude that these data are compatible with the hypothesis that severity of the condition and blood type are independent. At least these data, because of the close agreement between observed and expected frequencies, do not provide sufficient evidence to indicate a lack of independence between the two criteria of classification.

Degrees of Freedom   The student will note that the number of degrees of freedom is equal to the number of cell frequencies in Table 10.4.3 which could be filled in arbitrarily while maintaining the observed marginal totals. For example, given that 1320 subjects are free of the condition and that 543 of these have blood type A, 211 have blood type B, and 90 have blood type AB, we must enter 476 as the number with blood type O. This accounts for 3 degrees of freedom. On applying the same logic to the second row, we find that we can arbitrarily fill in only three cells, which accounts for 3 more degrees of freedom. When we attempt to fill in the cells of the third row, we find that these frequencies are already determined by the frequencies of the first two rows and, thus, the total degrees of freedom are 6.

Small Expected Frequencies   The problem of small expected frequencies discussed in the previous section may be encountered when analyzing the data of contingency tables. Although there is a lack of concensus on how to handle this problem, many authors currently follow the rule given by Cochran (10). He suggests that for contingency tables with more than 1 degree of freedom a minimum expectation of 1 is allowable if no more than 20 percent of the cells have expected frequencies of less than 5. To meet this rule, adjacent rows and/or adjacent columns may be combined when to do so is logical in light of other considerations. If X^2 is based on less than 30 degrees of freedom, expected frequencies as small as 2 can be tolerated. We did not experience the problem of small expected frequencies in Example 10.4.1, since they were all greater than 5.

TABLE 10.4.5

Nutritional Status and Academic Performance of 500 Elementary School Children

Academic performance	NUTRITIONAL STATUS		
	Poor	Good	Total
Poor	105	15	120
Satisfactory	80	300	380
Total	185	315	500

The 2 × 2 Contingency Table Sometimes each of two criteria of classification may be broken down into only two categories, or levels. When data are cross classified in this manner, the result is a contingency table consisting of two rows and two columns. Such a table is commonly referred to as a 2 × 2 table. When we apply the $(r - 1)(c - 1)$ rule for finding degrees of freedom to a 2 × 2 table the result is 1 degree of freedom. Let us illustrate this with an example.

Example 10.4.2

1. *Data* A sample of 500 elementary school children in a certain school system were cross-classified by nutritional status and academic performance. The results are shown in Table 10.4.5.

   The researchers wished to know if they could conclude that there is a relationship between nutritional status and academic performance. A chi-square test is appropriate for reaching a decision.

2. *Assumptions* The sample is a simple random sample.

3. *Hypotheses*

   H_0: Nutritional status and academic performance are independent.
   H_A: The two variables are not independent.

   Let $\alpha = .05$.

4. *Test Statistic* The test statistic is

$$X^2 = \sum_{i=1}^{k} \left[\frac{(O_i - E_i)^2}{E_i} \right]$$

5. *Distribution of the Test Statistic* When H_0 is true X^2 is distributed approximately as χ^2 with $(2 - 1)(2 - 1) = (1)(1) = 1$ degree of freedom.

6. *Decision Rule* Reject H_0 if the computed value of X^2 is equal to or greater than 3.841.

7. *Calculation of the Test Statistic* The value of X^2 may be computed by first calculating the expected cell frequencies in the manner discussed above. In

TABLE 10.4.6

A 2 × 2 Contingency Table

Second criterion of classification	FIRST CRITERION OF CLASSIFICATION		
	1	*2*	*Total*
1	a	b	$a + b$
2	c	d	$c + d$
Total	$a + c$	$b + d$	n

the case of a 2 × 2 contingency table, however, X^2 may be calculated by the following shortcut formula:

$$X^2 = \frac{n(ad - bc)^2}{(a + c)(b + d)(a + b)(c + d)} \quad \text{(10.4.1)}$$

where a, b, c, and d are the observed cell frequencies as shown in Table 10.4.6. For our illustrative example we have

$$X^2 = \frac{500[(105)(300) - (15)(80)]^2}{(185)(315)(120)(380)}$$
$$= 172.746$$

8. *Statistical Decision* Since 172.746 > 3.841, we reject H_0. For this test, $p < .005$.

9. *Conclusion* The researcher may conclude that there is a relationship between the two characteristics under study.

The problems of how to handle small expected frequencies and small total sample sizes may arise in the analysis of 2 × 2 contingency tables. Cochran (10) suggests that the χ^2 test should not be used if $n < 20$ or if $20 < n < 40$ and any expected frequency is less than 5. When $n \geq 40$ an expected cell frequency as small as 1 can be tolerated.

Yates' Correction The observed frequencies in a contingency table are discrete and thereby give rise to a discrete statistic, X^2, which is approximated by the χ^2 distribution which is continuous. Yates (17) in 1934 proposed a procedure for correcting for this in the case of 2 × 2 tables. The correction consists of subtracting half the total number of observations from the absolute value of the quantity $ad - bc$ before squaring. That is,

$$X^2_{corrected} = \frac{n(|ad - bc| - .5n)^2}{(a + c)(b + d)(a + b)(c + d)} \quad \text{(10.4.2)}$$

It is generally agreed that no correction is necessary for larger contingency tables. Although Yates' correction for 2 × 2 tables has been used extensively in the past, recent investigators, for example, Grizzle (18), Lancaster (19), Pearson (20), and Plackett (21), have questioned its use. The work of Grizzle, in particular, has strengthened the case against the use of the correction on the basis that it too often results in an overly conservative test; that is, the use of the correction too often leads to nonrejection of the null hypothesis. As a result some practitioners are recommending against its use. This seems to be a reasonable recommendation to follow.

We may, as a matter of interest, apply the correction to our current example. Using Equation 10.4.2 and the data from Table 10.4.5, we may compute

$$X^2_{\text{corrected}} = \frac{500[|(105)(300) - (15)(80)| - (.5)(500)]^2}{(185)(315)(120)(380)}$$

$$= 169.907$$

As might be expected, with a sample this large, the difference in the two results is not dramatic.

The characteristics of a chi-square test of independence that distinguish it from other chi-square tests are as follows:

1. Generally, a single sample is selected from a population of interest and the subjects or objects are cross-classified on the basis of the two variables of interest.

2. The rationale for calculating expected cell frequencies is based on the probability law which states that if two events (here the two criteria of classification) are independent, the probability of their joint occurrence is equal to the product of their individual probabilities.

3. The hypotheses and conclusions are stated in terms of the independence (or lack of independence) of two variables.

EXERCISES

In the exercises that follow perform the test at the indicated level of significance and determine the p value.

10.4.1 A sample of 250 physicians were cross-classified according to their specialty and the area of the country in which they practice. The results were as follows.

| Area of country | SPECIALTY | | | | |
	A	B	C	D	Total
North	20	18	12	17	67
South	6	22	15	13	56
East	4	6	14	11	35
West	10	19	23	40	92
Total	40	65	64	81	250

Do these data provide sufficient evidence to indicate a lack of independence between the two criteria of classification? Let $\alpha = .01$.

10.4.2 Five hundred employees of a firm that manufactures a product suspected of being associated with respiratory disorders were cross-classified by level of exposure to the product and whether or not they exhibited symptoms of respiratory disorders. The results are shown in this table.

	LEVEL OF EXPOSURE			
Symptoms present	High	Limited	No known exposure	Total
Yes	185	33	17	235
No	120	73	72	265
Total	305	106	89	500

Do these data provide sufficient evidence, at the .01 level of significance, to indicate a relationship between level of exposure and the presence of respiratory disorder symptoms?

10.4.3 Five hundred elementary school children were cross-classified by socioeconomic group and the presence or absence of a certain speech defect. The results were as follows.

	SOCIOECONOMIC GROUP				
Speech defect	Upper	Upper middle	Lower middle	Lower	Total
Present	8	24	32	27	91
Absent	42	121	138	108	409
Total	50	145	170	135	500

Are these data compatible with the hypothesis that the speech defect is unrelated to socioeconomic status?

10.4.4 A group of 350 adults who participated in a health survey were asked whether or not they were on a diet. The responses by sex are as follows.

	SEX		
	Male	Female	Total
On diet	14	25	39
Not on diet	159	152	311
Total	173	177	350

Do these data suggest that being on a diet is dependent on sex? Let $\alpha = .05$.

10.4.5 A sample of 500 college students participated in a study designed to evaluate the level of college students' knowledge of a certain group of common diseases. The following table shows the students classified by major field of study and level of knowledge of the group of diseases.

| | KNOWLEDGE OF DISEASES | | |
Major	Good	Poor	Total
Premedical	31	91	122
Other	19	359	378
Total	50	450	500

Do these data suggest that there is a relationship between knowledge of the group of diseases and major field of study of the college students from which the present sample was drawn? Let $\alpha = .05$.

10.4.6   The following table shows the results of a survey in which the subjects were a sample of 300 adults residing in a certain metropolitan area. Each subject was asked to indicate which of three policies they favored with respect to smoking in public places.

| Highest education level | POLICY FAVORED | | | | |
	No restrictions on smoking	Smoking allowed in designated areas only	No smoking at all	No opinion	Total
College graduate	5	44	23	3	75
High school graduate	15	100	30	5	150
Grade school graduate	15	40	10	10	75
Total	24	195	60	21	300

Can one conclude from these data that, in the sampled population, there is a relationship between level of education and attitude toward smoking in public places? Let $\alpha = .05$.

10.5   TESTS OF HOMOGENEITY

A characteristic of the examples and exercises presented in the last section is that, in each case, the total sample was assumed to have been drawn before the entities were classified according to the two criteria of classification. That is, the observed number of entities falling into each cell was determined after the sample was drawn. As a result, the row and column totals are chance quantities not under the control of the investigator. We think of the sample drawn under these conditions as a single sample drawn from a single population. On occasion, however, either row or column totals may be under the control of the investigator; that is, the investigator may specify that independent samples be drawn from each of several populations. In this case one set of marginal totals is said to be *fixed*, while the other set, corresponding to the criterion of classification applied to the samples, is *random*. The former procedure, as we have seen, leads to a chi-square test of independence. The latter situation leads to a chi-square *test of homogeneity*. The two situations not only involve different sampling procedures; they lead to different questions and null hypotheses. The test of independence is concerned with the question: Are the two criteria of classification independent? The homogeneity test is concerned with the question: Are the samples drawn from populations that are homogeneous with respect to some criterion of classification? In the latter case the null hypothesis states that the samples are drawn from the same population. Despite these differences in concept and sampling procedure, the two tests are mathematically identical, as we see when we consider the following example.

Example 10.5.1

1. *Data*   A researcher, studying the extent of drug usage among college students who were professed users of drugs, selected from this group a sample of 150 freshmen, 135 sophomores, 125 juniors, and 100 seniors. Each student completed a questionnaire in which he or she indicated the extent of his or her use of drugs as either experimental, casual, or moderate to heavy. The results are shown in Table 10.5.1. The manner of selecting the samples, that is, selecting a specified number from each population, has the effect of fixing the row totals of the table. Are these data compatible with the hypothesis that the four populations are homogeneous with respect to the extent of drug usage?

TABLE 10.5.1

Extent of Drug Usage Among 510 College Students Classified by Class Standing

Class standing	EXTENT OF DRUG USAGE			
	Experimental	*Casual*	*Moderate to heavy*	*Total*
Freshman	57	50	43	150
Sophomore	57	58	20	135
Junior	56	45	24	125
Senior	45	22	33	100
Total	215	175	120	510

2. *Assumptions* We assume that we have a simple random sample from each of the four populations of interest.

3. *Hypotheses*

H_0: The four populations (freshmen, sophomores, juniors, and seniors) are homogeneous with respect to extent of drug usage.

H_A: The four populations are not homogeneous with respect to extent of drug usage.

Let $\alpha = .05$.

4. *Test Statistic* The test statistic is the now familiar $X^2 = \sum[(O_i - E_i)^2/E_i]$.

5. *Distribution of Test Statistic* If H_0 is true X^2 is distributed approximately as χ^2 with $(4 - 1)(3 - 1) = (3)(2) = 6$ degrees of freedom.

6. *Decision Rule* Reject H_0 if the computed value of X^2 is equal to or greater than 12.592.

7. *Calculation of Test Statistic* We need expected frequencies for each of the cells in Table 10.5.1. If the populations are indeed homogeneous, or, equivalently, if the samples are all drawn from the same population, with respect to drug usage, our best estimate of the proportion in the combined population who have used drugs only experimentally is $215/510 = .4216$. By the same token, if the four populations are homogeneous, we interpret this probability as applying to each of the populations individually. For example, under the null hypothesis, $215/510$ is our best estimate of the probability that a student picked at random from the drug users will be an experimental user only. We would expect, then, to find $(215/510) \times 150 = 63.24$ out of 150 freshmen to be experimental users. Similarly, we would expect $(215/510) \times 135 = 56.91$ sophomores, $(215/510) \times 125 = 52.70$ juniors, and $(215/510) \times 100 = 42.16$ seniors to be experimental users.

We see again that the shortcut procedure of multiplying appropriate marginal totals and dividing by the grand total yields the expected frequencies for the cells. The expected frequencies computed in this manner, along with the observed frequencies, are shown in Table 10.5.2. The expected frequencies are enclosed in parentheses.

From the data in Table 10.5.2 we compute the following test statistic:

$$X^2 = \frac{(57 - 63.24)^2}{63.24} + \frac{(50 - 51.47)^2}{51.47} + \cdots + \frac{(33 - 23.53)^2}{23.53} = 19.4$$

8. *Statistical Decision* We find by consulting Table F that the probability of obtaining a value of X^2 as large or larger than 19.4 when the null hypothesis is true is less than .005. The decision then is to reject the null hypothesis.

9. *Conclusion* We conclude that the populations are not homogeneous with respect to extent of drug usage.

Table 10.5.2

Observed and Expected Frequencies, Example 10.5.1

Class standing	EXTENT OF DRUG USAGE			Total
	Experimental	Casual	Moderate to heavy	
Freshman	57(63.24)	50(51.47)	43(35.29)	150
Sophomore	57(56.91)	58(46.32)	20(31.76)	135
Junior	56(52.70)	45(42.89)	24(29.41)	125
Senior	45(42.16)	22(34.31)	33(23.53)	100
Total	215	175	120	510

The rules for small expected frequencies given in the previous section are applicable when carrying out a test of homogeneity.

When the chi-square test of homogeneity is used to test the null hypothesis that two populations are homogeneous, and when there are only two levels of the criterion of classification, the data may be displayed in a 2×2 contingency table. The analysis is identical to the analysis of 2×2 tables given in Section 10.4.

In summary, the chi-square test of homogeneity has the following characteristics:

1. Two or more populations are identified in advance and an independent sample is drawn from each.
2. Sample subjects or objects are placed in appropriate categories of the variable of interest.
3. The calculation of expected cell frequencies is based on the rationale that if the populations are homogeneous as stated in the null hypothesis, the best estimate of the probability that a subject or object will fall into a particular category of the variable of interest can be obtained by pooling the sample data.
4. The hypotheses and conclusions are stated in terms of homogeneity (with respect to the variable of interest) of populations.

The chi-square test of homogeneity for the two-sample case provides an alternative method for testing the null hypothesis that two population proportions are equal. In Section 6.6, it will be recalled, we learned to test H_0: $p_1 = p_2$ against H_A: $p_1 \neq p_2$ by means of the statistic

$$z = \frac{(\hat{p}_1 - \hat{p}_2) - (p_1 - p_2)}{\sqrt{\dfrac{\bar{p}(1 - \bar{p})}{n_1} + \dfrac{\bar{p}(1 - \bar{p})}{n_2}}}$$

where $\bar{p}$ is obtained by pooling the data of the two independent samples available for analysis.

Suppose, for example, that in a test of H_0: $p_1 = p_2$ against H_A: $p_1 \neq p_2$, the sample data were as follows: $n_1 = 100$, $\hat{p}_1 = .60$, $n_2 = 120$, $\hat{p}_2 = .40$. When we pool the sample data we have

$$\bar{p} = \frac{.60(100) + .40(120)}{100 + 120} = \frac{108}{220} = .4909$$

and

$$z = \frac{.60 - .40}{\sqrt{\dfrac{(.4909)(.5091)}{100} + \dfrac{(.4909)(.5091)}{120}}} = 2.95469$$

which is significant at the .05 level since it is greater than the critical value of 1.96.

If we wish to test the same hypothesis using the chi-square approach, our contingency table would be

| | Characteristic Present | | |
Sample	Yes	No	Total
1	60	40	100
2	48	72	120
Total	108	112	220

By Equation 10.4.1 we compute

$$X^2 = \frac{220[(60)(72) - (40)(48)]^2}{(108)(112)(100)(120)} = 8.7302$$

which is significant at the .05 level since it is greater than the critical value of 3.841. We see, therefore, that we reach the same conclusion by both methods. This is not surprising since, as explained in Section 10.2, $\chi^2_{(1)} = z^2$. We note that $8.7302 = (2.95469)^2$ and that $3.841 = (1.96)^2$.

Computer Analysis The computer may be used to advantage in calculating X^2 for tests of independence and tests of homogeneity. Figure 10.5.1 shows the computer printout for Example 10.5.1 when the MINITAB program for computing X^2 from contingency tables is used.

```
EXPECTED FREQUENCIES ARE PRINTED BELOW OBSERVED FREQUENCIES
        I   C1   I   C2   I   C3   I TOTALS
------- I ------- I ------- I ------- I -------
   1    I  57   I  50   I  43   I   150
        I  63.2 I  51.5 I  35.3 I
------- I ------- I ------- I ------- I -------
   2    I  57   I  58   I  20   I   135
        I  56.9 I  46.3 I  31.8 I
------- I ------- I ------- I ------- I -------
   3    I  56   I  45   I  24   I   125
        I  52.7 I  42.9 I  29.4 I
------- I ------- I ------- I ------- I -------
   4    I  45   I  22   I  33   I   100
        I  42.2 I  34.3 I  23.5 I
------- I ------- I ------- I ------- I -------
 TOTALS I  215  I  175  I  120  I   510

TOTAL CHI SQUARE =

        .61 +  .04 + 1.68 +
        .00 + 2.94 + 4.36 +

        .21 +  .10 + 1.00 +
        .19 + 4.42 + 3.81 +

          = 19.37

DEGREES OF FREEDOM = (4-1) X (3-1) = 6
```

FIGURE 10.5.1

Printout of Computer Analysis of Example 10.5.1 Using MINITAB

EXERCISES

In the exercises that follow perform the test at the indicated level of significance and determine the p value.

10.5.1 In a study of the dental-caries status of children in six communities with varying levels of fluoride in the water supply, a sample of 125 children was selected from each of the communities and given a dental examination. The following table shows the number of caries-free children in each sample.

Community	Number of children in sample	Number of caries-free children
A	125	38
B	125	8
C	125	30
D	125	44
E	125	64
F	125	32
Total	750	216

Are these data compatible with the hypothesis that the six populations are homogeneous with respect to the proportion of children who are free of dental caries?

10.5.2 In a study designed to investigate the relationship between nutritional status and ability to perform certain tasks, 400 experimental animals who had been trained to perform these tasks were divided into four groups. The four groups were fed different diets representing different levels of nutritional deprivation ranging from a well-balanced diet (treatment condition 1) to one highly deficient in essential nutrients (treatment condition 4). After the animals had been on the diets for a period of time, their ability to perform tasks of varying difficulty was tested. Some of the animals were able to perform only the simple tasks, some could perform tasks of moderate difficulty, while others could perform the more difficult tasks. The following table shows the 400 animals classified according to treatment condition and level of difficulty of task performed.

| *Difficulty of task performed* | **TREATMENT CONDITION** | | | | |
	1	*2*	*3*	*4*	*Total*
Simple	3	12	27	49	91
Moderately difficult	13	62	36	48	159
Difficult	84	26	37	3	150
Total	100	100	100	100	400

Should the researchers conclude that there is a lack of homogeneity among the four groups with respect to ability to perform tasks of varying difficulty?

10.5.3 A sample of hospital records was selected from the records of patients with each of five diagnoses. The records were evaluated as to appropriateness of length of stay. The results are shown in the following table.

| *Diagnosis* | **LENGTH OF STAY EVALUATION** | | | |
	Appropriate length of stay	*Understay*	*Overstay*	*Total*
I	31	6	13	50
II	49	2	4	55
III	55	2	3	60
IV	42	9	24	75
V	33	3	9	45
Total	210	22	53	285

Do these data suggest a lack of homogeneity among the diagnostic groups with respect to length of stay evaluation? Let $\alpha = .05$.

10.5.4 In an air pollution study, a random sample of 200 households was selected from each of two communities. A respondent in each household was asked whether or not anyone in the household was bothered by air pollution. The responses were as follows.

| | ANY MEMBER OF HOUSEHOLD BOTHERED BY AIR POLLUTION? | | |
Community	Yes	No	Total
I	43	157	200
II	81	119	200
Total	124	276	400

Can the researchers conclude that the two communities differ with respect to the variable of interest? Let $\alpha = .05$.

10.5.5 In a simple random sample of 250 industrial workers with cancer, researchers found that 102 had worked at jobs classified as "high exposure" with respect to suspected cancer-causing agents. Of the remainder, 84 had worked at "moderate exposure" jobs, and 64 had experienced no known exposure because of their jobs. In an independent simple random sample of 250 industrial workers from the same area who had no history of cancer, 31 worked in "high exposure" jobs, 60 worked in 'moderate exposure" jobs, and 159 worked in jobs involving no known exposure to suspected cancer-causing agents. Does it appear from these data that persons working in jobs that expose them to suspected cancer-causing agents have an increased risk of contracting cancer? Let $\alpha = .05$.

10.6 SUMMARY

In this chapter some uses of the versatile chi-square distribution are discussed. Chi-square goodness-of-fit tests applied to the normal, binomial, and Poisson distributions are presented. We see that the procedure consists of computing a statistic

$$X^2 = \sum \frac{(O_i - E_i)^2}{E_i}$$

which measures the discrepancy between the observed (O_i) and expected (E_i) frequencies of occurrence of values in certain discrete categories. When the appropriate null hypothesis is true, this quantity is distributed approximately as χ^2. When X^2 is greater than or equal to the tabulated value of χ^2 for some α, the null hypothesis is rejected at the α level of significance.

Tests of independence and tests of homogeneity are also discussed in this chapter. The tests are mathematically equivalent but conceptually different. Again, these tests essentially test the goodness-of-fit of observed data to expectation under hypotheses, respectively, of independence of two criteria of classifying the data, and the homogeneity of proportions among two or more groups.

REVIEW QUESTIONS AND EXERCISES

1. Explain how the chi-square distribution may be derived.

2. What are the mean and variance of the chi-square distribution?

3. Explain how the degrees of freedom are computed for the chi-square goodness-of-fit tests.

4. State Cochran's rule for small expected frequencies in goodness-of-fit tests.

5. How does one adjust for small expected frequencies?

6. What is a contingency table?

7. How are the degrees of freedom computed when an X^2 value is computed from a contingency table?

8. Explain the rationale behind the method of computing the expected frequencies in a test of independence.

9. Explain the difference between a test of independence and a test of homogeneity.

10. Explain the rationale behind the method of computing the expected frequencies in a test of homogeneity.

11. A study of 190 pregnancies yielded the following data on the relationship between hypertension of mother and a certain complication of pregnancy.

A certain complication of pregnancy	MOTHER HYPERTENSIVE		
	Yes	No	Total
Present	23	55	78
Absent	12	100	112
Total	35	155	190

Do these data provide sufficient evidence to indicate that the two conditions are not independent? Let $\alpha = .01$. What is the p value?

12. A sample of 150 chronic carriers of a certain antigen and a sample of 500 noncarriers revealed the following blood group distributions.

Blood group	Carriers	Noncarriers	Total
O	72	230	302
A	54	192	246
B	16	63	79
AB	8	15	23
Total	150	500	650

Can one conclude from these data that the two populations from which the samples were drawn differ with respect to blood group distribution? Let $\alpha = .05$. What is the p value for this test?

13. The following table shows 200 males classified according to social class and headache status.

Headache group	SOCIAL CLASS			Total
	A	B	C	
No headache (in previous year)	6	30	22	58
Simple headache	11	35	17	63
Unilateral headache (nonmigraine)	4	19	14	37
Migraine	5	25	12	42
Total	26	109	65	200

Do these data provide sufficient evidence to indicate that headache status and social class are related? Let $\alpha = .05$. What is the p value for this test?

14. The following is the frequency distribution of scores made on an aptitude test by 175 applicants to a physical therapy training facility.

Score	Number of applicants
10–14	3
15–19	8
20–24	13
25–29	17
30–34	19
35–39	25
40–44	28
45–49	20
50–54	18
55–59	12
60–64	8
65–69	4
Total	175

Do these data provide sufficient evidence to indicate that the population of scores is not normally distributed? Let $\alpha = .05$. What is the p value for this test?

15. Several studies have contributed evidence to support the hypothesis that handling or "gentling" rats during the early part of their lives has beneficial results. Suppose a study in which a sample of handled rats and a sample of unhandled rats compared with respect to overall health status were to yield the following results.

| | GROUP | | |
Health status	Handled	Unhandled	Total
High	37	23	60
Low	13	27	40
Total	50	50	100

Would these data support the hypothesis? Let $\alpha = .05$.

16. A local health department sponsored a VD information program that was open to high school juniors and seniors who ranged in age from 16 through 19 years. The program director believed that each age level was equally interested in knowing more about VD. Since each age level was about equally represented in the area served, she felt that equal interest in VD would be reflected by equal age level attendance at the program. The age breakdown of those attending was as follows.

Age	Number attending
16	26
17	50
18	44
19	40

Are these data incompatible with the program director's belief that students in the four age levels are equally interested in VD? Let $\alpha = .05$. What is the p value for this test?

17. A survey of children under 15 years of age residing in the inner city area of a large city were classified according to ethnic group and hemoglobin level. The results were as follows.

| Ethnic group | HEMOGLOBIN LEVEL (gms/100 ml) | | | |
	10.0 or greater	9.0–9.9	< 9.0	Total
A	80	100	20	200
B	99	190	96	385
C	70	30	10	110
Total	249	320	126	695

Do these data provide sufficient evidence to indicate, at the .05 level of significance, that the two variables are related? What is the p value for this test?

18. A sample of reported cases of mumps in preschool children showed the following distribution by age.

Age (years)	Number of cases
Under 1	6
1	20
2	35
3	41
4	48
Total	150

Test the hypothesis that cases occur with equal frequency in the five age categories. Let $\alpha = .05$. What is the p value for this test?

19. Each of a sample of 250 men drawn from a population of suspected joint disease victims was asked which of three symptoms bother him most. The same question was asked of a sample of 300 suspected women joint disease victims. The results were as follows.

Symptom by which bothered most	Men	Women
Morning stiffness	111	102
Nocturnal pain	59	73
Joint swelling	80	125
	250	300

Do these data provide sufficient evidence to indicate that the two populations are not homogeneous with respect to major symptoms? Let $\alpha = .05$. What is the p value for this test?

For each of the following situations, indicate whether a null hypothesis of homogeneity or a null hypothesis of independence is appropriate.

20. A researcher wished to compare the status of three communities with respect to immunity against polio in preschool children. A sample of preschool children was drawn from each of the three communities.

21. In a study of the relationship between smoking and respiratory illness, a random sample of adults were classified according to consumption of tobacco and extent of respiratory symptoms.

22. A physician who wished to know more about the relationship between smoking and birth defects studies the health records of a sample of mothers and their children, including stillbirths and spontaneously aborted fetuses where possible.

23. A health research team believes that the incidence of depression is higher among people with hypoglycemia than among people who do not suffer from this condition.

24. In a simple random sample of 200 patients undergoing therapy at a drug abuse treatment center, 60 percent belonged to ethnic group I. The remainder belonged to ethnic group II. In ethnic group I, 60 were being treated for alcohol abuse (A), 25 for marijuana abuse (B), and 20 for abuse of heroin, illegal methadone, or some other opioid (C). The remainder had abused barbiturates, cocaine, amphetamines, hallucinogens, or some other nonopioid besides marijuana (D). In ethnic group II the abused drug category and the numbers involved were as follows:

A(28)   B(32)   C(13)   D(the remainder)

Can one conclude from these data that there is a relationship between ethnic group and choice of drug to abuse? Let $\alpha = .05$ and find the p value.

25. The following data are available on a population of 1200 adults. The variables are as follows:

Sex (A):	1 = male, 0 = female
Smoking status (B):	0 = nonsmoker, 1 = smoker
Drinking level (C):	0 = nondrinker,
	1 = light to moderate drinker
	2 = heavy drinker
Symptoms of respiratory disease (D):	1 = present, 0 = absent
High blood pressure status (E):	1 = present, 0 = absent

Select a simple random sample of size 100 from this population and carry out an analysis to see if you can conclude that there is a relationship between smoking status and symptoms of respiratory disease. Let $\alpha = .05$ and determine the p value for your test. Compare your results with those of your classmates.

Subject	VARIABLE					Subject	VARIABLE				
	A	B	C	D	E		A	B	C	D	E
1.	1	1	1	0	0	2.	1	1	1	1	1
3.	1	0	1	1	0	4.	1	0	1	0	1
5.	1	0	1	0	1	6.	1	0	0	0	1
7.	1	1	2	1	0	8.	1	0	0	0	0
9.	0	0	1	0	0	10.	0	0	0	0	0
11.	1	1	1	1	0	12.	1	0	1	1	1
13.	0	0	2	1	1	14.	1	0	2	0	1
15.	0	1	0	1	1	16.	1	0	2	0	0

Subject	A	B	C	D	E	Subject	A	B	C	D	E
17.	1	0	2	0	0	18.	1	0	1	0	0
19.	1	0	1	0	0	20.	0	0	1	0	0
21.	1	0	2	1	0	22.	0	1	0	1	1
23.	1	1	0	1	1	24.	1	0	1	0	0
25.	1	1	2	0	1	26.	0	1	1	1	1
27.	1	0	0	0	0	28.	0	1	1	0	0
29.	1	1	0	1	0	30.	0	1	1	1	1
31.	1	1	2	1	1	32.	1	0	1	1	0
33.	1	1	1	1	1	34.	1	0	1	0	0
35.	1	1	0	0	1	36.	1	1	1	1	1
37.	0	0	0	0	0	38.	0	1	0	0	0
39.	0	1	0	1	0	40.	1	0	1	0	0
41.	0	1	1	1	1	42.	1	0	0	1	0
43.	0	1	2	1	1	44.	1	0	2	0	0
45.	1	0	0	1	0	46.	1	1	0	1	1
47.	1	0	1	0	0	48.	1	0	1	0	1
49.	0	0	1	0	1	50.	1	1	0	0	1
51.	0	1	1	1	1	52.	1	0	1	0	0
53.	0	0	2	0	0	54.	1	1	2	1	1
55.	1	1	0	0	1	56.	0	1	2	1	0
57.	0	0	2	0	1	58.	1	1	1	1	1
59.	1	0	1	1	0	60.	1	1	1	1	1
61.	0	0	0	1	0	62.	0	1	0	1	1
63.	1	0	0	0	0	64.	0	0	0	0	0
65.	0	1	0	1	1	66.	1	0	1	0	0
67.	0	1	1	1	1	68.	1	0	0	0	0
69.	0	1	0	0	1	70.	0	1	2	0	1
71.	0	1	1	1	0	72.	1	1	0	1	0
73.	1	0	2	0	0	74.	1	0	1	0	1
75.	1	0	0	1	0	76.	0	0	0	0	0
77.	0	0	0	0	0	78.	0	0	0	1	0
79.	1	1	2	1	1	80.	1	0	2	0	0
81.	1	1	1	0	0	82.	1	0	1	0	1
83.	1	0	2	0	1	84.	1	0	1	1	0
85.	1	0	2	0	0	86.	0	0	0	1	0
87.	1	0	2	0	0	88.	0	0	1	0	0
89.	1	1	0	1	1	90.	1	0	2	0	1
91.	1	0	0	0	1	92.	0	1	0	1	1
93.	1	1	1	1	0	94.	1	0	1	1	0
95.	0	1	1	1	1	96.	1	1	1	1	1
97.	1	1	0	0	1	98.	0	0	2	0	0
99.	1	1	0	1	1	100.	1	1	0	1	1
101.	1	0	0	0	0	102.	0	1	0	1	1
103.	0	1	1	1	1	104.	1	1	1	0	1
105.	1	0	1	0	0	106.	1	0	2	0	0

Subject	A	B	C	D	E	Subject	A	B	C	D	E
107.	0	0	0	0	0	108.	1	1	0	1	0
109.	0	0	1	1	0	110.	1	0	2	1	0
111.	1	0	0	0	0	112.	0	0	0	0	0
113.	1	0	1	0	1	114.	1	1	2	0	1
115.	1	0	1	0	0	116.	0	0	2	0	1
117.	0	1	0	1	0	118.	1	0	1	1	1
119.	1	0	2	0	0	120.	1	1	1	1	1
121.	1	0	0	0	0	122.	0	0	0	1	0
123.	0	1	2	1	1	124.	1	1	0	0	1
125.	1	1	2	1	0	126.	1	1	0	1	1
127.	1	1	1	0	0	128.	0	0	0	0	0
129.	0	1	2	1	1	130.	0	0	1	0	0
131.	1	0	0	0	0	132.	0	1	1	1	1
133.	1	1	0	1	1	134.	1	0	2	1	0
135.	0	0	0	0	0	136.	0	1	0	1	0
137.	1	1	1	1	1	138.	1	1	1	0	0
139.	0	0	2	0	0	140.	0	0	0	0	0
141.	1	0	2	0	0	142.	0	1	1	1	1
143.	1	0	1	0	1	144.	1	0	0	1	0
145.	1	1	0	1	0	146.	1	1	1	1	1
147.	1	1	0	1	1	148.	1	1	0	0	0
149.	1	1	0	0	0	150.	0	0	2	0	1
151.	0	0	2	1	1	152.	1	1	1	0	0
153.	0	0	0	0	0	154.	1	1	0	1	0
155.	1	1	2	1	1	156.	0	0	2	1	0
157.	0	0	0	0	0	158.	1	1	0	1	1
159.	1	1	2	1	1	160.	1	1	0	1	1
161.	1	0	1	0	1	162.	1	0	1	0	1
163.	1	0	2	0	0	164.	1	0	1	0	1
165.	0	1	0	0	0	166.	1	0	1	0	1
167.	1	0	0	1	0	168.	1	1	1	1	1
169.	1	0	2	0	1	170.	1	0	1	0	1
171.	0	1	0	0	1	172.	1	1	2	1	0
173.	0	1	2	1	1	174.	0	1	1	1	1
175.	0	0	1	1	1	176.	1	0	1	0	0
177.	1	1	1	1	0	178.	0	0	2	0	0
179.	1	0	0	1	0	180.	1	1	2	1	0
181.	0	0	1	0	1	182.	0	1	1	0	0
183.	0	0	1	0	0	184.	1	0	1	1	1
185.	1	0	0	0	0	186.	1	1	0	1	1
187.	1	0	1	0	0	188.	1	1	1	0	1
189.	1	0	2	0	0	190.	0	1	2	1	1
191.	0	0	0	0	0	192.	0	0	0	0	0
193.	1	1	0	1	1	194.	1	0	1	1	1
195.	0	0	2	0	1	196.	1	0	1	1	0

| Subject | VARIABLE | | | | | Subject | VARIABLE | | | | |
	A	B	C	D	E		A	B	C	D	E
197.	1	0	1	0	1	198.	1	1	0	1	0
199.	0	0	0	0	0	200.	1	0	2	0	1
201.	0	0	2	0	1	202.	1	0	2	1	0
203.	1	1	0	0	0	204.	1	0	1	0	0
205.	1	1	0	1	1	206.	1	1	0	1	1
207.	1	1	0	1	1	208.	0	1	2	1	1
209.	0	1	1	1	1	210.	1	0	2	0	1
211.	0	1	1	1	1	212.	1	1	1	0	0
213.	1	0	2	1	1	214.	0	0	1	0	0
215.	0	1	1	1	1	216.	0	0	0	0	0
217.	0	1	1	1	1	218.	1	0	2	0	1
219.	1	0	2	0	1	220.	1	0	2	0	1
221.	1	1	2	0	1	222.	1	0	0	0	0
223.	1	0	2	1	1	224.	1	0	1	1	1
225.	0	1	1	1	0	226.	1	0	1	0	1
227.	1	0	0	0	0	228.	1	1	2	1	1
229.	0	1	1	0	1	230.	1	1	1	1	1
231.	0	0	2	0	0	232.	1	0	1	0	0
233.	1	0	0	1	1	234.	0	1	0	1	1
235.	0	1	0	1	1	236.	1	0	1	0	1
237.	1	1	1	1	0	238.	1	0	1	0	0
239.	0	1	2	1	1	240.	0	0	1	0	0
241.	1	0	2	0	1	242.	1	0	0	1	0
243.	0	1	1	0	1	244.	0	1	1	1	1
245.	1	0	1	0	1	246.	0	0	0	0	0
247.	0	1	0	0	0	248.	0	0	2	0	1
249.	1	0	0	0	0	250.	0	1	1	1	1
251.	0	1	0	1	0	252.	0	0	1	1	1
253.	1	0	2	1	1	254.	0	1	2	1	1
255.	1	1	0	1	0	256.	1	0	1	0	0
257.	1	0	0	0	0	258.	1	0	0	0	0
259.	1	0	1	0	1	260.	0	1	0	1	0
261.	0	0	2	1	1	262.	1	1	1	0	1
263.	1	1	0	0	0	264.	1	0	1	0	0
265.	1	1	0	1	0	266.	0	0	0	0	0
267.	1	0	0	0	0	268.	0	0	0	0	0
269.	1	0	0	1	0	270.	1	0	1	0	0
271.	0	0	1	1	1	272.	1	0	0	0	1
273.	0	0	0	0	0	274.	1	0	1	0	1
275.	1	0	0	0	0	276.	1	0	1	0	1
277.	1	1	0	1	1	278.	1	0	1	0	0
279.	1	1	0	1	1	280.	1	1	1	1	1
281.	1	0	1	1	0	282.	0	0	0	1	0
283.	1	1	0	1	0	284.	1	1	0	1	0
285.	0	1	0	0	1	286.	1	0	0	0	0
287.	0	0	0	0	0	288.	1	1	1	1	0

Subject	VARIABLE					Subject	VARIABLE				
	A	B	C	D	E		A	B	C	D	E
289.	0	0	1	0	0	290.	1	0	1	0	1
291.	0	1	2	1	0	292.	0	1	0	1	0
293.	1	0	2	0	0	294.	1	1	2	0	1
295.	0	1	0	0	1	296.	0	0	1	1	0
297.	1	0	2	0	0	298.	1	0	1	1	1
299.	0	1	2	1	0	300.	1	0	0	0	0
301.	0	1	2	1	1	302.	1	1	1	1	1
303.	0	1	2	1	1	304.	1	1	0	0	0
305.	1	0	0	0	0	306.	1	1	0	1	1
307.	1	1	2	1	0	308.	1	0	1	0	1
309.	1	0	2	0	0	310.	0	1	1	0	0
311.	1	0	0	0	0	312.	1	0	2	1	1
313.	0	0	0	0	0	314.	0	1	1	1	1
315.	1	1	1	1	0	316.	1	0	0	0	0
317.	1	1	0	1	0	318.	0	0	2	0	0
319.	0	1	2	1	1	320.	0	0	2	0	0
321.	0	1	0	1	0	322.	1	0	1	1	0
323.	0	0	0	1	0	324.	1	1	1	0	0
325.	0	1	2	1	1	326.	0	0	0	0	0
327.	1	1	0	1	0	328.	1	1	2	1	1
329.	1	1	0	1	0	330.	1	0	0	0	0
331.	0	1	0	1	1	332.	0	1	1	0	0
333.	1	1	0	0	1	334.	1	0	1	0	1
335.	0	0	1	1	0	336.	0	0	0	0	0
337.	1	1	0	1	0	338.	1	0	2	0	1
339.	0	1	1	1	0	340.	0	1	1	1	1
341.	1	0	2	0	1	342.	0	1	2	1	1
343.	1	0	0	0	0	344.	0	1	1	1	1
345.	1	0	1	0	1	346.	1	1	1	1	1
347.	0	1	0	0	0	348.	1	1	0	0	0
349.	0	0	1	1	1	350.	1	1	1	1	1
351.	1	0	0	0	1	352.	0	0	1	0	0
353.	1	0	1	0	1	354.	0	1	0	1	0
355.	1	1	1	1	1	356.	1	0	2	0	1
357.	1	0	1	1	0	358.	0	0	1	1	0
359.	1	1	0	0	0	360.	1	0	1	0	0
361.	1	0	1	0	1	362.	1	1	0	1	1
363.	1	0	0	0	0	364.	1	1	1	1	1
365.	1	1	0	1	0	366.	0	1	1	1	1
367.	1	0	0	0	0	368.	1	1	1	0	1
369.	0	0	0	0	0	370.	1	1	2	1	0
371.	0	0	1	0	1	372.	1	0	0	1	0
373.	0	1	1	1	1	374.	0	1	0	1	1
375.	1	0	1	1	1	376.	0	0	0	0	0
377.	1	1	2	1	1	378.	1	0	1	0	0
379.	1	0	2	0	0	380.	1	0	2	0	1

	VARIABLE						VARIABLE				
Subject	A	B	C	D	E	Subject	A	B	C	D	E
381.	0	0	0	1	0	382.	1	0	0	0	0
383.	0	1	1	1	1	384.	0	1	2	0	1
385.	1	0	1	0	0	386.	0	0	0	0	0
387.	1	0	1	0	1	388.	1	1	2	0	1
389.	0	1	0	1	1	390.	1	0	1	1	1
391.	0	0	2	0	1	392.	0	0	2	1	1
393.	0	1	1	1	1	394.	1	0	1	0	1
395.	0	0	1	0	0	396.	0	0	1	0	0
397.	1	1	0	1	1	398.	0	0	1	0	1
399.	0	0	0	1	0	400.	0	0	0	0	0
401.	0	1	1	0	1	402.	1	0	2	0	0
403.	1	1	1	1	1	404.	1	1	1	1	1
405.	1	0	1	0	0	406.	1	1	2	1	1
407.	1	0	1	0	0	408.	0	0	0	0	0
409.	0	0	1	1	1	410.	1	1	0	1	1
411.	0	1	0	0	1	412.	1	0	2	1	1
413.	0	0	1	0	1	414.	1	0	2	0	0
415.	0	1	0	1	1	416.	0	0	1	0	1
417.	1	0	0	0	0	418.	1	0	0	0	0
419.	1	0	0	1	0	420.	1	1	1	1	0
421.	1	0	0	0	0	422.	1	0	0	0	0
423.	1	1	0	1	1	424.	1	0	1	1	0
425.	0	1	2	0	1	426.	1	1	2	1	0
427.	1	0	1	0	1	428.	1	1	0	1	0
429.	0	0	2	0	1	430.	1	1	1	1	0
431.	1	1	2	1	1	432.	1	0	1	0	1
433.	0	1	0	1	0	434.	0	0	1	0	0
435.	1	1	2	0	0	436.	0	1	1	1	0
437.	0	0	0	1	1	438.	0	0	0	0	0
439.	0	0	0	0	0	440.	1	0	0	0	0
441.	1	0	2	0	1	442.	1	1	0	1	0
443.	1	0	2	1	0	444.	0	1	1	1	0
445.	1	1	1	0	1	446.	0	1	0	0	0
447.	0	0	0	1	0	448.	1	1	1	1	0
449.	1	1	0	1	1	450.	1	0	0	0	0
451.	1	1	2	1	1	452.	1	1	2	1	1
453.	1	1	1	0	1	454.	0	1	1	1	0
455.	1	1	1	1	1	456.	0	1	1	1	1
457.	1	0	1	0	0	458.	1	1	0	1	1
459.	0	0	2	0	0	460.	1	0	0	0	0
461.	1	0	0	0	0	462.	1	1	1	0	1
463.	1	1	0	1	0	464.	0	1	1	0	1
465.	1	1	1	1	1	466.	1	0	0	0	0
467.	1	0	1	1	0	468.	0	1	2	1	1
469.	0	0	0	1	0	470.	1	0	2	0	0
471.	0	0	0	0	0	472.	1	0	0	0	0

Subject	VARIABLE A	B	C	D	E	Subject	VARIABLE A	B	C	D	E
473.	0	1	0	1	1	474.	0	0	0	1	0
475.	0	0	0	0	1	476.	1	0	1	0	1
477.	1	1	0	1	0	478.	1	1	1	1	1
479.	0	1	1	1	1	480.	1	0	1	0	0
481.	0	1	2	0	1	482.	1	1	2	0	0
483.	1	1	0	1	0	484.	0	0	1	0	0
485.	1	0	0	1	0	486.	1	0	0	0	0
487.	0	0	1	0	0	488.	1	1	2	1	1
489.	0	0	2	0	1	490.	1	1	0	1	1
491.	1	0	1	0	0	492.	0	1	1	1	1
493.	0	0	1	1	0	494.	1	1	1	0	1
495.	1	0	0	0	0	496.	0	0	0	1	0
497.	0	0	1	0	0	498.	1	0	0	0	0
499.	1	1	2	1	1	500.	0	0	1	0	0
501.	1	1	1	1	1	502.	0	0	0	0	0
503.	0	0	1	0	0	504.	1	1	2	1	1
505.	1	0	1	0	1	506.	0	1	1	0	0
507.	0	1	1	1	1	508.	1	1	2	1	0
509.	0	1	1	0	0	510.	1	1	2	1	1
511.	1	0	0	1	0	512.	1	1	1	1	1
513.	0	1	1	1	0	514.	0	0	0	1	0
515.	1	0	0	0	0	516.	1	1	2	1	1
517.	1	1	1	1	1	518.	0	0	0	0	0
519.	0	1	1	1	1	520.	1	1	0	0	1
521.	1	0	0	0	0	522.	1	0	1	1	0
523.	1	1	2	1	1	524.	0	1	2	0	1
525.	1	1	2	1	1	526.	1	1	1	1	1
527.	1	0	0	0	0	528.	0	1	1	0	0
529.	0	1	1	0	0	530.	1	0	2	0	1
531.	1	0	1	0	1	532.	1	0	0	0	0
533.	1	0	1	0	0	534.	0	1	1	1	0
535.	1	0	0	1	0	536.	1	1	1	1	0
537.	1	1	0	1	0	538.	1	1	0	1	0
539.	1	1	0	1	0	540.	1	0	2	0	0
541.	1	0	1	0	1	542.	1	1	1	0	1
543.	1	1	1	1	1	544.	1	0	1	0	1
545.	0	1	0	0	0	546.	0	1	1	1	1
547.	1	0	1	1	1	548.	0	0	1	0	0
549.	0	1	1	1	1	550.	1	1	1	1	1
551.	0	0	2	0	1	552.	0	1	1	1	0
553.	0	0	1	1	0	554.	0	1	2	0	0
555.	0	1	2	1	1	556.	1	0	1	0	0
557.	0	0	1	0	1	558.	0	1	2	0	0
559.	0	1	2	1	1	560.	1	0	0	0	0
561.	1	0	2	0	0	562.	1	0	1	0	0
563.	1	0	2	1	0	564.	1	0	1	0	1

Subject	A	B	C	D	E	Subject	A	B	C	D	E
565.	0	0	0	1	0	566.	1	1	0	1	1
567.	1	0	0	0	0	568.	0	0	0	0	0
569.	1	1	1	1	1	570.	0	1	0	1	1
571.	1	0	1	0	0	572.	1	1	0	1	0
573.	0	0	2	0	0	574.	0	1	0	1	0
575.	1	1	1	0	1	576.	0	0	0	0	0
577.	1	1	2	0	1	578.	1	0	0	1	0
579.	1	0	2	1	0	580.	0	0	0	0	0
581.	0	1	1	1	0	582.	1	0	0	0	0
583.	1	0	2	0	0	584.	1	1	1	1	1
585.	1	0	1	0	1	586.	1	1	0	1	0
587.	1	0	2	1	0	588.	0	0	2	0	0
589.	0	0	1	0	1	590.	0	1	2	1	1
591.	1	0	2	0	0	592.	1	0	0	0	0
593.	1	0	2	0	0	594.	1	1	0	0	0
595.	1	0	2	1	1	596.	1	1	1	1	1
597.	0	0	0	1	0	598.	1	0	1	0	0
599.	1	1	1	1	0	600.	1	0	0	0	0
601.	1	1	1	1	1	602.	1	1	2	1	0
603.	0	0	0	0	1	604.	0	1	2	0	1
605.	0	0	1	0	0	606.	1	0	1	0	0
607.	0	1	0	1	1	608.	1	1	2	1	1
609.	0	1	1	1	1	610.	1	0	1	0	0
611.	0	0	2	1	0	612.	1	0	0	1	0
613.	1	0	2	0	1	614.	0	0	0	0	0
615.	0	0	2	0	1	616.	1	0	0	0	0
617.	0	0	0	1	0	618.	0	0	0	0	0
619.	0	0	1	0	1	620.	1	1	1	0	0
621.	0	0	0	0	0	622.	1	0	1	0	0
623.	1	1	2	1	1	624	1	1	2	0	0
625.	0	0	0	1	0	626.	0	0	2	0	0
627.	0	0	2	1	1	628.	1	0	0	0	0
629.	1	0	2	0	0	630.	1	1	2	1	1
631.	1	0	1	0	0	632.	0	0	0	0	1
633.	1	1	0	1	0	634.	0	1	0	1	1
635.	1	0	1	1	0	636.	1	1	1	1	1
637.	1	1	2	1	1	638.	0	0	0	0	0
639.	1	1	0	1	1	640.	1	0	1	0	1
641.	0	1	0	0	1	642.	1	0	1	1	1
643.	1	1	2	1	1	644.	1	0	2	0	0
645.	0	0	2	0	1	646.	0	0	1	0	0
647.	0	0	1	0	0	648.	1	1	0	1	0
649.	1	1	0	1	1	650.	1	1	1	0	1
651.	1	0	2	0	1	652.	0	0	1	0	1
653.	0	0	0	1	0	654.	0	0	0	0	0
655.	0	1	1	0	1	656.	1	1	0	1	0

Subject	VARIABLE A	B	C	D	E	Subject	VARIABLE A	B	C	D	E
657.	1	1	1	1	0	658.	0	0	1	1	1
659.	1	1	1	1	1	660.	0	1	1	1	1
661.	1	1	1	0	0	662.	1	0	2	0	0
663.	0	0	0	0	0	664.	0	0	1	1	0
665.	0	0	1	0	0	666.	1	1	0	1	0
667.	0	0	1	0	1	668.	1	1	0	1	1
669.	1	0	0	0	0	670.	1	0	1	0	1
671.	1	0	2	0	0	672.	0	0	1	0	1
673.	1	0	1	1	1	674.	0	1	2	1	1
675.	0	0	0	0	0	676.	0	1	1	1	1
677.	1	0	1	0	1	678.	1	0	0	0	0
679.	1	0	1	1	0	680.	0	1	1	0	1
681.	0	0	0	1	0	682.	0	0	0	0	0
683.	1	0	0	0	0	684.	0	0	1	0	1
685.	1	0	2	0	0	686.	1	0	1	1	0
687.	1	1	0	1	1	688.	0	0	1	0	0
689.	1	1	0	0	1	690.	0	0	0	0	0
691.	0	1	1	1	1	692.	0	0	2	0	0
693.	0	0	2	0	0	694.	1	1	0	1	1
695.	0	0	2	0	0	696.	1	1	1	1	0
697.	1	0	1	1	0	698.	0	0	0	1	0
699.	0	0	1	0	1	700.	0	0	1	0	0
701.	0	0	0	0	0	702.	0	0	1	1	0
703.	1	1	0	1	0	704.	0	1	1	1	1
705.	1	0	2	0	1	706.	0	0	1	0	0
707.	1	0	1	0	1	708.	0	0	0	0	0
709.	1	0	1	1	0	710.	0	1	0	1	0
711.	0	0	0	0	0	712.	1	0	0	0	0
713.	1	1	1	0	0	714.	0	0	0	0	0
715.	1	0	2	0	0	716.	0	0	0	0	0
717.	1	1	1	0	1	718.	1	1	0	1	1
719.	1	1	2	1	1	720.	0	1	1	1	1
721.	0	0	0	1	1	722.	1	0	0	1	0
723.	1	0	2	0	1	724.	0	0	2	0	1
725.	1	0	0	0	0	726.	0	1	0	0	0
727.	1	1	0	1	1	728.	0	0	1	0	1
729.	0	0	2	0	0	730.	0	0	0	0	0
731.	1	1	2	1	0	732.	1	1	0	1	0
733.	1	0	2	1	0	734.	0	1	1	0	0
735.	1	1	0	1	0	736.	0	0	1	0	1
737.	1	0	0	0	0	738.	1	1	0	1	1
739.	0	0	2	0	1	740.	1	0	0	1	0
741.	1	1	0	1	1	742.	1	0	0	1	0
743.	0	0	2	0	1	744.	0	1	1	1	1
745.	1	0	2	0	0	746.	1	1	1	1	1
747.	1	0	1	1	1	748.	1	0	0	0	0

	Variable						**Variable**				
Subject	A	B	C	D	E	Subject	A	B	C	D	E
749.	1	1	0	1	0	750.	0	1	0	0	0
751.	1	1	1	0	1	752.	1	1	1	1	1
753.	1	1	2	1	0	754.	1	0	1	0	1
755.	0	0	2	0	1	756.	1	0	1	0	0
757.	0	1	1	1	1	758.	0	0	1	1	0
759.	0	1	2	0	1	760.	0	0	2	1	0
761.	1	1	2	1	1	762.	0	0	0	0	0
763.	1	0	2	0	1	764.	1	0	1	0	1
765.	0	1	1	1	1	766.	0	1	2	1	1
767.	1	1	1	0	0	768.	1	0	1	0	1
769.	1	0	1	1	1	770.	1	0	0	0	0
771.	1	0	0	0	0	772.	1	1	0	1	0
773.	1	0	0	0	0	774.	1	1	0	1	0
775.	1	0	0	0	0	776.	0	1	0	1	1
777.	1	0	1	1	1	778.	1	0	2	0	1
779.	0	1	0	1	1	780.	0	1	0	0	1
781.	1	0	0	0	0	782.	1	1	0	1	1
783.	0	1	1	1	1	784.	0	0	0	0	0
785.	1	1	1	1	1	786.	1	1	2	0	1
787.	0	0	1	0	1	788.	1	0	1	1	1
789.	1	1	0	1	0	790.	0	0	1	0	0
791.	1	1	1	1	0	792.	1	0	0	0	1
793.	1	1	0	1	0	794.	1	1	1	0	1
795.	1	1	2	1	1	796.	1	0	0	0	0
797.	1	0	0	1	0	798.	0	1	1	1	0
799.	1	0	0	1	0	800.	1	0	2	0	1
801.	0	0	2	0	1	802.	1	0	0	0	0
803.	0	0	1	0	1	804.	1	0	0	1	0
805.	1	0	0	0	1	806.	1	0	0	0	0
807.	1	0	1	0	1	808.	1	1	1	1	1
809.	1	1	1	1	0	810.	1	1	1	0	1
811.	1	1	1	1	1	812.	1	1	2	1	0
813.	1	0	1	0	1	814.	1	0	2	0	1
815.	0	0	1	1	1	816.	1	1	0	1	1
817.	1	0	2	0	0	818.	1	1	1	1	1
819.	0	0	0	0	0	820.	0	1	0	0	1
821.	1	0	2	0	0	822.	1	0	1	1	1
823.	1	0	2	1	1	824.	1	0	2	0	0
825.	0	1	0	1	0	826.	0	1	1	0	0
827.	0	0	2	0	1	828.	0	0	1	0	0
829.	1	0	2	0	0	830.	1	1	2	1	1
831.	0	1	1	1	1	832.	1	1	2	0	1
833.	0	1	1	1	0	834.	0	1	0	1	1
835.	0	1	0	1	1	836.	0	0	1	1	1
837.	1	0	0	0	0	838.	0	1	1	1	1
839.	1	0	0	0	0	840.	0	1	1	0	1
841.	0	0	0	1	0	842.	1	0	2	0	0

Subject	A	B	C	D	E	Subject	A	B	C	D	E
843.	0	1	2	1	1	844.	0	1	1	1	1
845.	0	0	0	0	0	846.	1	0	0	0	0
847.	1	1	0	1	1	848.	0	1	1	0	1
849.	1	0	1	0	0	850.	1	1	0	1	1
851.	1	0	0	0	0	852.	0	0	1	0	0
853.	0	1	1	1	1	854.	1	0	1	1	1
855.	0	1	1	1	1	856.	0	1	1	1	0
857.	1	0	1	0	0	858.	1	0	0	1	0
859.	1	1	0	1	0	860.	1	0	2	0	1
861.	1	1	2	0	0	862.	0	0	0	0	0
863.	0	1	1	1	1	864.	1	0	0	0	0
865.	1	0	0	0	0	866.	1	1	0	0	1
867.	1	1	0	1	1	868.	0	1	0	1	1
869.	1	0	2	1	1	870.	1	0	0	0	0
871.	1	1	0	1	0	872.	0	1	1	1	0
873.	0	0	2	1	0	874.	0	0	1	0	1
875.	1	1	1	1	1	876.	1	0	0	0	0
877.	1	0	2	0	0	878.	1	0	1	0	0
879.	1	0	0	0	0	880.	1	0	1	0	1
881.	0	0	2	1	0	882.	0	1	0	0	0
883.	1	0	0	0	0	884.	1	1	2	1	1
885.	0	1	0	0	1	886.	1	0	0	1	0
887.	0	0	1	0	1	888.	1	0	2	0	0
889.	0	0	1	1	0	890.	1	0	2	0	0
891.	0	0	0	0	1	892.	1	1	2	1	1
893.	0	0	1	0	1	894.	1	1	0	1	1
895.	0	1	2	0	0	896.	1	0	0	0	0
897.	0	1	1	1	1	898.	0	1	1	1	1
899.	0	0	0	0	0	900.	1	1	2	1	1
901.	1	0	0	1	0	902.	1	1	1	1	0
903.	1	0	1	0	1	904.	1	1	2	0	1
905.	0	1	0	1	0	906.	1	1	1	1	1
907.	1	1	1	1	0	908.	1	0	2	0	1
909.	1	0	2	1	0	910.	1	1	1	0	1
911.	1	0	2	0	0	912.	0	1	1	1	0
913.	0	0	1	0	0	914.	0	1	2	1	0
915.	1	0	1	0	1	916.	1	1	1	0	1
917.	1	0	1	0	1	918.	1	0	1	0	0
919.	0	0	1	1	0	920.	1	0	1	0	1
921.	1	1	0	1	1	922.	0	0	1	0	0
923.	1	1	1	1	1	924.	0	1	0	1	0
925.	1	0	1	0	0	926.	1	0	1	1	0
927.	1	1	1	0	0	928.	1	1	0	1	1
929.	1	1	1	1	1	930.	1	1	2	1	1
931.	1	1	0	1	1	932.	1	0	0	1	0
933.	1	0	0	0	0	934.	1	1	2	1	1
935.	0	1	0	0	0	936.	1	1	2	1	1

		VARIABLE						VARIABLE			
Subject	A	B	C	D	E	Subject	A	B	C	D	E
937.	1	1	2	1	1	938.	0	0	0	0	0
939.	1	0	0	0	0	940.	0	0	0	0	0
941.	0	0	2	0	0	942.	0	0	1	0	1
943.	1	0	0	1	0	944.	1	1	0	1	1
945.	1	0	2	1	1	946.	0	0	0	0	0
947.	1	0	1	0	0	948.	0	1	0	1	1
949.	1	0	2	0	0	950.	1	1	0	0	0
951.	0	0	1	0	1	952.	0	0	1	1	0
953.	1	1	1	0	1	954.	0	1	1	1	1
955.	1	1	2	1	1	956.	1	1	0	1	0
957.	1	0	1	0	0	958.	0	1	1	1	0
959.	0	0	0	0	0	960.	0	0	1	0	1
961.	1	0	1	0	0	962.	0	1	0	0	1
963.	1	1	0	1	1	964.	1	0	2	1	0
965.	0	0	1	0	1	966.	0	1	1	1	1
967.	0	1	1	1	1	968.	1	1	1	1	1
969.	1	1	0	1	1	970.	1	0	1	0	0
971.	0	0	2	0	0	972.	0	0	1	0	1
973.	1	0	1	1	1	974.	0	0	1	1	1
975.	0	1	0	0	0	976.	1	0	1	0	0
977.	1	0	1	0	1	978.	1	0	1	0	1
979.	1	0	0	0	0	980.	1	1	0	0	0
981.	1	0	1	0	1	982.	1	0	0	0	0
983.	1	0	0	0	0	984.	1	0	0	1	0
985.	0	1	1	1	0	986.	1	1	0	1	0
987.	1	0	2	1	0	988.	1	0	1	0	1
989.	1	0	2	0	1	990.	1	1	0	1	1
991.	1	0	0	0	0	992.	1	0	1	0	1
993.	0	1	1	1	1	994.	1	0	0	1	0
995.	1	1	2	0	1	996.	1	0	1	0	1
997.	1	1	1	1	0	998.	1	0	2	0	1
999.	1	0	0	0	0	1000.	1	1	1	1	1
1001.	1	1	0	0	0	1002.	1	0	0	0	1
1003.	1	1	0	1	0	1004.	0	0	1	0	1
1005.	1	0	0	1	0	1006.	0	0	2	0	0
1007.	1	0	1	1	1	1008.	0	0	0	0	0
1009.	1	1	2	1	0	1010.	0	0	2	0	1
1011.	1	0	1	1	1	1012.	0	0	0	0	0
1013.	0	1	0	1	0	1014.	1	0	1	0	1
1015.	0	0	1	0	1	1016.	1	1	1	0	1
1017.	1	0	1	1	1	1018.	0	1	0	1	1
1019.	0	1	0	1	0	1020.	1	0	1	0	0
1021.	0	0	0	0	0	1022.	1	0	0	0	0
1023.	1	1	1	1	1	1024.	0	0	2	0	1
1025.	0	1	0	1	1	1026.	1	0	0	0	0

	Variable						Variable				
Subject	*A*	*B*	*C*	*D*	*E*	*Subject*	*A*	*B*	*C*	*D*	*E*
1027.	1	1	0	1	0	1028.	1	0	2	0	0
1029.	1	1	1	1	1	1030.	1	1	1	0	0
1031.	0	1	0	0	0	1032.	0	1	0	1	1
1033.	1	1	0	1	0	1034.	0	0	1	1	0
1035.	1	0	0	1	0	1036.	1	0	0	0	0
1037.	1	1	1	1	1	1038.	1	0	2	0	1
1039.	0	0	1	0	0	1040.	0	0	0	0	0
1041.	1	0	1	0	1	1042.	1	0	2	0	0
1043.	1	1	1	0	0	1044.	0	0	0	1	0
1045.	1	0	2	1	0	1046.	1	1	1	1	1
1047.	0	1	1	1	1	1048.	1	0	1	0	0
1049.	1	0	0	0	0	1050.	0	0	0	0	0
1051.	0	1	0	1	1	1052.	0	0	0	0	1
1053.	1	0	2	1	0	1054.	0	0	0	0	0
1055.	1	1	2	0	1	1056.	1	1	0	1	1
1057.	0	0	2	0	1	1058.	1	0	1	0	0
1059.	1	1	0	1	0	1060.	1	0	1	1	1
1061.	1	0	2	0	0	1062.	1	0	0	0	0
1063.	0	0	1	0	1	1064.	1	0	2	0	0
1065.	1	1	1	1	0	1066.	1	0	1	0	0
1067.	0	0	0	1	0	1068.	1	1	2	1	1
1069.	0	0	0	0	0	1070.	1	0	1	0	0
1071.	0	1	1	1	1	1072.	1	1	0	1	0
1073.	1	0	1	0	1	1074.	1	1	1	0	1
1075.	1	0	1	1	0	1076.	1	1	2	1	0
1077.	0	0	1	1	0	1078.	1	0	2	0	1
1079.	1	0	2	0	1	1080.	1	1	2	1	1
1081.	1	0	1	0	0	1082.	0	0	0	0	0
1083.	0	0	0	1	0	1084.	0	0	0	0	0
1085.	1	1	1	1	1	1086.	1	0	0	0	0
1087.	0	0	1	0	1	1088.	1	0	1	0	1
1089.	1	1	0	0	1	1090.	1	1	2	0	0
1091.	1	1	1	1	0	1092.	0	1	2	1	1
1093.	1	0	0	0	1	1094.	1	1	0	1	1
1095.	1	1	0	1	1	1096.	0	1	0	1	1
1097.	0	0	0	1	0	1098.	1	0	0	1	0
1099.	1	0	1	0	1	1100.	0	1	2	1	1
1101.	1	0	0	0	0	1102.	1	0	1	1	1
1103.	0	1	1	0	1	1104.	1	0	0	0	0
1105.	0	0	0	0	0	1106.	1	0	2	0	0
1107.	1	0	0	1	0	1108.	1	1	0	0	1
1109.	0	1	1	1	1	1110.	1	1	0	1	0
1111.	1	0	0	1	0	1112.	1	0	0	0	0
1113.	1	1	2	1	1	1114.	0	0	1	0	0
1115.	0	0	0	0	0	1116.	1	1	0	1	1

| | **VARIABLE** | | | | | | **VARIABLE** | | | | |
Subject	A	B	C	D	E	Subject	A	B	C	D	E
1117.	0	1	0	1	1	1118.	1	0	1	0	0
1119.	1	0	0	0	0	1120.	1	0	0	0	0
1121.	1	0	0	1	0	1122.	1	1	1	1	1
1123.	0	1	0	0	1	1124.	0	0	2	1	0
1125.	1	0	0	0	0	1126.	0	1	2	0	1
1127.	0	1	0	1	0	1128.	0	0	0	0	0
1129.	1	1	2	1	0	1130.	0	0	0	0	0
1131.	0	0	0	0	0	1132.	1	0	0	0	1
1133.	1	1	0	1	0	1134.	1	0	2	0	1
1135.	1	1	0	0	1	1136.	0	0	1	1	1
1137.	1	0	1	0	1	1138.	1	1	2	1	1
1139.	0	0	2	0	1	1140.	1	0	1	0	0
1141.	0	1	1	1	0	1142.	0	0	0	1	0
1143.	1	1	0	1	0	1144.	0	1	1	1	1
1145.	1	0	1	0	0	1146.	1	0	0	1	0
1147.	1	0	1	0	0	1148.	0	0	0	0	0
1149.	1	1	1	1	0	1150.	1	0	1	0	1
1151.	0	1	1	1	1	1152.	1	1	0	0	0
1153.	1	0	0	0	0	1154.	0	0	0	1	0
1155.	1	0	0	0	0	1156.	0	0	0	1	0
1157.	1	0	1	0	0	1158.	1	1	1	0	1
1159.	1	0	1	0	1	1160.	1	1	0	1	0
1161.	1	1	2	1	1	1162.	0	0	0	0	0
1163.	0	1	2	1	1	1164.	0	1	1	0	1
1165.	1	1	1	1	1	1166.	1	0	1	0	0
1167.	1	0	1	0	0	1168.	0	0	1	0	0
1169.	1	1	2	1	1	1170.	1	0	1	1	0
1171.	1	1	1	1	0	1172.	1	0	1	0	1
1173.	0	0	2	0	0	1174.	1	0	1	0	1
1175.	1	0	1	1	0	1176.	1	1	0	1	0
1177.	1	0	2	1	1	1178.	0	1	0	1	0
1179.	1	0	2	0	1	1180.	0	0	1	0	0
1181.	1	0	0	0	0	1182.	1	0	1	0	1
1183.	0	1	0	1	1	1184.	1	0	0	1	0
1185.	0	0	1	0	1	1186.	1	1	2	0	0
1187.	0	1	0	1	1	1188.	1	1	2	1	1
1189.	0	1	0	1	1	1190.	1	0	2	0	1
1191.	1	0	1	1	0	1192.	1	1	0	1	1
1193.	0	0	0	0	0	1194.	1	0	1	0	1
1195.	0	1	1	1	1	1196.	1	0	2	0	1
1197.	1	1	1	1	0	1198.	0	0	1	0	1
1199.	1	0	1	0	0	1200.	1	1	0	0	0

26. Refer to Exercise 25. Select a simple random sample of size 100 from the population and carry out a test to see if you can conclude that there is a relationship between drinking status and high blood pressure status in the population. Let $\alpha = .05$ and determine the p value. Compare your results with those of your classmates.

27. Refer to Exercise 25. Select a simple random sample of size 100 from the population and carry out a test to see if you can conclude that there is a relationship between sex and smoking status in the population. Let $\alpha = .05$ and determine the p value. Compare your results with those of your classmates.

28. Refer to Exercise 25. Select a simple random sample of size 100 from the population and carry out a test to see if you can conclude that there is a relationship between sex and drinking level in the population. Let $\alpha = .05$ and find the p value. Compare your results with those of your classmates.

REFERENCES

References Cited

1. Karl Pearson, "On the Criterion That a Given System of Deviations from the Probable in the Case of a Correlated System of Variables Is Such That It Can Be Reasonably Supposed to Have Arisen from Random Sampling," *The London, Edinburgh and Dublin Philosophical Magazine and Journal of Science*, Fifth Series, *50* (1900), 157–175. Reprinted in *Karl Pearson's Early Statistical Papers*, Cambridge University Press, 1948.

2. H. O. Lancaster, *The Chi-Squared Distribution*, Wiley, New York, 1969.

3. Harald Cramer, *Mathematical Methods of Statistics*, Princeton University Press, Princeton, N.J., 1958.

4. Ram C. Dahiya and John Gurland, "Pearson Chi-Squared Test of Fit with Random Intervals," *Biometrika*, *59* (1972), 147–153.

5. G. S. Watson, "On Chi-Square Goodness-of-Fit Tests for Continuous Distributions," *Journal of the Royal Statistical Society*, B, *20* (1958), 44–72.

6. G. S. Watson, "The χ^2 Goodness-of-Fit Test for Normal Distributions," *Biometrika*, *44* (1957), 336–348.

7. G. S. Watson, "Some Recent Results in Chi-Square Goodness-of-Fit Tests," *Biometrics*, *15* (1959), 440–468.

8. G. R. Chase, "On the Chi-Square Test When the Parameters Are Estimated Independently of the Sample," *Journal of the American Statistical Association*, *67* (1972), 609–611.

9. William G. Cochran, "The χ^2 Test of Goodness of Fit," *Annals of Mathematical Statistics*, *23* (1952), 315–345.

10. William G. Cochran, "Some Methods for Strengthening the Common χ^2 Tests," *Biometrics*, *10* (1954), 417–451.

11. John T. Roscoe and Jackson A. Byars, "An Investigation of the Restraints with Respect to Sample Size Commonly Imposed on the Use of the Chi-Square Statistic," *Journal of the American Statistical Association*, *66* (1971), 755–759.

12. James K. Yarnold, "The Minimum Expectation in X^2 Goodness-of-Fit Tests and the Accuracy of Approximations for the Null Distribution," *Journal of the American Statistical Association*, *65* (1970), 864–886.

13. Merle W. Tate and Leon A. Hyer, "Significance Values for an Exact Multinomial Test and Accuracy of the Chi-Square Approximation," U.S. Department of Health, Education and Welfare, Office of Education, Bureau of Research, August 1969.

14. Malcolm J. Slakter, "Comparative Validity of the Chi-Square and Two Modified Chi-Square Goodness-of-Fit Tests for Small but Equal Expected Frequencies," *Biometrika*, *53* (1966), 619–23.

15. Malcolm J. Slakter, "A Comparison of the Pearson Chi-Square and Kolmogorov Goodness-of-Fit Tests with Respect to Validity," *Journal of the American Statistical Association*, *60* (1965), 854–858.

16. R. C. Lewontin and J. Felsenstein, "The Robustness of Homogeneity Tests in $2 \times N$ Tables," *Biometrics*, *21* (1965), 19–33.

17. F. Yates, "Contingency Tables Involving Small Numbers and the χ^2 Tests," *Journal of the Royal Statistical Society Supplement*, *1*, 1934 (Series B), 217–235.

18. J. E. Grizzle, "Continuity Correction in the χ^2 Test for 2×2 Tables," *The American Statistician*, *21* (October 1967), 28–32.

19. H. O. Lancaster, "The Combination of Probabilities Arising from Data in Discrete Distributions," *Biometrika*, *36* (1949), 370–382.

20. E. S. Pearson, "The Choice of Statistical Test Illustrated on the Interpretation of Data in a 2×2 Table," *Biometrika*, *34* (1947), 139–167.

21. R. L. Plackett, "The Continuity Correction in 2×2 Tables," *Biometrika*, *51* (1964), 427–338.

CHAPTER 11

Nonparametric and Distribution-Free Statistics

11.1 INTRODUCTION

The statistical inference procedures we have discussed up to this point, with one exception, are classified as *parametric statistics*. The one exception is our uses of chi-square: as a test of goodness-of-fit, and as a test of independence. These uses of chi-square come under the heading of *nonparametric statistics*.

The obvious question now is: What is the difference? In answer, let us recall the nature of the inferential procedures that we have categorized as *parametric*. In each case, our interest was focused on estimating or testing a hypothesis about one or more population parameters. Furthermore, central to these procedures was a knowledge of the functional form of the population from which were drawn the samples providing the basis for the inference.

An example of a parametric statistical test is the widely used *t* test. The most common uses of this test are for testing a hypothesis about a single population mean or the difference between two population means. One of the assumptions underlying the valid use of this test is that the sampled population or populations are at least approximately normally distributed.

As we shall learn, the procedures that we discuss in this chapter either are not concerned with population parameters or do not depend on knowledge of the sampled population. Strictly speaking, only those procedures that test hypotheses that are not statements about population parameters are classified as *nonparametric*, while those that make no assumption about the sampled population are called *distribution-free* procedures. Despite this distinction, it is customary to use the terms *nonparametric* and *distribution-free* interchangeably and to discuss the various procedures of both types under the heading of *nonparametric statistics*. We will follow this convention. This point is discussed by Kendall and Sundrum (1) and Gibbons (2).

The above discussion implies the following two advantages of nonparametric statistics.

1. They allow for the testing of hypotheses that are not statements about population parameter values. Some of the chi-square tests of goodness-of-fit and the tests of independence are examples of tests possessing this advantage.

2. Nonparametric tests may be used when the form of the sampled population is unknown.

Other advantages have been listed by several writers, for example, Gibbons (2), Blum and Fattu (3), Moses (4), and Siegel (5). In addition to the two already mentioned, the following are most frequently given.

3. Nonparametric procedures are computationally easier and consequently more quickly applied than parametric procedures. This can be a desirable feature in certain cases, but when time is not at a premium, it merits a low priority as a criterion for choosing a nonparametric test.

4. Nonparametric procedures may be applied when the data being analyzed consist merely of rankings or classifications. That is, the data may not be based on a measurement scale strong enough to allow the arithmetic operations necessary for carrying out parametric procedures. The subject of measurement scales is discussed in more detail in the next section.

Although nonparametric statistics enjoy a number of advantages, their disadvantages must also be recognized. Moses (4) has noted the following.

1. The use of nonparametric procedures with data that can be handled with a parametric procedure results in a waste of data.

2. The application of some of the nonparametric tests may be laborious for large samples.

In a general introductory textbook, space limitations prevent the presentation of more than a sampling of nonparametric procedures. Additional procedures discussed at an introductory or intermediate level may be found in the books by Conover (6), Siegel (5), Bradley (7), Kraft and Van Eeden (8), Tate and Clelland (9), Gibbons (10), Mosteller and Rourke (11), Pierce (12), Noether (13), Hollander and Wolfe (14), Daniel (15), Marascuilo and McSweeney (16), and Lehmann (17). More mathematicallly rigorous books have been written by Gibbons (2), Hajek (18), and Walsh (19, 20), Savage (21) has prepared a bibliography of nonparametric statistics.

11.2 MEASUREMENT SCALES

As was pointed out in the previous section, one of the advantages of nonparametric statistical procedures is that they can be used with data that are based on a weak measurement scale. To understand fully the meaning of this statement, it is necessary to know and understand the meaning of measurement and the various measurement scales most frequently used. This subject is discussed in considerable detail in the literature. Refer, in particular, to the writings of Stevens (22, 23).

Measurement   This may be defined as the assignment of numbers to objects or events according to a set of rules. The various measurement scales result from the fact that measurement may be carried out under different sets of rules.

The Nominal Scale   The lowest measurement scale is the *nominal scale*. As the name implies it consists of "naming" observations or classifying them into various mutually exclusive and collectively exhaustive categories. The practice of using numbers to distinguish among the various medical diagnoses constitutes measurement on a nominal scale. Other examples include such dichotomies as male–female, well–sick, under 65 years of age–65 and over, child–adult, and married–not married. As we have seen, under certain conditions the chi-square test may be used for testing hypotheses regarding the frequency of occurrence of observations in various nominal categories.

The Ordinal Scale   Whenever observations are not only different from category to category, but can be ranked according to some criterion, they are said to be measured on an ordinal scale. Convalescing patients may be characterized as unimproved, improved, and much improved. Individuals may be classified according to socioeconomic status as low, medium, or high. The intelligence of children may be above average, average, or below average. In each of these examples the members of any one category are all considered equal, but the members of one category are considered lower, worse, or smaller than those in another category, which in turn bears a similar relationship to another category. For example, a much improved patient is in better health than one classified as improved, while a patient who has improved is in better condition than one who has not improved. It is usually impossible to infer that the difference between members of one category, and the next adjacent category is equal to the difference between members of that category and the members of the category adjacent to it. The degree of improvement between unimproved and improved is probably not the same as that between improved and much improved. The implication is that if a finer breakdown were made resulting in more categories, these, too, could be ordered in a similar manner. The function of numbers assigned to ordinal data is to order (or rank) the observations from lowest to highest and, hence, the term ordinal. Numerous nonparametric statistical tests have been developed for use with ordinal data.

The Interval Scale   The *interval scale* is a more sophisticated scale than the nominal or ordinal in that with this scale it is not only possible to order measurements, but also the distance between any two measurements is known. We know, say, that the difference between a measurement of 20 and a measurement of 30 is equal to the difference between measurements of 30 and 40. The ability to do this implies the use of a unit distance and a zero point, both of which are arbitrary. The selected zero point is not a true zero in that it does not indicate a total absence of the quantity being measured. Perhaps the best example of an interval scale is provided by the way in which temperature is usually measured. The unit of measurement is the degree and the point of comparison is the arbitrarily chosen "zero degrees." The interval scale unlike the nominal and ordinal scales is a truly quantitative scale. When the interval scale of measurement has been achieved, and when the assumptions of the model are met, the usual parametric statistical procedures such as the t test and the F test may be used.

The Ratio Scale   The highest level of measurement is the *ratio scale*. This scale is characterized by the fact that equality of ratios as well as equality of intervals may be determined. Fundamental to the ratio scale is a true zero point. The measurement of such familiar traits as height, weight, and length makes use of the ratio scale. When the ratio scale of measurement has been achieved, any statistical procedure may be used provided that the assumptions of the particular model employed are met.

Many authors, for example Conover (6) and Siegel (5), indicate that different statistical tests require different measurement scales. Although this idea appears to be followed in practice, Anderson (24), Gaito (25), Lord (26), and Armstrong (27) present some interesting alternative points of view. The subject is also discussed by Borgatta and Bohrnstedt (28).

11.3   THE SIGN TEST

The familiar t test is not strictly valid for testing (a) the null hypothesis that a population mean is equal to some particular value, or (b) the null hypothesis that the mean of a population of differences between pairs of measurements is equal to zero unless it is known (or the researcher is willing to assume) that the relevant populations are normally distributed. Case b will be recognized as a situation that was analyzed by the paired comparisons test in Chapter 6. When the normality assumptions cannot be be made or when the data at hand are ranks rather than measurements on an interval or ratio scale, an alternative test must be sought. Although the t test is known to be rather insensitive to violations of the normality assumption, there are times when an alternative test is desirable.

A frequently used nonparametric test that does not depend on the assumptions of the t test, or measurement beyond the ordinal scale, is the *sign test*. This test focuses on the median rather than the mean as a measure of central tendency or location. The median and mean will be equal in symmetric distributions. The only assumption underlying the test is that the distribution of the variable of interest is continuous.

The sign test gets its name from the fact that pluses and minuses, rather than numerical values, provide the raw data used in the calculations. We illustrate the

TABLE 11.3.1

General Appearance Scores of 10 Mentally Retarded Girls

Girl	Score	Girl	Score
1	4	6	6
2	5	7	10
3	8	8	7
4	8	9	6
5	9	10	6

use of the sign test, first in the case of a single sample, and then by an example involving paired samples.

Example 11.3.1 Researchers wished to know if instruction in personal care and grooming would improve the appearance of mentally retarded girls.

1. *Data* In a school for the mentally retarded, 10 girls selected at random received special instruction in personal care and grooming. Two weeks after completion of the course of instruction the girls were interviewed by a nurse and a social worker who assigned each girl a score based on her general appearance. The investigators believed that at best the scores achieved the level of an ordinal scale. They felt that although a score of, say, 8 represented a better appearance than a score of 6, they were unwilling to say that the difference between scores of 6 and 8 was equal to the difference between, say, scores of 8 and 10; or that the difference between scores of 6 and 8 represented twice as much improvement as the difference between scores of 5 and 6. The scores are shown in Table 11.3.1. We wish to know if we can conclude that the median score of the population from which we assume this sample to have been drawn is different from 5.

2. *Assumptions* We assume that the measurements are taken on a continuous variable.

3. *Hypotheses*

$$H_0: \text{ The population median is 5.}$$

$$H_A: \text{ The population median is not 5.}$$

Let $\alpha = .05$.

4. *Test Statistic* The test statistic for the sign test is either the observed number of plus signs or the observed number of minus signs. The nature of the alternative hypothesis determines which of these test statistics is appropriate.

In a given test, any one of the following alternative hypotheses is possible:

$$H_A: P(+) > P(-) \qquad \text{one-sided alternative}$$
$$H_A: P(+) < P(-) \qquad \text{one-sided alternative}$$
$$H_A: P(+) \neq P(-) \qquad \text{two-sided alternative}$$

If the alternative hypothesis is

$$H_A: P(+) > P(-)$$

a sufficiently small number of minus signs causes rejection of H_0. The test statistic is the number of minus signs. Similarly, if the alternative hypothesis is

$$H_A: P(+) < P(-)$$

a sufficiently small number of plus signs causes rejection of H_0. The test statistic is the number of plus signs. If the alternative hypothesis is

$$H_A: P(+) \neq P(-)$$

either a sufficiently small number of plus signs or a sufficiently small number of minus signs causes rejection of the null hypothesis. We may take as the test statistic the less frequently occurring sign.

5. *Distribution of the Test Statistic* As a first step in determining the nature of the test statistic, let us examine the data in Table 11.3.1 to determine which scores lie above and which ones lie below the hypothesized median of 5. If we assign a plus sign to those scores that lie above the hypothesized median and a minus to those that fall below, we have the results shown in Table 11.3.2.

If the null hypothesis were true, that is, if the median were, in fact, 5, we would expect the numbers of scores falling above and below 5 to be approximately equal. This line of reasoning suggests an alternative way in which we could have stated the null hypothesis, namely that the probability of a plus is equal to the probability of a minus, and these probabilities are

TABLE 11.3.2

Scores Above ($+$) and Below ($-$) the Hypothesized Median Based on Data of Example 11.3.1

Girl	1	2	3	4	5	6	7	8	9	10
Score relative to hypothesized median	−	0	+	+	+	+	+	+	+	+

each equal to .5. Stated symbolically, the hypothesis would be

$$H_0: P(+) = P(-) = .5$$

In other words, we would expect about the same number of plus signs as minus signs in Table 11.3.2 when H_0 is true. A look at Table 11.3.2 reveals a preponderance of pluses; specifically, we observe eight pluses, one minus, and one zero, which was assigned to the score that fell exactly on the median. The usual procedure for handling zeros is to eliminate them from the analysis and reduce n, the sample size, accordingly. If we follow this procedure our problem reduces to one consisting of nine observations of which eight are plus and one is minus.

Since the number of pluses and minuses is not the same, we wonder if the distribution of signs is sufficiently disproportionate to cast doubt on our hypothesis. Stated another way, we wonder if this small a number of minuses could have come about by chance alone when the null hypothesis is true; or if the number is so small that something other than chance (that is, a false null hypothesis) is responsible for the results.

Based on what we learned in Chapter 3, it seems reasonable to conclude that the observations in Table 11.3.2 constitute a set of n independent random variables from the Bernoulli population with parameter, p. If we let k = the test statistic, the sampling distribution of k is the binominal probability distribution with parameter $p = .5$ if the null hypothesis is true.

6. *Decision Rule*   The decision rule depends on the alternative hypothesis.

For $H_A: P(+) > P(-)$, reject H_0 if, when H_0 is true, the probability of observing k or fewer minus signs is less than or equal to α.

For $H_A: P(+) < P(-)$, reject H_0 if the probability of observing, when H_0 is true, k or fewer plus signs is equal to or less than α.

For $H_A: P(+) \neq P(-)$, reject H_0 if (given that H_0 is true) the probability of obtaining a value of k as extreme as or more extreme than was actually computed is equal to or less than $\alpha/2$.

7. *Calculation of Test Statistic*   We may determine the probability of observing x or fewer minus signs when given a sample of size n and parameter p by evaluating the following expression:

$$P(k \leq x|n, p) = \sum_{k=0}^{x} \binom{n}{k} p^k q^{n-k} \qquad (11.3.1)$$

For our example we would compute

$$\binom{9}{0}(.5)^0(.5)^{9-0} + \binom{9}{1}(.5)^1(.5)^{9-1} = .00195 + .01758 = .0195$$

8. *Statistical Decision*   In Table A we find

$$P(k \leq 1|9, .5) = .0195$$

With a two-sided test either a very small number of minuses or a very small number of pluses would cause rejection of the null hypothesis. Since, in our example, there are fewer minuses, we focus our attention on minuses rather than pluses. By setting α equal to .05, we are saying that if the number of minuses is so small that the probability of observing this few or fewer is less than .025 (half of α), we will reject the null hypothesis. The probability we have computed, .0195, is less than .025. We, therefore, reject the null hypothesis.

9. *Conclusion*   We conclude that the median score is not 5. The p value for this test is $2(.0195) = .0390$.

When the data to be analyzed consist of observations in matched pairs and the assumptions underlying the t test are not met, or the measurement scale is weak, the sign test may be employed to test the null hypothesis that the median difference is 0. An alternative way of stating the null hypothesis is

$$P(X_i > Y_i) = P(X_i < Y_i) = .5$$

One of the matched scores, say Y_i, is subtracted from the other score, X_i. If Y_i is less than X_i, the sign of the difference is $+$, and if Y_i is greater than X_i, the sign of the difference is $-$. If the median difference is 0, we would expect a pair picked at random to be just as likely to yield a $+$ as a $-$ when the subtraction is performed. We may state the null hypothesis, then, as

$$H_0: P(+) = P(-) = .5$$

In a random sample of matched pairs we would expect the number of $+$'s and $-$'s to be about equal. If there are too many $+$'s or too many $-$'s than can be accounted for by chance alone when the null hypothesis is true, we will entertain some doubt about the truth of our null hypothesis. By means of the sign test, we can decide how many of one sign constitutes more than can be accounted for by chance alone.

Example 11.3.2   A dental research team wished to know if teaching people how to brush their teeth would be beneficial.

1. *Data*   Twelve pairs of patients seen in a dental clinic were obtained by carefully matching on such factors as age, sex, intelligence, and initial oral hygiene scores. One member of each pair received instruction on how to brush the teeth and on other oral hygiene matters. Six months later all 24 subjects were examined and assigned an oral hygiene score by a dental hygienist unaware of which subjects had received the instruction. A low score indicates a high level of oral hygiene. The results are shown in Table 11.3.3.

2. *Assumptions*   We assume that the population of differences between pairs of scores is a continuous variable.

TABLE 11.3.3

Oral Hygiene Scores of 12 Subjects Receiving Oral Hygiene Instruction (X_i) and 12 Subjects Not Receiving Instruction (Y_i)

	SCORE	
Pair number	Instructed (X_i)	Not instructed (Y_i)
1	1.5	2.0
2	2.0	2.0
3	3.5	4.0
4	3.0	2.5
5	3.5	4.0
6	2.5	3.0
7	2.0	3.5
8	1.5	3.0
9	1.5	2.5
10	2.0	2.5
11	3.0	2.5
12	2.0	2.5

3. *Hypotheses* If the instruction produces a beneficial effect, this fact would be reflected in the scores assigned to the members of each pair. If we take the differences $X_i - Y_i$ we would expect to observe more $-$'s than $+$'s if instruction had been beneficial, since a low score indicates a higher level of oral hygiene. If, in fact, instruction is beneficial, the median of the hypothetical population of all such differences would be less than 0, that is, negative. If, on the other hand, instruction has no effect, the median of this population would be zero. The null and alternate hypotheses, then, are

H_0:   the median of the differences is zero $[P(+) = P(-)]$

H_A: the median of the differences is negative $[P(+) < P(-)]$

Let α be .05.

4. *Test Statistic* The test statistic is $k = 2$ the number of plus signs.

5. *Distribution of the Test Statistic* The sampling distribution of k is the binomial distribution with parameters n and .5 if H_0 is true.

6. *Decision Rule* Reject H_0 if $P(k \leq 2|11, .5) \leq .05$.

7. *Calculation of the Test Statistic* As will be seen, the procedure here is identical to the single sample procedure once the score differences have been obtained for each pair. Performing the subtractions and observing signs yields the results shown in Table 11.3.4.

     The nature of the hypothesis indicates a one-sided test so that all of $\alpha = .05$ is associated with the rejection region which consists of all values of

TABLE 11.3.4

Signs of Differences ($X_i - Y_i$) in Oral Hygiene Scores of 12 Subjects Instructed (X_i) and 12 Matched Subjects Not Instructed (Y_i)

Pair	1	2	3	4	5	6	7	8	9	10	11	12
Sign of score differences	−	0	−	+	−	−	−	−	−	−	+	−

k (where k is equal to the number of + signs) for which the probability of obtaining that many or fewer pluses due to chance alone when H_0 is true is equal to or less than .05. We see in Table 11.3.4 that the experiment yielded one zero, two pluses, and nine minuses. When we eliminate the zero, the effective sample size is $n = 11$ with two pluses and nine minuses. In other words, since a "small" number of plus signs will cause rejection of the null hypothesis, the value of our test statistic is $k = 2$.

8. *Statistical Decision* We want to know the probability of obtaining no more than two pluses out of eleven tries when the null hypothesis is true. As we have seen, the answer is obtained by evaluating the appropriate binomial expression. In this example we find

$$P(k \le 2|11, .5) = \sum_{k=0}^{2} \binom{11}{k}(.5)^k(.5)^{11-k}$$

By consulting Table A, we find this probability to be .0327. Since .0327 is less than .05, we must reject H_0.

9. *Conclusion* We conclude that the median difference is negative. That is, we conclude that the instruction was beneficial. For this test, $p = .0327$.

As has been demonstrated, the sign test may be used with a single sample or with two samples in which each member of one sample is matched with a member of the other sample to form a sample of matched pairs. We have also seen that the alternative hypothesis may lead to either a one-sided or a two-sided test. In either case we concentrate on the less frequently occurring sign and calculate the probability of obtaining that few or fewer of that sign.

We use the least frequently occurring sign as our test statistic because the binomial probabilities in Table A are "less than or equal to" probabilities. By using the least frequently occurring sign we can obtain the probability we need directly from Table A without having to do any subtracting. If the probabilities in Table A were "greater than or equal to" probabilities, which are often found in tables of the binomial distribution, we would use the more frequently occurring sign as our test statistic in order to take advantage of the convenience of obtaining the desired probability directly from the table without having to do any subtracting. In fact, we could, in our present examples, use the more frequently occurring sign as our test statistic, but since Table A contains "less than or equal to" probabilities we would have to perform a subtraction operation to obtain the desired probability. As an illustration, consider the last example. If we use as our test statistic the most frequently occurring sign it is 9, the number of minuses. The desired probability, then, is the probability of 9 or more minuses, when $n = 11$, and $p = .5$. That is, we want

$$P(k \geq 9|11, .5)$$

Since, however, Table A contains "less than or equal to" probabilities, we must obtain this probability by subtraction. That is,

$$
\begin{aligned}
P(k \geq 9|11, .5) &= 1 - P(k \leq 8|11, .5) \\
&= 1 - .9673 \\
&= .0327
\end{aligned}
$$

which is the result obtained previously.

Sample Size We saw in Chapter 4 that when the sample size is large and when p is close to .5, the binomial distribution may be approximated by the normal distribution. The rule of thumb used was that the normal approximation is appropriate when both np and nq are greater than 5. When $p = .5$, as was hypothesized in our two examples, a sample of size 12 would satisfy the rule of thumb. Following this guideline, one could use the normal approximation when the sign test is used to test the null hypothesis that the median or median difference is 0 and n is equal to or greater than 12. Gibbons (2) concurs in this recommendation, although Siegel (5) recommends that n be, at least, 25. Since the procedure involves approximating a continuous distribution by a discrete distribution, the continuity correction of .5 is generally used. The test statistic then is

$$z = \frac{(k \pm .5) - .5n}{.5\sqrt{n}} \tag{11.3.2}$$

which is compared with the value of z from the unit normal distribution corresponding to the chosen level of significance. In Equation 11.3.2, $k + .5$ is used when $k < n/2$ and $k - .5$ is used when $k > n/2$.

In addition to the references already cited, the sign test is discussed in considerable detail by Dixon and Mood (29).

EXERCISES

11.3.1 A random sample of 15 student nurses were given a test to measure their level of authoritarianism with the following results.

Student number	1	2	3	4	5	6	7	8	9	10	11	12
Authoritarianism score	75	90	85	110	115	95	132	74	82	104	88	124
Student number	13	14	15									
Authoritarianism score	110	76	98									

Test at the .05 level of significance, the null hypothesis that the median score for the population sampled is 100. Determine the p value.

11.3.2 We want to know if a particular remotivation program is effective with physical therapy patients. A group of 20 patients seen in a physical therapy clinic were given a test, designed to measure their level of motivation, prior to participation in an experimental remotivation program. At the end of the program the patients were retested. The before and after scores were as follows.

Patient	SCORE Before	After	Patient	SCORE Before	After
1	10	15	11	16	24
2	8	10	12	10	23
3	5	10	13	15	25
4	14	14	14	5	15
5	15	25	15	24	20
6	22	20	16	20	25
7	17	20	17	14	24
8	10	22	18	10	23
9	8	16	19	15	25
10	20	18	20	14	25

What do you conclude from these data? Let $\alpha = .05$. A high score indicates a high level of motivation. Determine the p value.

11.3.3 A sample of 15 patients suffering from asthma participated in an experiment to study the effect of a new treatment on pulmonary function. Among the various measurements recorded were those of forced expiratory volume (liters) in 1 second (FEV_1) before and after application of the treatment. The results were as follows.

Subject	Before	After	Subject	Before	After
1	1.69	1.69	9	2.58	2.44
2	2.77	2.22	10	1.84	4.17
3	1.00	3.07	11	1.89	2.42
4	1.66	3.35	12	1.91	2.94
5	3.00	3.00	13	1.75	3.04
6	.85	2.74	14	2.46	4.62
7	1.42	3.61	15	2.35	4.42
8	2.82	5.14			

On the basis of these data, can one conclude that the treatment is effective in increasing the FEV_1 level? Let $\alpha = .05$ and find the p value.

11.4   THE MEDIAN TEST

The sign test for two samples given in the preceding section required that the samples be related. It is frequently of interest to make inferences based on two samples that are not related, that is, samples that are independent. To test the null hypothesis that two samples are drawn from populations with equal means, one employs the t or z statistic (depending on the sample sizes and/or whether the population variances are known or unknown) when the conditions for these parametric tests are met. A nonparametric counterpart is provided by the median test that may be used to test the null hypothesis that two independent samples have been drawn from populations with equal medians. The test, attributed mainly to Mood (30) and Westenberg (31), is also discussed by Brown and Mood (32) and Moses (4) as well as in several other references already cited.

Example 11.4.1   Do urban and rural male junior high school students differ with respect to their level of mental health?

1.   *Data*   Members of a random sample of 12 male students from a rural junior high school and an independent random sample of 16 male students from an urban junior high school were given a test to measure their level of mental health. The results are shown in Table 11.4.1.
     To determine if we can conclude that there is a difference we perform a hypothesis test. Suppose we choose a .05 level of significance.

TABLE 11.4.1
Level of Mental Health Scores of Junior High Boys

	SCHOOL		
Urban	*Rural*	*Urban*	*Rural*
35	29	25	50
26	50	27	37
27	43	45	34
21	22	46	31
27	42	33	
38	47	26	
23	42	46	
25	32	41	

2. *Assumptions* The assumptions underlying the test are (a) the samples are selected independently and at random from their respective populations, (b) the populations are of the same form, differing only in location, and (c) the variable of interest is continuous. The level of measurement must be, at least, ordinal. The two samples do not have to be of equal size.

3. *Hypotheses*

$$H_0: M_U = M_R$$

$$H_A: M_U \neq M_R$$

Let $\alpha = .05$. M_U is the median score of the sampled population of urban students and M_R is the median score of the sampled population of rural students.

4. *Test Statistic* As will be shown in the discussion that follows, the test statistic is X^2 as computed, for example, by Equation 10.4.1 for a 2×2 contingency table.

5. *Distribution of the Test Statistic* When H_0 is true and the assumptions are met, X^2 is distributed approximately as χ^2 with 1 degree of freedom.

6. *Decision Rule* Reject H_0 if the computed value of χ^2 is ≥ 3.841 (since $\alpha = .05$).

7. *Calculation of Test Statistic* The first step in calculating the test statistic is to compute the common median of the two samples combined. This is done by arranging the observations in ascending order and, since the total number of observations is even, obtaining the mean of the two middle numbers. For our example the median is $(33 + 34)/2 = 33.5$.

We now determine for each group the number of observations falling above and below the common median. The resulting frequencies are arranged in a 2×2 table. For the present example we obtain Table 11.4.2.

If the two samples are, in fact, from populations with the same median, we would expect about one half the scores in each sample to be above the combined median and about one half to be below. If the conditions relative to sample size and expected frequencies for a 2×2 contingency table as discussed in Chapter 10 are met, the chi-square test with 1 degree of freedom may be used to test the null hypothesis of equal population medians. For our

TABLE 11.4.2

Level of Mental Health Scores of Junior High School Boys

	Urban	Rural	Total
Number of scores above median	6	8	14
Number of scores below median	10	4	14
Total	16	12	28

example we have, by Formula 10.4.1,

$$X^2 = \frac{28[(6)(4) - (8)(10)]^2}{(16)(12)(14)(14)} = 2.33$$

8. *Statistical Decision*   Since 2.33 < 3.841, the critical value of χ^2 with $\alpha = .05$ and 1 degree of freedom, we are unable to reject the null hypothesis on the basis of these data.

9. *Conclusion*   We conclude that the two samples may have been drawn from populations with equal medians. Since 2.33 < 2.706, we have $p > .10$.

Sometimes one or more observed values will be exactly equal to the computed median and, hence, will fall neither above nor below the median. We note that if $n_1 + n_2$ is odd, at least one value will always be exactly equal to the median. This raises the question of what to do with observations of this kind. Siegel (5) recommends dropping them from the analysis if $n_1 + n_2$ is large and there are only a few values that fall at the combined median, or dichotomizing the scores into those that exceed the median and those that do not, in which case the observations that equal the median will be counted in the second category. Alternative procedures are suggested by Senders (33) and Hays and Winkler (34). The problem is discussed in considerable detail by Bradley (7).

Before we use the chi-square test, it should be determined that the necessary conditions for a 2 × 2 contingency table discussed in Chapter 10 are met. If they are not, a procedure known as Fisher's exact test (35) may be used. This test is discussed and illustrated by Daniel (15).

The median test extends logically to the case where it is desired to test the null hypothesis that $k \geq 3$ samples are from populations with equal medians. For this test a 2 × k contingency table may be constructed by using the frequencies that fall above and below the median computed from combined samples. If conditions as to sample size and expected frequencies are met, X^2 may be computed and compared with the critical χ^2 with $k - 1$ degrees of freedom. Senders (33) and Conover (6) give numerical examples for the case where more than two samples are involved.

EXERCISES

11.4.1   Fifteen patient records from each of two hospitals were reviewed and assigned a score designed to measure level of care. The scores were as follows:

Hospital A: 99, 85, 73, 98, 83, 88, 99, 80, 74, 91, 80, 94, 94, 98, 80

Hospital B: 78, 74, 69, 79, 57, 78, 79, 68, 59, 91, 89, 55, 60, 55, 79

Would you conclude, at the .05 level of significance, that the two population medians are different? Determine the p value.

11.4.2 The following serum albumin values were obtained from 17 normal and 13 hospitalized subjects.

SERUM ALBUMIN (g/100 ml)

Normal Subjects		Hospitalized Subjects	
2.4	3.0	1.5	3.1
3.5	3.2	2.0	1.3
3.1	3.5	3.4	1.5
4.0	3.8	1.7	1.8
4.2	3.9	2.0	2.0
3.4	4.0	3.8	1.5
4.5	3.5	3.5	
5.0	3.6		
2.9			

Would you conclude at the .05 level of significance that the medians of the two populations sampled are different? Determine the p value.

11.5 THE MANN–WHITNEY TEST

The median test discussed in the preceding section does not make full use of all the information present in the two samples when the variable of interest is measured on at least an ordinal scale. By reducing an observation's information content to merely that of whether or not it falls above or below the common median is a waste of information. If, for testing the desired hypothesis, there is available a procedure that makes use of more of the information inherent in the data, that procedure should be used if possible. Such a nonparametric procedure that can often be used instead of the median test is the Mann–Whitney test (36). Since this test is based on the ranks of the observations it utilizes more information than does the median test.

The assumptions underlying the Mann–Whitney test are as follows:

1. The two samples, of size n and m, respectively, available for analysis have been independently and randomly drawn from their respective populations.

2. The measurement scale is at least ordinal.

3. If the populations differ at all, they differ only with respect to their medians.

When these assumptions are met we may test the null hypothesis that the two populations have equal medians against either of the three possible alternatives: (1) the populations do not have equal medians (two-sided test), (2) the median of

TABLE 11.5.1

Hemoglobin Determinations (grams) for 25 Laboratory Animals

Exposed animals (X)	Unexposed animals (Y)
14.4	17.4
14.2	16.2
13.8	17.1
16.5	17.5
14.1	15.0
16.6	16.0
15.9	16.9
15.6	15.0
14.1	16.3
15.3	16.8
15.7	
16.7	
13.7	
15.3	
14.0	

population 1 is larger than the median of population 2 (one-sided test), or (3) the median of population 1 is smaller than the median of population 2 (one-sided test). If the two populations are symmetric, so that within each population the mean and median are the same, the conclusions we reach regarding the two population medians will also apply to the two population means. The following example illustrates the use of the Mann–Whitney test.

Example 11.5.1 A researcher designed an experiment to assess the effects of prolonged inhalation of cadmium oxide.

1. *Data* Fifteen laboratory animals served as experimental subjects, while 10 similar animals served as controls. The variable of interest was hemoglobin level following the experiment. The results are shown in Table 11.5.1. We wish to know if we can conclude that prolonged inhalation of cadmium oxide reduces hemoglobin level.

2. *Assumptions* We presume that the assumptions of the Mann–Whitney test are met.

3. *Hypotheses* The null and alternative hypotheses are as follows:

$$H_0: M_X \geq M_Y$$

$$H_A: M_X < M_Y$$

TABLE 11.5.2

Original Data and Ranks, Example 11.5.1

X	Rank	Y	Rank
13.7	1		
13.8	2		
14.0	3		
14.1	4.5		
14.1	4.5		
14.2	6		
14.4	7		
		15.0	8.5
		15.0	8.5
15.3	10.5		
15.3	10.5		
15.6	12		
15.7	13		
15.9	14		
		16.0	15
		16.2	16
		16.3	17
16.5	18		
16.6	19		
16.7	20		
		16.8	21
		16.9	22
		17.1	23
		17.4	24
		17.5	25
Total		$\overline{145}$	

where M_X is the median of a population of animals exposed to cadmium oxide and M_Y is the median of a population of animals not exposed to the substance. Suppose we let $\alpha = .05$.

4. *Test Statistic*   To compute the test statistic we combine the two samples and rank all observations from smallest to largest while keeping track of the sample to which each observation belongs. Tied observations are assigned a rank equal to the mean of the rank positions for which they are tied. The results of this step are shown in Table 11.5.2.

The test statistic is

$$T = S - \frac{n(n+1)}{2} \tag{11.5.1}$$

where n is the number of sample X observations and S is the sum of the ranks assigned to the sample observations from the population of X values. The choice of which sample's values we label X is arbitrary.

5. *Distribution of Test Statistic* Critical values from the distribution of the test statistic are given in Table J for various levels of α.

6. *Decision Rule* If the median of the X population is, in fact, smaller than the median of the Y population, as specified in the alternative hypothesis, we would expect (for equal sample sizes) the sum of the ranks assigned to the observations from the X population to be smaller than the sum of the ranks assigned to the observations from the Y population. The test statistic is based on this rationale in such a way that a sufficiently small value of T will cause rejection of H_0: $M_X \geq M_Y$.

For our example, the decision rule is:

Reject H_0: $M_X \geq M_Y$ if the computed T is less than w_α, where w_α is the critical value of T obtained by entering Appendix Table J with n, the number of X observations; m, the number of Y observations; and α, the chosen level of significance.

If we use the Mann–Whitney procedure to test

$$H_0: M_X \leq M_Y$$

against

$$H_A: M_X > M_Y$$

sufficiently large values of T will cause rejection so that the decision rule is:

Reject H_0: $M_X \leq M_Y$ if computed T is greater than $w_{1-\alpha}$, where $W_{1-\alpha} = nm - W_\alpha$.

For the two-sided test situation with

$$H_0: M_X = M_Y$$
$$H_A: M_X \neq M_Y$$

Computed values of T that are either sufficiently large or sufficiently small will cause rejection of H_0: the decision rule for this case, then, is:

Reject H_0: $M_X = M_Y$ if the computed value of T is either less than $w_{\alpha/2}$ or greater than $w_{1-\alpha/2}$, where $w_{\alpha/2}$ is the critical value of T for n, m, and $\alpha/2$ given in Appendix Table J, and $w_{1-\alpha/2} = nm - w_{\alpha/2}$.

7. *Calculation of Test Statistic* For our present example we have, as shown in Table 11.5.2, $S = 145$, so that

$$T = 145 - \frac{15(15 + 1)}{2} = 25$$

8. *Statistical Decision* When we enter Table J with $n = 15$, $m = 10$, and $\alpha = .05$, we find the critical value of w_α to be 45. Since $25 < 45$, we reject H_0.

9. *Conclusion* We conclude that M_X is smaller than M_Y. This leads to the conclusion that prolonged inhalation of cadmium oxide does reduce the hemoglobin level.

Since $22 < 25 < 30$, we have for this test $.005 > p > .001$.

When either n or m is greater than 20 we cannot use Appendix Table J to obtain critical values for the Mann–Whitney test. When this is the case we may compute

$$z = \frac{T - mn/2}{\sqrt{nm(n + m + 1)/12}} \tag{11.5.2}$$

and compare the result, for significance, with critical values of the unit normal distribution.

If a large proportion of the observations are tied, a correction factor proposed by Noether (37) may be used.

EXERCISES

11.5.1 The following are the antibody responses of subjects receiving a booster dose of one of two types of rabies vaccine.

Type 1 (X):   1.25, 5.30, 1.70, 1.00, 8.50, 3.75, 8.10, 2.25, 5.6, 7.85

Type 2 (Y):   .57, 3.90, 8.20, 1.20, 1.70, 1.00, 4.55, 5.20, 2.16, 1.90, 4.6

Can we conclude on the basis of these data that the two types of vaccine differ in their effects? Let $\alpha = .05$, and find the p value.

11.5.2 The following are the lengths of times (in minutes) spent in the operating room by 20 subjects undergoing the same operative procedure. Eleven of the subjects were patients in hospital A and 9 were in hospital B.

Hospital A: 35, 30, 33, 39, 41, 29, 30, 36, 45, 40, 31

Hospital B: 45, 38, 42, 50, 48, 51, 32, 37, 46

On the basis of these data can one conclude that, for the same operative procedure, patients in hospital B tend to be in the operating room longer than patients in hospital A? Let $\alpha = .01$ and find the p value.

11.5.3 The following weights (kg) were obtained on a sample of 12-year-old male schoolchildren in a developing country. The children resided in two different geographic areas.

Area A	Area B	Area A	Area B
54.1	36.3	58.9	26.8
33.8	31.4	45.7	25.1
56.2	51.9	55.3	29.8
32.0	51.8	50.8	26.1
33.7	46.4	35.2	
58.6	36.2		

Can one conclude from these data that, on the average, 12-year-old males from area A differ in weight from 12-year-old males from area B? Let $\alpha = .05$ and find the p value.

11.6 THE KOLMOGOROV–SMIRNOV GOODNESS-OF-FIT TEST

When one wishes to know how well the distribution of sample data conforms to some theoretical distribution, a test known as the Kolmogorov–Smirnov goodness-of-fit test provides an alternative to the chi-square goodness-of-fit test discussed in Chapter 10. The test gets its name from A. Kolmogorov and N. V. Smirnov, two Russian mathematicians who introduced two closely related tests in the 1930s.

Kolmogorov's work (38) is concerned with the one-sample case as discussed here. Smirnov's work (39) deals with the case involving two samples in which interest centers on testing the hypothesis that the distributions of the two parent populations are identical. The test for the first situation is frequently referred to as the Kolmogorov–Smirnov one-sample test. The test for the two-sample case, commonly referred to as the Kolmogorov–Smirnov two-sample test, will not be discussed here. Those interested in this topic should refer to Conover (6), Gibbons (2, 10), and Daniel (15).

In using the Kolmogorov–Smirnov goodness-of-fit test a comparison is made between some theoretical cumulative distribution function, $F_T(x)$, and a sample cumulative distribution function, $F_S(x)$. The sample is a random sample from a population with unknown cumulative distribution function $F(x)$. It will be recalled (Section 3.2) that a cumulative distribution function gives the probability that X is equal to or less than a particular value, x. That is, by means of the sample cumulative distribution function, $F_S(x)$, we may determine $P(X \leq x)$. If there is close agreement between the theoretical and sample cumulative distributions, the hypothesis that the sample was drawn from the population with the specified cumulative distribution function, $F_T(x)$, is supported. If, however, there is a discrepancy between the theoretical and observed cumulative distribution functions too great to be attributed to chance alone, when H_0 is true, the hypothesis is rejected.

The difference between the theoretical cumulative distribution function, $F_T(x)$, and the sample cumulative distribution function, $F_S(x)$, is measured by the statistic D, which is the greatest vertical distance between $F_S(x)$ and $F_T(x)$. When a two-sided test is appropriate, that is, when the hypotheses are

$$H_0: F(x) = F_T(x) \qquad \text{for all } x \text{ from } -\infty \text{ to } +\infty$$
$$H_A: F(x) \neq F_T(x) \qquad \text{for at least one } x$$

the test statistic is

$$D = \sup_x |F_S(x) - F_T(x)| \tag{11.6.1}$$

which is read, "D equals the supremum, (greatest) over all x, of the absolute value of the difference $F_S(x)$ minus $F_T(x)$."

The null hypothesis is rejected at the α level of significance if the computed value of D exceeds the value shown in Table K for $1 - \alpha$ (two sided) and the sample size n. Tests in which the alternative is one sided are possible. Numerical examples are given by Gibbons (10) and Goodman (40).

The assumptions underlying the Kolmogorov–Smirnov test include the following.

1. The sample is a random sample.
2. The hypothesized distribution $F_T(x)$ is continuous.

Noether (41) has shown that when values of D are based on a discrete theoretical distribution, the test is conservative. When the test is used with discrete data, then, the investigator should bear in mind that the true probability of committing a type I error is, at most equal to α, the stated level of significance. The test is also conservative if one or more parameters have to be estimated from sample data.

Example 11.6.1

1. *Data* Fasting blood glucose determinations made on 36 nonobese, apparently healthy, adult males are shown in Table 11.6.1. We wish to know if we may conclude that these data are not from a normally distributed population with a mean of 80 and a standard deviation of 6.

2. *Assumptions* The sample available is a simple random sample from a continuous population distribution.

3. *Hypotheses* The appropriate hypotheses are

$$H_0: F(x) = F_T(x) \qquad \text{for all } x \text{ from } -\infty \text{ to } +\infty$$
$$H_A: F(x) \neq F_T(x) \qquad \text{for at least one } x$$

Let $\alpha = .05$.

TABLE 11.6.1

Fasting Blood Glucose Values (mg / 100 ml) for 36 Nonobese, Apparently Healthy, Adult Males

75	92	80	80	84	72
84	77	81	77	75	81
80	92	72	77	78	76
77	86	77	92	80	78
68	78	92	68	80	81
87	76	80	87	77	86

4. *Test Statistic*   See Equation 11.6.1.

5. *Distribution of Test Statistic*   Critical values of the test statistic for selected values of α are given in Table K.

6. *Decision Rule*   Reject H_0 if the computed value of D exceeds .221, the critical value of D for $n = 36$ and $\alpha = .05$.

7. *Calculation of Test Statistic*   Our first step is to compute values of $F_S(x)$ as shown in Table 11.6.2.

We obtain values of $F_T(x)$ by first converting each observed value of x to a value of the unit normal variable, z. From Appendix Table C we then find the area between $-\infty$ and z. From these areas we are able to compute values of $F_T(x)$. The procedure, which is similar to that used to obtain expected relative frequencies in the chi-square goodness-of-fit test is summarized in Table 11.6.3.

TABLE 11.6.2

Values of $F_S(x)$ for Example 11.6.1

x	Frequency	Cumulative frequency	$F_S(x)$
68	2	2	.0556
72	2	4	.1111
75	2	6	.1667
76	2	8	.2222
77	6	14	.3889
78	3	17	.4722
80	6	23	.6389
81	3	26	.7222
84	2	28	.7778
86	2	30	.8333
87	2	32	.8889
92	4	36	1.0000
	$\overline{36}$		

TABLE 11.6.3

Steps in Calculation of $F_T(x)$ for Example 11.6.1

x	$z = (x - 80)/6$	$F_T(x)$
68	-2.00	.0228
72	-1.33	.0918
75	$-.83$	.2033
76	$-.67$	.2514
77	$-.50$	.3085
78	$-.33$	.3707
80	$.00$	.5000
81	$.17$	.5675
84	$.67$	.7486
86	1.00	.8413
87	1.17	.8790
92	2.00	.9772

The test statistic D may be computed algebraically, or it may be determined graphically by actually measuring the largest vertical distance between the curves of $F_S(x)$ and $F_T(x)$ on a graph. The graphs of the two distributions are shown in Figure 11.6.1.

Examination of the graphs of $F_S(x)$ and $F_T(x)$ reveals that $D \approx .16 = (.72 - .56)$. Now let us compute the value of D algebraically. The possible values of $|F_S(x) - F_T(x)|$ are shown in Table 11.6.4. This table shows that the exact value of D is .1547.

8. *Statistical Decision* Reference to Table K reveals that a computed D of .1547 is not significant at any reasonable level. Therefore, we are not willing to reject H_0.

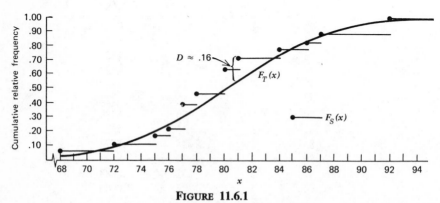

FIGURE 11.6.1

$F_S(x)$ and $F_T(x)$ for Example 11.6.1

TABLE 11.6.4

Calculation of $|F_S(x) - F_T(x)|$ for Example 11.6.1.

x	$F_S(x)$	$F_T(x)$	$|F_S(x) - F_T(x)|$
68	.0556	.0228	.0328
72	.1111	.0918	.0193
75	.1667	.2033	.0366
76	.2222	.2514	.0292
77	.3889	.3085	.0804
78	.4722	.3707	.1015
80	.6389	.5000	.1389
81	.7222	.5675	.1547
84	.7778	.7486	.0292
86	.8333	.8413	.0080
87	.8889	.8790	.0099
92	1.0000	.9772	.0228

9. *Conclusion* The sample may have come from the specified distribution. Since we have a two-sided test, and since .1547 < .174, we have $p > .20$.

The reader should be aware that in determining the value of D *it is not always sufficient to compute and choose from the possible values of $|F_S(x) - F_T(x)|$. The largest vertical distance between $F_S(x)$ and $F_T(x)$ may not occur at an observed value, x, but at some other value of X.* Such a situation is illustrated in Figure 11.6.2. We see that if only values of $|F_S(x) - F_T(x)|$ at the left endpoints of the horizontal bars are considered we would incorrectly compute D as $|.2 - .4| = .2$. One can see by examining the graph, however, that the largest vertical distance between $F_S(x)$ and $F_T(x)$ occurs at the right endpoint of the horizontal bar originating at the point corresponding to $x = .4$, and the correct value of D is $|.5 - .2| = .3$.

One can determine the correct value of D algebraically by computing, in addition to the differences $|F_S(x) - F_T(x)|$, the differences $F_S(x_{i-1}) - F_T(x_i)|$ for all values of $i = 1, 2, \ldots, r + 1$, where r = the number of different values of x and $F_S(x_0) = 0$. The correct value of the test statistic will then be

$$D = \underset{1 \le i \le r}{\text{maximum}} \left\{ \text{maximum} \left[|F_S(x_i) - F_T(x_i)|, |F_S(x_{i-1}) - F_T(x_i)|\right] \right\}$$

$$(11.6.2)$$

The advantages and disadvantages of the Kolmogorov–Smirnov goodness-of-fit test in comparison with the chi-square test have been discussed by Goodman

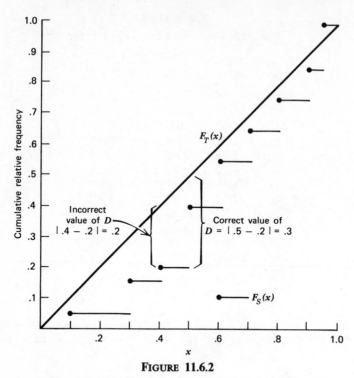

FIGURE 11.6.2

Graph of Fictitious Data Showing Correct Calculation of D

(40), Massey (42), Birnbaum (43), and Slakter (44). The following are some important points of comparison.

1. The Kolmogorov–Smirnov test does not require that the observations be grouped as is the case with the chi-square test. The consequence of this difference is that the Kolmogorov–Smirnov test makes use of all the information present in a set of data.

2. The Kolmogorov–Smirnov test can be used with any size sample. It will be recalled that certain minimum sample sizes are required for the use of the chi-square test.

3. As has been noted the Kolmogorov–Smirnov test is not applicable when parameters have to be estimated from the sample. The chi-square test may be used in these situations by reducing the degrees of freedom by 1 for each parameter estimated.

4. The problem of the assumption of a continuous theoretical distribution has already been mentioned.

EXERCISES

11.6.1 The weights at autopsy of the brains of 25 adults suffering from a certain disease were as follows.

Weight of brain (grams)

859	1073	1041	1166	1117
962	1051	1064	1141	1202
973	1001	1016	1168	1255
904	1012	1002	1146	1233
920	1039	1086	1140	1348

Can one conclude from these data that the sampled population is not normally distributed with a mean of 1050 and a standard deviation of 50? Determine the p value for this test.

11.6.2 IQs of a sample of 30 adolescents arrested for drug abuse in a certain metropolitan jurisdiction were as follows.

IQ

95	100	91	106	109	110
98	104	97	100	107	119
92	106	103	106	105	112
101	91	105	102	101	110
101	95	102	104	107	118

Do these data provide sufficient evidence that the sampled population of IQ scores is not normally distributed with a mean of 105 and a standard deviation of 10? Determine the p value.

11.6.3 For a sample of apparently normal subjects who served as controls in an experiment, the following systolic blood pressure readings were recorded at the beginning of the experiment.

162	177	151	167
130	154	179	146
147	157	141	157
153	157	134	143
141	137	151	161

Can one conclude on the basis of these data that the population of blood pressures from which the sample was drawn is not normally distributed with $\mu = 150$ and $\sigma = 12$? Determine the p value.

11.7 THE KRUSKAL–WALLIS ONE-WAY ANALYSIS OF VARIANCE BY RANKS

In Chapter 7 we discuss how one-way analysis of variance may be used to test the null hypothesis that several population means are equal. When the assumptions underlying this technique are not met, that is, when the populations from which the samples are drawn are not normally distributed with equal variances, or when the data for analysis consist only of ranks, a nonparametric alternative to the one-way analysis of variance may be used to test the hypothesis of equal location parameters. As was pointed out in Section 11.4, the median test may be extended to accommodate the situation involving more than two groups. A deficiency of this test, however, is the fact that it uses only a small amount of the information available, that is, whether or not the observations are above or below a single number, the median of the combined samples. Several nonparametric analogs to analysis of variance are available that use more information by taking into account the magnitude of each observation relative to the magnitude of every other observation. Perhaps the best known of these procedures is the Kruskal–Wallis one-way analysis of variance by ranks (45).

The application of the test involves the following steps.

1. The $n_1, n_2, \ldots, n_k$ observations from the k groups are combined into a single series of size n and arranged in order of magnitude from smallest to largest. The observations are then replaced by ranks from 1, which is assigned to the smallest observation, to n, which is assigned to the largest observation. When two or more observations have the same value, each observation is given the mean of the ranks for which it is tied.

2. The ranks assigned to observations in each of the k groups are added separately to give k rank sums.

3. The test statistic

$$H = \frac{12}{n(n+1)} \sum_{j=1}^{k} \frac{R_j^2}{n_j} - 3(n+1) \qquad (11.7.1)$$

is computed.

In Equation 11.7.1

$$k = \text{the number of groups}$$
$$n_j = \text{the number of observations in the } j\text{th group}$$
$$n = \text{the number of observations in all groups combined}$$
$$R_j = \text{the sum of the ranks in the } j\text{th group}$$

4. When there are three groups and five or fewer observations in each group, the significance of the computed H is determined by consulting Appendix Table

TABLE 11.7.1

Reaction Time in Seconds of 13
Experimental Animals

	GROUP	
I	*II*	*III*
17	8	2
20	7	5
40	9	4
31	8	3
35		

L. When there are more than five observations in one or more of the groups, H is compared with tabulated values of χ^2 with $k - 1$ degrees of freedom. The adequacy of the chi-square approximation for small samples is discussed by Gabriel and Lachenbruch (46).

Example 11.7.1

1. *Data*   The effects of two drugs on reaction time to a certain stimulus were studied in three groups of experimental animals. Group III served as a control while the animals in group I were treated with drug A and those in group II were treated with drug B prior to the application of the stimulus. Table 11.7.1 shows the reaction times in seconds of the 13 animals.

    Can we conclude that the three populations represented by the three samples differ with respect to reaction time? We can so conclude if we can reject the null hypothesis that the three populations do not differ in their reaction times.

2. *Assumptions*   The samples are independent random samples from their respective populations. The measurement scale employed is at least ordinal. The distributions of the values in the sampled populations are identical except for the possibility that one or more of the populations are composed of values that tend to be larger than those of the other populations.

3. *Hypotheses*

    H_0:   The population distributions are all identical.

    H_A: At least one of the populations tends to exhibit
    larger values than at least one of the other populations.

    Let $\alpha = .01$.

4. *Test Statistic*   See Equation 11.7.1.

TABLE 11.7.2

The Data of Table 11.7.1 Replaced by Ranks

	GROUP	
I	*II*	*III*
9	6.5	1
10	5	4
13	8	3
11	6.5	2
12		
$R_1 = 55$	$R_2 = 26$	$R_3 = 10$

5. *Distribution of Test Statistic*   Critical values of H for various sample sizes and α levels are given in Table L.

6. *Decision Rule*   The null hypothesis will be rejected if the computed value of H is so large that the probability of obtaining a value that large or larger when H_0 is true is equal to or less than the chosen significance level, α.

7. *Calculation of Test Statistic*   When the three samples are combined into a single series and ranked, the table of ranks shown in Table 11.7.2 may be constructed.

   The null hypothesis implies that the observations in the three samples constitute a single sample of size 13 from a single population. If this is true, we would expect the ranks to be well distributed among the three groups. Consequently, we would expect the total sum of ranks to be divided among the three groups in proportion to group size. Departures from these conditions are reflected in the magnitude of the test statistic H.

   From the data in Table 11.7.2 and Equation 11.7.1 we obtain

$$H = \frac{12}{13(13 + 1)} \left[\frac{(55)^2}{5} + \frac{(26)^2}{4} + \frac{(10)^2}{4} \right] - 3(13 + 1)$$
$$= 10.68$$

8. *Statistical Decision*   Table L shows that when the n_j's are 5, 4, and 4, the probability of obtaining a value of $H \geq 10.68$ is less than .009. The null hypothesis can be rejected at the .01 level of significance.

9. *Conclusion*   We conclude that there is a difference in the average reaction time among the three populations. For this test, $p < .009$.

It will be noted that the two tied values in sample II were each assigned the rank of 6.5. We may adjust the value of H for this tie by dividing it by

$$1 - \frac{\sum T}{n^3 - n} \tag{11.7.2}$$

where $T = t^3 - t$. The letter t is used to designate the number of tied observations in a group of tied values. In our present example there is only one group of tied values but, in general, there may be several groups of tied values resulting in several values of T. Since there were only two tied observations in our one group of ties, we have $T = 2^3 - 2 = 6$ and $\sum T = 6$, so that Expression 11.7.2 is

$$1 - \frac{6}{13^3 - 13} = .9973$$

and

$$\frac{H}{1 - \dfrac{\sum T}{n^3 - n}} = \frac{10.68}{.9973} = 10.71$$

which, of course, is also significant at the .01 level.

As is the case here, the effect of the adjustment for ties is usually negligible. Note also that the effect of the adjustment is to increase H, so that if the unadjusted H is significant at the chosen level, there is no need to apply the adjustment.

Now let us illustrate the procedure when there are more than three groups and, at least, one of the n_j's is greater than 5.

Example 11.7.2   Table 11.7.3 shows the net book value of equipment capital per bed for a sample of hospitals from each of five types of hospitals. We wish to determine, by means of the Kruskal–Wallis test, if we can conclude that the average net book value of equipment capital per bed differs among the five types

TABLE 11.7.3
Net Book Value of Equipment per Bed by Hospital Type

		TYPE HOSPITAL		
A	B	C	D	E
$1735(11)	$5260(35)	$2790(20)	$3475(26)	$6090(40)
1520(2)	4455(28)	2400(12)	3115(22)	6000(38)
1476(1)	4480(29)	2655(16)	3050(21)	5894(37)
1688(7)	4325(27)	2500(13)	3125(23)	5705(36)
1702(10)	5075(32)	2755(19)	3275(24)	6050(39)
2667(17)	5225(34)	2592(14)	3300(25)	6150(41)
1575(4)	4613(30)	2601(15)	2730(18)	5110(33)
1602(5)	4887(31)	1648(6)		
1530(3)		1700(9)		
1698(8)				
$R_1 = 68$	$R_2 = 246$	$R_3 = 124$	$R_4 = 159$	$R_5 = 264$

of hospital. The ranks of the 41 values, along with the sum of ranks for each group, are shown in the table.

From the sums of the ranks we compute

$$H = \frac{12}{41(41 + 1)} \left[\frac{(68)^2}{10} + \frac{(246)^2}{8} + \frac{(124)^2}{9} + \frac{(159)^2}{7} + \frac{(264)^2}{7} \right]$$

$$- 3(41 + 1)$$

$$= 36.39$$

Reference to Table F with $k-1 = 4$ degrees of freedom indicates that the probability of obtaining a value of H as large or larger than 36.39, due to chance alone, when there is no difference among the groups is less than .005. We conclude, then, that there is a difference among the five populations with respect to the average value of the variable of interest.

EXERCISES

For the following exercises, perform the test at the indicated level of significance and determine the p value.

11.7.1 The following table shows the pesticide residue levels (ppb) in blood samples from four populations of human subjects. Use the Kruskal–Wallis test to test at the .05 level of significance the null hypothesis that there is no difference among the populations with respect to average level of pesticide residue.

	POPULATION		
A	B	C	D
10	4	15	7
37	35	5	11
12	32	10	10
31	19	12	8
11	33	6	2
9	18	6	5
44	11	9	4
12	7	11	5
15	32	9	2
42	17	14	6
23	8	15	3

11.7.2 The following are average daily charges made to patients hospitalized for a certain surgical procedure by samples of hospitals located in three different areas of the country.

	AREA	
I	*II*	*III*
$80.75	$58.63	$84.21
78.15	72.70	101.76
85.40	64.20	107.74
71.94	62.50	115.30
82.05	63.24	126.15

Can we conclude at the .05 level of significance that the three areas differ with respect to the average daily charges?

11.7.3 Refer to the data in Table 7.2.1 (Example 7.2.1) and apply the Kruskal–Wallis test. Compare the results with those obtained when the F test was used.

11.7.4 Refer to Exercise 7.2.1 and apply the Kruskal–Wallis Test.

11.7.5 Refer to Exercise 7.2.2 and apply the Kruskal–Wallis Test.

11.7.6 Refer to Exercise 7.2.3 and apply the Kruskal–Wallis Test.

11.7.7 Refer to Exercise 7.2.4 and apply the Kruskal–Wallis Test.

11.7.8 Hepatic γ-glutamyl transpeptidase (GGTP) activity was measured in 22 patients undergoing percutaneous liver biopsy. The results were as follows:

Subject	Diagnosis	Hepatic GGTP (μ/g protein)
1	Normal liver	27.7
2	Primary biliary cirrhosis	45.9
3	Alcoholic liver disease	85.3
4	Primary biliary cirrhosis	39.0
5	Normal liver	25.8
6	Persistent hepatitis	39.6
7	Chronic active hepatitis	41.8
8	Alcoholic liver disease	64.1
9	Persistent hepatitis	41.1
10	Persistent hepatitis	35.3
11	Alcoholic liver disease	71.5
12	Primary biliary cirrhosis	40.9
13	Normal liver	38.1
14	Primary biliary cirrhosis	40.4
15	Primary biliary cirrhosis	34.0
16	Alcoholic liver disease	74.4
17	Alcoholic liver disease	78.2
18	Persistent hepatitis	32.6
19	Chronic active hepatitis	46.3
20	Normal liver	39.6
21	Chronic active hepatitis	52.7
22	Chronic active hepatitis	57.2

Can we conclude from these sample data that the average population GGTP level differs among the five diagnostic groups? Let $\alpha = .05$ and find the p value.

11.8 THE FRIEDMAN TWO-WAY
ANALYSIS OF VARIANCE BY RANKS

Just as we may on occasion have need of a nonparametric analog to the parametric one-way analysis of variance, we may also find it necessary to analyze the data in a two-way classification by nonparametric methods analogous to the two-way analysis of variance. Such a need may arise because the assumptions necessary for parametric analysis of variance are not met, because the measurement scale employed is weak, or because results are needed in a hurry. A test frequently employed under these circumstances is the Friedman two-way analysis of variance by ranks (47, 48). This test is appropriate whenever the data are measured on, at least, an ordinal scale and can be meaningfully arranged in a two-way classification as is given for the randomized block experiment discussed in Chapter 7. The following example illustrates this procedure.

Example 11.8.1 A physical therapist conducted a study to compare three models of low-volt electrical stimulators.

1. *Data*   Nine physical therapists were asked to rank them in order of preference. A rank of 1 indicates first preference. The results are shown in Table 11.8.1.

    We wish to know if we can conclude that the models are not preferred equally.

2. *Assumptions*   The observations appearing in a given block are independent of the observations appearing in each of the other blocks, and within each block measurement on at least an ordinal scale is achieved.

TABLE 11.8.1

Physical Therapists' Rankings of Three Models of Low-Volt Electrical Stimulators

	MODEL		
Therapist	*A*	*B*	*C*
1	2	3	1
2	2	3	1
3	2	3	1
4	1	3	2
5	3	2	1
6	1	2	3
7	2	3	1
8	1	3	2
9	1	3	2
R_j	15	25	14

3. *Hypotheses*   In general, the hypotheses are

H_0:   The treatments all have identical effects.

H_A: At least one treatment tends to yield larger observations than at least one of the other treatments.

For our present example we state the hypotheses as follows:

H_0:   The three models are equally preferred.

H_A: The three models are not equally preferred.

Let $\alpha = .05$.

4. *Test Statistic*   By means of the Friedman test we will be able to determine if it is reasonable to assume that the columns of ranks have been drawn from the same population. If the null hypothesis is true we would expect the observed distribution of ranks within any column to be the result of chance factors and, hence, we would expect the numbers 1, 2, and 3 to occur with approximately the same frequency in each column. If, on the other hand, the null hypothesis is false (that is, the models are not equally preferred), we would expect a preponderance of relatively high (or low) ranks in at least one column. This condition would be reflected in the sums of the ranks. The Friedman test will tell us whether or not the observed sums of ranks are so discrepant that it is not likely they are a result of chance when H_0 is true.

Since the data already consist of rankings within blocks (rows), our first step is to sum the ranks within each column (treatment). These sums are the R_j's shown in Table 11.8.1. A test statistic, denoted by Friedman as χ_r^2 is computed as follows:

$$\chi_r^2 = \frac{12}{nk(n+1)} \sum_{j=1}^{k} (R_j)^2 - 3n(k+1) \qquad \textbf{(11.8.1)}$$

where n = the number of rows (blocks) and k = the number of columns (treatments).

5. *Distribution of Test Statistic*   Critical values for various values of n and k are given in Table M.

6. *Decision Rule*   Reject H_0 if the probability of obtaining (when H_0 is true) a value of χ_r^2 as large or larger than actually computed is less than or equal to α.

7. *Calculation of Test Statistic*   Using the data in Table 11.8.1 and Equation 11.8.1, we compute

$$\chi_r^2 = \frac{12}{9(3)(3+1)} \left[(15)^2 + (25)^2 + (14)^2 \right] - 3(9)(3+1)$$

$$= 8.222$$

TABLE 11.8.2

Percent Decrease in Salivary Flow of Experimental Animals Following Different Dose Levels of Atropine

Animal number	DOSE LEVEL			
	A	B	C	D
1	29(1)	48(2)	75(3)	100(4)
2	72(2)	30(1)	100(3.5)	100(3.5)
3	70(1)	100(4)	86(2)	96(3)
4	54(2)	35(1)	90(3)	99(4)
5	5(1)	43(3)	32(2)	81(4)
6	17(1)	40(2)	76(3)	81(4)
7	74(1)	100(3)	100(3)	100(3)
8	6(1)	34(2)	60(3)	81(4)
9	16(1)	39(2)	73(3)	79(4)
10	52(2)	34(1)	88(3)	96(4)
11	8(1)	42(3)	31(2)	79(4)
12	29(1)	47(2)	72(3)	99(4)
13	71(1)	100(3.5)	97(2)	100(3.5)
14	7(1)	33(2)	58(3)	79(4)
15	68(1)	99(4)	84(2)	93(3)
16	70(2)	30(1)	99(3.5)	99(3.5)
R_j	20	36.5	44	59.5

8. *Statistical Decision*   When we consult Appendix Table Ma, we find that the probability of obtaining a value of χ_r^2 as large as 8.222 due to chance alone, when the null hypothesis is true, is .016. We are able, therefore, to reject the null hypothesis.

9. *Conclusion*   We conclude that the three models of low-volt electrical stimulator are not equally preferred. For this test, $p = .016$.

When the original data consist of measurements on an interval or a ratio scale instead of ranks, the measurements are assigned ranks based on their relative magnitudes within blocks. If ties occur each value is assigned the mean of the ranks for which it is tied.

When the values of k and or n exceed those given in Table M of the Appendix, the critical value of χ_r^2 is obtained by consulting the χ^2 table (Table F) with the chosen α and $k - 1$ degrees of freedom.

Example 11.8.2   Table 11.8.2 shows the responses, in percent decrease in salivary flow, of 16 experimental animals following different dose levels of atropine. The ranks (in parentheses) and the sum of the ranks are also given in the table. We wish to see if we may conclude that the different dose levels produce different responses. That is, we test the null hypothesis of no difference in response among the four dose levels.

From the data we compute

$$\chi_r^2 = \frac{12}{16(4)(4+1)} \left[(20)^2 + (36.5)^2 + (44)^2 + (59.5)^2 \right] - 3(16)(4+1)$$

$$= 30.32$$

Reference to Table F indicates that with $k - 1 = 3$ degrees of freedom the probability of getting a value of χ_r^2 as large as 30.32 due to change alone is, when H_0 is true, less than .005. We reject the null hypothesis and conclude that the different dose levels do produce different responses.

EXERCISES

For the following exercises perform the test at the indicated level of significance and determine the p value.

11.8.1 The following table shows the scores made by nine randomly selected student nurses on final examinations in three subject areas.

Student number	SUBJECT AREA Fundamentals	Physiology	Anatomy
1	98	95	77
2	95	71	79
3	76	80	91
4	95	81	84
5	83	77	80
6	99	70	93
7	82	80	87
8	75	72	81
9	88	81	83

Test the null hypothesis that student nurses constituting the population from which the above sample was drawn perform equally well in all three subject areas against the alternative hypothesis that they perform better in, at least, one area. Let $\alpha = .05$.

11.8.2 Fifteen randomly selected physical therapy students were given the following instructions: "Assume that you will marry a person with one of the following handicaps (the handicaps were listed and designated by the letters A to J). Rank these handicaps from 1 to 10 according to your first, second, third (and so on) choice of a handicap for your marriage partner." The results are shown in the following table.

Student	HANDICAP									
number	A	B	C	D	E	F	G	H	I	J
1	1	3	5	9	8	2	4	6	7	10
2	1	4	5	7	8	2	3	6	9	10
3	2	3	7	8	9	1	4	6	5	10
4	1	4	7	8	9	2	3	6	5	10
5	1	4	7	8	10	2	3	6	5	9
6	2	3	7	9	8	1	4	5	6	10
7	2	4	6	9	8	1	3	7	5	10
8	1	5	7	9	10	2	3	4	6	8
9	1	4	5	7	8	2	3	6	9	10
10	2	3	6	8	9	1	4	7	5	10
11	2	4	5	8	9	1	3	7	6	10
12	2	3	6	8	10	1	4	5	7	9
13	3	2	6	9	8	1	4	7	5	10
14	2	5	7	8	9	1	3	4	6	10
15	2	3	6	7	8	1	5	4	9	10

Test the null hypothesis of no preference for handicaps against the alternative that some handicaps are preferred over others. Let $\alpha = .05$.

11.8.3 Refer to Example 7.3.1. Work the example using the Friedman test and compare the results with those obtained using the parametric two-way analysis of variance.

11.8.4 Refer to Exercise 7.3.1. Work this exercise by using the Friedman test.

11.8.5 Use the Friedman test to work Exercise 7.3.2.

11.8.6 Use the Friedman test to work Exercise 7.3.3.

11.8.7 Use the Friedman test to work Exercise 7.3.4.

11.8.8 Ten subjects with exercise-induced asthma participated in an experiment to compare the protective effect of a drug administered in four dose levels. Saline was used as a control. The variable of interest was change in FEV_1 after administration of the drug or saline. The results were as follows.

Subject	Saline	DOSE LEVEL OF DRUG (mg/ml)			
		2	10	20	40
1	−.68	−.32	−.14	−.21	−.32
2	−1.55	−.56	−.31	−.21	−.16
3	−1.41	−.28	−.11	−.08	−.83
4	−.76	−.56	−.24	−.41	−.08
5	−.48	−.25	−.17	−.04	−.18
6	−3.12	−1.99	−1.22	−.55	−.75
7	−1.16	−.88	−.87	−.54	−.84
8	−1.15	−.31	−.18	−.07	−.09
9	−.78	−.24	−.39	−.11	−.51
10	−2.12	−.35	−.28	+.11	−.41

Can one conclude on the basis of these data that different dose levels have different effects? Let $\alpha = .05$ and find the p value.

11.9 THE SPEARMAN RANK CORRELATION COEFFICIENT

Several nonparametric measures of correlation are available to the researcher. Refer to Kendall (49), Kruskal (50), and Hotelling and Pabst (51) for detailed discussions of the various methods.

A frequently used procedure that is attractive because of the simplicity of the calculations involved is due to Spearman (52). The measure of correlation computed by this method is called the Spearman rank correlation coefficient and is designated by r_s. This procedure makes use of the two sets of ranks which may be assigned to the sample values of X and Y, the independent and continuous variables of a bivariate distribution. The usually tested hypotheses and their alternatives are as follows.

(1)　H_0: X and Y are mutually independent.
　　 H_A: X and Y are not mutually independent.

(2)　H_0: X and Y are mutually independent.
　　 H_A: There is a tendency for large values of X and large
　　　　 values of Y to be paired together.

(3)　H_0: X and Y are mutually independent.
　　 H_A: There is a tendency for large values of X to be
　　　　 paired with small values of Y.

The hypotheses specified in no. 1 lead to a two-sided test and are used when it is desired to detect any departure from independence. The one-sided tests indicated by nos. 2 and 3 are used, respectively, when investigators wish to know if they can conclude that the variables are directly or inversely correlated.

The hypothesis-testing procedure involves the following steps.

1. Rank the values of X from 1 to n (the number of pairs of values of X and Y in the sample). Rank the values of Y from 1 to n.

2. Compute d_i for each pair of observations by subtracting the rank of Y_i from the rank of X_i.

3. Square each d_i and compute $\sum d_i^2$, the sum of the squared values.

4. Compute

$$r_s = 1 - \frac{6 \sum d_i^2}{n(n^2 - 1)} \tag{11.9.1}$$

5. If n is between 4 and 30 compare the computed value of r_s with the critical values, r_s^*, of Table N. For the two-sided test, H_0 is rejected at the α significance level if r_s is greater than r_s^* or less than $-r_s^*$ where r_s^* is at the intersection of the column headed $\alpha/2$ and the row corresponding to n. For the one-sided test with H_A specifying direct correlation, H_0 is rejected at the α significance level if r_s is greater than r_s^* for α and n. The null hypothesis is rejected at the α significance level in the other one-sided test if r_s is less than $-r_s^*$ for α and n.

6. If n is greater than 30, one may compute

$$z = r_s\sqrt{n - 1} \tag{11.9.2}$$

and use Table C to obtain critical values.

7. Tied observations present rather a problem. Glasser and Winter (53) point out that the use of Table N is strictly valid only when the data do not contain any ties (unless some random procedure for breaking ties is employed). In practice, however, the table is frequently used after some other method for handling ties has been employed. If the number of ties is large, a correction for ties found in Siegel (5) may be employed. The correction factor for ties is

$$T = \frac{t^3 - t}{12} \tag{11.9.3}$$

where t = the number of observations that are tied for some particular rank. When this correction factor is used r_s is computed from

$$r_s = \frac{\sum x^2 + \sum y^2 - \sum d_i^2}{2\sqrt{\sum x^2 \sum y^2}} \tag{11.9.4}$$

instead of from Equation 11.9.1.

In Equation 11.9.4

$$\sum x^2 = \frac{n^3 - n}{12} - \sum T_x$$

$$\sum y^2 = \frac{n^3 - n}{12} - \sum T_y$$

$T_x =$ the sum of the values of T for the various tied ranks in X, and

$T_y =$ the sum of the values of T for the various tied ranks in Y

Gibbons (2), as well as Siegel (5), points out that unless the number of ties is excessive the correction makes very little difference in the value of r_s. When the number of ties is small, we can follow the usual procedure of assigning

TABLE 11.9.1

Age and EEG Output Value for 20 Subjects

Subject number	Age (X)	EEG output value (Y)
1	20	98
2	21	75
3	22	95
4	24	100
5	27	99
6	30	65
7	31	64
8	33	70
9	35	85
10	38	74
11	40	68
12	42	66
13	44	71
14	46	62
15	48	69
16	51	54
17	53	63
18	55	52
19	58	67
20	60	55

the tied observations the mean of the ranks for which they are tied and proceed with steps 2 to 6 above. Edgington (54) discusses the problem of ties in some detail.

Example 11.9.1

1. *Data* In a study of the relationship between age and the EEG, data were collected on 20 subjects between the ages of 20 and 60 years. Table 11.9.1 shows the age and a particular EEG output value for each of the 20 subjects. The investigator wishes to know if it can be concluded that this particular EEG output is inversely correlated with age.

2. *Assumptions* We presume that the sample available for analysis is a simple random sample and that both X and Y are measured on at least the ordinal scale.

3. *Hypotheses*

   H_0: This EEG output and age are mutually independent.
   H_A: There is a tendency for this EEG output to decrease with age.

Suppose we let $\alpha = .05$.

4. *Test Statistic*   See Equation 11.9.1.

5. *Distribution of Test Statistic*   Critical values of the test statistic are given in Table N.

6. *Decision Rule*   For the present test we will reject H_0 if the computed value of r_s is less than $-.3789$.

7. *Calculation of Test Statistic*   When the X and Y values are ranked we have the results shown in Table 11.9.2. The d_i, d_i^2 and $\sum d_i^2$ are shown in the same table.

Substitution of the data from Table 11.9.2 into Equation 11.9.1 gives

$$r_s = 1 - \frac{6(2340)}{20\left[(20)^2 - 1\right]} = -.76$$

8. *Statistical Decision*   Since our computed $r_s = -.76$ is less than the critical r_s^* we reject the null hypothesis.

9. *Conclusion*   We conclude that the two variables are inversely related. Since $-.76 < -.6586$, we have for this test $p < .001$.

TABLE 11.9.2

Ranks for Data of Example 11.9.1

Subject number	Rank (X)	Rank (Y)	d_i	d_i^2
1	1	18	-17	289
2	2	15	-13	169
3	3	17	-14	196
4	4	20	-16	256
5	5	19	-14	196
6	6	7	-1	1
7	7	6	1	1
8	8	12	-4	16
9	9	16	-7	49
10	10	14	-4	16
11	11	10	1	1
12	12	8	4	16
13	13	13	0	0
14	14	4	10	100
15	15	11	4	16
16	16	2	14	196
17	17	5	12	144
18	18	1	17	289
19	19	9	10	100
20	20	3	17	289

$$\sum d_i^2 = 2340$$

Let us now illustrate the procedure for a sample with $n > 30$ and some tied observations.

Example 11.9.2 In Table 11.9.3 are shown the ages and concentrations (ppm) of a certain mineral in the tissue of 35 subjects on whom autopsies were performed as part of a large research project.

The ranks, d_i, d_i^2, and $\sum d_i^2$ are shown in Table 11.9.4. Let us test, at the .05 level of significance, the null hypothesis that X and Y are mutually independent against the two-sided alternative that they are not mutually independent.

From the data in Table 11.9.4 we compute

$$r_s = 1 - \frac{6(1788.5)}{35[35^2 - 1]} = .75$$

To test the significance of r_s we compute

$$z = .75\sqrt{35 - 1} = 4.37$$

Since 4.37 is greater than $z = 3.09$, $p < 2(.001) = .002$, and we reject H_0 and conclude that the two variables under study are not mutually independent.

TABLE 11.9.3

Age and Mineral Concentration (ppm) in Tissue of 35 Subjects

Subject number	Age (X)	Mineral concentration (Y)	Subject number	Age (X)	Mineral concentration (Y)
1	82	169.63	19	50	4.48
2	85	48.94	20	71	46.93
3	83	41.16	21	54	30.91
4	64	63.95	22	62	34.27
5	82	21.09	23	47	41.44
6	53	5.40	24	66	109.88
7	26	6.33	25	34	2.78
8	47	4.26	26	46	4.17
9	37	3.62	27	27	6.57
10	49	4.82	28	54	61.73
11	65	108.22	29	72	47.59
12	40	10.20	30	41	10.46
13	32	2.69	31	35	3.06
14	50	6.16	32	75	49.57
15	62	23.87	33	50	5.55
16	33	2.70	34	76	50.23
17	36	3.15	35	28	6.81
18	53	60.59			

TABLE 11.9.4

Ranks for Data of Example 11.9.2

Subject number	Rank (X)	Rank (Y)	d_i	d_i^2	Subject number	Rank (X)	Rank (Y)	d_i	d_i^2
1	32.5	35	−2.5	6.25	19	17	9	8	64.00
2	35	27	8	64.00	20	28	25	3	9.00
3	34	23	11	121.00	21	21.5	21	.5	.25
4	25	32	−7	49.00	22	23.5	22	1.5	2.25
5	32.5	19	13.5	182.25	23	13.5	24	−10.5	110.25
6	19.5	11	8.5	72.25	24	27	34	−7	49.00
7	1	14	−13	169.00	25	6	3	3	9.00
8	13.5	8	5.5	30.25	26	12	7	5	25.00
9	9	6	3	9.00	27	2	15	−13	169.00
10	15	10	5	25.00	28	21.5	31	−9.5	90.25
11	26	33	−7	49.00	29	29	26	3	9.00
12	10	17	−7	49.00	30	11	18	−7	49.00
13	4	1	3	9.00	31	7	4	3	9.00
14	17	13	4	16.00	32	30	28	2	4.00
15	23.5	20	3.5	12.25	33	17	12	5	25.00
16	5	2	3	9.00	34	31	29	2	4.00
17	8	5	3	9.00	35	3	16	−13	169.00
18	19.5	30	−10.5	110.25					

$$\sum d_i^2 = 1788.5$$

For comparative purposes let us correct for ties using Equation 11.9.3 and then compute r_s by Equation 11.9.4.

In the rankings of X we had six groups of ties that were broken by assigning the values 13.5, 17, 19.5, 21.5, 23.5, and 32.5. In five of the groups two observations tied, and in one group three observations tied. We, therefore, compute five values of

$$T_x = \frac{2^3 - 2}{12} = \frac{6}{12} = .5$$

and one value of

$$T_x = \frac{3^3 - 3}{12} = \frac{24}{12} = 2$$

From these computations, we have $\sum T_x = 5(.5) + 2 = 4.5$, so that

$$\sum x^2 = \frac{35^3 - 35}{12} - 4.5 = 3565.5$$

Since no ties occurred in the Y rankings, we have $\sum T_y = 0$ and

$$\sum y^2 = \frac{35^3 - 35}{12} - 0 = 3570.0$$

From Table 11.9.4 we have $\sum d_i^2 = 1788.5$. From these data we may now compute by Equation 11.9.4

$$r_s = \frac{3565.5 + 3570.0 - 1788.5}{2\sqrt{(3565.5)(3570)}} = .75$$

We see that in this case the correction for ties does not make any difference in the value of r_s.

EXERCISES

For the following exercises perform the test at the indicated level of significance and determine the p value.

11.9.1 The following table shows 15 randomly selected geographic areas ranked by population density and age-adjusted death rate. Can we conclude at the .05 level of significance that population density and age-adjusted death rate are not mutually independent?

	RANK BY	
Area	Population density (X)	Age-adjusted death rate (Y)
1	8	10
2	2	14
3	12	4
4	4	15
5	9	11
6	3	1
7	10	12
8	5	7
9	6	8
10	14	5
11	7	6
12	1	2
13	13	9
14	15	3
15	11	13

11.9.2 The following table shows 10 communities ranked by DMF teeth per 100 children and fluoride concentration in ppm in the public water supply.

| | RANK BY | |
| | DMF teeth per 100 | Fluoride concentration |
Community	Children (X)	(Y)
1	8	1
2	9	3
3	7	4
4	3	9
5	2	8
6	4	7
7	1	10
8	5	6
9	6	5
10	10	2

Do these data provide sufficient evidence to indicate that the number of DMF teeth per 100 children tends to decrease as fluoride concentration increases? Let $\alpha = .05$.

11.9.3 Refer to Example 8.7.1. Compute r_s and test the null hypothesis that the two methods are mutually independent. Compare the results with those obtained by using the methods of Chapter 8.

11.9.4 Compute r_s for the data of Exercise 8.7.3 and test the null hypothesis that the two variables are mutually independent. Compare the results with those obtained using the methods of Chapter 8.

11.9.5 Compute r_s for the data of Exercise 8.7.4. Do the data provide sufficient evidence to indicate that CO concentration increases as traffic volume increases?

11.9.6 Compute r_s for the data of Exercise 8.7.5 and test the null hypothesis that the two variables are mutually independent.

11.9.7 Seventeen patients with a history of congestive heart failure participated in a study to assess the effects of exercise on various bodily functions. During a period of exercise the following data were collected on the percent change in plasma norepinephrine (Y) and the percent change in oxygen consumption (X).

Subject	X	Y
1	500	525
2	475	130
3	390	325
4	325	190
5	325	90
6	205	295
7	200	180
8	75	74
9	230	420

(Table continued)

Subject	X	Y
10	50	60
11	175	105
12	130	148
13	76	75
14	200	250
15	174	102
16	201	151
17	125	130

On the basis of these data can one conclude that there is an association between the two variables? Let $\alpha = .05$.

11.10 SUMMARY

This chapter is concerned with nonparametric statistical tests. These tests may be used either when the assumptions underlying the parametric tests are not realized or when the data to be analyzed are measured on a scale too weak for the arithmetic procedures necessary for the parametric tests.

The four measurement scales—nominal, ordinal, interval, and ratio—are defined and illustrated.

Finally, seven nonparametric tests are described and illustrated. Except for the Kolmogorov–Smirnov goodness-of-fit test, each test provides a nonparametric alternative to a well-known parametric test. There are a number of other nonparametric tests available, many of which are described and illustrated in the references cited at the end of this chapter.

REVIEW QUESTIONS AND EXERCISES

1. Define nonparametric statistics.
2. What is meant by the term distribution-free statistical tests?
3. What are the advantages of using nonparametric statistical tests?
4. What are some of the disadvantages of the nonparametric tests?
5. Define the word measurement.
6. List in order of sophistication and describe the four measurement scales.

7. Describe a situation in your particular area of interest where each of the following tests could be used. Use real or realistic data and test an appropriate hypothesis using each test.
   (a) The sign test
   (b) The median test
   (c) The Mann–Whitney test
   (d) The Kolmogorov–Smirnov goodness-of-fit test
   (e) The Kruskal–Wallis one-way analysis of variance by ranks
   (f) The Friedman two-way analysis of variance by ranks
   (g) The Spearman rank correlation coefficient

8. The following are the ranks of the ages (X) of 20 surgical patients and the dose (Y) of an analgesic agent required to block one spinal segment.

Rank of age in years (X)	Rank of dose requirement (Y)	Rank of age in years (X)	Rank of dose requirement (Y)
1	1	11	13
2	7	12	5
3	2	13	11
4	4	14	16
5	6	15	20
6	8	16	18
7	3	17	19
8	15	18	17
9	9	19	10
10	12	20	14

   Compute r_s and test (two-sided) for significance. Let $\alpha = .05$. Determine the p value for this test.

9. The following pulmonary function data were collected on children with muscular dystrophy before and after a period of respiratory therapy. Scores are expressed as percent of predicted normal values for height, weight, and body surface measurement.

Forced Vital Capacity (Liters)
Before: 74 65 84 89 84 65 78 86 83 82
After: 79 78 100 92 104 70 81 84 85 90

   Use the sign test to determine whether one should conclude that the therapy is effective. Let $\alpha = .05$. What is the p value?

10. Three methods of reducing skin bacterial load by bathing were compared. Bacteria counts were made on the right foot of subjects before and after treatment. The variable of interest was percent reduction of bacteria. Twenty-seven nursing student volunteers participated in the experiment. The three methods of bathing the foot were whirlpool agitation, spraying, and soaking. The results were as follows:

Whirlpool		Spraying		Soaking	
91	80	18	16	6	10
87	92	22	15	6	12
88	81	20	26	8	5
84	93	29	19	9	9
86		25		13	

Can one conclude on the basis of these data that the three methods are not equally effective? Let $\alpha = .05$. What is the p value for this test?

11. Ten subjects with bronchial asthma participated in an experiment to evaluate the relative effectiveness of three drugs. The following table shows the change in FEV_1 (forced expired volume in 1 second) values (expressed as liters) two hours after drug administration.

	DRUG		
Subject	A	B	C
1	.00	.13	.26
2	.04	.17	.23
3	.02	.20	.21
4	.02	.27	.19
5	.04	.11	.36
6	.03	.18	.25
7	.05	.21	.32
8	.02	.23	.38
9	.00	.24	.30
10	.12	.08	.30

Are these data sufficient to indicate a difference in drug effectiveness? Let $\alpha = .05$. What is the p value for this test?

12. Sera from two groups of subjects following streptococcal infection were assayed for neutralizing antibodies to streptolysin O (ASO). The results were as follows.

ASO (MEASURED IN TODD UNITS)

Group A	Group B
324	558
275	108
349	291
604	863
566	303
810	640
340	358
295	503
357	646
580	689
344	250
655	540
380	630
503	190
314	

Do these data provide sufficient evidence to indicate a difference in population medians? Let $\alpha = .05$. What is the p-value for this test? Use both the median test and the Mann–Whitney test and compare the results.

13. The following are the $PaCO_2$ (mm Hg) values in 16 patients with bronchopulmonary disease:

$$39, 40, 45, 48, 49, 56, 60, 75, 42, 48, 32, 37, 32, 33, 33, 36$$

Use the Kolmogorov–Smirnov test to test the null hypothesis that $PaCO_2$ values in the sampled population are normally distributed with $\mu = 44$ and $\sigma = 12$.

14. The following table shows the caloric intake (cal/day/kg) and oxygen consumption, VO_2 (ml/min/kg) in 10 infants.

Caloric intake (X)	VO₂ (Y)
50	7.0
70	8.0
90	10.5
120	11.0
40	9.0
100	10.8
150	12.0
110	10.0
75	9.5
160	11.9

Test the null hypothesis that the two variables are mutually independent against the alternative that they are directly related. Let $\alpha = .05$. What is the p value for this test?

REFERENCES

References Cited

1. M. G. Kendall and R. M. Sundrum, "Distribution-Free Methods and Order Properties," *Review of the International Statistical Institute, 21:3* (1953), 124–134.

2. Jean Dickinson Gibbons, *Nonparametric Statistical Inference*, McGraw-Hill, New York, 1971.

3. J. R. Blum and N. A. Fattu, "Nonparametric Methods," *Review of Educational Research, 24* (1954), 467–487.

4. L. E. Moses, "Nonparametric Statistics for Psychological Research," *Psychological Bulletin, 49* (1952), 122–143.

5. Sidney Siegel, *Nonparametric Statistics for the Behaviorial Sciences*, McGraw-Hill, New York, 1956.

6. W. J. Conover, *Practical Nonparametric Statistics*, Second Edition, Wiley, New York, 1980.

7. James V. Bradley, *Distribution-Free Statistics*, Prentice-Hall, Englewood Cliffs, N.J., 1968.

8. Charles H. Kraft and Constance Van Eeden, *A Nonparametric Introduction to Statistics*, Macmillan, New York, 1968.

9. Merle W. Tate and Richard C. Clelland, *Nonparametric and Shortcut Statistics in the Social, Biological, and Medical Sciences*, Interstate Printers and Publishers, Danville, Ill., 1957.

10. Jean Dickinson Gibbons, *Nonparametric Methods for Quantitative Analysis*, Holt, Rinehart and Winston, New York, 1976.

11. Frederick Mosteller and Robert E. K. Rourke, *Sturdy Statistics: Nonparametric and Order Statistics*, Reading, Mass., Addison-Wesley, 1973.

12. Albert Pierce, *Fundamentals of Nonparametric Statistics*, Belmont, Cal., Dickensen, 1970.

13. G. E. Noether, *Introduction to Statistics: A Fresh Approach*, Boston, Houghton Mifflin, 1971.

14. Myles Hollander and Douglas A. Wolfe, *Nonparametric Statistical Methods*, Wiley, New York, 1973.

15. Wayne W. Daniel, *Practical Nonparametric Statistics*, Houghton Mifflin, Boston, 1978.

16. Leonard A. Marascuilo and Maryellen McSweeney, *Nonparametric and Distribution-Free Methods for the Social Sciences*, Brooks/Cole, Monterey, Calif., 1977.

17. E. L. Lehmann, *Nonparametrics, Statistical Methods Based on Ranks*, Holden-Day, San Francisco, 1975.

18. Jaroslav Hajek, *A Course in Nonparametric Statistics*, Holden-Day, San Francisco, 1969.

19. John Edward Walsh, *Handbook of Nonparametric Statistics*, Van Nostrand, Princeton, N.J., 1962.

20. John E. Walsh, *Handbook of Nonparametric Statistics*, II, Van Nostrand, Princeton, N.J., 1965.

21. I. R. Savage, *Bibliography of Nonparametric Statistics*, Harvard University Press, Cambridge, Mass. (1962).

22. S. S. Stevens, "On the Theory of Scales of Measurement," *Science*, *103* (1946), 677–680.

23. S. S. Stevens, "Mathematics, Measurement and Psychophysics," in S. S. Stevens (editor), *Handbook of Experimental Psychology*, Wiley, New York, 1951.

24. N. H. Anderson, "Scales and Statistics: Parametric and Nonparametric," *Psychological Bulletin*, *58* (1961), 305–316.

25. J. Gaito, "Nonparametric Methods in Psychological Research," *Psychological Reports*, *5* (1959), 115–125.

26. F. M. Lord, "On the Statistical Treatment of Football Numbers," *American Psychologist*, *8* (1953), 750–751.

27. Gordon D. Armstrong, "Parametric Statistics and Ordinal Data: A Pervasive Misconception," *Nursing Research*, *30* (1981), 60–62.

28. Edgar F. Borgatta and George W. Bohrnstedt, "Level of Measurement Once Over Again," *Sociological Methods & Research*, *9* (1980), 147–160.

29. W. J. Dixon and A. M. Mood, "The Statistical Sign Test," *Journal of the American Statistical Association*, *41* (1946), 557–566.

30. A. M. Mood, *Introduction to the Theory of Statistics*, McGraw-Hill, New York, 1950.

31. J. Westenberg, "Significance Test for Median and Interquartile Range in Samples from Continuous Populations of Any Form," *Proceedings Koninklijke Nederlandse Akademie Van Wetenschappen*, *51* (1948), 252–261.

32. G. W. Brown and A. M. Mood, "On Median Tests for Linear Hypotheses," *Proceedings of the Second Berkeley Symposium on Mathematical Statistics and Probability*, University of California Press, Berkeley, 1951, pp. 159–166.

33. V. L. Senders, *Measurement and Statistics*, Oxford University Press, New York, 1958.

34. William L. Hays and Robert L. Winkler, *Statistics: Probability, Inference, and Decision*, Holt, Rinehart and Winston, New York, 1971.

35. Ronald A. Fisher, *Statistical Methods for Research Workers*, Thirteenth Edition, Hafner, New York, 1958.

36. H. B. Mann and D. R. Whitney, "On a Test of Whether One of Two Random Variables Is Stochastically Larger Than the Other," *Annals of Mathematical Statistics*, *18* (1947), 50–60.

37. Gottfried E. Noether, *Elements of Nonparametric Statistics*, Wiley, New York, 1967.

38. A. N. Kolmogorov, "Sulla Determinazione Empirical di una Legge di Distribuizione," *Giornale dell' Istitute Italiano degli Altuari*, *4* (1933), 83–91.

39. N. V. Smirnov, "Estimate of Deviation Between Empirical Distribution Functions in Two Independent Samples," (in Russian) *Bulletin Moscow University*, *2* (1939), 3–16.

40. L. A. Goodman, "Kolmogorov-Smirnov Tests for Psychological Research," *Psychological Bulletin*, *51* (1954), 160–168.

41. Gottfried E. Noether, *Elements of Nonparametric Statistics*, Wiley, New York, 1967.

42. F. J. Massey, "The Kolmogorov-Smirnov Test for Goodness of Fit," *Journal of the American Statistical Association*, *46* (1951), 68–78.

43. Z. W. Birnbaum, "Numerical Tabulation of the Distribution of Kolmogorov's Statistic for Finite Sample Size," *Journal of the American Statistical Association*, *47* (1952), 425–441.

44. M. J. Slakter, "A Comparison of the Pearson Chi-Square and Kolmogorov Goodness-of-Fit Tests with Respect to Validity," *Journal of the American Statistical Association*, *60* (1965), 854–58.

45. W. H. Kruskal and W. A. Wallis, "Use of Ranks in One-Criterion Analysis of Variance," *Journal of the American Statistical Association 47* (1952), 583–621; errata, ibid., *48* (1953), 907–911.

46. K. R. Gabriel and P. A. Lachenbruch, "Nonparametric ANOVA in Small Samples: A Monte Carlo Study of the Adequacy of the Asymptotic Approximation," *Biometrics*, *25* (1969), 593–596.

47. M. Friedman, "The Use of Ranks to Avoid the Assumption of Normality Implicit in the Analysis of Variance," *Journal of the American Statistical Association*, *32* (1937), 675–701.

48. M. Friedman, "A Comparison of Alternative Tests of Significance for the Problem of *m* Rankings," *Annals of Mathematical Statistics*, *11* (1940), 86–92.

49. M. G. Kendall, *Rank Correlation Methods*, Second Edition, Hafner, New York, 1955.

50. W. H. Kruskal, "Ordinal Measures of Association," *Journal of the American Statistical Association*, *53* (1958), 814–861.

51. Harold Hotelling and Margaret R. Pabst, "Rank Correlation and Tests of Significance Involving No Assumption of Normality," *Annals of Mathematical Statistics*, *7* (1936), 29–43.

52. C. Spearman, "The Proof and Measurement of Association Between Two Things," *American Journal of Psychology*, *15* (1904), 72–101.

53. G. J. Glasser and R. F. Winter, "Critical Values of the Coefficient of Rank Correlation for Testing the Hypothesis of Independence," *Biometrika*, *48* (1961), 444–448.

54. Eugene S. Edgington, *Statistical Inference: The Distribution-Free Approach*, McGraw-Hill, New York, 1969.

Additional References: Journal Articles
 1. Olive Jean Dunn, "Multiple Comparisons Using Rank Sums," *Technometrics*, *6* (1964), 241–252.

 2. B. J. McDonald and W. A. Thompson, "Rank Sum Multiple Comparisons in One- and Two-Way Classifications," *Biometrika*, *54* (1967), 487–498.

 3. Wayne W. Daniel and Carol E. Coogler, "Some Quick and Easy Statistical Procedures for the Physical Therapist," *Physical Therapy*, *54* (1974), 135–140.

 4. Wayne W. Daniel and Beaufort B. Longest, "Some Practical Statistical Procedures," *Journal of Nursing Administration*, *5* (1975), 23–27.

 5. Wayne W. Daniel, *On Nonparametric and Robust Tests for Dispersion: A Selected Bibliography*, Vance Bibliographies, Monticello, Ill., December 1979.

 6. Wayne W. Daniel, *Goodness-of-fit: A Selected Bibliography for the Statistician and the Researcher*, Vance Bibliographies, Monticello, Ill., May 1980.

CHAPTER 12

Vital Statistics

12.1 INTRODUCTION

The private physician arrives at a diagnosis and treatment plan for an individual patient by means of a case history, a physical examination, and various laboratory tests. The community may be thought of as a living complex organism for which the public health team is the physician. To carry out this role satisfactorily the public health team must also make use of appropriate tools and techniques for evaluating the health status of the community. Traditionally these tools and techniques have consisted of the community's vital statistics, which include the counts of births, deaths, illnesses, and the various rates and ratios that may be computed from them.

In succeeding sections we give some of the more useful and widely used rates and ratios. Before proceeding, however, let us distinguish between the terms *rate* and *ratio* by defining each as follows.

1. *Rate* Although there are some exceptions, the term rate usually is reserved to refer to those calculations that imply the probability of the occurrence of some event. A rate is expressed in the form

$$\left(\frac{a}{a+b}\right)k \tag{12.1.1}$$

where

$$a = \text{the frequency with which an event has occurred during}$$
$$\text{some specified period of time}$$
$$a + b = \text{the number of persons exposed to the risk of the event}$$
$$\text{during the same period of time}$$
$$k = \text{some number such as 10, 100, 1000, 10,000, or 100,000}$$

As indicated by Expression 12.1.1 the numerator of a rate is a component part of the denominator. The purpose of the multiplier, k, called the base, is to avoid results involving the very small numbers that may arise in the calculation of rates and to facilitate comprehension of the rate. The value chosen for k will depend on the magnitudes of the numerator and denominator.

2. *Ratio*   A ratio is a fraction of the form

$$\left(\frac{c}{d}\right)k \tag{12.1.2}$$

where k is some base as already defined and both c and d refer to the frequency of occurrence of some event or item. In the case of a ratio, as opposed to a rate, the numerator is not a component part of the denominator. We can speak, for example, of the person–doctor ratio or the person–hospital-bed ratio of a certain geographical area. The values of k most frequently used in ratios are 1 and 100.

12.2   DEATH RATES AND RATIOS

The rates and ratios discussed in this section are concerned with the occurrence of death. Death rates express the relative frequency of the occurrence of death within some specified interval of time in a specific population. The denominator of a death rate is referred to as the population at risk. The numerator represents only those deaths that occurred in the population specified by the denominator.

1. *Annual Crude Death Rate*   The annual crude death rate is defined as

$$\frac{\text{total number of deaths during year (January 1 to December 31)}}{\text{total population as of July 1}} \cdot k$$

where the value of k is usually chosen as 1000. This is the most widely used rate for measuring the overall health of a community. To compare the crude death rates of two communities is hazardous, unless it is known that the communities are comparable with respect to the many characteristics, other than health conditions, that influence the death rate. Variables that enter into the picture include age, race, sex, and socioeconomic status. When two populations must be compared on the basis of death rates, adjustments may be made to reconcile the population differences with respect to these variables. The same precautions should be exercised when comparing the annual death rates for the same community for two different years.

2. *Annual Specific Death Rates*   It is usually more meaningful and enlightening to observe the death rates of small, well-defined subgroups of the total

population. Rates of this type are called *specific death rates* and are defined as

$$\frac{\text{total number of deaths in a specific subgroup during a year}}{\text{total population in the specific subgroup as of July 1}} \cdot k$$

where k is usually equal to 1000. Subgroups for which specific death rates may be computed include those groups that may be distinguished on the basis of sex, race, and age. Specific rates may be computed for two or more characteristics simultaneously. For example, we may compute the death rate for white males, thus obtaining a race–sex specific rate. Cause-specific death rates may also be computed by including in the numerator only those deaths due to a particular cause of death, say cancer, heart disease, or accidents. Because of the small fraction that results, the base, k, for a cause-specific rate is usually 100,000 or 1,000,000.

3. *Adjusted or Standardized Death Rates* As we have already pointed out, the usefulness of the crude death rate is restricted by the fact that it does not reflect the composition of the population with respect to certain characteristics by which it is influenced. We have seen that by means of specific death rates various segments of the population may be investigated individually. If, however, we attempt to obtain an overall impression of the health of a population by looking at individual specific death rates, we are soon overwhelmed by their great number.

What is wanted is a single figure that measures the forces of mortality in a population while holding constant one or more of the compositional factors such as age, race, or sex. Such a figure, called an *adjusted death rate*, is available. It is most commonly obtained by what is known as the *direct method* of adjustment. The method consists essentially of applying to a *standard population* specific rates observed in the population of interest. From the resulting expected numbers we may compute an overall rate that tells us what the rate for the population of interest would be if that population had the same composition as the standard population. This method is not restricted to the computation of adjusted death rates only, but it may be used to obtain other adjusted rates, for example, an adjusted birth rate. If two or more populations are adjusted in this manner, they are then directly comparable on the basis of the adjustment factors. Opinions differ as to what population should be used as the standard. The population of the United States as of the last decennial census is frequently used. For adjustment calculations a population of 1,000,000, reflecting the composition of the standard population and called the *standard million*, is usually used. In the following example we illustrate the direct method of adjustment to obtain an age-adjusted death rate.

Example 12.2.1 The 1970 crude death rate for Georgia was 9.1 deaths per 1000 population. Let us obtain an age-adjusted death rate for Georgia by using

TABLE 12.2.1

Calculation of Age-Adjusted Death Rate for Georgia, 1970, by Direct Method

1	2	3	4	5	6
				Standard population based on	Number of expected deaths in
			Age-specific death rates	U.S. population	standard
Age (years)	Population[a]	Deaths[a]	(per 100,000)	1970[b]	population
0 to 4	424,600	2,483	584.8	84,416	494
5 to 14	955,000	449	47.0	200,508	94
15 to 24	863,000	1,369	158.6	174,406	277
25 to 34	608,100	1,360	223.6	122,569	274
35 to 44	518,400	2,296	442.9	113,614	503
45 to 54	486,400	4,632	952.3	114,265	1,088
55 to 64	384,400	7,792	2,027.1	91,480	1,854
65 to 74	235,900	9,363	3,969.1	61,195	2,429
75 and over	132,900	12,042	9,060.9	37,547	3,402
Total	4,608,700	41,786[c]		1,000,000	10,415

[a] *Georgia Vital and Morbidity Statistics* 1970, Georgia Department of Public Health, Atlanta, Georgia.
[b] *1970 Census of Population*, PC(1)-B1, Table 49.
[c] Excludes 44 deaths at unknown age.

the 1970 United States census as the standard population. In other words, we want a death rate that could have been expected in Georgia if the age composition of the Georgia population had been the same as that of the United States in 1970. The data necessary for the calculations are shown in Table 12.2.1.

The procedure for calculating an age-adjusted death rate by the direct method consists of the following steps.

1. The population of interest is listed (column 2) according to age group (column 1).

2. The deaths in the population of interest are listed (column 3) by age group.

3. The age-specific death rates (column 4) for each age group are calculated by dividing column 3 by column 2 and multiplying by 100,000.

4. The standard population (column 5) is listed by age group.

5. The expected number of deaths in the standard population for each group (column 6) is computed by multiplying column 4 by column 5 and dividing by 100,000. The entries in column 6 are the deaths that would be expected in the standard population if the persons in this population had been exposed to the same risk of death experienced by the population being adjusted.

6. The entries in column 6 are summed to obtain the total number of expected deaths in the standard population.

7. The age-adjusted death rate is computed in the same manner as a crude death rate. That is, the age-adjusted death rate is equal to

$$\frac{\text{total number of expected deaths}}{\text{total standard population}} \cdot 1000$$

In the present example we have an age-adjusted death rate of

$$\frac{10{,}415}{1{,}000{,}000} \cdot 1000 = 10.4$$

We see, then, that the crude death rate has been increased from 9.1 per 1000 to 10.4 per 1000 by adjusting the 1970 population of Georgia to the age distribution of the standard population. This increase in the death rate following adjustment reflects the fact that in 1970 the population of Georgia was slightly younger than the population of the United States as a whole. For example, only 8 percent of the Georgia population was 65 years of age or older whereas 10 percent of the United States population was in that age group.

4. *Maternal Mortality Rate*   This rate is defined as

$$\frac{\text{deaths from all puerperal causes during a year}}{\text{total live births during the year}} \cdot k$$

where k is taken as 1000 or 100,000. The preferred denominator for this rate is the number of women who were pregnant during the year. This denominator, however, is impossible to determine.

A death from a puerperal cause is a death that can be ascribed to some phase of childbearing. Because of the decline in the maternal mortality rate in the United States, it is more convenient to use $k = 100{,}000$. In some countries, however, $k = 1000$ results in a more convenient rate. The decline in the maternal mortality rate in this country also has had the effect of reducing its usefulness as a discriminator among communities with varying qualities of medical care and health facilities.

Some limitations of the maternal mortality rate include the following.

(a) Fetal deaths are not included in the denominator. This results in an inflated rate, since a mother can die from a puerperal cause without producing a live birth.

(b) A maternal death can be counted only once, although twins or larger multiple births may have occurred. Such cases cause the denominator to be too large and, hence, there is a too small rate.

(c) Under-registration of live births, which result in a too small denominator, causes the rate to be too large.

(d) A maternal death may occur in a year later than the year in which the birth occurred. Although there are exceptions, in most cases the transfer of maternal deaths will balance out in a given year.

5. *Infant Mortality Rate*   This rate is defined as

$$\frac{\text{number of deaths under 1 year of age during a year}}{\text{total number of live births during the year}} \cdot k$$

where k is generally taken as 1000. Use and interpretation of this rate must be made in light of its limitations which are similar to those that characterize the maternal mortality rate. Many of the infants who die in a given calendar year were born during the previous year; and, similarly, many children born in a given calendar year will die during the following year. In populations with a stable birthrate this does not pose a serious problem. In periods of rapid change, however, some adjustment should be made. One way to make an adjustment is to allocate the infant deaths to the calendar year in which the infants were born before computing the rate.

6. *Neonatal Mortality Rate*   In an effort to better understand the nature of infant deaths, rates for ages less than a year are frequently computed. Of these, the one most frequently computed is the *neonatal mortality rate*, which is defined as

$$\frac{\text{number of deaths under 28 days of age during a year}}{\text{total number of live births during the year}} \cdot k$$

where $k = 1000$.

7. *Fetal Death Rate*   This rate is defined as

$$\frac{\text{total number of fetal deaths during a year}}{\text{total deliveries during the year}} \cdot k$$

where k is usually taken to be 1000. A fetal death is defined as a product of conception that shows no sign of life after complete birth. There are several problems associated with the use and interpretation of this rate. There is variation among reporting areas with respect to the duration of gestation. Some areas report all fetal deaths regardless of length of gestation while others have a minimum gestation period that must be reached before reporting is required. Another objection to the fetal death rate is that it does not take into account the extent to which a community is trying to reproduce. The ratio to be considered next has been proposed to overcome this objection.

8. *Fetal Death Ratio* This ratio is defined as

$$\frac{\text{total number of fetal deaths during a year}}{\text{total number of live births during the year}} \cdot k$$

where k is taken as 100 or 1000.

Some authorities have suggested that the number of fetal deaths as well as live births be included in the denominator in an attempt to include all pregnancies in the computation of the ratio. The objection to this suggestion rests on the incompleteness of fetal death reporting.

9. *Perinatal Mortality Rate* · Since fetal deaths occurring late in pregnancy and neonatal deaths frequently have the same underlying causes, it has been suggested that the two be combined to obtain what is known as the *perinatal mortality rate*. This rate is computed as

$$\frac{\begin{array}{c}(\text{number of fetal deaths of 28 weeks or more}) \\ + (\text{infant deaths under 7 days})\end{array}}{\begin{array}{c}(\text{number of fetal deaths of 28 weeks or more}) \\ + (\text{number of live births})\end{array}} \cdot k$$

where $k = 1000$.

10. *Cause-of-Death Ratio* This ratio is defined as

$$\frac{\text{number of deaths due to a specific disease during a year}}{\text{total number of deaths due to all causes during the year}} \cdot k$$

where $k = 100$. This index is used to measure the relative importance of a given cause of death. It should be used with caution in comparing one community with another. A higher cause-of-death ratio in one community than that in another may be because the first community has a low mortality from other causes.

11. *Proportional Mortality Ratio* This index has been suggested as a single measure for comparing the overall health conditions of different communities. It is defined as

$$\frac{\text{number of deaths of persons 50 years of age and older}}{\text{total number of deaths}} \cdot k$$

where $k = 100$. The specified class is usually an age group such as 50 years and over, or a cause of death category, such as accidents. The proportional mortality ratio exhibits certain defects as noted by Linder and Grove (1).

EXERCISES

12.2.1 The following annual data were reported for a certain geographic area.

	Total	NUMBER White	Nonwhite
Estimated population			
as of July 1	597,500	361,700	235,800
Total live births	12,437	6,400	6,037
Immature births	1,243	440	803
Fetal deaths:			
Total	592	365	227
Under 20 weeks gestation	355	269	86
20 to 27 weeks gestation	103	42	61
28 weeks and more	123	49	74
Unknown length of gestation	11	5	6
Deaths			
Total all ages	6,219	3,636	2,583
Under 1 year	267	97	170
Under 28 days	210	79	131
Deaths from immaturity	16	12	4
Maternal Deaths	2	—	2
Cause of death			
Malignant neoplasms	948	626	322
Ischaemic heart disease	1,697	1,138	559

Source: *Georgia Vital and Morbidity Statistics 1970*, Georgia Department of Public Health, Atlanta, p. 47.

From these data compute the following rates and ratios: (a) crude death rate, (b) race-specific death rates for white and nonwhite, (c) maternal mortality rate, (d) infant mortality rate, (e) neonatal mortality rate, (f) fetal death ratio, (g) cause of death ratios for malignant neoplasms and ischaemic heart disease.

12.2.2 The following table shows the deaths and estimated population by age for the state of Georgia for 1971. Use these data to compute the age-adjusted death rate for Georgia, 1971. Use the same standard population that was used in Example 12.2.1.

Age (years)	Estimated population	Deaths
0 to 4	423,700	2,311
5 to 14	947,900	480
15 to 24	891,300	1,390
25 to 34	623,700	1,307
35 to 44	520,000	2,137
45 to 54	494,200	4,640

(Table continued)

Age (years)	Estimated population	Deaths
55 to 64	388,600	7,429
65 to 74	243,000	9,389
75 and over	136,000	12,411
Total	4,668,400	41,494[a]

Source: Statistics Section, Office of Evaluation and Research, Georgia Department of Human Resources, Atlanta.

[a] Excludes 42 deaths at unknown age.

12.3   MEASURES OF FERTILITY

The term *fertility* as used by American demographers refers to the actual bearing of children as opposed to the capacity to bear children, for which phenomenon the term *fecundity* is used. A knowledge of the "rate" of childbearing in a community is important to the health worker in planning services and facilities for mothers, infants, and children. The following are the six basic measures of fertility.

1. *Crude Birth Rate*   This rate is the most widely used of the fertility measures. It is obtained from

$$\frac{\text{total number of live births during a year}}{\text{total population as of July 1}} \cdot k$$

where $k = 1000$. For an illustration of the computation of this and the other five rates, see Table 12.3.1.

2. *General Fertility Rate*   This rate is defined as

$$\frac{\text{number of live births during a year}}{\text{total number of women of childbearing age}} \cdot k$$

where $k = 1000$, and the childbearing age is usually defined as ages 15 through 44 or ages 15 through 49. The attractive feature of this rate, when compared to the crude birth rate, is the fact that the denominator approximates the number of persons actually exposed to the risk of bearing a child.

3. *Age-Specific Fertility Rate*   Since the rate of childbearing is not uniform throughout the childbearing ages, a rate that permits the analysis of fertility rates for shorter maternal age intervals is desirable. The rate used is the

TABLE 12.3.1
Illustration of Procedures for Computing Six Basic Measures of Fertility, for Georgia, 1970

1	2	3	4	5	6	7
Age of woman (years)	Number of women in population[a]	Number of births to women of specified age[b]	Age-specific birthrate per 1000 women	Standard population based on U.S. population 1970[c]	Expected births	Cumulative fertility rate
15 to 19	220,100	21,790	99.0	193,762	19,182	495.0
20 to 24	209,500	37,051	176.9	173,583	30,707	1,379.5
25 to 29	170,100	22,135	130.1	140,764	18,313	2,030.0
30 to 34	139,100	9,246	66.5	119,804	7,967	2,362.5
35 to 39	135,400	3,739	27.6	116,925	3,227	2,500.5
40 to 49	261,700	1,044	4.0	255,162	1,021	2,540.5
Total	1,135,900	95,005		1,000,000	80,417	

Computation of six basic rates:

(1) Crude birth rate = total births divided by total population
$$= (95,584/4,608,700)(1000) = 21.$$

(2) General fertility rate = $(95,584/1,135,900)(1000) = 84.1$.

(3) Age-specific fertility rates = entries in column 3 divided by entries in column 2 multiplied by 1000 for each age group. Results appear in column 4.

(4) Total fertility rate = the sum of each age-specific rate multiplied by the age interval width = $(99.0)(5) + (176.9)(5) + (130.1)(5) + (66.5)(5) + (27.6)(5) + (4.0)(10) = 2,540.5$.

(5) Cumulative fertility rate = age-specific birth rate multiplied by age interval width cumulated by age. See column 7.

(6) Standardized general fertility rate = $(80,417/1,000,000)(1000) = 80.4$.

[a] Statistics Section, Office of Evaluation and Research, Georgia Department of Human Resources, Atlanta.

[b] *Georgia Vital and Morbidity Statistics 1970*, Georgia Department of Public Health, Atlanta.

[c] 1970 Census of Population, PC(1)-B1.

age-specific fertility rate which is defined as

$$\frac{\text{number of births to women of a certain age in a year}}{\text{total number of women of the specified age}} \cdot k$$

where $k = 1000$. Age-specific rates may be computed for single years of age, or any age interval. Rates for five-year age groups are the ones most frequently computed. Specific fertility rates may be computed also for other population subgroups such as those defined by race, socioeconomic status, and various demographic characteristics.

4. *Total Fertility Rate* If the age-specific fertility rates for all ages are added and multiplied by the interval into which the ages were grouped, the result is called the *total fertility rate*. The resulting figure is an estimate of the number of children a cohort of 1000 women would have if, during their reproductive years, they reproduced at the rates represented by the age-specific fertility rates from which the total fertility rate is computed.

5. *Cumulative Fertility Rate* The cumulative fertility rate is computed in the same manner as the total fertility rate except that the adding process can terminate at the end of any desired age group. The numbers in column 7 of Table 12.3.1 are the cumulative fertility rates through the ages indicated in column 1. The final entry in the cumulative fertility rate column is the total fertility rate.

6. *Standardized Fertility Rate* Just as the crude death rate may be standardized or adjusted, so may we standardize the general fertility rate. The procedure is identical to that discussed in Section 12.2 for adjusting the crude death rate. The necessary computations for computing the age-standardized fertility rate are shown in Table 12.3.1.

EXERCISES

12.3.1 The data in the following table are for the state of Georgia for the year 1971.

Age of woman (years)	Number of women in population	Number of births to women of specified age
15 to 19	225,200	21,834
20 to 24	217,600	35,997
25 to 29	173,400	21,670
30 to 34	143,300	8,935
35 to 39	134,100	3,464
40 to 49	267,800	925[a]

Source: Statistics Section, Office of Evaluation and Research, Georgia Department of Human Resources, Atlanta.

[a] May include some births to women over 49 years of age.

From the above data compute the following rates:

(a) Age-specific fertility rates for each age group.
(b) Total fertility rate.
(c) Cumulative fertility rate through each age group.
(d) General fertility rate standardized by age.

Use the standard population shown in Table 12.3.1.

12.3.2 There were a total of 95,546 live births in Georgia in 1971. The estimated total population as of July 1, 1971 was 4,668,400, and the number of women between the ages of 15 and 49 was 1,161,400. Use these data to compute:
(a) The crude birth rate.
(b) The general fertility rate.

12.4 MEASURES OF MORBIDITY

Another area that concerns the health worker who is analyzing the health of a community is *morbidity*. Data for the study of the morbidity of a community are not, as a rule, as readily available and complete as are the data on births and deaths because of incompleteness of reporting and differences among states with regard to laws requiring the reporting of diseases. The two rates most frequently used in the study of diseases in a community are the *incidence rate* and the *prevalence rate*.

1. *Incidence Rate*   This rate is defined as

$$\frac{\text{total number of new cases of a specific disease during a year}}{\text{total population as of July 1}} \cdot k$$

where the value of k depends on the magnitude of the numerator. A base of 1000 is used when convenient, but 100 can be used for the more common diseases, and 10,000 or 100,000 can be used for those less common or rare. This rate, which measures the degree to which new cases are occurring in the community, is useful in helping determine the need for initiation of preventive measures. It is a meaningful measure for both chronic and acute diseases.

2. *Prevalence Rate*   Although it is referred to as a rate, the *prevalence rate* is really a ratio, since it is computed from

$$\frac{\text{total number of cases, new or old, existing at a point in time}}{\text{total population at that point in time}} \cdot k$$

where the value of k is selected by the same criteria as for the incidence rate. This rate is especially useful in the study of chronic diseases, but it may also be computed for acute diseases.

3. *Case-Fatality Ratio*   This ratio is useful in determining how well the treatment program for a certain disease is succeeding. It is defined as

$$\frac{\text{total number of deaths due to a disease}}{\text{total number of cases due to the disease}} \cdot k$$

where $k = 100$. The period of time covered is arbitrary, depending on the nature of the disease, and it may cover several years for an endemic disease.

Note that this ratio can be interpreted as the probability of dying following contraction of the disease in question and, as such, reveals the seriousness of the disease.

4. *Immaturity Ratio*   This ratio is defined as

$$\frac{\text{number of live births under 2500 grams during a year}}{\text{total number of live births during the year}} \cdot k$$

where $k = 100$.

5. *Secondary Attack Rate*   This rate measures the occurrence of a contagious disease among susceptible persons who have been exposed to a primary case and is defined as

$$\frac{\begin{array}{c}\text{number of additional cases among contacts of a}\\ \text{primary case within the maximum incubation period}\end{array}}{\text{total number of susceptible contacts}} \cdot k$$

where $k = 100$. This rate is used to measure the spread of infection and is usually applied to closed groups such as a household or classroom where it can be reasonably assumed that all members were, indeed, contacts.

12.5 SUMMARY

This chapter is concerned with the computation and interpretation of various rates and ratios that are useful in studying the health of a community. More specifically, we discuss the more important rates and ratios relating to births, deaths, and morbidity. Individuals who wish to continue their reading in this area should consult the references.

REFERENCES

References Cited
1. Forest E. Linder and Robert D. Grove, *Vital Statistics Rates in the United States, 1900–1940*, United States Government Printing Office, Washington, D.C., 1947.

Other References
1. Donald J. Bogue, *Principles of Demography*, Wiley, New York, 1969.
2. John P. Fox, Carrie E. Hall, and Lila R. Elveback, *Epidemiology*, Macmillan, New York, 1970, Chapter 7.
3. Bernard G. Greenberg, "Biostatistics," in Hugh Rodman Leavell and E. Gurney Clark, *Preventive Medicine*, Third Edition, McGraw-Hill, New York, 1965.
4. Mortimer Spiegelman, *Introduction to Demography*, Revised Edition, Harvard University Press, Cambridge, Massachusetts, 1968.

APPENDIX

651

TABLE A

Cumulative Binomial Probability Distribution

$$P(X \le x|n,p) = \sum_{X=0}^{x} \binom{n}{x} p^x q^{n-x}$$

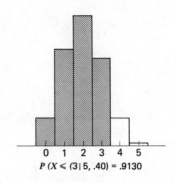

$P(X \le (3|5, .40) = .9130$

$n = 5$

p x	.01	.02	.03	.04	.05	.06	.07	.08	.09	.10
0	.9510	.9039	.8587	.8154	.7738	.7339	.6957	.6591	.6240	.5905
1	.9990	.9962	.9915	.9852	.9774	.9681	.9575	.9456	.9326	.9185
2	1.0000	.9999	.9997	.9994	.9988	.9980	.9969	.9955	.9937	.9914
3	1.0000	1.0000	1.0000	1.0000	1.0000	.9999	.9999	.9998	.9997	.9995
4	1.0000	1.0000	1.0000	1.0000	1.0000	1.0000	1.0000	1.0000	1.0000	1.0000

p x	.11	.12	.13	.14	.15	.16	.17	.18	.19	.20
0	.5584	.5277	.4984	.4704	.4437	.4182	.3939	.3707	.3487	.3277
1	.9035	.8875	.8708	.8533	.8352	.8165	.7973	.7776	.7576	.7373
2	.9888	.9857	.9821	.9780	.9734	.9682	.9625	.9563	.9495	.9421
3	.9993	.9991	.9987	.9983	.9978	.9971	.9964	.9955	.9945	.9933
4	1.0000	1.0000	1.0000	.9999	.9999	.9999	.9999	.9998	.9998	.9997
5	1.0000	1.0000	1.0000	1.0000	1.0000	1.0000	1.0000	1.0000	1.0000	1.0000

p x	.21	.22	.23	.24	.25	.26	.27	.28	.29	.30
0	.3077	.2887	.2707	.2536	.2373	.2219	.2073	.1935	.1804	.1681
1	.7167	.6959	.6749	.6539	.6328	.6117	.5907	.5697	.5489	.5282
2	.9341	.9256	.9164	.9067	.8965	.8857	.8743	.8624	.8499	.8369
3	.9919	.9903	.9886	.9866	.9844	.9819	.9792	.9762	.9728	.9692
4	.9996	.9995	.9994	.9992	.9990	.9988	.9986	.9983	.9979	.9976
5	1.0000	1.0000	1.0000	1.0000	1.0000	1.0000	1.0000	1.0000	1.0000	1.0000

p x	.31	.32	.33	.34	.35	.36	.37	.38	.39	.40
0	.1564	.1454	.1350	.1252	.1160	.1074	.0992	.0916	.0845	.0778
1	.5077	.4875	.4675	.4478	.4284	.4094	.3907	.3724	.3545	.3370
2	.8234	.8095	.7950	.7801	.7648	.7491	.7330	.7165	.6997	.6826
3	.9653	.9610	.9564	.9514	.9460	.9402	.9340	.9274	.9204	.9130
4	.9971	.9966	.9961	.9955	.9947	.9940	.9931	.9921	.9910	.9898
5	1.0000	1.0000	1.0000	1.0000	1.0000	1.0000	1.0000	1.0000	1.0000	1.0000

TABLE A (*Continued*)

n = 5 (Continued)

p / x	.41	.42	.43	.44	.45	.46	.47	.48	.49	.50
0	.0715	.0656	.0602	.0551	.0503	.0459	.0418	.0380	.0345	.0312
1	.3199	.3033	.2871	.2714	.2562	.2415	.2272	.2135	.2002	.1875
2	.6651	.6475	.6295	.6114	.5931	.5747	.5561	.5375	.5187	.5000
3	.9051	.8967	.8879	.8786	.8688	.8585	.8478	.8365	.8247	.8125
4	.9884	.9869	.9853	.9835	.9815	.9794	.9771	.9745	.9718	.9688
5	1.0000	1.0000	1.0000	1.0000	1.0000	1.0000	1.0000	1.0000	1.0000	1.0000

n = 6

p / x	.01	.02	.03	.04	.05	.06	.07	.08	.09	.10
0	.9415	.8858	.8330	.7828	.7351	.6899	.6470	.6064	.5679	.5314
1	.9985	.9943	.9875	.9784	.9672	.9541	.9392	.9227	.9048	.8857
2	1.0000	.9998	.9995	.9988	.9978	.9962	.9942	.9915	.9882	.9841
3	1.0000	1.0000	1.0000	1.0000	.9999	.9998	.9997	.9995	.9992	.9987
4	1.0000	1.0000	1.0000	1.0000	1.0000	1.0000	1.0000	1.0000	1.0000	.9999
5	1.0000	1.0000	1.0000	1.0000	1.0000	1.0000	1.0000	1.0000	1.0000	1.0000

p / x	.11	.12	.13	.14	.15	.16	.17	.18	.19	.20
0	.4970	.4644	.4336	.4046	.3771	.3513	.3269	.3040	.2824	.2621
1	.8655	.8444	.8224	.7997	.7765	.7528	.7287	.7044	.6799	.6554
2	.9794	.9739	.9676	.9605	.9527	.9440	.9345	.9241	.9130	.9011
3	.9982	.9975	.9966	.9955	.9941	.9925	.9906	.9884	.9859	.9830
4	.9999	.9999	.9998	.9997	.9996	.9995	.9993	.9990	.9987	.9984
5	1.0000	1.0000	1.0000	1.0000	1.0000	1.0000	1.0000	1.0000	1.0000	.9999
6	1.0000	1.0000	1.0000	1.0000	1.0000	1.0000	1.0000	1.0000	1.0000	1.0000

p / x	.21	.22	.23	.24	.25	.26	.27	.28	.29	.30
0	.2431	.2252	.2084	.1927	.1780	.1642	.1513	.1393	.1281	.1176
1	.6308	.6063	.5820	.5578	.5339	.5104	.4872	.4644	.4420	.4202
2	.8885	.8750	.8609	.8461	.8306	.8144	.7977	.7804	.7626	.7443
3	.9798	.9761	.9720	.9674	.9624	.9569	.9508	.9443	.9372	.9295
4	.9980	.9975	.9969	.9962	.9954	.9944	.9933	.9921	.9907	.9891
5	.9999	.9999	.9999	.9998	.9998	.9997	.9996	.9995	.9994	.9993
6	1.0000	1.0000	1.0000	1.0000	1.0000	1.0000	1.0000	1.0000	1.0000	1.0000

p / x	.31	.32	.33	.34	.35	.36	.37	.38	.39	.40
0	.1079	.0989	.0905	.0827	.0754	.0687	.0625	.0568	.0515	.0467
1	.3988	.3780	.3578	.3381	.3191	.3006	.2828	.2657	.2492	.2333
2	.7256	.7064	.6870	.6672	.6471	.6268	.6063	.5857	.5650	.5443
3	.9213	.9125	.9031	.8931	.8826	.8714	.8596	.8473	.8343	.8208
4	.9873	.9852	.9830	.9805	.9777	.9746	.9712	.9675	.9635	.9590
5	.9991	.9989	.9987	.9985	.9982	.9978	.9974	.9970	.9965	.9959
6	1.0000	1.0000	1.0000	1.0000	1.0000	1.0000	1.0000	1.0000	1.0000	1.0000

TABLE A (*Continued*)

n = 6 (Continued)

p x	.41	.42	.43	.44	.45	.46	.47	.48	.49	.50
0	.0422	.0381	.0343	.0308	.0277	.0248	.0222	.0198	.0176	.0156
1	.2181	.2035	.1895	.1762	.1636	.1515	.1401	.1293	.1190	.1094
2	.5236	.5029	.4823	.4618	.4415	.4214	.4015	.3820	.3627	.3437
3	.8067	.7920	.7768	.7610	.7447	.7280	.7107	.6930	.6748	.6562
4	.9542	.9490	.9434	.9373	.9308	.9238	.9163	.9083	.8997	.8906
5	.9952	.9945	.9937	.9927	.9917	.9905	.9892	.9878	.9862	.9844
6	1.0000	1.0000	1.0000	1.0000	1.0000	1.0000	1.0000	1.0000	1.0000	1.0000

n = 7

p x	.01	.02	.03	.04	.05	.06	.07	.08	.09	.10
0	.9321	.8681	.8080	.7514	.6983	.6485	.6017	.5578	.5168	.4783
1	.9980	.9921	.9829	.9706	.9556	.9382	.9187	.8974	.8745	.8503
2	1.0000	.9997	.9991	.9980	.9962	.9937	.9903	.9860	.9807	.9743
3	1.0000	1.0000	1.0000	.9999	.9998	.9996	.9993	.9988	.9982	.9973
4	1.0000	1.0000	1.0000	1.0000	1.0000	1.0000	1.0000	.9999	.9999	.9998
5	1.0000	1.0000	1.0000	1.0000	1.0000	1.0000	1.0000	1.0000	1.0000	1.0000

p x	.11	.12	.13	.14	.15	.16	.17	.18	.19	.20
0	.4423	.4087	.3773	.3479	.3206	.2951	.2714	.2493	.2288	.2097
1	.8250	.7988	.7719	.7444	.7166	.6885	.6604	.6323	.6044	.5767
2	.9669	.9584	.9487	.9380	.9262	.9134	.8995	.8846	.8687	.8520
3	.9961	.9946	.9928	.9906	.9879	.9847	.9811	.9769	.9721	.9667
4	.9997	.9996	.9994	.9991	.9988	.9983	.9978	.9971	.9963	.9953
5	1.0000	1.0000	1.0000	1.0000	.9999	.9999	.9999	.9998	.9997	.9996
6	1.0000	1.0000	1.0000	1.0000	1.0000	1.0000	1.0000	1.0000	1.0000	1.0000

p x	.21	.22	.23	.24	.25	.26	.27	.28	.29	.30
0	.1920	.1757	.1605	.1465	.1335	.1215	.1105	.1003	.0910	.0824
1	.5494	.5225	.4960	.4702	.4449	.4204	.3965	.3734	.3510	.3294
2	.8343	.8159	.7967	.7769	.7564	.7354	.7139	.6919	.6696	.6471
3	.9606	.9539	.9464	.9383	.9294	.9198	.9095	.8984	.8866	.8740
4	.9942	.9928	.9912	.9893	.9871	.9847	.9819	.9787	.9752	.9712
5	.9995	.9994	.9992	.9989	.9987	.9983	.9979	.9974	.9969	.9962
6	1.0000	1.0000	1.0000	1.0000	.9999	.9999	.9999	.9999	.9998	.9998
7	1.0000	1.0000	1.0000	1.0000	1.0000	1.0000	1.0000	1.0000	1.0000	1.0000

TABLE A (*Continued*)

$n = 7$ (Continued)

p x	.31	.32	.33	.34	.35	.36	.37	.38	.39	.40
0	.0745	.0672	.0606	.0546	.0490	.0440	.0394	.0352	.0314	.0280
1	.3086	.2887	.2696	.2513	.2338	.2172	.2013	.1863	.1721	.1586
2	.6243	.6013	.5783	.5553	.5323	.5094	.4866	.4641	.4419	.4199
3	.8606	.8466	.8318	.8163	.8002	.7833	.7659	.7479	.7293	.7102
4	.9668	.9620	.9566	.9508	.9444	.9375	.9299	.9218	.9131	.9037
5	.9954	.9945	.9935	.9923	.9910	.9895	.9877	.9858	.9836	.9812
6	.9997	.9997	.9996	.9995	.9994	.9992	.9991	.9989	.9986	.9984
7	1.0000	1.0000	1.0000	1.0000	1.0000	1.0000	1.0000	1.0000	1.0000	1.0000

p x	.41	.42	.43	.44	.45	.46	.47	.48	.49	.50
0	.0249	.0221	.0195	.0173	.0152	.0134	.0117	.0103	.0090	.0078
1	.1459	.1340	.1228	.1123	.1024	.0932	.0847	.0767	.0693	.0625
2	.3983	.3771	.3564	.3362	.3164	.2973	.2787	.2607	.2433	.2266
3	.6906	.6706	.6502	.6294	.6083	.5869	.5654	.5437	.5219	.5000
4	.8937	.8831	.8718	.8598	.8471	.8337	.8197	.8049	.7895	.7734
5	.9784	.9754	.9721	.9684	.9643	.9598	.9549	.9496	.9438	.9375
6	.9981	.9977	.9973	.9968	.9963	.9956	.9949	.9941	.9932	.9922
7	1.0000	1.0000	1.0000	1.0000	1.0000	1.0000	1.0000	1.0000	1.0000	1.0000

$n = 8$

p x	.01	.02	.03	.04	.05	.06	.07	.08	.09	.10
0	.9227	.8508	.7837	.7214	.6634	.6096	.5596	.5132	.4703	.4305
1	.9973	.9897	.9777	.9619	.9428	.9208	.8965	.8702	.8423	.8131
2	.9999	.9996	.9987	.9969	.9942	.9904	.9853	.9789	.9711	.9619
3	1.0000	1.0000	.9999	.9998	.9996	.9993	.9987	.9978	.9966	.9950
4	1.0000	1.0000	1.0000	1.0000	1.0000	1.0000	.9999	.9999	.9997	.9996
5	1.0000	1.0000	1.0000	1.0000	1.0000	1.0000	1.0000	1.0000	1.0000	1.0000

p x	.11	.12	.13	.14	.15	.16	.17	.18	.19	.20
0	.3937	.3596	.3282	.2992	.2725	.2479	.2252	.2044	.1853	.1678
1	.7829	.7520	.7206	.6889	.6572	.6256	.5943	.5634	.5330	.5033
2	.9513	.9392	.9257	.9109	.8948	.8774	.8588	.8392	.8185	.7969
3	.9929	.9903	.9871	.9832	.9786	.9733	.9672	.9603	.9524	.9437
4	.9993	.9990	.9985	.9979	.9971	.9962	.9950	.9935	.9917	.9896
5	1.0000	.9999	.9999	.9998	.9998	.9997	.9995	.9993	.9991	.9988
6	1.0000	1.0000	1.0000	1.0000	1.0000	1.0000	1.0000	1.0000	.9999	.9999
7	1.0000	1.0000	1.0000	1.0000	1.0000	1.0000	1.0000	1.0000	1.0000	1.0000

TABLE A (*Continued*)

$n = 8$ (Continued)

x \ p	.21	.22	.23	.24	.25	.26	.27	.28	.29	.30
0	.1517	.1370	.1236	.1113	.1001	.0899	.0806	.0722	.0646	.0576
1	.4743	.4462	.4189	.3925	.3671	.3427	.3193	.2969	.2756	.2553
2	.7745	.7514	.7276	.7033	.6785	.6535	.6282	.6027	.5772	.5518
3	.9341	.9235	.9120	.8996	.8862	.8719	.8567	.8406	.8237	.8059
4	.9871	.9842	.9809	.9770	.9727	.9678	.9623	.9562	.9495	.9420
5	.9984	.9979	.9973	.9966	.9958	.9948	.9936	.9922	.9906	.9887
6	.9999	.9998	.9998	.9997	.9996	.9995	.9994	.9992	.9990	.9987
7	1.0000	1.0000	1.0000	1.0000	1.0000	1.0000	1.0000	1.0000	.9999	.9999
8	1.0000	1.0000	1.0000	1.0000	1.0000	1.0000	1.0000	1.0000	1.0000	1.0000

x \ p	.31	.32	.33	.34	.35	.36	.37	.38	.39	.40
0	.0514	.0457	.0406	.0360	.0319	.0281	.0248	.0218	.0192	.0168
1	.2360	.2178	.2006	.1844	.1691	.1548	.1414	.1289	.1172	.1064
2	.5264	.5013	.4764	.4519	.4278	.4042	.3811	.3585	.3366	.3154
3	.7874	.7681	.7481	.7276	.7064	.6847	.6626	.6401	.6172	.5941
4	.9339	.9250	.9154	.9051	.8939	.8820	.8693	.8557	.8414	.8263
5	.9866	.9841	.9813	.9782	.9747	.9707	.9664	.9615	.9561	.9502
6	.9984	.9980	.9976	.9970	.9964	.9957	.9949	.9939	.9928	.9915
7	.9999	.9999	.9999	.9998	.9998	.9997	.9996	.9996	.9995	.9993
8	1.0000	1.0000	1.0000	1.0000	1.0000	1.0000	1.0000	1.0000	1.0000	1.0000

x \ p	.41	.42	.43	.44	.45	.46	.47	.48	.49	.50
0	.0147	.0128	.0111	.0097	.0084	.0072	.0062	.0053	.0046	.0039
1	.0963	.0870	.0784	.0705	.0632	.0565	.0504	.0448	.0398	.0352
2	.2948	.2750	.2560	.2376	.2201	.2034	.1875	.1724	.1581	.1445
3	.5708	.5473	.5238	.5004	.4770	.4537	.4306	.4078	.3854	.3633
4	.8105	.7938	.7765	.7584	.7396	.7202	.7001	.6795	.6584	.6367
5	.9437	.9366	.9289	.9206	.9115	.9018	.8914	.8802	.8682	.8555
6	.9900	.9883	.9864	.9843	.9819	.9792	.9761	.9728	.9690	.9648
7	.9992	.9990	.9988	.9986	.9983	.9980	.9976	.9972	.9967	.9961
8	1.0000	1.0000	1.0000	1.0000	1.0000	1.0000	1.0000	1.0000	1.0000	1.0000

$n = 9$

x \ p	.01	.02	.03	.04	.05	.06	.07	.08	.09	.10
0	.9135	.8337	.7602	.6925	.6302	.5730	.5204	.4722	.4279	.3874
1	.9966	.9869	.9718	.9522	.9288	.9022	.8729	.8417	.8088	.7748
2	.9999	.9994	.9980	.9955	.9916	.9862	.9791	.9702	.9595	.9470
3	1.0000	1.0000	.9999	.9997	.9994	.9987	.9977	.9963	.9943	.9917
4	1.0000	1.0000	1.0000	1.0000	1.0000	.9999	.9998	.9997	.9995	.9991
5	1.0000	1.0000	1.0000	1.0000	1.0000	1.0000	1.0000	1.0000	1.0000	.9999
6	1.0000	1.0000	1.0000	1.0000	1.0000	1.0000	1.0000	1.0000	1.0000	1.0000

TABLE A (*Continued*)

$n = 9$ (Continued)

x \ p	.11	.12	.13	.14	.15	.16	.17	.18	.19	.20
0	.3504	.3165	.2855	.2573	.2316	.2082	.1869	.1676	.1501	.1342
1	.7401	.7049	.6696	.6343	.5995	.5652	.5315	.4988	.4670	.4362
2	.9327	.9167	.8991	.8798	.8591	.8371	.8139	.7895	.7643	.7382
3	.9883	.9842	.9791	.9731	.9661	.9580	.9488	.9385	.9270	.9144
4	.9986	.9979	.9970	.9959	.9944	.9925	.9902	.9875	.9842	.9804
5	.9999	.9998	.9997	.9996	.9994	.9991	.9987	.9983	.9977	.9969
6	1.0000	1.0000	1.0000	1.0000	1.0000	.9999	.9999	.9998	.9998	.9997
7	1.0000	1.0000	1.0000	1.0000	1.0000	1.0000	1.0000	1.0000	1.0000	1.0000

x \ p	.21	.22	.23	.24	.25	.26	.27	.28	.29	.30
0	.1199	.1069	.0952	.0846	.0751	.0665	.0589	.0520	.0458	.0404
1	.4066	.3782	.3509	.3250	.3003	.2770	.2548	.2340	.2144	.1960
2	.7115	.6842	.6566	.6287	.6007	.5727	.5448	.5171	.4898	.4628
3	.9006	.8856	.8696	.8525	.8343	.8151	.7950	.7740	.7522	.7297
4	.9760	.9709	.9650	.9584	.9511	.9429	.9338	.9238	.9130	.9012
5	.9960	.9949	.9935	.9919	.9900	.9878	.9851	.9821	.9787	.9747
6	.9996	.9994	.9992	.9990	.9987	.9983	.9978	.9972	.9965	.9957
7	1.0000	1.0000	.9999	.9999	.9999	.9999	.9998	.9997	.9997	.9996
8	1.0000	1.0000	1.0000	1.0000	1.0000	1.0000	1.0000	1.0000	1.0000	1.0000

x \ p	.31	.32	.33	.34	.35	.36	.37	.38	.39	.40
0	.0355	.0311	.0272	.0238	.0207	.0180	.0156	.0135	.0117	.0101
1	.1788	.1628	.1478	.1339	.1211	.1092	.0983	.0882	.0790	.0705
2	.4364	.4106	.3854	.3610	.3373	.3144	.2924	.2713	.2511	.2318
3	.7065	.6827	.6585	.6338	.6089	.5837	.5584	.5331	.5078	.4826
4	.8885	.8748	.8602	.8447	.8283	.8110	.7928	.7738	.7540	.7334
5	.9702	.9652	.9596	.9533	.9464	.9388	.9304	.9213	.9114	.9006
6	.9947	.9936	.9922	.9906	.9888	.9867	.9843	.9816	.9785	.9750
7	.9994	.9993	.9991	.9989	.9986	.9983	.9979	.9974	.9969	.9962
8	1.0000	1.0000	1.0000	.9999	.9999	.9999	.9999	.9998	.9998	.9997
9	1.0000	1.0000	1.0000	1.0000	1.0000	1.0000	1.0000	1.0000	1.0000	1.0000

x \ p	.41	.42	.43	.44	.45	.46	.47	.48	.49	.50
0	.0087	.0074	.0064	.0054	.0046	.0039	.0033	.0028	.0023	.0020
1	.0628	.0558	.0495	.0437	.0385	.0338	.0296	.0259	.0225	.0195
2	.2134	.1961	.1796	.1641	.1495	.1358	.1231	.1111	.1001	.0898
3	.4576	.4330	.4087	.3848	.3614	.3386	.3164	.2948	.2740	.2539
4	.7122	.6903	.6678	.6449	.6214	.5976	.5735	.5491	.5246	.5000
5	.8891	.8767	.8634	.8492	.8342	.8183	.8015	.7839	.7654	.7461
6	.9710	.9666	.9617	.9563	.9502	.9436	.9363	.9283	.9196	.9102
7	.9954	.9945	.9935	.9923	.9909	.9893	.9875	.9855	.9831	.9805
8	.9997	.9996	.9995	.9994	.9992	.9991	.9989	.9986	.9984	.9980
9	1.0000	1.0000	1.0000	1.0000	1.0000	1.0000	1.0000	1.0000	1.0000	1.0000

TABLE A (*Continued*)

$n = 10$

p x	.01	.02	.03	.04	.05	.06	.07	.08	.09	.10
0	.9044	.8171	.7374	.6648	.5987	.5386	.4840	.4344	.3894	.3487
1	.9957	.9838	.9655	.9418	.9139	.8824	.8483	.8121	.7746	.7361
2	.9999	.9991	.9972	.9938	.9885	.9812	.9717	.9599	.9460	.9298
3	1.0000	1.0000	.9999	.9996	.9990	.9980	.9964	.9942	.9912	.9872
4	1.0000	1.0000	1.0000	1.0000	.9999	.9998	.9997	.9994	.9990	.9984
5	1.0000	1.0000	1.0000	1.0000	1.0000	1.0000	1.0000	1.0000	.9999	.9999
6	1.0000	1.0000	1.0000	1.0000	1.0000	1.0000	1.0000	1.0000	1.0000	1.0000

p x	.11	.12	.13	.14	.15	.16	.17	.18	.19	.20
0	.3118	.2785	.2484	.2213	.1969	.1749	.1552	.1374	.1216	.1074
1	.6972	.6583	.6196	.5816	.5443	.5080	.4730	.4392	.4068	.3758
2	.9116	.8913	.8692	.8455	.8202	.7936	.7659	.7372	.7078	.6778
3	.9822	.9761	.9687	.9600	.9500	.9386	.9259	.9117	.8961	.8791
4	.9975	.9963	.9947	.9927	.9901	.9870	.9832	.9787	.9734	.9672
5	.9997	.9996	.9994	.9990	.9986	.9980	.9973	.9963	.9951	.9936
6	1.0000	1.0000	.9999	.9999	.9999	.9998	.9997	.9996	.9994	.9991
7	1.0000	1.0000	1.0000	1.0000	1.0000	1.0000	1.0000	1.0000	.9999	.9999
8	1.0000	1.0000	1.0000	1.0000	1.0000	1.0000	1.0000	1.0000	1.0000	1.0000

p x	.21	.22	.23	.24	.25	.26	.27	.28	.29	.30
0	.0947	.0834	.0733	.0643	.0563	.0492	.0430	.0374	.0326	.0282
1	.3464	.3185	.2921	.2673	.2440	.2222	.2019	.1830	.1655	.1493
2	.6474	.6169	.5863	.5558	.5256	.4958	.4665	.4378	.4099	.3828
3	.8609	.8413	.8206	.7988	.7759	.7521	.7274	.7021	.6761	.6496
4	.9601	.9521	.9431	.9330	.9219	.9096	.8963	.8819	.8663	.8497
5	.9918	.9896	.9870	.9839	.9803	.9761	.9713	.9658	.9596	.9527
6	.9988	.9984	.9979	.9973	.9965	.9955	.9944	.9930	.9913	.9894
7	.9999	.9998	.9998	.9997	.9996	.9994	.9993	.9990	.9988	.9984
8	1.0000	1.0000	1.0000	1.0000	1.0000	1.0000	.9999	.9999	.9999	.9999
9	1.0000	1.0000	1.0000	1.0000	1.0000	1.0000	1.0000	1.0000	1.0000	1.0000

p x	.31	.32	.33	.34	.35	.36	.37	.38	.39	.40
0	.0245	.0211	.0182	.0157	.0135	.0115	.0098	.0084	.0071	.0060
1	.1344	.1206	.1080	.0965	.0860	.0764	.0677	.0598	.0527	.0464
2	.3566	.3313	.3070	.2838	.2616	.2405	.2206	.2017	.1840	.1673
3	.6228	.5956	.5684	.5411	.5138	.4868	.4600	.4336	.4077	.3823
4	.8321	.8133	.7936	.7730	.7515	.7292	.7061	.6823	.6580	.6331
5	.9449	.9363	.9268	.9164	.9051	.8928	.8795	.8652	.8500	.8338
6	.9871	.9845	.9815	.9780	.9740	.9695	.9644	.9587	.9523	.9452
7	.9980	.9975	.9968	.9961	.9952	.9941	.9929	.9914	.9897	.9877
8	.9998	.9997	.9997	.9996	.9995	.9993	.9991	.9989	.9986	.9983
9	1.0000	1.0000	1.0000	1.0000	1.0000	1.0000	1.0000	.9999	.9999	.9999
10	1.0000	1.0000	1.0000	1.0000	1.0000	1.0000	1.0000	1.0000	1.0000	1.0000

TABLE A (*Continued*)

$n = 10$ (Continued)

p x	.41	.42	.43	.44	.45	.46	.47	.48	.49	.50
0	.0051	.0043	.0036	.0030	.0025	.0021	.0017	.0014	.0012	.0010
1	.0406	.0355	.0309	.0269	.0233	.0201	.0173	.0148	.0126	.0107
2	.1517	.1372	.1236	.1111	.0996	.0889	.0791	.0702	.0621	.0547
3	.3575	.3335	.3102	.2877	.2660	.2453	.2255	.2067	.1888	.1719
4	.6078	.5822	.5564	.5304	.5044	.4784	.4526	.4270	.4018	.3770
5	.8166	.7984	.7793	.7593	.7384	.7168	.6943	.6712	.6474	.6230
6	.9374	.9288	.9194	.9092	.8980	.8859	.8729	.8590	.8440	.8281
7	.9854	.9828	.9798	.9764	.9726	.9683	.9634	.9580	.9520	.9453
8	.9979	.9975	.9969	.9963	.9955	.9946	.9935	.9923	.9909	.9893
9	.9999	.9998	.9998	.9997	.9997	.9996	.9995	.9994	.9992	.9990
10	1.0000	1.0000	1.0000	1.0000	1.0000	1.0000	1.0000	1.0000	1.0000	1.0000

$n = 11$

p x	.01	.02	.03	.04	.05	.06	.07	.08	.09	.10
0	.8953	.8007	.7153	.6382	.5688	.5063	.4501	.3996	.3544	.3138
1	.9948	.9805	.9587	.9308	.8981	.8618	.8228	.7819	.7399	.6974
2	.9998	.9988	.9963	.9917	.9848	.9752	.9630	.9481	.9305	.9104
3	1.0000	1.0000	.9998	.9993	.9984	.9970	.9947	.9915	.9871	.9815
4	1.0000	1.0000	1.0000	1.0000	.9999	.9997	.9995	.9990	.9983	.9972
5	1.0000	1.0000	1.0000	1.0000	1.0000	1.0000	1.0000	.9999	.9998	.9997
6	1.0000	1.0000	1.0000	1.0000	1.0000	1.0000	1.0000	1.0000	1.0000	1.0000

p x	.11	.12	.13	.14	.15	.16	.17	.18	.19	.20
0	.2775	.2451	.2161	.1903	.1673	.1469	.1288	.1127	.0985	.0859
1	.6548	.6127	.5714	.5311	.4922	.4547	.4189	.3849	.3526	.3221
2	.8880	.8634	.8368	.8085	.7788	.7479	.7161	.6836	.6506	.6174
3	.9744	.9659	.9558	.9440	.9306	.9154	.8987	.8803	.8603	.8389
4	.9958	.9939	.9913	.9881	.9841	.9793	.9734	.9666	.9587	.9496
5	.9995	.9992	.9988	.9982	.9973	.9963	.9949	.9932	.9910	.9883
6	1.0000	.9999	.9999	.9998	.9997	.9995	.9993	.9990	.9986	.9980
7	1.0000	1.0000	1.0000	1.0000	1.0000	1.0000	.9999	.9999	.9998	.9998
8	1.0000	1.0000	1.0000	1.0000	1.0000	1.0000	1.0000	1.0000	1.0000	1.0000

p x	.21	.22	.23	.24	.25	.26	.27	.28	.29	.30
0	.0748	.0650	.0564	.0489	.0422	.0364	.0314	.0270	.0231	.0198
1	.2935	.2667	.2418	.2186	.1971	.1773	.1590	.1423	.1270	.1130
2	.5842	.5512	.5186	.4866	.4552	.4247	.3951	.3665	.3390	.3127
3	.8160	.7919	.7667	.7404	.7133	.6854	.6570	.6281	.5989	.5696
4	.9393	.9277	.9149	.9008	.8854	.8687	.8507	.8315	.8112	.7897
5	.9852	.9814	.9769	.9717	.9657	.9588	.9510	.9423	.9326	.9218
6	.9973	.9965	.9954	.9941	.9924	.9905	.9881	.9854	.9821	.9784
7	.9997	.9995	.9993	.9991	.9988	.9984	.9979	.9973	.9966	.9957
8	1.0000	1.0000	.9999	.9999	.9999	.9998	.9998	.9997	.9996	.9994
9	1.0000	1.0000	1.0000	1.0000	1.0000	1.0000	1.0000	1.0000	1.0000	1.0000

TABLE A (*Continued*)

$n = 11$ (Continued)

x \ p	.31	.32	.33	.34	.35	.36	.37	.38	.39	.40
0	.0169	.0144	.0122	.0104	.0088	.0074	.0062	.0052	.0044	.0036
1	.1003	.0888	.0784	.0690	.0606	.0530	.0463	.0403	.0350	.0302
2	.2877	.2639	.2413	.2201	.2001	.1814	.1640	.1478	.1328	.1189
3	.5402	.5110	.4821	.4536	.4256	.3981	.3714	.3455	.3204	.2963
4	.7672	.7437	.7193	.6941	.6683	.6419	.6150	.5878	.5603	.5328
5	.9099	.8969	.8829	.8676	.8513	.8339	.8153	.7957	.7751	.7535
6	.9740	.9691	.9634	.9570	.9499	.9419	.9330	.9232	.9124	.9006
7	.9946	.9933	.9918	.9899	.9878	.9852	.9823	.9790	.9751	.9707
8	.9992	.9990	.9987	.9984	.9980	.9974	.9968	.9961	.9952	.9941
9	.9999	.9999	.9999	.9998	.9998	.9997	.9996	.9995	.9994	.9993
10	1.0000	1.0000	1.0000	1.0000	1.0000	1.0000	1.0000	1.0000	1.0000	1.0000

x \ p	.41	.42	.43	.44	.45	.46	.47	.48	.49	.50
0	.0030	.0025	.0021	.0017	.0014	.0011	.0009	.0008	.0006	.0005
1	.0261	.0224	.0192	.0164	.0139	.0118	.0100	.0084	.0070	.0059
2	.1062	.0945	.0838	.0740	.0652	.0572	.0501	.0436	.0378	.0327
3	.2731	.2510	.2300	.2100	.1911	.1734	.1567	.1412	.1267	.1133
4	.5052	.4777	.4505	.4236	.3971	.3712	.3459	.3213	.2974	.2744
5	.7310	.7076	.6834	.6586	.6331	.6071	.5807	.5540	.5271	.5000
6	.8879	.8740	.8592	.8432	.8262	.8081	.7890	.7688	.7477	.7256
7	.9657	.9601	.9539	.9468	.9390	.9304	.9209	.9105	.8991	.8867
8	.9928	.9913	.9896	.9875	.9852	.9825	.9794	.9759	.9718	.9673
9	.9991	.9988	.9986	.9982	.9978	.9973	.9967	.9960	.9951	.9941
10	.9999	.9999	.9999	.9999	.9998	.9998	.9998	.9997	.9996	.9995
11	1.0000	1.0000	1.0000	1.0000	1.0000	1.0000	1.0000	1.0000	1.0000	1.0000

$n = 12$

x \ p	.01	.02	.03	.04	.05	.06	.07	.08	.09	.10
0	.8864	.7847	.6938	.6127	.5404	.4759	.4186	.3677	.3225	.2824
1	.9938	.9769	.9514	.9191	.8816	.8405	.7967	.7513	.7052	.6590
2	.9998	.9985	.9952	.9893	.9804	.9684	.9532	.9348	.9134	.8891
3	1.0000	.9999	.9997	.9990	.9978	.9957	.9925	.9880	.9820	.9744
4	1.0000	1.0000	1.0000	.9999	.9998	.9996	.9991	.9984	.9973	.9957
5	1.0000	1.0000	1.0000	1.0000	1.0000	1.0000	.9999	.9998	.9997	.9995
6	1.0000	1.0000	1.0000	1.0000	1.0000	1.0000	1.0000	1.0000	1.0000	.9999
7	1.0000	1.0000	1.0000	1.0000	1.0000	1.0000	1.0000	1.0000	1.0000	1.0000

TABLE A (*Continued*)

$n = 12$ (Continued)

x \ p	.11	.12	.13	.14	.15	.16	.17	.18	.19	.20
0	.2470	.2157	.1880	.1637	.1422	.1234	.1069	.0924	.0798	.0687
1	.6133	.5686	.5252	.4834	.4435	.4055	.3696	.3359	.3043	.2749
2	.8623	.8333	.8023	.7697	.7358	.7010	.6656	.6298	.5940	.5583
3	.9649	.9536	.9403	.9250	.9078	.8886	.8676	.8448	.8205	.7946
4	.9935	.9905	.9867	.9819	.9761	.9690	.9607	.9511	.9400	.9274
5	.9991	.9986	.9978	.9967	.9954	.9935	.9912	.9884	.9849	.9806
6	.9999	.9998	.9997	.9996	.9993	.9990	.9985	.9979	.9971	.9961
7	1.0000	1.0000	1.0000	1.0000	.9999	.9999	.9998	.9997	.9996	.9994
8	1.0000	1.0000	1.0000	1.0000	1.0000	1.0000	1.0000	1.0000	1.0000	.9999
9	1.0000	1.0000	1.0000	1.0000	1.0000	1.0000	1.0000	1.0000	1.0000	1.0000

x \ p	.21	.22	.23	.24	.25	.26	.27	.28	.29	.30
0	.0591	.0507	.0434	.0371	.0317	.0270	.0229	.0194	.0164	.0138
1	.2476	.2224	.1991	.1778	.1584	.1406	.1245	.1100	.0968	.0850
2	.5232	.4886	.4550	.4222	.3907	.3603	.3313	.3037	.2775	.2528
3	.7674	.7390	.7096	.6795	.6488	.6176	.5863	.5548	.5235	.4925
4	.9134	.8979	.8808	.8623	.8424	.8210	.7984	.7746	.7496	.7237
5	.9755	.9696	.9626	.9547	.9456	.9354	.9240	.9113	.8974	.8822
6	.9948	.9932	.9911	.9887	.9857	.9822	.9781	.9733	.9678	.9614
7	.9992	.9989	.9984	.9979	.9972	.9964	.9953	.9940	.9924	.9905
8	.9999	.9999	.9998	.9997	.9996	.9995	.9993	.9990	.9987	.9983
9	1.0000	1.0000	1.0000	1.0000	1.0000	.9999	.9999	.9999	.9998	.9998
10	1.0000	1.0000	1.0000	1.0000	1.0000	1.0000	1.0000	1.0000	1.0000	1.0000

x \ p	.31	.32	.33	.34	.35	.36	.37	.38	.39	.40
0	.0116	.0098	.0082	.0068	.0057	.0047	.0039	.0032	.0027	.0022
1	.0744	.0650	.0565	.0491	.0424	.0366	.0315	.0270	.0230	.0196
2	.2296	.2078	.1876	.1687	.1513	.1352	.1205	.1069	.0946	.0834
3	.4619	.4319	.4027	.3742	.3467	.3201	.2947	.2704	.2472	.2253
4	.6968	.6692	.6410	.6124	.5833	.5541	.5249	.4957	.4668	.4382
5	.8657	.8479	.8289	.8087	.7873	.7648	.7412	.7167	.6913	.6652
6	.9542	.9460	.9368	.9266	.9154	.9030	.8894	.8747	.8589	.8418
7	.9882	.9856	.9824	.9787	.9745	.9696	.9641	.9578	.9507	.9427
8	.9978	.9972	.9964	.9955	.9944	.9930	.9915	.9896	.9873	.9847
9	.9997	.9996	.9995	.9993	.9992	.9989	.9986	.9982	.9978	.9972
10	1.0000	1.0000	1.0000	.9999	.9999	.9999	.9999	.9998	.9998	.9997
11	1.0000	1.0000	1.0000	1.0000	1.0000	1.0000	1.0000	1.0000	1.0000	1.0000

TABLE A (*Continued*)

n = 12 (Continued)

p\x	.41	.42	.43	.44	.45	.46	.47	.48	.49	.50
0	.0018	.0014	.0012	.0010	.0008	.0006	.0005	.0004	.0003	.0002
1	.0166	.0140	.0118	.0099	.0083	.0069	.0057	.0047	.0039	.0032
2	.0733	.0642	.0560	.0487	.0421	.0363	.0312	.0267	.0227	.0193
3	.2047	.1853	.1671	.1502	.1345	.1199	.1066	.0943	.0832	.0730
4	.4101	.3825	.3557	.3296	.3044	.2802	.2570	.2348	.2138	.1938
5	.6384	.6111	.5833	.5552	.5269	.4986	.4703	.4423	.4145	.3872
6	.8235	.8041	.7836	.7620	.7393	.7157	.6911	.6657	.6396	.6128
7	.9338	.9240	.9131	.9012	.8883	.8742	.8589	.8425	.8249	.8062
8	.9817	.9782	.9742	.9696	.9644	.9585	.9519	.9445	.9362	.9270
9	.9965	.9957	.9947	.9935	.9921	.9905	.9886	.9863	.9837	.9807
10	.9996	.9995	.9993	.9991	.9989	.9986	.9983	.9979	.9974	.9968
11	1.0000	1.0000	1.0000	.9999	.9999	.9999	.9999	.9999	.9998	.9998
12	1.0000	1.0000	1.0000	1.0000	1.0000	1.0000	1.0000	1.0000	1.0000	1.0000

n = 13

p\x	.01	.02	.03	.04	.05	.06	.07	.08	.09	.10
0	.8775	.7690	.6730	.5882	.5133	.4474	.3893	.3383	.2935	.2542
1	.9928	.9730	.9436	.9068	.8646	.8186	.7702	.7206	.6707	.6213
2	.9997	.9980	.9938	.9865	.9755	.9608	.9422	.9201	.8946	.8661
3	1.0000	.9999	.9995	.9986	.9969	.9940	.9897	.9837	.9758	.9658
4	1.0000	1.0000	1.0000	.9999	.9997	.9993	.9987	.9976	.9959	.9935
5	1.0000	1.0000	1.0000	1.0000	1.0000	.9999	.9999	.9997	.9995	.9991
6	1.0000	1.0000	1.0000	1.0000	1.0000	1.0000	1.0000	1.0000	.9999	.9999
7	1.0000	1.0000	1.0000	1.0000	1.0000	1.0000	1.0000	1.0000	1.0000	1.0000

p\x	.11	.12	.13	.14	.15	.16	.17	.18	.19	.20
0	.2198	.1898	.1636	.1408	.1209	.1037	.0887	.0758	.0646	.0550
1	.5730	.5262	.4814	.4386	.3983	.3604	.3249	.2920	.2616	.2336
2	.8349	.8015	.7663	.7296	.6920	.6537	.6152	.5769	.5389	.5017
3	.9536	.9391	.9224	.9033	.8820	.8586	.8333	.8061	.7774	.7473
4	.9903	.9861	.9807	.9740	.9658	.9562	.9449	.9319	.9173	.9009
5	.9985	.9976	.9964	.9947	.9925	.9896	.9861	.9817	.9763	.9700
6	.9998	.9997	.9995	.9992	.9987	.9981	.9973	.9962	.9948	.9930
7	1.0000	1.0000	.9999	.9999	.9998	.9997	.9996	.9994	.9991	.9988
8	1.0000	1.0000	1.0000	1.0000	1.0000	1.0000	1.0000	.9999	.9999	.9998
9	1.0000	1.0000	1.0000	1.0000	1.0000	1.0000	1.0000	1.0000	1.0000	1.0000

TABLE A (*Continued*)

$n = 13$ (Continued)

x \ p	.21	.22	.23	.24	.25	.26	.27	.28	.29	.30
0	.0467	.0396	.0334	.0282	.0238	.0200	.0167	.0140	.0117	.0097
1	.2080	.1846	.1633	.1441	.1267	.1111	.0971	.0846	.0735	.0637
2	.4653	.4301	.3961	.3636	.3326	.3032	.2755	.2495	.2251	.2025
3	.7161	.6839	.6511	.6178	.5843	.5507	.5174	.4845	.4522	.4206
4	.8827	.8629	.8415	.8184	.7940	.7681	.7411	.7130	.6840	.6543
5	.9625	.9538	.9438	.9325	.9198	.9056	.8901	.8730	.8545	.8346
6	.9907	.9880	.9846	.9805	.9757	.9701	.9635	.9560	.9473	.9376
7	.9983	.9976	.9968	.9957	.9944	.9927	.9907	.9882	.9853	.9818
8	.9998	.9996	.9995	.9993	.9990	.9987	.9982	.9976	.9969	.9960
9	1.0000	1.0000	.9999	.9999	.9999	.9998	.9997	.9996	.9995	.9993
10	1.0000	1.0000	1.0000	1.0000	1.0000	1.0000	1.0000	1.0000	.9999	.9999
11	1.0000	1.0000	1.0000	1.0000	1.0000	1.0000	1.0000	1.0000	1.0000	1.0000

x \ p	.31	.32	.33	.34	.35	.36	.37	.38	.39	.40
0	.0080	.0066	.0055	.0045	.0037	.0030	.0025	.0020	.0016	.0013
1	.0550	.0473	.0406	.0347	.0296	.0251	.0213	.0179	.0151	.0126
2	.1815	.1621	.1443	.1280	.1132	.0997	.0875	.0765	.0667	.0579
3	.3899	.3602	.3317	.3043	.2783	.2536	.2302	.2083	.1877	.1686
4	.6240	.5933	.5624	.5314	.5005	.4699	.4397	.4101	.3812	.3530
5	.8133	.7907	.7669	.7419	.7159	.6889	.6612	.6327	.6038	.5744
6	.9267	.9146	.9012	.8865	.8705	.8532	.8346	.8147	.7935	.7712
7	.9777	.9729	.9674	.9610	.9538	.9456	.9365	.9262	.9149	.9023
8	.9948	.9935	.9918	.9898	.9874	.9846	.9813	.9775	.9730	.9679
9	.9991	.9988	.9985	.9980	.9975	.9968	.9960	.9949	.9937	.9922
10	.9999	.9999	.9998	.9997	.9997	.9995	.9994	.9992	.9990	.9987
11	1.0000	1.0000	1.0000	1.0000	1.0000	1.0000	.9999	.9999	.9999	.9999
12	1.0000	1.0000	1.0000	1.0000	1.0000	1.0000	1.0000	1.0000	1.0000	1.0000

x \ p	.41	.42	.43	.44	.45	.46	.47	.48	.49	.50
0	.0010	.0008	.0007	.0005	.0004	.0003	.0003	.0002	.0002	.0001
1	.0105	.0088	.0072	.0060	.0049	.0040	.0033	.0026	.0021	.0017
2	.0501	.0431	.0370	.0316	.0269	.0228	.0192	.0162	.0135	.0112
3	.1508	.1344	.1193	.1055	.0929	.0815	.0712	.0619	.0536	.0461
4	.3258	.2997	.2746	.2507	.2279	.2065	.1863	.1674	.1498	.1334
5	.5448	.5151	.4854	.4559	.4268	.3981	.3701	.3427	.3162	.2905
6	.7476	.7230	.6975	.6710	.6437	.6158	.5873	.5585	.5293	.5000
7	.8886	.8736	.8574	.8400	.8212	.8012	.7800	.7576	.7341	.7095
8	.9621	.9554	.9480	.9395	.9302	.9197	.9082	.8955	.8817	.8666
9	.9904	.9883	.9859	.9830	.9797	.9758	.9713	.9662	.9604	.9539
10	.9983	.9979	.9973	.9967	.9959	.9949	.9937	.9923	.9907	.9888
11	.9998	.9998	.9997	.9996	.9995	.9993	.9991	.9989	.9986	.9983
12	1.0000	1.0000	1.0000	1.0000	1.0000	1.0000	.9999	.9999	.9999	.9999
13	1.0000	1.0000	1.0000	1.0000	1.0000	1.0000	1.0000	1.0000	1.0000	1.0000

TABLE A (*Continued*)

$n = 14$

x \ p	.01	.02	.03	.04	.05	.06	.07	.08	.09	.10
0	.8687	.7536	.6528	.5647	.4877	.4205	.3620	.3112	.2670	.2288
1	.9916	.9690	.9355	.8941	.8470	.7963	.7436	.6900	.6368	.5846
2	.9997	.9975	.9923	.9833	.9699	.9522	.9302	.9042	.8745	.8416
3	1.0000	.9999	.9994	.9981	.9958	.9920	.9864	.9786	.9685	.9559
4	1.0000	1.0000	1.0000	.9998	.9996	.9990	.9980	.9965	.9941	.9908
5	1.0000	1.0000	1.0000	1.0000	1.0000	.9999	.9998	.9996	.9992	.9985
6	1.0000	1.0000	1.0000	1.0000	1.0000	1.0000	1.0000	1.0000	.9999	.9998
7	1.0000	1.0000	1.0000	1.0000	1.0000	1.0000	1.0000	1.0000	1.0000	1.0000

x \ p	.11	.12	.13	.14	.15	.16	.17	.18	.19	.20
0	.1956	.1670	.1423	.1211	.1028	.0871	.0736	.0621	.0523	.0440
1	.5342	.4859	.4401	.3969	.3567	.3193	.2848	.2531	.2242	.1979
2	.8061	.7685	.7292	.6889	.6479	.6068	.5659	.5256	.4862	.4481
3	.9406	.9226	.9021	.8790	.8535	.8258	.7962	.7649	.7321	.6982
4	.9863	.9804	.9731	.9641	.9533	.9406	.9259	.9093	.8907	.8702
5	.9976	.9962	.9943	.9918	.9885	.9843	.9791	.9727	.9651	.9561
6	.9997	.9994	.9991	.9985	.9978	.9968	.9954	.9936	.9913	.9884
7	1.0000	.9999	.9999	.9998	.9997	.9995	.9992	.9988	.9983	.9976
8	1.0000	1.0000	1.0000	1.0000	1.0000	.9999	.9999	.9998	.9997	.9996
9	1.0000	1.0000	1.0000	1.0000	1.0000	1.0000	1.0000	1.0000	1.0000	1.0000

x \ p	.21	.22	.23	.24	.25	.26	.27	.28	.29	.30
0	.0369	.0309	.0258	.0214	.0178	.0148	.0122	.0101	.0083	.0068
1	.1741	.1527	.1335	.1163	.1010	.0874	.0754	.0648	.0556	.0475
2	.4113	.3761	.3426	.3109	.2811	.2533	.2273	.2033	.1812	.1608
3	.6634	.6281	.5924	.5568	.5213	.4864	.4521	.4187	.3863	.3552
4	.8477	.8235	.7977	.7703	.7415	.7116	.6807	.6490	.6168	.5842
5	.9457	.9338	.9203	.9051	.8883	.8699	.8498	.8282	.8051	.7805
6	.9848	.9804	.9752	.9690	.9617	.9533	.9437	.9327	.9204	.9067
7	.9967	.9955	.9940	.9921	.9897	.9868	.9833	.9792	.9743	.9685
8	.9994	.9992	.9989	.9984	.9978	.9971	.9962	.9950	.9935	.9917
9	.9999	.9999	.9998	.9998	.9997	.9995	.9993	.9991	.9988	.9983
10	1.0000	1.0000	1.0000	1.0000	1.0000	.9999	.9999	.9999	.9998	.9998
11	1.0000	1.0000	1.0000	1.0000	1.0000	1.0000	1.0000	1.0000	1.0000	1.0000

TABLE A (*Continued*)

n = 14 (Continued)

x \ p	.31	.32	.33	.34	.35	.36	.37	.38	.39	.40
0	.0055	.0045	.0037	.0030	.0024	.0019	.0016	.0012	.0010	.0008
1	.0404	.0343	.0290	.0244	.0205	.0172	.0143	.0119	.0098	.0081
2	.1423	.1254	.1101	.0963	.0839	.0729	.0630	.0543	.0466	.0398
3	.3253	.2968	.2699	.2444	.2205	.1982	.1774	.1582	.1405	.1243
4	.5514	.5187	.4862	.4542	.4227	.3920	.3622	.3334	.3057	.2793
5	.7546	.7276	.6994	.6703	.6405	.6101	.5792	.5481	.5169	.4859
6	.8916	.8750	.8569	.8374	.8164	.7941	.7704	.7455	.7195	.6925
7	.9619	.9542	.9455	.9357	.9247	.9124	.8988	.8838	.8675	.8499
8	.9895	.9869	.9837	.9800	.9757	.9706	.9647	.9580	.9503	.9417
9	.9978	.9971	.9963	.9952	.9940	.9924	.9905	.9883	.9856	.9825
10	.9997	.9995	.9994	.9992	.9989	.9986	.9981	.9976	.9969	.9961
11	1.0000	.9999	.9999	.9999	.9999	.9998	.9997	.9997	.9995	.9994
12	1.0000	1.0000	1.0000	1.0000	1.0000	1.0000	1.0000	1.0000	1.0000	.9999
13	1.0000	1.0000	1.0000	1.0000	1.0000	1.0000	1.0000	1.0000	1.0000	1.0000

x \ p	.41	.42	.43	.44	.45	.46	.47	.48	.49	.50
0	.0006	.0005	.0004	.0003	.0002	.0002	.0001	.0001	.0001	.0001
1	.0066	.0054	.0044	.0036	.0029	.0023	.0019	.0015	.0012	.0009
2	.0339	.0287	.0242	.0203	.0170	.0142	.0117	.0097	.0079	.0065
3	.1095	.0961	.0839	.0730	.0632	.0545	.0468	.0399	.0339	.0287
4	.2541	.2303	.2078	.1868	.1672	.1490	.1322	.1167	.1026	.0898
5	.4550	.4246	.3948	.3656	.3373	.3100	.2837	.2585	.2346	.2120
6	.6645	.6357	.6063	.5764	.5461	.5157	.4852	.4549	.4249	.3953
7	.8308	.8104	.7887	.7656	.7414	.7160	.6895	.6620	.6337	.6047
8	.9320	.9211	.9090	.8957	.8811	.8652	.8480	.8293	.8094	.7880
9	.9788	.9745	.9696	.9639	.9574	.9500	.9417	.9323	.9218	.9102
10	.9951	.9939	.9924	.9907	.9886	.9861	.9832	.9798	.9759	.9713
11	.9992	.9990	.9987	.9983	.9978	.9973	.9966	.9958	.9947	.9935
12	.9999	.9999	.9999	.9998	.9997	.9997	.9996	.9994	.9993	.9991
13	1.0000	1.0000	1.0000	1.0000	1.0000	1.0000	1.0000	1.0000	1.0000	.9999
14	1.0000	1.0000	1.0000	1.0000	1.0000	1.0000	1.0000	1.0000	1.0000	1.0000

n = 15

x \ p	.01	.02	.03	.04	.05	.06	.07	.08	.09	.10
0	.8601	.7386	.6333	.5421	.4633	.3953	.3367	.2863	.2430	.2059
1	.9904	.9647	.9270	.8809	.8290	.7738	.7168	.6597	.6035	.5490
2	.9996	.9970	.9906	.9797	.9638	.9429	.9171	.8870	.8531	.8159
3	1.0000	.9998	.9992	.9976	.9945	.9896	.9825	.9727	.9601	.9444
4	1.0000	1.0000	.9999	.9998	.9994	.9986	.9972	.9950	.9918	.9873
5	1.0000	1.0000	1.0000	1.0000	.9999	.9999	.9997	.9993	.9987	.9978
6	1.0000	1.0000	1.0000	1.0000	1.0000	1.0000	1.0000	.9999	.9998	.9997
7	1.0000	1.0000	1.0000	1.0000	1.0000	1.0000	1.0000	1.0000	1.0000	1.0000

TABLE A (*Continued*)

$n = 15$ (Continued)

x \ p	.11	.12	.13	.14	.15	.16	.17	.18	.19	.20
0	.1741	.1470	.1238	.1041	.0874	.0731	.0611	.0510	.0424	.0352
1	.4969	.4476	.4013	.3583	.3186	.2821	.2489	.2187	.1915	.1671
2	.7762	.7346	.6916	.6480	.6042	.5608	.5181	.4766	.4365	.3980
3	.9258	.9041	.8796	.8524	.8227	.7908	.7571	.7218	.6854	.6482
4	.9813	.9735	.9639	.9522	.9383	.9222	.9039	.8833	.8606	.8358
5	.9963	.9943	.9916	.9879	.9832	.9773	.9700	.9613	.9510	.9389
6	.9994	.9990	.9985	.9976	.9964	.9948	.9926	.9898	.9863	.9819
7	.9999	.9999	.9998	.9996	.9994	.9990	.9986	.9979	.9970	.9958
8	1.0000	1.0000	1.0000	1.0000	.9999	.9999	.9998	.9997	.9995	.9992
9	1.0000	1.0000	1.0000	1.0000	1.0000	1.0000	1.0000	1.0000	.9999	.9999
10	1.0000	1.0000	1.0000	1.0000	1.0000	1.0000	1.0000	1.0000	1.0000	1.0000

x \ p	.21	.22	.23	.24	.25	.26	.27	.28	.29	.30
0	.0291	.0241	.0198	.0163	.0134	.0109	.0089	.0072	.0059	.0047
1	.1453	.1259	.1087	.0935	.0802	.0685	.0583	.0495	.0419	.0353
2	.3615	.3269	.2945	.2642	.2361	.2101	.1863	.1645	.1447	.1268
3	.6105	.5726	.5350	.4978	.4613	.4258	.3914	.3584	.3268	.2969
4	.8090	.7805	.7505	.7190	.6865	.6531	.6190	.5846	.5500	.5155
5	.9252	.9095	.8921	.8728	.8516	.8287	.8042	.7780	.7505	.7216
6	.9766	.9702	.9626	.9537	.9434	.9316	.9183	.9035	.8870	.8689
7	.9942	.9922	.9896	.9865	.9827	.9781	.9726	.9662	.9587	.9500
8	.9989	.9984	.9977	.9969	.9958	.9944	.9927	.9906	.9879	.9848
9	.9998	.9997	.9996	.9994	.9992	.9989	.9985	.9979	.9972	.9963
10	1.0000	1.0000	.9999	.9999	.9999	.9998	.9998	.9997	.9995	.9993
11	1.0000	1.0000	1.0000	1.0000	1.0000	1.0000	1.0000	1.0000	.9999	.9999
12	1.0000	1.0000	1.0000	1.0000	1.0000	1.0000	1.0000	1.0000	1.0000	1.0000

x \ p	.31	.32	.33	.34	.35	.36	.37	.38	.39	.40
0	.0038	.0031	.0025	.0020	.0016	.0012	.0010	.0008	.0006	.0005
1	.0296	.0248	.0206	.0171	.0142	.0117	.0096	.0078	.0064	.0052
2	.1107	.0962	.0833	.0719	.0617	.0528	.0450	.0382	.0322	.0271
3	.2686	.2420	.2171	.1940	.1727	.1531	.1351	.1187	.1039	.0905
4	.4813	.4477	.4148	.3829	.3519	.3222	.2938	.2668	.2413	.2173
5	.6916	.6607	.6291	.5968	.5643	.5316	.4989	.4665	.4346	.4032
6	.8491	.8278	.8049	.7806	.7548	.7278	.6997	.6705	.6405	.6098
7	.9401	.9289	.9163	.9023	.8868	.8698	.8513	.8313	.8098	.7869
8	.9810	.9764	.9711	.9649	.9578	.9496	.9403	.9298	.9180	.9050
9	.9952	.9938	.9921	.9901	.9876	.9846	.9810	.9768	.9719	.9662
10	.9991	.9988	.9984	.9978	.9972	.9963	.9953	.9941	.9925	.9907
11	.9999	.9998	.9997	.9996	.9995	.9994	.9991	.9989	.9985	.9981
12	1.0000	1.0000	1.0000	1.0000	.9999	.9999	.9999	.9998	.9998	.9997
13	1.0000	1.0000	1.0000	1.0000	1.0000	1.0000	1.0000	1.0000	1.0000	1.0000

TABLE A (*Continued*)

n = 15 (Continued)

p / x	.41	.42	.43	.44	.45	.46	.47	.48	.49	.50
0	.0004	.0003	.0002	.0002	.0001	.0001	.0001	.0001	.0000	.0000
1	.0042	.0034	.0027	.0021	.0017	.0013	.0010	.0008	.0006	.0005
2	.0227	.0189	.0157	.0130	.0107	.0087	.0071	.0057	.0046	.0037
3	.0785	.0678	.0583	.0498	.0424	.0359	.0303	.0254	.0212	.0176
4	.1948	.1739	.1546	.1367	.1204	.1055	.0920	.0799	.0690	.0592
5	.3726	.3430	.3144	.2869	.2608	.2359	.2125	.1905	.1699	.1509
6	.5786	.5470	.5153	.4836	.4522	.4211	.3905	.3606	.3316	.3036
7	.7626	.7370	.7102	.6824	.6535	.6238	.5935	.5626	.5314	.5000
8	.8905	.8746	.8573	.8385	.8182	.7966	.7735	.7490	.7233	.6964
9	.9596	.9521	.9435	.9339	.9231	.9110	.8976	.8829	.8667	.8491
10	.9884	.9857	.9826	.9789	.9745	.9695	.9637	.9570	.9494	.9408
11	.9975	.9968	.9960	.9949	.9937	.9921	.9903	.9881	.9855	.9824
12	.9996	.9995	.9993	.9991	.9989	.9986	.9982	.9977	.9971	.9963
13	1.0000	1.0000	.9999	.9999	.9999	.9998	.9998	.9997	.9996	.9995
14	1.0000	1.0000	1.0000	1.0000	1.0000	1.0000	1.0000	1.0000	1.0000	1.0000

n = 16

p / x	.01	.02	.03	.04	.05	.06	.07	.08	.09	.10
0	.8515	.7238	.6143	.5204	.4401	.3716	.3131	.2634	.2211	.1853
1	.9891	.9601	.9182	.8673	.8108	.7511	.6902	.6299	.5711	.5147
2	.9995	.9963	.9887	.9758	.9571	.9327	.9031	.8688	.8306	.7892
3	1.0000	.9998	.9989	.9968	.9930	.9868	.9779	.9658	.9504	.9316
4	1.0000	1.0000	.9999	.9997	.9991	.9981	.9962	.9932	.9889	.9830
5	1.0000	1.0000	1.0000	1.0000	.9999	.9998	.9995	.9990	.9981	.9967
6	1.0000	1.0000	1.0000	1.0000	1.0000	1.0000	.9999	.9999	.9997	.9995
7	1.0000	1.0000	1.0000	1.0000	1.0000	1.0000	1.0000	1.0000	1.0000	.9999
8	1.0000	1.0000	1.0000	1.0000	1.0000	1.0000	1.0000	1.0000	1.0000	1.0000

p / x	.11	.12	.13	.14	.15	.16	.17	.18	.19	.20
0	.1550	.1293	.1077	.0895	.0743	.0614	.0507	.0418	.0343	.0281
1	.4614	.4115	.3653	.3227	.2839	.2487	.2170	.1885	.1632	.1407
2	.7455	.7001	.6539	.6074	.5614	.5162	.4723	.4302	.3899	.3518
3	.9093	.8838	.8552	.8237	.7899	.7540	.7164	.6777	.6381	.5981
4	.9752	.9652	.9529	.9382	.9209	.9012	.8789	.8542	.8273	.7982
5	.9947	.9918	.9880	.9829	.9765	.9685	.9588	.9473	.9338	.9183
6	.9991	.9985	.9976	.9962	.9944	.9920	.9888	.9847	.9796	.9733
7	.9999	.9998	.9996	.9993	.9989	.9984	.9976	.9964	.9949	.9930
8	1.0000	1.0000	.9999	.9999	.9998	.9997	.9996	.9993	.9990	.9985
9	1.0000	1.0000	1.0000	1.0000	1.0000	1.0000	.9999	.9999	.9998	.9998
10	1.0000	1.0000	1.0000	1.0000	1.0000	1.0000	1.0000	1.0000	1.0000	1.0000

TABLE A (*Continued*)

					$n = 16$ (Continued)					

$\begin{matrix}&p\\x&\end{matrix}$	.21	.22	.23	.24	.25	.26	.27	.28	.29	.30
0	.0230	.0188	.0153	.0124	.0100	.0081	.0065	.0052	.0042	.0033
1	.1209	.1035	.0883	.0750	.0635	.0535	.0450	.0377	.0314	.0261
2	.3161	.2827	.2517	.2232	.1971	.1733	.1518	.1323	.1149	.0994
3	.5582	.5186	.4797	.4417	.4050	.3697	.3360	.3041	.2740	.2459
4	.7673	.7348	.7009	.6659	.6302	.5940	.5575	.5212	.4853	.4499
5	.9008	.8812	.8595	.8359	.8103	.7831	.7542	.7239	.6923	.6598
6	.9658	.9568	.9464	.9342	.9204	.9049	.8875	.8683	.8474	.8247
7	.9905	.9873	.9834	.9786	.9729	.9660	.9580	.9486	.9379	.9256
8	.9979	.9970	.9959	.9944	.9925	.9902	.9873	.9837	.9794	.9743
9	.9996	.9994	.9992	.9988	.9984	.9977	.9969	.9959	.9945	.9929
10	.9999	.9999	.9999	.9998	.9997	.9996	.9994	.9992	.9989	.9984
11	1.0000	1.0000	1.0000	1.0000	1.0000	.9999	.9999	.9999	.9998	.9997
12	1.0000	1.0000	1.0000	1.0000	1.0000	1.0000	1.0000	1.0000	1.0000	1.0000

$\begin{matrix}&p\\x&\end{matrix}$	.31	.32	.33	.34	.35	.36	.37	.38	.39	.40
0	.0026	.0021	.0016	.0013	.0010	.0008	.0006	.0005	.0004	.0003
1	.0216	.0178	.0146	.0120	.0098	.0079	.0064	.0052	.0041	.0033
2	.0856	.0734	.0626	.0533	.0451	.0380	.0319	.0266	.0222	.0183
3	.2196	.1953	.1730	.1525	.1339	.1170	.1018	.0881	.0759	.0651
4	.4154	.3819	.3496	.3187	.2892	.2613	.2351	.2105	.1877	.1666
5	.6264	.5926	.5584	.5241	.4900	.4562	.4230	.3906	.3592	.3288
6	.8003	.7743	.7469	.7181	.6881	.6572	.6254	.5930	.5602	.5272
7	.9119	.8965	.8795	.8609	.8406	.8187	.7952	.7702	.7438	.7161
8	.9683	.9612	.9530	.9436	.9329	.9209	.9074	.8924	.8758	.8577
9	.9908	.9883	.9852	.9815	.9771	.9720	.9659	.9589	.9509	.9417
10	.9979	.9972	.9963	.9952	.9938	.9921	.9900	.9875	.9845	.9809
11	.9996	.9995	.9993	.9990	.9987	.9983	.9977	.9970	.9962	.9951
12	1.0000	.9999	.9999	.9999	.9998	.9997	.9996	.9995	.9993	.9991
13	1.0000	1.0000	1.0000	1.0000	1.0000	1.0000	1.0000	.9999	.9999	.9999
14	1.0000	1.0000	1.0000	1.0000	1.0000	1.0000	1.0000	1.0000	1.0000	1.0000

TABLE A (*Continued*)

n = 16 (Continued)

p / x	.41	.42	.43	.44	.45	.46	.47	.48	.49	.50
0	.0002	.0002	.0001	.0001	.0001	.0001	.0000	.0000	.0000	.0000
1	.0026	.0021	.0016	.0013	.0010	.0008	.0006	.0005	.0003	.0003
2	.0151	.0124	.0101	.0082	.0066	.0053	.0042	.0034	.0027	.0021
3	.0556	.0473	.0400	.0336	.0281	.0234	.0194	.0160	.0131	.0106
4	.1471	.1293	.1131	.0985	.0853	.0735	.0630	.0537	.0456	.0384
5	.2997	.2720	.2457	.2208	.1976	.1759	.1559	.1374	.1205	.1051
6	.4942	.4613	.4289	.3971	.3660	.3359	.3068	.2790	.2524	.2272
7	.6872	.6572	.6264	.5949	.5629	.5306	.4981	.4657	.4335	.4018
8	.8381	.8168	.7940	.7698	.7441	.7171	.6889	.6596	.6293	.5982
9	.9313	.9195	.9064	.8919	.8759	.8584	.8393	.8186	.7964	.7728
10	.9766	.9716	.9658	.9591	.9514	.9426	.9326	.9214	.9089	.8949
11	.9938	.9922	.9902	.9879	.9851	.9817	.9778	.9732	.9678	.9616
12	.9988	.9984	.9979	.9973	.9965	.9956	.9945	.9931	.9914	.9894
13	.9998	.9998	.9997	.9996	.9994	.9993	.9990	.9987	.9984	.9979
14	1.0000	1.0000	1.0000	1.0000	.9999	.9999	.9999	.9999	.9998	.9997
15	1.0000	1.0000	1.0000	1.0000	1.0000	1.0000	1.0000	1.0000	1.0000	1.0000

n = 17

p / x	.01	.02	.03	.04	.05	.06	.07	.08	.09	.10
0	.8429	.7093	.5958	.4996	.4181	.3493	.2912	.2423	.2012	.1668
1	.9877	.9554	.9091	.8535	.7922	.7283	.6638	.6005	.5396	.4818
2	.9994	.9956	.9866	.9714	.9497	.9218	.8882	.8497	.8073	.7618
3	1.0000	.9997	.9986	.9960	.9912	.9836	.9727	.9581	.9397	.9174
4	1.0000	1.0000	.9999	.9996	.9988	.9974	.9949	.9911	.9855	.9779
5	1.0000	1.0000	1.0000	1.0000	.9999	.9997	.9993	.9985	.9973	.9953
6	1.0000	1.0000	1.0000	1.0000	1.0000	1.0000	.9999	.9998	.9996	.9992
7	1.0000	1.0000	1.0000	1.0000	1.0000	1.0000	1.0000	1.0000	1.0000	.9999
8	1.0000	1.0000	1.0000	1.0000	1.0000	1.0000	1.0000	1.0000	1.0000	1.0000

p / x	.11	.12	.13	.14	.15	.16	.17	.18	.19	.20
0	.1379	.1138	.0937	.0770	.0631	.0516	.0421	.0343	.0278	.0225
1	.4277	.3777	.3318	.2901	.2525	.2187	.1887	.1621	.1387	.1182
2	.7142	.6655	.6164	.5676	.5198	.4734	.4289	.3867	.3468	.3096
3	.8913	.8617	.8290	.7935	.7556	.7159	.6749	.6331	.5909	.5489
4	.9679	.9554	.9402	.9222	.9013	.8776	.8513	.8225	.7913	.7582
5	.9925	.9886	.9834	.9766	.9681	.9577	.9452	.9305	.9136	.8943
6	.9986	.9977	.9963	.9944	.9917	.9882	.9837	.9780	.9709	.9623
7	.9998	.9996	.9993	.9989	.9983	.9973	.9961	.9943	.9920	.9891
8	1.0000	.9999	.9999	.9998	.9997	.9995	.9992	.9988	.9982	.9974
9	1.0000	1.0000	1.0000	1.0000	1.0000	.9999	.9999	.9998	.9997	.9995
10	1.0000	1.0000	1.0000	1.0000	1.0000	1.0000	1.0000	1.0000	1.0000	.9999
11	1.0000	1.0000	1.0000	1.0000	1.0000	1.0000	1.0000	1.0000	1.0000	1.0000

TABLE A (*Continued*)

$n = 17$ (Continued)

x \ p	.21	.22	.23	.24	.25	.26	.27	.28	.29	.30
0	.0182	.0146	.0118	.0094	.0075	.0060	.0047	.0038	.0030	.0023
1	.1004	.0849	.0715	.0600	.0501	.0417	.0346	.0286	.0235	.0193
2	.2751	.2433	.2141	.1877	.1637	.1422	.1229	.1058	.0907	.0774
3	.5073	.4667	.4272	.3893	.3530	.3186	.2863	.2560	.2279	.2019
4	.7234	.6872	.6500	.6121	.5739	.5357	.4977	.4604	.4240	.3887
5	.8727	.8490	.8230	.7951	.7653	.7339	.7011	.6671	.6323	.5968
6	.9521	.9402	.9264	.9106	.8929	.8732	.8515	.8279	.8024	.7752
7	.9853	.9806	.9749	.9680	.9598	.9501	.9389	.9261	.9116	.8954
8	.9963	.9949	.9930	.9906	.9876	.9839	.9794	.9739	.9674	.9597
9	.9993	.9989	.9984	.9978	.9969	.9958	.9943	.9925	.9902	.9873
10	.9999	.9998	.9997	.9996	.9994	.9991	.9987	.9982	.9976	.9968
11	1.0000	1.0000	1.0000	.9999	.9999	.9998	.9998	.9997	.9995	.9993
12	1.0000	1.0000	1.0000	1.0000	1.0000	1.0000	1.0000	1.0000	.9999	.9999
13	1.0000	1.0000	1.0000	1.0000	1.0000	1.0000	1.0000	1.0000	1.0000	1.0000

x \ p	.31	.32	.33	.34	.35	.36	.37	.38	.39	.40
0	.0018	.0014	.0011	.0009	.0007	.0005	.0004	.0003	.0002	.0002
1	.0157	.0128	.0104	.0083	.0067	.0054	.0043	.0034	.0027	.0021
2	.0657	.0556	.0468	.0392	.0327	.0272	.0225	.0185	.0151	.0123
3	.1781	.1563	.1366	.1188	.1028	.0885	.0759	.0648	.0550	.0464
4	.3547	.3222	.2913	.2622	.2348	.2094	.1858	.1640	.1441	.1260
5	.5610	.5251	.4895	.4542	.4197	·.3861	.3535	.3222	.2923	.2639
6	.7464	.7162	.6847	.6521	.6188	.5848	.5505	.5161	.4818	.4478
7	.8773	.8574	.8358	.8123	.7872	.7605	.7324	.7029	.6722	.6405
8	.9508	.9405	.9288	.9155	.9006	.8841	.8659	.8459	.8243	.8011
9	.9838	.9796	.9746	.9686	.9617	.9536	.9443	.9336	.9216	.9081
10	.9957	.9943	.9926	.9905	.9880	.9849	.9811	.9766	.9714	.9652
11	.9991	.9987	.9983	.9977	.9970	.9960	.9949	.9934	.9916	.9894
12	.9998	.9998	.9997	.9996	.9994	.9992	.9989	.9985	.9981	.9975
13	1.0000	1.0000	1.0000	.9999	.9999	.9999	.9998	.9998	.9997	.9995
14	1.0000	1.0000	1.0000	1.0000	1.0000	1.0000	1.0000	1.0000	1.0000	.9999
15	1.0000	1.0000	1.0000	1.0000	1.0000	1.0000	1.0000	1.0000	1.0000	1.0000

TABLE A (*Continued*)

$n = 17$ (Continued)

p x	.41	.42	.43	.44	.45	.46	.47	.48	.49	.50
0	.0001	.0001	.0001	.0001	.0000	.0000	.0000	.0000	.0000	.0000
1	.0016	.0013	.0010	.0008	.0006	.0004	.0003	.0002	.0002	.0001
2	.0100	.0080	.0065	.0052	.0041	.0032	.0025	.0020	.0015	.0012
3	.0390	.0326	.0271	.0224	.0184	.0151	.0123	.0099	.0080	.0064
4	.1096	.0949	.0817	.0699	.0596	.0505	.0425	.0356	.0296	.0245
5	.2372	.2121	.1887	.1670	.1471	.1288	.1122	.0972	.0838	.0717
6	.4144	.3818	.3501	.3195	.2902	.2623	.2359	.2110	.1878	.1662
7	.6080	.5750	.5415	.5079	.4743	.4410	.4082	.3761	.3448	.3145
8	.7762	.7498	.7220	.6928	.6626	.6313	.5992	.5665	.5333	.5000
9	.8930	.8764	.8581	.8382	.8166	.7934	.7686	.7423	.7145	.6855
10	.9580	.9497	.9403	.9295	.9174	.9038	.8888	.8721	.8538	.8338
11	.9867	.9835	.9797	.9752	.9699	.9637	.9566	.9483	.9389	.9283
12	.9967	.9958	.9946	.9931	.9914	.9892	.9866	.9835	.9798	.9755
13	.9994	.9992	.9989	.9986	.9981	.9976	.9969	.9960	.9950	.9936
14	.9999	.9999	.9998	.9998	.9997	.9996	.9995	.9993	.9991	.9988
15	1.0000	1.0000	1.0000	1.0000	1.0000	1.0000	.9999	.9999	.9999	.9999
16	1.0000	1.0000	1.0000	1.0000	1.0000	1.0000	1.0000	1.0000	1.0000	1.0000

$n = 18$

p x	.01	.02	.03	.04	.05	.06	.07	.08	.09	.10
0	.8345	.6951	.5780	.4796	.3972	.3283	.2708	.2229	.1831	.1501
1	.9862	.9505	.8997	.8393	.7735	.7055	.6378	.5719	.5091	.4503
2	.9993	.9948	.9843	.9667	.9419	.9102	.8725	.8298	.7832	.7338
3	1.0000	.9996	.9982	.9950	.9891	.9799	.9667	.9494	.9277	.9018
4	1.0000	1.0000	.9998	.9994	.9985	.9966	.9933	.9884	.9814	.9718
5	1.0000	1.0000	1.0000	.9999	.9998	.9995	.9990	.9979	.9962	.9936
6	1.0000	1.0000	1.0000	1.0000	1.0000	1.0000	.9999	.9997	.9994	.9988
7	1.0000	1.0000	1.0000	1.0000	1.0000	1.0000	1.0000	1.0000	.9999	.9998
8	1.0000	1.0000	1.0000	1.0000	1.0000	1.0000	1.0000	1.0000	1.0000	1.0000

p x	.11	.12	.13	.14	.15	.16	.17	.18	.19	.20
0	.1227	.1002	.0815	.0662	.0536	.0434	.0349	.0281	.0225	.0180
1	.3958	.3460	.3008	.2602	.2241	.1920	.1638	.1391	.1176	.0991
2	.6827	.6310	.5794	.5287	.4797	.4327	.3881	.3462	.3073	.2713
3	.8718	.8382	.8014	.7618	.7202	.6771	.6331	.5888	.5446	.5010
4	.9595	.9442	.9257	.9041	.8794	.8518	.8213	.7884	.7533	.7164
5	.9898	.9846	.9778	.9690	.9581	.9449	.9292	.9111	.8903	.8671
6	.9979	.9966	.9946	.9919	.9882	.9833	.9771	.9694	.9600	.9487
7	.9997	.9994	.9989	.9983	.9973	.9959	.9940	.9914	.9880	.9837
8	1.0000	.9999	.9998	.9997	.9995	.9992	.9987	.9980	.9971	.9957
9	1.0000	1.0000	1.0000	1.0000	.9999	.9999	.9998	.9996	.9994	.9991
10	1.0000	1.0000	1.0000	1.0000	1.0000	1.0000	1.0000	.9999	.9999	.9998
11	1.0000	1.0000	1.0000	1.0000	1.0000	1.0000	1.0000	1.0000	1.0000	1.0000

TABLE A (*Continued*)

$n = 18$ (Continued)

x \ p	.21	.22	.23	.24	.25	.26	.27	.28	.29	.30
0	.0144	.0114	.0091	.0072	.0056	.0044	.0035	.0027	.0021	.0016
1	.0831	.0694	.0577	.0478	.0395	.0324	.0265	.0216	.0176	.0142
2	.2384	.2084	.1813	.1570	.1353	.1161	.0991	.0842	.0712	.0600
3	.4586	.4175	.3782	.3409	.3057	.2728	.2422	.2140	.1881	.1646
4	.6780	.6387	.5988	.5586	.5187	.4792	.4406	.4032	.3671	.3327
5	.8414	.8134	.7832	.7512	.7174	.6824	.6462	.6093	.5719	.5344
6	.9355	.9201	.9026	.8829	.8610	.8370	.8109	.7829	.7531	.7217
7	.9783	.9717	.9637	.9542	.9431	.9301	.9153	.8986	.8800	.8593
8	.9940	.9917	.9888	.9852	.9807	.9751	.9684	.9605	.9512	.9404
9	.9986	.9980	.9972	.9961	.9946	.9927	.9903	.9873	.9836	.9790
10	.9997	.9996	.9994	.9991	.9988	.9982	.9975	.9966	.9954	.9939
11	1.0000	.9999	.9999	.9998	.9998	.9997	.9995	.9993	.9990	.9986
12	1.0000	1.0000	1.0000	1.0000	1.0000	.9999	.9999	.9999	.9998	.9997
13	1.0000	1.0000	1.0000	1.0000	1.0000	1.0000	1.0000	1.0000	1.0000	1.0000

x \ p	.31	.32	.33	.34	.35	.36	.37	.38	.39	.40
0	.0013	.0010	.0007	.0006	.0004	.0003	.0002	.0002	.0001	.0001
1	.0114	.0092	.0073	.0058	.0046	.0036	.0028	.0022	.0017	.0013
2	.0502	.0419	.0348	.0287	.0236	.0193	.0157	.0127	.0103	.0082
3	.1432	.1241	.1069	.0917	.0783	.0665	.0561	.0472	.0394	.0328
4	.2999	.2691	.2402	.2134	.1886	.1659	.1451	.1263	.1093	.0942
5	.4971	.4602	.4241	.3889	.3550	.3224	.2914	.2621	.2345	.2088
6	.6889	.6550	.6202	.5849	.5491	.5133	.4776	.4424	.4079	.3743
7	.8367	.8122	.7859	.7579	.7283	.6973	.6651	.6319	.5979	.5634
8	.9280	.9139	.8981	.8804	.8609	.8396	.8165	.7916	.7650	.7368
9	.9736	.9671	.9595	.9506	.9403	.9286	.9153	.9003	.8837	.8653
10	.9920	.9896	.9867	.9831	.9788	.9736	.9675	.9603	.9520	.9424
11	.9980	.9973	.9964	.9953	.9938	.9920	.9898	.9870	.9837	.9797
12	.9996	.9995	.9992	.9989	.9986	.9981	.9974	.9966	.9956	.9942
13	.9999	.9999	.9999	.9998	.9997	.9996	.9995	.9993	.9990	.9987
14	1.0000	1.0000	1.0000	1.0000	1.0000	.9999	.9999	.9999	.9998	.9998
15	1.0000	1.0000	1.0000	1.0000	1.0000	1.0000	1.0000	1.0000	1.0000	1.0000

TABLE A (*Continued*)

n = 18 (Continued)

p \ x	.41	.42	.43	.44	.45	.46	.47	.48	.49	.50
0	.0001	.0001	.0000	.0000	.0000	.0000	.0000	.0000	.0000	.0000
1	.0010	.0008	.0006	.0004	.0003	.0002	.0002	.0001	.0001	.0001
2	.0066	.0052	.0041	.0032	.0025	.0019	.0015	.0011	.0009	.0007
3	.0271	.0223	.0182	.0148	.0120	.0096	.0077	.0061	.0048	.0038
4	.0807	.0687	.0582	.0490	.0411	.0342	.0283	.0233	.0190	.0154
5	.1849	.1628	.1427	.1243	.1077	.0928	.0795	.0676	.0572	.0481
6	.3418	.3105	.2807	.2524	.2258	.2009	.1778	.1564	.1368	.1189
7	.5287	.4938	.4592	.4250	.3915	.3588	.3272	.2968	.2678	.2403
8	.7072	.6764	.6444	.6115	.5778	.5438	.5094	.4751	.4409	.4073
9	.8451	.8232	.7996	.7742	.7473	.7188	.6890	.6579	.6258	.5927
10	.9314	.9189	.9049	.8893	.8720	.8530	.8323	.8098	.7856	.7597
11	.9750	.9693	.9628	.9551	.9463	.9362	.9247	.9117	.8972	.8811
12	.9926	.9906	.9882	.9853	.9817	.9775	.9725	.9666	.9598	.9519
13	.9983	.9978	.9971	.9962	.9951	.9937	.9921	.9900	.9875	.9846
14	.9997	.9996	.9994	.9993	.9990	.9987	.9983	.9977	.9971	.9962
15	1.0000	.9999	.9999	.9999	.9999	.9998	.9997	.9996	.9995	.9993
16	1.0000	1.0000	1.0000	1.0000	1.0000	1.0000	1.0000	1.0000	.9999	.9999
17	1.0000	1.0000	1.0000	1.0000	1.0000	1.0000	1.0000	1.0000	1.0000	1.0000

n = 19

p \ x	.01	.02	.03	.04	.05	.06	.07	.08	.09	.10
0	.8262	.6812	.5606	.4604	.3774	.3086	.2519	.2051	.1666	.1351
1	.9847	.9454	.8900	.8249	.7547	.6829	.6121	.5440	.4798	.4203
2	.9991	.9939	.9817	.9616	.9335	.8979	.8561	.8092	.7585	.7054
3	1.0000	.9995	.9978	.9939	.9868	.9757	.9602	.9398	.9147	.8850
4	1.0000	1.0000	.9998	.9993	.9980	.9956	.9915	.9853	.9765	.9648
5	1.0000	1.0000	1.0000	.9999	.9998	.9994	.9986	.9971	.9949	.9914
6	1.0000	1.0000	1.0000	1.0000	1.0000	.9999	.9998	.9996	.9991	.9983
7	1.0000	1.0000	1.0000	1.0000	1.0000	1.0000	1.0000	.9999	.9999	.9997
8	1.0000	1.0000	1.0000	1.0000	1.0000	1.0000	1.0000	1.0000	1.0000	1.0000

p \ x	.11	.12	.13	.14	.15	.16	.17	.18	.19	.20
0	.1092	.0881	.0709	.0569	.0456	.0364	.0290	.0230	.0182	.0144
1	.3658	.3165	.2723	.2331	.1985	.1682	.1419	.1191	.0996	.0829
2	.6512	.5968	.5432	.4911	.4413	.3941	.3500	.3090	.2713	.2369
3	.8510	.8133	.7725	.7292	.6841	.6380	.5915	.5451	.4995	.4551
4	.9498	.9315	.9096	.8842	.8556	.8238	.7893	.7524	.7136	.6733
5	.9865	.9798	.9710	.9599	.9463	.9300	.9109	.8890	.8643	.8369
6	.9970	.9952	.9924	.9887	.9837	.9772	.9690	.9589	.9468	.9324
7	.9995	.9991	.9984	.9974	.9959	.9939	.9911	.9874	.9827	.9767
8	.9999	.9998	.9997	.9995	.9992	.9986	.9979	.9968	.9953	.9933
9	1.0000	1.0000	1.0000	.9999	.9999	.9998	.9996	.9993	.9990	.9984
10	1.0000	1.0000	1.0000	1.0000	1.0000	1.0000	.9999	.9999	.9998	.9997
11	1.0000	1.0000	1.0000	1.0000	1.0000	1.0000	1.0000	1.0000	1.0000	1.0000

TABLE A (*Continued*)

$n = 19$ (Continued)

x \ p	.21	.22	.23	.24	.25	.26	.27	.28	.29	.30
0	.0113	.0089	.0070	.0054	.0042	.0033	.0025	.0019	.0015	.0011
1	.0687	.0566	.0465	.0381	.0310	.0251	.0203	.0163	.0131	.0104
2	.2058	.1778	.1529	.1308	.1113	.0943	.0795	.0667	.0557	.0462
3	.4123	.3715	.3329	.2968	.2631	.2320	.2035	.1776	.1542	.1332
4	.6319	.5900	.5480	.5064	.4654	.4256	.3871	.3502	.3152	.2822
5	.8071	.7749	.7408	.7050	.6677	.6295	.5907	.5516	.5125	.4739
6	.9157	.8966	.8751	.8513	.8251	.7968	.7664	.7343	.7005	.6655
7	.9693	.9604	.9497	.9371	.9225	.9059	.8871	.8662	.8432	.8180
8	.9907	.9873	.9831	.9778	.9713	.9634	.9541	.9432	.9306	.9161
9	.9977	.9966	.9953	.9934	.9911	.9881	.9844	.9798	.9742	.9674
10	.9995	.9993	.9989	.9984	.9977	.9968	.9956	.9940	.9920	.9895
11	.9999	.9999	.9998	.9997	.9995	.9993	.9990	.9985	.9980	.9972
12	1.0000	1.0000	1.0000	.9999	.9999	.9999	.9998	.9997	.9996	.9994
13	1.0000	1.0000	1.0000	1.0000	1.0000	1.0000	1.0000	1.0000	.9999	.9999
14	1.0000	1.0000	1.0000	1.0000	1.0000	1.0000	1.0000	1.0000	1.0000	1.0000

x \ p	.31	.32	.33	.34	.35	.36	.37	.38	.39	.40
0	.0009	.0007	.0005	.0004	.0003	.0002	.0002	.0001	.0001	.0001
1	.0083	.0065	.0051	.0040	.0031	.0024	.0019	.0014	.0011	.0008
2	.0382	.0314	.0257	.0209	.0170	.0137	.0110	.0087	.0069	.0055
3	.1144	.0978	.0831	.0703	.0591	.0495	.0412	.0341	.0281	.0230
4	.2514	.2227	.1963	.1720	.1500	.1301	.1122	.0962	.0821	.0696
5	.4359	.3990	.3634	.3293	.2968	.2661	.2373	.2105	.1857	.1629
6	.6294	.5927	.5555	.5182	.4812	.4446	.4087	.3739	.3403	.3081
7	.7909	.7619	.7312	.6990	.6656	.6310	.5957	.5599	.5238	.4878
8	.8997	.8814	.8611	.8388	.8145	.7884	.7605	.7309	.6998	.6675
9	.9595	.9501	.9392	.9267	.9125	.8965	.8787	.8590	.8374	.8139
10	.9863	.9824	.9777	.9720	.9653	.9574	.9482	.9375	.9253	.9115
11	.9962	.9949	.9932	.9911	.9886	.9854	.9815	.9769	.9713	.9648
12	.9991	.9988	.9983	.9977	.9969	.9959	.9946	.9930	.9909	.9884
13	.9998	.9998	.9997	.9995	.9993	.9991	.9987	.9983	.9977	.9969
14	1.0000	1.0000	.9999	.9999	.9999	.9998	.9998	.9997	.9995	.9994
15	1.0000	1.0000	1.0000	1.0000	1.0000	1.0000	1.0000	1.0000	.9999	.9999
16	1.0000	1.0000	1.0000	1.0000	1.0000	1.0000	1.0000	1.0000	1.0000	1.0000

TABLE A (*Continued*)

n = 19 (Continued)

x \ p	.41	.42	.43	.44	.45	.46	.47	.48	.49	.50
0	.0000	.0000	.0000	.0000	.0000	.0000	.0000	.0000	.0000	.0000
1	.0006	.0005	.0004	.0003	.0002	.0001	.0001	.0001	.0001	.0000
2	.0043	.0033	.0026	.0020	.0015	.0012	.0009	.0007	.0005	.0004
3	.0187	.0151	.0122	.0097	.0077	.0061	.0048	.0037	.0029	.0022
4	.0587	.0492	.0410	.0340	.0280	.0229	.0186	.0150	.0121	.0096
5	.1421	.1233	.1063	.0912	.0777	.0658	.0554	.0463	.0385	.0318
6	.2774	.2485	.2213	.1961	.1727	.1512	.1316	.1138	.0978	.0835
7	.4520	.4168	.3824	.3491	.3169	.2862	.2570	.2294	.2036	.1796
8	.6340	.5997	.5647	.5294	.4940	.4587	.4238	.3895	.3561	.3238
9	.7886	.7615	.7328	.7026	.6710	.6383	.6046	.5701	.5352	.5000
10	.8960	.8787	.8596	.8387	.8159	.7913	.7649	.7369	.7073	.6762
11	.9571	.9482	.9379	.9262	.9129	.8979	.8813	.8628	.8425	.8204
12	.9854	.9817	.9773	.9720	.9658	.9585	.9500	.9403	.9291	.9165
13	.9960	.9948	.9933	.9914	.9891	.9863	.9829	.9788	.9739	.9682
14	.9991	.9988	.9984	.9979	.9972	.9964	.9954	.9940	.9924	.9904
15	.9999	.9998	.9997	.9996	.9995	.9993	.9990	.9987	.9983	.9978
16	1.0000	1.0000	1.0000	.9999	.9999	.9999	.9999	.9998	.9997	.9996
17	1.0000	1.0000	1.0000	1.0000	1.0000	1.0000	1.0000	1.0000	1.0000	1.0000

n = 20

x \ p	.01	.02	.03	.04	.05	.06	.07	.08	.09	.10
0	.8179	.6676	.5438	.4420	.3585	.2901	.2342	.1887	.1516	.1216
1	.9831	.9401	.8802	.8103	.7358	.6605	.5869	.5169	.4516	.3917
2	.9990	.9929	.9790	.9561	.9245	.8850	.8390	.7879	.7334	.6769
3	1.0000	.9994	.9973	.9926	.9841	.9710	.9529	.9294	.9007	.8670
4	1.0000	1.0000	.9997	.9990	.9974	.9944	.9893	.9817	.9710	.9568
5	1.0000	1.0000	1.0000	.9999	.9997	.9991	.9981	.9962	.9932	.9887
6	1.0000	1.0000	1.0000	1.0000	1.0000	.9999	.9997	.9994	.9987	.9976
7	1.0000	1.0000	1.0000	1.0000	1.0000	1.0000	1.0000	.9999	.9998	.9996
8	1.0000	1.0000	1.0000	1.0000	1.0000	1.0000	1.0000	1.0000	1.0000	.9999
9	1.0000	1.0000	1.0000	1.0000	1.0000	1.0000	1.0000	1.0000	1.0000	1.0000

TABLE A *(Continued)*

$n = 20$ (Continued)

p\x	.11	.12	.13	.14	.15	.16	.17	.18	.19	.20
0	.0972	.0776	.0617	.0490	.0388	.0306	.0241	.0189	.0148	.0115
1	.3376	.2891	.2461	.2084	.1756	.1471	.1227	.1018	.0841	.0692
2	.6198	.5631	.5080	.4550	.4049	.3580	.3146	.2748	.2386	.2061
3	.8290	.7873	.7427	.6959	.6477	.5990	.5504	.5026	.4561	.4114
4	.9390	.9173	.8917	.8625	.8298	.7941	.7557	.7151	.6729	.6296
5	.9825	.9740	.9630	.9493	.9327	.9130	.8902	.8644	.8357	.8042
6	.9959	.9933	.9897	.9847	.9781	.9696	.9591	.9463	.9311	.9133
7	.9992	.9986	.9976	.9962	.9941	.9912	.9873	.9823	.9759	.9679
8	.9999	.9998	.9995	.9992	.9987	.9979	.9967	.9951	.9929	.9900
9	1.0000	1.0000	.9999	.9999	.9998	.9996	.9993	.9989	.9983	.9974
10	1.0000	1.0000	1.0000	1.0000	1.0000	.9999	.9999	.9998	.9996	.9994
11	1.0000	1.0000	1.0000	1.0000	1.0000	1.0000	1.0000	1.0000	.9999	.9999
12	1.0000	1.0000	1.0000	1.0000	1.0000	1.0000	1.0000	1.0000	1.0000	1.0000

p\x	.21	.22	.23	.24	.25	.26	.27	.28	.29	.30
0	.0090	.0069	.0054	.0041	.0032	.0024	.0018	.0014	.0011	.0008
1	.0566	.0461	.0374	.0302	.0243	.0195	.0155	.0123	.0097	.0076
2	.1770	.1512	.1284	.1085	.0913	.0763	.0635	.0526	.0433	.0355
3	.3690	.3289	.2915	.2569	.2252	.1962	.1700	.1466	.1256	.1071
4	.5858	.5420	.4986	.4561	.4148	.3752	.3375	.3019	.2685	.2375
5	.7703	.7343	.6965	.6573	.6172	.5765	.5357	.4952	.4553	.4164
6	.8929	.8699	.8442	.8162	.7858	.7533	.7190	.6831	.6460	.6080
7	.9581	.9464	.9325	.9165	.8982	.8775	.8545	.8293	.8018	.7723
8	.9862	.9814	.9754	.9680	.9591	.9485	.9360	.9216	.9052	.8867
9	.9962	.9946	.9925	.9897	.9861	.9817	.9762	.9695	.9615	.9520
10	.9991	.9987	.9981	.9972	.9961	.9945	.9926	.9900	.9868	.9829
11	.9998	.9997	.9996	.9994	.9991	.9986	.9981	.9973	.9962	.9949
12	1.0000	1.0000	.9999	.9999	.9998	.9997	.9996	.9994	.9991	.9987
13	1.0000	1.0000	1.0000	1.0000	1.0000	1.0000	.9999	.9999	.9998	.9997
14	1.0000	1.0000	1.0000	1.0000	1.0000	1.0000	1.0000	1.0000	1.0000	1.0000

TABLE A (*Continued*)

n = 20 (Continued)

x \ p	.31	.32	.33	.34	.35	.36	.37	.38	.39	.40
0	.0006	.0004	.0003	.0002	.0002	.0001	.0001	.0001	.0001	.0000
1	.0060	.0047	.0036	.0028	.0021	.0016	.0012	.0009	.0007	.0005
2	.0289	.0235	.0189	.0152	.0121	.0096	.0076	.0060	.0047	.0036
3	.0908	.0765	.0642	.0535	.0444	.0366	.0300	.0245	.0198	.0160
4	.2089	.1827	.1589	.1374	.1182	.1011	.0859	.0726	.0610	.0510
5	.3787	.3426	.3082	.2758	.2454	.2171	.1910	.1671	.1453	.1256
6	.5695	.5307	.4921	.4540	.4166	.3803	.3453	.3118	.2800	.2500
7	.7409	.7078	.6732	.6376	.6010	.5639	.5265	.4892	.4522	.4159
8	.8660	.8432	.8182	.7913	.7624	.7317	.6995	.6659	.6312	.5956
9	.9409	.9281	.9134	.8968	.8782	.8576	.8350	.8103	.7837	.7553
10	.9780	.9721	.9650	.9566	.9468	.9355	.9225	.9077	.8910	.8725
11	.9931	.9909	.9881	.9846	.9804	.9753	.9692	.9619	.9534	.9435
12	.9982	.9975	.9966	.9955	.9940	.9921	.9898	.9868	.9833	.9790
13	.9996	.9994	.9992	.9989	.9985	.9979	.9972	.9963	.9951	.9935
14	.9999	.9999	.9999	.9998	.9997	.9996	.9994	.9991	.9988	.9984
15	1.0000	1.0000	1.0000	1.0000	1.0000	.9999	.9999	.9998	.9998	.9997
16	1.0000	1.0000	1.0000	1.0000	1.0000	1.0000	1.0000	1.0000	1.0000	1.0000

x \ p	.41	.42	.43	.44	.45	.46	.47	.48	.49	.50
0	.0000	.0000	.0000	.0000	.0000	.0000	.0000	.0000	.0000	.0000
1	.0004	.0003	.0002	.0002	.0001	.0001	.0001	.0000	.0000	.0000
2	.0028	.0021	.0016	.0012	.0009	.0007	.0005	.0004	.0003	.0002
3	.0128	.0102	.0080	.0063	.0049	.0038	.0029	.0023	.0017	.0013
4	.0423	.0349	.0286	.0233	.0189	.0152	.0121	.0096	.0076	.0059
5	.1079	.0922	.0783	.0660	.0553	.0461	.0381	.0313	.0255	.0207
6	.2220	.1959	.1719	.1499	.1299	.1119	.0958	.0814	.0688	.0577
7	.3804	.3461	.3132	.2817	.2520	.2241	.1980	.1739	.1518	.1316
8	.5594	.5229	.4864	.4501	.4143	.3793	.3454	.3127	.2814	.2517
9	.7252	.6936	.6606	.6264	.5914	.5557	.5196	.4834	.4474	.4119
10	.8520	.8295	.8051	.7788	.7507	.7209	.6896	.6568	.6229	.5881
11	.9321	.9190	.9042	.8877	.8692	.8489	.8266	.8024	.7762	.7483
12	.9738	.9676	.9603	.9518	.9420	.9306	.9177	.9031	.8867	.8684
13	.9916	.9893	.9864	.9828	.9786	.9735	.9674	.9603	.9520	.9423
14	.9978	.9971	.9962	.9950	.9936	.9917	.9895	.9867	.9834	.9793
15	.9996	.9994	.9992	.9989	.9985	.9980	.9973	.9965	.9954	.9941
16	.9999	.9999	.9999	.9998	.9997	.9996	.9995	.9993	.9990	.9987
17	1.0000	1.0000	1.0000	1.0000	1.0000	.9999	.9999	.9999	.9999	.9998
18	1.0000	1.0000	1.0000	1.0000	1.0000	1.0000	1.0000	1.0000	1.0000	1.0000

TABLE A (*Continued*)

$n = 25$

p x	.01	.02	.03	.04	.05	.06	.07	.08	.09	.10
0	.7778	.6035	.4670	.3604	.2774	.2129	.1630	.1244	.0946	.0718
1	.9742	.9114	.8280	.7358	.6424	.5527	.4696	.3947	.3286	.2712
2	.9980	.9868	.9620	.9235	.8729	.8129	.7466	.6768	.6063	.5371
3	.9999	.9986	.9938	.9835	.9659	.9402	.9064	.8649	.8169	.7636
4	1.0000	.9999	.9992	.9972	.9928	.9850	.9726	.9549	.9314	.9020
5	1.0000	1.0000	.9999	.9996	.9988	.9969	.9935	.9877	.9790	.9666
6	1.0000	1.0000	1.0000	1.0000	.9998	.9995	.9987	.9972	.9946	.9905
7	1.0000	1.0000	1.0000	1.0000	1.0000	.9999	.9998	.9995	.9989	.9977
8	1.0000	1.0000	1.0000	1.0000	1.0000	1.0000	1.0000	.9999	.9998	.9995
9	1.0000	1.0000	1.0000	1.0000	1.0000	1.0000	1.0000	1.0000	1.0000	.9999
10	1.0000	1.0000	1.0000	1.0000	1.0000	1.0000	1.0000	1.0000	1.0000	1.0000

p x	.11	.12	.13	.14	.15	.16	.17	.18	.19	.20
0	.0543	.0409	.0308	.0230	.0172	.0128	.0095	.0070	.0052	.0038
1	.2221	.1805	.1457	.1168	.0931	.0737	.0580	.0454	.0354	.0274
2	.4709	.4088	.3517	.3000	.2537	.2130	.1774	.1467	.1204	.0982
3	.7066	.6475	.5877	.5286	.4711	.4163	.3648	.3171	.2734	.2340
4	.8669	.8266	.7817	.7332	.6821	.6293	.5759	.5228	.4708	.4207
5	.9501	.9291	.9035	.8732	.8385	.7998	.7575	.7125	.6653	.6167
6	.9844	.9757	.9641	.9491	.9305	.9080	.8815	.8512	.8173	.7800
7	.9959	.9930	.9887	.9827	.9745	.9639	.9505	.9339	.9141	.8909
8	.9991	.9983	.9970	.9950	.9920	.9879	.9822	.9748	.9652	.9532
9	.9998	.9996	.9993	.9987	.9979	.9965	.9945	.9917	.9878	.9827
10	1.0000	.9999	.9999	.9997	.9995	.9991	.9985	.9976	.9963	.9944
11	1.0000	1.0000	1.0000	1.0000	.9999	.9998	.9997	.9994	.9990	.9985
12	1.0000	1.0000	1.0000	1.0000	1.0000	1.0000	.9999	.9999	.9998	.9996
13	1.0000	1.0000	1.0000	1.0000	1.0000	1.0000	1.0000	1.0000	1.0000	.9999
14	1.0000	1.0000	1.0000	1.0000	1.0000	1.0000	1.0000	1.0000	1.0000	1.0000

TABLE A (*Continued*)

$n = 25$ (Continued)

p x	.21	.22	.23	.24	.25	.26	.27	.28	.29	.30
0	.0028	.0020	.0015	.0010	.0008	.0005	.0004	.0003	.0002	.0001
1	.0211	.0162	.0123	.0093	.0070	.0053	.0039	.0029	.0021	.0016
2	.0796	.0640	.0512	.0407	.0321	.0252	.0196	.0152	.0117	.0090
3	.1987	.1676	.1403	.1166	.0962	.0789	.0642	.0519	.0417	.0332
4	.3730	.3282	.2866	.2484	.2137	.1826	.1548	.1304	.1090	.0905
5	.5675	.5184	.4701	.4233	.3783	.3356	.2956	.2585	.2245	.1935
6	.7399	.6973	.6529	.6073	.5611	.5149	.4692	.4247	.3817	.3407
7	.8642	.8342	.8011	.7651	.7265	.6858	.6435	.6001	.5560	.5118
8	.9386	.9212	.9007	.8772	.8506	.8210	.7885	.7535	.7162	.6769
9	.9760	.9675	.9569	.9440	.9287	.9107	.8899	.8662	.8398	.8106
10	.9918	.9883	.9837	.9778	.9703	.9611	.9498	.9364	.9205	.9022
11	.9976	.9964	.9947	.9924	.9893	.9852	.9801	.9736	.9655	.9558
12	.9994	.9990	.9985	.9977	.9966	.9951	.9931	.9904	.9870	.9825
13	.9999	.9998	.9996	.9994	.9991	.9986	.9979	.9970	.9957	.9940
14	1.0000	1.0000	.9999	.9999	.9998	.9997	.9995	.9992	.9988	.9982
15	1.0000	1.0000	1.0000	1.0000	1.0000	.9999	.9999	.9998	.9997	.9995
16	1.0000	1.0000	1.0000	1.0000	1.0000	1.0000	1.0000	1.0000	.9999	.9999
17	1.0000	1.0000	1.0000	1.0000	1.0000	1.0000	1.0000	1.0000	1.0000	1.0000

p x	.31	.32	.33	.34	.35	.36	.37	.38	.39	.40
0	.0001	.0001	.0000	.0000	.0000	.0000	.0000	.0000	.0000	.0000
1	.0011	.0008	.0006	.0004	.0003	.0002	.0002	.0001	.0001	.0001
2	.0068	.0051	.0039	.0029	.0021	.0016	.0011	.0008	.0006	.0004
3	.0263	.0207	.0162	.0126	.0097	.0074	.0056	.0043	.0032	.0024
4	.0746	.0610	.0496	.0400	.0320	.0255	.0201	.0158	.0123	.0095
5	.1656	.1407	.1187	.0994	.0826	.0682	.0559	.0454	.0367	.0294
6	.3019	.2657	.2321	.2013	.1734	.1483	.1258	.1060	.0886	.0736
7	.4681	.4253	.3837	.3439	.3061	.2705	.2374	.2068	.1789	.1536
8	.6361	.5943	.5518	.5092	.4668	.4252	.3848	.3458	.3086	.2735
9	.7787	.7445	.7081	.6700	.6303	.5896	.5483	.5067	.4653	.4246
10	.8812	.8576	.8314	.8025	.7712	.7375	.7019	.6645	.6257	.5858
11	.9440	.9302	.9141	.8956	.8746	.8510	.8249	.7964	.7654	.7323
12	.9770	.9701	.9617	.9515	.9396	.9255	.9093	.8907	.8697	.8462
13	.9917	.9888	.9851	.9804	.9745	.9674	.9588	.9485	.9363	.9222
14	.9974	.9964	.9950	.9931	.9907	.9876	.9837	.9788	.9729	.9656
15	.9993	.9990	.9985	.9979	.9971	.9959	.9944	.9925	.9900	.9868
16	.9998	.9998	.9996	.9995	.9992	.9989	.9984	.9977	.9968	.9957
17	1.0000	1.0000	.9999	.9999	.9998	.9997	.9996	.9994	.9992	.9988
18	1.0000	1.0000	1.0000	1.0000	1.0000	.9999	.9999	.9999	.9998	.9997
19	1.0000	1.0000	1.0000	1.0000	1.0000	1.0000	1.0000	1.0000	1.0000	.9999
20	1.0000	1.0000	1.0000	1.0000	1.0000	1.0000	1.0000	1.0000	1.0000	1.0000

TABLE A (*Continued*)

$n = 25$ (Continued)

x \ p	.41	.42	.43	.44	.45	.46	.47	.48	.49	.50
0	.0000	.0000	.0000	.0000	.0000	.0000	.0000	.0000	.0000	.0000
1	.0000	.0000	.0000	.0000	.0000	.0000	.0000	.0000	.0000	.0000
2	.0003	.0002	.0002	.0001	.0001	.0000	.0000	.0000	.0000	.0000
3	.0017	.0013	.0009	.0007	.0005	.0003	.0002	.0002	.0001	.0001
4	.0073	.0055	.0042	.0031	.0023	.0017	.0012	.0009	.0006	.0005
5	.0233	.0184	.0144	.0112	.0086	.0066	.0050	.0037	.0028	.0020
6	.0606	.0495	.0401	.0323	.0258	.0204	.0160	.0124	.0096	.0073
7	.1308	.1106	.0929	.0773	.0639	.0523	.0425	.0342	.0273	.0216
8	.2407	.2103	.1823	.1569	.1340	.1135	.0954	0795	.0657	.0539
9	.3849	.3465	.3098	.2750	.2424	.2120	.1840	.1585	.1354	.1148
10	.5452	.5044	.4637	.4235	.3843	.3462	.3098	.2751	.2426	.2122
11	.6971	.6603	.6220	.5826	.5426	.5022	.4618	.4220	.3829	.3450
12	.8203	.7920	.7613	.7285	.6937	.6571	.6192	.5801	.5402	.5000
13	.9059	.8873	.8664	.8431	.8173	.7891	.7587	.7260	.6914	.6550
14	.9569	.9465	.9344	.9203	.9040	.8855	.8647	.8415	.8159	.7878
15	.9829	.9780	.9720	.9647	.9560	.9457	.9337	.9197	.9036	.8852
16	.9942	.9922	.9897	.9866	.9826	.9778	.9719	.9648	.9562	.9461
17	.9983	.9977	.9968	.9956	.9942	.9923	.9898	.9868	.9830	.9784
18	.9996	.9994	.9992	.9988	.9984	.9977	.9969	.9959	.9945	.9927
19	.9999	.9999	.9998	.9997	.9996	.9995	.9992	.9989	.9985	.9980
20	1.0000	1.0000	1.0000	1.0000	.9999	.9999	.9998	.9998	.9997	.9995
21	1.0000	1.0000	1.0000	1.0000	1.0000	1.0000	1.0000	1.0000	.9999	.9999
22	1.0000	1.0000	1.0000	1.0000	1.0000	1.0000	1.0000	1.0000	1.0000	1.0000

TABLE B

Cumulative Poisson Distribution $P(X \leq x|\lambda)$. **1000 Times the Probability of** x **or Fewer Occurrences of Event That Has Average Number of Occurrences Equal to** λ

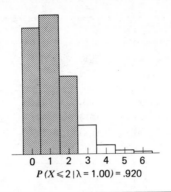

$P(X \leqslant 2|\lambda = 1.00) = .920$

x \ λ	.02	.04	.06	.08	.10	.15	.20	.25
0	980	961	942	923	905	861	819	779
1	1000	999	998	997	995	990	982	974
2		1000	1000	1000	1000	999	999	998
3						1000	1000	1000

x \ λ	.30	.35	.40	.45	.50	.55	.60	.65
0	741	705	670	638	607	577	549	522
1	963	951	938	925	910	894	878	861
2	996	994	992	989	986	982	977	972
3	1000	1000	999	999	998	998	997	996
4			1000	1000	1000	1000	1000	999
5								1000

x \ λ	.70	.75	.80	.85	.90	.95	1.0	1.1
0	497	472	449	427	407	387	368	333
1	844	827	809	791	772	754	736	699
2	966	959	953	945	937	929	920	900
3	994	993	991	989	987	984	981	974
4	999	999	999	998	998	997	996	995
5	1000	1000	1000	1000	1000	1000	999	999
6							1000	1000

TABLE B (*Continued*)

x \ λ	1.2	1.3	1.4	1.5	1.6	1.7	1.8	1.9
0	301	273	247	223	202	183	165	150
1	663	627	592	558	525	493	463	434
2	879	857	833	809	783	757	731	704
3	966	957	946	934	921	907	891	875
4	992	989	986	981	976	970	964	956
5	998	998	997	996	994	992	990	987
6	1000	1000	999	999	999	998	997	997
7			1000	1000	1000	1000	999	999
8							1000	1000

x \ λ	2.0	2.2	2.4	2.6	2.8	3.0	3.2	3.4
0	135	111	091	074	061	050	041	033
1	406	355	308	267	231	199	171	147
2	677	623	570	518	469	423	380	340
3	857	819	779	736	692	647	603	558
4	947	928	904	877	848	815	781	744
5	983	975	964	951	935	916	895	871
6	995	993	988	983	976	966	955	942
7	999	998	997	995	992	988	983	977
8	1000	1000	999	999	998	997	994	992
9			1000	1000	999	999	998	997
10					1000	1000	1000	999
11								1000

x \ λ	3.6	3.8	4.0	4.2	4.4	4.6	4.8	5.0
0	027	022	018	015	012	010	008	007
1	126	107	092	078	066	056	048	040
2	303	269	238	210	185	163	143	125
3	515	473	433	395	359	326	294	265
4	706	668	629	590	551	513	476	440
5	844	816	785	753	720	686	651	616
6	927	909	889	867	844	818	791	762
7	969	960	949	936	921	905	887	867
8	988	984	979	972	964	955	944	932
9	996	994	992	989	985	980	975	968
10	999	998	997	996	994	992	990	986
11	1000	999	999	999	998	997	996	995
12		1000	1000	1000	999	999	999	998
13					1000	1000	1000	999
14								1000

TABLE B (*Continued*)

x \ λ	5.2	5.4	5.6	5.8	6.0	6.2	6.4	6.6
0	006	005	004	003	002	002	002	001
1	034	029	024	021	017	015	012	010
2	109	095	082	072	062	054	046	040
3	238	213	191	170	151	134	119	105
4	406	373	342	313	285	259	235	213
5	581	546	512	478	446	414	384	355
6	732	702	670	638	606	574	542	511
7	845	822	797	771	744	716	687	658
8	918	903	886	867	847	826	803	780
9	960	951	941	929	916	902	886	869
10	982	977	972	965	957	949	939	927
11	993	990	988	984	980	975	969	963
12	997	996	995	993	991	989	986	982
13	999	999	998	997	996	995	994	992
14	1000	999	999	999	999	998	997	997
15		1000	1000	1000	999	999	999	999
16					1000	1000	1000	999
17								1000

x \ λ	6.8	7.0	7.2	7.4	7.6	7.8	8.0	8.5
0	001	001	001	001	001	000	000	000
1	009	007	006	005	004	004	003	002
2	034	030	025	022	019	016	014	009
3	093	082	072	063	055	048	042	030
4	192	173	156	140	125	112	100	074
5	327	301	276	253	231	210	191	150
6	480	450	420	392	365	338	313	256
7	628	599	569	539	510	481	453	386
8	755	729	703	676	648	620	593	523
9	850	830	810	788	765	741	717	653
10	915	901	887	871	854	835	816	763
11	955	947	937	926	915	902	888	849
12	978	973	967	961	954	945	936	909
13	990	987	984	980	976	971	966	949
14	996	994	993	991	989	986	983	973
15	998	998	997	996	995	993	992	986
16	999	999	999	998	998	997	996	993
17	1000	1000	999	999	999	999	998	997
18			1000	1000	1000	1000	999	999
19							1000	999
20								1000

TABLE B (*Continued*)

λ / x	9.0	9.5	10.0	10.5	11.0	11.5	12.0	12.5
1	001	001	000	000	000	000	000	000
2	006	004	003	002	001	001	001	000
3	021	015	010	007	005	003	002	002
4	055	040	029	021	015	011	008	005
5	116	089	067	050	038	028	020	015
6	207	165	130	102	079	060	046	035
7	324	269	220	179	143	114	090	070
8	456	392	333	279	232	191	155	125
9	587	522	458	397	341	289	242	201
10	706	645	583	521	460	402	347	297
11	803	752	697	639	579	520	462	406
12	876	836	792	742	689	633	576	519
13	926	898	864	825	781	733	682	628
14	959	940	917	888	854	815	772	725
15	978	967	951	932	907	878	844	806
16	989	982	973	960	944	924	899	869
17	995	991	986	978	968	954	937	916
18	998	996	993	988	982	974	963	948
19	999	998	997	994	991	986	979	969
20	1000	999	998	997	995	992	988	983
21		1000	999	999	998	996	994	991
22			1000	999	999	998	997	995
23				1000	1000	999	999	998
24						1000	999	999
25							1000	999
26								1000

TABLE B (*Continued*)

x \ λ	13.0	13.5	14.0	14.5	15	16	17	18
3	001	001	000	000	000	000	000	000
4	004	003	002	001	001	000	000	000
5	011	008	006	004	003	001	001	000
6	026	019	014	010	008	004	002	001
7	054	041	032	024	018	010	005	003
8	100	079	062	048	037	022	013	007
9	166	135	109	088	070	043	026	015
10	252	211	176	145	118	077	049	030
11	353	304	260	220	185	127	085	055
12	463	409	358	311	268	193	135	092
13	573	518	464	413	363	275	201	143
14	675	623	570	518	466	368	281	208
15	764	718	669	619	568	467	371	287
16	835	798	756	711	664	566	468	375
17	890	861	827	790	749	659	564	469
18	930	908	883	853	819	742	655	562
19	957	942	923	901	875	812	736	651
20	975	965	952	936	917	868	805	731
21	986	980	971	960	947	911	861	799
22	992	989	983	976	967	942	905	855
23	996	994	991	986	981	963	937	899
24	998	997	995	992	989	978	959	932
25	999	998	997	996	994	987	975	955
26	1000	999	999	998	997	993	985	972
27		1000	999	999	998	996	991	983
28			1000	999	999	998	995	990
29				1000	1000	999	997	994
30						999	999	997
31						1000	999	998
32							1000	999
33								1000

TABLE B (*Continued*)

x \ λ	19	20	21	22	23	24	25
6	001	000	000	000	000	000	000
7	002	001	000	000	000	000	000
8	004	002	001	001	000	000	000
9	009	005	003	002	001	000	000
10	018	011	006	004	002	001	001
11	035	021	013	008	004	003	001
12	061	039	025	015	009	005	003
13	098	066	043	028	017	011	006
14	150	105	072	048	031	020	012
15	215	157	111	077	052	034	022
16	292	221	163	117	082	056	038
17	378	297	227	169	123	087	060
18	469	381	302	232	175	128	092
19	561	470	384	306	238	180	134
20	647	559	471	387	310	243	185
21	725	644	558	472	389	314	247
22	793	721	640	556	472	392	318
23	849	787	716	637	555	473	394
24	893	843	782	712	635	554	473
25	927	888	838	777	708	632	553
26	951	922	883	832	772	704	629
27	969	948	917	877	827	768	700
28	980	966	944	913	873	823	763
29	988	978	963	940	908	868	818
30	993	987	976	959	936	904	863
31	996	992	985	973	956	932	900
32	998	995	991	983	971	953	929
33	999	997	994	989	981	969	950
34	999	999	997	994	988	979	966
35	1000	999	998	996	993	987	978
36		1000	999	998	996	992	985
37			999	999	997	995	991
38			1000	999	999	997	994
39				1000	999	998	997
40					1000	999	998
41						999	999
42						1000	999
43							1000

TABLE C

Normal Curve Areas $P(z \leq z_0)$ Entries in the Body of the Table are Areas Between $-\infty$ and z

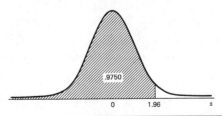

.9750

z	-0.09	-0.08	-0.07	-0.06	-0.05	-0.04	-0.03	-0.02	-0.01	0.00	z
-3.80	.0001	.0001	.0001	.0001	.0001	.0001	.0001	.0001	.0001	.0001	-3.80
-3.70	.0001	.0001	.0001	.0001	.0001	.0001	.0001	.0001	.0001	.0001	-3.70
-3.60	.0001	.0001	.0001	.0001	.0001	.0001	.0001	.0001	.0002	.0002	-3.60
-3.50	.0002	.0002	.0002	.0002	.0002	.0002	.0002	.0002	.0002	.0002	-3.50
-3.40	.0002	.0003	.0003	.0003	.0003	.0003	.0003	.0003	.0003	.0003	-3.40
-3.30	.0003	.0004	.0004	.0004	.0004	.0004	.0004	.0005	.0005	.0005	-3.30
-3.20	.0005	.0005	.0005	.0006	.0006	.0006	.0006	.0006	.0007	.0007	-3.20
-3.10	.0007	.0007	.0008	.0008	.0008	.0008	.0009	.0009	.0009	.0010	-3.10
-3.00	.0010	.0010	.0011	.0011	.0011	.0012	.0012	.0013	.0013	.0013	-3.00
-2.90	.0014	.0014	.0015	.0015	.0016	.0016	.0017	.0018	.0018	.0019	-2.90
-2.80	.0019	.0020	.0021	.0021	.0022	.0023	.0023	.0024	.0025	.0026	-2.80
-2.70	.0026	.0027	.0028	.0029	.0030	.0031	.0032	.0033	.0034	.0035	-2.70
-2.60	.0036	.0037	.0038	.0039	.0040	.0041	.0043	.0044	.0045	.0047	-2.60
-2.50	.0048	.0049	.0051	.0052	.0054	.0055	.0057	.0059	.0060	.0062	-2.50
-2.40	.0064	.0066	.0068	.0069	.0071	.0073	.0075	.0078	.0080	.0082	-2.40
-2.30	.0084	.0087	.0089	.0091	.0094	.0096	.0099	.0102	.0104	.0107	-2.30
-2.20	.0110	.0113	.0116	.0119	.0122	.0125	.0129	.0132	.0136	.0139	-2.20
-2.10	.0143	.0146	.0150	.0154	.0158	.0162	.0166	.0170	.0174	.0179	-2.10
-2.00	.0183	.0188	.0192	.0197	.0202	.0207	.0212	.0217	.0222	.0228	-2.00
-1.90	.0233	.0239	.0244	.0250	.0256	.0262	.0268	.0274	.0281	.0287	-1.90
-1.80	.0294	.0301	.0307	.0314	.0322	.0329	.0336	.0344	.0351	.0359	-1.80
-1.70	.0367	.0375	.0384	.0392	.0401	.0409	.0418	.0427	.0436	.0446	-1.70
-1.60	.0455	.0465	.0475	.0485	.0495	.0505	.0516	.0526	.0537	.0548	-1.60
-1.50	.0559	.0571	.0582	.0594	.0606	.0618	.0630	.0643	.0655	.0668	-1.50
-1.40	.0681	.0694	.0708	.0721	.0735	.0749	.0764	.0778	.0793	.0808	-1.40
-1.30	.0823	.0838	.0853	.0869	.0885	.0901	.0918	.0934	.0951	.0968	-1.30
-1.20	.0985	.1003	.1020	.1038	.1056	.1075	.1093	.1112	.1131	.1151	-1.20
-1.10	.1170	.1190	.1210	.1230	.1251	.1271	.1292	.1314	.1335	.1357	-1.10
-1.00	.1379	.1401	.1423	.1446	.1469	.1492	.1515	.1539	.1562	.1587	-1.00
-0.90	.1611	.1635	.1660	.1685	.1711	.1736	.1762	.1788	.1814	.1841	-0.90
-0.80	.1867	.1894	.1922	.1949	.1977	.2005	.2033	.2061	.2090	.2119	-0.80
-0.70	.2148	.2177	.2206	.2236	.2266	.2296	.2327	.2358	.2389	.2420	-0.70
-0.60	.2451	.2483	.2514	.2546	.2578	.2611	.2643	.2676	.2709	.2743	-0.60
-0.50	.2776	.2810	.2843	.2877	.2912	.2946	.2981	.3015	.3050	.3085	-0.50
-0.40	.3121	.3156	.3192	.3228	.3264	.3300	.3336	.3372	.3409	.3446	-0.40
-0.30	.3483	.3520	.3557	.3594	.3632	.3669	.3707	.3745	.3783	.3821	-0.30
-0.20	.3859	.3897	.3936	.3974	.4013	.4052	.4090	.4129	.4168	.4207	-0.20
-0.10	.4247	.4286	.4325	.4364	.4404	.4443	.4483	.4522	.4562	.4602	-0.10
0.00	.4641	.4681	.4721	.4761	.4801	.4840	.4880	.4920	.4960	.5000	0.00

APPENDIX

TABLE C (*Continued*)

z	0.00	0.01	0.02	0.03	0.04	0.05	0.06	0.07	0.08	0.09	z
0.00	.5000	.5040	.5080	.5120	.5160	.5199	.5239	.5279	.5319	.5359	0.00
0.10	.5398	.5438	.5478	.5517	.5557	.5596	.5636	.5675	.5714	.5753	0.10
0.20	.5793	.5832	.5871	.5910	.5948	.5987	.6026	.6064	.6103	.6141	0.20
0.30	.6179	.6217	.6255	.6293	.6331	.6368	.6406	.6443	.6480	.6517	0.30
0.40	.6554	.6591	.6628	.6664	.6700	.6736	.6772	.6808	.6844	.6879	0.40
0.50	.6915	.6950	.6985	.7019	.7054	.7088	.7123	.7157	.7190	.7224	0.50
0.60	.7257	.7291	.7324	.7357	.7389	.7422	.7454	.7486	.7517	.7549	0.60
0.70	.7580	.7611	.7642	.7673	.7704	.7734	.7764	.7794	.7823	.7852	0.70
0.80	.7881	.7910	.7939	.7967	.7995	.8023	.8051	.8078	.8106	.8133	0.80
0.90	.8159	.8186	.8212	.8238	.8264	.8289	.8315	.8340	.8365	.8389	0.90
1.00	.8413	.8438	.8461	.8485	.8508	.8531	.8554	.8577	.8599	.8621	1.00
1.10	.8643	.8665	.8686	.8708	.8729	.8749	.8770	.8790	.8810	.8830	1.10
1.20	.8849	.8869	.8888	.8907	.8925	.8944	.8962	.8980	.8997	.9015	1.20
1.30	.9032	.9049	.9066	.9082	.9099	.9115	.9131	.9147	.9162	.9177	1.30
1.40	.9192	.9207	.9222	.9236	.9251	.9265	.9279	.9292	.9306	.9319	1.40
1.50	.9332	.9345	.9357	.9370	.9382	.9394	.9406	.9418	.9429	.9441	1.50
1.60	.9452	.9463	.9474	.9484	.9495	.9505	.9515	.9525	.9535	.9545	1.60
1.70	.9554	.9564	.9573	.9582	.9591	.9599	.9608	.9616	.9625	.9633	1.70
1.80	.9641	.9649	.9656	.9664	.9671	.9678	.9686	.9693	.9699	.9706	1.80
1.90	.9713	.9719	.9726	.9732	.9738	.9744	.9750	.9756	.9761	.9767	1.90
2.00	.9772	.9778	.9783	.9788	.9793	.9798	.9803	.9808	.9812	.9817	2.00
2.10	.9821	.9826	.9830	.9834	.9838	.9842	.9846	.9850	.9854	.9857	2.10
2.20	.9861	.9864	.9868	.9871	.9875	.9878	.9881	.9884	.9887	.9890	2.20
2.30	.9893	.9896	.9898	.9901	.9904	.9906	.9909	.9911	.9913	.9916	2.30
2.40	.9918	.9920	.9922	.9925	.9927	.9929	.9931	.9932	.9934	.9936	2.40
2.50	.9938	.9940	.9941	.9943	.9945	.9946	.9948	.9949	.9951	.9952	2.50
2.60	.9953	.9955	.9956	.9957	.9959	.9960	.9961	.9962	.9963	.9964	2.60
2.70	.9965	.9966	.9967	.9968	.9969	.9970	.9971	.9972	.9973	.9974	2.70
2.80	.9974	.9975	.9976	.9977	.9977	.9978	.9979	.9979	.9980	.9981	2.80
2.90	.9981	.9982	.9982	.9983	.9984	.9984	.9985	.9985	.9986	.9986	2.90
3.00	.9987	.9987	.9987	.9988	.9988	.9989	.9989	.9989	.9990	.9990	3.00
3.10	.9990	.9991	.9991	.9991	.9992	.9992	.9992	.9992	.9993	.9993	3.10
3.20	.9993	.9993	.9994	.9994	.9994	.9994	.9994	.9995	.9995	.9995	3.20
3.30	.9995	.9995	.9995	.9996	.9996	.9996	.9996	.9996	.9996	.9997	3.30
3.40	.9997	.9997	.9997	.9997	.9997	.9997	.9997	.9997	.9997	.9998	3.40
3.50	.9998	.9998	.9998	.9998	.9998	.9998	.9998	.9998	.9998	.9998	3.50
3.60	.9998	.9998	.9999	.9999	.9999	.9999	.9999	.9999	.9999	.9999	3.60
3.70	.9999	.9999	.9999	.9999	.9999	.9999	.9999	.9999	.9999	.9999	3.70
3.80	.9999	.9999	.9999	.9999	.9999	.9999	.9999	.9999	.9999	.9999	3.80

Table D

Random Digits

85967	73152	14511	85285	36009	95892	36962	67835	63314	50162
07483	51453	11649	86348	76431	81594	95848	36738	25014	15460
96283	01898	61414	83525	04231	13604	75339	11730	85423	60698
49174	12074	98551	37895	93547	24769	09404	76548	05393	96770
97366	39941	21225	93629	19574	71565	33413	56087	40875	13351
90474	41469	16812	81542	81652	45554	27931	93994	22375	00953
28599	64109	09497	76235	41383	31555	12639	00619	22909	29563
25254	16210	89717	65997	82667	74624	36348	44018	64732	93589
28785	02760	24359	99410	77319	73408	58993	61098	04393	48245
84725	86576	86944	93296	10081	82454	76810	52975	10324	15457
41059	66456	47679	66810	15941	84602	14493	65515	19251	41642
67434	41045	82830	47617	36932	46728	71183	36345	41404	81110
72766	68816	37643	19959	57550	49620	98480	25640	67257	18671
92079	46784	66125	94932	64451	29275	57669	66658	30818	58353
29187	40350	62533	73603	34075	16451	42885	03448	37390	96328
74220	17612	65522	80607	19184	64164	66962	82310	18163	63495
03786	02407	06098	92917	40434	60602	82175	04470	78754	90775
75085	55558	15520	27038	25471	76107	90832	10819	56797	33751
09161	33015	19155	11715	00551	24909	31894	37774	37953	78837
75707	48992	64998	87080	39333	00767	45637	12538	67439	94914
21333	48660	31288	00086	79889	75532	28704	62844	92337	99695
65626	50061	42539	14812	48895	11196	34335	60492	70650	51108
84380	07389	87891	76255	89604	41372	10837	66992	93183	56920
46479	32072	80083	63868	70930	89654	05359	47196	12452	38234
59847	97197	55147	76639	76971	55928	36441	95141	42333	67483
31416	11231	27904	57383	31852	69137	96667	14315	01007	31929
82066	83436	67914	21465	99605	83114	97885	74440	99622	87912
01850	42782	39202	18582	46214	99228	79541	78298	75404	63648
32315	89276	89582	87138	16165	15984	21466	63830	30475	74729
59388	42703	55198	80380	67067	97155	34160	85019	03527	78140
58089	27632	50987	91373	07736	20436	96130	73483	85332	24384
61705	57285	30392	23660	75841	21931	04295	00875	09114	32101
18914	98982	60199	99275	41967	35208	30357	76772	92656	62318
11965	94089	34803	48941	69709	16784	44642	89761	66864	62803
85251	48111	80936	81781	93248	67877	16498	31924	51315	79921
66121	96986	84844	93873	46352	92183	51152	85878	30490	15974
53972	96642	24199	58080	35450	03482	66953	49521	63719	57615
14509	16594	78883	43222	23093	58645	60257	89250	63266	90858
37700	07688	65533	72126	23611	93993	01848	03910	38552	17472
85466	59392	72722	15473	73295	49759	56157	60477	83284	56367
52969	55863	42312	67842	05673	91878	82738	36563	79540	61935
42744	68315	17514	02878	97291	74851	42725	57894	81434	62041
26140	13336	67726	61876	29971	99294	96664	52817	90039	53211
95589	56319	14563	24071	06916	59555	18195	32280	79357	04224
39113	13217	59999	49952	83021	47709	53105	19295	88318	41626
41392	17622	18994	98283	07249	52289	24209	91139	30715	06604
54684	53645	79246	70183	87731	19185	08541	33519	07223	97413
89442	61001	36658	57444	95388	36682	38052	46719	09428	94012
36751	16778	54888	15357	68003	43564	90976	58904	40512	07725
98159	02564	21416	74944	53049	88749	02865	25772	89853	88714

TABLE E

Percentiles of the _t_ Distribution

$P(t_{10} \leqslant 2.2281) = .975$

d. f.	$t_{.90}$	$t_{.95}$	$t_{.975}$	$t_{.99}$	$t_{.995}$
1	3.078	6.3138	12.706	31.821	63.657
2	1.886	2.9200	4.3027	6.965	9.9248
3	1.638	2.3534	3.1825	4.541	5.8409
4	1.533	2.1318	2.7764	3.747	4.6041
5	1.476	2.0150	2.5706	3.365	4.0321
6	1.440	1.9432	2.4469	3.143	3.7074
7	1.415	1.8946	2.3646	2.998	3.4995
8	1.397	1.8595	2.3060	2.896	3.3554
9	1.383	1.8331	2.2622	2.821	3.2498
10	1.372	1.8125	2.2281	2.764	3.1693
11	1.363	1.7959	2.2010	2.718	3.1058
12	1.356	1.7823	2.1788	2.681	3.0545
13	1.350	1.7709	2.1604	2.650	3.0123
14	1.345	1.7613	2.1448	2.624	2.9768
15	1.341	1.7530	2.1315	2.602	2.9467
16	1.337	1.7459	2.1199	2.583	2.9208
17	1.333	1.7396	2.1098	2.567	2.8982
18	1.330	1.7341	2.1009	2.552	2.8784
19	1.328	1.7291	2.0930	2.539	2.8609
20	1.325	1.7247	2.0860	2.528	2.8453
21	1.323	1.7207	2.0796	2.518	2.8314
22	1.321	1.7171	2.0739	2.508	2.8188
23	1.319	1.7139	2.0687	2.500	2.8073
24	1.318	1.7109	2.0639	2.492	2.7969
25	1.316	1.7081	2.0595	2.485	2.7874
26	1.315	1.7056	2.0555	2.479	2.7787
27	1.314	1.7033	2.0518	2.473	2.7707
28	1.313	1.7011	2.0484	2.467	2.7633
29	1.311	1.6991	2.0452	2.462	2.7564
30	1.310	1.6973	2.0423	2.457	2.7500
35	1.3062	1.6896	2.0301	2.438	2.7239
40	1.3031	1.6839	2.0211	2.423	2.7045
45	1.3007	1.6794	2.0141	2.412	2.6896
50	1.2987	1.6759	2.0086	2.403	2.6778
60	1.2959	1.6707	2.0003	2.390	2.6603
70	1.2938	1.6669	1.9945	2.381	2.6480
80	1.2922	1.6641	1.9901	2.374	2.6388
90	1.2910	1.6620	1.9867	2.368	2.6316
100	1.2901	1.6602	1.9840	2.364	2.6260
120	1.2887	1.6577	1.9799	2.358	2.6175
140	1.2876	1.6558	1.9771	2.353	2.6114
160	1.2869	1.6545	1.9749	2.350	2.6070
180	1.2863	1.6534	1.9733	2.347	2.6035
200	1.2858	1.6525	1.9719	2.345	2.6006
∞	1.282	1.645	1.96	2.326	2.576

TABLE F

Percentiles of the Chi-Square Distribution

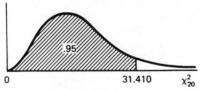

$$P(\chi^2_{20} \leqslant 31.410) = .95$$

d. f.	$\chi^2_{.005}$	$\chi^2_{.025}$	$\chi^2_{.05}$	$\chi^2_{.90}$	$\chi^2_{.95}$	$\chi^2_{.975}$	$\chi^2_{.99}$	$\chi^2_{.995}$
1	.0000393	.000982	.00393	2.706	3.841	5.024	6.635	7.879
2	.0100	.0506	.103	4.605	5.991	7.378	9.210	10.597
3	.0717	.216	.352	6.251	7.815	9.348	11.345	12.838
4	.207	.484	.711	7.779	9.488	11.143	13.277	14.860
5	.412	.831	1.145	9.236	11.070	12.832	15.086	16.750
6	.676	1.237	1.635	10.645	12.592	14.449	16.812	18.548
7	.989	1.690	2.167	12.017	14.067	16.013	18.475	20.278
8	1.344	2.180	2.733	13.362	15.507	17.535	20.090	21.955
9	1.735	2.700	3.325	14.684	16.919	19.023	21.666	23.589
10	2.156	3.247	3.940	15.987	18.307	20.483	23.209	25.188
11	2.603	3.816	4.575	17.275	19.675	21.920	24.725	26.757
12	3.074	4.404	5.226	18.549	21.026	23.336	26.217	28.300
13	3.565	5.009	5.892	19.812	22.362	24.736	27.688	29.819
14	4.075	5.629	6.571	21.064	23.685	26.119	29.141	31.319
15	4.601	6.262	7.261	22.307	24.996	27.488	30.578	32.801
16	5.142	6.908	7.962	23.542	26.296	28.845	32.000	34.267
17	5.697	7.564	8.672	24.769	27.587	30.191	33.409	35.718
18	6.265	8.231	9.390	25.989	28.869	31.526	34.805	37.156
19	6.844	8.907	10.117	27.204	30.144	32.852	36.191	38.582
20	7.434	9.591	10.851	28.412	31.410	34.170	37.566	39.997
21	8.034	10.283	11.591	29.615	32.671	35.479	38.932	41.401
22	8.643	10.982	12.338	30.813	33.924	36.781	40.289	42.796
23	9.260	11.688	13.091	32.007	35.172	38.076	41.638	44.181
24	9.886	12.401	13.848	33.196	36.415	39.364	42.980	45.558
25	10.520	13.120	14.611	34.382	37.652	40.646	44.314	46.928
26	11.160	13.844	15.379	35.563	38.885	41.923	45.642	48.290
27	11.808	14.573	16.151	36.741	40.113	43.194	46.963	49.645
28	12.461	15.308	16.928	37.916	41.337	44.461	48.278	50.993
29	13.121	16.047	17.708	39.087	42.557	45.722	49.588	52.336
30	13.787	16.791	18.493	40.256	43.773	46.979	50.892	53.672
35	17.192	20.569	22.465	46.059	49.802	53.203	57.342	60.275
40	20.707	24.433	26.509	51.805	55.758	59.342	63.691	66.766
45	24.311	28.366	30.612	57.505	61.656	65.410	69.957	73.166
50	27.991	32.357	34.764	63.167	67.505	71.420	76.154	79.490
60	35.535	40.482	43.188	74.397	79.082	83.298	88.379	91.952
70	43.275	48.758	51.739	85.527	90.531	95.023	100.425	104.215
80	51.172	57.153	60.391	96.578	101.879	106.629	112.329	116.321
90	59.196	65.647	69.126	107.565	113.145	118.136	124.116	128.299
100	67.328	74.222	77.929	118.498	124.342	129.561	135.807	140.169

APPENDIX

Table G

Percentiles of the F Distribution
$$F_{.995}$$

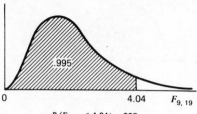

$$0 \qquad\qquad 4.04 \qquad F_{9,\,19}$$

$P(F_{9,\,19} \leqslant 4.04) = .995$

Denominator Degrees of Freedom	Numerator Degrees of Freedom								
	1	2	3	4	5	6	7	8	9
1	16211	20000	21615	22500	23056	23437	23715	23925	24091
2	198.5	199.0	199.2	199.2	199.3	199.3	199.4	199.4	199.4
3	55.55	49.80	47.47	46.19	45.39	44.84	44.43	44.13	43.88
4	31.33	26.28	24.26	23.15	22.46	21.97	21.62	21.35	21.14
5	22.78	18.31	16.53	15.56	14.94	14.51	14.20	13.96	13.77
6	18.63	14.54	12.92	12.03	11.46	11.07	10.79	10.57	10.39
7	16.24	12.40	10.88	10.05	9.52	9.16	8.89	8.68	8.51
8	14.69	11.04	9.60	8.81	8.30	7.95	7.69	7.50	7.34
9	13.61	10.11	8.72	7.96	7.47	7.13	6.88	6.69	6.54
10	12.83	9.43	8.08	7.34	6.87	6.54	6.30	6.12	5.97
11	12.23	8.91	7.60	6.88	6.42	6.10	5.86	5.68	5.54
12	11.75	8.51	7.23	6.52	6.07	5.76	5.52	5.35	5.20
13	11.37	8.19	6.93	6.23	5.79	5.48	5.25	5.08	4.94
14	11.06	7.92	6.68	6.00	5.56	5.26	5.03	4.86	4.72
15	10.80	7.70	6.48	5.80	5.37	5.07	4.85	4.67	4.54
16	10.58	7.51	6.30	5.64	5.21	4.91	4.69	4.52	4.38
17	10.38	7.35	6.16	5.50	5.07	4.78	4.56	4.39	4.25
18	10.22	7.21	6.03	5.37	4.96	4.66	4.44	4.28	4.14
19	10.07	7.09	5.92	5.27	4.85	4.56	4.34	4.18	4.04
20	9.94	6.99	5.82	5.17	4.76	4.47	4.26	4.09	3.96
21	9.83	6.89	5.73	5.09	4.68	4.39	4.18	4.01	3.88
22	9.73	6.81	5.65	5.02	4.61	4.32	4.11	3.94	3.81
23	9.63	6.73	5.58	4.95	4.54	4.26	4.05	3.88	3.75
24	9.55	6.66	5.52	4.89	4.49	4.20	3.99	3.83	3.69
25	9.48	6.60	5.46	4.84	4.43	4.15	3.94	3.78	3.64
26	9.41	6.54	5.41	4.79	4.38	4.10	3.89	3.73	3.60
27	9.34	6.49	5.36	4.74	4.34	4.06	3.85	3.69	3.56
28	9.28	6.44	5.32	4.70	4.30	4.02	3.81	3.65	3.52
29	9.23	6.40	5.28	4.66	4.26	3.98	3.77	3.61	3.48
30	9.18	6.35	5.24	4.62	4.23	3.95	3.74	3.58	3.45
40	8.83	6.07	4.98	4.37	3.99	3.71	3.51	3.35	3.22
60	8.49	5.79	4.73	4.14	3.76	3.49	3.29	3.13	3.01
120	8.18	5.54	4.50	3.92	3.55	3.28	3.09	2.93	2.81
∞	7.88	5.30	4.28	3.72	3.35	3.09	2.90	2.74	2.62

TABLE G (*Continued*)

Denominator Degrees of Freedom	Numerator Degrees of Freedom									
	10	12	15	20	24	30	40	60	120	∞
1	24224	24426	24630	24836	24940	25044	25148	25253	25359	25465
2	199.4	199.4	199.4	199.4	199.5	199.5	199.5	199.5	199.5	199.5
3	43.69	43.39	43.08	42.78	42.62	42.47	42.31	42.15	41.99	41.83
4	20.97	20.70	20.44	20.17	20.03	19.89	19.75	19.61	19.47	19.32
5	13.62	13.38	13.15	12.90	12.78	12.66	12.53	12.40	12.27	12.14
6	10.25	10.03	9.81	9.59	9.47	9.36	9.24	9.12	9.00	8.88
7	8.38	8.18	7.97	7.75	7.65	7.53	7.42	7.31	7.19	7.08
8	7.21	7.01	6.81	6.61	6.50	6.40	6.29	6.18	6.06	5.95
9	6.42	6.23	6.03	5.83	5.73	5.62	5.52	5.41	5.30	5.19
10	5.85	5.66	5.47	5.27	5.17	5.07	4.97	4.86	4.75	4.64
11	5.42	5.24	5.05	4.86	4.76	4.65	4.55	4.44	4.34	4.23
12	5.09	4.91	4.72	4.53	4.43	4.33	4.23	4.12	4.01	3.90
13	4.82	4.64	4.46	4.27	4.17	4.07	3.97	3.87	3.76	3.65
14	4.60	4.43	4.25	4.06	3.96	3.86	3.76	3.66	3.55	3.44
15	4.42	4.25	4.07	3.88	3.79	3.69	3.58	3.48	3.37	3.26
16	4.27	4.10	3.92	3.73	3.64	3.54	3.44	3.33	3.22	3.11
17	4.14	3.97	3.79	3.61	3.51	3.41	3.31	3.21	3.10	2.98
18	4.03	3.86	3.68	3.50	3.40	3.30	3.20	3.10	2.99	2.87
19	3.93	3.76	3.59	3.40	3.31	3.21	3.11	3.00	2.89	2.78
20	3.85	3.68	3.50	3.32	3.22	3.12	3.02	2.92	2.81	2.69
21	3.77	3.60	3.43	3.24	3.15	3.05	2.95	2.84	2.73	2.61
22	3.70	3.54	3.36	3.18	3.08	2.98	2.88	2.77	2.66	2.55
23	3.64	3.47	3.30	3.12	3.02	2.92	2.82	2.71	2.60	2.48
24	3.59	3.42	3.25	3.06	2.97	2.87	2.77	2.66	2.55	2.43
25	3.54	3.37	3.20	3.01	2.92	2.82	2.72	2.61	2.50	2.38
26	3.49	3.33	3.15	2.97	2.87	2.77	2.67	2.56	2.45	2.33
27	3.45	3.28	3.11	2.93	2.83	2.73	2.63	2.52	2.41	2.29
28	3.41	3.25	3.07	2.89	2.79	2.69	2.59	2.48	2.37	2.25
29	3.38	3.21	3.04	2.86	2.76	2.66	2.56	2.45	2.33	2.21
30	3.34	3.18	3.01	2.82	2.73	2.63	2.52	2.42	2.30	2.18
40	3.12	2.95	2.78	2.60	2.50	2.40	2.30	2.18	2.06	1.93
60	2.90	2.74	2.57	2.39	2.29	2.19	2.08	1.96	1.83	1.69
120	2.71	2.54	2.37	2.19	2.09	1.98	1.87	1.75	1.61	1.43
∞	2.52	2.36	2.19	2.00	1.90	1.79	1.67	1.53	1.36	1.00

Table G (*Continued*)

$$F_{.99}$$

Denominator Degrees of Freedom	Numerator Degrees of Freedom								
	1	2	3	4	5	6	7	8	9
1	4052	4999.5	5403	5625	5764	5859	5928	5981	6022
2	98.50	99.00	99.17	99.25	99.30	99.33	99.36	99.37	99.39
3	34.12	30.82	29.46	28.71	28.24	27.91	27.67	27.49	27.35
4	21.20	18.00	16.69	15.98	15.52	15.21	14.98	14.80	14.66
5	16.26	13.27	12.06	11.39	10.97	10.67	10.46	10.29	10.16
6	13.75	10.92	9.78	9.15	8.75	8.47	8.26	8.10	7.98
7	12.25	9.55	8.45	7.85	7.46	7.19	6.99	6.84	6.72
8	11.26	8.65	7.59	7.01	6.63	6.37	6.18	6.03	5.91
9	10.56	8.02	6.99	6.42	6.06	5.80	5.61	5.47	5.35
10	10.04	7.56	6.55	5.99	5.64	5.39	5.20	5.06	4.94
11	9.65	7.21	6.22	5.67	5.32	5.07	4.89	4.74	4.63
12	9.33	6.93	5.95	5.41	5.06	4.82	4.64	4.50	4.39
13	9.07	6.70	5.74	5.21	4.86	4.62	4.44	4.30	4.19
14	8.86	6.51	5.56	5.04	4.69	4.46	4.28	4.14	4.03
15	8.68	6.36	5.42	4.89	4.56	4.32	4.14	4.00	3.89
16	8.53	6.23	5.29	4.77	4.44	4.20	4.03	3.89	3.78
17	8.40	6.11	5.18	4.67	4.34	4.10	3.93	3.79	3.68
18	8.29	6.01	5.09	4.58	4.25	4.01	3.84	3.71	3.60
19	8.18	5.93	5.01	4.50	4.17	3.94	3.77	3.63	3.52
20	8.10	5.85	4.94	4.43	4.10	3.87	3.70	3.56	3.46
21	8.02	5.78	4.87	4.37	4.04	3.81	3.64	3.51	3.40
22	7.95	5.72	4.82	4.31	3.99	3.76	3.59	3.45	3.35
23	7.88	5.66	4.76	4.26	3.94	3.71	3.54	3.41	3.30
24	7.82	5.61	4.72	4.22	3.90	3.67	3.50	3.36	3.26
25	7.77	5.57	4.68	4.18	3.85	3.63	3.46	3.32	3.22
26	7.72	5.53	4.64	4.14	3.82	3.59	3.42	3.29	3.18
27	7.68	5.49	4.60	4.11	3.78	3.56	3.39	3.26	3.15
28	7.64	5.45	4.57	4.07	3.75	3.53	3.36	3.23	3.12
29	7.60	5.42	4.54	4.04	3.73	3.50	3.33	3.20	3.09
30	7.56	5.39	4.51	4.02	3.70	3.47	3.30	3.17	3.07
40	7.31	5.18	4.31	3.83	3.51	3.29	3.12	2.99	2.89
60	7.08	4.98	4.13	3.65	3.34	3.12	2.95	2.82	2.72
120	6.85	4.79	3.95	3.48	3.17	2.96	2.79	2.66	2.56
∞	6.63	4.61	3.78	3.32	3.02	2.80	2.64	2.51	2.41

TABLE G (*Continued*)

Denominator Degrees of Freedom	Numerator Degrees of Freedom									
	10	12	15	20	24	30	40	60	120	∞
1	6056	6106	6157	6209	6235	6261	6287	6313	6339	6366
2	99.40	99.42	99.43	99.45	99.46	99.47	99.47	99.48	99.49	99.50
3	27.23	27.05	26.87	26.69	26.60	26.50	26.41	26.32	26.22	26.13
4	14.55	14.37	14.20	14.02	13.93	13.84	13.75	13.65	13.56	13.46
5	10.05	9.89	9.72	9.55	9.47	9.38	9.29	9.20	9.11	9.02
6	7.87	7.72	7.56	7.40	7.31	7.23	7.14	7.06	6.97	6.88
7	6.62	6.47	6.31	6.16	6.07	5.99	5.91	5.82	5.74	5.65
8	5.81	5.67	5.52	5.36	5.28	5.20	5.12	5.03	4.95	4.86
9	5.26	5.11	4.96	4.81	4.73	4.65	4.57	4.48	4.40	4.31
10	4.85	4.71	4.56	4.41	4.33	4.25	4.17	4.08	4.00	3.91
11	4.54	4.40	4.25	4.10	4.02	3.94	3.86	3.78	3.69	3.60
12	4.30	4.16	4.01	3.86	3.78	3.70	3.62	3.54	3.45	3.36
13	4.10	3.96	3.82	3.66	3.59	3.51	3.43	3.34	3.25	3.17
14	3.94	3.80	3.66	3.51	3.43	3.35	3.27	3.18	3.09	3.00
15	3.80	3.67	3.52	3.37	3.29	3.21	3.13	3.05	2.96	2.87
16	3.69	3.55	3.41	3.26	3.18	3.10	3.02	2.93	2.84	2.75
17	3.59	3.46	3.31	3.16	3.08	3.00	2.92	2.83	2.75	2.65
18	3.51	3.37	3.23	3.08	3.00	2.92	2.84	2.75	2.66	2.57
19	3.43	3.30	3.15	3.00	2.92	2.84	2.76	2.67	2.58	2.49
20	3.37	3.23	3.09	2.94	2.86	2.78	2.69	2.61	2.52	2.42
21	3.31	3.17	3.03	2.88	2.80	2.72	2.64	2.55	2.46	2.36
22	3.26	3.12	2.98	2.83	2.75	2.67	2.58	2.50	2.40	2.31
23	3.21	3.07	2.93	2.78	2.70	2.62	2.54	2.45	2.35	2.26
24	3.17	3.03	2.89	2.74	2.66	2.58	2.49	2.40	2.31	2.21
25	3.13	2.99	2.85	2.70	2.62	2.54	2.45	2.36	2.27	2.17
26	3.09	2.96	2.81	2.66	2.58	2.50	2.42	2.33	2.23	2.13
27	3.06	2.93	2.78	2.63	2.55	2.47	2.38	2.29	2.20	2.10
28	3.03	2.90	2.75	2.60	2.52	2.44	2.35	2.26	2.17	2.06
29	3.00	2.87	2.73	2.57	2.49	2.41	2.33	2.23	2.14	2.03
30	2.98	2.84	2.70	2.55	2.47	2.39	2.30	2.21	2.11	2.01
40	2.80	2.66	2.52	2.37	2.29	2.20	2.11	2.02	1.92	1.80
60	2.63	2.50	2.35	2.20	2.12	2.03	1.94	1.84	1.73	1.60
120	2.47	2.34	2.19	2.03	1.95	1.86	1.76	1.66	1.53	1.38
∞	2.32	2.18	2.04	1.88	1.79	1.70	1.59	1.47	1.32	1.00

TABLE G (*Continued*)
$$F_{.975}$$

Denominator Degrees of Freedom	Numerator Degrees of Freedom								
	1	2	3	4	5	6	7	8	9
1	647.8	799.5	864.2	899.6	921.8	937.1	948.2	956.7	963.3
2	38.51	39.00	39.17	39.25	39.30	39.33	39.36	39.37	39.39
3	17.44	16.04	15.44	15.10	14.88	14.73	14.62	14.54	14.47
4	12.22	10.65	9.98	9.60	9.36	9.20	9.07	8.98	8.90
5	10.01	8.43	7.76	7.39	7.15	6.98	6.85	6.76	6.68
6	8.81	7.26	6.60	6.23	5.99	5.82	5.70	5.60	5.52
7	8.07	6.54	5.89	5.52	5.29	5.12	4.99	4.90	4.82
8	7.57	6.06	5.42	5.05	4.82	4.65	4.53	4.43	4.36
9	7.21	5.71	5.08	4.72	4.48	4.32	4.20	4.10	4.03
10	6.94	5.46	4.83	4.47	4.24	4.07	3.95	3.85	3.78
11	6.72	5.26	4.63	4.28	4.04	3.88	3.76	3.66	3.59
12	6.55	5.10	4.47	4.12	3.89	3.73	3.61	3.51	3.44
13	6.41	4.97	4.35	4.00	3.77	3.60	3.48	3.39	3.31
14	6.30	4.86	4.24	3.89	3.66	3.50	3.38	3.29	3.21
15	6.20	4.77	4.15	3.80	3.58	3.41	3.29	3.20	3.12
16	6.12	4.69	4.08	3.73	3.50	3.34	3.22	3.12	3.05
17	6.04	4.62	4.01	3.66	3.44	3.28	3.16	3.06	2.98
18	5.98	4.56	3.95	3.61	3.38	3.22	3.10	3.01	2.93
19	5.92	4.51	3.90	3.56	3.33	3.17	3.05	2.96	2.88
20	5.87	4.46	3.86	3.51	3.29	3.13	3.01	2.91	2.84
21	5.83	4.42	3.82	3.48	3.25	3.09	2.97	2.87	2.80
22	5.79	4.38	3.78	3.44	3.22	3.05	2.93	2.84	2.76
23	5.75	4.35	3.75	3.41	3.18	3.02	2.90	2.81	2.73
24	5.72	4.32	3.72	3.38	3.15	2.99	2.87	2.78	2.70
25	5.69	4.29	3.69	3.35	3.13	2.97	2.85	2.75	2.68
26	5.66	4.27	3.67	3.33	3.10	2.94	2.82	2.73	2.65
27	5.63	4.24	3.65	3.31	3.08	2.92	2.80	2.71	2.63
28	5.61	4.22	3.63	3.29	3.06	2.90	2.78	2.69	2.61
29	5.59	4.20	3.61	3.27	3.04	2.88	2.76	2.67	2.59
30	5.57	4.18	3.59	3.25	3.03	2.87	2.75	2.65	2.57
40	5.42	4.05	3.46	3.13	2.90	2.74	2.62	2.53	2.45
60	5.29	3.93	3.34	3.01	2.79	2.63	2.51	2.41	2.33
120	5.15	3.80	3.23	2.89	2.67	2.52	2.39	2.30	2.22
∞	5.02	3.69	3.12	2.79	2.57	2.41	2.29	2.19	2.11

TABLE G (*Continued*)

Denominator Degrees of Freedom	Numerator Degrees of Freedom									
	10	12	15	20	24	30	40	60	120	∞
1	968.6	976.7	984.9	993.1	997.2	1001	1006	1010	1014	1018
2	39.40	39.41	39.43	39.45	39.46	39.46	39.47	39.48	39.49	39.50
3	14.42	14.34	14.25	14.17	14.12	14.08	14.04	13.99	13.95	13.90
4	8.84	8.75	8.66	8.56	8.51	8.46	8.41	8.36	8.31	8.26
5	6.62	6.52	6.43	6.33	6.28	6.23	6.18	6.12	6.07	6.02
6	5.46	5.37	5.27	5.17	5.12	5.07	5.01	4.96	4.90	4.85
7	4.76	4.67	4.57	4.47	4.42	4.36	4.31	4.25	4.20	4.14
8	4.30	4.20	4.10	4.00	3.95	3.89	3.84	3.78	3.73	3.67
9	3.96	3.87	3.77	3.67	3.61	3.56	3.51	3.45	3.39	3.33
10	3.72	3.62	3.52	3.42	3.37	3.31	3.26	3.20	3.14	3.08
11	3.53	3.43	3.33	3.23	3.17	3.12	3.06	3.00	2.94	2.88
12	3.37	3.28	3.18	3.07	3.02	2.96	2.91	2.85	2.79	2.72
13	3.25	3.15	3.05	2.95	2.89	2.84	2.78	2.72	2.66	2.60
14	3.15	3.05	2.95	2.84	2.79	2.73	2.67	2.61	2.55	2.49
15	3.06	2.96	2.86	2.76	2.70	2.64	2.59	2.52	2.46	2.40
16	2.99	2.89	2.79	2.68	2.63	2.57	2.51	2.45	2.38	2.32
17	2.92	2.82	2.72	2.62	2.56	2.50	2.44	2.38	2.32	2.25
18	2.87	2.77	2.67	2.56	2.50	2.44	2.38	2.32	2.26	2.19
19	2.82	2.72	2.62	2.51	2.45	2.39	2.33	2.27	2.20	2.13
20	2.77	2.68	2.57	2.46	2.41	2.35	2.29	2.22	2.16	2.09
21	2.73	2.64	2.53	2.42	2.37	2.31	2.25	2.18	2.11	2.04
22	2.70	2.60	2.50	2.39	2.33	2.27	2.21	2.14	2.08	2.00
23	2.67	2.57	2.47	2.36	2.30	2.24	2.18	2.11	2.04	1.97
24	2.64	2.54	2.44	2.33	2.27	2.21	2.15	2.08	2.01	1.94
25	2.61	2.51	2.41	2.30	2.24	2.18	2.12	2.05	1.98	1.91
26	2.59	2.49	2.39	2.28	2.22	2.16	2.09	2.03	1.95	1.88
27	2.57	2.47	2.36	2.25	2.19	2.13	2.07	2.00	1.93	1.85
28	2.55	2.45	2.34	2.23	2.17	2.11	2.05	1.98	1.91	1.83
29	2.53	2.43	2.32	2.21	2.15	2.09	2.03	1.96	1.89	1.81
30	2.51	2.41	2.31	2.20	2.14	2.07	2.01	1.94	1.87	1.79
40	2.39	2.29	2.18	2.07	2.01	1.94	1.88	1.80	1.72	1.64
60	2.27	2.17	2.06	1.94	1.88	1.82	1.74	1.67	1.58	1.48
120	2.16	2.05	1.94	1.82	1.76	1.69	1.61	1.53	1.43	1.31
∞	2.05	1.94	1.83	1.71	1.64	1.57	1.48	1.39	1.27	1.00

TABLE G (*Continued*)

$$F_{.95}$$

Denominator Degrees of Freedom	Numerator Degrees of Freedom								
	1	2	3	4	5	6	7	8	9
1	161.4	199.5	215.7	224.6	230.2	234.0	236.8	238.9	240.5
2	18.51	19.00	19.16	19.25	19.30	19.33	19.35	19.37	19.38
3	10.13	9.55	9.28	9.12	9.01	8.94	8.89	8.85	8.81
4	7.71	6.94	6.59	6.39	6.26	6.16	6.09	6.04	6.00
5	6.61	5.79	5.41	5.19	5.05	4.95	4.88	4.82	4.77
6	5.99	5.14	4.76	4.53	4.39	4.28	4.21	4.15	4.10
7	5.59	4.74	4.35	4.12	3.97	3.87	3.79	3.73	3.68
8	5.32	4.46	4.07	3.84	3.69	3.58	3.50	3.44	3.39
9	5.12	4.26	3.86	3.63	3.48	3.37	3.29	3.23	3.18
10	4.96	4.10	3.71	3.48	3.33	3.22	3.14	3.07	3.02
11	4.84	3.98	3.59	3.36	3.20	3.09	3.01	2.95	2.90
12	4.75	3.89	3.49	3.26	3.11	3.00	2.91	2.85	2.80
13	4.67	3.81	3.41	3.18	3.03	2.92	2.83	2.77	2.71
14	4.60	3.74	3.34	3.11	2.96	2.85	2.76	2.70	2.65
15	4.54	3.68	3.29	3.06	2.90	2.79	2.71	2.64	2.59
16	4.49	3.63	3.24	3.01	2.85	2.74	2.66	2.59	2.54
17	4.45	3.59	3.20	2.96	2.81	2.70	2.61	2.55	2.49
18	4.41	3.55	3.16	2.93	2.77	2.66	2.58	2.51	2.46
19	4.38	3.52	3.13	2.90	2.74	2.63	2.54	2.48	2.42
20	4.35	3.49	3.10	2.87	2.71	2.60	2.51	2.45	2.39
21	4.32	3.47	3.07	2.84	2.68	2.57	2.49	2.42	2.37
22	4.30	3.44	3.05	2.82	2.66	2.55	2.46	2.40	2.34
23	4.28	3.42	3.03	2.80	2.64	2.53	2.44	2.37	2.32
24	4.26	3.40	3.01	2.78	2.62	2.51	2.42	2.36	2.30
25	4.24	3.39	2.99	2.76	2.60	2.49	2.40	2.34	2.28
26	4.23	3.37	2.98	2.74	2.59	2.47	2.39	2.32	2.27
27	4.21	3.35	2.96	2.73	2.57	2.46	2.37	2.31	2.25
28	4.20	3.34	2.95	2.71	2.56	2.45	2.36	2.29	2.24
29	4.18	3.33	2.93	2.70	2.55	2.43	2.35	2.28	2.22
30	4.17	3.32	2.92	2.69	2.53	2.42	2.33	2.27	2.21
40	4.08	3.23	2.84	2.61	2.45	2.34	2.25	2.18	2.12
60	4.00	3.15	2.76	2.53	2.37	2.25	2.17	2.10	2.04
120	3.92	3.07	2.68	2.45	2.29	2.17	2.09	2.02	1.96
∞	3.84	3.00	2.60	2.37	2.21	2.10	2.01	1.94	1.88

TABLE G (*Continued*)

Denominator Degrees of Freedom	Numerator Degrees of Freedom									
	10	12	15	20	24	30	40	60	120	∞
1	241.9	243.9	245.9	248.0	249.1	250.1	251.1	252.2	253.3	254.3
2	19.40	19.41	19.43	19.45	19.45	19.46	19.47	19.48	19.49	19.50
3	8.79	8.74	8.70	8.66	8.64	8.62	8.59	8.57	8.55	8.53
4	5.96	5.91	5.86	5.80	5.77	5.75	5.72	5.69	5.66	5.63
5	4.74	4.68	4.62	4.56	4.53	4.50	4.46	4.43	4.40	4.36
6	4.06	4.00	3.94	3.87	3.84	3.81	3.77	3.74	3.70	3.67
7	3.64	3.57	3.51	3.44	3.41	3.38	3.34	3.30	3.27	3.23
8	3.35	3.28	3.22	3.15	3.12	3.08	3.04	3.01	2.97	2.93
9	3.14	3.07	3.01	2.94	2.90	2.86	2.83	2.79	2.75	2.71
10	2.98	2.91	2.85	2.77	2.74	2.70	2.66	2.62	2.58	2.54
11	2.85	2.79	2.72	2.65	2.61	2.57	2.53	2.49	2.45	2.40
12	2.75	2.69	2.62	2.54	2.51	2.47	2.43	2.38	2.34	2.30
13	2.67	2.60	2.53	2.46	2.42	2.38	2.34	2.30	2.25	2.21
14	2.60	2.53	2.46	2.39	2.35	2.31	2.27	2.22	2.18	2.13
15	2.54	2.48	2.40	2.33	2.29	2.25	2.20	2.16	2.11	2.07
16	2.49	2.42	2.35	2.28	2.24	2.19	2.15	2.11	2.06	2.01
17	2.45	2.38	2.31	2.23	2.19	2.15	2.10	2.06	2.01	1.96
18	2.41	2.34	2.27	2.19	2.15	2.11	2.06	2.02	1.97	1.92
19	2.38	2.31	2.23	2.16	2.11	2.07	2.03	1.98	1.93	1.88
20	2.35	2.28	2.20	2.12	2.08	2.04	1.99	1.95	1.90	1.84
21	2.32	2.25	2.18	2.10	2.05	2.01	1.96	1.92	1.87	1.81
22	2.30	2.23	2.15	2.07	2.03	1.98	1.94	1.89	1.84	1.78
23	2.27	2.20	2.13	2.05	2.01	1.96	1.91	1.86	1.81	1.76
24	2.25	2.18	2.11	2.03	1.98	1.94	1.89	1.84	1.79	1.73
25	2.24	2.16	2.09	2.01	1.96	1.92	1.87	1.82	1.77	1.71
26	2.22	2.15	2.07	1.99	1.95	1.90	1.85	1.80	1.75	1.69
27	2.20	2.13	2.06	1.97	1.93	1.88	1.84	1.79	1.73	1.67
28	2.19	2.12	2.04	1.96	1.91	1.87	1.82	1.77	1.71	1.65
29	2.18	2.10	2.03	1.94	1.90	1.85	1.81	1.75	1.70	1.64
30	2.16	2.09	2.01	1.93	1.89	1.84	1.79	1.74	1.68	1.62
40	2.08	2.00	1.92	1.84	1.79	1.74	1.69	1.64	1.58	1.51
60	1.99	1.92	1.84	1.75	1.70	1.65	1.59	1.53	1.47	1.39
120	1.91	1.83	1.75	1.66	1.61	1.55	1.50	1.43	1.35	1.25
∞	1.83	1.75	1.67	1.57	1.52	1.46	1.39	1.32	1.22	1.00

TABLE G (*Continued*)

$$F_{.90}$$

Denominator Degrees of Freedom	Numerator Degrees of Freedom								
	1	2	3	4	5	6	7	8	9
1	39.86	49.50	53.59	55.83	57.24	58.20	58.91	59.44	59.86
2	8.53	9.00	9.16	9.24	9.29	9.33	9.35	9.37	9.38
3	5.54	5.46	5.39	5.34	5.31	5.28	5.27	5.25	5.24
4	4.54	4.32	4.19	4.11	4.05	4.01	3.98	3.95	3.94
5	4.06	3.78	3.62	3.52	3.45	3.40	3.37	3.34	3.32
6	3.78	3.46	3.29	3.18	3.11	3.05	3.01	2.98	2.96
7	3.59	3.26	3.07	2.96	2.88	2.83	2.78	2.75	2.72
8	3.46	3.11	2.92	2.81	2.73	2.67	2.62	2.59	2.56
9	3.36	3.01	2.81	2.69	2.61	2.55	2.51	2.47	2.44
10	3.29	2.92	2.73	2.61	2.52	2.46	2.41	2.38	2.35
11	3.23	2.86	2.66	2.54	2.45	2.39	2.34	2.30	2.27
12	3.18	2.81	2.61	2.48	2.39	2.33	2.28	2.24	2.21
13	3.14	2.76	2.56	2.43	2.35	2.28	2.23	2.20	2.16
14	3.10	2.73	2.52	2.39	2.31	2.24	2.19	2.15	2.12
15	3.07	2.70	2.49	2.36	2.27	2.21	2.16	2.12	2.09
16	3.05	2.67	2.46	2.33	2.24	2.18	2.13	2.09	2.06
17	3.03	2.64	2.44	2.31	2.22	2.15	2.10	2.06	2.03
18	3.01	2.62	2.42	2.29	2.20	2.13	2.08	2.04	2.00
19	2.99	2.61	2.40	2.27	2.18	2.11	2.06	2.02	1.98
20	2.97	2.59	2.38	2.25	2.16	2.09	2.04	2.00	1.96
21	2.96	2.57	2.36	2.23	2.14	2.08	2.02	1.98	1.95
22	2.95	2.56	2.35	2.22	2.13	2.06	2.01	1.97	1.93
23	2.94	2.55	2.34	2.21	2.11	2.05	1.99	1.95	1.92
24	2.93	2.54	2.33	2.19	2.10	2.04	1.98	1.94	1.91
25	2.92	2.53	2.32	2.18	2.09	2.02	1.97	1.93	1.89
26	2.91	2.52	2.31	2.17	2.08	2.01	1.96	1.92	1.88
27	2.90	2.51	2.30	2.17	2.07	2.00	1.95	1.91	1.87
28	2.89	2.50	2.29	2.16	2.06	2.00	1.94	1.90	1.87
29	2.89	2.50	2.28	2.15	2.06	1.99	1.93	1.89	1.86
30	2.88	2.49	2.28	2.14	2.05	1.98	1.93	1.88	1.85
40	2.84	2.44	2.23	2.09	2.00	1.93	1.87	1.83	1.79
60	2.79	2.39	2.18	2.04	1.95	1.87	1.82	1.77	1.74
120	2.75	2.35	2.13	1.99	1.90	1.82	1.77	1.72	1.68
∞	2.71	2.30	2.08	1.94	1.85	1.77	1.72	1.67	1.63

TABLE G (*Continued*)

Denominator Degrees of Freedom	Numerator Degrees of Freedom									
	10	12	15	20	24	30	40	60	120	∞
1	60.19	60.71	61.22	61.74	62.00	62.26	62.53	62.79	63.06	63.33
2	9.39	9.41	9.42	9.44	9.45	9.46	9.47	9.47	9.48	9.49
3	5.23	5.22	5.20	5.18	5.18	5.17	5.16	5.15	5.14	5.13
4	3.92	3.90	3.87	3.84	3.83	3.82	3.80	3.79	3.78	3.76
5	3.30	3.27	3.24	3.21	3.19	3.17	3.16	3.14	3.12	3.10
6	2.94	2.90	2.87	2.84	2.82	2.80	2.78	2.76	2.74	2.72
7	2.70	2.67	2.63	2.59	2.58	2.56	2.54	2.51	2.49	2.47
8	2.54	2.50	2.46	2.42	2.40	2.38	2.36	2.34	2.32	2.29
9	2.42	2.38	2.34	2.30	2.28	2.25	2.23	2.21	2.18	2.16
10	2.32	2.28	2.24	2.20	2.18	2.16	2.13	2.11	2.08	2.06
11	2.25	2.21	2.17	2.12	2.10	2.08	2.05	2.03	2.00	1.97
12	2.19	2.15	2.10	2.06	2.04	2.01	1.99	1.96	1.93	1.90
13	2.14	2.10	2.05	2.01	1.98	1.96	1.93	1.90	1.88	1.85
14	2.10	2.05	2.01	1.96	1.94	1.91	1.89	1.86	1.83	1.80
15	2.06	2.02	1.97	1.92	1.90	1.87	1.85	1.82	1.79	1.76
16	2.03	1.99	1.94	1.89	1.87	1.84	1.81	1.78	1.75	1.72
17	2.00	1.96	1.91	1.86	1.84	1.81	1.78	1.75	1.72	1.69
18	1.98	1.93	1.89	1.84	1.81	1.78	1.75	1.72	1.69	1.66
19	1.96	1.91	1.86	1.81	1.79	1.76	1.73	1.70	1.67	1.63
20	1.94	1.89	1.84	1.79	1.77	1.74	1.71	1.68	1.64	1.61
21	1.92	1.87	1.83	1.78	1.75	1.72	1.69	1.66	1.62	1.59
22	1.90	1.86	1.81	1.76	1.73	1.70	1.67	1.64	1.60	1.57
23	1.89	1.84	1.80	1.74	1.72	1.69	1.66	1.62	1.59	1.55
24	1.88	1.83	1.78	1.73	1.70	1.67	1.64	1.61	1.57	1.53
25	1.87	1.82	1.77	1.72	1.69	1.66	1.63	1.59	1.56	1.52
26	1.86	1.81	1.76	1.71	1.68	1.65	1.61	1.58	1.54	1.50
27	1.85	1.80	1.75	1.70	1.67	1.64	1.60	1.57	1.53	1.49
28	1.84	1.79	1.74	1.69	1.66	1.63	1.59	1.56	1.52	1.48
29	1.83	1.78	1.73	1.68	1.65	1.62	1.58	1.55	1.51	1.47
30	1.82	1.77	1.72	1.67	1.64	1.61	1.57	1.54	1.50	1.46
40	1.76	1.71	1.66	1.61	1.57	1.54	1.51	1.47	1.42	1.38
60	1.71	1.66	1.60	1.54	1.51	1.48	1.44	1.40	1.35	1.29
120	1.65	1.60	1.55	1.48	1.45	1.41	1.37	1.32	1.26	1.19
∞	1.60	1.55	1.49	1.42	1.38	1.34	1.30	1.24	1.17	1.00

Table H

Percentage Points of the Studentized Range for 2 through 20 Treatments
Upper 5% points

Error df	2	3	4	5	6	7	8	9	10
1	17.97	26.98	32.82	37.08	40.41	43.12	45.40	47.36	49.07
2	6.08	8.33	9.80	10.88	11.74	12.44	13.03	13.54	13.99
3	4.50	5.91	6.82	7.50	8.04	8.48	8.85	9.18	9.46
4	3.93	5.04	5.76	6.29	6.71	7.05	7.35	7.60	7.83
5	3.64	4.60	5.22	5.67	6.03	6.33	6.58	6.80	6.99
6	3.46	4.34	4.90	5.30	5.63	5.90	6.12	6.32	6.49
7	3.34	4.16	4.68	5.06	5.36	5.61	5.82	6.00	6.16
8	3.26	4.04	4.53	4.89	5.17	5.40	5.60	5.77	5.92
9	3.20	3.95	4.41	4.76	5.02	5.24	5.43	5.59	5.74
10	3.15	3.88	4.33	4.65	4.91	5.12	5.30	5.46	5.60
11	3.11	3.82	4.26	4.57	4.82	5.03	5.20	5.35	5.49
12	3.08	3.77	4.20	4.51	4.75	4.95	5.12	5.27	5.39
13	3.06	3.73	4.15	4.45	4.69	4.88	5.05	5.19	5.32
14	3.03	3.70	4.11	4.41	4.64	4.83	4.99	5.13	5.25
15	3.01	3.67	4.08	4.37	4.59	4.78	4.94	5.08	5.20
16	3.00	3.65	4.05	4.33	4.56	4.74	4.90	5.03	5.15
17	2.98	3.63	4.02	4.30	4.52	4.70	4.86	4.99	5.11
18	2.97	3.61	4.00	4.28	4.49	4.67	4.82	4.96	5.07
19	2.96	3.59	3.98	4.25	4.47	4.65	4.79	4.92	5.04
20	2.95	3.58	3.96	4.23	4.45	4.62	4.77	4.90	5.01
24	2.92	3.53	3.90	4.17	4.37	4.54	4.68	4.81	4.92
30	2.89	3.49	3.85	4.10	4.30	4.46	4.60	4.72	4.82
40	2.86	3.44	3.79	4.04	4.23	4.39	4.52	4.63	4.73
60	2.83	3.40	3.74	3.98	4.16	4.31	4.44	4.55	4.65
120	2.80	3.36	3.68	3.92	4.10	4.24	4.36	4.47	4.56
∞	2.77	3.31	3.63	3.86	4.03	4.17	4.29	⁻4.39	4.47

Error df	11	12	13	14	15	16	17	18	19	20
1	50.59	51.96	53.20	54.33	55.36	56.32	57.22	58.04	58.83	59.56
2	14.39	14.75	15.08	15.38	15.65	15.91	16.14	16.37	16.57	16.77
3	9.72	9.95	10.15	10.35	10.52	10.69	10.84	10.98	11.11	11.24
4	8.03	8.21	8.37	8.52	8.66	8.79	8.91	9.03	9.13	9.23
5	7.17	7.32	7.47	7.60	7.72	7.83	7.93	8.03	8.12	8.21
6	6.65	6.79	6.92	7.03	7.14	7.24	7.34	7.43	7.51	7.59
7	6.30	6.43	6.55	6.66	6.76	6.85	6.94	7.02	7.10	7.17
8	6.05	6.18	6.29	6.39	6.48	6.57	6.65	6.73	6.80	6.87
9	5.87	5.98	6.09	6.19	6.28	6.36	6.44	6.51	6.58	6.64
10	5.72	5.83	5.93	6.03	6.11	6.19	6.27	6.34	6.40	6.47
11	5.61	5.71	5.81	5.90	5.98	6.06	6.13	6.20	6.27	6.33
12	5.51	5.61	5.71	5.80	5.88	5.95	6.02	6.09	6.15	6.21
13	5.43	5.53	5.63	5.71	5.79	5.86	5.93	5.99	6.05	6.11
14	5.36	5.46	5.55	5.64	5.71	5.79	5.85	5.91	5.97	6.03
15	5.31	5.40	5.49	5.57	5.65	5.72	5.78	5.85	5.90	5.96

TABLE H (*Continued*)

Error df	11	12	13	14	15	16	17	18	19	20
16	5.26	5.35	5.44	5.52	5.59	5.66	5.73	5.79	5.84	5.90
17	5.21	5.31	5.39	5.47	5.54	5.61	5.67	5.73	5.79	5.84
18	5.17	5.27	5.35	5.43	5.50	5.57	5.63	5.69	5.74	5.79
19	5.14	5.23	5.31	5.39	5.46	5.53	5.59	5.65	5.70	5.75
20	5.11	5.20	5.28	5.36	5.43	5.49	5.55	5.61	5.66	5.71
24	5.01	5.10	5.18	5.25	5.32	5.38	5.44	5.49	5.55	5.59
30	4.92	5.00	5.08	5.15	5.21	5.27	5.33	5.38	5.43	5.47
40	4.82	4.90	4.98	5.04	5.11	5.16	5.22	5.27	5.31	5.36
60	4.73	4.81	4.88	4.94	5.00	5.06	5.11	5.15	5.20	5.24
120	4.64	4.71	4.78	4.84	4.90	4.95	5.00	5.04	5.09	5.13
∞	4.55	4.62	4.68	4.74	4.80	4.85	4.89	4.93	4.97	5.01

Upper 1% Points

Error df	2	3	4	5	6	7	8	9	10
1	90.03	135.0	164.3	185.6	202.2	215.8	227.2	237.0	245.6
2	14.04	19.02	22.29	24.72	26.63	28.20	29.53	30.68	31.69
3	8.26	10.62	12.17	13.33	14.24	15.00	15.64	16.20	16.69
4	6.51	8.12	9.17	9.96	10.58	11.10	11.55	11.93	12.27
5	5.70	6.98	7.80	8.42	8.91	9.32	9.67	9.97	10.24
6	5.24	6.33	7.03	7.56	7.97	8.32	8.61	8.87	9.10
7	4.95	5.92	6.54	7.01	7.37	7.68	7.94	8.17	8.37
8	4.75	5.64	6.20	6.62	6.96	7.24	7.47	7.68	7.86
9	4.60	5.43	5.96	6.35	6.66	6.91	7.13	7.33	7.49
10	4.48	5.27	5.77	6.14	6.43	6.67	6.87	7.05	7.21
11	4.39	5.15	5.62	5.97	6.25	6.48	6.67	6.84	6.99
12	4.32	5.05	5.50	5.84	6.10	6.32	6.51	6.67	6.81
13	4.26	4.96	5.40	5.73	5.98	6.19	6.37	6.53	6.67
14	4.21	4.89	5.32	5.63	5.88	6.08	6.26	6.41	6.54
15	4.17	4.84	5.25	5.56	5.80	5.99	6.16	6.31	6.44
16	4.13	4.79	5.19	5.49	5.72	5.92	6.08	6.22	6.35
17	4.10	4.74	5.14	5.43	5.66	5.85	6.01	6.15	6.27
18	4.07	4.70	5.09	5.38	5.60	5.79	5.94	6.08	6.20
19	4.05	4.67	5.05	5.33	5.55	5.73	5.89	6.02	6.14
20	4.02	4.64	5.02	5.29	5.51	5.69	5.84	5.97	6.09
24	3.96	4.55	4.91	5.17	5.37	5.54	5.69	5.81	5.92
30	3.89	4.45	4.80	5.05	5.24	5.40	5.54	5.65	5.76
40	3.82	4.37	4.70	4.93	5.11	5.26	5.39	5.50	5.60
60	3.76	4.28	4.59	4.82	4.99	5.13	5.25	5.36	5.45
120	3.70	4.20	4.50	4.71	4.87	5.01	5.12	5.21	5.30
∞	3.64	4.12	4.40	4.60	4.76	4.88	4.99	5.08	5.16

TABLE H (*Continued*)

Error df	11	12	13	14	15	16	17	18	19	20
1	253.2	260.0	266.2	271.8	277.0	281.8	286.3	290.4	294.3	298.0
2	32.59	33.40	34.13	34.81	35.43	36.00	36.53	37.03	37.50	37.95
3	17.13	17.53	17.89	18.22	18.52	18.81	19.07	19.32	19.55	19.77
4	12.57	12.84	13.09	13.32	13.53	13.73	13.91	14.08	14.24	14.40
5	10.48	10.70	10.89	11.08	11.24	11.40	11.55	11.68	11.81	11.93
6	9.30	9.48	9.65	9.81	9.95	10.08	10.21	10.32	10.43	10.54
7	8.55	8.71	8.86	9.00	9.12	9.24	9.35	9.46	9.55	9.65
8	8.03	8.18	8.31	8.44	8.55	8.66	8.76	8.85	8.94	9.03
9	7.65	7.78	7.91	8.03	8.13	8.23	8.33	8.41	8.49	8.57
10	7.36	7.49	7.60	7.71	7.81	7.91	7.99	8.08	8.15	8.23
11	7.13	7.25	7.36	7.46	7.56	7.65	7.73	7.81	7.88	7.95
12	6.94	7.06	7.17	7.26	7.36	7.44	7.52	7.59	7.66	7.73
13	6.79	6.90	7.01	7.10	7.19	7.27	7.35	7.42	7.48	7.55
14	6.66	6.77	6.87	6.96	7.05	7.13	7.20	7.27	7.33	7.39
15	6.55	6.66	6.76	6.84	6.93	7.00	7.07	7.14	7.20	7.26
16	6.46	6.56	6.66	6.74	6.82	6.90	6.97	7.03	7.09	7.15
17	6.38	6.48	6.57	6.66	6.73	6.81	6.87	6.94	7.00	7.05
18	6.31	6.41	6.50	6.58	6.65	6.73	6.79	6.85	6.91	6.97
19	6.25	6.34	6.43	6.51	6.58	6.65	6.72	6.78	6.84	6.89
20	6.19	6.28	6.37	6.45	6.52	6.59	6.65	6.71	6.77	6.82
24	6.02	6.11	6.19	6.26	6.33	6.39	6.45	6.51	6.56	6.61
30	5.85	5.93	6.01	6.08	6.14	6.20	6.26	6.31	6.36	6.41
40	5.69	5.76	5.83	5.90	5.96	6.02	6.07	6.12	6.16	6.21
60	5.53	5.60	5.67	5.73	5.78	5.84	5.89	5.93	5.97	6.01
120	5.37	5.44	5.50	5.56	5.61	5.66	5.71	5.75	5.79	5.83
∞	5.23	5.29	5.35	5.40	5.45	5.49	5.54	5.57	5.61	5.65

TABLE I

Transformation of r to z

The Body of the Table Contains Values of $z = .5[\ln(1 + r) / (1 - r)] = \tanh^{-1}r$ for Corresponding Values of r, the Correlation Coefficient.

r	.00	.01	.02	.03	.04	.05	.06	.07	.08	.09
.0	.00000	.01000	.02000	.03001	.04002	.05004	.06007	.07012	.08017	.09024
.1	.10034	.11045	.12058	.13074	.14093	.15114	.16139	.17167	.18198	.19234
.2	.20273	.21317	.22366	.23419	.24477	.25541	.26611	.27686	.28768	.29857
.3	.30952	.32055	.33165	.34283	.35409	.36544	.37689	.38842	.40006	.41180
.4	.42365	.43561	.44769	.45990	.47223	.48470	.49731	.51007	.52298	.53606
.5	.54931	.56273	.57634	.59014	.60415	.61838	.63283	.64752	.66246	.67767
.6	.69315	.70892	.72500	.74142	.75817	.77530	.79281	.81074	.82911	.84795
.7	.86730	.88718	.90764	.92873	.95048	.97295	.99621	1.02033	1.04537	1.07143
.8	1.09861	1.12703	1.15682	1.18813	1.22117	1.25615	1.29334	1.33308	1.37577	1.42192
.9	1.47222	1.52752	1.58902	1.65839	1.73805	1.83178	1.94591	2.09229	2.29756	2.64665

TABLE J
Quantiles of the Mann–Whitney Test Statistic

n	p	m = 2	3	4	5	6	7	8	9	10	11	12	13	14	15	16	17	18	19	20
2	.001	0	0	0	0	0	0	0	0	0	0	0	0	0	0	0	0	0	0	0
	.005	0	0	0	0	0	0	0	0	0	0	0	0	0	0	0	0	0	1	1
	.01	0	0	0	0	0	0	0	0	0	0	0	1	1	1	1	1	1	2	2
	.025	0	0	0	0	0	0	1	1	1	1	2	2	2	2	2	3	3	3	3
	.05	0	0	0	1	1	1	2	2	2	2	3	3	4	4	4	4	5	5	5
	.10	0	1	1	2	2	2	3	3	4	4	5	5	5	6	6	7	7	8	8
3	.001	0	0	0	0	0	0	0	0	0	0	0	0	0	0	0	1	1	1	1
	.005	0	0	0	0	0	0	0	1	1	1	2	2	2	3	3	3	3	4	4
	.01	0	0	0	0	0	1	1	2	2	2	3	3	3	4	4	5	5	5	6
	.025	0	0	1	1	2	2	3	3	4	4	5	5	6	6	7	7	8	8	9
	.05	0	1	1	2	3	3	4	5	5	6	6	7	8	8	9	10	10	11	12
	.10	1	2	2	3	4	5	6	6	7	8	9	10	11	11	12	13	14	15	16
4	.001	0	0	0	0	0	0	0	0	1	1	1	2	2	2	3	3	4	4	4
	.005	0	0	0	0	1	1	2	2	3	3	4	4	5	6	6	7	7	8	9
	.01	0	0	1	1	2	2	3	4	4	5	6	6	7	8	8	9	10	10	11
	.025	0	1	2	2	3	4	5	5	6	7	8	9	10	11	12	12	13	14	15
	.05	0	1	3	3	4	5	6	7	8	9	10	11	12	13	15	16	17	18	19
	.10	1	2	4	5	6	7	8	10	11	12	13	14	16	17	18	19	21	22	23
5	.001	0	0	0	0	0	0	1	2	2	3	3	4	4	5	6	6	7	8	8
	.005	0	0	0	1	2	2	3	4	5	6	7	8	8	9	10	11	12	13	14
	.01	0	0	1	2	3	4	5	6	7	8	9	10	11	12	13	14	15	16	17
	.025	0	1	2	3	4	6	7	8	9	10	12	13	14	15	16	18	19	20	21
	.05	1	2	3	5	6	7	9	10	12	13	14	16	17	19	20	21	23	24	26
	.10	2	3	5	6	8	9	11	13	14	16	18	19	21	23	24	26	28	29	31

Table J (Continued)

n	p	m = 2	3	4	5	6	7	8	9	10	11	12	13	14	15	16	17	18	19	20
6	.001	0	0	0	0	0	0	2	3	4	5	5	6	7	8	9	10	11	12	13
	.005	0	0	1	2	3	4	5	6	7	8	10	11	12	13	14	16	17	18	19
	.01	0	0	2	3	4	5	7	8	9	10	12	13	14	16	17	19	20	21	23
	.025	0	2	3	4	6	7	9	11	12	14	15	17	18	20	22	23	25	26	28
	.05	1	3	4	6	8	9	11	13	15	17	18	20	22	24	26	27	29	31	33
	.10	2	4	6	8	10	12	14	16	18	20	22	24	26	28	30	32	35	37	39
7	.001	0	0	0	0	1	2	3	4	6	7	8	9	10	11	12	14	15	16	17
	.005	0	0	1	2	4	5	7	8	10	11	13	14	16	17	19	20	22	23	25
	.01	0	1	2	4	5	7	8	10	12	13	15	17	18	20	22	24	25	27	29
	.025	0	2	4	6	7	9	11	13	15	17	19	21	23	25	27	29	31	33	35
	.05	1	3	5	7	9	12	14	16	18	20	22	25	27	29	31	34	36	38	40
	.10	2	5	7	9	12	14	17	19	22	24	27	29	32	34	37	39	42	44	47
8	.001	0	0	0	1	2	3	5	6	7	9	10	12	13	15	16	18	19	21	22
	.005	0	0	2	3	5	7	8	10	12	14	16	18	19	21	23	25	27	29	31
	.01	0	1	3	5	7	8	10	12	14	16	18	21	23	25	27	29	31	33	35
	.025	1	3	5	7	9	11	14	16	18	20	23	25	27	30	32	35	37	39	42
	.05	2	4	6	9	11	14	16	19	21	24	27	29	32	34	37	40	42	45	48
	.10	3	6	8	11	14	17	20	23	25	28	31	34	37	40	43	46	49	52	55
9	.001	0	0	0	2	3	4	6	8	9	11	13	15	16	18	20	22	24	26	27
	.005	0	1	2	4	6	8	10	12	14	17	19	21	23	25	28	30	32	34	37
	.01	0	2	4	6	8	10	12	15	17	19	22	24	27	29	32	34	37	39	41
	.025	1	3	5	8	11	13	16	18	21	24	27	29	32	35	38	40	43	46	49
	.05	2	5	7	10	13	16	19	22	25	28	31	34	37	40	43	46	49	52	55
	.10	3	6	10	13	16	19	23	26	29	32	36	39	42	46	49	53	56	59	63
10	.001	0	0	1	2	4	6	7	9	11	13	15	18	20	22	24	26	28	30	33
	.005	0	1	3	5	7	10	12	14	17	19	22	25	27	30	32	35	38	40	43
	.01	0	2	4	7	9	12	14	17	20	23	25	28	31	34	37	39	42	45	48
	.025	1	4	6	9	12	15	18	21	24	27	30	34	37	40	43	46	49	53	56
	.05	2	5	8	12	15	18	21	25	28	32	35	38	42	45	49	52	56	59	63
	.10	4	7	11	14	18	22	25	29	33	37	40	44	48	52	55	59	63	67	71

TABLE J (Continued)

n	p	m = 2	3	4	5	6	7	8	9	10	11	12	13	14	15	16	17	18	19	20
11	.001	0	0	1	3	5	7	9	11	13	16	18	21	23	25	28	30	33	35	38
	.005	0	1	3	6	8	11	14	17	19	22	25	28	31	34	37	40	43	46	49
	.01	0	2	5	8	10	13	16	19	23	26	29	32	35	38	42	45	48	51	54
	.025	1	4	7	10	14	17	20	24	27	31	34	38	41	45	48	52	56	59	63
	.05	2	6	9	13	17	20	24	28	32	35	39	43	47	51	55	58	62	66	70
	.10	4	8	12	16	20	24	28	32	37	41	45	49	53	58	62	66	70	74	79
12	.001	0	0	1	3	5	8	10	13	15	18	21	24	26	29	32	35	38	41	43
	.005	0	2	4	7	10	13	16	19	22	25	28	32	35	38	42	45	48	52	55
	.01	0	3	6	9	12	15	18	22	25	29	32	36	39	43	47	50	54	57	61
	.025	2	5	8	12	15	19	23	27	30	34	38	42	46	50	54	58	62	66	70
	.05	3	6	10	14	18	22	27	31	35	39	43	48	52	56	61	65	69	73	78
	.10	5	9	13	18	22	27	31	36	40	45	50	54	59	64	68	73	78	82	87
13	.001	0	0	2	4	6	9	12	15	18	21	24	27	30	33	36	39	43	46	49
	.005	0	2	4	8	11	14	18	21	25	28	32	35	39	43	46	50	54	58	61
	.01	1	3	6	10	13	17	21	24	28	32	36	40	44	48	52	56	60	64	68
	.025	2	5	9	13	17	21	25	29	34	38	42	46	51	55	60	64	68	73	77
	.05	3	7	12	16	20	25	29	34	38	43	48	52	57	62	66	71	76	81	85
	.10	5	10	14	19	24	29	34	39	44	49	54	59	64	69	75	80	85	90	95
14	.001	0	0	2	4	7	10	13	16	20	23	26	30	33	37	40	44	47	51	55
	.005	0	2	5	8	12	16	19	23	27	31	35	39	43	47	51	55	59	64	68
	.01	1	3	7	11	14	18	23	27	31	35	39	44	48	52	57	61	66	70	74
	.025	2	6	10	14	18	23	27	32	37	41	46	51	56	60	65	70	75	79	84
	.05	4	8	12	17	22	27	32	37	42	47	52	57	62	67	72	78	83	88	93
	.10	5	11	16	21	26	32	37	42	48	53	59	64	70	75	81	86	92	98	103
15	.001	0	0	2	5	8	11	15	18	22	25	29	33	37	41	44	48	52	56	60
	.005	0	3	6	9	13	17	21	25	30	34	38	43	47	52	56	61	65	70	74
	.01	1	4	8	12	16	20	25	29	34	38	43	48	52	57	62	67	71	76	81
	.025	2	6	11	15	20	25	30	35	40	45	50	55	60	65	71	76	81	86	91
	.05	4	8	13	19	24	29	35	40	45	51	56	62	67	73	78	84	89	95	101
	.10	6	11	17	23	28	34	40	46	52	58	64	69	75	81	87	93	99	105	111

Table J (Continued)

n	p	m = 2	3	4	5	6	7	8	9	10	11	12	13	14	15	16	17	18	19	20
16	.001	0	0	3	6	9	12	16	20	24	28	32	36	40	44	49	53	57	61	66
	.005	0	3	6	10	14	19	23	28	32	37	42	46	51	56	61	66	71	75	80
	.01	1	4	8	13	17	22	27	32	37	42	47	52	57	62	67	72	77	83	88
	.025	2	7	12	16	22	27	32	38	43	48	54	60	65	71	76	82	87	93	99
	.05	4	9	15	20	26	31	37	43	49	55	61	66	72	78	84	90	96	102	108
	.10	6	12	18	24	30	37	43	49	55	62	68	75	81	87	94	100	107	113	120
17	.001	0	1	3	6	10	14	18	22	26	30	35	39	44	48	53	58	62	67	71
	.005	0	3	7	11	16	20	25	30	35	40	45	50	55	61	66	71	76	82	87
	.01	1	5	9	14	19	24	29	34	39	45	50	56	61	67	72	78	83	89	94
	.025	3	7	12	18	23	29	35	40	46	52	58	64	70	76	82	88	94	100	106
	.05	4	10	16	21	27	34	40	46	52	58	65	71	78	84	90	97	103	110	116
	.10	7	13	19	26	32	39	46	53	59	66	73	80	86	93	99	107	114	121	128
18	.001	0	1	4	7	11	15	19	24	28	33	38	43	47	52	57	62	67	72	77
	.005	0	3	7	12	17	22	27	32	38	43	48	54	59	65	71	76	82	88	93
	.01	1	5	10	15	20	25	31	37	42	48	54	60	66	71	77	83	89	95	101
	.025	3	8	13	19	25	31	37	43	49	56	62	68	75	81	87	94	100	107	113
	.05	5	11	17	23	29	36	42	49	56	62	69	76	83	89	96	103	110	117	124
	.10	7	14	21	28	35	42	49	56	63	70	78	85	92	99	107	114	121	129	136
19	.001	0	1	4	8	12	16	21	26	30	35	41	46	51	56	61	67	72	78	83
	.005	1	4	8	13	18	23	29	34	40	46	52	58	64	70	75	82	88	94	100
	.01	2	5	10	16	21	27	33	39	45	51	57	64	70	76	83	89	95	102	108
	.025	3	8	14	20	26	33	39	46	53	59	66	73	79	86	93	100	107	114	120
	.05	5	11	18	24	31	38	45	52	59	66	73	81	88	95	102	110	117	124	131
	.10	8	15	22	29	37	44	52	59	67	74	82	90	98	105	113	121	129	136	144
20	.001	0	1	4	8	13	17	22	27	33	38	43	49	55	60	66	71	77	83	89
	.005	1	4	9	14	19	25	31	37	43	49	55	61	68	74	80	87	93	100	106
	.01	2	6	11	17	23	29	35	41	48	54	61	68	74	81	88	94	101	108	115
	.025	3	9	15	21	28	35	42	49	56	63	70	77	84	91	99	106	113	120	128
	.05	5	12	19	26	33	40	48	55	63	70	78	85	93	101	108	116	124	131	139
	.10	8	16	23	31	39	47	55	63	71	79	87	95	103	111	120	128	136	144	152

TABLE K

Quantiles of the Kolmogorov Test Statistic

One-Sided Test					
$p = .90$	.95	.975	.99	.995
Two-Sided Test					
$p = .80$	.90	.95	.98	.99
$n = 1$	.900	.950	.975	.990	.995
2	.684	.776	.842	.900	.929
3	.565	.636	.708	.785	.829
4	.493	.565	.624	.689	.734
5	.447	.509	.563	.627	.669
6	.410	.468	.519	.577	.617
7	.381	.436	.483	.538	.576
8	.358	.410	.454	.507	.542
9	.339	.387	.430	.480	.513
10	.323	.369	.409	.457	.489
11	.308	.352	.391	.437	.468
12	.296	.338	.375	.419	.449
13	.285	.325	.361	.404	.432
14	.275	.314	.349	.390	.418
15	.266	.304	.338	.377	.404
16	.258	.295	.327	.366	.392
17	.250	.286	.318	.355	.381
18	.244	.279	.309	.346	.371
19	.237	.271	.301	.337	.361
20	.232	.265	.294	.329	.352
21	.226	.259	.287	.321	.344
22	.221	.253	.281	.314	.337
23	.216	.247	.275	.307	.330
24	.212	.242	.269	.301	.323
25	.208	.238	.264	.295	.317
26	.204	.233	.259	.290	.311
27	.200	.229	.254	.284	.305
28	.197	.225	.250	.279	.300
29	.193	.221	.246	.275	.295
30	.190	.218	.242	.270	.290
31	.187	.214	.238	.266	.285
32	.184	.211	.234	.262	.281
33	.182	.208	.231	.258	.277
34	.179	.205	.227	.254	.273
35	.177	.202	.224	.251	.269
36	.174	.199	.221	.247	.265
37	.172	.196	.218	.244	.262
38	.170	.194	.215	.241	.258
39	.168	.191	.213	.238	.255
40	.165	.189	.210	.235	.252
Approximation for $n > 40$	$\dfrac{1.07}{\sqrt{n}}$	$\dfrac{1.22}{\sqrt{n}}$	$\dfrac{1.36}{\sqrt{n}}$	$\dfrac{1.52}{\sqrt{n}}$	$\dfrac{1.63}{\sqrt{n}}$

TABLE L

Critical Values of the Kruskal–Wallis Test Statistic

Sample Sizes			Critical Value	α
n_1	n_2	n_3		
2	1	1	2.7000	.500
2	2	1	3.6000	.200
2	2	2	4.5714	.067
			3.7143	.200
3	1	1	3.2000	.300
3	2	1	4.2857	.100
			3.8571	.133
3	2	2	5.3572	.029
			4.7143	.048
			4.5000	.067
			4.4643	.105
3	3	1	5.1429	.043
			4.5714	.100
			4.0000	.129
3	3	2	6.2500	.011
			5.3611	.032
			5.1389	.061
			4.5556	.100
			4.2500	.121
3	3	3	7.2000	.004
			6.4889	.011
			5.6889	.029
			5.6000	.050
			5.0667	.086
			4.6222	.100
4	1	1	3.5714	.200
4	2	1	4.8214	.057
			4.5000	.076
			4.0179	.114
4	2	2	6.0000	.014
			5.3333	.033
			5.1250	.052
			4.4583	.100
			4.1667	.105
4	3	1	5.8333	.021
			5.2083	.050
			5.0000	.057
			4.0556	.093
			3.8889	.129

APPENDIX

TABLE L (*Continued*)

Sample Sizes			Critical	
n_1	n_2	n_3	Value	α
4	3	2	6.4444	.008
			6.3000	.011
			5.4444	.046
			5.4000	.051
			4.5111	.098
			4.4444	.102
4	3	3	6.7455	.010
			6.7091	.013
			5.7909	.046
			5.7273	.050
			4.7091	.092
			4.7000	.101
4	4	1	6.6667	.010
			6.1667	.022
			4.9667	.048
			4.8667	.054
			4.1667	.082
			4.0667	.102
4	4	2	7.0364	.006
			6.8727	.011
			5.4545	.046
			5.2364	.052
			4.5545	.098
			4.4455	.103
4	4	3	7.1439	.010
			7.1364	.011
			5.5985	.049
			5.5758	.051
			4.5455	.099
			4.4773	.102
4	4	4	7.6538	.008
			7.5385	.011
			5.6923	.049
			5.6538	.054
			4.6539	.097
			4.5001	.104
5	1	1	3.8571	.143
5	2	1	5.2500	.036
			5.0000	.048
			4.4500	.071
			4.2000	.095
			4.0500	.119

TABLE L (*Continued*)

Sample Sizes			Critical Value	α
n_1	n_2	n_3		
5	2	2	6.5333	.008
			6.1333	.013
			5.1600	.034
			5.0400	.056
			4.3733	.090
			4.2933	.122
5	3	1	6.4000	.012
			4.9600	.048
			4.8711	.052
			4.0178	.095
			3.8400	.123
5	3	2	6.9091	.009
			6.8218	.010
			5.2509	.049
			5.1055	.052
			4.6509	.091
			4.4945	.101
5	3	3	7.0788	.009
			6.9818	.011
			5.6485	.049
			5.5152	.051
			4.5333	.097
			4.4121	.109
5	4	1	6.9545	.008
			6.8400	.011
			4.9855	.044
			4.8600	.056
			3.9873	.098
			3.9600	.102
5	4	2	7.2045	.009
			7.1182	.010
			5.2727	.049
			5.2682	.050
			4.5409	.098
			4.5182	.101
5	4	3	7.4449	.010
			7.3949	.011
			5.6564	.049
			5.6308	.050

TABLE L (*Continued*)

Sample Sizes			Critical Value	α
n_1	n_2	n_3		
			4.5487	.099
			4.5231	.103
5	4	4	7.7604	.009
			7.7440	.011
			5.6571	.049
			5.6176	.050
			4.6187	.100
			4.5527	.102
5	5	1	7.3091	.009
			6.8364	.011
			5.1273	.046
			4.9091	.053
			4.1091	.086
			4.0364	.105
5	5	2	7.3385	.010
			7.2692	.010
			5.3385	.047
			5.2462	.051
			4.6231	.097
			4.5077	.100
5	5	3	7.5780	.010
			7.5429	.010
			5.7055	.046
			5.6264	.051
			4.5451	.100
			4.5363	.102
5	5	4	7.8229	.010
			7.7914	.010
			5.6657	.049
			5.6429	.050
			4.5229	.099
			4.5200	.101
5	5	5	8.0000	.009
			7.9800	.010
			5.7800	.049
			5.6600	.051
			4.5600	.100
			4.5000	.102

TABLE M*a*

Exact Distribution of χ_r^2 for Tables with from 2 to 9 Sets of Three Ranks
($k = 3$; $n = 2, 3, 4, 5, 6, 7, 8, 9$)
P Is the Probability of Obtaining a Value of χ_r^2 as Great as or Greater than the Corresponding Value of χ_r^2

$n = 2$		$n = 3$		$n = 4$		$n = 5$	
χ_r^2	P	χ_r^2	P	χ_r^2	P	χ_r^2	P
0	1.000	.000	1.000	.0	1.000	.0	1.000
1	.833	0.667	.944	.5	.931	.4	.954
3	.500	2.000	.528	1.5	.653	1.2	.691
4	.167	2.667	.361	2.0	.431	1.6	.522
		4.667	1.94	3.5	.273	2.8	.367
		6.000	.028	4.5	.125	3.6	.182
				6.0	.069	4.8	.124
				6.5	.042	5.2	.093
				8.0	.0046	6.4	.039
						7.6	.024
						8.4	.0085
						10.0	.00077

$n = 6$		$n = 7$		$n = 8$		$n = 9$	
.00	1.000	.000	1.000	.00	1.000	.000	1.000
0.33	.956	.286	.964	.25	.967	.222	.971
1.00	.740	.857	.768	.75	.794	.667	.814
1.33	.570	1.143	.620	1.00	.654	.889	.865
2.33	.430	2.000	.486	1.75	.531	1.556	.569
3.00	.252	2.571	.305	2.25	.355	2.000	.398
4.00	.184	3.429	.237	3.00	.285	2.667	.328
4.33	.142	3.714	.192	3.25	.236	2.889	.278
5.33	.072	4.571	.112	4.00	.149	3.556	.187
6.33	.052	5.429	.085	4.75	.120	4.222	.154
7.00	.029	6.000	.052	5.25	.079	4.667	.107
8.33	.012	7.143	.027	6.25	.047	5.556	.069
9.00	.0081	7.714	.021	6.75	.038	6.000	.057
9.33	.0055	8.000	.016	7.00	.030	6.222	.048
10.33	.0017	8.857	.0084	7.75	.018	6.889	.031
12.00	.00013	10.286	.0036	9.00	.0099	8.000	.019
		10.571	.0027	9.25	.0080	8.222	.016
		11.143	.0012	9.75	.0048	8.667	.010
		12.286	.00032	10.75	.0024	9.556	.0060
		14.000	.000021	12.00	.0011	10.667	.0035
				12.25	.00086	10.889	.0029
				13.00	.00026	11.556	.0013
				14.25	.000061	12.667	.00066
				16.00	.0000036	13.556	.00035
						14.000	.00020
						14.222	.000097
						14.889	.000054
						16.222	.000011
						18.000	.0000006

TABLE Mb

Exact Distribution of χ_r^2 for Tables with from 2 to 4 Sets of Four Ranks
($k = 4$; $n = 2, 3, 4$)
P Is the Probability of Obtaining a Value of χ_r^2 as Great as or Greater than the Corresponding Value of χ_r^2

$n = 2$		$n = 3$		$n = 4$			
χ_r^2	P	χ_r^2	P	χ_r^2	P	χ_r^2	P
.0	1.000	.2	1.000	.0	1.000	5.7	.141
.6	.958	.6	.958	.3	.992	6.0	.105
1.2	.834	1.0	.910	.6	.928	6.3	.094
1.8	.792	1.8	.727	.9	.900	6.6	.077
2.4	.625	2.2	.608	1.2	.800	6.9	.068
3.0	.542	2.6	.524	1.5	.754	7.2	.054
3.6	.458	3.4	.446	1.8	.677	7.5	.052
4.2	.375	3.8	.342	2.1	.649	7.8	.036
4.8	.208	4.2	.300	2.4	.524	8.1	.033
5.4	.167	5.0	.207	2.7	.508	8.4	.019
6.0	.042	5.4	.175	3.0	.432	8.7	.014
		5.8	.148	3.3	.389	9.3	.012
		6.6	.075	3.6	.355	9.6	.0069
		7.0	.054	3.9	.324	9.9	.0062
		7.4	.033	4.5	.242	10.2	.0027
		8.2	.017	4.8	.200	10.8	.0016
		9.0	.0017	5.1	.190	11.1	.00094
				5.4	.158	12.0	.000072

TABLE N

Critical Values of the Spearman Test Statistic. Approximate Upper-tail Critical Values r_s^*, Where $P(r_s > r_s^*) \leq \alpha$, $n = 4(1)30$ Significance Level, α.

n	.001	.005	.010	.025	.050	.100
4	—	—	—	—	.8000	.8000
5	—	—	.9000	.9000	.8000	.7000
6	—	.9429	.8857	.8286	.7714	.6000
7	.9643	.8929	.8571	.7450	.6786	.5357
8	.9286	.8571	.8095	.7143	.6190	.5000
9	.9000	.8167	.7667	.6833	.5833	.4667
10	.8667	.7818	.7333	.6364	.5515	.4424
11	.8364	.7545	.7000	.6091	.5273	.4182
12	.8182	.7273	.6713	.5804	.4965	.3986
13	.7912	.6978	.6429	.5549	.4780	.3791
14	.7670	.6747	.6220	.5341	.4593	.3626
15	.7464	.6536	.6000	.5179	.4429	.3500
16	.7265	.6324	.5824	.5000	.4265	.3382
17	.7083	.6152	.5637	.4853	.4118	.3260
18	.6904	.5975	.5480	.4716	.3994	.3148
19	.6737	.5825	.5333	.4579	.3895	.3070
20	.6586	.5684	.5203	.4451	.3789	.2977
21	.6455	.5545	.5078	.4351	.3688	.2909
22	.6318	.5426	.4963	.4241	.3597	.2829
23	.6186	.5306	.4852	.4150	.3518	.2767
24	.6070	.5200	.4748	.4061	.3435	.2704
25	.5962	.5100	.4654	.3977	.3362	.2646
26	.5856	.5002	.4564	.3894	.3299	.2588
27	.5757	.4915	.4481	.3822	.3236	.2540
28	.5660	.4828	.4401	.3749	.3175	.2490
29	.5567	.4744	.4320	.3685	.3113	.2443
30	.5479	.4665	.4251	.3620	.3059	.2400

Note: The corresponding lower-tail critical value for r_s is $-r_s^*$.

ACKNOWLEDGMENTS FOR TABLES

D. From *A Million Random Digits with 100,000 Normal Deviates* by The Rand Corporation, The Free Press, Glencoe, Ill., 1955. Reprinted by permission.

E. Reproduced from *Documenta Geigy, Scientific Tables*, Seventh Edition, 1970, courtesy of CIBA-Geigy Limited, Basle, Switzerland.

F. From A. Hald and S. A. Sinkbaek, "A Table of Percentage Points of the χ^2 Distribution," *Skandinavisk Aktuarietidskrift*, *33* (1950), 168–175. Used by permission.

G. From *Biometrika Tables for Statisticians*, Third Edition, Vol. I, Bentley House, London, 1970. Reprinted by permission.

H. From *Biometrika Tables for Statisticians*, Third Edition, Vol. I, Bentley House, London, 1970. Used by permission.

J. Adapted from L. R. Verdooren, "Extended Tables of Critical Values for Wilcoxon's Test Statistic," *Biometrika*, *50* (1963), 177–186. Used by permission of the author and E. S. Pearson on behalf of the Biometrika Trustees. The adaptation is due to W. J. Conover, *Practical Nonparametric Statistics*, New York: John Wiley, 1971, 384–388.

K. From L. H. Miller, "Table of Percentage Points of Kolmogorov Statistics," *Journal of the American Statistical Association*, *51* (1956), 111–121. Reprinted by permission of the American Statistical Association. The table as reprinted here follows the format found in W. J. Conover, *Practical Nonparametric Statistics*, © 1971, by John Wiley & Sons, Inc.

L. From W. H. Kruskal and W. A. Wallis, "Use of Ranks in One-Criterion Analysis of Variance," *Journal of the American Statistical Association*, *47* (1952), 583–621; errata, ibid., *48* (1953), 907–911. Reprinted by permission of the American Statistical Association.

M. From M. Friedman, "The Use of Ranks to Avoid the Assumption of Normality Implicit in the Analysis of Variance," *Journal of the American Statistical Association*, *32* (1937), 675–701. Reprinted by permission.

N. From Gerald J. Glasser and Robert F. Winter, "Critical Values of the Coefficient of Rank Correlation for Testing the Hypothesis of Independence," *Biometrika 48* (1961), 444–448. Used by permission. The table as reprinted here contains corrections given in W. J. Conover, *Practical Nonparametric Statistics*, © 1971, by John Wiley & Sons, Inc.

ANSWERS TO ODD-NUMBERED EXERCISES

Chapter 1

1.5.1. Suggested class intervals: 55–59, 60–64, . . . , 80–84.

1.5.3. Suggested class intervals: 80–89, 90–99, . . . , 140–149.

1.5.5. Suggested class intervals: 0–2, 3–5, 6–8, 9–11, 12–14, 15–17.

1.5.7.

Class interval	Frequency	Relative frequency
110–139	8	.0516
140–169	16	.1032
170–199	46	.2968
200–229	49	.3161
230–259	26	.1677
260–289	9	.0581
290–319	1	.0065
	155	1.0000

1.6.1. (a) 65.1 (b) 65 (c) 65.

1.6.3. (a) 9.4 (b) 11.

1.6.5. Chicken.

1.7.1. (a) 16 (b) 19.66 (c) 4.4.

1.7.3. (a) 12 (b) 17.83 (c) 4.2 (d) 44.68.

1.7.5. (a) 45.1667 (b) 54.5 (c) 47 (d) 391.7879 (e) 19.7936.

1.9.1. (a) 66.9 (b) 66.6 (c) 65–69 (d) 46.45 (e) 6.8.

1.9.3. (a) 117.33 (b) 116.46 (c) 110–119 (d) 217.26 (e) 14.7.

1.9.5. (a) 6.53 (b) 5.85 (c) 15.1182 (d) 3.89.

1.9.7. $\bar{x} = 203.8548$, Median $= 204.0918$, $s^2 = 1368.4126$, $s = 36.9921$.

Review Exercises

21. $\bar{x} = 23.07$ Median $= 22$ $s^2 = 54.92$ $s = 7.4$.

23. $\bar{x} = 9.42$ Median $= 9.45$ $s^2 = 2.90$ $s = 1.7$.

25. $\bar{x} = 35.23$ Median $= 35.64$ $s^2 = 168.0360$ $s = 12.96$.

27. $\bar{x} = 5.4593$ Median $= 5.205$, $s^2 = 2.085715$ $s = 1.4442004$.

Chapter 2

2.4.1. (a) 160 (b) 26 (c) 688 (d) 308 (e) 1025 (f) 263.

2.5.1. (a) 30 (b) 210 (c) 30,240 (d) 15 (e) 35 (f) 252 (g) 56 (h) 126 (i) 10.

2.5.3. 120.

2.5.5. 6.

2.5.7. 4.

2.5.9. 560.

2.5.11. 5040.

2.6.1. (a) .16 (b) .03 (c) .67 (d) .30 (e) .995 (f) .26.

2.6.3. (a) .45 (b) .41 (c) .10 (d) .04.

2.6.5. .95.

2.6.7. .301.

Review Exercises

11. (a) 30 (b) 54 (c) 33 (d) 85 (e) 100 (f) 77 (g) 7 (h) 80.

13. .64.

15. .49.

17. .75.

19. .21.

Chapter 3

3.3.1. (a) .1484 (b) .8915 (c) .1085 (d) .2012.

3.3.3. (a) .9729 (b) .9095 (c) .7827 (d) .0271.

3.3.5. (a) .001 (b) .027 (c) .972 (d) .271 (e) .972 (f) .729.

3.4.1. (a) .176 (b) .384 (c) .440 (d) .427.

3.4.3. (a) .105 (b) .032 (c) .007 (d) .440.

3.4.5. (a) .086 (b) .946 (c) .463 (d) .664 (e) .026.

3.6.1. .4236.

3.6.3. .2912.

3.6.5. .0099.

3.6.7. .95.

3.6.9. .901.

3.6.11. -2.54.

3.6.13. 1.77.

3.6.15. 1.32.

3.6.17. (a) .1151 (b) .2119 (c) .6730 (d) .1012.

3.6.19. (a) .7486 (b) .0228 (c) .4772 (d) .0228.

3.6.21. (a) .6826 (b) .2743 (c) .2743 (d) .3446 (e) .8833.

Review Exercises

15. .1719.

17. (a) .0916 (b) .0905 (c) .9095 (d) .1845 (e) .2502.

19. (a) .762 (b) .238 (c) .065.

21. (a) .0668 (b) .6247 (c) .6826.

23. (a) .0013 (b) .0668 (c) .8931.

25. 57.10.

27. (a) 64.75 (b) 118.45 (c) 130.15 (d) 131.8.

29. 14.90.

31. 10.6.

Chapter 4

4.4.1. (a) .9772 (b) .9544 (c) .0082 (d) .0082 (e) 11.51% (f) 94,520.

4.4.3. (a) .1814 (b) .8016 (c) .0643.

4.4.5. (a) .5 (b) .7333 (c) .9772.

4.4.7. $\mu_{\bar{x}} = 5$; $\sigma_{\bar{x}}^2 = 3$.

4.4.9. (a) .0853 (b) .0104 (c) .7973.

4.5.1. .1379.

4.5.3. .0038.

4.5.5. .0139.

4.6.1. (a) .0808 (b) .8384 (c) .2005.

4.6.3. .0823.

4.6.5. (a) .1539 (b) .3409 (c) .5230.

4.7.1. .008.

4.7.3. .0520.

Review Exercises

13. .8664.

15. .0011.

17. .0082.

19. .8882.

21. .0019.

23. Normally distributed.

25. .0166.

27. 25, 1.4.

Chapter 5

5.2.1. (a) 88, 92 (b) 87, 93 (c) 86, 94.

5.2.3. (a) 7.63, 8.87 (b) 7.51, 8.99 (c) 7.28, 9.22.

5.2.5. 1576.125, 1919.125.

5.3.1. 5.76, 8.24; 5.46, 8.54; 4.76, 9.24.

5.3.3. 69.58, 76.42; 68.87, 77.13; 67.41, 78.59.

5.3.5. .00964, .02160.

5.4.1. (a) -2.0, 13.0 (b) -3.5, 14.5 (c) -6.3, 17.3.

5.4.3. (a) 4, 10 (b) 3.5, 10.5 (c) 2.3, 11.7.

5.4.5. -4.33, 7.33; -5.50, 8.50; -7.86, 10.86. Use 40 df.

5.4.7. 2.1, 4.5; 1.8, 4.8; 1.3, 4.8.

5.4.9. 24.7, 33.2.

5.5.1. .04, .12; .03, .13; .01, .15.

5.5.3. .14, .34; .12, .36; .08, .40.

5.6.1. $-.02$, .10; $-.03$, .11; $-.06$, .14.

5.6.3. $-.07$, .19; $-.09$, .21; $-.14$, .26.

5.7.1. 27, 16.

5.7.3. 19.

5.8.1. 683, 1068.

5.8.3. 385, 289.

5.9.1. $8.40 < \sigma^2 < 18.54$; $2.90 < \sigma < 4.31$.

5.9.3. $630{,}307.86 < \sigma^2 < 1{,}878{,}027.08$; $793.92 < \sigma < 1{,}370.41$.

5.9.5. $1.37 < \sigma^2 < 4.35$; $1.17 < \sigma < 2.09$.

5.9.7. $170.98503 < \sigma^2 < 630.65006$.

5.10.1. $1.48 < (\sigma_2^2/\sigma_1^2) < 9.76$.

5.10.3. $.49 < (\sigma_1^2/\sigma_2^2) < 2.95$.

5.10.5. $.90 < (\sigma_1^2/\sigma_2^2) < 3.52$.

5.10.7. $5.1263398 < (\sigma_U^2/\sigma_N^2) < 60.298059$.

Review Exercises

13. $\bar{x} = 79.87$ $s^2 = 28.1238$ $s = 5.30$ $76.93, 82.81$.

15. $\hat{p} = .30$ $.19, .41$.

17. $\hat{p}_1 = .20$ $\hat{p}_2 = .54$ $.26, .42$.

19. $\hat{p} = .90$ $.87, .93$.

21. $\bar{x} = 19.23$ $s^2 = 20.2268$ $16.01, 22.45$.

23. $-12.219, -7.215$.

Chapter 6

6.2.1. Yes, $z = 3$ $p = .0013$.

6.2.3. Yes, $z = 2.67$ $p = .0038$.

6.2.5. Yes, $z = -5.73$ $p < 0.0001$

6.2.7. No, $t = -1.5$ $.05 < p < .10$

6.2.9. Yes, $z = 3.08$ $p = .0010$

6.2.11. $z = 4$, $p < .0001$

6.2.13. $t = .1271$, $p > .20$.

6.2.15. $z = -4.18$. Reject H_0. $p < .0001$.

6.3.1. Yes, $z = -11.91$ $p < .0001$.

6.3.3. $s_p^2 = 29.30$, $t = -3.62$ Reject H_0 $p < 2(.005) = .01$.

6.3.5. Do not reject H_0 $z = 1.40$ $p = 2(.0808) = .1616$.

6.3.7. $s_p^2 = 5421.25$ $t = -6.66$ Reject H_0 $p < 2(.005) = .010$.

6.3.9. $z = 3.39$ Reject H_0 $p = 2(1 - .9997) = .0006$.

6.3.11. $t = -3.3567$ $p < .01$.

6.4.1. Yes, $t = 16.63$ $p < .005$.

6.4.3. Reject H_0 $t = 3.46$; $.1, .5$ $p < 2(.005) = .010$.

6.4.5. $t = 7.801$. $p < .01$.

6.5.1. Reject H_0 $z = 3.26$ $p < .0010$.

6.5.3. Yes, $z = -8.94$ $p < .0010$.

6.5.5. $z = -2.64$. $p = .0041$.

6.6.1. $\bar{p} = .66$ $z = 2.45$ Reject H_0.
$p = .0142$.

6.6.3. $\bar{p} = .27$ $z = 2.90$ Reject H_0.
$\bar{p} = .0038$.

6.7.1. Yes, $\chi^2 = 54$ $p < .005$.

6.7.3. $\chi^2 = 6.75$ Do not reject H_0 $p > .05$ (two-sided test).

6.7.5. $\chi^2 = 28.8$ Do not reject H_0 $p > .10$.

6.7.7. $\chi^2 = 22.036$. $.10 > p > .05$.

6.8.1. No   V.R. = 2.08 $p > .10$.

6.8.3. No   V.R. = 1.83 $p > .10$.

6.8.5. Reject H_0 V.R. = 4 $.01 < p < .025$.

6.8.7. V.R. = 2.1417. $p > .10$.

Review Exercises

19. $\bar{d} = .40$ $s_d^2 = .2871$ $s_d = .54$ $t = 2.869$ $.005 < p < .01$.

21. $z = 1.095$ $.1379 > p > .1357$.

23. $t = 3.873$ $p < .005$.

25. $\bar{d} = 11.49$ $s_d^2 = 256.6790$ $s_d = 16.02$ $t = 2.485$ $.025 > p > .01$.

Chapter 7

7.2.1. Yes   V.R. = 6.03 $p < .005$.

7.2.3. Yes   V.R. = 67.80 $p < .005$.

7.2.5. No   V.R. = 4.70 $.025 > p > .01$.

7.2.7. $\bar{x}_A - \bar{x}_C$ and $\bar{x}_B - \bar{x}_C$ are significant.

7.2.9. $\bar{x}_D - \bar{x}_E$ and $\bar{x}_A - \bar{x}_E$ are significant.

7.2.11. V.R. = 11.5808, $p < .005$. $\bar{x}_A - \bar{x}_C$ significant; $\bar{x}_B - \bar{x}_C$ significant;
$\bar{x}_B - \bar{x}_D$ significant.

7.3.1. Yes   V.R. = 13.17 $p < .005$.

7.3.3. Yes   V.R. = 30.22 $p < .005$.

7.3.5. V.R. = 276.64 $p < .005$.

7.4.1. (a) V.R. $(A) = 38.78$, V.R. $(B) = 21.57$, V.R. $(AB) = 2.24$
(b) $\alpha < .03$ $p(A) < .005$ $p(B) < .005$ $.10 > p(AB) > .05$.

7.4.3. V.R. $(A) = 5.84$, V.R. $(B) = 44.34$, V.R. $(AB) = 30.09$

$.025 > p(A) > .01$ $p(B) < .005$ $p(AB) < .005$.

7.4.5. Since $83.02 > 2.81$, reject H_0 of equal treatments effects. $p < .005$.
Since $7.60 > 2.42$, reject H_0 of equal physical therapy programs effects.
$p < .005$. Since $5.88 > 1.89$, reject H_0 of no interaction.
$p < .005$. Critical values of F obtained by linear interpolation.

Review Exercises

3. V.R. $= 8.042$ Reject H_0 $p \geq .005$.

5. V.R. $= .825$ Do not reject H_0 $p > .10$.

7. V.R. $(A) = 6.325$ $.005 < p(A) < .01$ $p(B) < .005$ $.01 > p(AB) > .005$; V.R. $(B) = 38.856$; V.R. $(AB) = 4.970$.

9. V.R. $= 14.4364$, $p < .005$.

11. V.R. $= 6.32049$, $.01 > p > .005$.

13. V.R. $= 3.1187$, $.05 > p > .025$. No significant differences among individual pairs of means.

15. V.R. $(A) = 29.4021$, $p < .005$;
V.R. $(B) = 31.4898$, $p < .005$; V.R. $(AB) = 7.11596$, $p < .005$.

17. 499.5, 9, 166.5, 61.1667, 2.8889, 57.6346, $< .005$.

19. (a) Completely randomized (b) 3 (c) 30 (d) No, because $1.0438 < 3.35$.

Chapter 8

8.3.1. $\hat{y} = 7.05 + 4.09x$.

8.3.3. $\hat{y} = 1.36 - .008x$.

8.3.5. $\hat{y} = 106.67 - 1.54x$.

8.4.1. (a) .90 (b) V.R. $= 103.41$ ($p < .005$) (c) $t = 10.17$ ($p < .01$)
(f) 3.21, 4.97.

8.4.3. (a) .94 (b) V.R. $= 152.42$ ($p < .005$) (c) $t = -12.29$ ($p < .01$)
(f) $-.009$, $-.007$.

8.4.5. (a) .78 (b) V.R. $= 28.91$ ($p < .005$) (c) $t = -5.38$ ($p < .01$)
(f) -2.2, $-.88$.

8.5.1. (a) $15.23 \pm (2.2010)(1.3558)(.2774)$
(b) $15.23 \pm (2.2010)(1.3558)(1.0377)$.

8.5.3. (a) $.96 \pm (2.2622)(.0341)(.3162)$
(b) $.96 \pm (2.2622)(.0341)(1.0488)$.

8.5.5. (a) $98.97 \pm (2.3060)(2.6004)(.3210)$
(b) $98.97 \pm (2.3060)(2.6004)(1.0503)$.

8.5.7. (a) $\hat{y} = 30.79 + .7116(90) = 94.834$ (b) 67.755, 121.913.

8.7.1.   (b) .90   (c) $t = 7.44$ ($p < .01$)   .719, .966.
       (Linear interpolation used.)

8.7.3.   (b) .93   (c) $t = 7.16$ ($p < .01$)   .725, .983.
       (Linear interpolation used.)

8.7.5.   (b) .54   (c) $t = 3.08$ ($p < .01$)   .184, .771.
       (Linear interpolation used.)

Review Exercises

17. $r = -.74$ $t = -4.67$

19. $\hat{y} = 19.41 + .9895x$ $r^2 = .9994$ $t = 144.89$.

Source	SS	d.f.	MS	V.R.
Regression	1364.4880	1	1364.4880	20,992.123
Residual	.8453	13	.0650	
Total	1365.3333	14		

21. $\hat{y} = 1.2714 + .8533x$ $r^2 = .6878$ $t = 5.35$.

Source	SS	d.f.	MS	V.R.
Regression	1.6498	1	1.6498	28.64
Residual	.7489	13	.0576	
Total	2.3987	14		

23. $\hat{y} = 61.8819 + .509687x$; V.R. $= 4.285$; $.10 > p > .05$; $t = 2.07$; $.10 > p > .05$; Approximate 95% confidence interval for ρ: $-.03$, .79; 110.3022; 87.7773, 132.8271.

25. $\hat{y} = 37.4559 + .0798579x$; V.R. $= 73.957$; $p < .005$; $t = 8.6013$; $p < .01$; Approximate 95% confidence interval for ρ: .80, .99; 40.6150, 42.2826.

Chapter 9

9.3.1. $\hat{y} = 4639.56 - 244.43x_1 - 422.93x_2$.

9.3.3. $\hat{y} = 13.45 + 4.02x_1 + 2.81x_2$.

9.3.5. $\hat{y} = -422.00 + 11.17x_1 - .63x_2$.

9.4.1.   (a) .93.

(b)

Source	SS	d.f.	MS	V.R.	p
Regression	8,247,389.5	2	4,123,694.80	49.57	< .005
Residual	582,308.5	7	83,186.93		
Total	8,829,698.0	9			

       (c) $t(b_1) = -3.45$, $t(b_2) = -3.22$ ($.01 < p < .02$ for both tests.)

9.4.3. (a) .67.

(b)

Source	SS	d.f.	MS	V.R.	p
Regression	452.56	2	226.28	7.05	$.01 < p < .025$
Residual	224.70	7	32.10		
Total	677.26	9			

(c) $t(b_1) = 3.75$; ($p < .01$); $t(b_2) = 2.04$, $(.05 < p < .10)$

9.4.5. (a) .31.

(b)

Source	SS	d.f.	MS	V.R.	p
Regression	17,023.01	2	8,511.505	4.89	$.01 < p < .025$
Residual	38,276.99	22	1,739.86		
Total	55,300.00	24			

(c) $t(b_1) = 3.05$, ($p < .01$); $t(b_2) = -.67$, ($p > .20$).

9.5.1. (a) $2148.62 \pm (2.3646)(288.42)$

$$\times \sqrt{\tfrac{1}{10} + (.060319)(.8)^2 + (.207378)(-.7)^2 \atop +2(-.086740)(.8)(-.7)}.$$

(b) Add 1 under the radical.

9.5.3. (a) $50.41 \pm (2.3646)(5.67)$

$$\times \sqrt{\tfrac{1}{10} + (.035757)(-1.99)^2 + (.059206)(.44)^2 \atop +2(.022857)(-1.99)(.44)}.$$

(b) Add 1 under the radical.

9.5.5. (a) $532.90 \pm 2.0739(41.71)$

$$\times \sqrt{\tfrac{1}{25} + (.007678)(-.52)^2 + (.000506)(-2.12)^2 \atop +2(-.000002)(-.52)(-2.12)}.$$

(b) Add 1 under the radical.

9.6.1. $\hat{y} = -11.6 + .920x_1 + 17.3x_2$; $t_1 = 9.96$; $t_2 = 7.51$.
$R^2 = .864$. For hypertensive subjects: $\hat{y} = 5.7 + .920x_1$.
For normal subjects: $\hat{y} = -11.6 + .920x_1$.
Since $t = 7.51 > 2.0930$, reject H_0 that $\beta_2 = 0$. Since $7.51 > 2.8609$,
$p < 2(.005) = .01$. $17.334 \pm 2.0930(2.309)$; 12.50, 22.17.

9.6.3. $\hat{y} = 257 - 2.01x_1 = 75.9x_2 + 73.5x_3 - .754x_4 - 2.02x_5$.
$t_1 = -5.18$ $t_2 = 2.80$ $t_3 = 2.82$ $t_4 = -1.56$ $t_5 = -4.12$
$R^2 = .907$.

For area A: $\hat{y} = 332.9 - 2.764x_1$
For area B: $\hat{y} = 330.5 - 4.03x_1$
For area C: $\hat{y} = 257 - 2.01x_1$
H_0: $\beta_2 = 0$: $t = 2.80 > 2.7239$, $p < .01$.   Reject H_0.
H_0: $\beta_3 = 0$: $t = 2.82 > 2.7239$, $p < .01$.   Reject H_0.
H_0: $\beta_4 = 0$: $t = -1.56$.   Do not reject H_0 since $-1.56 > -1.6896$.
$p > .10$.
H_0: $\beta_5 = 0$: $t = -4.12$.   Reject H_0 since $-4.12 < -2.7239$. $p < .01$.

9.7.1. (a) $b_0 = 36.5323$, $b_1 = 6.4758$, $b_2 = -.5749$, $R_{y.12} = .9976$,
   $F = 933$ ($p < .005$)
   (b) $r_{y1.2} = .505$, $t = 1.755$ $(.20 > p > .10)$
   $r_{y2.1} = -.084$, $t = -.253$ ($p > .20$)
   $r_{12.y} = .902$, $t = 6.268$ ($p < .01$).

9.7.3. (a) $b_0 = -422.00$, $b_1 = 11.17$, $b_2 = -.63$, $R_{y.12} = .5548$
   $F = 4.89$ ($.01 < p < .025$)
   (b) $r_{y1.2} = .546$, $t = 3.056$ ($p < .01$)
   $r_{y2.1} = -.142$, $t = -.673$ ($p > .20$)
   $r_{12.y} = .078$, $t = .367$ ($p > .20$).

9.7.5. $\hat{y} = 13.500946 - 7.1312752x_1 + 9.0677309x_2 + .017812535x_3 - .35758245x_3$ $R^2 = .967005$; $R = .9834$.

Review Exercises

7. $R = .3496$ $F = .83$ ($p > .10$).

9. (a) $\hat{y} = 11.43 + 1.26x_1 + 3.11x_2$
   (b) $R^2 = .92$.

(c)

Source	SS	d.f.	MS	V.R.	p
Regression	1827.004659	2	913.50	69.048	< .005
Residual	158.728641	12	13.23		
	1985.7333	14			

(d) $\hat{y} = 11.43 + 1.26(10) + 3.11(5) = 39.56$.

11. (a) $\hat{y} = -126.487 + .176285x_1 - 1.56304x_2 + 1.5745x_3 + 1.62902x_4$

(b)

Source	SS	d.f.	MS	V.R.	p
Regression	30873.80	4	7718.440	13.655	< .005
Residual	5774.92	10	577.492		
	36648.72	14			

(c) $t_1 = 4.3967$; $t_2 = -.77684$; $t_3 = 3.53284$; $t_4 = 2.59102$

(d) $R^2_{y.1234} = .8424255$; $R_{y.1234} = .911784$.

13. $\hat{y} = -111.6 + 80.387x_1 + 159.40x_2$

Coefficient	t
-111.6	$-.97$
80.387	15.53
159.40	2.12
$R^2 = .918$	

For males, $\hat{y} = 42.8 + 80.387x_1$

For females, $\hat{y} = -111.6 + 80.387x_1$

For $t = 2.12$, $.05 > p > .02$.

Chapter 10

10.3.1. $X^2 = 2.072$ $p > .005$.

10.3.3. $X^2 = 3.417$ $p > .10$.

10.3.5. $X^2 = 2.21$ $p > .10$.

10.4.1. $X^2 = 27.272$ $p < .005$.

10.4.3. $X^2 = .765$ $p > .10$.

10.4.5. $X^2 = 42.579$ $p < .005$.

10.5.1. $X^2 = 65.855$ $p < .005$.

10.5.3. $X^2 = 32.754$ $p < .005$.

10.5.5. $X^2 = 82.373$ with 2 d.f.   Reject H_0. $p < .005$.

Review Exercises

11. $X^2 = 10.7827$ $p < .005$.

13. $X^2 = 3.4505$ $p > .10$.

15. $X^2 = 8.1667$ $p < .005$.

17. $X^2 = 67.8015$ $p < .005$.

19. $X^2 = 7.2577$ $.05 > p > .025$.

21. Independence.

23. Homogeneity.

Chapter 11

11.3.1. $P = .3036$ $p = .3036$.

11.3.3. $P(x \leq 2|13, .5) = .0112$. Since $.0112 < .05$, reject H_0. $p = .0112$.

11.4.1. $X^2 = 16.13$ $p < .005$.

11.5.1. $T = 71$ $p > .20$.

11.5.3. $T = 151 - 11(12)/2 = 85$. $W_{.025} = 27$, $W_{.975} = 11(10) - 27 = 83$.
Reject H_0. Since $83 < 85 < 87$, $.05 > p > .02$.

11.6.1. $D = .3241$ $p < .01$.

11.6.3. $D = .1319$ $p > .20$.

11.7.1. $H = 19.55$ $p < .005$.

11.7.3. $H = 23.0375$ $p < .005$.

11.7.5. $H = 13.86$ $p < .005$.

11.7.7. $H = 14.75$ $.01 > p > .005$.

11.8.1. $\chi_r^2 = 8.67$ $p = .01$.

11.8.3. $\chi_r^2 = 13.74$ $p < .005$.

11.8.5. $\chi_r^2 = 11.1$ $p = .00094$.

11.8.7. $\chi_r^2 = 7.28$ $p \approx .054$.

11.9.1. $r_S = -0.07$ $p > .20$.

11.9.3. r_S (corrected for ties) $= .95$. $p < .002$.

11.9.5. $r_S = .99$ $p < .001$.

11.9.7. $r_S = .6826$.
Since $.6826 > .4853$, reject H_0. $.002 < p < .010$.

Review Exercises

9. $P(X \le 1|10, .5) = 1 - .9893 = .0107$ $p = .09127$.

11. $\chi_r^2 = 16.2$ $p < .005$.

13. $D = .1587$ $p > .20$.

Chapter 12

12.2.1. (a) 10.4 (b) 10.1, 11.0 (c) 16.1 (d) 21.5 (e) 16.9 (f) 47.6
(g) 15.2, 27.3.

12.3.1. (a) 97.0, 165.4, 125.0, 62.4, 5.8, 3.5
(b) 2413 (c) 485; 1312; 1937; 2249; 2378; 2413.
(d) 76.5.

Index